MW01098343

THERMODYNAMICS

Fundamentals for Applications

Other Books by the Authors

J. P. O'Connell

Computer Calculations for Multicomponent Vapor-Liquid Equilibria (coauthor)

Computer Calculations for Multicomponent Vapor-Liquid and Liquid-Liquid Equilibria (coauthor)

The Properties of Gases and Liquids (coauthor of 5th edition)

J. M. Haile

Molecular-Based Study of Fluids (coeditor)

Chemical Engineering Applications of Molecular Simulation (editor)

Molecular Dynamics Simulation

Technical Style: Technical Writing in a Digital Age

Lectures in Thermodynamics: Heat and Work

Analysis of Data

THERMODYNAMICS

Fundamentals for Applications

J. P. O'Connell
University of Virginia

and

J. M. Haile
Macatea Productions

CAMBRIDGE UNIVERSITY PRESS
Cambridge, New York, Melbourne, Madrid, Cape Town, Singapore, São Paulo

Cambridge University Press
40 West 20th Street, New York, NY 10011-4211, USA

www.cambridge.org
Information on this title: www.cambridge.org/9780521582063

First published 2005

Printed in the United States of America

A catalog record for this book is available from the British Library.

Library of Congress Cataloging in Publication Data

O'Connell, John P. (John Paul)
Thermodynamics : fundamentals for applications / John P. O'Connell, J. M. Haile.
p. cm.
Includes bibliographical references and index.
ISBN 0-521-58206-7 (alk. paper)
1. Thermodynamics. I. Haile, J. M. II. Title.

QC311.O3 2004
536'.7–dc22 2004057542

ISBN-13 978-0-521-58206-3 hardback
ISBN-10 0-521-58206-7 hardback

To Verna and Tricia

If it were easy … it cannot be educational.
In education, as elsewhere, the broad
primrose path leads to a nasty place.

Alfred North Whitehead
"The Aims of Education," in
Alfred North Whitehead, An Anthology,
F. S. C. Northrop and M. W. Gross, eds.,
Macmillan, New York, 1953, p. 90.

remarkable things
occur in accordance with Nature,
the cause of which is unknown;
others occur contrary to Nature,
which are produced by skill
for the benefit of mankind.

Mechanica, Aristotle (384–322 BCE)

Many scholars doubt that the *Mechanica,*
the oldest known textbook on engineering,
was written by Aristotle. Perhaps it was written
by Straton of Lampsacus (a.k.a. Strato Physicus,
died c. 270 BCE), who was a graduate student
under Aristotle and who eventually succeeded
Theophrastus as head of the Peripatetic school.

CONTENTS

PREFACE

Thermodynamics is fundamental and applicable to all technical endeavors. Its two brief laws provide a complete basis for establishing the states of pure substances and their mixtures. It shows us the directions in which those states tend to change when systems are prodded by external forces. It provides a secure foundation for scientific investigations into all forms of matter. It reveals constraints on interconversions of heat and work, on separations of components from solutions, and on ultimate extents of chemical reactions. It can guide screening for feasibility of alternative processes, and when a design has been selected, it can contribute to the optimization of that design.

Although thermodynamics describes natural phenomena, those descriptions are in fact products of creative, systematic, human minds. Nature unfolds without any explicit reference to energy, entropy, or fugacity; these are unnatural concepts created by humans. Nevertheless, the complexities observed in Nature can be organized by appealing to thermodynamic methodology. With proper understanding, generalized thermodynamic techniques can be used to deal effectively with many aspects of reality. But to gain that understanding, thermodynamics must be studied in a systematic way that uncovers its structure and economy.

Thermodynamic ideas originated almost 200 years ago, but the subject continues to evolve. Although some claim that "there is nothing new in thermodynamics," scholars still find challenges in its abstractness, rigor, and universality. They debate the "best" ways to phrase its basic principles and to identify the limits of its application. In addition, much current work seeks to extend thermodynamics into new domains of technological development. For example, modern computers enable us to test models at every level of complexity. As a result, thermodynamics is now being used to an unprecedented extent as a basis for creating models that can correlate and predict natural phenomena. In many applications, experimental data are being replaced by modeling and simulation of thermodynamic properties.

But in spite of all this, we have come to the realization that the full power of thermodynamics can be used to advantage only after its foundations are fully assimilated. So in this text we concentrate on fundamentals, rather than modeling, with the belief that a deeper knowledge of the basics will enhance your ability to combine them with models when you need to apply thermodynamics to practical situations.

Over our years of teaching, we have identified three common attitudes that many students bring to an advanced study of thermodynamics. One is "I don't like this stuff, it's too abstract, it's not engineering, so I'll get by as best I can." This attitude springs from frustration with an earlier exposure to the material—an incomplete or misleading experience, often exacerbated by a weak background that cannot support the development of a sound logical structure. These frustrations must be confronted and relieved, if the student is to become self-reliant with thermodynamic concepts and proficient in their use. Self-reliance and proficiency require care, maturity, and intelligence. For these students, a major objective of an advanced course is to overcome confusion and antipathy from earlier exposures, while fully integrating the concepts, knowledge, and procedures. This is a demanding but essential exercise because only then can a learner see full relevance, make prudent applications, and have a satisfying learning experience.

A second attitude stands diametrically opposed to the first: "This stuff is ok; it's just calculus plus reasoning, I could learn it if I really wanted to." With this attitude, students read the text and, feeling quite comfortable, turn the page; but later they have difficulty applying what they've read. Their knowledge is superficial. Perhaps these students delude themselves because the logic of thermodynamics seems relatively straightforward compared to the subtleties encountered when trying to apply that logic in realistic situations. To combat this attitude, we have tried to go well beyond simple derivations of relevant thermodynamic relations; you may find the ratio of words-to-equations much higher here than in many texts. We do not avoid discussing the exceptions, the special cases, the constraints, the limiting behaviors that must be addressed to reach deeper understandings. Nor do we demure from making subjective judgments about relative importance, about issues to be confronted in choosing from among alternatives, about the levels of approximation that can apply to the problem at hand. These kinds of issues constitute a large part of engineering practice, and thermodynamics provides a rich and varied environment in which to develop appreciation for and skill in dealing with such issues—thermodynamics can foster the development of sound engineering judgment.

A third attitude that some students bring to a graduate course can be characterized by thoughts like these: "Where is the equation to solve this problem? Where is the example that shows me how to solve this problem? Where is the solution manual that shows me whether my numerical answer is correct?" Too many students confuse problem solving with studying; too many think the objective in solving a problem is to get a number, so they can move on to the next problem. Too many believe the book, or some book somewhere, has all the answers. In writing this text, we have made determined efforts to subvert these attitudes: we present relatively few worked examples and our examples are not like the problems at the ends of chapters. Many of our problems are not answered with a number, and in many others, the student is to explain or compare or discuss the numbers that have been obtained. Following the advice of master teachers and our own experience, we do not provide a solution manual for the problems.

In solving problems, we expect students to always begin by articulating a general statement or equation that can be expected to apply to nearly any situation; examples include the first and second laws. We call these the *always true*, and we expect students to memorize a small select set of *always true* relations, for at least one of them will apply to any situation. Hence, thermodynamics always gives a starting point for problem solving. In contrast, students are not expected to memorize equations that represent models—except, of course, for very simple ones, like the ideal gas. This approach encourages students to deploy the full power of thermodynamic deduction. It also motivates students to distinguish models from the few really important concepts that apply in most situations. When encountering this approach for the first time, some students become impatient or disdainful. But most grow to appreciate that there are clear rules for distinguishing those few *always true* relations that are to be memorized from those many *always true* relations that can always be derived when needed, just as there are clear rules for distinguishing the *always true* from models, which are always special cases.

But problem solving is only one part of an engineering education, and somehow, in the continuing efforts to convert engineering students into effective problem solvers, more important and rewarding aspects of learning have been ignored: the delight of discovery, the satisfaction of grasping an intricate and convoluted argument, the stimulation from contributing to a wide-ranging enquiry in a scholarly atmosphere. With much of the earlier computational burden now relieved by computers and software, it might be expected that the imbalance of problem solving against deeper understandings would be redressed. But the reality is otherwise, for too often in today's scholasticism, computers have been adopted as tools for even more elaborate problem solving, widening the gulf between computation and thought. This book is an attempt, however modest, to bridge that gulf.

The material in this book is developed for beginning graduate students in chemical engineering, and the needs of those students are, in our view, best served if we focus on macroscopic thermodynamics. In this book, models and molecular concepts are confined to illustrations and brief discussions; nevertheless, studied thoroughly, this material alone is sufficient for a full semester course. Alternatively, to create time to study contemporary applications from other sources, certain chapters (such as 0–6) could be covered less thoroughly or selected sections (some in Chapters 8, 9, 11, and 12) might be omitted. However, because we build a logical structure in a systematic fashion, familiarity with the content of the early chapters is essential if you are to fully comprehend the development and applications presented in later chapters.

Students best overcome misconceptions and grow to reliable, efficient practice when studying with a master teacher. Such study can be enhanced by a textbook that stimulates deeper explorations of the material. Our goal is that this text will stimulate students and their instructors to dig deeper, so they begin to appreciate the distinctive structure of thermodynamics, to become effective in its use, to enrich their vision of Nature's unity and diversity, and to enhance their professional proficiency.

ACKNOWLEDGMENTS

We are grateful to our many students and colleagues for their encouragement and enthusiasm: their feedback helped us improve earlier drafts of this text and stimulated us to persevere. Special thanks to Professor E. Dendy Sloan, who provided support and a place to write, while one of us spent a sabbatical at Colorado School of Mines. Thanks also to J. Mitchell Haile, who willingly provided all manner of technical support as we thrashed our way through dozens of computers and scores of software packages in the course of the writing.

The page design, composition, and layout for this book were done by Macatea Productions (www.macatea.com).

THERMODYNAMICS
Fundamentals for Applications

0

INTRODUCTION

You are a member of a group assigned to experimentally determine the behavior of certain mixtures that are to be used in a new process. Your first task is to make a 1000-ml mixture that is roughly equimolar in isopropanol and water; then you will determine the exact composition to within ±0.002 mole fraction. Your equipment consists of a 1000-ml volumetric flask, assorted pipettes and graduated cylinders, a thermometer, a barometer, a library, and a brain. You measure 300 ml of water and stir it into 700 ml of alcohol—Oops!—the meniscus falls below the 1000-ml line. Must have been careless. You repeat the procedure: same result. Something doesn't seem right.

At the daily meeting it quickly becomes clear that other members of the group are also perplexed. For example, Leia reports that she's getting peculiar results with the isopropanol-methyl(ethyl)ketone mixtures: her volumes are *greater* than the sum of the pure component volumes. Meanwhile, Luke has been measuring the freezing points of water in ethylene glycol and he claims that the freezing point of the 50% mixture is well *below* the freezing points of both pure water and pure glycol. Then Han interrupts to say that 50:50 mixtures of benzene and hexafluorobenzene freeze at temperatures *higher* than either pure component.

These conflicting results are puzzling; can they all be true? To keep the work going efficiently, the group needs to deal with the phenomena in an orderly way. Furthermore, you want to understand what's happening in these mixtures so that next time you won't be surprised.

0.1 NATURAL PHENOMENA

These kinds of phenomena affect the course of chemical engineering practice. As chemical engineers we create new processes for new products and refurbish old processes to meet new specifications. Those processes may involve mixing, separation, chemical reaction, heat transfer, and mass transfer. To make homemade ice cream we mix fluids, promote heat transfer, and induce a phase change, without worrying much about efficiency or reproducibility. But to design an economical process that makes ice cream in a consistent and efficient manner, we must have quantitative knowledge of the properties and phase behavior of pure substances and their mixtures.

The acquisition of that knowledge appears to be an overwhelming task. An essentially infinite number of mixtures can be formed from the more than 22,000,000 pure substances now identified by the *Chemical Abstracts Registry*, a large number of properties must be studied, and an extensive range of operating variables must be explored. We will never be able to measure the properties needed for all possible mixtures over all required conditions. Theory is of limited help: our inability to create a detailed quantum mechanical description of matter, coupled with our ignorance of intermolecular forces, prevents our computing from first principles all the property values we may need. Is there anything we can do?

The most successful approach combines classical thermodynamics with modeling. Classical thermodynamics provides a grand scheme for organizing our knowledge of chemical systems, including reaction and phase equilibria. Thermodynamics provides rigorous relations among quantities, thereby reducing the amount of experiment that must be done and providing tests for consistency. Thermodynamics establishes necessary and sufficient conditions for the occurrence of vapor-liquid, liquid-liquid, and solid-fluid equilibria; further, thermodynamics identifies directions for mass transfer and chemical reactions. Thermodynamics allows us to determine how a situation will respond to changes in temperature, pressure, and composition. Thermodynamics identifies bounds: What is the least amount of heat and work that must be expended on a given process? What is the best yield we can obtain from a chemical reaction?

Thermodynamics carries us a long way toward the solution of a problem, but it doesn't carry us to the end because thermodynamics *itself* involves no numbers. To get numbers we must either do experiments or do some more fundamental theory, such as statistical mechanics or molecular simulation. With the demand for property values far exceeding both the predictive power of theory and the range of experiment, we use modeling to interpolate and extrapolate the limited available data.

This book is intended to help you master the concepts and tools of modern thermodynamic analysis. To achieve that goal, we will review fundamentals, especially those that pertain to mixtures, reaction equilibria, and phase equilibria: the objective is to solidify your grounding in the essentials. In most undertakings the first step is the most difficult, and yet, without the essentials, we haven't a clue as to how to start. A virtue of thermodynamics is that it *always* gives us a starting point for an analysis. But to pursue the rest of an analysis intelligently, you must choose models that are appropriate for your problem, taking into account the advantages and limitations that they offer. Finally, to complete an analysis efficiently and effectively, you must have experience. This book tries to instruct you in *how* to perform thermodynamic analyses and provides opportunities for you to practice that procedure. The program begins in Chapter 1, but before embarking we use the rest of this introduction to clarify some misconceptions you may have obtained from previous exposures to the subject.

0.2 THERMODYNAMICS, SCIENCE, AND ENGINEERING

Chemical engineering thermodynamics balances science and engineering. But when the subject is studied, that balance can be easily upset either in favor of a "practical" study that ignores scientifically-based generality, consistency, and constraint, or in favor of a "scientific" study that ignores practical motivation and utility. Beyond the introductory level, such unbalanced approaches rarely promote facility with the material. To clarify this issue, we use this section to distinguish the development of science from the practice of engineering.

Legend has it that a falling apple inspired Newton's theory of gravitation. More likely the theory was the culmination of much thinking and several observations, of which the last perhaps involved an apple. Once his theory was tested in various situations and found satisfactory, it became known as a universal law. Newton's encounter with an apple may or may not have happened, but nevertheless the story conveys the most common method of discovery. This method, in which a few particular observations are extended to a single broad generality, is called *induction*. The method is summarized schematically on the left side of Figure 0.1. (For more on the role of induction in scientific discovery, see Polya [1].)

The law of gravitation illustrates the principal goal of science: to identify, organize, codify, and compress a large amount of information into a concise statement. Another example is Maxwell's proposal that electricity and magnetism can be described by the same set of differential equations. Still another example occurs in linear transport the-

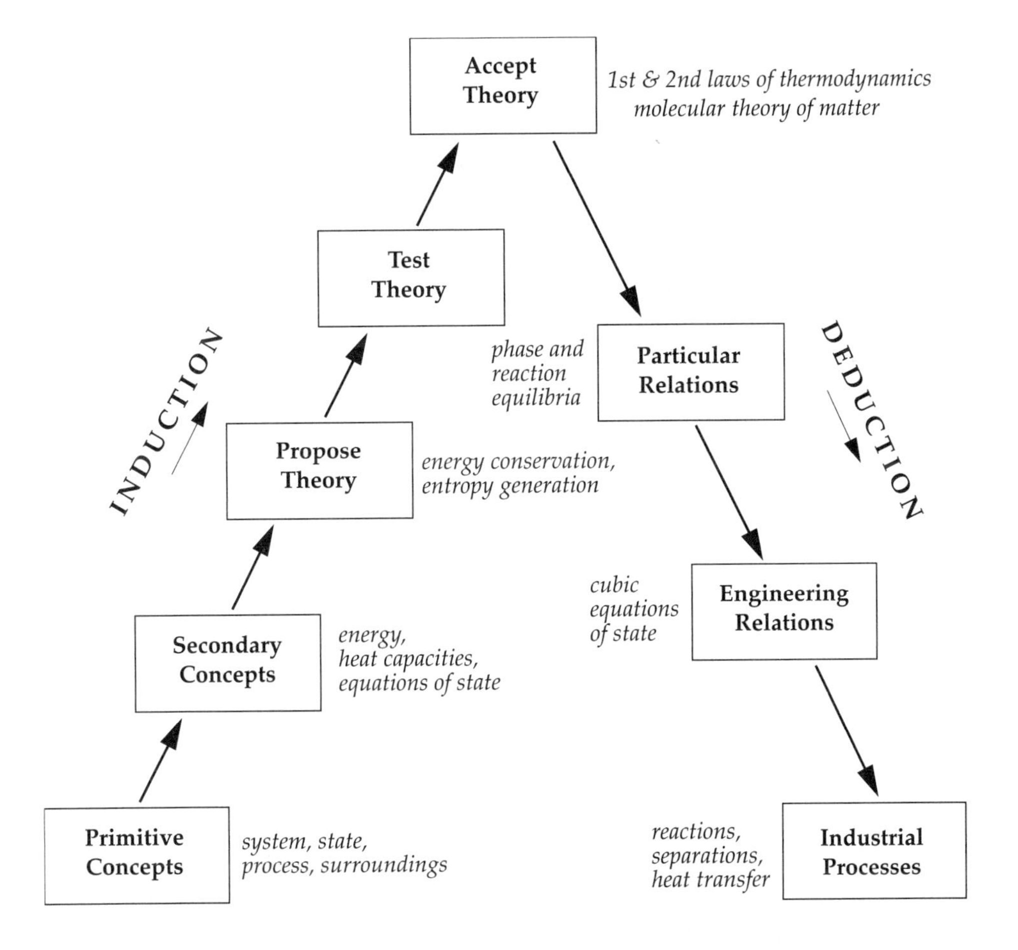

Figure 0.1 Schematic of the principal ways in which science and engineering are practiced. Science proceeds mainly by *induction* from primitive concepts to general theories. From those generalities engineering proceeds by *deduction* to create new processes and products.

ory in which Newton's law of viscosity, Fourier's law of heat conduction, Fick's law of mass transfer, and Ohm's law of electrical conduction all collapse into a single form

$$\text{flux} = -\text{coefficient} \times \text{gradient} \tag{0.2.1}$$

Such generalizations can also be found in thermodynamics; for example, Gibbs described phase equilibrium using the thermodynamics originally developed to analyze heat engines and other thermal processes. These examples illustrate that the more highly developed a scientific discipline, the fewer, broader and more powerful its laws, so that one general goal of science is to make efficient use of brain power [2],

$$\text{science} \Leftrightarrow \text{the economy of thought} \tag{0.2.2}$$

The practice of engineering is an activity distinct from the development of science. A well-engineered product or process accomplishes its allotted task through simple design, easy operation, moderate cost, infrequent maintenance, and long life: one well-engineered product was the original Volkswagen Beetle. These attributes of design, operation, and maintenance all contribute to an efficient use of resources; i.e.,

$$\text{engineering} \Leftrightarrow \text{the economy of resources} \tag{0.2.3}$$

Engineering practice is not science, but economic insights from science contribute to the economical use of resources: the general theories and laws produced by the minds of scientists become tools in the hands of engineers. But because those theories and laws are so general (to achieve economy of thought), we must first reduce them to forms appropriate to our situation. This method, in which a generality is reduced to apply to a particular case, is called *deduction*; it is the primary way by which engineers use science. This use is illustrated on the right side of Figure 0.1.

The broad generalities of science are of such overwhelming importance that they deserve a handy and memorable name: we call them the things that are *always true*. An example is the statement of conservation of mass. Conservation of mass represents economy of thought because it applies to any situation that does not involve nuclear reactions. But to actually use it, we must deduce the precise form that pertains to our problem: What substances are involved? What are the input and output streams? Is the situation a transient or steady state?

Besides the generalities of natural phenomena, science produces another set of things that are always true: definitions. Definitions promote clear thinking as science pushes along its path toward new generalities. By construction, definitions are always true and therefore they are important to engineering analysis. Ignoring definitions leads to fuzzy analysis and ambiguous communications. While there is much science in thermodynamics, engineers rarely study thermodynamics for the sake of its science. Instead, we must confront the science because articulating an always true serves as a crucial step in every thermodynamic analysis.

As you use this book to restudy thermodynamics, you may realize that your earlier experience with the subject was more like the left-hand side (uphill) of Figure 0.1. It may not have been clear that your goal was to reach the top, so that everything you did afterwards could be downhill (right-hand side). You may even have tried to "tunnel through" to applications, meaning you may have memorized particular formulae and used them without serious regard for their origins or limitations. It is true that formulae must be used, but we should apply their most general and reliable forms, being sensitive to what they can and cannot say about a particular situation.

In this text our goal is to enable you to deduce those methods and relations that pertain to particular applications. We develop fundamentals in an uphill approach, and we apply those fundamentals in a downhill fashion, taking advantage of any knowledge you may already have and attempting to include all the essentials in an accessible way. Throughout, we include sample applications appropriate to the level of learning you should have achieved, and we exhort you to develop facility with the material through repetition, practice, and extension.

To become proficient with thermodynamics and reach deep levels of understanding, you must have not only ability. In addition, you must adapt to alternative ways of thinking, make a commitment to learning, and exercise your new skills through personal reflection, interactive conversation, and problem solving. In this way you, your classmates, and your instructor can all benefit from your efforts.

0.3 WHY THERMODYNAMICS IS CHALLENGING

In this section we cite two stumbling blocks that often hinder a study of thermodynamics: its scope and its abstract nature. Both can lead to frustration, but in this book we try to offer strategies that help you minimize your frustrations with the material.

0.3.1 Large Number of Relations

In studying thermodynamics, it is easy to be overwhelmed by the large number of mathematical relations. Those relations may be algebraic, such as equations of state, or they may be differential, such as the Maxwell relations. The number is large because many variables are needed to describe natural phenomena and because additional variables have been created by humans to achieve economy of thought. To keep the material under control, it must be organized in ways that are sensible rather than arbitrary. Numerous relations may arise in the search for economy of thought, but in studying a subject we should economize resources, such as brain power, by appealing to orderliness and relative importance.

As an example, consider these four properties: temperature T, pressure P, volume V, and entropy S. For a system of constant mass we can use these four properties to form twelve common first derivatives:

$$\left(\frac{\partial P}{\partial T}\right)_V, \left(\frac{\partial P}{\partial T}\right)_S, \left(\frac{\partial V}{\partial T}\right)_P, \left(\frac{\partial V}{\partial T}\right)_S, \left(\frac{\partial S}{\partial T}\right)_P, \left(\frac{\partial S}{\partial T}\right)_V$$

$$\left(\frac{\partial P}{\partial S}\right)_V, \left(\frac{\partial P}{\partial S}\right)_T, \left(\frac{\partial V}{\partial P}\right)_S, \left(\frac{\partial V}{\partial P}\right)_T, \left(\frac{\partial S}{\partial V}\right)_P, \left(\frac{\partial S}{\partial V}\right)_T$$

How shall we organize these derivatives? We choose an engineering approach in which we group them according to relative importance; that is, we declare as most important those derivatives that convey the most useful information. If we do this, we obtain a hierarchy of derivatives ranked from most useful to least useful.

The hierarchy can be constructed from the simple rules presented in Chapter 3, but for now we merely note that such rankings can easily be found. So, of the twelve derivatives involving T, P, V, and S, three are very useful, six are moderately useful, and three are rarely used by engineers. Consequently, in an engineering study of those twelve derivatives, you should devote your effort to the most important nine—a savings of 25%. Moreover, by developing such patterns and using them repetitively, we hope to help you grapple with the material in systematic and successful ways.

0.3.2 Abstraction in Thermodynamic Properties

Thermodynamic abstraction takes two forms. One occurs in *conceptuals*—quantities such as entropy, chemical potential, and fugacity—which are often presented as arbitrarily defined concepts having only tenuous contacts to reality. Abstraction, it is true, is a prevalent feature of engineering thermodynamics; but it cannot be otherwise, for abstraction serves vital functions. Through the mechanism of conceptual properties, abstraction achieves economy of thought by providing simple expressions for the constraints that Nature imposes on phenomena. Moreover, through simplification, abstraction achieves economy of resources by providing means for identifying and separating important quantities from unimportant details.

Non-measurable concepts repel engineers—people who like to get their hands on things. But to use conceptuals effectively, we must appreciate why they have been invented and understand how they connect to reality. So in presenting abstract quantities, we will not only provide formal definitions, but we will also rationalize their forms relative to alternatives and offer interpretations that provide physical meaning.

In addition to physical interpretations, we will also try to reduce the level of abstraction by appealing to molecular theory. It is true that thermodynamics can be developed in a logical and self-contained way without introducing molecules, and in fact the subject is often taught in that way. But such a presentation may be a disservice to today's students who are familiar and comfortable with molecules. Whenever we can, we use molecular theory to provide physical interpretations, to simplify explanations, to generalize results, and to stimulate insight into macroscopic phenomena.

0.3.3 Abstraction in Thermodynamic Modeling

The second abstraction occurs in *modeling*. In science and engineering, progress often involves isolating the dominant elements from a complex situation—a cutting away of undergrowth to reveal more clearly both forest and trees. Although abstract models are not real, without them we would be overwhelmed by the complexities of reality. Moreover, even when an abstraction—call it an idealization—does not precisely represent part of a real situation, the idealization might serve as a basis for systematic learning and later analysis.

One such strategy separates reality into ideal and correction terms. For thermodynamic properties this separation often takes an additive form

$$\text{real} = \text{ideal} + \text{correction} \tag{0.3.1}$$

This pattern appears in the virial equation of state, in correlations of gas properties based on residual properties, and in correlations of liquid mixture properties based on excess properties. Another separation of reality takes a multiplicative form,

$$\text{real} = \text{ideal} \times \text{correction} \tag{0.3.2}$$

This pattern is used to correlate gas volumes in terms of the compressibility factor, to correlate gas phase fugacities in terms of fugacity coefficients, and to correlate liquid mixture fugacities in terms of activity coefficients.

According to a traditional engineering view, much of the abstraction in thermodynamics can be eliminated if we avoid its scientific foundations and discuss only its practical applications. Alternatively, according to a traditional scientific view, when we combine modeling with thermodynamics to enhance its usefulness, we spoil its beauty and logical consistency. In this text we intend to strike a middle ground between these conflicting views. We seek to preserve and exploit the subject's logic, but we will also combine the scientific formalism with engineering modeling because we intend to actually apply the science to realistic situations.

0.4 THE ROLE OF THERMODYNAMIC MODELING

In § 0.1 we noted that pure thermodynamics is not generally sufficient to solve engineering problems. Thermodynamics provides numerous relations among such quantities as temperature, pressure, heat capacities, and chemical potentials, but to obtain numerical values for those quantities, we must rely on experimental data—thermodynamics itself provides no numbers.

But reliable experiments are expensive and time-consuming to perform, and consequently we rarely have enough data to satisfy engineering needs. So we contrive models to extend the range of validity of data. At the present time, successful models usually have some basis in molecular theory. As suggested by Figure 0.2, modern model building involves an interplay among thermodynamics, molecular theory, molecular simulation, and experiment: thermodynamics identifies quantities that are important in a particular application, molecular theory provides mathematical forms for representing those quantities, while molecular simulation and experiment provide data for obtaining values of parameters in the mathematical forms.

The resulting models may be used in various applications, including chemical reaction equilibria, which is important to chemical reactor design, and phase equilibria, which arises in distillation, solvent extraction, and crystallization. But in addition to such traditional applications, thermodynamic models may also be used to help solve many other engineering problems, such as those involving surface and interfacial phenomena, supercritical extraction, hazardous waste removal, polymer and composite material development, and biological processing.

No single book could provide a complete description of all the topics—fundamentals, experiments, modeling, and applications—implied by Figure 0.2. In this book we choose to emphasize fundamental thermodynamics (Parts I, II, and III) and calculations for systems having multiple phases and reactions (Part IV); these topics arise in many common applications. Since we cannot possibly cover everything, we will concentrate on the fundamentals and illustrate their use in enough applications so you

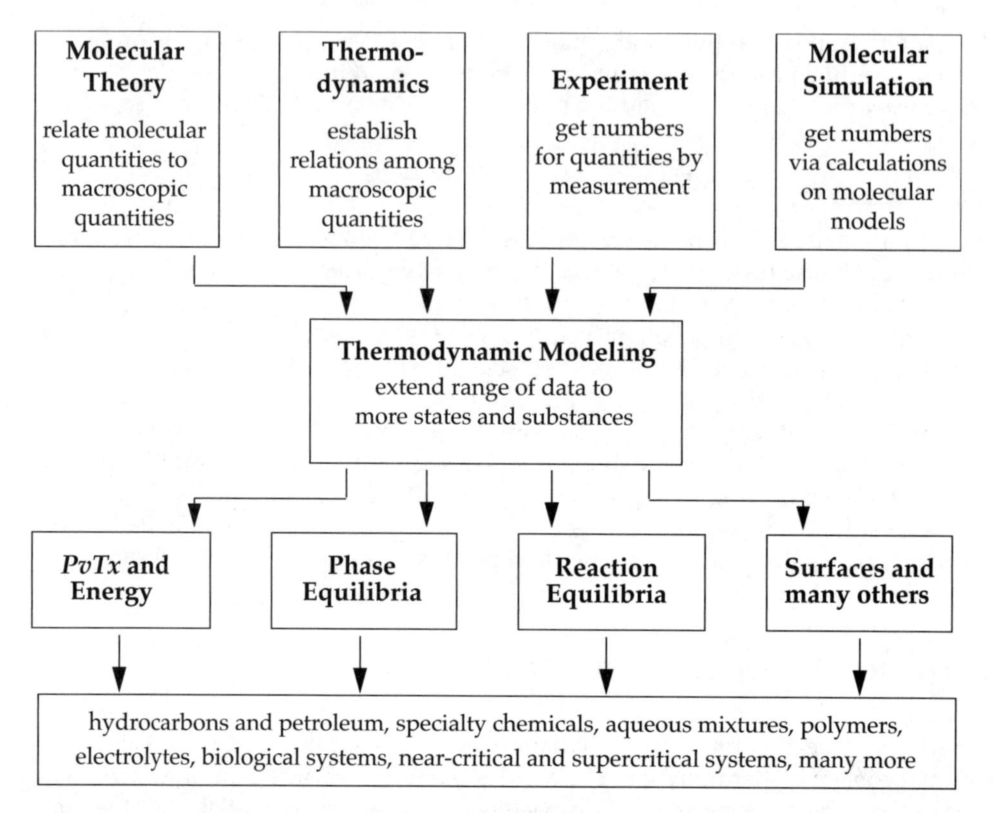

Figure 0.2 By combining molecular theory, thermodynamics, experimental data, and molecular simulation, thermodynamic modeling simplifies and extends descriptions of physical and chemical properties. This contributes to the reliable and accurate design, optimization, and operation of engineering processes and equipment. Note the distinction between molecular models used in molecular simulation and macroscopic models used in thermodynamics.

can learn how they are applied. As a result, you should be able to take advantage of thermodynamics in situations that are not covered explicitly here. Truly fundamental concepts are permanent and universal, it is only the applications that go in and out of style.

LITERATURE CITED

[1] G. Polya, *Mathematics and Plausible Reasoning*, vol. I, *Induction and Analogy in Mathematics*, Princeton University Press, Princeton, NJ, 1954.

[2] E. Mach, *The Science of Mechanics*, 3rd pbk. ed., T. J. McCormack (transl.), The Open Court Publishing Company, LaSalle, IL, 1974.

PART I

THE BASICS

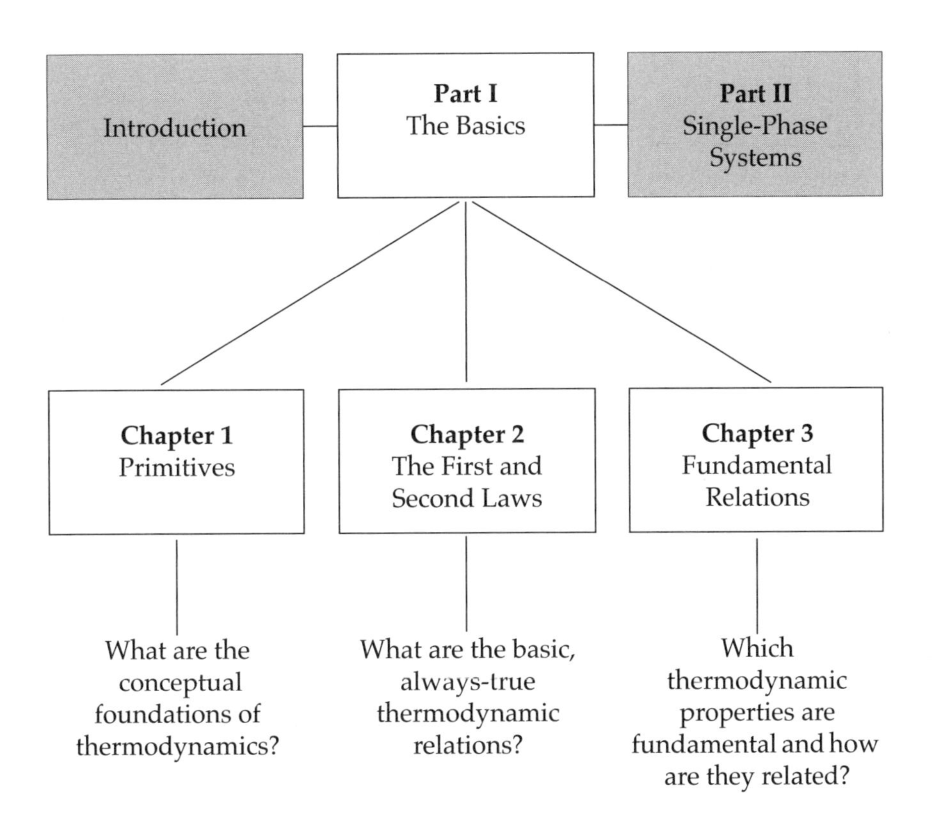

1

PRIMITIVES

In this chapter we review elementary concepts that are used to describe Nature. These concepts are so basic that we call them *primitives,* for everything in later chapters builds on these ideas. You have probably encountered this material before, but our presentation may be new to you. The chapter is divided into primitive things (§ 1.1), primitive quantities (§ 1.2), primitive changes (§ 1.3), and primitive analyses (§ 1.4).

1.1 PRIMITIVE THINGS

Every thermodynamic analysis focuses on a *system*—what you're talking about. The system occupies a definite region in space: it may be composed of one homogeneous phase or many disparate parts. When we start an analysis, we must properly and explicitly identify the system; otherwise, our analysis will be vague and perhaps misleading. In some situations there is only one correct identification of the system; in other situations, several correct choices are possible, but some may simplify an analysis more than others.

A system can be described at either of two levels: a *macroscopic* description pertains to a system sufficiently large to be perceived by human senses; a *microscopic* description pertains to individual molecules and how those molecules interact with one another. Thermodynamics applies to macroscopic entities; nevertheless, we will occasionally appeal to microscopic descriptions to interpret macroscopic phenomena. Both levels contain primitive things.

1.1.1 Macroscopic Things

Beyond the system lies the rest of the universe, which we call the *surroundings*. Actually, the surroundings include only that part of the universe close enough to affect the system in some way. For example, in studying how air in a balloon responds to being moved from a cool room to a warm one, we might choose the air in the balloon to be the system and choose the air in the warmer room to be the surroundings. If the universe beyond the room does not affect the balloon, then objects and events outside the room can be ignored.

An *interaction* is a means by which we can cause a change in the system while we remain in the surroundings; that is, an action in the surroundings will cause a response in the system only if the proper interaction exists. Interactions are of two types: thermal and nonthermal. A *nonthermal* interaction connects some variable x in the system to a variable y in the surroundings. This means that x and y are not independent; instead, they are coupled by a relation of the form

$$F(x, y) = 0 \tag{1.1.1}$$

Each nonthermal interaction involves a force that tends to change something about the system. Of most concern to us will be the nonthermal interaction in which a mechanical force deforms the system volume. In this case, the system volume is x in (1.1.1) and the surroundings have volume y. When the system volume increases, the volume of the surroundings necessarily decreases, and vice versa. One of these variables, typically the system variable x, is chosen to measure the extent of the interaction; this variable is called the *interaction coordinate.*

When two or more nonthermal interactions are established, the choice of interaction coordinates must be done carefully, to ensure that the coordinates are mutually independent. That is, each interaction coordinate must be capable of being manipulated while all others are held fixed. Such coordinates are called *generalized coordinates,* the interaction corresponding to a generalized coordinate is said to be *conjugate* to its coordinate, and each conjugate interaction is said to be *orthogonal* to every other interaction [1–3]. As suggested by Figure 1.1, many orthogonal interactions are possible; examples (with their conjugate coordinates) are mechanical interactions (volume), chemical interactions (composition), gravitational interactions (position relative to a mass), and electrical interactions (position relative to a charge).

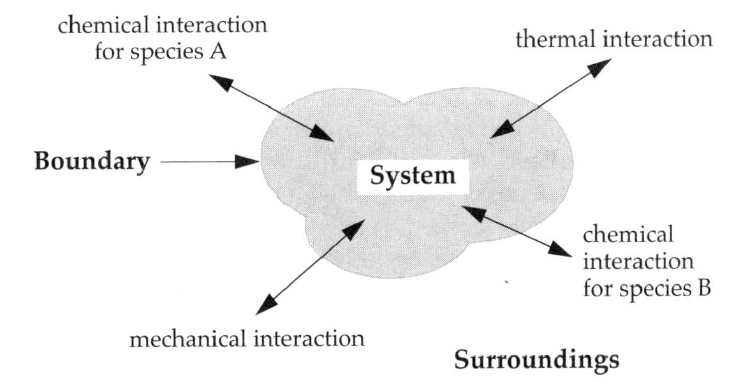

Figure 1.1 A system may engage in several kinds of orthogonal interactions with its surroundings. Examples include mechanical interactions, by which a force acts to change some coordinate of the system; chemical interactions, by which amounts of species change either by chemical reaction or by diffusion across boundaries; and thermal interactions, by which the system responds to a temperature difference across the boundary.

Table 1.1 Examples of boundaries between systems and surroundings

Boundary	Constraints on interactions
Open	Any interaction is possible
Closed	Impenetrable by matter, but other kinds of interactions can occur
Semipermeable	Penetrable by some chemical species, but not by others; all other interactions are possible
Insulated	Thermal interactions are not possible, but nonthermal interactions can occur
Rigid	Boundary cannot be mechanically deformed
Isolated	No interactions can occur

Besides nonthermal interactions, the system and surroundings may be connected through a *thermal* interaction. The thermal interaction causes a change in the system by means of a difference in hotness and coldness, which is measured by a temperature difference between system and surroundings. The thermal interaction distinguishes thermodynamics from other branches of science: when the thermal interaction is unimportant or irrelevant, some other branch of knowledge can be applied. For example, in predicting the motions of bodies in the solar system, the interactions are gravitational and classical mechanics describes the motion. For the behavior of electrons in molecules, the interactions are electromagnetic and quantum mechanics applies.

Boundaries separate a system from its surroundings, and the nature of the boundary may limit how the system interacts with its surroundings. Therefore the location and nature of the boundary must be carefully and completely articulated to successfully analyze a system. Boundaries are usually physical entities, such as walls, but they can be chosen to be imaginary. Common boundaries are listed in Table 1.1.

1.1.2 Microscopic Things

Molecular theory asserts that all matter is composed of molecules, with molecules made up of one or more atoms. What evidence do we have for the existence of molecules? That is, why do we believe that matter is ultimately composed of lumps, rather than being continuous on all scales? (For a review of the nineteenth-century debate on the discrete vs. continuous universe, see Nye [4].) One piece of evidence is the law of definite proportions: the elements of the periodic table combine in discrete amounts to form compounds. Another piece of evidence is obtained by shining X rays on a crystalline solid: the resulting diffraction pattern is an array of discrete points, not a continuous spectrum. More evidence is provided by Brownian motion; see Figure 1.2.

Molecules themselves exhibit certain primitive characteristics: (a) they have size and shape, (b) they exert forces on one another, and (c) they are in constant motion at high velocities. Molecules vary in size according to the number and kind of constituent atoms: an argon atom has a "diameter" of about $3.4(10^{-10})$ m; a fully extended octane chain (C_8H_{18}) is about $10(10^{-10})$ m long; the double helix of human DNA (a polymer) is about $20(10^{-10})$ m thick and, when extended, is about 0.04 m long [5].

These microscopic sizes imply that huge numbers of molecules make up a macroscopic chunk of matter: there are about as many molecules in one living cell as there are cells in one common domestic cat [6].

The size and shape of a molecule constitute its *molecular structure*, which is a primary aspect of molecular identity. But identity may not be conserved: in the absence of chemical reactions, identity is preserved at the molecular level, but when reactions do occur, identity is preserved only at the atomic level. Molecular structure results from forces acting among constituent atoms. These forces are of two types: (a) chemical forces, which are caused by sharing of electrons and are the primary determinants of structure, and (b) physical forces, which are mainly electrostatic. Molecular structure is dynamic, not static, because the atoms in a molecule are continually moving about stable positions: the structure ascribed to a molecule is really a time-average over a distribution. In large molecules the structure may be an average over several different "sub-structures" that are formed when groups of atoms rearrange themselves relative to other parts of the molecule. Such rearrangements occur, for example, as internal rotations in alkanes and folding motions in proteins. Molecular structure and its distribution can be distorted by changes in temperature and pressure.

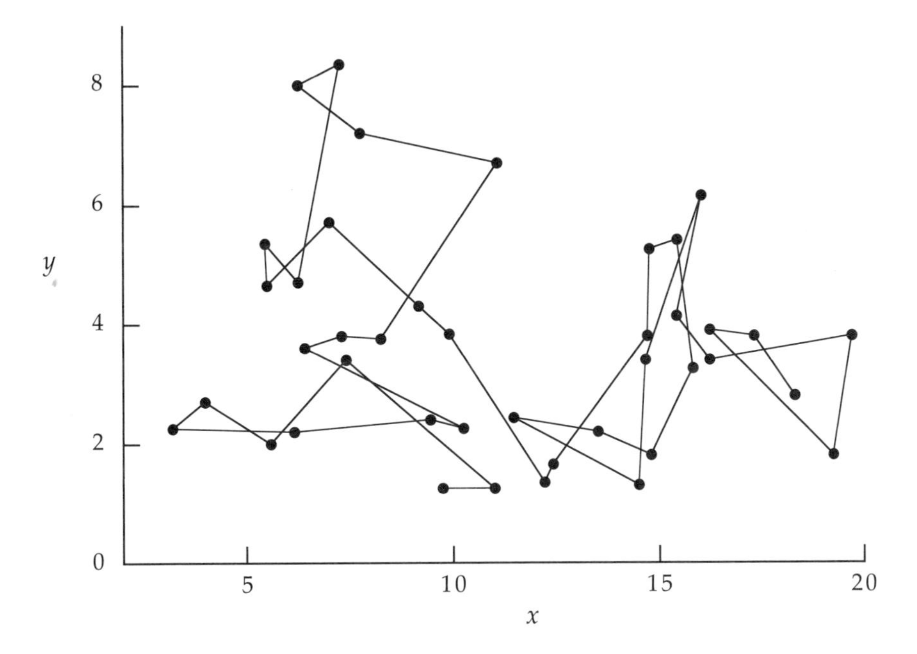

Figure 1.2 One piece of evidence for the existence of molecules is Brownian motion: a small macroscopic particle suspended in a medium will exhibit irregular trajectories caused by the particle colliding with molecules of the medium. The trajectories shown here are from Perrin [7], in which a mastic grain of $1.06(10^{-6})$ m diameter was suspended in a liquid. The dots represent positions of the grain observed at intervals of 30 seconds, with the positions projected onto a horizontal plane (orthogonal to the force of gravity). The straight lines indicate the order of observations; but otherwise, they have no physical significance. (Units on the axes are arbitrary.) Note that this image is incomplete because it is a two-dimensional projection from a three-dimensional phenomenon.

Besides forces acting among atoms on one molecule (*intramolecular forces*), there are also *intermolecular forces* acting between molecules. Such forces depend on distances between molecular centers and, in nonspherical molecules, on the relative orientations of the molecules. When molecules are widely separated, as in a gas, intermolecular forces are small; see Figure 1.3. If we squeeze the gas, it may condense to form a liquid; evidently, when molecules are pushed moderately close together they attract one another. But if we squeeze on the condensate, the liquid resists strongly: when molecules are close together they repel one another. This behavior is typical.

Even a superficial knowledge of molecular structure and intermolecular forces may help us explain why some substances behave as they do. For example, at ambient conditions the chain molecule n-decane $C_{10}H_{22}$ is a liquid, while the double-ring molecule naphthalene $C_{10}H_8$ is solid. This difference is not caused by the small difference in molecular masses—these substances have similar boiling points and critical points. Rather, it is caused by the difference in molecular structure. Differences in structure cause differences in molecular flexibility and in the ability of molecules to pack. Such differences lead to different temperatures at which molecular kinetic energies overcome intermolecular potential energies thereby allowing molecular centers to move enough to produce phase changes; for example, solids melt and liquids vaporize.

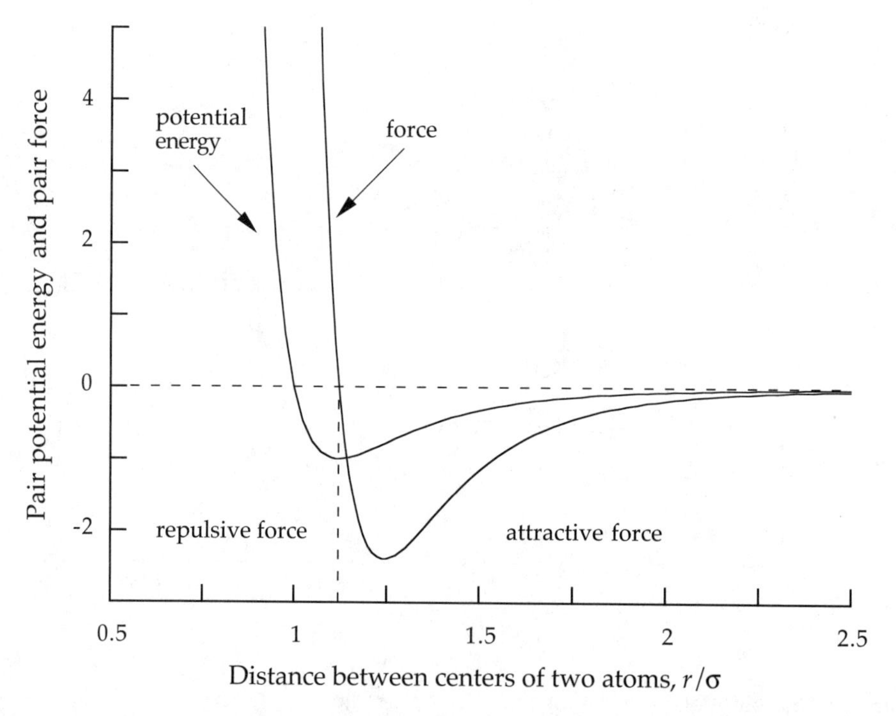

Figure 1.3 Schematic of the potential energy and force acting between two spherical molecules, such as those of argon. When two molecules are far apart, they do not interact, so both the force and the potential energy are zero. When the molecules are close together, their electron clouds are distorted, causing a strong repulsive force. At intermediate separations, the molecules attract one another. Here the scales on ordinate and abscissa are dimensionless. On the abscissa, distances have been divided by σ, which is related to the atomic diameter. On the ordinate, energies were divided by the magnitude of the minimum energy u_{min}, while dimensionless forces were computed as $F\sigma/u_{min}$.

According to kinetic theory, molecules in liquids and gases are continually moving. We see this in Brownian motion, and in some cases, we can sense molecular diffusion: when a bottle is opened, we can soon decide whether it contained ammonia or perfume. Further, molecular motion serves as the mechanism for the thermal interaction.

1.2 PRIMITIVE QUANTITIES

Once we have identified the system, its boundaries, and its interactions with the surroundings, we must describe the condition of the system. This description involves certain quantities, called *properties*, whose values depend only on the current condition. We take properties to be macroscopic concepts; microscopically, there are additional quantities, such as bond lengths, force constants, and multipole moments, that describe molecular structure and define intermolecular forces. These microscopic quantities are not properties, but they contribute to the values taken by properties.

In thermodynamics, we assume properties are continuous and differentiable. These assumptions cannot be rigorously confirmed because sufficient experiments cannot be done to verify them; nevertheless, they allow us to invoke the mathematical limit for transforming discretely distributed data into continuous functions. They seem to fail only in special cases, such as at critical points. These mathematical assumptions are so significant that they could be considered fundamental laws.

1.2.1 Generalized Forces

Recall from § 1.1.1 that we impose changes on a system via thermal and nonthermal interactions. In the case of nonthermal interactions, changes are caused by forces. Common forces and their conjugate nonthermal interactions are listed in Table 1.2. A force has the following characteristics:

(a) It causes or can cause a change in the condition of a system; the change results in a modification of the value of a generalized coordinate.

(b) It can be measured by a balancing procedure; that is, an unknown force is measurable by finding a calibrated standard that stops the action of the unknown force.

Table 1.2 Common macroscopic interactions

Interaction	Generalized coordinate	Conjugate force
Mechanical	System volume	Pressure
Gravitational	Position of a mass	Gravitational field
Interfacial	Area of boundary	Interfacial tension
Chemical	Species mole number	Chemical potential
Electrical	Position of electric charge	Electric field

In classical mechanics forces are said to be *conservative* if they can be written as the negative gradient of some potential energy function. An example is the force F_g exerted on an object in a gravitational field of potential energy E_p,

$$F_g = -\frac{dE_p}{dz} = -\frac{d(mgz)}{dz} = -mg \tag{1.2.1}$$

Here m is the mass of the object, g is the gravitational acceleration, and z is the distance the object's center lies from the center of the field. The negative sign indicates an attractive force and we recognize the result as Newton's second law.

We can extend this idea to thermodynamics by defining any force to be conservative if it is proportional to some thermodynamic potential function differentiated with respect to a generalized coordinate. Under this definition, the forces cited in Table 1.2 are all conservative. A particular example is the pressure involved in the mechanical interaction; in Chapter 2 we will find that

$$P = -\left(\frac{\partial U}{\partial V}\right)_{NS} \tag{1.2.2}$$

where S is the entropy. Here the internal energy U serves as the thermodynamic potential function that connects the generalized coordinate V to its conjugate force P. One of our goals is to identify thermodynamic potential functions for computationally convenient choices of generalized coordinates and their conjugate forces.

Besides conservative forces, there are other forces that are not conjugate to a generalized coordinate through a derivative of some potential function. All such forces are said to be *dissipative*, because they add to the amount of energy needed to change a state; ultimately, that extra energy is dissipated as heat. Common examples are frictional forces that must be overcome whenever one part of a system moves relative to other parts. All real macroscopic forces have dissipative components, and one of the goals of thermodynamics is to account for any energy dissipated as heat.

For the thermal interaction, the force is sometimes identified as the temperature with its generalized coordinate being the entropy [8]. Such an identification provides an obvious and appealing symmetry because it makes thermal interactions appear to be structurally analogous to nonthermal interactions; however, we prefer not to make such an identification because for all known nonthermal interactions the generalized coordinate can be measured, whereas entropy cannot. In this book we will consider only mechanical, gravitational, interfacial, and chemical forces plus the thermal interactions; others will not be used.

1.2.2 Equilibrium and State

The condition of a system is said to be an *equilibrium* one when all forces are in balance and the thermal interaction is not acting, either because it is blocked or because temperatures are the same on both sides of the boundary. These restrictions apply not only to interactions across system boundaries, but also to interactions between system parts. At equilibrium, macroscopic properties do not change with time nor with macroscopic position within a uniform portion of the system. Equilibrium conditions differ from *steady state* conditions. During steady states, net interactions are constant with time, while at equilibrium net interactions are not merely constant, but zero. Moreover, when equilibrium conditions are disturbed by a small interaction, the system tends to resist the interaction; that is, a small disturbance from equilibrium causes only a small bounded change in the system's condition. This is called *Le Chatelier's principle.*

Equilibrium is an idealized concept because everything in the universe is apparently changing on some time-scale (the scales range from femtoseconds to eons). The concept is useful when changes occur on time-scales that are unimportant to the observer. For example, a system may have corroding boundaries or its contents may be decomposing because of electromagnetic radiation (visible or ultraviolet light, for example); it may be expanding via chemical explosion or collapsing under glacial weight. In any situation, we must identify those interactions that occur over the time-scale of our application. "Equilibrium" is said to exist when those interactions are brought into balance. If other interactions are long-lived compared to the time-scale of interest and if, during that time-scale, those interactions have little effect on the system's condition, then those interactions can be ignored.

By stipulating values for a certain number of properties, we establish the condition of the system: the thermodynamic *state*. The number of properties needed depends on such things as the number of parts of the system and the number of chemical species in each part. This issue will be addressed in Chapter 3. When only a few properties are sufficient to identify the state, it may be useful to construct a *state diagram* by plotting independent properties on mutually orthogonal coordinate axes. The dimensionality of this diagram equals the number of properties needed to identify the state.

We say a state is *well-defined* when sufficient property values are specified to locate a system on its state diagram. If, in a well-defined state, the system is at equilibrium, then the condition is said to be an *equilibrium state*. Consequently, all equilibrium states are well-defined, but well-defined states need not be equilibrium states. In fact, a well-defined state may not be physically realizable—it may be thermodynamically unstable or hypothetical or an idealization. For example, many well-defined states of an ideal gas cannot be realized in a laboratory; nevertheless, thermodynamic analyses can be performed on such hypothetical systems.

Since by definition properties depend only on the state, properties are called *state functions*. State functions have convenient mathematical attributes. For example, in the calculus they form exact differentials (see Appendix A); this means that if a system is changed from state 1 to state 2, then the change in any state function F is computed merely by forming the difference

$$\Delta F = F_2 - F_1 \tag{1.2.3}$$

For specified initial (1) and final (2) states, the value of the change ΔF is *always* the same, regardless of how state 2 is produced from state 1. Examples of measurable state functions include temperature, pressure, volume, heat capacity, and number of moles. Properties constitute an important set of primitives, for without state functions, there would be no thermodynamics.

1.2.3 Extensive and Intensive Properties

Thermodynamic properties can be classified in various ways. One classification divides properties into two kinds: extensive and intensive. *Extensive* properties are those whose experimental values *must* be obtained by a measurement that encompasses the entire system, either directly or indirectly. An indirect measurement would apply to systems of disparate parts; measurements would be performed on all the parts and the results added to obtain the total property for the system. Examples include the total volume, the total amount of material, and the total internal energy.

Intensive properties are those whose experimental values *can* be obtained either by inserting a probe at discrete points into the system or (equivalently) by extracting a sample from the system. If the system is composed of disparate parts, values for intensive properties may differ in different parts. Examples of intensive properties are the temperature, pressure, density, and internal energy per mole.

Redlich [2] suggests a simple thought-experiment that allows us to distinguish extensive properties from intensive ones. Let our system be in an equilibrium state, for which values of properties can be assigned, and imagine replicating the system (fancifully, run it through a duplicating machine), while keeping the original state undisturbed. Our new system is now a composite of the original plus the replica. *Extensive* properties are those whose values in the composite *differ* from those in the original system, while *intensive* properties are those whose values are the *same* in both the composite and the original.

These operational distinctions between extensive and intensive avoid ambiguities that can occur in other definitions. Some of those definitions merely say that extensive properties are proportional to the amount of material N in the system, while intensive properties are independent of N. Other definitions are more specific by identifying extensive properties to be those that are homogeneous of degree one in N, while intensive properties are of degree zero (see Appendix A).

But these definitions can lead to ambiguities, especially when we must interpret certain partial derivatives that often arise in thermodynamics. For example, is the system pressure P extensive? Some definitions suggest that P does not change with N, and for a pure substance it is true that

$$\left(\frac{\partial P}{\partial N}\right)_{Tv} = 0 \tag{1.2.4}$$

where $v = V/N$ is the molar volume. That is, here $P = P(T, v)$ does not change when material is added to the system because the container volume V must increase to keep the molar volume v constant. However, it is also true that

$$\left(\frac{\partial P}{\partial N}\right)_{TV} \neq 0 \tag{1.2.5}$$

where the quantity held fixed is the container volume V. In fact, for a pure ideal gas,

$$\left(\frac{\partial P}{\partial N}\right)_{TV} = \frac{RT}{V} \neq 0 \tag{1.2.6}$$

because for an ideal gas $P = NRT/V$. That is, P increases when we increase the amount of an ideal gas while T and container volume V remain fixed. The lesson here is that an intensive property (such as P) may or may not respond to a change in N, depending on which quantities are held fixed when N is changed.

Any extensive property can be made intensive by dividing it by the total amount of material in the system; however, not all extensive properties are proportional to the amount of material. For example, the interfacial area between the system and its boundary satisfies our definition of an extensive property, but this area changes not only when we change the amount of material but also when we merely change the shape of the system. Further, although some intensive properties can be made extensive by multiplying by the amount of material, temperature and pressure cannot be made extensive.

In this book we restrict ourselves to extensive properties that are homogeneous of degree one in the amount of material. Specifically, for a multicomponent system containing component mole numbers $N_1, N_2, \ldots$, we will use only those extensive properties F that are related to their intensive analogs f by

$$F(p_1, p_2, N_1, N_2, \ldots) = Nf(p_1, p_2, x_1, x_2, \ldots) \tag{1.2.7}$$

Here p_1 and p_2 are any two independent intensive properties, the $x_i = N_i/N$ are mole fractions, and $N = \Sigma N_i$. Therefore, if we fix values for p_1 and p_2 while doubling all mole numbers, then values for all extensive properties F double. However, we do not expect that (1.2.7) is either necessary or sufficient for identifying extensive properties.

One motivation for distinguishing extensive from intensive is that the intensive thermodynamic state does not depend on the amount of material. The same intensive state can be attained in a hot toddy the size of a tea cup or the size of a swimming pool. This means we can perform a single analysis using intensive variables, but then apply the results to various systems of different sizes.

1.2.4 Measurables and Conceptuals

Thermodynamic analyses are also helped by another classification of properties: one that distinguishes measurables from conceptuals. *Measurables* are properties whose values can be determined directly from an experiment; these are the properties of ultimate interest because they can be monitored and controlled in an industrial setting. Examples are temperature, pressure, total volume, mole fraction, surface area, and electric charge. *Conceptuals* are properties whose values cannot be obtained directly from experiment; their values must be obtained by some mathematical procedure applied to measurables. (In some cases we can contrive special experimental situations so that a *change* in a conceptual can be measured.) Conceptuals simplify thermodynamic analyses; for example, conceptuals often simplify those basic equations that describe Nature's constraints on a system or process. The common conceptuals are energy, entropy, the Gibbs energy, chemical potential, fugacity, and activity coefficient.

Conceptuals play an intermediate role in engineering practice; they are a means to an end. For example, assume we are to diagnose and correct a process (perhaps a distillation column) that is behaving abnormally (improper product concentration in the overhead). To document the abnormality, we collect data on certain measurables (say temperature, pressure, and composition). We translate these measurements into values for conceptuals (such as energies and fugacities) and perform an analysis that reveals the source of the abnormality (perhaps insufficient heat supplied). Then using relations between conceptuals and measurables, we formulate a strategy for correcting the problem; the strategy is implemented via measurables and interactions.

1.3 PRIMITIVE CHANGES

The engineer's task is not merely to describe the current thermodynamic state of a system; an engineer must also anticipate how that state will respond when conditions in the surroundings change. A related problem is also important; that is, an engineer may need to decide how to manipulate conditions in the surroundings to produce a desired change in the system. For example, consider a vapor in equilibrium with an equimolar mixture of ethanol and water initially at 1 bar. Say we want to increase the pressure to 10 bar, while preserving the two phases and the equimolar composition in the liquid. The thermodynamic problem is to identify the new temperature and new vapor composition, but the engineering problem is to identify the valve settings needed to achieve the desired final state. Any time a system moves from one equilibrium state to another, the change is called a *process*. Processes include all kinds of physical changes, which are typically monitored by changes in temperature, pressure, composition, and phase; moreover, processes can also include chemical changes—changes in molecular identities—which occur during chemical reactions.

Possible processes are limited by the nature of system boundaries and by conditions in the surroundings. The kinds of processes allowed by particular boundaries are listed in Table 1.3. Often we cause a particular process to occur by bringing the system into contact with a *reservoir* that forces a particular system property to remain constant. Common reservoirs include the thermal (or heat) reservoir, which maintains the system at a constant temperature (an isothermal process), and the mechanical reservoir, which imposes its pressure on the system (isobaric process).

We will find it useful to identify certain limiting cases of processes. To facilitate the discussion, we introduce the following notation. Let Δ represent the net total of all

Table 1.3 Typical boundaries and reservoirs with their corresponding processes

Boundary or reservoir	Process
Closed	Constant mass
Thermally insulated	Adiabatic
Rigid	Constant volume (isometric)
Closed and rigid	Constant density (isochoric)
Closed, rigid, insulated	Constant energy
Heat reservoir	Constant temperature (isothermal)
Mechanical reservoir	Constant pressure (isobaric)

driving forces acting on a system, and let δ be the differential analog of Δ. In general, the driving forces can be divided into two types: *external* forces Δ_{ext} that act across system boundaries and *internal* forces Δ_{int} that act within the system but between different parts of it. As a result, we can write

$$\Delta = \Delta_{ext} + \Delta_{int} \tag{1.3.1}$$

Moreover, any driving force may be composed of both conservative and dissipative components; we let $\mathcal{F}$ represent all dissipative components of the driving forces.

We first define the *static limit* of any process as that produced when all net driving forces are removed,

$$\lim_{\Delta \to 0} (\text{process}) = \lim_{\substack{\Delta_{ext} \to 0 \\ \Delta_{int} \to 0}} (\text{process}) = \begin{pmatrix} \text{equilibrium} \\ \text{state} \end{pmatrix} \tag{1.3.2}$$

This means that in the static limit, we expect any process to degenerate to an equilibrium state: a physically realizable point on a state diagram. But note that to achieve equilibrium, all external and internal driving forces must be zero. In general, an equilibrium state is not obtained by taking only the external driving forces to zero; for example, an isolated system need not be at equilibrium, nor need its state even be well-defined.

In some (troublesome) situations, taking all external driving forces to zero does result in a well-defined state, but the presence of internal driving forces precludes equilibrium. These states can often be identified by administering a small disturbance. For example, by careful addition, we may create a supersaturated solution of sugar in water. When all net external driving forces are brought to zero, the state is well-defined: the solution is a single liquid phase at a definite temperature, pressure, and composition. However, this well-defined state is not at equilibrium; in supersaturated solutions there exist internal driving forces tending to produce a new phase, although this tendency is kinetically limited. But if we disturb the solution, perhaps by adding a small crystal of sugar, those internal driving forces are relieved by rapid formation of solid sugar.

If, instead of taking all driving forces to zero, we make them differential, then we say the process is *quasi-static*,

$$\lim_{\Delta \to \delta} (\text{process}) = \begin{pmatrix} \text{quasi-static} \\ \text{process} \end{pmatrix} \tag{1.3.3}$$

Differential driving forces produce a differential process; however, we can contrive a finite process by stringing together a sequence of quasi-static steps. From an equilibrium state (a point on a state diagram) we use differential driving forces to take a step, then we let the system relax back to equilibrium. This new equilibrium condition locates a new point on a state diagram. Repeating the sequence (differential step + relaxation to equilibrium) many times, we generate a series of points that represent a

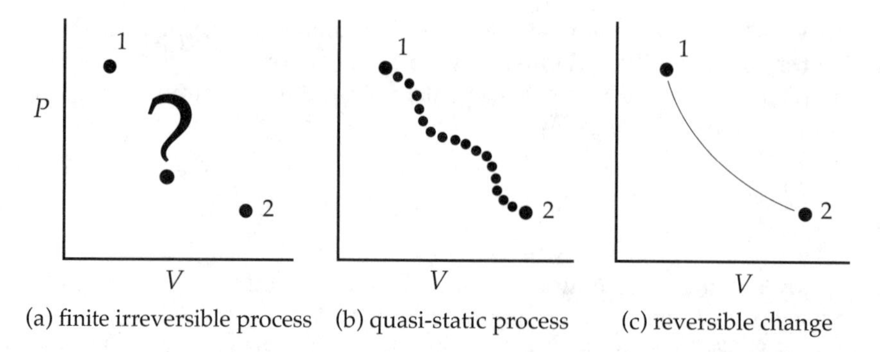

Figure 1.4 Comparison of changes of state as represented on a state (PV) diagram for a pure, one-phase substance. During an (a) irreversible process, intermediate states are unknown and unknowable; during a (b) quasi-static process, the system moves in small discrete steps between identifiable equilibrium states; during a (c) reversible change, every intermediate state is a well-defined equilibrium state.

process path on a state diagram. Such a quasi-static process is illustrated schematically in the middle panel of Figure 1.4.

Even though a quasi-static process is driven differentially, the driving forces may still contain dissipative components. These components may arise because some properties have finite differences across boundaries or they may arise from differential effects accumulated over a finite process. If we could remove all dissipative components $\mathcal{F}$, so the process would be driven only by conservative forces, then the change of state would be *reversible*. This reversible limit can be expressed as

$$\lim_{\mathcal{F} \to 0} (\text{process}) = \begin{pmatrix} \text{reversible} \\ \text{change} \end{pmatrix} \tag{1.3.4}$$

Formally, this limit is sufficient to define a reversible change, but in practice the dissipative components $\mathcal{F}$ can be made to vanish only by simultaneously making the total driving force Δ vanish. To remind ourselves of this, we rewrite (1.3.4) in the form

$$\lim_{\substack{\mathcal{F} \to 0 \\ \Delta \to 0}} (\text{process}) = \begin{pmatrix} \text{reversible} \\ \text{change} \end{pmatrix} \tag{1.3.5}$$

To the degree that a reversible change is viewed as a process, analogous to a quasi-static process, the following distinction occurs: if the dissipative forces can be made to vanish, $\mathcal{F} \to 0$, then the driving forces must also vanish, $\Delta \to 0$; however, the converse is not necessarily true. That is, if $\Delta \to 0$, then we may or may not also have $\mathcal{F} \to 0$. In other words, a reversible change has quasi-static characteristics, but a quasi-static process need not be reversible [9]. Since, in the reversible limit, all driving forces are taken to zero, every state visited during a reversible change is an equilibrium state; hence, a reversible change can be represented by a continuous line on a state diagram.

Now we address the apparent contradiction between the limit in (1.3.5) and that in (1.3.2): both have $\Delta \to 0$, but with different results. The resolution is that the static limit in (1.3.2) can describe a real process, while the reversible limit in (1.3.5) is an idealization. That is, a reversible "process" is not really a process at all [10], it is only a continuous sequence of equilibrium states on a state diagram. We emphasize this distinction by calling the reversible limit a *reversible change*, not a reversible "process."

In a reversible change, no energy is used to overcome dissipative forces, so a reversible path from initial state 1 to final state 2 can also be traversed in the opposite direction, returning both system and surroundings to their initial conditions. The equilibrium states visited during the process 2-1 are identical to those visited during 1-2, just in reverse order. Although the reversible change is an unrealizable idealization, it is useful because (i) it allows calculations to be done using only system properties and (ii) it provides bounds on energy requirements for a process.

All real processes are in fact *irreversible*: they proceed in a finite time and are not a continuous string of equilibrium states. Typically, an irreversible process involves a stage during which the state of the system *cannot* be identified, as in the top part of Figure 1.5. Irreversible processes are driven by macroscopic property gradients across system boundaries, so that in practice no real change can be reversed without causing some change in the surroundings. That is, irreversible processes involve dissipative forces, such as friction and turbulence, which must be overcome to return the system to any previous state. The magnitudes of dissipative forces depend on system state and on the magnitudes of property gradients; these determine the *degree* of irreversibility. Strongly irreversible processes are less efficient than weakly irreversible ones. Often, highly irreversible processes are driven by large gradients, which make the process proceed quickly: fast processes are usually more irreversible than slow ones. But process speed may not correlate with gradient size; for example, if a boundary poses a large resistance, then even a slow process may require a large driving force.

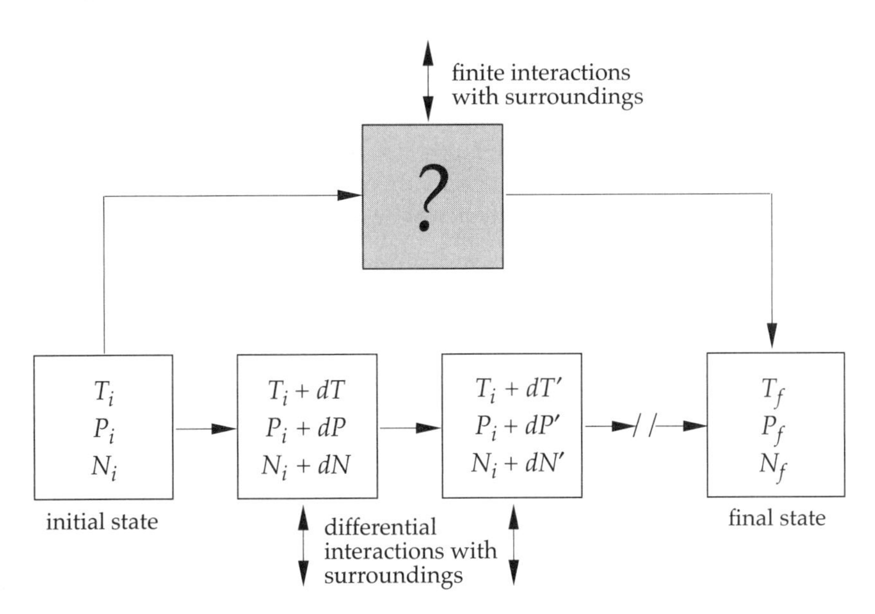

Figure 1.5 Schematic of a quasi-static process compared with a finite irreversible process. The system initially in a state having properties T_i, P_i, and N_i is to be changed to a final state having T_f, P_f, and N_f. In the finite irreversible process (*top*) the system passes through intermediate states that are undefined. During the quasi-static process (*bottom*) the change occurs in differential stages; at the end of each stage the system is allowed to relax to an intermediate state that is well-defined. In both processes, overall changes in state functions, such as $\Delta T = T_f - T_i$ and $\Delta P = P_f - P_i$, are the same.

1.4 PRIMITIVE ANALYSES

In later chapters, much of our attention will focus on analyzing how a system responds to a process. The primitive stages of an analysis lead to a sketch or diagram that helps us visualize the system and the processes acting on it. We divide such sketches into two general classes: one for closed systems (§ 1.4.1), the other for open systems (§ 1.4.2). For closed systems, no further primitive concepts apply, and a thermodynamic analysis proceeds as described in Chapter 2. But for open systems, the sketch can be enhanced by invoking one additional primitive concept: equations that represent system inventories. These equations are discussed in § 1.4.2.

1.4.1 Closed-System Analyses: Two-Picture Problems

When processes are applied to closed systems, we can usually identify the system state at two or more different times. The diagrams in Figure 1.4 and schematics in Figure 1.5 are of this type; in those examples, we know the initial and final states of the system. Intermediate states may be knowable (reversible) or unknowable (irreversible); nevertheless, the identities of two states may be sufficient to allow us to analyze the change. We call these situations "two-picture" problems because the primitive analysis leads us to sketches representing two (or more) system states.

1.4.2 Open-System Analyses: One-Picture Problems

When streams are flowing through an open system, a primitive analysis leads us to represent the situation by a single sketch, perhaps like that in Figure 1.6. We call this a "one-picture" problem. In these situations we can extend the primitive analysis to include equations that represent inventories on selected quantities. We develop those equations here.

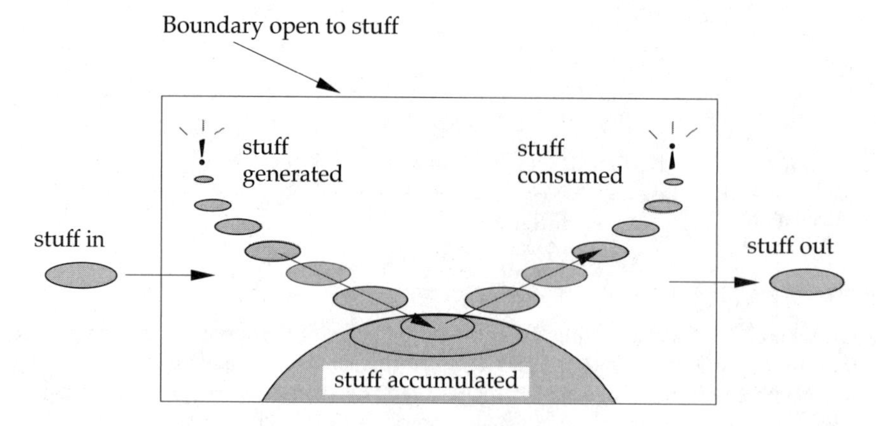

Figure 1.6 Schematic representation of terms appearing in the stuff equation (1.4.1). The amount of stuff accumulated in a system may change because of stuff added to the system, stuff removed from the system, stuff generated in the system, or stuff consumed in the system.

For the system in Figure 1.6, the boundary allows transfer of some quantity which, for generality, we call *stuff*. By identifying all ways by which the amount of stuff may change, we obtain a general balance equation, which we call the *stuff equation* [11],

$$\begin{pmatrix}\text{Rate of}\\ \text{stuff into}\\ \text{system by}\\ \text{interactions}\end{pmatrix} - \begin{pmatrix}\text{Rate of}\\ \text{stuff out}\\ \text{of system by}\\ \text{interactions}\end{pmatrix} + \begin{pmatrix}\text{Rate of}\\ \text{generation}\\ \text{of stuff}\\ \text{in system}\end{pmatrix} - \begin{pmatrix}\text{Rate of}\\ \text{consumption}\\ \text{of stuff}\\ \text{in system}\end{pmatrix} = \begin{pmatrix}\text{Rate of}\\ \text{accumulation}\\ \text{of stuff}\\ \text{in system}\end{pmatrix} \tag{1.4.1}$$

In general the stuff equation is a differential equation and its accumulation term can be positive, negative, or zero; that is, the amount of stuff in the system may increase, decrease, or remain constant with time. In a particular situation several kinds of stuff may need to be inventoried; examples include molecules, energy, and entropy.

The stuff equation applies to both conserved and non-conserved quantities. Conserved quantities can be neither created nor destroyed; so, for such quantities the stuff equation reduces to a *general conservation principle*

$$\begin{pmatrix}\text{Rate of conserved}\\ \text{stuff into}\\ \text{system by}\\ \text{interactions}\end{pmatrix} - \begin{pmatrix}\text{Rate of conserved}\\ \text{stuff out}\\ \text{of system by}\\ \text{interactions}\end{pmatrix} = \begin{pmatrix}\text{Rate of}\\ \text{accumulation of}\\ \text{conserved stuff}\\ \text{in system}\end{pmatrix} \tag{1.4.2}$$

One important conservation principle is provided by molecular theory: atoms are conserved parcels of matter. (We ignore subatomic processes such as fission or fusion and consider only changes that do not modify the identities of atoms.) At the macroscopic level this conservation principle is the mass or *material balance*

$$\begin{pmatrix}\text{Rate of}\\ \text{material into}\\ \text{system by}\\ \text{interactions}\end{pmatrix} - \begin{pmatrix}\text{Rate of}\\ \text{material out}\\ \text{of system by}\\ \text{interactions}\end{pmatrix} = \begin{pmatrix}\text{Rate of}\\ \text{accumulation}\\ \text{of material}\\ \text{in system}\end{pmatrix} \tag{1.4.3}$$

If, instead of total material, the inventory is to be conducted on chemical species (moles), then (1.4.3) continues to apply, so long as chemical reactions are not occurring. If reactions occur, then mole numbers may change and (1.4.1) would apply rather than (1.4.3). So, in the absence of nuclear processes, mass is *always* conserved, but moles are generally conserved only in the absence of chemical reactions.

If during a process the rates of accumulation, generation, and consumption are all zero, then the process is said to be in *steady state* with respect to transfer of that particular stuff. In such cases the general differential balance (1.4.1) reduces to a simple algebraic equation

$$\begin{pmatrix}\text{Steady rate}\\ \text{of stuff into}\\ \text{system}\end{pmatrix} - \begin{pmatrix}\text{Steady rate}\\ \text{of stuff out}\\ \text{of system}\end{pmatrix} = 0 \tag{1.4.4}$$

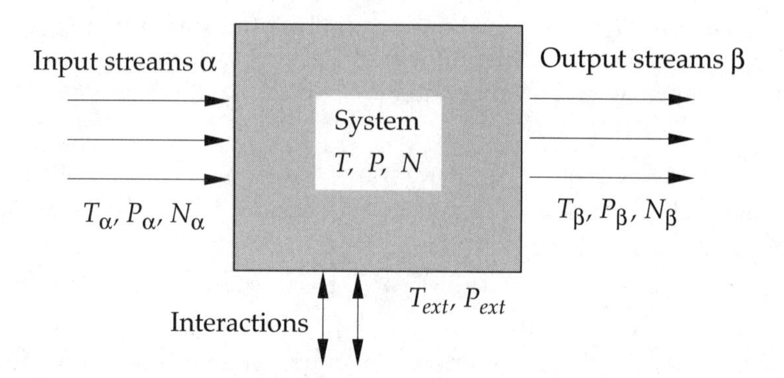

Figure 1.7 Schematic of a one-picture situation in which an open system exchanges mass with its surroundings via input and output streams. In addition, the system exchanges energy with its surroundings via thermal and nonthermal interactions. T_{ext} is the temperature and P_{ext} is the pressure on the external side of the boundary at the point where energy transfers occur.

If the process is not a steady state, then it is a *transient*, and the system either gains (rate of accumulation > 0) or loses (rate of accumulation < 0) stuff over time. In the analysis of any real process, the appropriate form (1.4.1)–(1.4.3) must be identified and integrated. For some processes the integration can readily be done analytically, such as for steady states (1.4.4), but others may require elaborate numerical treatments.

The one-picture approach generalizes to situations in which mass and energy enter and leave the system simultaneously, as in Figure 1.7. Mass may enter the system through any number of input streams and leave through additional output streams. Each stream may have its own temperature, pressure, composition, and flow rate. Further, energy may also be transferred to and from the system via thermal and nonthermal interactions with the surroundings. In such situations, we can write a stuff equation for each molecular species and a separate, independent stuff equation for energy, as we shall see in Chapter 2.

1.5 SUMMARY

We have reviewed the primitive things, quantities, changes, and analyses that form the basis for thermodynamics as it is developed in this book. Whenever possible we have offered definitions of the primitives, but in every case we moved beyond simple definitions: we tried to show *why* each primitive is important, and we tried to clarify subtleties that often surround certain primitives.

At the macroscopic level, primitive things include the system and the boundary that separates the system from its surroundings. Macroscopic things also include the thermal and nonthermal interactions by which we stand in the surroundings and either measure something in the system or cause a change in the system. Macroscopic things are composed of microscopic things—molecules, atoms, and the forces that act among them. Although classical thermodynamics is a purely macroscopic discipline, we will, when it is economical to do so, use molecular arguments to help explain macroscopically observed behavior. Moreover, molecular theory is now used as a basis for developing many thermodynamic models; to use those models properly, we need some appreciation of molecular theory.

Primitive quantities include generalized forces, the concepts of equilibrium and state, and ways to classify properties. The ideas surrounding force, equilibrium, and state are absolutely crucial because they identify those situations which are amenable to thermodynamic analysis. We will have much more to say about these concepts; for example, we want to devise quantitative ways for identifying the state of a system and for deciding whether the system is at equilibrium. Although classifications of properties are not crucial, the classifications—extensive and intensive or measurable and conceptual—facilitate our development and study of the subject.

Changes in a system state are caused by interactions, and we focused on the distinction between reversible changes and irreversible processes. The importance of this distinction cannot be overemphasized because its implications seem to often be misunderstood. The implications can contribute to engineering practice; for example, calculations for reversible changes require values only for differences in system properties, but calculations for irreversible processes require values for quantities of both system and surroundings. Consequently, calculations for reversible changes are nearly always easier than those for irreversible processes. We prefer easy calculations.

Although reversible changes are idealizations—real processes are *always* irreversible—they can be useful. In some situations the value of a quantity computed for a reversible change is *exactly* the same as that for an irreversible process, so we calculate the quantity using the reversible change. In other situations the values computed for a reversible change *bound* the values for the irreversible process, and those bounds may contribute to an engineering design or to the operation of a production facility. In still other situations, an efficiency for a real irreversible process may be known relative to that for a reversible change; then, we compute quantities for the reversible change and apply the efficiency factor to obtain the value for the real process.

These uses are important to a proper *application* of thermodynamics in real situations. But in addition, the distinction between reversible and irreversible lies at the core of the *science* of thermodynamics; for example, what happens to the energy that is wasted in irreversible processes? This is a purely thermodynamic question.

Finally, we discussed the primitive steps in beginning an analysis that will determine how a system responds to processes. Those primitive steps culminate either in a two-picture diagram for closed systems or in a one-picture diagram for open systems. In addition, for open systems we identified forms of a general balance equation that apply to any kind of stuff that may cross system boundaries. With all these primitive concepts in place, we can begin the uphill development of thermodynamics.

LITERATURE CITED

[1] O. Redlich, "Fundamental Thermodynamics Since Carathéodory," *Rev. Mod. Phys.*, **40**, 556 (1968).

[2] O. Redlich, *Thermodynamics: Fundamentals, Applications*, Elsevier, Amsterdam, 1976, ch. 1.

[3] O. Redlich, "The Basis of Thermodynamics," in *A Critical Review of Thermodynamics*, E. B. Stuart, A. J. Brainard, and B. Gal-Or (eds.), Mono Book Corp., Baltimore, 1970, p. 439.

[4] M. J. Nye, *Molecular Reality: A Perspective on the Scientific Work of Jean Perrin*, Macdonald, London, 1972.

[5] G. S. Kutter, *The Universe and Life*, Jones and Barlett, Boston, 1987, pp. 266, 290, 372.

[6] S. Vogel, *Life's Devices*, Princeton University Press, Princeton, NJ, 1988.

[7] J. Perrin, *Atoms*, D. L. Hammick (transl.), van Nostrand, New York, 1916, p. 115.

[8] S. K. Ma, *Statistical Mechanics*, World Scientific, Philadelphia, 1985, ch. 2.

[9] J. Kestin, *A Course in Thermodynamics*, vol. 1, revised printing, Hemisphere Publishing Corporation, New York, 1979, ch. 4, pp. 133–34.

[10] A. Sommerfeld, *Thermodynamics and Statistical Mechanics*, Academic Press, New York, 1955, p. 19.

[11] The name *stuff equation* for the general balance equation is not ours, but we are embarrassed to report we don't know who originated the idea.

[12] B. E. Poling, J. M. Prausnitz, and J. P. O'Connell, *Properties of Gases and Liquids*, 5th ed., McGraw-Hill, New York, 2001.

[13] Yu. Ya. Fialkov, *The Extraordinary Properties of Ordinary Solutions*, MIR, Moscow, 1985, p. 19.

PROBLEMS

1.1 For each of the following situations, identify (i) the system, (ii) the boundaries, (iii) the surroundings, and (iv) the kinds of interactions that can occur. Do *not* attempt to solve the problem.

(a) Hot coffee is placed in a vacuum bottle and the top is sealed. Estimate the temperature of the coffee after 4 hours.

(b) A can of your favorite beverage, initially at room temperature, is placed in a freezer. How long must the can remain there to cool the liquid to 40°F?

(c) A bottle of soda is capped when the pressure of carbonation is 0.20 MPa. How long before the pressure has dropped to 0.11 MPa?

(d) Each tire on a car is charged with air to 0.20 MPa. The car is then driven for 300 km at an average speed of 100 km/h. Estimate the tire pressure at the end of the trip.

(e) If the price of electric power is $0.10 per kWh in Denver, what is the cost of heating 500 cm^3 of water from 300 K to boiling on an electric stove in Denver?

(f) At the end of the trip in part (d), a pinhole leak develops in the car's radiator and coolant is being lost at the rate of 3 l/hr. Is the leaking coolant vapor or liquid? Ten minutes later, has the engine temperature increased or decreased?

(g) Tabitha the Untutored put her birthday balloon near a sunny window and, for the next few days, observed interesting behavior: each afternoon the balloon was closer to the ceiling than it was in the morning, and each day its maximum height was less than the day before. What was the maximum temperature of the balloon each day, and how many days passed before the balloon failed to rise from the floor?

1.2 For each situation in Problem 1.1, discuss how you would use abstraction (i.e., simplifying assumptions) to make the *system* amenable to analysis. Do not attempt to abstract the *process*. Estimate the order of magnitude of the error introduced by each simplification.

1.3 For each situation in Problem 1.1, cite the *process* involved. What abstraction (i.e., simplifying assumptions) could you use to make each process amenable to analysis? Would your abstraction make the estimate for the desired quantity too large or too small? What additional data would you need to solve each problem?

1.4 For each process in Problem 1.1, cite those aspects that are dissipative.

1.5 How would you determine whether the thermodynamic state of a system depended on the shape of its boundary? If you found that it did, what would be the consequences?

1.6 If energy is a conceptual and not measurable, then what is being measured in kilowatt-hours by that circular device (with the rotating disc) on the exterior of most American houses?

1.7 Using only what you know at this moment, and without referring to any resource, estimate the diameter of one water molecule. Clearly state any assumptions made and estimate the uncertainty in your answer.

1.8 According to kinetic theory, the root-mean-square (rms) velocity of an atom in a monatomic fluid is related to the absolute temperature by $v_{rms} = (3kT/m)^{1/2}$ where m is the mass of one atom, k is the Boltzmann constant, $k = R/N_A = 1.381(10^{-23})$ J/(K molecule), and N_A is Avogadro's number. Compute the rms velocity (in km/hr) for one argon atom at 300 K.

1.9 At atmospheric pressure aqueous mixtures of simple alcohols exhibit the following kinds of phase behavior. Explain these using molecular forces and structure.

(a) Methanol and water mix in all proportions and do not exhibit an azeotrope.

(b) Ethanol and water mix in all proportions and form an azeotrope when the mixture is nearly pure ethanol.

(c) Normal propanol mixes with water in all proportions, as does isopropanol, and both mixtures form azeotropes near the equimolar composition. The n-propanol azeotrope has a higher concentration of water than does the isopropanol azeotrope.

(d) Normal butanol and isobutanol are each only partially miscible in water; however, at pressures above ambient, each butanol mixes with water in all proportions and each exhibits an azeotrope.

(e) 2-methyl-2-propanol and trimethylmethanol each mix with water in all proportions and form azeotropes at compositions near pure water. The 2-methyl-2-propanol azeotrope has a higher concentration of alcohol than does the trimethylmethanol azeotrope.

1.10 Following are the melting (T_m), boiling (T_b), and critical (T_c) temperatures for benzene, cyclohexane, decane, and naphthalene. Explain the trends in terms of molecular structure and forces. Data from [12].

Substance	Mol. wt.	T_m(°C)	T_b(°C)	T_c(°C)
Benzene	78.1	5.5	80.1	288.9
Cyclohexane	84.2	6.5	80.6	280.3
Naphthalene	128.2	78.2	128.0	475.2
Decane	142.3	–29.7	174.1	344.5

1.11 Following are the melting (T_m), boiling (T_b), and critical (T_c) temperatures of the normal alkanes from C_1 to C_{10}. Explain the trends in terms of molecular structure and forces. Data from [12].

Substance	Mol. wt.	T_m(K)	T_b(K)	T_c(K)
Methane	16.	40.7	111.7	190.6
Ethane	30.1	90.4	184.6	305.3
Propane	44.1	91.4	231.0	369.8
Butane	58.1	134.8	272.7	425.1
Pentane	72.	143.4	309.2	469.7
Hexane	86.2	177.8	341.9	507.6
Heptane	100.2	182.6	371.6	540.2
Octane	114.2	216.4	389.8	568.7
Nonane	128.3	219.7	424.0	594.6
Decane	142.3	243.5	447.3	617.7

1.12 Following are the melting (T_m) and boiling (T_b) temperatures of selected hydrides from Group VI of the periodic table. Explain the trends in terms of molecular structure and forces. Data from [13].

Substance	Formula	Mol. wt.	T_m(°C)	T_b(°C)
Hydrogen telluride	H_2Te	130	–51	–4
Hydrogen selenide	H_2Se	81	–61	–42
Hydrogen sulfide	H_2S	34	–82	–61
Hydrogen oxide	H_2O	18	0	100

1.13 The ideal-gas equation of state, $PV = NRT$, applies to certain pure gases and their mixtures. Consider such a mixture confined in a closed, rigid vessel, and containing N_i moles of each component i. Determine the expression that would allow you to compute how the pressure would change when a small amount of pure substance i is added to the mixture at fixed temperature; that is, find the expression for the following partial derivative,

$$\left(\frac{\partial P}{\partial N_i}\right)_{TVN_{j \neq i}}$$

1.14 Consider a binary gas mixture that obeys the virial equation of state,

$$\frac{Pv}{RT} = 1 + \frac{[x_1^2 B_{11}(T) + 2x_1 x_2 B_{12}(T) + x_2^2 B_{22}(T)]}{v}$$

Here, $v = V/N$, each x_i is a mole fraction for one component, $x_i = N_i/(N_1 + N_2)$, and the B_{ij} are called second virial coefficients; they are intensive properties that depend only on temperature. Derive the expression for the partial derivative

$$\left(\frac{\partial P}{\partial N_1}\right)_{TVN_2}$$

which has a physical interpretation analogous to that given for the partial derivative in Problem 1.13.

1.15 Use the stuff equation for mass to obtain equations for the following situations.

(a) A mixing device is steadily fed material through streams 1 and 2, and it steadily discharges material through stream 3. The flow rates $\dot{N}_1$ and $\dot{N}_3$ are known; write an equation for the feed rate of stream 2.

(b) A mixing device is steadily fed by streams 1 and 2 for a duration τ; this device has no discharge. If the feed rates $\dot{N}_1$ and $\dot{N}_2$ are known, write an equation for the total moles accumulated.

(c) A mixing device is fed by streams 1 and 2 for a duration τ; however, the feed rates are not steady, but are proportional to time, t,

$$\dot{N}_1 = a_1 t \qquad \text{and} \qquad \dot{N}_2 = a_2 t$$

If the constants a_1 and a_2 are known, write an equation for the change in the moles accumulated.

(d) A chemical reactor is steadily fed by two streams: one feeds pure reactant A at rate $\dot{N}_A$, the other feeds pure reactant B at rate $\dot{N}_B$. In the reactor, A and B combine to form product C (i.e., A + B $\rightarrow$ C), and C is discharged at a steady rate. Given the fractional conversion of A, $\alpha_A = (N_{Ain} - N_{Aout})/N_{Ain}$, write an equation for $\dot{N}_C$, the rate at which C is discharged.

2

THE FIRST AND SECOND LAWS

Much of thermodynamics concerns the causes and consequences of changing the state of a system. For example, you may be confronted with a polymerization process that converts esters to polyesters for the textile industry, or you may need a process that removes heat from a chemical reactor to control the reaction temperature and thereby control the rate of reaction. You may need a process that pressurizes a petroleum feed to a flash distillation unit, or you may need a process that recycles plastic bottles into garbage bags. In these and a multitude of other such situations, a system is to be subjected to a process that converts an initial state into some final state.

Changes of state are achieved by processes that force the system and its surroundings to exchange material or energy or both. Energy may be exchanged directly as heat and work; energy is also carried by any material that enters or leaves a system. A change of state may involve not only changes in measurables, such as T and P, but it may also involve phase changes and chemical reactions. To design and operate such processes we must be able to predict and control material and energy transfers.

Thermodynamics helps us determine energy transfers that accompany a change of state. To compute those energetic effects, we can choose from two basic strategies, as illustrated in Figure 2.1. In the first strategy we directly compute the heat and work that accompany a process. But to perform such calculations, we must know the process path that the system follows from the initial to the final state. That is, heat and work are process-dependent quantities. Unfortunately, the path can be properly characterized only for reversible changes. All real processes are irreversible and rarely do we know enough about the process to be able to directly compute heat and work.

In the second strategy we avoid direct computations of heat and work by reformulating our problem in terms of thermodynamic state functions. State functions depend on the condition of the system, not on the process; for example, changes in state functions are determined only from the initial and final states of the system (see Figure 2.2). State functions simplify thermodynamic calculations because they allow us to analyze ill-defined real processes in terms of well-defined changes of state. So long as the initial and final system states are the same, then we can compute changes in state functions along any computationally convenient path.

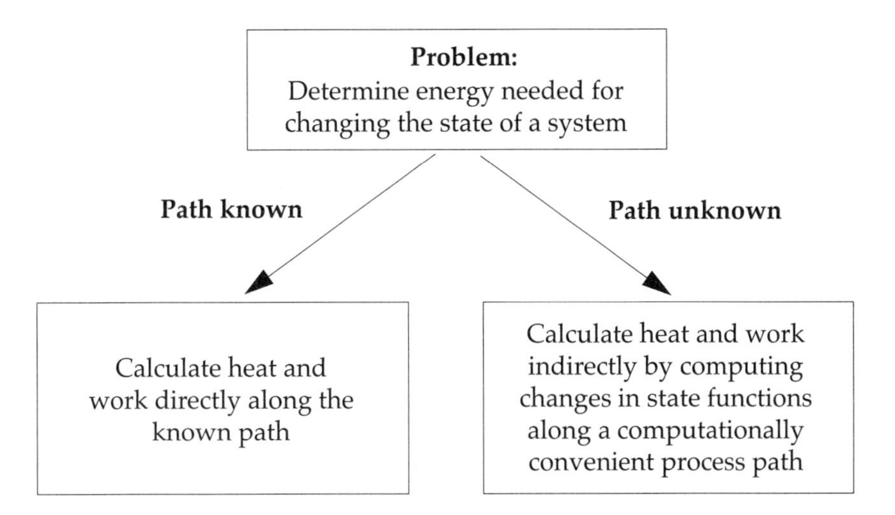

Figure 2.1 Thermodynamics provides two basic strategies for computing energy requirements associated with changes of state in closed systems. When we know the process path then heat and work effects can be computed directly. But more often the process path is not known, and then we compute changes in state functions along a *convenient* path.

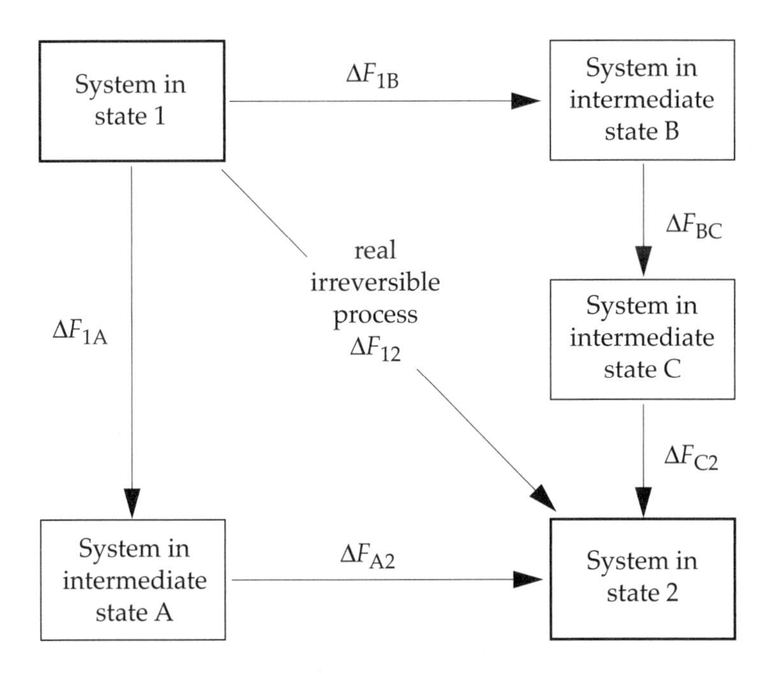

Figure 2.2 A principal value of thermodynamics is that process diagrams, like this one, commute: along any indirect, computationally convenient process-path, the change in any state function F is the same as its change along the direct irreversible path: $\Delta F_{12} = \Delta F_{1A} + \Delta F_{A2} = \Delta F_{1B} + \Delta F_{BC} + \Delta F_{C2}$.

In this chapter we develop expressions that relate heat and work to state functions: those relations constitute the first and second laws of thermodynamics. We begin by reviewing basic concepts about work (§ 2.1); that discussion leads us to the first law (§ 2.2) for closed systems. Our development follows the ideas of Redlich [1]. Then we rationalize the second law (§ 2.3) for closed systems, basing our arguments on those originally devised by Carathéodory [2–4]. Finally, by straightforward applications of the stuff equations introduced in § 1.4, we extend the first and second laws to open systems (§ 2.4).

2.1 WORK

In this section we review those general features of work that lead to the first law of thermodynamics. We start with fundamental ideas about mechanical work (§ 2.1.1), then consider the work that causes a change of system volume (§ 2.1.2), and we offer an example (§ 2.1.3). Finally, we discuss experimental observations about adiabatic work that serve as the foundation for the first law (§ 2.1.4).

2.1.1 Work to Displace a System

As our system, consider a macroscopic object, say a table. To move the table a distance x, we must apply a force F and thereby expend an amount of work W. When the force is exerted in the direction of the motion (F is parallel to x), then the work is given by

$$W = \int F \, dx \tag{2.1.1}$$

When F and x are not parallel, we replace the integrand in (2.1.1) with the component of F that is parallel to x; this may be found by forming the vector dot product between the force vector and a unit vector parallel to x. For a differential change in position dx, we need only a small amount of work δW,

$$\delta W = F \, dx \tag{2.1.2}$$

Before going further we choose a sign convention for W. In this book we consistently make quantities positive when they are added to the system; therefore, we use

$W > 0$, when work is done on the system, and

$W < 0$, when the system does work on its surroundings.

This choice is arbitrary, so when studying thermodynamics you must take care to identify the convention that applies to the material at hand.

Notions of work can be illustrated by considering mechanical situations in which we want either to change the position of our system in the earth's gravitational field or to change the velocity of our system. For gravitational effects, the force is given by Newton's second law

$$F = mg \tag{2.1.3}$$

where m is the system mass and g is the gravitational acceleration. Over most elevations of interest g is a constant, independent of the system's distance from the center of the earth. Therefore, using (2.1.3) in (2.1.1) gives the work required to change the system height from z_1 to z_2,

$$W = mg\int_{z_1}^{z_2} dz = mg(z_2 - z_1) \equiv \Delta E_p \tag{2.1.4}$$

where z is measured along a line from the center of the earth to the system's center of mass. This work is usually identified as a change in gravitational potential energy ΔE_p, which depends on a variable Δz that is external to the system. Our sign convention implies that when $z_2 > z_1$ we must do work to elevate the system, $W > 0$, and then the external potential energy increases.

Now consider the work needed to change a system's velocity. Our system has mass m and is initially moving with velocity v. To change the velocity we must exert a force, which is given by Newton's second law written in the form

$$F = m\frac{dv}{dt} \tag{2.1.5}$$

For a change in velocity from v_1 to v_2, we substitute (2.1.5) into (2.1.1),

$$W = m\int_{x_1}^{x_2} \frac{dv}{dt}\,dx = m\int_{t_1}^{t_2} \frac{dv}{dt}\frac{dx}{dt}\,dt \tag{2.1.6}$$

$$= m\int_{v_1}^{v_2} v\,dv = \frac{m}{2}(v_2^2 - v_1^2) \equiv \Delta E_k \tag{2.1.7}$$

This work is usually identified as the change in kinetic energy, ΔE_k, which depends on a quantity Δv^2 that is external to the system. Our sign convention implies that to increase a system's velocity, we must exert a force parallel to that velocity, so $W > 0$, and hence the external kinetic energy increases. The external kinetic and potential energies sum to the total external energy E_{ext}, so their changes obey

$$\Delta E_{ext} = \Delta E_k + \Delta E_p \tag{2.1.8}$$

Substituting an energy change for an amount of work encompasses important concepts that are easily overlooked: First, changes in a particular form of energy (a conceptual) can sometimes be interpreted as a particular kind of work (a measurable). That is, we establish a relation between an abstract quantity and one that has physical meaning. Second, these forms of energy are defined only as changes; however, by defining a particular position or velocity to be zero, we can create (apparently) absolute values for E_p or E_k. The difference forms that occur in (2.1.4), (2.1.7), and (2.1.8) allow us to introduce arbitrary reference points for conceptuals: in (2.1.4) values of position can be measured relative to *any* arbitrary height, while in (2.1.7), velocities can be measured relative to *any* convenient frame of motion.

2.1.2 Work Accompanying a Volume Change

In addition to the mechanical forms of work discussed above, there are many other forms. For example, work is involved in electrical charging that results from a current flow, in changes of surface area that are opposed by surface tension, and in magnetization caused by a magnetic field. Such forms are all equivalent to mechanical work. However, in the thermodynamics of fluids, the most common form is the mechanical work that deforms the system boundary and thereby changes the system volume.

If, during such a deformation, a force acts on a uniform segment of the boundary, then we can multiply and divide (2.1.2) by the uniform segment area A and write,

$$\delta W = -\frac{F}{A} d(xA) = -P_{ext}\, dV \tag{2.1.9}$$

Here V is the system volume and P_{ext} is the surrounding (external) pressure exerted on the boundary to produce the deformation. The pressure P_{ext} is *always* positive. The negative sign in (2.1.9) is chosen so that when the system volume decreases ($dV < 0$), the work is positive, and when the volume increases ($dV > 0$), the work is negative. For a finite deformation from V_1 to V_2,

$$W = -\int_{V_1}^{V_2} P_{ext}\, dV \qquad \textit{always true} \tag{2.1.10}$$

This equation is always true because it is a definition: it defines the work done when V changes by a finite amount.

Although (2.1.10) is always true, it is only useful when we know how P_{ext} and V are related during the process that deforms our system. Rarely do we have such a relation, and even if we did, it would apply only to particular situations because P_{ext} is a process variable. Generally, we prefer to use the system pressure P rather than the external pressure P_{ext}, but P is often undefined during an irreversible process, so (2.1.10) must be used, perhaps with an estimate for $P_{ext}(V)$. If the system pressure P is defined, then it can be related to V through an equation of state, but this would help us compute the work only if we knew how P and P_{ext} were related. In a process, the exact relation between P and P_{ext} is determined by the behavior of the boundary, but at least we can make the following general observations.

In § 1.2.1 we noted that any force acting on a system generally decomposes into conservative and dissipative components. For a pressure that deforms the system volume, the conservative component equals the system pressure P, so the general relation between P and P_{ext} can be written as

$$P_{ext} = P \pm \mathcal{P} \tag{2.1.11}$$

Here $\mathcal{P}$ is the pressure needed to overcome dissipative forces. Since dissipative forces *always* act to oppose the tendency for change, the sign on $\mathcal{P}$ is determined by the prevailing force. For example, to compress the system, we must have $P_{ext} > P$, and the dissipative pressure opposes the compression by increasing the pressure that must be overcome by P_{ext},

$$P_{ext} = P + \mathcal{P} \qquad \textit{for compression} \tag{2.1.12}$$

However, in an expansion, $P_{ext} < P$, and now the dissipative pressure opposes the expansion by reducing the pressure that is to overcome P_{ext},

$$P_{ext} = P - \mathcal{P} \qquad \textit{for expansion} \tag{2.1.13}$$

Equations (2.1.12) and (2.1.13) are written for finite irreversible processes. For a quasi-static process, the dissipative pressure is a differential quantity $d\mathcal{P}$. Moreover, for a reversible change, dissipative components vanish ($\mathcal{P} = 0$), and $P = P_{ext}$. Then (2.1.9) gives the *reversible* work,

$$\delta W_{rev} = -P\,dV \tag{2.1.14}$$

Likewise, for a finite reversible change (2.1.14) becomes

$$W_{rev} = -\int_{V_1}^{V_2} P\,dV \tag{2.1.15}$$

Equations (2.1.14) and (2.1.15) are idealizations that are never obeyed exactly by real systems. A reversible change is not a realizable process, it is merely a sequence of equilibrium states on a state diagram (see § 1.3 and the Example in § 2.1.3).

If two states can be connected by both a reversible change and an irreversible process, then we can relate the reversible work to the irreversible work. Substituting the expression for P_{ext} (2.1.11) into the definition of work (2.1.9), we find

$$\delta W_{irr} = \delta W_{rev} + \delta W_{lost} \tag{2.1.16}$$

The quantity δW_{lost} is called the *lost work*; it is the energy needed to overcome dissipative forces,

$$\delta W_{lost} \equiv -\mathcal{P}dV \geq 0 \tag{2.1.17}$$

The lost work is zero for a reversible change; otherwise, it is *always* positive. For example, an irreversible compression has $dV < 0$, $\mathcal{P} > 0$ by (2.1.12), and hence $\delta W_{lost} > 0$. Similarly, an irreversible expansion has $dV > 0$, $\mathcal{P} < 0$ by (2.1.13), so again we have $\delta W_{lost} > 0$. Rearranging (2.1.16) we find*

$$\delta W_{irr} - \delta W_{rev} = \delta W_{lost} \geq 0 \tag{2.1.18}$$

* When the other sign convention is chosen for the work, the lost work is defined by $\delta W_{lost} = \delta W_{rev} - \delta W_{irr}$, so in both sign conventions the lost work is positive or zero.

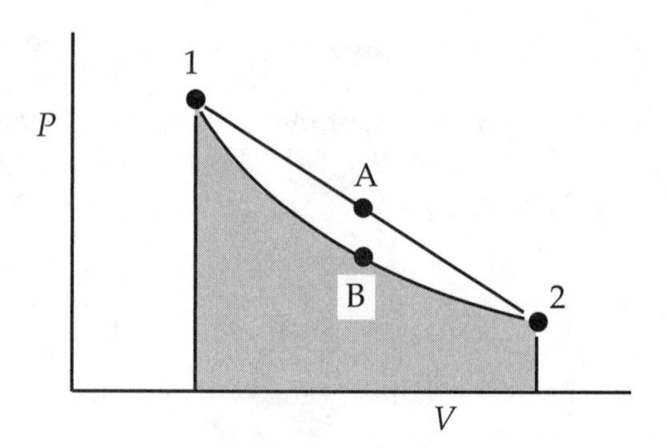

Figure 2.3 For reversible changes the work associated with a volume change is measured by the area under the process path on a P-V diagram. Here the area under path 1A2 differs from that under path 1B2 and therefore the amount of work differs for the two processes: in general, work is a process-dependent quantity, not a state function.

Consequently, we always have

$$\delta W_{irr} \geq \delta W_{rev} \tag{2.1.19}$$

This means that a reversible change is more efficient than any irreversible process between the same two states. During a compression the irreversible work done to the system is larger than W_{rev} because part of W_{irr} is wasted in overcoming dissipative forces that oppose the compression. Likewise, during an expansion the irreversible work done by the system is less than W_{rev} because part of W_{irr} must overcome dissipative forces that oppose the expansion. Lost work measures irreversibilities: high irreversibilities imply large values for W_{lost}. But W_{lost} is a process variable, so it cannot be computed solely from system properties; values for W_{lost} must be either measured or estimated.

Even if a change of state were reversible, direct computation of the work would still require us to know how the system pressure changes during the process; that is, to evaluate the integral in (2.1.15) we must have a quantitative form for the integrand $P(V)$. The integral represents an area on a state diagram, plotted in terms of pressure and volume. Obviously the magnitude of that area depends not only on the initial and final states, $P(V_1)$ and $P(V_2)$, but also on the process-path that connects them. So, work does not form exact differentials and its value depends on the process, as well as on the initial and final states of the system. See Figure 2.3 and the Example in § 2.1.3.

Such process dependence complicates analyses because every time we encounter a different variation of P with V, we must reanalyze the entire situation to find the work. Moreover, if the process is irreversible, then either P is undefined during the process or the variation of P with V is unknown. Either situation prevents us from computing the work directly and solely in terms of system properties. Life can be simpler when we can deal with state functions.

2.1.3 Example

How does the work for an irreversible process differ from that for a quasi-static process and for a reversible change?

Consider a 5-cm ID cylinder fitted with a double-headed piston, as in Figure 2.4. One end of the cylinder is loaded with 0.01 moles of methane; the other end is charged with 0.02 moles of air, initially at 1.4 bar. There is friction at the contact points between the piston heads and cylinder walls. The air chamber is fitted with a pressure gauge and two vents to the surrounding atmosphere. One vent is a large ID line fitted with a

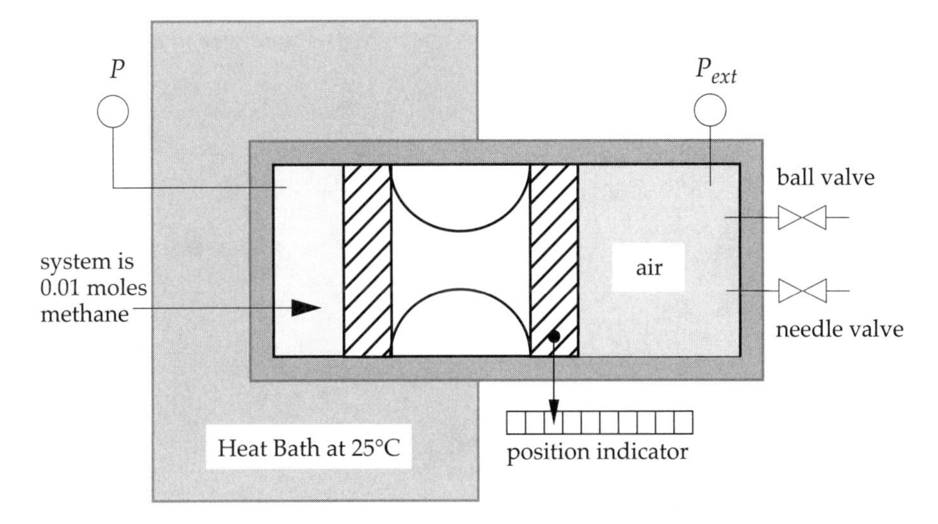

Figure 2.4 Schematic diagram of the double-headed piston-cylinder apparatus. To move the piston, any pressure imbalance between the methane and air chambers must overcome friction at the surfaces of contact between the piston heads and cylinder walls.

ball valve; the other is a small ID line fitted with a needle valve. The atmospheric pressure is constant at 1 bar. Attached to the piston is a position indicator, by which we can determine the volumes of both the methane and the air. The methane chamber is immersed in a heat bath that is maintained at 25°C.

System. We identify the system as the 0.01 moles of methane. When the system is at equilibrium, its temperature is 25°C. Initially the methane pressure is $P_1 = 1.4$ bar, and we want to decrease that value to $P_4 = 1.1$ bar.

Process 1: *Finite stepwise irreversible expansion.* From the initial conditions, we want to expand the methane to 1.1 bar. To create a pressure imbalance across the piston, we vent the air chamber. In this first example, air is removed by cycling the ball valve open and shut three times. During each cycle enough air is vented to reduce P by 0.1 bar. The vent line is so large that each drop in P_{ext} is nearly instantaneous. Between each cycle the system is given time to reestablish equilibrium, as indicated by steady readings on the pressure gauges and the position indicator.

This expansion can be illustrated by a "process" diagram on which we plot the air pressure measured by the gauge (which is external to the methane system) and the methane volume, determined from the position of the piston. On such a diagram the process is approximately a decreasing staircase. The area under the curve on the process diagram gives the magnitude of the work done by the expanding methane,

$$W = -\int_{V_1}^{V_4} P_{ext}\, dV = -\sum_{i=2}^{4} P_{ext_i}(V_i - V_{i-1}) \tag{2.1.20}$$

Since the system volume increases ($V_4 > V_1$), the work is negative: the expanding methane does work on its surroundings. Each cycling of the valve produces a step-

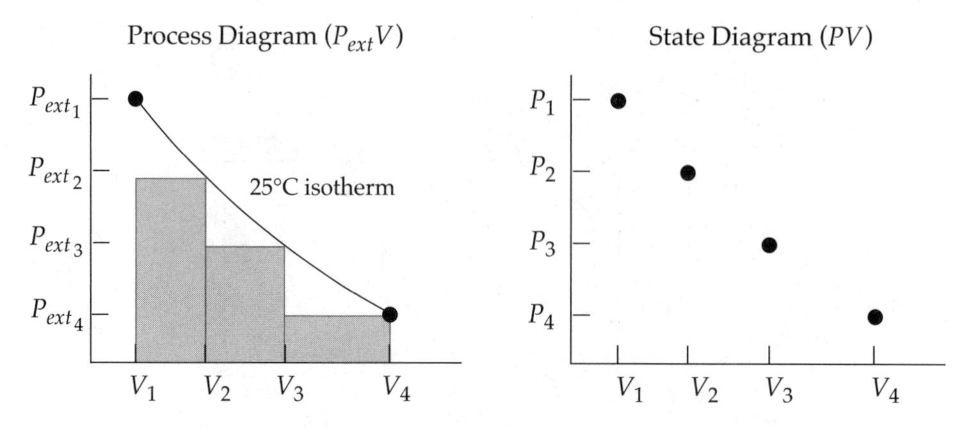

Figure 2.5 Process 1 on a process diagram (*left*) and on a state diagram (*right*)

change in P_{ext}, and the integral in (2.1.20) accumulates the areas of the three shaded rectangles shown on the left in Figure 2.5.

If we want to show a state diagram for the methane during the expansion, we can only plot the four points at which the system is at equilibrium. (See the right side of Figure 2.5.) We cannot draw a process path on the state diagram because during each expansion cycle the methane is not in any well-defined state: the process is irreversible. This means that the state diagram cannot be used to evaluate the work. In particular, a smooth curve connecting the four points does not represent the process path, and the area under that curve would *not* be the work done: $W \neq -\int P dV$.

Process 2: *Quasi-static irreversible expansion.* From the initial conditions, we now expand the methane by just barely opening the needle valve, slowly venting air. The process path is a continuous curve on the process diagram, but the curve is not smooth; even though the air is vented continuously, the piston does not move continuously because of friction between the piston and cylinder walls. During any movement the total force acting on the piston is the algebraic sum of contributions from the methane, the air, and friction,

$$F_{total} = F_{methane} - F_{air} - F_{sf} - F_{kf} = (P_{methane} - P_{air})A - F_{sf} - F_{kf} \quad (2.1.21)$$

where A is the cross-sectional area of the piston. The frictional force is composed of two components: a static part (sf) and a kinetic part (kf). Kinetic friction is zero when the piston is stationary, while static friction is present whenever there is a tendency to change the piston's velocity. To move the piston from rest, sufficient air must be removed so that the imbalance ΔP is large enough to overcome static friction, where

$$\Delta P = P_{methane} - P_{air} \quad (2.1.22)$$

To keep the piston moving once it starts, ΔP must exceed the combined effects of static and kinetic friction. But as the piston moves, $P_{methane}$ decreases while P_{air} increases, because the moving piston compresses air faster than air is being vented. So for sufficiently slow venting, the movement of the piston causes ΔP to decrease and frictional forces stop the motion. Therefore, the methane volume is first constant while P_{ext} decreases, then the piston moves, increasing both V and P_{ext}. When P_{ext} is large

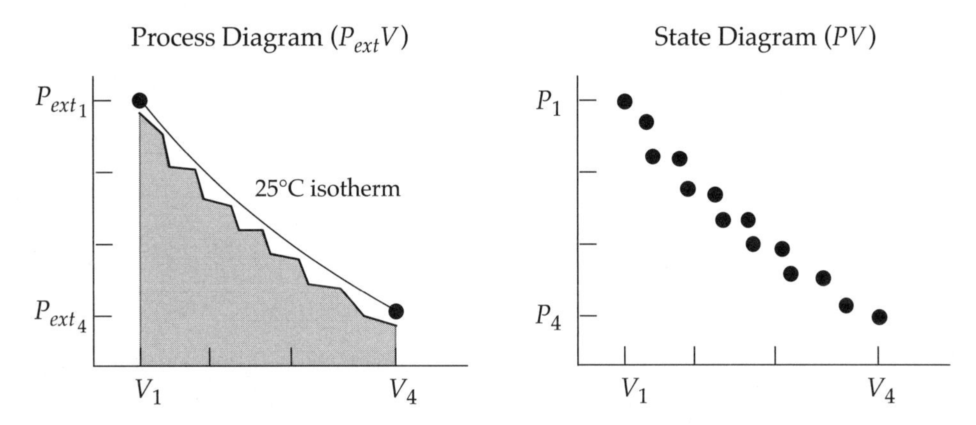

Figure 2.6 Process 2 on a process diagram (*left*) and on a state diagram (*right*)

enough, the piston stops, and the cycle repeats. We assume that after each cycle, the piston remains stationary long enough for the methane to reach an equilibrium state at which its pressure P is defined. The process appears as in Figure 2.6.

Since this process is driven by small pressure imbalances, the process is essentially quasi-static. On the state diagram (in Figure 2.6) the process path is the sequence of P-V points read from the pressure gauge and position indicator when the system passes though the intermediate stationary states. But even though the process is driven by small pressure differences, dissipative forces are present and the process is irreversible. The work is still the area under the path on the process diagram. But the work would not be given by the area under a smooth curve on the state diagram.

Process 3: *Reversible change.* To convert Process 2 into the reversible change shown in Figure 2.7, we must remove any friction. We lubricate the piston-cylinder interface; however, we cannot remove all friction and, consequently, a reversible change cannot be attained. But by removing as much friction as possible and venting the air slowly, the discontinuities in the process path in Figure 2.6 could be made smaller and the corresponding points on the state diagram would be more numerous and closer

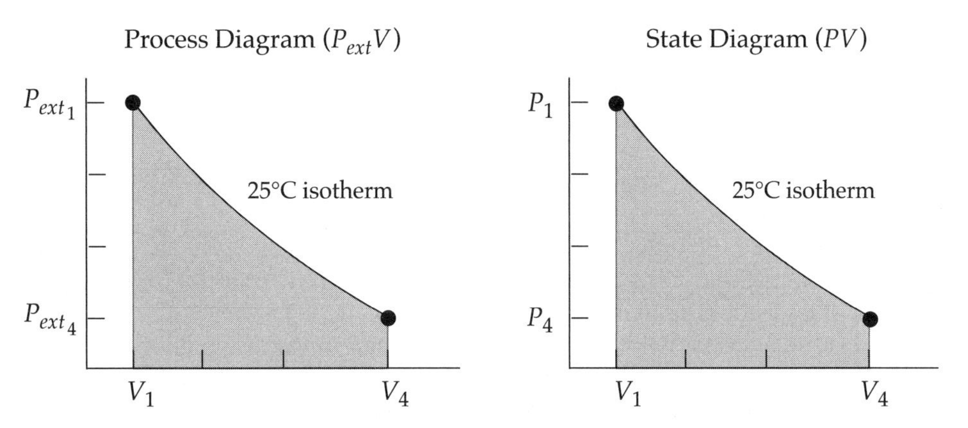

Figure 2.7 In the limiting case of a reversible change (Process 3), the process path would appear the same on both a process diagram (*left*) and a state diagram (*right*).

together. Extrapolating to the limit of no friction, the process paths on the two diagrams would coincide: the expansion would follow the 25°C isotherm on both diagrams. Then, both would appear as in Figure 2.7 and the reversible work could be computed from either via

$$W = -\int_{V_1}^{V_4} P_{ext}\, dV = -\int_{V_1}^{V_4} P dV \tag{2.1.23}$$

2.1.4 Adiabatic Work

Consider a pure gas held in the cylinder of a piston-cylinder apparatus; the cylinder is thermally insulated from its surroundings. We take the gas to be our system, which is initially at equilibrium at temperature T_1, volume V_1, and total number of moles N. Experiment shows that specifying these three quantities fixes the thermodynamic state of pure systems. From this initial state we place a single large weight on top of the piston, exerting a constant pressure (P_{ext}) on the gas and changing its volume. The system is closed so N is unchanged and the system is insulated so the process is adiabatic. At the end of the process the system is allowed to relax to equilibrium at its final volume V_n; the temperature is then measured and found to have *increased*, $T_n > T_1$. The work used in the process is given by (2.1.10) as

$$W_A = -\int_{V_1}^{V_n} P_{ext}\, dV = -P_{ext}(V_n - V_1) \tag{2.1.24}$$

This is a real irreversible process, and so (2.1.10) rather than (2.1.15) applies. We label this work with subscript A to indicate that a particular adiabatic process was used.

Now we repeat the experiment using a different adiabatic process B. The system is still closed, and the initial and final states are still $[T_1, V_1]$ and $[T_n, V_n]$, but we use a sequence of steps with various weights, so the volume changes in a different way; hence, the degree of irreversibility differs from that in process A. In general, to achieve the required final state $[T_n, V_n]$ we may have to use some combination of compressions and expansions. The work required for this second process is

$$W_B = -\int_{V_1}^{V_n} P_{ext}\, dV = -\sum_{i=1}^{n-1} P_{ext_i}(V_{i+1} - V_i) \tag{2.1.25}$$

Remarkably, we obtain the *same* value as we found for the first adiabatic process; that is, experimentally we find

$$W_A = W_B \tag{2.1.26}$$

This means that, although the integrands in (2.1.24) and (2.1.25) differ for the two processes, the areas under the two curves are the same.

We now repeat the same total change of state many times using all manner of irreversible processes. For example, besides applying a pressure, we might pass a current through an electrical resistor inserted into the gas (electrical work), or we might rotate a paddle-wheel mounted in the gas (mechanical work). Some of these experiments might even approximate reversible changes. In all cases, so long as the system has the same contents, remains closed, and has the same thermally insulated boundary, then we find that the *same* amount of work is needed to change from the initial state $[T_1, V_1]$ to the final state $[T_n, V_n]$. That is, experiment shows that the adiabatic work done on or by closed systems is *independent* of the process.

The work to displace a system or change its velocity can be viewed as a change in external energy (§ 2.1.1). Similarly, the adiabatic work can be interpreted as a change in an energy; we call it the *internal energy U*. Therefore, using W_{ad} for the adiabatic work, we write

$$\Delta U = W_{ad} \qquad \textit{always true} \qquad (2.1.27)$$

The internal energy is an extensive conceptual (nonmeasurable) property of a system. It is called *internal* because its value is determined by the system's state as characterized by system properties such as temperature T, pressure P, and number of moles N. This distinguishes U from the external energy (2.1.8), which is related to measurables determined by the external position or velocity of the system. Macroscopically, internal energy can be interpreted as the means by which energy is stored in the mass of a system. A microscopic interpretation of U is given in § 2.2.3.

Experiments analogous to those just described were first performed by Joule in the 1840s [5]. Those experiments accomplished several things: they fully discredited the old caloric theory of heat (a theory that considered heat to be transported by movement of a substance called *caloric*), they demonstrated that a temperature change can occur without heat transfer, and they provided a numerical conversion factor between equivalent amounts of heat and work. However for us, Joule's most important result leads to (2.1.27).

2.2 THE FIRST LAW

Consider a change of state that can be accomplished both adiabatically and nonadiabatically; we want to extend our analysis to include the nonadiabatic paths. That is, we repeat the experiments of § 2.1.4 using the same closed system and the same initial and final states $[T_1, V_1]$ and $[T_n, V_n]$; however, we remove the thermal insulation. So the difference compared to the experiments in § 2.1.4 is that now the system and its surroundings are in contact via two interactions: the thermal interaction and a force (a nonthermal interaction) that changes the system volume. As in § 2.1.4, we perform a series of experiments in which we use many different irreversible processes to cause the same change of state. The results from these experiments show that, for such nonadiabatic processes, the work computed from (2.1.10) is always greater than the adiabatic value. Moreover the various nonadiabatic processes give values of the work that differ from one another: nonadiabatic work is a path function.

2.2.1 Heat

For the experiments just described, which were performed between the same initial and final states, the difference between the adiabatic (W_{ad}) and nonadiabatic (W_{nad}) work must be the energy transferred through the thermal interaction. We call this energy the *heat Q*, which is an extensive, measurable, process variable,

$$Q \equiv W_{ad} - W_{nad} \tag{2.2.1}$$

Although this definition of Q may differ from ones more familiar to you, in fact (2.2.1) possesses all the attributes normally associated with heat transfer. It is important because it provides a precise quantitative relation that is generally useful.

Equation (2.2.1) provides a means for determining Q by measuring work: for a given process between states 1 and 2, Q may be determined by measuring the work required by the process, and then measuring the work required by any adiabatic process between the same two states. If we want a value for reversible heat transfer, the nonadiabatic process must be reversible; however, the value of work for the adiabatic process is independent of reversibility. When state 2 cannot be reached adiabatically from state 1, then instead of measuring the adiabatic work W_{12}, we would measure the adiabatic work for the opposite process (from state 2 to state 1) W_{21}. Then (2.1.27) allows us to compute W_{12} by

$$W_{ad12} = -W_{ad21} \tag{2.2.2}$$

Note that the definition of heat in (2.2.1) does not involve temperature. Temperature is a property that measures "hotness," the intensity of heat; it does not measure a quantity of heat, which is not a property. Temperature and heat are related only in situations in which a temperature difference is allowed to affect a system: a temperature difference can cause heat transfer via the thermal interaction. But heat transfer is not necessary to change a temperature, nor does temperature necessarily change as a result of a heat transfer.

Just as for work, we must choose a sign convention for heat. We use

$Q > 0$, if heat is transferred to the system from the surroundings, and

$Q < 0$, if heat is transferred from the system to the surroundings.

Since the nonadiabatic work is a path function and the rhs of (2.2.1) is a linear combination of a path function (W_{nad}) and a change in a state function (W_{ad}), heat must also be a path function. It is not a property of the system. Like work, heat is energy that can be identified only as it crosses the system boundary.

2.2.2 The First Law for Closed Systems

What we have accomplished thus far can now be collected and condensed into the first law for closed systems. We begin with differential processes and state the first law in two parts. In the first part, we identify the adiabatic work as the change in a state function, the internal energy,

Part 1, Law 1 $\quad dU = \delta W_{ad} \quad$ *closed systems* $\quad$ (2.2.3)

Then we substitute (2.2.3) into the differential form of (2.2.1) to obtain

Part 2, Law 1 $\quad dU = \delta Q + \delta W \quad$ *closed systems* $\quad$ (2.2.4)

These two equations represent the first law for differential processes acting on closed systems; they are always true.

When a change of state can be performed both reversibly and irreversibly, the change in U must be the same for both. Then we can write

$$dU = \delta Q + \delta W = \delta Q_{rev} + \delta W_{rev} = \delta Q_{rev} - PdV \tag{2.2.5}$$

Even though $\delta Q \neq \delta Q_{rev}$ and $\delta W \neq \delta W_{rev}$ the sum $(\delta Q + \delta W)$ *must* have the same value as the sum $(\delta Q_{rev} + \delta W_{rev})$. This suggests that some results computed for reversible changes can be applied to real (irreversible) processes.

Because heat and work do not form exact differentials, we have written δQ and δW rather than dQ and dW. That is, δQ represents a small *amount* of heat, while dU represents a differential *change* in internal energy. Integrating δQ produces a finite amount of heat Q, while integrating dU produces a finite change in internal energy ΔU. Then for a finite process, we can integrate (2.2.4) to obtain

$$\Delta U = Q + W \tag{2.2.6}$$

Equation (2.2.6) makes the remarkable assertion that the algebraic combination of two path functions *always* yields a change in a state function: that is, between two specified state points, the value of ΔU is always the same, regardless of the values of Q and W used to cause the change of state.

Heat and work are not properties of either the system or the surroundings; they exist only during the interaction that carries them across the boundary. However, for certain special processes Q and W are separately related to changes in state functions. We have already seen that if *no* thermal interaction exists, then the adiabatic work equals ΔU and it can be calculated assuming a reversible change. Likewise, if *only* a thermal interaction connects the system to the surroundings (the process is *workfree*), then the heat transferred equals ΔU,

$$\Delta U_{wf} = Q_{wf} \tag{2.2.7}$$

If the center of our closed system undergoes changes in position or velocity during a process, then we must allow for possible changes in external energy. In such cases the first law (2.2.6) becomes

$$\Delta U + \Delta E_k + \Delta E_p = Q + W \tag{2.2.8}$$

The combined internal and external energy of a system is called the *total energy* E

$$\Delta E \equiv \Delta U + \Delta E_k + \Delta E_p \tag{2.2.9}$$

However, in many situations ΔE_k and ΔE_p are either identically zero or they are negligible compared to the magnitude of ΔU, so (2.2.6) is usually sufficient for our needs. Note that like E_k and E_p, we cannot ascribe absolute values to E or U; we can only obtain changes in their values or values relative to some arbitrarily chosen reference state. In the latter case, the value at a reference state is often set to zero (e.g., $U_{ref} = 0$), so that tables *appear* to contain absolute values. This is done for computational convenience because, in practice, we only need values for changes and in such calculations, the value at the reference state cancels [for example, $\Delta U = (U_2 - U_{ref}) - (U_1 - U_{ref}) = (U_2 - U_1)$].

In many applications the quantities we can actually measure or manipulate are the heat and work effects on the external side of the system boundary. We call these Q_{ext} and W_{ext}; they would be measured at a point on the boundary at which the surroundings have temperature T_{ext} and pressure P_{ext}. These external heat and work effects would differ from the heat and work effects felt by the system whenever the system boundary possesses a finite mass that could store energy. In such cases, the second part of the first law for closed systems generalizes (via the stuff equation (1.4.1) and Figure 1.7) to

$$dU + dU_b = \delta Q + \delta W + dU_b = \delta Q_{ext} + \delta W_{ext} \tag{2.2.10}$$

Here dU_b represents a change in internal energy of the boundary, and we have assumed changes in kinetic and potential energies are negligible. The advantage to (2.2.10) is that it explicitly contains those process variables that might be used to manipulate the system's state.

2.2.3 Molecular Interpretations

The internal energy U is a macroscopic property that represents the mechanism by which energy is stored in a system. Microscopically, energy is stored in the kinetic and potential energies of individual molecules. These molecular energies differ from the external kinetic and potential energies, which are associated with the center of mass of the entire system. In a static system, changes in external energy are zero; nevertheless, the molecules possess kinetic energy because they are continually moving, and they contain potential energy because molecules exert forces on one another. Consequently, the internal energy is viewed as being composed of two molecular contributions,

$$u = \frac{U}{\mathcal{N}} = \langle u_k \rangle + \langle u_p \rangle \tag{2.2.11}$$

Here u is the intensive internal energy, $\mathcal{N}$ is the number of molecules, u_k is the molecular kinetic energy, and u_p is the molecular potential energy; the angle brackets indicate an average over all molecules. Note that in molecular theory we can write an equation that represents the absolute internal energy, but in thermodynamics we cannot.

Consider a system of $\mathcal{N}$ spherical atoms (such as those of argon), each having a mass m and some velocity v_i. The atomic velocities differ, but we can form an average velocity (which would be zero for a static system) and an average molecular kinetic energy (which is always greater than zero),

$$\langle u_k \rangle = \frac{m}{2\mathcal{N}} \sum_{i=1}^{\mathcal{N}} v_i^2 \tag{2.2.12}$$

Recall that one mole of gas has $\mathcal{N} \approx 10^{23}$ molecules, so it is not practical for us to know the velocities of the molecules; nevertheless, relations like (2.2.12) can offer insight into the meanings of thermodynamic quantities. The many different velocities contained in (2.2.12) form a *distribution*; for a static system at equilibrium, that distribution is a Gaussian.

The potential portion of the internal energy decomposes into several parts, and those parts generally make very different contributions to changes in the internal energy of a system. The most important part is the configurational internal energy $\langle u_c \rangle$ which results from forces acting between different molecules. You are familiar with some of these forces; they include hydrogen bonding and forces arising from interactions between dipole moments. A second part of $\langle u_p \rangle$ arises from vibrational and rotational motions of atoms within individual molecules (intramolecular interactions). These energies may contribute to ΔU at high densities when neighboring molecules inhibit the motions of one another. When these effects are important, they are usually combined into a change in configurational energy $\Delta\langle u_c \rangle$. The third contribution to $\langle u_p \rangle$ originates from electronic energies associated with chemical bonds; however, in non-reacting systems at common temperatures, changes in these energies do not contribute to ΔU. Therefore, the molecular expression (2.2.11) for a change in internal energy is usually written as the sum of kinetic and configurational contributions,

$$\Delta u = \Delta\langle u_k \rangle + \Delta\langle u_c \rangle \tag{2.2.13}$$

When heat and work cross a boundary and enter a system, those energies are immediately parceled out among the molecules—some goes to change molecular kinetic energies and the rest goes to change molecular configurational energies. We do not talk about heat or work "in" a system because to the molecules it is all just so much energy: the molecules do not know whether that energy came from some heat effect or from some work mode. Further, in thermodynamics the state is specified by values for a certain number of properties (such as T, P, and N), but in statistical mechanics the state is specified by just two things: the molecular energies *available* to the system and the *distribution* of molecular energies among those available. When we change the state, we may be changing the available energies, or the distribution, or both. In fact, in statistical mechanics we can show that work modes change the available energies while heat effects change the distribution.

After a change of state, the system relaxes to equilibrium; this means that the molecules must properly distribute themselves among the available energies that are allowed to that state. (Note that thermodynamics gives us no information as to how long such a relaxation process might take.) Each equilibrium state has a unique set of available energies and a unique distribution among those energies, independent of the process by which the state is attained. During a reversible change, the system is driven by differential forces from one equilibrium state to the next, and the molecules continuously readjust themselves to maintain the correct distribution among available energies. Consequently, the macroscopic state is always well-defined and the change can be represented by a continuous curve on a state diagram. But during an irreversible process, the available energies and the distribution are out of balance and are not those of any well-defined state: intermediate states are unknown and unknowable. The state becomes knowable only when an irreversible process is complete and the molecules have achieved the correct distribution among available energies.

2.3 THE SECOND LAW

Recall that a principal goal of this chapter is to relate path functions to state functions. One realization of that goal is provided by the first law (2.2.6): the algebraic sum of heat and work produces a change in a state function—the internal energy. But can we be more direct? Is there a way to directly relate work and heat to state functions, without combining the two through the first law? For the reversible work caused by a volume change, we know that

$$\delta W_{rev} = -PdV \tag{2.3.1}$$

So, although δW_{rev} depends on path, dividing by P gives a change in a state function,

$$\frac{\delta W_{rev}}{P} = -dV \tag{2.3.2}$$

That is, the reciprocal pressure is an integrating factor that converts δW_{rev} into an exact differential. (Integrating factors are discussed in Appendix A.)

However, it is not immediately obvious how to convert δQ into an exact differential. We might tentatively guess that a form analogous to (2.3.2) can be found; that is, perhaps there is another integrating factor λ such that

$$\frac{\delta Q_{rev}}{\lambda} = dS \tag{2.3.3}$$

Here S stands for a new state function and the identity of the integrating factor λ is yet to be discovered. The objectives of this section are to develop (2.3.3) and identify λ.

2.3.1 Entropy and Thermodynamic Temperature

Our presentation of the second law is based on the rigorous development by Carathéodory [2]. Carathéodory's approach has been described in detail by Chandrasekhar [3] and Kestin [4], so we need only outline the arguments here.

We begin by introducing the idea of *inaccessible* states. Consider a closed system containing a pure single phase; this system has only one interaction with its surroundings—the work mode that can change the volume. Therefore, the system can undergo only adiabatic processes. For a pure substance at constant mass, we need only two properties to identify the state; we choose volume V and internal energy U. Experimentally, we find that for reversible changes of volume, our system follows a unique curve on a UV-state diagram, as in Figure 2.8. That curve is called a *reversible adiabat*; note that its slope is negative,

$$\left(\frac{\partial U}{\partial V}\right)_{Q=0} < 0 \tag{2.3.4}$$

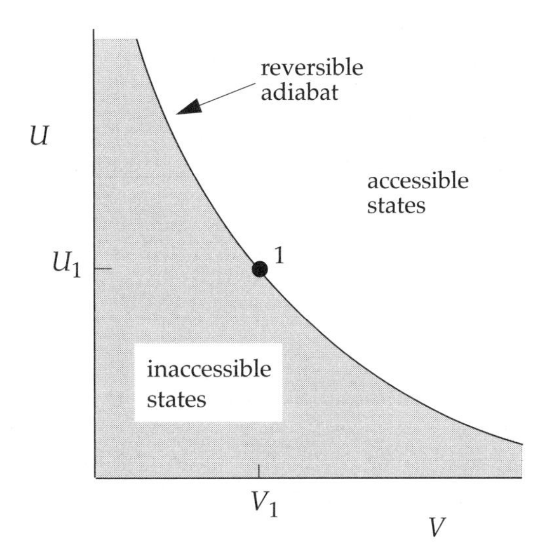

Figure 2.8 Schematic plot of states accessible via adiabatic processes in closed systems. From the initial state 1 only states on the line can be reached by reversible adiabatic volume changes. States above the reversible adiabat can be reached only by processes that include irreversible adiabatic volume changes. States below the reversible adiabat cannot be reached by *any* adiabatic volume change.

From state 1 any reversible adiabatic volume change leaves the system somewhere on the curve shown in Figure 2.8. Further, each reversible adiabat is unique; that is, reversible adiabats do not intersect. If they did, then it would be possible to find two different values of the adiabatic work for the same change of state; this would violate the first law.

If we want to move from state 1 in Figure 2.8 to states not on the reversible adiabat, then we find experimentally that we must use an irreversible process. However, using an irreversible adiabatic process does not allow us to reach every other state on the diagram. From state 1, we can only reach states *above* the reversible adiabat by means of some irreversible adiabatic process; those states are said to be *accessible.* To reach states below the line (shaded region in Figure 2.8.), we must use some nonadiabatic process; that is, we must transfer heat. This means that a particular asymmetry exists among the states that are accessible using adiabatic processes.

The existence of states that are inaccessible to adiabatic processes was shown by Carathéodory to be necessary and sufficient for the existence of an integrating factor that converts δQ_{rev} into an exact differential [2–4]. From the calculus we know that for differential equations in two independent variables, an integrating factor *always* exists; in fact, an infinite number of integrating factors exist. Experimentally, we find that for pure one-phase substances, only two independent intensive properties are needed to identify a thermodynamic state. So for the experimental situation we have described, we can write δQ_{rev} as a function of two variables and choose the integrating factor. The simplest choice is to identify the integrating factor as the positive absolute thermodynamic temperature $\lambda = T$. Then (2.3.3) becomes

$$dS \equiv \frac{\delta Q_{rev}}{T} \tag{2.3.5}$$

The new state function S is named the *entropy;* it is an extensive, conceptual property and has dimensions of (energy/absolute temperature). Although we have discussed the development of (2.3.5) in terms of pure substances, which require only two properties to identify the state, Carathéodory proved that this entire development extends in a straightforward (but tedious) way to systems requiring any number of properties. Consequently, the definition (2.3.5) is completely general [2–4].

Microscopically, the absolute temperature is proportional to the average kinetic energies of the molecules. For a substance such as argon, whose molecules are spheres, the molecular kinetic energy in (2.2.12) is caused by translational motion,

$$\frac{3kT}{2} = \langle u_k \rangle = \frac{m}{2\mathcal{N}} \sum_{i=1}^{\mathcal{N}} \mathrm{v}_i^2 \tag{2.3.6}$$

Here k is Boltzmann's constant, m is the mass of one molecule, $\mathcal{N}$ is the total number of molecules present, v_i is the velocity of molecule i, and the angle brackets represent an average over molecules, as in (2.2.12). Boltzmann's constant is related to the gas constant R by $k = RN_A$, where N_A is Avogadro's number; therefore, k can be interpreted as the molecular gas constant. Note that temperature is simply proportional to the kinetic contribution to the molecular internal energy in (2.2.12).

Absolute temperature is defined by (2.3.6), but that definition applies only when $\mathcal{N}$ is large enough for there to be a statistically meaningful distribution of molecular velocities so that a reliable average can be determined. This means temperature is a macroscopic property; an individual molecule does not have a temperature, it has velocity and kinetic energy. A change in temperature measures the work needed to change the time-average molecular velocities. In adiabatic processes the temperature changes, even though no heat is transferred, because when work is done on or by a system, the average molecular velocities must change.

2.3.2 The Second Law for Closed Systems

What we have accomplished so far can now be collected into the second law for closed systems. Analogous to the first law, we state the second law in two parts. First is the definition of the entropy, which relates a path function to a new state function,

Part 1, Law 2 $$dS \equiv \frac{\delta Q_{rev}}{T} \qquad \textit{closed systems} \tag{2.3.5}$$

The second part prescribes the observed limits on the directions of adiabatic processes in closed systems (i.e., it identifies those states that are accessible and inaccessible by adiabatic processes),

Part 2, Law 2

$$dS > 0, \text{ for irreversible adiabatic processes}$$
$$dS = 0, \text{ for reversible adiabatic changes of state} \tag{2.3.7}$$
$$dS < 0, \text{ for changes of state that cannot be realized adiabatically}$$

The second part of the second law (2.3.7) divides a state diagram into three parts (a line and two areas), as illustrated on the TS diagram in Figure 2.9. We emphasize that the second law (2.3.7) does *not* preclude the system entropy S from decreasing; in fact,

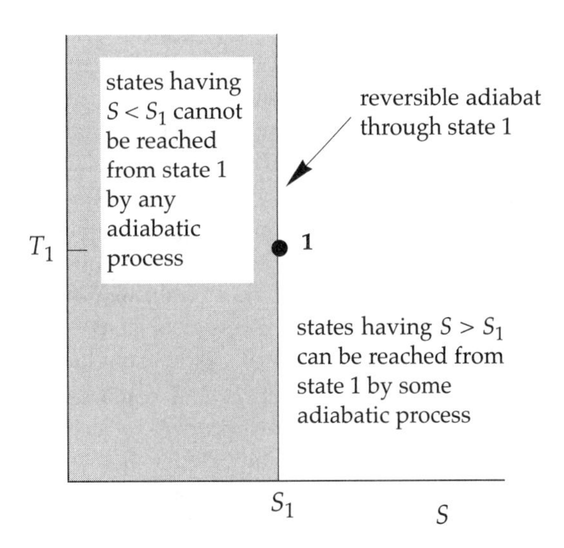

Figure 2.9 Graphical interpretation of the second law (2.3.7) for closed systems. For any closed system initially at state 1, adiabatic processes can only move the system to states having entropy $S_2 \geq S_1$; on this diagram, all such states lie on or to the right of the vertical line through state 1. Shaded region here corresponds to the shaded region in Figure 2.8.

the entropy must decrease whenever heat is removed from a system during a work-free process. Let us compare the second law (2.3.7) with the first law (2.2.4): the first law asserts that energy is a conserved quantity, but in contrast, the second law asserts that entropy may *not* be conserved; entropy is created during irreversible processes.

The traditional formulation of the second law is given by (2.3.5) and (2.3.7); however, there is an alternative that may be useful, especially for open systems. Again, the statement is in two parts: a definition of entropy plus an assertion that entropy is not conserved because we now explicitly include entropy changes in the boundary. The definition takes the form of the stuff equation (1.4.1) with Figure 1.7,

$$dS + dS_b \equiv \frac{\delta Q_{ext}}{T_{ext}} + dS_{gen} \tag{2.3.8}$$

Here Q_{ext} is the amount of heat crossing at the outside boundary between the system and its surroundings, T_{ext} is the temperature at the (external) point of heat transfer, and S_b is the entropy of the boundary. If we interpret (2.3.8) as a stuff equation, then the lhs is the accumulation term, the first term on the rhs is an interaction term, and dS_{gen} represents entropy generated in the system and its boundaries. The sign convention for Q_{ext} is the same as that for any quantity crossing a boundary, see the Example in § 2.3.3. Note that (2.3.8) contains no work term. The form of the second law in (2.3.8) is completely general: it applies to open and closed systems undergoing any kind of change of state. A particular advantage to (2.3.8) is that it explicitly contains Q_{ext}, so it may be used to identify those portions of the boundary at which irreversibilities occur.

The generation term in (2.3.8) accounts for entropy created by dissipative forces. The second part of the second law states that this generation term is always either positive or zero:

$$dS_{gen} \begin{cases} = 0 & \text{for reversible changes} \\ > 0 & \text{for irreversible processes} \end{cases} \tag{2.3.9}$$

For reversible changes the generation term is zero because dissipative forces are not present (see § 1.3); consequently, entropy is conserved during a reversible change. Further, for a reversible change, (2.3.8) and (2.3.9) reduce to the definition in (2.3.5). If a change of state is adiabatic and reversible, then the first term on the rhs of (2.3.8) is zero and the remainder, combined with (2.3.9), reduces to the equality in (2.3.7).

In practice, the second law form (2.3.8) is useful only when the generation term can be estimated, typically through some measured or estimated efficiency factor. Normally, changes in entropy are computed by integrating over changes in measurables, as we show in Part II of this book. But in any case, we would combine (2.3.5) and (2.3.8) to obtain S_{gen}. For the special case of a reversible change ($dS_{gen} = 0$), (2.3.8) provides a useful relation among property changes and heat flows.

But aside from these practical considerations, (2.3.8) may offer some additional insight over (2.3.5). For example, (2.3.8) shows that the system entropy can increase, decrease, or remain constant, depending on the relative sizes of the two terms on the rhs. More importantly, (2.3.8) helps clarify the nonconservative nature of entropy: when a process drives a system through unidentifiable states, then the generation term in (2.3.8) will be positive and entropy is created in the system and its boundary. Consequently, for a specified value of T_{ext} and a given change of state, $\delta Q_{irr} \le \delta Q_{rev}$, which means $\delta W_{irr} \ge \delta W_{rev}$ [this is consistent with the sign of the lost work given in (2.1.18)]. Therefore for the specified change of state, either the magnitude of the work actually produced is *less* than that obtained from the reversible change or the magnitude of the work actually required is *more* than that needed for the reversible change.

To say this another way, entropy is created through the action of dissipative forces that are wasteful because they convert some energy into heat, reducing the amount available for performing useful work. Consequently, we try to control process efficiency by minimizing the generation term: to increase efficiency, decrease dS_{gen}. We attain maximum efficiency when $dS_{gen} = 0$; however, this means all steps would be reversible, which is impossible. Different kinds of irreversibilities are produced by different property gradients, so it is natural to ask which gradients—temperature, pressure, or composition—create the largest entropies. In many applications, temperature gradients are most wasteful. Fortunately, temperature gradients are usually the ones most easily modified and controlled; for example, air and water streams fed to power-plant boilers and to multiple-effect evaporators are routinely preheated by hot exit streams to reduce temperature gradients, providing considerable gains in efficiency.

When two states can be connected by both a reversible change and an irreversible process, we can combine the first and second laws to show that dS_{gen} can never be negative (see Problem 2.9). The result is

$$T dS_{gen} = \delta W_{lost} + \left(1 - \frac{T}{T_{ext}}\right) \delta Q_{irr} \tag{2.3.10}$$

We want to deduce the sign of dS_{gen}. We know that $\delta W_{lost} \ge 0$, so we need to show that the heat-transfer term in (2.3.10) cannot be negative. There are three cases.

(a) Consider an irreversible process that adds heat to the system. We have $\delta W_{lost} > 0$ while $\delta Q_{irr} > 0$ and $T_{ext} > T$, so both terms on the rhs of (2.3.10) are positive. Hence, $dS_{gen} > 0$.

(b) Now consider an irreversible process that removes heat from the system. We still have $\delta W_{lost} > 0$ while $\delta Q_{irr} < 0$ and $T_{ext} < T$, so both terms on the rhs of (2.3.10) are again positive. Hence, we still have $dS_{gen} > 0$.

(c) For any reversible change, $\delta W_{lost} = 0$ and $T_{ext} = T \pm dT$, so both terms on the rhs are zero; hence, $dS_{gen} = 0$.

Equation (2.3.10) shows that in closed systems, entropy can be generated in two general ways. First, as already discussed in § 2.1.2, the lost work δW_{lost} is the energy needed to overcome dissipative forces that act to oppose a mechanical process. Second, the heat-transfer term in (2.3.10) contributes when a finite temperature difference irreversibly drives heat across system boundaries. This second term is zero in two important special cases: (a) for adiabatic processes, $\delta Q_{irr} = 0$, and (b) for processes in which heat is driven by a differential temperature difference, $T_{ext} = T \pm dT$. In both of these special cases, (2.3.10) reduces to

$$TdS_{gen} = \delta W_{lost} \tag{2.3.11}$$

This special form, rather than (2.3.10), is more often presented in textbooks. However, (2.3.10) makes clear that even in a workfree process we expect entropy to be generated because of irreversible heat transfer. In fact, when the heat transfer term is not zero, it is usually larger that the lost work term.

2.3.3 Example

How do the sign and magnitude of Q_{ext} differ from those for Q?

In the first part of the second law (2.3.8), we introduced a heat transfer term Q_{ext} that represents the amount of heat entering or leaving a system at the external side of its boundary. This is in contrast to the heat Q, which is the amount of heat that actually enters or leaves the system. These two terms, Q_{ext} and Q, may differ, depending on the amount of entropy generated in the boundary; however, in many applications, the two will have similar magnitudes.

Now consider the signs of Q and Q_{ext}. We have adopted a consistent sign convention for all quantities that cross a boundary: anything entering a system is positive, anything leaving a system is negative (§ 2.2.1). So we expect the sign of Q_{ext} to be the same as that for Q.

But note that our convention means that the sign depends on the identity of the system. When dealing with Q_{ext} this can lead to confusion because to compute a value for Q_{ext} we might reverse the identities of system and surroundings. Such a reversal may enable us to take advantage of quantities in the surroundings whose values we know, so that a value for Q_{ext} can be computed.

To keep clear the proper sign for Q_{ext} it may be helpful to imagine yourself as an observer located in the "surroundings" to which the current step in your calculation applies. (This is the natural location for an engineer; that is, we typically stand in the surroundings and try to measure or manipulate quantities in the system.) If the heat

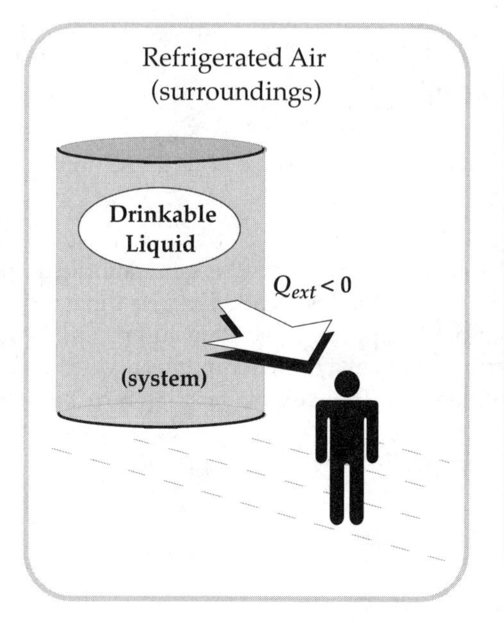

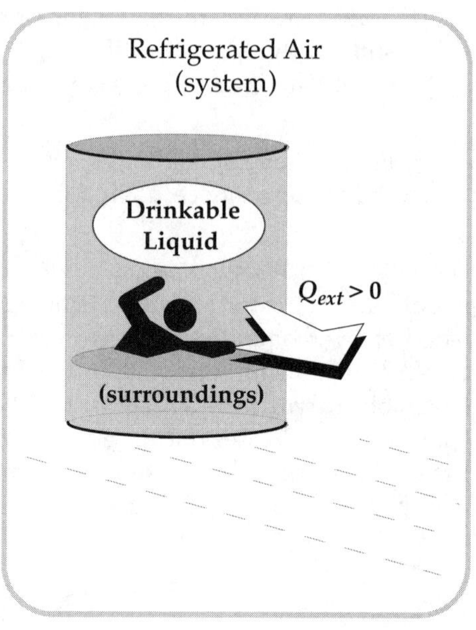

Figure 2.10 Schematics of a beverage cooling in a refrigerator. (*left*) The system is the drinkable liquid, and the engineer stands in the surroundings, which is the air inside the refrigerator. (*right*) The roles of system and surroundings are reversed compared to those at left. In both cases, heat is removed from the beverage and the engineer is in the surroundings, but the sign of Q_{ext} is determined by our choice for the system.

Q_{ext} is coming toward you, then it is leaving the system and it is negative; if heat is moving away from you, then it is entering the system and it is positive.

To have an example, consider cooling a can of beverage in a refrigerator, as in Figure 2.10. On the left in Figure 2.10 we identify the following equivalences:

system	⇔	beverage
boundary	⇔	can
surroundings	⇔	air in refrigerator

We always place the observer in the surroundings, which on the left in Figure 2.10 is the refrigerated air (i.e., inside the refrigerator). Then, since the beverage is being cooled, heat moves toward the observer, so $Q_{ext} < 0$.

However, sometimes calculations are simplified if we choose the refrigerated air as the system, rather than the beverage. Then we would identify these equivalences:

system	⇔	air in refrigerator
boundary	⇔	can
surroundings	⇔	beverage

The observer is always in the surroundings, which now is the beverage, as on the right in Figure 2.10. We are still cooling the beverage, but now heat moves away the observer, so $Q_{ext} > 0$. Therefore, the sign of Q_{ext} on the right in Figure 2.10 differs from that on the left in Figure 2.10, although the processes in both are exactly the same. The signs of quantities crossing boundaries depend on the identity of the system.

2.4 THERMODYNAMIC STUFF EQUATIONS

In § 2.2 and 2.3 we presented the first and second laws for closed systems. In practice these would apply to such situations as those batch processes in which the amount of material in the system is constant over the period of interest. But many production facilities are operated with material and energy entering and leaving the system. To analyze such situations, we must extend the first and second laws to open systems. The extensions are obtained by straightforward applications of the stuff equations cited in § 1.4. We begin by clarifying our notation (§ 2.4.1), then we write stuff equations for material (§ 2.4.2), for energy (§ 2.4.3), and for entropy (§ 2.4.4). These three stuff equations are always true and must be satisfied by any process, and therefore they can be used to test whether a proposed process is thermodynamically feasible (§ 2.4.5).

2.4.1 Notation

In the sections that follow we will repeatedly encounter the sum $(U + PV)$, so it will be convenient to replace that sum with a single symbol H. Hence, we define

$$H \equiv U + PV \tag{2.4.1}$$

Later, we will find that H is more than a notational convenience. For example, we already know that U is a state function, and it is simple to show (Problem 2.26) that (PV) is also a state function; consequently, H is a state function. It is named the *enthalpy*: an extensive, conceptual property of any system. It has dimensions of energy.

For the special case in which a closed system undergoes a reversible isobaric change of state, we can assign a physical interpretation to dH. In such cases, (2.4.1) gives

$$dH = dU + PdV = dU - \delta W_{rev} = \delta Q_{rev} \qquad \textit{isobaric} \tag{2.4.2}$$

That is, for reversible isobaric changes of state, the enthalpy change of the system is the same as the heat transferred to or from the system. Unfortunately, for other processes acting on closed systems, no such simple interpretation applies; nevertheless, we will find the enthalpy to be a useful conceptual for both closed and open systems.

In the remainder of § 2.4 we will be concerned with open systems in which both mass and energy can enter and leave the system. We adopt the following notation. At any instant, the system has total number of moles N, total energy $E = Ne$, and total entropy $S = Ns$. Material can enter the system through any number of feed streams $\alpha = 1, 2, \ldots$ and leave through any number of discharge streams $\beta = 1, 2, \ldots$. Energy may enter and leave the system by heat transfer Q, through work modes, and by material entering and leaving the system.

The possible work modes are of two kinds: (a) those that deform the boundary W_b and (b) those that cross the boundary W_{sh}. The latter includes wires carrying electrical current and rotating shafts or reciprocating pistons for performing mechanical work; these modes are called *shaft work*, hence the subscript *sh* on W. Common mechanical devices that produce or consume shaft work include pumps and compressors, which

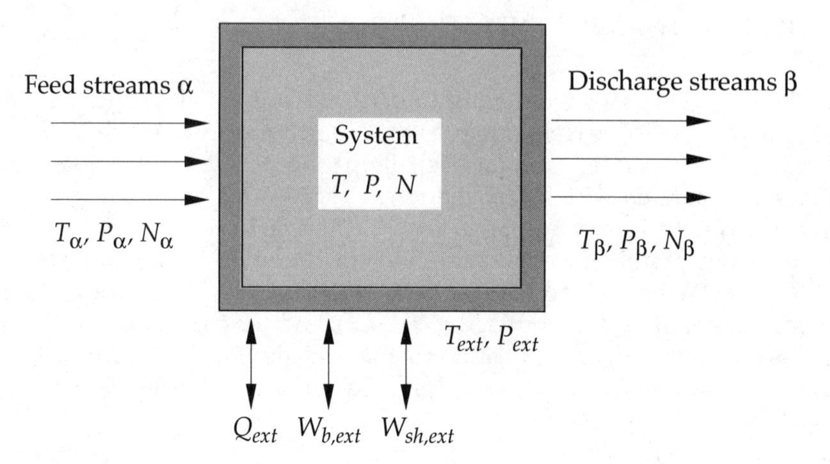

Figure 2.11 Schematic of a system open to mass transfer through streams α and β, heat transfer through any number of conduits, and work effects via any number of interactions that either deform the boundary or cross the boundary. The subscripts *ext* on Q and W indicate that their values may differ on the inside and outside of the boundary. The thickness of the boundary has been enlarged for emphasis.

do work on the system, so $W_{sh} > 0$, and turbines, which are driven by the system, so $W_{sh} < 0$. However, we caution that shaft work is a generic term for any work that crosses a boundary; it is not limited to electrical and mechanical modes and it does not necessarily involve a physical shaft. But in any case, the effects of shaft work can always be made equivalent to those of some mechanical device that does have a rotating shaft. Our notation is summarized in Figure 2.11.

2.4.2 Material Balance

In § 2.4 we restrict our attention to systems in which no chemical reactions occur; then, over any small time interval dt, the change in the number of moles in the system is given by the overall material balance (1.4.3), which we write as

$$dN = \sum_{\alpha} dN_{\alpha} - \sum_{\beta} dN_{\beta} \tag{2.4.3}$$

If the feed and discharge streams flow at steady state, then no change occurs in the amount of material in the system, and (2.4.3) reduces to

$$\sum_{\alpha} dN_{\alpha} - \sum_{\beta} dN_{\beta} = 0 \tag{2.4.4}$$

We will sometimes write balance equations, such as (2.4.3) and (2.4.4), explicitly in terms of flow *rates* of material and energy, with a flow rate indicated by an over-dot. For example, in terms of molar flow rates, (2.4.3) becomes

$$\dot{N}\,dt = \sum_{\alpha} \dot{N}_{\alpha}\,dt - \sum_{\beta} \dot{N}_{\beta}\,dt \tag{2.4.5}$$

where $\dot{N} = dN/dt$. Since the time interval dt is common to all terms, (2.4.5) simplifies to

$$\dot{N} = \sum_{\alpha} \dot{N}_{\alpha} - \sum_{\beta} \dot{N}_{\beta} \tag{2.4.6}$$

2.4.3 First Law for Open Systems (Energy Balance)

At any instant, the system in Figure 2.11 has total energy $E = Ne$, where e represents the combined internal and external energies,

$$e = u + e_k + e_p \tag{2.4.7}$$

Here u is the molar internal energy, e_k is the molar kinetic energy, and e_p is the molar potential energy of the system. In many cases, changes in the kinetic and potential energies are zero or are negligible, and then e in (2.4.7) is merely the internal energy u.

Equation (2.4.7) gives the total energy in the system at any instant; now consider how that value might change. We have already cited heat transfer Q, work effects that deform the boundary W_b, and shaft work W_{sh} (see Figure 2.11). Besides these, the system energy may change because of material crossing the boundary. Material flowing in any stream α (or β) has internal energy U_α, kinetic energy $m_\alpha \text{v}_\alpha^2/2$ due to its motion relative to the system, and potential energy $m_\alpha z_\alpha g$ due to its position relative to a reference elevation. Therefore, each stream can have

$$N_\alpha e_\alpha = N_\alpha u_\alpha + \frac{1}{2} m_\alpha \text{v}_\alpha^2 + m_\alpha z_\alpha g \tag{2.4.8}$$

In addition, there is work associated with making each chunk of material flow through the system. Specifically, each volume element of any stream (α or β) contributes to the flow by deforming the volume element ahead of it, thereby doing work of the usual ($P\,dV$) form. So for any stream (α or β)

$$\delta W_i^{flow} = \pm P_i\,dV_i = \pm P_i v_i\,dN_i \qquad i = \alpha, \beta \tag{2.4.9}$$

The sign of this work term is positive for streams α entering the system and negative for streams β leaving the system.

Finally, we want to be completely general at this point, so we also consider the possibility that energy in the boundary may change during a process. We let $E_b = N_b e_b$ represent, at any instant, the total energy in the boundary. This energy can be decomposed into internal, kinetic, and potential energies, just as in (2.4.7). In many applications the boundary energy E_b is negligible, not because e_b is small, but because N_b is small.

Collecting all possible ways by which energy can be exchanged between the system and its surroundings, we obtain the total energy balance (first law) for open systems

$$d(Ne) + dE_b = \sum_{\alpha}(e_\alpha + P_\alpha v_\alpha)dN_\alpha - \sum_{\beta}(e_\beta + P_\beta v_\beta)dN_\beta + \delta Q + \delta W_b + \delta W_{sh} \tag{2.4.10}$$

Note that the total differential $d(Ne)$ on the lhs differs from the terms dN_α and dN_β on the rhs; that is, (2.4.10) allows for the possibility that the amount of material in the system can change during a process. This most general form of the first law is *always true.* However in many situations, this general form simplifies because some contributions are zero or are negligible compared to other contributions. For easy reference, we collect many of its useful forms here.

Closed systems. For closed systems, all $dN_\alpha = 0$, all $dN_\beta = 0$, while N and N_b are constant, so the general form (2.4.10) reduces to

$$N de + N_b de_b = \delta Q + \delta W \tag{2.4.11}$$

We have written W for $W_b + W_{sh}$. This closed-system form simplifies further in these special situations:

(a) Negligible external energy and negligible boundary mass,

$$dU = \delta Q + \delta W \tag{2.4.12}$$

(b) No thermal interaction (adiabatic process),

$$dU = \delta W = \delta W_{rev} \tag{2.4.13}$$

(c) No work modes (workfree process),

$$dU = \delta Q = \delta Q_{rev} \tag{2.4.14}$$

Open systems. For open systems the kinetic and potential energy terms are usually negligible compared to the internal energy terms; then, $e_i = u_i$ for the system and for each stream. Hence, the general form (2.4.10) simplifies to

$$d(Nu) + dU_b = \sum_{\alpha} h_\alpha dN_\alpha - \sum_{\beta} h_\beta dN_\beta + \delta Q + \delta W_b + \delta W_{sh} \tag{2.4.15}$$

Note we have introduced the enthalpy (2.4.1) for each stream. For steady-state flows of mass and energy, $d(Ne) = 0$, $dU_b = 0$, and

$$\frac{dN_\alpha}{dt} = \dot{N}_\alpha = \text{constant for all streams } \alpha \text{ and } \beta \tag{2.4.16}$$

$$\frac{\delta Q}{dt} = \dot{Q} = \text{constant for all heat conduits} \tag{2.4.17}$$

$$\frac{\delta W_{sh}}{dt} = \dot{W}_{sh} = \text{constant for all shaft work modes} \tag{2.4.18}$$

$$\frac{\delta W_b}{dt} = \dot{W}_b = 0 \tag{2.4.19}$$

At steady state, (2.4.19) applies because we cannot contrive a way to continuously deform the boundary. Under the restrictions (2.4.16)–(2.4.19), the general energy balance simplifies to

$$\sum_\alpha h_\alpha \dot{N}_\alpha - \sum_\beta h_\beta \dot{N}_\beta + \dot{Q} + \dot{W}_{sh} = 0 \qquad \textit{steady state} \tag{2.4.20}$$

For steady-state workfree processes, (2.4.20) shows that the heat transferred can be computed from the enthalpy change between inlets and outlets; common applications include steady-state heat exchangers. For steady-state adiabatic processes, (2.4.20) shows that the shaft work can be obtained from the enthalpy change; these situations arise in adiabatic pumps, turbines, and compressors.

2.4.4 The Second Law for Open Systems (Entropy Balance)

In addition to material and energy balances, we may also perform an entropy balance on the system in Figure 2.11. But since entropy is not conserved (entropy can be generated in the system and its boundary), we must appeal to a more general form of the stuff equation, namely (1.4.1). The balance can be written as a generalization of the first part of the second law (2.3.8), in which terms are now included to account for entropy carried by the streams:

$$d(Ns) + dS_b = \sum_\alpha s_\alpha dN_\alpha - \sum_\beta s_\beta dN_\beta + \frac{\delta Q_{ext}}{T_{ext}} + dS_{gen} \tag{2.4.21}$$

This general form of the second law is *always true*. Note that (2.4.21) contains no work effects—the system entropy is not affected by work interactions. Further, note that the second part of the second law (2.3.9) still applies, so the generation term in (2.4.21) must be positive or zero. In contrast, the system entropy S may increase, decrease, or remain constant. For closed systems (2.4.21) reduces to (2.3.8).

If the process is a steady state, then no net change occurs in the amount of entropy in the system or boundary and (2.4.21) reduces to

$$\sum_\alpha s_\alpha dN_\alpha - \sum_\beta s_\beta dN_\beta + \frac{\delta Q_{ext}}{T_{ext}} + dS_{gen} = 0 \qquad \textit{steady state} \qquad (2.4.22)$$

Further, if the boundary is well-insulated, then we are left with

$$\sum_\alpha s_\alpha dN_\alpha - \sum_\beta s_\beta dN_\beta + dS_{gen} = 0 \qquad \textit{steady state, adiabat} \qquad (2.4.23)$$

Finally, if a reversible change occurs, then $dS_{gen} = 0$, and (2.4.23) for an insulated steady-flow system reduces to

$$\sum_\alpha s_\alpha dN_\alpha - \sum_\beta s_\beta dN_\beta = 0 \qquad \textit{steady state, reversible adiabat} \qquad (2.4.24)$$

Rate forms of (2.4.22)–(2.4.24), analogous to (2.4.20) can easily be written.

2.4.5 Feasibility Analyses

The material, energy, and entropy balances presented in § 2.4.2–2.4.4 must be obeyed by *any* process that does not involve chemical or nuclear reactions. Consequently, they can be used to help troubleshoot problems that may arise in many process operations; they may also be used to test the thermodynamic feasibility of a process that may be proposed during a design project. To be feasible a process must satisfy

(a) conservation of matter (2.4.3),

(b) the first law (2.4.10), and

(c) the second law (2.4.21).

The conservation equations for matter (2.4.3) and energy (2.4.10) provide constraints on quantities, and therefore they allow us to test for consistency in the specifications of a proposed process. For example, with these conservation laws we may be able to test whether the proposed outputs of matter or energy are consistent with the proposed inputs. However, the entropy balance (2.4.21) is not a conservation law, and therefore it does not provide a check on quantities or consistency. Instead, it provides a constraint on the direction of a proposed process. Some proposed processes can be performed in both forward and reverse directions, but many others can be performed in only one way. In the latter cases, the entropy balance can be used to identify the ranges of operating variables (temperatures, pressure, flow rates) that must be used to make a proposed process proceed in the desired direction.

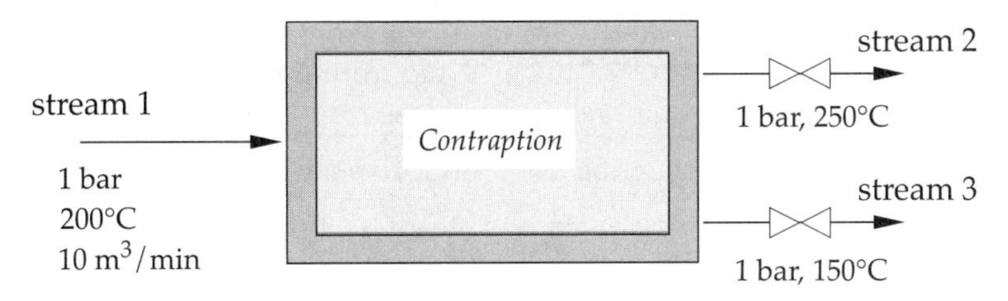

Figure 2.12 Schematic of a device intended to convert a feed stream (1) into two discharge streams, one hotter (2), the other cooler (3) than the feed

Although the general stuff equations (2.4.3), (2.4.10), and (2.4.21) are always true, they may not always be useful. To be useful, sufficient information must be available from calculations or measurements. Specifically, to test whether a process satisfies the first law, we must have either (a) complete specifications of the initial and final states of the system, or (b) values for both the heat and the work. To test whether the second law is satisfied, we must know the value for the heat; then we would use (2.4.21) to compute dS_{gen} and determine whether the second part of the second law is obeyed.

But when these criteria are met, the thermodynamic stuff equations are powerful and versatile. In particular, they can be implemented without knowing the detailed mechanisms by which a proposed process is to accomplish its task. This occurs because the first and second laws establish equivalences between process variables (Q and W) and changes in system variables (such as u, h, and s).

2.4.6 Example

How do we use the thermodynamic stuff equations to test the feasibility of a proposed process?

Dr. Emmett Brown has built a mysterious contraption which is housed in an insulated container; no wires or shafts penetrate the container walls. The device is supplied with steam at 1 bar, 200°C, and a steady rate of 10 m^3/min. The device splits the feed into two streams, which leave the device at steady flow rates. Doc Brown claims that stream 2 leaves the device at 1 bar, 250°C while stream 3 leaves at 1 bar, 150°C. A schematic is shown in Figure 2.12. Relevant thermodynamic properties of each stream are given in Table 2.1. Can this device perform as advertised?

Table 2.1 Properties of steam at 1 bar; from steam tables in [6]

	Stream 1	**Stream 2**	**Stream 3**
T (°C)	200.	250.	150.
v (m^3/kg)	2.172	2.406	1.936
u (kJ/kg)	2658.	2733.	2582.
h (kJ/kg)	2875.	2974.	2776.
s (kJ/kg K)	7.834	8.032	7.613

In Table 2.1 the values for u and s are relative to zero values at the reference state, which was chosen to be 0°C and 1.013 bar. Note from the table that each stream has values for $Pv = h - u$ that are small relative to those for u and h.

System. We choose the system to be the contraption and the steam within.

Step 1: *Apply conservation of mass.* Steady-state mass flow through the device must obey

$$\dot{m}_1 = \dot{m}_2 + \dot{m}_3 = 4.6 \text{ kg/min} \tag{2.4.25}$$

Note that this does not necessarily mean that the flow rates of streams 2 and 3 will be the same.

Step 2: *Apply the first law.* For a steady flow situation with negligible changes in external kinetic and potential energies, the first-law form (2.4.20) applies,

$$\sum_{\alpha} h_\alpha \dot{m}_\alpha - \sum_{\beta} h_\beta \dot{m}_\beta + \dot{Q} + \dot{W}_{sh} = 0 \tag{2.4.26}$$

Here we have used a mass basis rather than a mole basis. Recall that the index α runs over inlet streams, while index β runs over outlets. In this problem, the device is insulated, no shaft work crosses the boundary, and we presume the boundary itself has negligible mass. Then (2.4.26) reduces to

$$h_1 \dot{m}_1 - h_2 \dot{m}_2 - h_3 \dot{m}_3 = 0 \tag{2.4.27}$$

Solving (2.4.25) and (2.4.27) simultaneously with enthalpies from Table 2.1, we find

$$\dot{m}_2 = \dot{m}_3 = \frac{1}{2} \dot{m}_1 = 2.3 \text{ kg/min} \tag{2.4.28}$$

This means that, to satisfy the first law, valves must be adjusted so that streams 2 and 3 leave the device at the same mass flow rate.

Step 3: *Apply the second law.* For a system flowing at steady state through an insulated enclosure, the second law in the form (2.4.23) applies

$$\sum_{\alpha} s_\alpha \dot{m}_\alpha - \sum_{\beta} s_\beta \dot{m}_\beta + \dot{S}_{gen} = 0 \tag{2.4.29}$$

where, again, we have used a mass basis. The entropy generation term must always be positive or zero; therefore, we must have

$$\dot{S}_{gen} = s_2 \dot{m}_2 + s_3 \dot{m}_3 - s_1 \dot{m}_1 \geq 0 \tag{2.4.30}$$

Using the mass flow rates from steps 1 and 2 and the entropy values from Table 2.1, we find

$$\dot{S}_{gen} = -0.053 \text{ kJ/(min K)} \tag{2.4.31}$$

So even though the first law might be satisfied, the second law would be violated; the contraption will not function as planned. Can you contrive modifications of the process that would make the desired initial and final states feasible?

Comment. Note that we have made a definitive statement about the feasibility of a proposed process *without* knowing details about the process itself. We are able to do so because the first and second laws effectively replace process-dependent heat and work effects with process-independent changes in state functions.

2.5 SUMMARY

In this chapter we have developed the first and second laws for closed and open systems. For closed systems both laws are motivated by the desire to relate the process variables Q and W to changes in system properties. To emphasize this common theme, we have stated each law in two parts: part 1 defines a new state function (either U or S) and part 2 imposes limitations on how the new state function changes with certain changes of state. For closed systems, the first law asserts that an exact differential (dU) is obtained from the algebraic sum of δQ and δW, while the second law asserts that an exact differential (dS) is obtained by applying an integrating factor to δQ_{rev}. If a quantity forms an exact differential, then it is a system property, and changes in its value are not affected by the process that connects two states.

For open systems, the first and second laws are particular forms of the general stuff equation presented in § 1.4. The first law represents an energy balance on a system, and it asserts that energy is a conserved quantity. Similarly, the second law represents an entropy balance, but the second law asserts that entropy is not conserved: through the actions of dissipative forces, entropy is created (but never consumed) during any irreversible process.

The first and second laws are concise statements of constraints that Nature imposes on energy transfers involved in any process. As such, they can be used in several ways to obtain quantitative information about processes that connect a given initial state to a given final state. A typical engineering question is to determine the amounts of Q and W needed for a proposed change of state. If the process path is known or can be reliably estimated, then Q and W may be computed directly, without recourse to the laws of thermodynamics. But in most situations, the process path is unknown and in fact unknowable; then, our strategy would be to invoke the first and second laws, so that we may perform the analysis solely in terms of system-dependent state functions. Variations on this problem are also common; for example, we may know an initial state and need to identify the final state that would result when known amounts of Q and W cross the boundary.

Other applications of the laws include feasibility analyses, such as in the Example of § 2.4.6. In these situations we usually know the initial and final states that we want; the question is whether it is thermodynamically possible to start from the given initial state and achieve the desired final state. If the answer is no, then we consider whether adjustments can be made in either the initial or final state to obtain a change that is thermodynamically possible. If the answer is yes, then we may proceed to a detailed design of a process itself, including an assessment of economic feasibility.

But although the first and second laws meet our objective of relating Q and W to system properties, that objective has been obtained at a price. The price is that, while the first and second laws have identified new system properties, U and S, those new properties are conceptuals, not measurables. To obtain full benefit from the first and second laws, we must relate U and S to measurables—preferably measurable operating variables such as temperature, pressure, and composition. And so, the first and second laws have certainly achieved the economy of thought characteristic of science, but before we can apply those laws in an engineering setting, we must establish relations between conceptuals and measurables.

LITERATURE CITED

[1] O. Redlich, *Thermodynamics: Fundamentals, Applications*, Elsevier, Amsterdam, 1976.

[2] C. Carathéodory, "Untersuchungen über die Grundlagen der Thermodynamik," *Math. Ann.*, **67**, 355 (1909).

[3] S. Chandrasekhar, *An Introduction to the Study of Stellar Structure*, Dover, New York, 1958, ch. 1.

[4] J. Kestin, *A Course in Thermodynamics*, vol. 1, revised printing, Hemisphere Publishing Corporation, New York, 1979.

[5] For a readable discussion of Joule's experiments, see T. W. Chalmers, *Historical Researches*, Dawsons of Pall Mall, London, 1968.

[6] L. Haar, J. S. Gallagher, and G. S. Kell, *NBS/NRC Steam Tables*, Hemisphere Publishing Corp., New York, 1984.

[7] J. Rifkin and T. Howard, *Entropy: A New World View*, Bantam Books, New York, 1980, p. 57.

PROBLEMS

2.1 Argon is held in a vertical piston-cylinder apparatus. The piston has a diameter of 5 cm and a weight-pan is attached to the piston shaft.

(a) If atmospheric pressure is 1 bar and the gas is at 1.5 bar, what is the mass of piston and pan?

(b) How much work is done when a 50-kg mass is placed on the pan and the gas is compressed isothermally at 300 K?

(c) Approximately how much work is done if 50 kg of sand are placed on the pan 1 grain at a time while the gas is kept at 300 K?

(d) Repeat parts (b) and (c) for processes that are adiabatic and begin at 300 K. For such processes, PV^{γ} is a constant where $\gamma = 1.7$ for ambient argon.

2.2 For every workfree process between the same initial and final states, show that the heat effect is the same, regardless of how the heat is transferred.

2.3 For any workfree isothermal process on a closed system, show that $q = T\Delta s$.

2.4 In some of Joule's experiments, work was done on water held in an adiabatic calorimeter. The work was done by a rotating paddle, driven by falling weights. Assume the volume of the water remains constant during these experiments.

(a) In one experiment a 25-kg mass was allowed to fall 20 times through a height of 2 m; what was the maximum amount of work done?

(b) If a 25-kg mass were fired into the calorimeter and brought to a standstill, what should its initial velocity be to accomplish the same effect as in (a)?

(c) If the calorimeter held 1.2 kg of water and if process (a) caused the water temperature to rise from 288 to 290 K, what is the numerical value for the factor that connects temperature rise to work under these conditions?

2.5 In some of his experiments, Joule used electrical work rather than mechanical work. To achieve the same effect as in Problem 2.4(a), for what duration would electrical work have to be provided to the calorimeter, if the current originated from a 100-volt battery and it encountered a 1000-ohm resistance?

2.6 Steam flows at 2.5 kg/s through a turbine, generating electricity at the rate of 1 MW. The inlet velocity of the steam is 100 m/s and the outlet velocity is 30 m/s; the inlet is located 30 m above the outlet. Of the total power generated, estimate the fraction contributed by the change in kinetic energy of the steam and the fraction contributed by the change in potential energy.

2.7 Fill in the missing entries (...) for the signs in the following table. Here ΔU, Q, and W apply to a closed system. (Note that "0" indicates a value of zero.)

	Sign of ΔU	**Sign of W**	**Sign of Q**
(a)	0	+	...
(b)	0	...	+
(c)	+	...	0
(d)	−	+	...
(e)	...	+	−
(f)	...	+	−
(g)	−	0	...

2.8 One mole of hydrogen initially at 10 bar, 300 K expands reversibly to 1 bar, 500 K. The expansion is carried out along a straight-line path on a Pv diagram. Determine Q and W. If necessary, assume $Pv = RT$, $(\partial u/\partial T)_v = 5R/2$, and $(\partial u/\partial v)_T = 0$.

2.9 Consider a closed system that can change from state 1 to state 2 via both a reversible change and an irreversible process. Assume boundary effects are negligible. Combine the first law and the expression for lost work (2.1.18) with the second-law forms (2.3.5) and (2.3.8) to derive (2.3.10).

2.10 Consider one mole of a gas that obeys $Pv = RT$ and has $(\partial u/\partial T)_v = 3R/2$ with $(\partial u/\partial v)_T = 0$. The gas undergoes a reversible change of state, so the first law

$$\delta q_{rev} = du - \delta w_{rev} \tag{P2.10.1}$$

can be written

$$\delta q_{rev} = \left(\frac{\partial u}{\partial T}\right)_v dT + P dv \tag{P2.10.2}$$

(a) Determine whether or not (P2.10.2) is an exact differential.

(b) For the special case of a reversible adiabatic change, (P2.10.2) becomes

$$0 = \left(\frac{\partial u}{\partial T}\right)_v dT + P dv \tag{P2.10.3}$$

Find the equation for the process path on a Tv diagram by solving the differential equation (P2.10.3) using exact differentials. If (P2.10.3) is not exact, you must find an integrating factor.

(c) If we replace δq_{rev} with the definition of the entropy, then (P2.10.2) is

$$ds = \frac{1}{T}\left(\frac{\partial u}{\partial T}\right)_v dT + \frac{P}{T} dv \tag{P2.10.4}$$

Determine whether or not this expression is an exact differential.

2.11 (a) Show that there are only two situations in which isentropic processes can occur on closed systems: either the process is reversible and adiabatic or the process removes heat from the system. Are all adiabatic processes isentropic?

(b) For open systems, show that it is possible for the system entropy to decrease or remain unchanged, even when heat enters the system. For the isentropic case, give the conditions that must apply to the inlet and outlet streams.

2.12 For a certain process, one mole of neon is needed at 300 K, 1 bar; the gas is available at 500 K, 3 bar. Determine whether or not the required change of state can be accomplished adiabatically. If necessary, assume neon obeys $Pv = RT$ and has $(\partial u/\partial T)_v = 3R/2$ with $(\partial u/\partial v)_T = 0$.

2.13 Assume carbon monoxide obeys $Pv = RT$, $(\partial u/\partial T)_v = 5R/2$, and $(\partial u/\partial v)_T = 0$. On a Pv diagram, sketch the states accessible from 20°C, 1 bar by (a) reversible adiabats and (b) real adiabatic processes. Are any states inaccessible by adiabatic processes?

2.14 When a closed system is compressed isothermally, we expect to remove heat from the system.

(a) Use the first and second laws to confirm this.

(b) Cite a situation in which this would *not* be true.

(c) When heat must be removed, show that more heat must be removed from a real compression than from a reversible one.

2.15 The Second Law is sometimes stated as *The total energy of the universe is constant, but the total entropy is continually increasing and must ultimately reach a maximum.*

(a) Is all of this always true? If not, what other statements should be made?

(b) Does this statement automatically imply an ultimate "heat death" of the universe where all heterogeneities of matter and energy are eliminated? Explain.

2.16 Rifkin and Howard [7] quote Bertrand Russell as saying *Whenever there is a great deal of energy in one region and very little in a neighboring region, energy tends to travel from the one region to the other, until equality is established. This whole process may be described as a tendency towards democracy.*

(a) Does energy always flow from a high "concentration" to a low one? If not, give an example in which it does not.

(b) This might imply that democracy is an equilibrium state and suggests that it will be the case in which energy is evenly distributed. Do you believe it? What about the distribution of entropy?

(c) Do you think such concepts can really be applied to human affairs?

(d) On what length scale might Russell's arguments become inexact?

2.17 One analysis of manufacturing efficiency can be made by determining wasted energy through an analysis of any heat and work effects together with the changes of state that the materials undergo. To illustrate, consider a process that changes a closed system's internal energy by Δu, entropy by Δs, and volume by Δv; no chemical reactions or changes of composition occur. The only energy input is heat q from condensing steam at the temperature T_S. The energy outputs are heat and volumetric work to the environment, which is at fixed T_o and P_o. Then, inefficiencies in the process can be measured by the extra amount of heat required due to irreversibilities. Use the first and second laws to show that this extra heat is given by

$$q_{lost} = q - \frac{T_S}{T_S - T_o}(\Delta u + P_o \Delta v - T_o \Delta s) \tag{P2.17.1}$$

Give an example of an industry where this analysis might be usable. Suggest how the values of Δu, Δs, and Δv might be determined.

2.18 A fluid is leaking steadily through a well-insulated valve at the end of a pipe.

(a) Do any *always true* relations connect conceptual properties of the fluid just upstream of the valve with those just downstream? If so, write them.

(b) Are there any *always true* relations that connect measurables just upstream of the valve with those just downstream? If so, write them.

2.19 A "heat engine" is any cyclic device that takes heat from a high-temperature reservoir, does useful work, and expels unused heat to a low-temperature reservoir. For a specified amount of heat into the engine, show that any real (i.e., irreversible) heat engine always produces less useful work than would a reversible heat engine operating between the same two reservoirs.

2.20 In analyzing real processes, when is the entropy balance helpful as opposed to merely being an additional equation with an additional unknown?

2.21 A "heat pump" is any cyclic device that uses work from the surroundings to move heat from a low-temperature reservoir to a high-temperature reservoir. For a specified amount of work, show that any real heat pump always removes less heat from the low-temperature reservoir than would a reversible heat pump between the same two reservoirs.

2.22 For the following situations, write appropriate forms of the mass and energy balances, but include only terms that are nonzero. *The problems are not to be solved.*

(a) Steam flows steadily through a horizontal, insulated nozzle. Find the diameter of the outlet that gives no change in velocity.

(b) A battery-driven toy runs until it stops. How much energy was in the battery at the start?

(c) Two metal blocks initially at different temperatures make contact in an insulated container. How much heat was transferred? From which block?

(d) Steam drives a turbine to steadily generate electricity. There are two steam outlets. What is the state of the steam in the second outlet?

(e) A 100-watt incandescent light bulb is turned on. What is the temperature of the glass surface after 10 minutes?

2.23 A closed insulated vessel having rigid walls is divided into two compartments by a membrane. One compartment is loaded with a fluid at state 1; the other compartment is evacuated. The membrane ruptures, allowing the fluid to fill the vessel. Show that the final state of the fluid (2) must have $u_2 = u_1$ but $s_2 > s_1$, regardless of the nature of the fluid. (This process is called a Joule expansion.)

2.24 A well-insulated cylinder, having a volume of 1 m^3, is initially filled with 1 kmole of helium at 300 K. A valve on the cylinder is opened, allowing the pressure to fall rapidly to 1 bar; then the valve is closed. After a period of time, a gauge reads 92.86 K for the temperature of helium in the cylinder. Discuss whether this temperature reading could be correct.

2.25 A rigid insulated vessel is divided into two compartments: one contains a fluid at T_1, P_1 and the other is under vacuum. The compartments are connected by a pipe fitted with a pressure relief valve; the relief valve bursts. You, as the engineer responsible for the unit, examine the system two hours later.

(a) What always true relations exist to connect initial conceptual properties to the final conceptuals?

(b) What always true relations exist to connect initial measurables to final measurables? Are these enough to determine whether pressure and temperature gauges on the vessel have been damaged?

2.26 In (2.4.1) we defined the enthalpy H to be the sum $(U + PV)$. We already know that U is a state function.

(a) Without using H, prove that the product PV is also a state function.

(b) Prove that the sum of *any* two state functions is also a state function.

3

FUNDAMENTAL RELATIONS

In the previous chapter we accomplished our first objective: we showed how the process variables heat and work are related to changes in system properties, the internal energy U and the entropy S. Those relations are provided by the first and second laws. Now our problem is to learn how to compute changes in U and S. Since U and S cannot be obtained directly from experiment, we must first relate ΔU and ΔS to measurable state functions, particularly temperature, pressure, volume, composition, and heat capacities. When we can establish such relations, our strategy in a process analysis can take the path on the left branch of the diagram shown in Figure 3.1.

Unfortunately, ΔU and ΔS are not always simply related to measurables, nor are ΔU and ΔS always directly related to convenient changes of state. So to ease conceptual and computational difficulties, we create additional state functions. Then we must establish how ΔU and ΔS are related to these new state functions and, in turn, how changes in the new functions are related to measurables. In these situations, our strategy follows the right branch of the diagram in Figure 3.1. In this chapter we develop relations that allow us to follow both strategies represented in the figure.

Our long-term goal is to be able to analyze processes, and since processes cause changes in system states, we begin by discussing the conditions that must be satisfied to characterize a state (§ 3.1). Then we introduce new conceptual state functions (§ 3.2) and show how they respond to changes in temperature, pressure, volume, and composition (§ 3.3 and § 3.4). Next we summarize those differential relations that enable us to use measurables to compute changes in conceptuals (§ 3.5); the relevant measurables include heat capacities, volumetric equations of state, and perhaps results from phase equilibrium experiments.

Lastly, we combine the first and second laws to obtain explicit expressions for the reversible heat and reversible work (§ 3.6 and § 3.7). Those expressions are general in that they apply to mixtures of any number of components in open or closed systems; however, as with everything done in Part I of this book, the expressions apply only to a single homogeneous phase. The expressions for Q_{rev} and W_{rev} given in § 3.7 complete the program outlined in Figure 3.1.

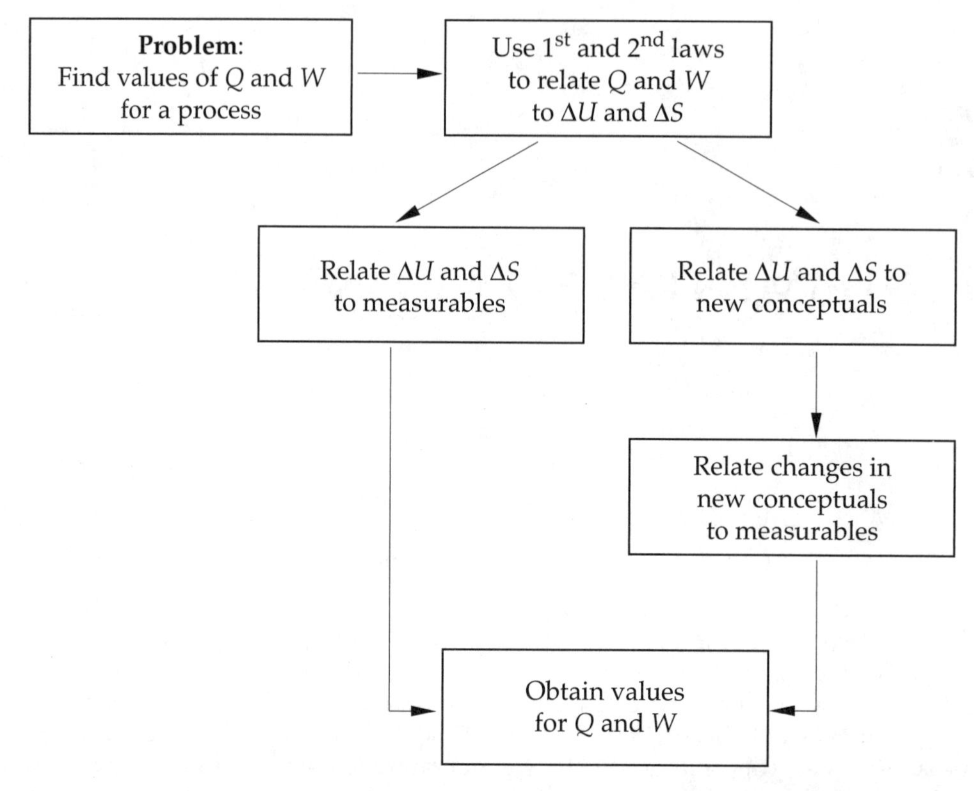

Figure 3.1 A thermodynamic analysis of a process usually proceeds by using the first and second laws to relate path functions to changes in conceptual properties. Then values of property changes are computed either (i) directly, by relating them to measurables (*left*), or (ii) indirectly, by first relating those changes to still other conceptual properties, and in turn, relating those conceptuals to measurables (*right*).

3.1 STATE OF SINGLE HOMOGENEOUS PHASES

One objective of thermodynamics is to analyze how the state (§ 1.2.2) responds when a system undergoes a process or sequence of processes (§ 1.3). In this section we address two important questions that naturally arise concerning relations between process and state: In § 3.1.1 we determine the minimum number of interactions that are required to change the state and in § 3.1.2 we determine the number of property values required to identify the final state. We restrict our attention here to multicomponent systems forming a single homogeneous phase; the generalizations to multiphase and reacting systems are considered in Chapters 9 and 10.

We distinguish between intensive state and extensive state. The *intensive* state can be identified solely in terms of intensive properties, and therefore it does not involve amounts of material. In contrast, identification of an *extensive* state must include a value for at least one extensive property, usually either the total amount of material or the total volume. Often only intensive states are needed to perform process analyses, while extensive states are usually needed to perform process designs.

3.1.1 Number of Interactions to Change a State

To change a thermodynamic state, we stand in the surroundings and apply interactions that cross the boundary. So we would like to know the number of orthogonal interactions that are available for changing the extensive state,

$$\mathcal{V}_{max} \equiv \begin{pmatrix} \text{number of available} \\ \text{orthogonal interactions} \end{pmatrix} \tag{3.1.1}$$

For a mixture of C components, there are C independent mole numbers, each of which could be manipulated through its own interaction. In addition, most systems of interest have the thermal interaction plus a work interaction that can change the system's volume. Therefore, in most cases the maximum number of orthogonal interactions will be given by

$$\mathcal{V}_{max} = C + 2 \qquad \textit{single, homogeneous phase} \tag{3.1.2}$$

If other orthogonal work modes are present, such as electrical or surface work, then the number on the rhs of (3.1.2) would increase accordingly.

The value given by (3.1.2) represents the maximum number of orthogonal interactions. However, the actual number will be less when external constraints are imposed. An *external constraint* blocks or controls an interaction so that it is not available for manipulating the system. For example, we might insulate the system to block the thermal interaction. Let S_{ext} count the number of external constraints imposed on interactions. Then, to manipulate the state, we would have

$$\mathcal{V} = \mathcal{V}_{max} - S_{ext} = C + 2 - S_{ext} \qquad \textit{single, homogeneous phase} \tag{3.1.3}$$

A special case of (3.1.3) occurs when we block all interactions that would change the amounts of components in a closed system. Then $S_{ext} = C$, and (3.1.3) reduces to

$$\mathcal{V} = 2 \qquad \textit{single, homogeneous phase and } C \textit{ known amounts} \tag{3.1.4}$$

So we have only two interactions available to manipulate the state. This result is *Duhem's theorem* applied to a single homogeneous phase. The extension of (3.1.4) to multiphase systems is developed in Chapter 9.

To change an intensive state, we have two possibilities. (a) We might fix the amounts of all components, so (3.1.4) applies. Then we can change the intensive state using the thermal interaction or the PV work mode or both. (b) We might want to change the composition. But we cannot directly manipulate a mole fraction, we can only change a composition by changing amounts of components, so (3.1.3) applies. Therefore (3.1.3) generally gives the number of available interactions for changing both extensive and intensive states. Note that it is possible to change the extensive state without changing the intensive state.

3.1.2 Number of Properties to Identify an Equilibrium State

After a change is finished, the system relaxes to an equilibrium state; that state is identified by giving values for properties, so we need to know how many property values are required. Experiment shows that a modest number of properties are sufficient to identify the state; that is, only a few properties are *independent*. To test for a complete set of independent properties, we specify values for $\mathcal{F}_{ex}$ properties $\{p_i, i = 1, 2, \dots, \mathcal{F}_{ex}\}$. If the value of each property p_i can be freely manipulated, while the value of any other property not in the set cannot be freely manipulated, then the $\mathcal{F}_{ex}$ properties $\{p_i\}$ form a complete set of independent properties. This implies that any property F is related to the properties p_i through some function ψ,

$$F = \psi(p_1, p_2, \dots, p_{\mathcal{F}_{ex}}) \tag{3.1.5}$$

A relation such as (3.1.5) is called an *equation of state*. The obvious question now is, What must be the value of $\mathcal{F}_{ex}$?

Our initial guess is likely to be that $\mathcal{F}_{ex} = \mathcal{V}$, which would mean that the number of properties needed to identify the extensive state is the same as the number of interactions available for manipulating the extensive state. But, in fact, $\mathcal{F}_{ex}$ may differ from $\mathcal{V}$ because of constraints. There are competing effects from two kinds of constraints.

(a) External constraints were introduced in § 3.1.1. But although external constraints reduce the number of available interactions during a state change, once equilibrium is established, external constraints do not affect the number of properties needed to identify the final state. It is true that an external constraint may couple two otherwise independent properties while a process is being carried out, but that coupling does not apply to the equilibrium state. For example, consider a pure fluid in an isolated system; hence, $\mathcal{S}_{ext} = 3$ and (3.1.3) gives $\mathcal{V} = 0$. That is, no interactions are available to manipulate an isolated system. Nevertheless, $\mathcal{F}_{ex} \neq 0$; that is, an essentially infinite number of states can be isolated, so we still need some number of properties to identify the particular equilibrium state confined to an isolated system.

(b) Internal constraints are those imposed by Nature through such mechanisms as multiphase and reaction equilibria. Internal constraints couple otherwise independent properties, thereby reducing the total number needed to identify equilibrium states. Let $\mathcal{S}$ represent the number of internal constraints, then the number of independent properties $\mathcal{F}_{ex}$ needed to identify the extensive state is given by

$$\mathcal{F}_{ex} = \mathcal{V}_{max} - \mathcal{S} \tag{3.1.6}$$

Using (3.1.2) for the usual situations of interest, we have

$$\mathcal{F}_{ex} = C + 2 - \mathcal{S} \qquad \textit{single, homogeneous phase} \tag{3.1.7}$$

where at least one of the $\mathcal{F}_{ex}$ properties must be extensive. The only internal constraints available to homogeneous one-phase fluids are those that occur at vapor-liquid critical points. Vapor-liquid critical points are one-phase situations having $\mathcal{S} = 2$. Then a pure fluid would have $\mathcal{F}_{ex} = 1$: we need only the amount of material to identify the extensive state of a pure fluid at its critical point.

To identify intensive states, we need $(C-1)$ independent mole fractions, rather than C independent mole numbers. So the number of independent properties needed to identify an intensive state is

$$\mathcal{F} = \mathcal{F}_{ex} - 1 \tag{3.1.8}$$

Hence,

$$\mathcal{F} = C + 1 - S \qquad \textit{single, homogeneous phase} \tag{3.1.9}$$

where all $\mathcal{F}$ properties must be intensive. The quantity $\mathcal{F}$ is often called the number of *degrees of freedom*. The Gibbs phase rule extends (3.1.9) to multiphase systems.

3.1.3 Proper Counting

We must take care to avoid misusing or misinterpreting the values given by $\mathcal{V}$ and $\mathcal{F}_{ex}$. Here are three common pitfalls to avoid.

(a) Do not confuse $\mathcal{V}$ with $\mathcal{F}_{ex}$. $\mathcal{V}$ is the number of orthogonal interactions needed to manipulate a system, while $\mathcal{F}_{ex}$ is the number of independent properties needed to identify an extensive state after a process is completed and equilibrium is established. External constraints reduce the number of interactions available for manipulation, while internal constraints reduce the number of properties required for identification. The numbers $\mathcal{F}_{ex}$ and $\mathcal{F}$ play crucial roles in testing whether thermodynamic problems are well-posed, that is, whether the number of knowns is sufficient to allow us to compute values for unknowns.

(b) A second pitfall is to assume that when you have established values for $\mathcal{F}$ independent variables, then you have *uniquely* defined the intensive state. This may not be so: an equation of state may not be monotone in its independent variables. For example, some properties of some pure fluids pass through extrema, as in Figure 3.2. Such

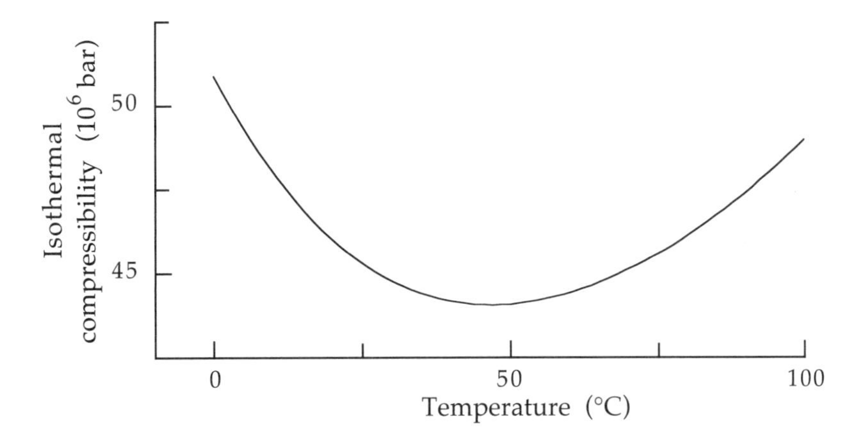

Figure 3.2 Isothermal compressibility κ_T (§ 3.3.3) of pure liquid water at 1 atm. For pure liquid water, $\mathcal{F} = 2$. Nevertheless, if we were to specify that $P = 1$ atm and that water had an isothermal compressibility $\kappa_T = 46(10^{-6})/\text{bar}$, we still could not uniquely identify the intensive state because κ_T for pure liquid water is not monotone in temperature. Data taken from [1].

extrema are common in fluid mixtures. For example, many mixtures in vapor-liquid equilibrium exhibit *azeotropes*: maxima or minima in isothermal plots of pressure vs. mole fraction and in isobaric plots of temperature vs. mole fraction.

(c) A third pitfall can occur when exercising the option of *replacement*. Note that $\mathcal{F}$ tells us only the *number* of properties needed as arguments in an equation of state; so long as those arguments are independent and intensive, we are free to choose their identities. But whether a particular property is legitimate depends on the identities of the other properties to be used. For example, assume we have an equation of state in the form $P = f(T, v)$; then we might replace the molar volume v with the density ρ and use the equation in the form $P = f'(T, \rho)$. However, we cannot keep v and replace temperature T with density ρ; that is, $P \neq f''(\rho, v)$ because a one-to-one correspondence exists between ρ and v (specifically, $\rho = 1/v$). So, when you replace one argument with another, you should confirm that the new arguments are mutually independent.

3.2 FUNDAMENTAL EQUATIONS

Often we can simplify an analysis by combining the first and second laws to eliminate heat and work in favor of changes in state functions. Such replacements yield the *fundamental equations* of thermodynamics. These equations allow us to determine the effects of state changes without requiring us to evaluate heat and work. In what follows, we first present the forms for closed systems (§ 3.2.1) and then give those for open systems (§ 3.2.2).

3.2.1 Closed Systems

Consider a closed homogeneous system that has negligible boundary mass and that has only two interactions with its surroundings: the thermal interaction and one mechanical work mode that can alter the system volume. Through these interactions the system is subjected to some differential process that changes the state. Since U is a state function, dU is unaffected by the reversibility of the process; so,

$$dU = dU_{rev} = dU_{irr} \tag{3.2.1}$$

Substituting the first law (2.2.4), we have

$$dU = \delta Q_{rev} + \delta W_{rev} = \delta Q_{irr} + \delta W_{irr} \tag{3.2.2}$$

For the reversible change we write $\delta Q_{rev} = TdS$ and $\delta W_{rev} = -PdV$. So making these substitutions leaves

$$dU = TdS - PdV = \delta Q_{irr} + \delta W_{irr} \tag{3.2.3}$$

or simply

$$dU = TdS - PdV \tag{3.2.4}$$

which is the *fundamental equation* for closed systems. We emphasize that (3.2.4) applies to any process regardless of reversibility. However, for irreversible processes ($T\,dS$) is not the heat transferred nor is ($-P\,dV$) the work done; in fact, as we discussed in § 2.3.2, $\delta Q_{irr} < (T\,dS)$ and $\delta W_{irr} > (-P\,dV)$. This means that, in irreversible processes, heat and work are distributed between ($T\,dS$) and ($-P\,dV$) in some unknown way; that distribution depends on the degree of irreversibility.

According to (3.1.9), we need two independent intensive properties from which to construct an equation of state for a pure single phase. The fundamental equation (3.2.4) implies that if we want to use the internal energy as the dependent variable, then we should use S and V as independent variables,*

$$U = U(S, V) \tag{3.2.5}$$

That is, S and V are the "natural" or "canonical" variables when we choose an equation of state to be explicit in U. Of course, we could express U in any pair of independent intensive quantities; for example, we could use $U(T, P)$ or $U(S, P)$ or $U(T, V)$, etc. But $U(S, V)$ is the natural choice because if we knew the function $U(S, V)$ for our system, that knowledge would be sufficient to determine values for the remaining properties in the fundamental equation. To do so, we would merely need to evaluate derivatives,

$$T = \left(\frac{\partial U}{\partial S}\right)_{VN} \quad \text{and} \quad -P = \left(\frac{\partial U}{\partial V}\right)_{SN} \tag{3.2.6}$$

However, if we had some other functional representation for U, we would not have sufficient information to compute the remaining properties in (3.2.4). For example, say we had the function $U(S, P)$. Then to obtain the volume V, instead of differentiating, we would have to integrate the second differential equation appearing in (3.2.6), and to evaluate that integral, we would need an integration constant; that is, we would need a value for U at some volume V. Consequently, $U(S, P)$ is not a complete description of our system and this is why we say that $U(S, V)$ is "fundamental."

But for engineering use, S and V are not convenient independent variables; S, for example, is not measurable at all and V may not be easy to control in a laboratory or industrial situation. We would prefer to use easily measured and controlled properties as independent variables; in particular, we would like to use T and P. But if we merely replace S and V in (3.2.5) with T and P, so we have $U(T, P)$, then we will have lost information and made subsequent computations of ΔU more complicated. Hence, if we want to replace S and V as independent variables but preserve the fundamental nature of the equation of state, then we must also change the dependent variable U. This can be done via Legendre transforms.

Legendre transformation is a mathematical technique for exchanging one independent variable for another in a function; see Appendix A. One consequence of such transformations is that, not only do we obtain a new independent variable, but we also obtain a new function. Legendre transforms have the structure

* For closed systems, relations among thermodynamics properties can be developed using extensive properties (such as U and V) or intensive properties (such as u and v). We usually use extensive properties.

$$\begin{pmatrix}\text{new}\\ \text{function}\end{pmatrix} = \begin{pmatrix}\text{old}\\ \text{function}\end{pmatrix} - \begin{pmatrix}\text{old}\\ \text{variable}\end{pmatrix}\begin{pmatrix}\text{new}\\ \text{variable}\end{pmatrix} \qquad (3.2.7)$$

where

$$\begin{pmatrix}\text{new}\\ \text{variable}\end{pmatrix} = \frac{\partial(\text{old function})}{\partial(\text{old variable})} \qquad (3.2.8)$$

Because of (3.2.8), one independent variable cannot be replaced by any arbitrarily chosen variable. For transformations of the fundamental equation, this means that the product of the old and new variables must have dimensions of energy.

In the fundamental equation (3.2.4) let us choose to replace V with its conjugate variable, the pressure P. Then the Legendre transform is

$$H = U - (-PV) = U + PV \qquad (3.2.9)$$

The new function H defined by this transform is the *enthalpy*, previously introduced in § 2.4.1. It is an extensive, conceptual state function and has dimensions of energy. Forming the total differential of (3.2.9) and substituting (3.2.4) for dU, we obtain

$$dH = TdS + VdP \qquad (3.2.10)$$

So, S and P are the canonical variables for H, and (3.2.10) is a form of the fundamental equation in which S and P are independent.

We obtain another form of the fundamental equation if we replace S with T in the original form (3.2.4). Therefore we introduce the Legendre transform

$$A = U - TS \qquad (3.2.11)$$

which defines another new conceptual state function, the *Helmholtz energy* A. Forming the total differential of (3.2.11) and using (3.2.4) to eliminate dU we find

$$dA = -SdT - PdV \qquad (3.2.12)$$

A fourth form of the fundamental equation can be obtained by applying a double Legendre transform to U,

$$G = U - (-PV) - TS = H - TS \qquad (3.2.13)$$

which defines still another new conceptual state function, the *Gibbs energy* G. Forming the total differential of (3.2.13) and substituting (3.2.10) for dH leaves

$$dG = -SdT + VdP \qquad (3.2.14)$$

The Helmholtz and Gibbs energies are both extensive, conceptual state functions having dimensions of energy. Unfortunately, only in special cases do the changes ΔA and ΔG have physical interpretations.

For closed systems, (3.2.4), (3.2.10), (3.2.12), and (3.2.14) are the four forms of the fundamental equation. For easy reference, we collect them together here:

$$dU = TdS - PdV \tag{3.2.4}$$

$$dH = TdS + VdP \tag{3.2.10}$$

$$dA = -SdT - PdV \tag{3.2.12}$$

$$dG = -SdT + VdP \tag{3.2.14}$$

Each of these is always true; they are four equivalent, though different, ways of conveying the same information. No one is any more basic than another. You must decide which is most appropriate for the problem at hand. The choice depends on which set of independent variables (S, V), (S, P), (T, V), or (T, P) best simplifies your problem. For example, in analyzing multiphase systems, the Gibbs energy (3.2.14) is often used because temperature and pressure are usually the variables most easily measured or controlled. But in developing models for PvT equations of state, the Helmholtz energy (3.2.12) is often used because we prefer to write those equations in the form $P = P(T, v)$. This preference usually simplifies the development, especially in models for multiphase systems wherein different values of the molar volume v can give the same pressure P. Note the distinction between analyzing experimental data (T and P are convenient) and developing theoretical models (T and v are convenient).

The new properties H, A, and G are conceptuals, as are S and U. Unfortunately, these new conceptuals are not amenable to physical interpretation, except in special situations. One special case is an open-system, such as in § 2.4.3, where we found that the enthalpy accounts for energy (flow work plus internal energy) entering and leaving the system via the mass in flowing streams. Another special case is the reversible isobaric change of state on closed systems, for then (3.2.10) reduces to $dH = \delta Q_{rev}$, as we showed in (2.4.2). Similarly, for a reversible isothermal change, (3.2.12) reduces to

$$dA = -PdV = \delta W_{rev} \qquad \textit{fixed T, closed system} \tag{3.2.15}$$

Since a reversible change provides the maximum (minimum) amount of work for a given expansion (compression), the change in Helmholtz energy provides a bound on the work associated with an isothermal process.

3.2.2 Open Systems

We now extend the fundamental equation to systems that can exchange mass with their surroundings. Through such systems may pass any number of components {1, 2, 3, ... }, for which we write the complete set of mole numbers as $\{N_1, N_2, N_3, \ldots\}$. We want to construct an extensive equation of state that provides the internal energy in terms of its canonical variables. But for an open system, the extensive internal energy U depends not only on S and V but also on the numbers of moles of each component present, so we write

$$U = U(S, V, N_1, N_2, \ldots) \tag{3.2.16}$$

Since U is a state function, its total differential is

$$dU = \left(\frac{\partial U}{\partial S}\right)_{VN} dS + \left(\frac{\partial U}{\partial V}\right)_{SN} dV + \sum_i \left(\frac{\partial U}{\partial N_i}\right)_{SVN_{j\neq i}} dN_i \tag{3.2.17}$$

Here $N = \Sigma N_i$ is the total number of moles present and the notation $N_{j\neq i}$ means that in taking the derivative, all mole numbers are held fixed except that of component i. Using (3.2.6) for the coefficients in (3.2.17) we find

$$dU = TdS - PdV + \sum_i \left(\frac{\partial U}{\partial N_i}\right)_{SVN_{j\neq i}} dN_i \tag{3.2.18}$$

Equation (3.2.18) is the first form of the fundamental equation for open systems. In the case of a reversible change, each term in (3.2.18) has a simple physical interpretation: $(T\,dS)$ is the heat crossing system boundaries; $(-P\,dV)$ is the work that alters the system volume; and $(\partial U/\partial N_i)\,dN_i$ is related to the work that causes component i to diffuse across system boundaries. For irreversible processes no such simple interpretations apply; nevertheless, since the lhs is an exact differential, (3.2.18) is valid regardless of whether a change of state is reversible. In a similar fashion we can extend each of the other forms of the fundamental equation to open systems. The results are

$$dH = TdS + VdP + \sum_i \left(\frac{\partial H}{\partial N_i}\right)_{SPN_{j\neq i}} dN_i \tag{3.2.19}$$

$$dA = -SdT - PdV + \sum_i \left(\frac{\partial A}{\partial N_i}\right)_{TVN_{j\neq i}} dN_i \tag{3.2.20}$$

$$dG = -SdT + VdP + \sum_i \left(\frac{\partial G}{\partial N_i}\right)_{TPN_{j\neq i}} dN_i \tag{3.2.21}$$

It is remarkable that in these four forms of the fundamental equation, the partial derivatives wrt N_i are numerically equal; that is,

$$\left(\frac{\partial U}{\partial N_i}\right)_{SVN_{j\neq i}} = \left(\frac{\partial H}{\partial N_i}\right)_{SPN_{j\neq i}} = \left(\frac{\partial A}{\partial N_i}\right)_{TVN_{j\neq i}} = \left(\frac{\partial G}{\partial N_i}\right)_{TPN_{j\neq i}} \tag{3.2.22}$$

It is therefore convenient to give these four derivatives a common symbol $\overline{G}_i$ and a special name—the *chemical potential*. We use the symbol $\overline{G}_i$ for reasons that will become obvious in § 3.4; the name chemical potential arises from processes described in Chapter 7. This choice implies that of the four derivatives in (3.2.22), we take that involving the Gibbs energy to be the defining relation for the chemical potential:

$$\overline{G}_i \equiv \left(\frac{\partial G}{\partial N_i}\right)_{TPN_{j\neq i}} \tag{3.2.23}$$

The chemical potential is an intensive conceptual state function and has dimensions of (energy / mole). It is closely related to the reversible work needed to add to the system a small amount of component *i*, when the addition is done with temperature, pressure, and all other mole numbers held fixed. (This statement is proved in § 3.7.3.) For a pure substance (3.2.23) simplifies to

$$\overline{G}_{\text{pure i}}(T, P) = \left(\frac{\partial G}{\partial N}\right)_{TP} = \left(\frac{\partial (Ng)}{\partial N}\right)_{TP} = g(T, P) \tag{3.2.24}$$

For pure substances, the chemical potential is merely the molar Gibbs energy.

For multicomponent open systems, then, the four extensive forms of the fundamental equation, (3.2.18)–(3.2.21), can be written as

$$dU = TdS - PdV + \sum_i \overline{G}_i\, dN_i \tag{3.2.25}$$

$$dH = TdS + VdP + \sum_i \overline{G}_i\, dN_i \tag{3.2.26}$$

$$dA = -SdT - PdV + \sum_i \overline{G}_i\, dN_i \tag{3.2.27}$$

$$dG = -SdT + VdP + \sum_i \overline{G}_i\, dN_i \tag{3.2.28}$$

3.2.3 Integrated Forms

The differential forms of the fundamental equation for open systems can be integrated over a change in the amount of material, yielding an integrated form for each equation. When our system is a mixture, we can change the amount N_i of each component *i* by the same factor *c*: $N_i \rightarrow cN_i$. The integration over the change is simply done if we remember that intensive properties (such as *T*, *P*, and $\overline{G}_i$) are independent of the number of moles present, while the total properties *S*, *U*, *H*, *A*, and *G* are homogeneous of degree one in the mole numbers. As a result, Euler's theorem for homogeneous functions applies (see Appendix A) and we can immediately write for (3.2.25)

$$U = TS - PV + \sum_i \bar{G}_i N_i \tag{3.2.29}$$

Similarly, the other open system forms (3.2.26)–(3.2.28) integrate to

$$H = TS + \sum_i \bar{G}_i N_i \tag{3.2.30}$$

$$A = -PV + \sum_i \bar{G}_i N_i \tag{3.2.31}$$

$$G = \sum_i \bar{G}_i N_i \tag{3.2.32}$$

Note that for a pure substance, (3.2.32) is the same as (3.2.24). These forms of the fundamental equation are consistent with the Legendre transforms that define H (3.2.9), A (3.2.11), and G (3.2.13).

3.3 RESPONSE TO A CHANGE IN *T*, *P*, OR *V*

In this and the next section we consider how properties in closed systems respond to changes in measurable state functions. Each such response is given by a partial derivative, and we are particularly interested in how conceptuals respond to changes in measurables because several of those derivatives are measurable, even though the conceptuals themselves are not.

We can consider any property (a state function) to be expressible as some function of temperature and pressure,

$$F = F(T, P) \tag{3.3.1}$$

Here F could be any of the extensive properties V, U, H, S, A, or G. Then the total differential of F gives rise to two partial derivatives,

$$dF = \left(\frac{\partial F}{\partial T}\right)_{PN} dT + \left(\frac{\partial F}{\partial P}\right)_{TN} dP \tag{3.3.2}$$

where subscript N means all mole numbers are held fixed. Alternatively, we could consider F to be expressible as some function of temperature and volume,

$$F = F(T, V) \tag{3.3.3}$$

Table 3.1 Classification of thermodynamic derivatives, with classes ranked by engineering importance. Here C_i represents a conceptual and M_i represents a measurable.

Class	Relative importance	Form
I	Most important	$\left(\frac{\partial M_1}{\partial M_2}\right)_{M_3}$
II	Second in importance	$\left(\frac{\partial C_1}{\partial M_2}\right)_{M_3}$
III	Third in importance	$\left(\frac{\partial M_1}{\partial M_2}\right)_{C_3}$ and $\left(\frac{\partial C_1}{\partial M_2}\right)_{C_3}$
IV	Least important	$\left(\frac{\partial C_1}{\partial C_2}\right)_{M_3}$ and $\left(\frac{\partial C_1}{\partial C_2}\right)_{C_3}$

where F could now be P, U, H, S, A, or G. Then the total differential involves two other partial derivatives,

$$dF = \left(\frac{\partial F}{\partial T}\right)_{VN} dT + \left(\frac{\partial F}{\partial V}\right)_{TN} dV \tag{3.3.4}$$

With these four kinds of partial derivatives and many dependent properties to consider, a huge number of partial derivatives can be formed. Fortunately, only a few have simple and useful forms; we are not interested here in the complicated or rarely used ones. We judge the importance of derivatives based on whether the dependent, independent, and held-fixed variables are conceptuals or measurables. Our classification scheme is summarized in Table 3.1.

3.3.1 Temperature Changes

In this section we present those class I and class II derivatives that show how properties respond to changes in temperature. First, we consider the effects of temperature changes on two measurables—pressure and volume; then we describe the effects on internal energy, enthalpy, and entropy; and finally, we present the effects on Gibbs and Helmholtz energies.

Response of P and v to changes in T. The response of pressure to a constant-volume change in temperature defines the *thermal pressure coefficient*, γ_v,

$$\gamma_v \equiv \left(\frac{\partial P}{\partial T}\right)_v \tag{3.3.5}$$

while the fractional response of volume to an isobaric change in temperature defines the *volume expansivity*, α,

$$\alpha \equiv \frac{1}{V}\left(\frac{\partial V}{\partial T}\right)_{PN} \tag{3.3.6}$$

Both of these class I derivatives are intensive measurable state functions. The thermal pressure coefficient is the slope of an isomet on a PT diagram and is positive for both liquids and gases. But γ_v-values for liquids are much greater than those for gases; representative values are given in Table 3.2.

The volume expansivity α is usually positive; that is, most materials expand on heating. For low-density gases, $\alpha \approx 1/T$ and it decreases with increasing pressure. In contrast, liquids have values that are roughly an order of magnitude smaller than $1/T$ and they are nearly constant over modest changes of temperature and pressure. The expansivity α of water is anomalous: it is negative at atmospheric pressure and temperatures below 4°C. Moreover, α for water is not monotone with either isobaric changes in temperature nor with isothermal changes in pressure.

Table 3.2 Thermodynamic response functions of air[a] compared to those of liquid water[b]

Property	Air	Saturated liquid water
Molecular weight	29	18
Temperature, T (K)	300	293.15
Pressure, P (bar)	1	0.023
Density, ρ (g/cm^3)	0.0012	1
Adiabatic compressibility, κ_s (bar^{-1})	0.72	45.6(10)$^{-6}$
Isothermal compressibility, κ_T (bar^{-1})	1	45.9(10)$^{-6}$
Isobaric heat capacity, c_p (J/mol K)	29	75.3
Isometric heat capacity, c_v (J/mol K)	21	74.8
Thermal pressure coefficient, γ_v (bar/K)	0.0033	4.6
Volume expansivity, α (K^{-1})	0.0033	21.(10^{-5})

a. Properties of air were computed assuming an ideal gas, except value for c_p taken from Vargaftik [2].

b. Properties for water taken from Rowlinson and Swinton [3].

Response of U, H, and S to changes in T. The response of the internal energy to an isometric change in T and that of the enthalpy to an isobaric change in T define the isometric and isobaric *heat capacities*,

$$C_v \equiv \left(\frac{\partial U}{\partial T}\right)_{VN} \tag{3.3.7}$$

$$C_p \equiv \left(\frac{\partial H}{\partial T}\right)_{PN} \tag{3.3.8}$$

These class II derivatives are extensive measurable state functions. Both C_v and C_p are always positive, so U (H) *always* increases with isometric (isobaric) increases in T. The heat capacities are experimentally accessible by measuring the temperature change that accompanies addition of a small amount of energy (such as heat) to a system at constant volume, to yield C_v, or reversibly at constant pressure, to yield C_p; that is,

$$C_v = \lim_{\Delta T \to 0} \left(\frac{\delta Q}{\Delta T}\right)_{VN} \tag{3.3.9}$$

$$C_p = \lim_{\Delta T \to 0} \left(\frac{\delta Q_{rev}}{\Delta T}\right)_{PN} \tag{3.3.10}$$

The heat capacities are sensitive to changes in T, generally they increase with increasing T. But, except near the gas-liquid critical point, they are weak functions of P and V.

Applying the definitions (3.3.7) and (3.3.8) to the fundamental equations (3.2.4) for dU and (3.2.10) for dH, respectively, we obtain the following expressions for the response of entropy to changes in temperature:

$$\left(\frac{\partial S}{\partial T}\right)_{VN} = \frac{C_v}{T} \tag{3.3.11}$$

$$\left(\frac{\partial S}{\partial T}\right)_{PN} = \frac{C_p}{T} \tag{3.3.12}$$

These class II derivatives are important because each gives the response of a conceptual to a change in state, with the response given solely in terms of measurables. Since C_p and C_v are positive, S must *always* increase with both isometric and isobaric increases in T.

Response of G and A to changes in T. From the forms of the fundamental equation (3.2.12) for dA and (3.2.14) for dG, we obtain the following temperature derivatives:

$$\left(\frac{\partial G}{\partial T}\right)_{PN} = \left(\frac{\partial A}{\partial T}\right)_{VN} = -S \tag{3.3.13}$$

Although these are class II derivatives, they are not generally useful for obtaining the response to a change in temperature, because the entropy is not directly measurable. However, if S can be obtained from a heat capacity via (3.3.11) or (3.3.12), then (3.3.13) can be integrated to obtain ΔG or ΔA. But (3.3.13) is more likely to be used to obtain expressions for S when the temperature dependence of G or A is known or can be estimated.

More useful are the Gibbs-Helmholtz equations, in which the temperature derivative of G/T is related to H and that of A/T is related to U. To derive the first of these, start with the Legendre transform that defines G,

$$G = H - TS \tag{3.2.13}$$

and substitute (3.3.13) for S,

$$G = H + T\left(\frac{\partial G}{\partial T}\right)_{PN} \tag{3.3.14}$$

This is a linear, first-order differential equation in the independent variables T and P and it can be solved by finding an integrating factor (see Appendix A). Equivalently, we multiply (3.3.14) by $1/T^2$ and rearrange to obtain

$$\frac{1}{T}\left(\frac{\partial G}{\partial T}\right)_{PN} - \frac{G}{T^2} = -\frac{H}{T^2} \tag{3.3.15}$$

Now we realize that

$$\frac{\partial}{\partial T}\left(\frac{G}{T}\right)_{PN} = \frac{1}{T}\left(\frac{\partial G}{\partial T}\right)_{PN} - \frac{G}{T^2} \tag{3.3.16}$$

So we substitute (3.3.15) into (3.3.16) and find

$$\frac{\partial}{\partial T}\left(\frac{G}{T}\right)_{PN} = -\frac{H}{T^2} \tag{3.3.17}$$

This is the *Gibbs-Helmholtz equation* for G; it provides the response of (G/T) to changes in temperature. By an analogous procedure, we can derive a second Gibbs-Helmholtz equation that gives the response of (A/T) to changes in T,

$$\frac{\partial}{\partial T}\left(\frac{A}{T}\right)_{VN} = -\frac{U}{T^2} \tag{3.3.18}$$

3.3.2 Example

How do we compute the response of the Gibbs energy to a finite isobaric change in temperature?

For a finite change in temperature, at fixed pressure, the corresponding change in the Gibbs energy g is formally obtained by integrating (3.3.13). But to perform that integration, we must know how the entropy s depends on T and P; this is rarely known, so (3.3.13) is little used. Alternatively, we may integrate the Gibbs-Helmholtz equation (3.3.17); for a change from T_1 to T_2, we obtain

$$\frac{g(T_2, P)}{RT_2} - \frac{g(T_1, P)}{RT_1} = -\int_{T_1}^{T_2} \frac{h(T)}{RT^2}\, dT \tag{3.3.19}$$

The rhs can be evaluated using an integration by parts, but a less direct attack is more economical. We start by writing the Legendre transform for g as

$$\Delta\left(\frac{g}{RT}\right) = \Delta\left(\frac{h}{RT}\right) - \frac{\Delta s}{R} \tag{3.3.20}$$

Then

$$\Delta\left(\frac{g}{RT}\right) = \frac{h(T_1)}{R}\Delta\left(\frac{1}{T}\right) + \frac{\Delta h}{RT_2} - \frac{\Delta s}{R} \tag{3.3.21}$$

An expression for Δh can be obtained by integrating the definition of c_p (3.3.8),

$$\Delta h = \int_{T_1}^{T_2} c_p(T)\, dT \tag{3.3.22}$$

Similarly, Δs can be obtained by integrating (3.3.12),

$$\frac{\Delta s}{R} = \int_{T_1}^{T_2} \frac{c_p(T)}{RT}\, dT \tag{3.3.23}$$

Substituting (3.3.22) and (3.3.23) into (3.3.21) gives

$$\frac{g(T_2, P)}{RT_2} - \frac{g(T_1, P)}{RT_1} = \frac{h(T_1, P)}{R}\left(\frac{1}{T_2} - \frac{1}{T_1}\right) + \frac{1}{RT_2}\int_{T_1}^{T_2} c_p(T)\, dT - \int_{T_1}^{T_2} \frac{c_p(T)}{RT}\, dT \tag{3.3.24}$$

where the integrals are to be evaluated at fixed P. If c_p is assumed constant, independent of T, then the integrals in (3.3.24) can be immediately evaluated. Otherwise, the temperature dependence of c_p is usually represented by some simple polynomial.

Note that (3.3.24) gives only $\Delta(g/RT)$, not Δg itself; even if we have values for T_2, T_1, and $\Delta(g/RT)$, we still cannot solve algebraically for Δg. Further, note that (3.3.24) contains $h(T_1, P)$, not Δh; hence, the value computed for $\Delta(g/RT)$ depends on the reference state at which the enthalpy is set to zero. In spite of these limitations, (3.3.24) is useful because the quantity g/RT arises naturally in many applications, such as descriptions of chemical reaction equilibria.

3.3.3 Pressure Changes

We first consider how volume responds to changes in P, then we consider how the conceptuals G, H, and S each respond.

Response of v to changes in P. Changes of volume in response to changes in pressure are given by the compressibilities. Two are in common use: one for isothermal changes κ_T and the other for reversible adiabatic changes κ_s,

$$\kappa_T \equiv -\frac{1}{v}\left(\frac{\partial v}{\partial P}\right)_T \tag{3.3.25}$$

$$\kappa_s \equiv -\frac{1}{v}\left(\frac{\partial v}{\partial P}\right)_s \tag{3.3.26}$$

Both compressibilities are intensive measurable state functions, though κ_T is proportional to a class I derivative, while κ_s is proportional to one of class III. Because volume decreases with increasing pressure, these definitions contain negative signs to make the compressibilities positive. Besides PvT experiments, κ_s can also be obtained from measurements of the speed of sound. The reciprocal isothermal compressibility is called the *bulk modulus*.

At the gas-liquid critical point κ_T diverges. Otherwise, values of the compressibilities are large for gases, but small and nearly constant for liquids; sample values are given in Table 3.2 and Figure 3.2. The idealizations

$$\kappa_T = 0 \quad \text{and} \quad \alpha = 0 \tag{3.3.27}$$

define an *incompressible substance* and are reliable approximations for normal liquids and solids over modest changes of state. The incompressible fluid is a simplification much used in fluid mechanics.

The isothermal compressibility (3.3.25), the thermal pressure coefficient (3.3.5), and the volume expansivity (3.3.6) satisfy a triple product rule (Appendix A):

$$\left(\frac{\partial P}{\partial T}\right)_v \left(\frac{\partial v}{\partial P}\right)_T \left(\frac{\partial T}{\partial v}\right)_P = -1 \tag{3.3.28}$$

Specifically,

$$\alpha = \gamma_v \kappa_T \tag{3.3.29}$$

Moreover, we can show (Problem 3.11) that the ratio of the compressibilities equals the ratio of the heat capacities,

$$\frac{\kappa_T}{\kappa_s} = \frac{C_p}{C_v} \tag{3.3.30}$$

and we can show that the difference in heat capacities always obeys

$$C_p - C_v = \frac{TV\alpha^2}{\kappa_T} > 0 \tag{3.3.31}$$

The inequality is always true because stable phases must have $\kappa_T > 0$, as we shall prove in Chapter 8.

Response of *G*, *H*, and *S* to changes in *P*. The fundamental equation provides three important relations for pressure derivatives of conceptuals. The first, obtained from (3.2.14) for dG, is

$$\left(\frac{\partial G}{\partial P}\right)_{TN} = V \tag{3.3.32}$$

This is an important class II derivative because it gives a response of the Gibbs energy directly and solely in terms of the measurables P, V, and T. Since $V > 0$, G must *always* increase with an isothermal increase in pressure.

Another pressure derivative is hidden in (3.2.14); it is one of the *Maxwell relations*. Recall from the calculus (Appendix A) that a function of two variables, such as $G(T, P)$, forms an exact total differential if its second cross-partial derivatives are equal; that is, if

$$\frac{\partial}{\partial P}\left[\left(\frac{\partial G}{\partial T}\right)_{PN}\right]_{TN} = \frac{\partial}{\partial T}\left[\left(\frac{\partial G}{\partial P}\right)_{TN}\right]_{PN} \tag{3.3.33}$$

But, from the development of the fundamental equation, we already know that G is a state function; therefore, (3.3.33) must be satisfied. Further, the fundamental equation (3.2.14) gives the two inner derivatives in (3.3.33): (3.3.13) for the inner T-derivative and (3.3.32) for the inner P-derivative. Therefore, on putting (3.3.13) into the lhs of (3.3.33) and (3.3.32) into the rhs, we find

$$\left(\frac{\partial S}{\partial P}\right)_{TN} = -\left(\frac{\partial V}{\partial T}\right)_{PN} = -V\alpha \quad \textit{a Maxwell relation} \tag{3.3.34}$$

This is an important class II derivative, because it gives the response of the entropy directly and solely in terms of P, V, and T. An analogous Maxwell relation can be derived from each of the other three forms of the fundamental equation, (3.2.4),

(3.2.10), and (3.2.12); however, those from (3.2.4) and (3.2.10) are unimportant class III derivatives. The Maxwell relation from (3.2.12) is given in § 3.3.4.

The third important pressure derivative gives the response of the enthalpy to isothermal changes in pressure. From the fundamental equation (3.2.10) and the Maxwell relation (3.3.34) we have

$$\left(\frac{\partial H}{\partial P}\right)_{TN} = T\left(\frac{\partial S}{\partial P}\right)_{TN} + V = -T\left(\frac{\partial V}{\partial T}\right)_{PN} + V = V(1-\alpha T) \tag{3.3.35}$$

For gases $\alpha \approx 1/T$, while for liquids v is small, so in both cases the molar enthalpy h is little affected by isothermal changes in pressure.

3.3.4 Volume Changes

There are no important class I derivatives which provide a response to changes in volume; however, three class II derivatives are important. One is given by the fundamental equation (3.2.12) for dA,

$$\left(\frac{\partial A}{\partial V}\right)_{TN} = -P \tag{3.3.36}$$

which relates a response of the Helmholtz energy to a measurable. The second is the Maxwell relation that arises from (3.2.12). Its derivation is exactly analogous to that given above for (3.3.34). The result is

$$\left(\frac{\partial S}{\partial V}\right)_{TN} = \left(\frac{\partial P}{\partial T}\right)_{v} = \gamma_v \qquad \textit{a Maxwell relation} \tag{3.3.37}$$

where γ_v is the thermal pressure coefficient. The importance of (3.3.37) is equal to that of the other Maxwell relation given in (3.3.34).

The third gives the response of the internal energy to an isothermal change in volume. It is derived from the fundamental equation (3.2.4) and the Maxwell relation (3.3.37) using a procedure analogous to that used for (3.3.35); the result is

$$\left(\frac{\partial U}{\partial V}\right)_{TN} = T\left(\frac{\partial P}{\partial T}\right)_{v} - P = T\gamma_v - P \tag{3.3.38}$$

For gases $\gamma_v \approx P/T$, so U is nearly independent of changes in volume. For liquids, we usually find $T\gamma_v > P$, and then U increases with isothermal increases in volume.

3.4 RESPONSE TO A CHANGE IN MOLE NUMBER

In the previous two sections we presented those simple derivative relations that characterize changes of state in closed systems or systems of constant composition. But engineering practice is more often concerned with open multicomponent systems—systems of variable composition. In those situations the behavior of our system is affected by the kinds and amounts of components that are present.

Multicomponent systems offer an extraordinary range of diverse behaviors. For example, in one-phase systems, property values of mixtures are often intermediate among those of the pure components; but equally often, values pass through extrema with composition, as they do for salts in water, for many polymer blends, and for biochemicals in solvents. In multiphase systems, different phases typically have different compositions and we exploit the spontaneous mass transfer between phases in such separation processes as distillation, extraction, crystallization, osmosis, and detergency. Furthermore, chemical reactions necessarily involve mixtures and thermodynamics controls the direction as well as the extent of reactions. In addition to using reactions to produce new products, reactions are important in cooking, combustion, and biological processes. The thermodynamics of multicomponent systems is central to chemical engineering practice.

In this section we consider how thermodynamic properties are affected by changes in the amounts of components. Such changes promote a response that is governed by the partial molar properties. In what follows, we apply the calculus and define certain useful quantities, but no new thermodynamics is introduced.

3.4.1 Partial Molar Properties

Consider any extensive property F for a mixture that contains C components whose mole numbers are $\{N_1, N_2, \ldots, N_C\}$. The mixture is a single homogeneous phase with no internal constraints, so (3.1.7) indicates that F depends on $(C + 2)$ independent variables:

$$F = F(T, P, N_1, N_2, \ldots, N_C) \tag{3.4.1}$$

Note that the list of independent variables contains both intensive and extensive properties; this is legitimate because F is extensive.

Now let the intensive analog of F be $f = F/N$, where N is the total number of moles in the mixture. In special cases (revealed in Chapters 4 and 5) F can be computed by a mole-fraction average of the pure-component properties $f_{\text{pure i}}$. But in general

$$f \approx \sum_i x_i f_{\text{pure i}} \tag{3.4.2}$$

is only an approximation that is sometimes correct and other times wrong. Tests of (3.4.2) are given in Figure 3.3 for estimating the molar volumes of two liquid mixtures. For water-ethanol, the simple average (3.4.2) produces mixture volumes within 1%, over the entire composition range. However, for mixtures of benzene and carbon tetrachloride, the volumes provided by (3.4.2) are in error by about 10% over a substantial range of compositions. For the volume, the approximation (3.4.2) can be seriously wrong because the forces acting among molecules in a mixture may not be simple averages of the forces acting among the same molecules in pure substances.

Since (3.4.2) is not generally obeyed, the question arises, What property of each component should be mole-fraction averaged to obtain the mixture value for f? Note that, analogous to (3.4.1), we can write

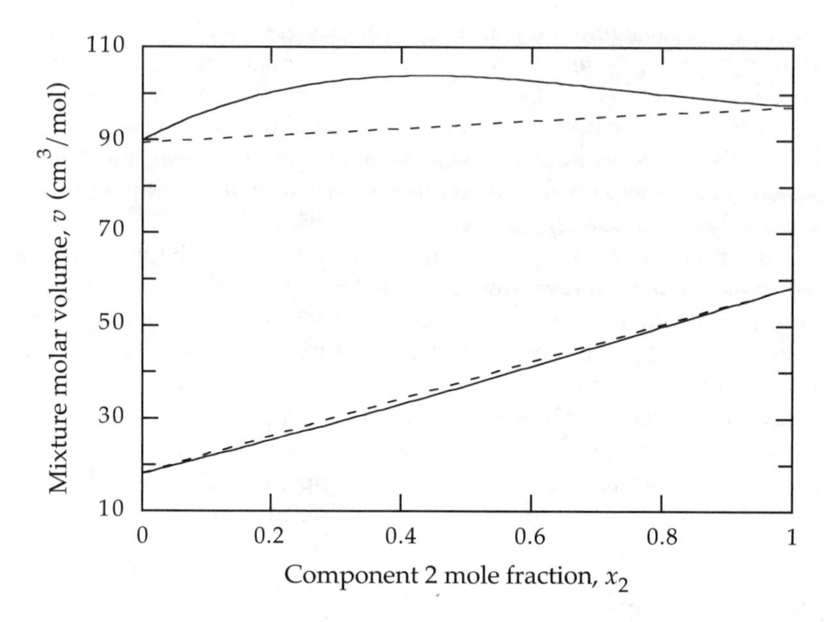

Figure 3.3 Tests of estimating mixture volumes by mole-fraction averaging the pure component volumes. The broken straight lines are the mole-fraction averages of the pure volumes, as computed via (3.4.2). The solid lines are the true mixture volumes taken from [4]. Benzene(1)-carbon tetrachloride(2) liquid mixtures (*top*) are at 25°C, 1 atm. The water(1)-ethanol(2) liquid mixtures (*bottom*) are at 20°C, 1 atm.

$$F = Nf(T, P, x_1, x_2, \ldots, x_{C-1}) \tag{3.4.3}$$

This means the extensive property F is homogeneous of order one in the total number of moles N. This homogeneity gives to extensive quantities a number of desirable attributes, which are developed in Appendix A. One rigorous consequence is

$$f = \sum_i x_i \bar{F}_i \qquad \textit{always true} \tag{3.4.4}$$

where

$$\bar{F}_i \equiv \left(\frac{\partial F}{\partial N_i}\right)_{TPN_{j \neq i}} \tag{3.4.5}$$

The derivative operator appearing in (3.4.5) is called the *partial molar derivative*, and the quantity $\bar{F}_i$ defined by (3.4.5) is called the *partial molar* F for component i. It is the partial molar property that can *always* be mole-fraction averaged to obtain the mixture property F. Note, however, that $\bar{F}_i$ is itself a property of the mixture, not a property of pure i; partial molar properties depend on temperature, pressure, and composition. We emphasize that the definition (3.4.5) demands that F be extensive and that the properties held fixed can *only* be temperature, pressure, and all other mole numbers except N_i. Partial molar properties are intensive state functions; they may be either measurable or conceptual depending on the identity of F.

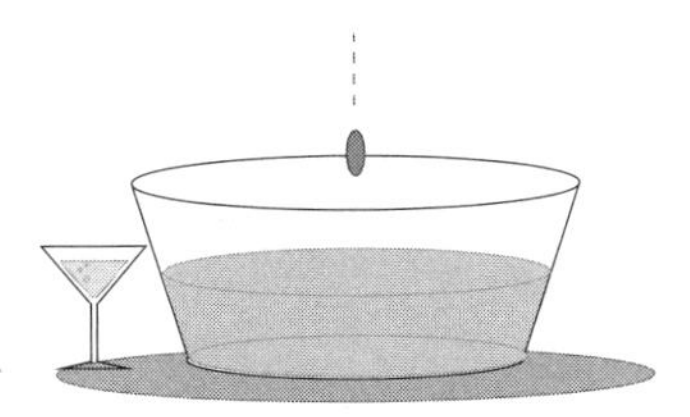

one 6-oz can frozen orange juice (undiluted)
one 6-oz can frozen lemonade (undiluted)
one 46-oz can pineapple juice
two qts. ginger ale (chilled)
(or one qt. ginger ale + one qt. champagne)

Combine first three ingredients and allow flavors to blend for 3–4 hours in a refrigerator. At serving time, pour mixture over ice in a large punch bowl. Add remaining ingredients and stir gently. Serves 10–12.

Figure 3.4 For this mixture, the partial molar volume for water can be determined, according to (3.4.5), by measuring how the total volume changes when a small amount of water is added to the equilibrium mixture with T, P, and the amounts of all other components fixed. (*Our thanks to Verna O'Connell for this recipe.)*

The definition (3.4.5) is amenable to a physical interpretation; for example, let F be the mixture volume V. According to (3.4.5), the partial molar volume can be obtained by fixing the state at a particular T, P, and composition and measuring an initial value for V. After adding a small amount of component i, while maintaining the values of T, P, and all other $N_{j\neq i}$, we measure the volume again. The ratio of the volume change to the amount of i added, $\Delta V/\Delta N_i$, is approximately $\bar{V}_i$; the approximation becomes exact as we decrease the amount of i added. See Figure 3.4. A partial molar property may be positive or negative, depending on whether F increases or decreases when a small amount of i is added.

For a pure substance, the sum in (3.4.4) contains only one term and we have

$$\bar{F}_{\text{pure 1}} = f_{\text{pure 1}} \tag{3.4.6}$$

That is, for a single component the partial molar property is merely the pure molar property. Hence, in the pure-fluid limit each isothermal-isobaric curve for a partial molar property (plotted against mole fraction) coincides with the value for the mixture property, as in Figure 3.5.

The partial derivative is a linear operator; therefore, the partial molar derivative (3.4.5) may be applied to all those expressions given in § 3.2, producing partial molar versions of the fundamental equations. In particular, when we apply the partial molar derivative to the integrated forms (3.2.29)–(3.2.31) of the fundamental equations, we obtain the following important relations among partial molar properties:

$$\bar{U}_i = T\bar{S}_i - P\bar{V}_i + \bar{G}_i \tag{3.4.7}$$

$$\bar{H}_i = T\bar{S}_i + \bar{G}_i \tag{3.4.8}$$

$$\bar{A}_i = -P\bar{V}_i + \bar{G}_i \tag{3.4.9}$$

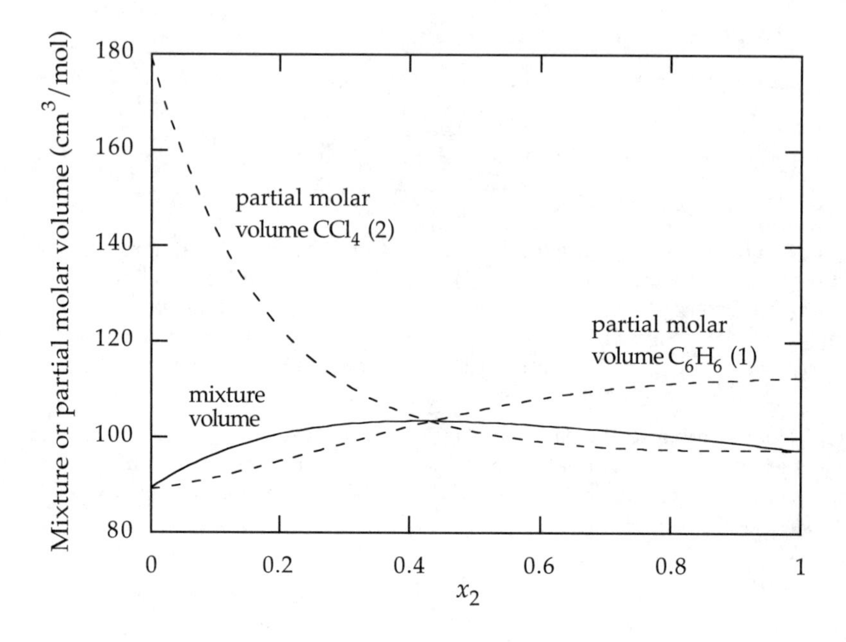

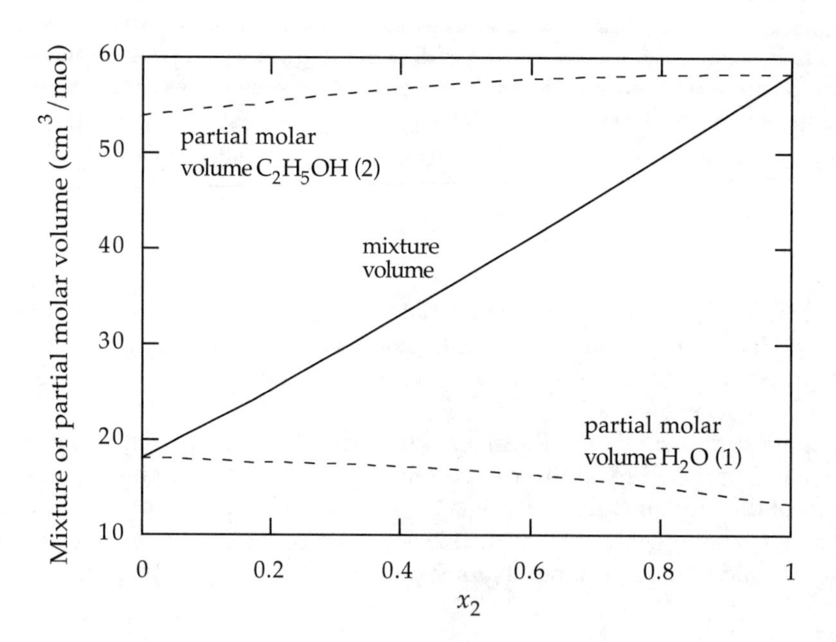

Figure 3.5 Molar volumes (solid lines) and partial molar volumes (broken lines) for binary liquid mixtures. *Top* is for benzene(1)-carbon tetrachloride(2) mixtures at 25°C, 1 atm. *Bottom* is for water(1)-ethanol(2) mixtures at 20°C, 1 atm. Note that if the partial molar volume of one component in a binary increases, then by the Gibbs-Duhem equation (3.4.13), the partial molar volume of the other component *must* decrease. Values computed from data in [4].

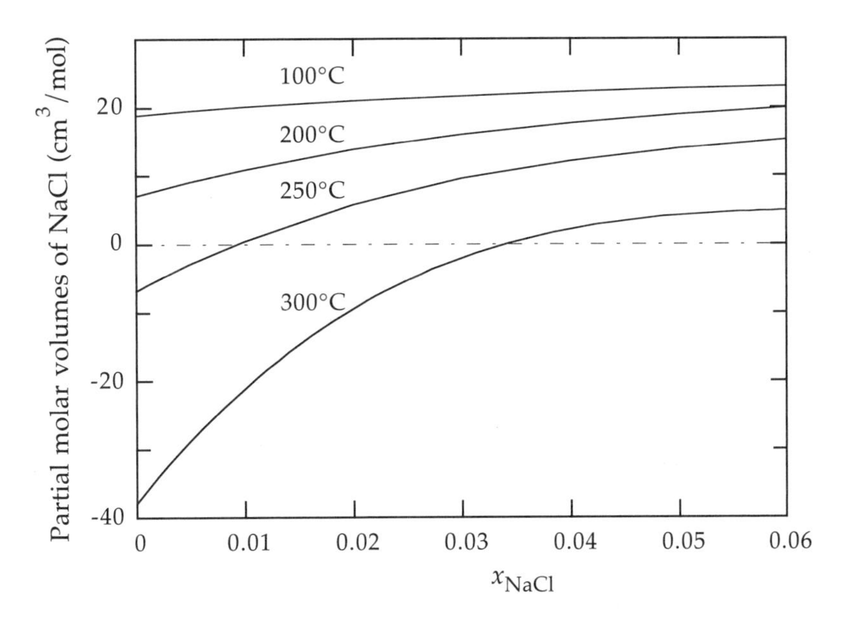

Figure 3.6 Partial molar volumes are usually positive, but they can be negative, as are these for NaCl in aqueous solutions, all at 200 bar. Computed from data in Pitzer et al. [5].

Figure 3.5 shows how partial molar volumes change with composition (a) for water-ethanol mixtures, wherein the $\bar{V}_i$ are weak functions of composition, and (b) for benzene-carbon tetrachloride mixtures, wherein the $\bar{V}_i$ are strong functions of composition. Partial molar volumes are usually positive as in Figure 3.5, but some are negative, indicating that the mixture contracts when a particular component is added. This generally happens for the partial molar volumes of "heavy" solutes when the mixture is near the critical point of the solvent. An example is the partial molar volume of NaCl in aqueous solution, shown in Figure 3.6. The negative values occur when attractive forces between solute (NaCl) and solvent (H_2O) molecules are strong enough to cause the mixture volume to decrease. Negative partial molar volumes indicate that interactions between unlike molecules (NaCl-H_2O) are stronger than those between solvent molecules (H_2O-H_2O).

3.4.2 Gibbs-Duhem Equations

Besides (3.4.4), another attribute of partial molar properties, also derived in Appendix A, is that they obey a set of relations known as Gibbs-Duhem equations. For the generic extensive property $F(T, P, \{N\})$, the general form of the *Gibbs-Duhem equation* is

$$\sum_i^C x_i \, d\bar{F}_i - \left(\frac{\partial f}{\partial T}\right)_{Px} dT - \left(\frac{\partial f}{\partial P}\right)_{Tx} dP = 0 \qquad (3.4.10)$$

On these derivatives the subscript x means that the composition is held fixed. For a mixture of C components, the Gibbs-Duhem equation (3.4.10) establishes a single relation among T, P, and the partial molar properties $\bar{F}_i$. That is, the intensive quantity f depends on only $(C + 1)$ independent intensive properties, as required by (3.1.9).

For isothermal-isobaric processes (3.4.10) reduces to a relation among the partial molar quantities themselves,

$$\sum_i^C x_i \, d\bar{F}_i = 0 \qquad \textit{fixed T and P} \qquad (3.4.11)$$

And in a binary mixture (3.4.11) further simplifies to

$$x_1 d\bar{F}_1 + x_2 \, d\bar{F}_2 = 0 \qquad \textit{fixed T and P} \qquad (3.4.12)$$

Since T and P are fixed in (3.4.12), an obvious choice is to use a mole fraction as the independent variable, then (3.4.12) can be written as

$$x_1 \left(\frac{\partial \bar{F}_1}{\partial x_1} \right)_{TP} = -x_2 \left(\frac{\partial \bar{F}_2}{\partial x_1} \right)_{TP} = x_2 \left(\frac{\partial \bar{F}_2}{\partial x_2} \right)_{TP} \qquad (3.4.13)$$

The last equality is valid because a binary has $dx_1 = -dx_2$. The simple form of the Gibbs-Duhem equation (3.4.13) says that in a binary at fixed T and P, if $\bar{F}_1$ increases as x_1 increases, then $\bar{F}_2$ *must* simultaneously decrease. This behavior can be seen in the partial molar volumes plotted in Figure 3.5; for example, in the water-ethanol mixtures, $\bar{V}_{etoh}$ increases with x_2 while simultaneously $\bar{V}_{hoh}$ decreases.

3.4.3 Chemical Potential

Note that the chemical potential $\bar{G}_i$, defined by (3.2.23), has the structure of (3.4.5); that is, the chemical potential is the partial molar Gibbs energy. This is why we use the partial-molar notation for the chemical potential: the notation reminds us that the chemical potential has mathematical and physical characteristics in common with other partial molar properties. For example, the integrated form of dG in (3.2.32) is consistent with the mole-fraction average (3.4.4) and the pure-fluid chemical potential (3.2.24) is consistent with (3.4.6) for the molar Gibbs energy. The chemical potential plays a central role in phase equilibria and chemical reaction equilibria; therefore, we will need to know how $\bar{G}_i$ responds to changes of state.

The response of G to a change in T is given by (3.3.13), while the response to a change in P is given by (3.3.32). Consider first the pressure derivative,

$$\left(\frac{\partial G}{\partial P} \right)_{TN} = V \qquad (3.3.32)$$

For mixtures, this derivative must be evaluated with all mole numbers fixed, and we remind ourselves of that by the subscript N. Now apply the partial molar derivative in (3.4.5) to both sides of (3.3.32); we obtain

$$\frac{\partial}{\partial N_i}\left[\left(\frac{\partial G}{\partial P}\right)_{TN}\right]_{TPN_{j\neq i}} = \overline{V}_i \qquad (3.4.14)$$

But G is a state function, so we can interchange the order of differentiation on the lhs, identify the resulting inner derivative as the chemical potential, and write

$$\left(\frac{\partial \overline{G}_i}{\partial P}\right)_{Tx} = \overline{V}_i \qquad (3.4.15)$$

Note that we now indicate constant composition (subscript x) because the chemical potential is intensive. Repeating these steps for the temperature derivative (3.3.13), we find

$$\left(\frac{\partial \overline{G}_i}{\partial T}\right)_{Px} = -\overline{S}_i \qquad (3.4.16)$$

Moreover, a Gibbs-Helmholtz equation relates the chemical potential to the partial molar enthalpy,

$$\left[\frac{\partial}{\partial T}\left(\frac{\overline{G}_i}{RT}\right)\right]_{Px} = -\frac{\overline{H}_i}{RT^2} \qquad (3.4.17)$$

In a mixture the chemical potentials of all components are not independent; rather, they are related through the Gibbs-Duhem equation. So letting $f = g$ in (3.4.10),

$$\sum_i^C x_i\, d\overline{G}_i - \left(\frac{\partial g}{\partial T}\right)_{Px} dT - \left(\frac{\partial g}{\partial P}\right)_{Tx} dP = 0 \qquad (3.4.18)$$

or

$$\sum_i^C x_i\, d\overline{G}_i = -s\, dT + v\, dP \qquad (3.4.19)$$

and for isothermal-isobaric processes,

$$\sum_i^C x_i\, d\overline{G}_i = 0 \qquad \textit{fixed T and P} \qquad (3.4.20)$$

In a mixture, the chemical potentials for all components cannot change freely in response to a change of state; rather, they must change so as to satisfy (3.4.19) or (3.4.20). Consequently, if we have a correlation that estimates (C –1) chemical potentials, then the last may be computed from the Gibbs-Duhem equation. Alternatively, if correlations are available to estimate all C chemical potentials for a mixture, then the Gibbs-Duhem equation can be used to test whether the correlations are thermodynamically consistent.

3.5 DIFFERENTIAL RELATIONS BETWEEN CONCEPTUALS AND MEASURABLES

With results from previous sections we can develop differential relations that enable us to compute conceptuals from measurables. We consider five conceptuals: U, H, S, A, and G. Recall we cannot obtain absolute values for these properties, we can compute only changes in their values caused by a change of state. Fortunately, values for changes ΔU, ΔH, ΔS, ΔA, and ΔG are sufficient for our needs.

First let us identify the measurables we need to carry out a computation.

(a) To account for temperature changes (§ 3.3.1), we need heat capacities in the form of either $C_p(T, P, N_1, N_2, \dots)$ or $C_v(T, V, N_1, N_2, \dots)$.

(b) To account for pressure or volume changes (§ 3.3.3 and 3.3.4), we need some volumetric equation of state for the measurables $\{T, V, P, N_1, N_2, \dots\}$,

$$F(P, V, T, N_1, N_2, \dots) = 0 \tag{3.5.1}$$

(c) To account for changes in composition, we need expressions for certain partial molar properties (§ 3.4). Usually these are obtained from (a) or (b) or both.

Volumetric equations of state (3.5.1) typically take one of two forms, either a pressure-explicit form,

$$P = P(T, v, x_1, x_2, \dots) \tag{3.5.2}$$

or a volume-explicit form

$$V = Nv(T, P, x_1, x_2, \dots) \tag{3.5.3}$$

Therefore our strategy differs somewhat depending on which of these describes our mixture to the desired accuracy and with minimum complexity. The pressure-explicit form is more general, so (3.5.2) is more commonly encountered, but (3.5.3) is usually more computationally convenient.

3.5.1 When T, P, and $\{N\}$ Are Independent

When temperature and pressure are the independent variables, the shortest route to the conceptuals is via the enthalpy and the entropy. So consider

$$H = H(T, P, N_1, N_2, \dots) \tag{3.5.4}$$

for which the total differential is

$$dH = \left(\frac{\partial H}{\partial T}\right)_{PN} dT + \left(\frac{\partial H}{\partial P}\right)_{TN} dP + \sum_i \left(\frac{\partial H}{\partial N_i}\right)_{TPN_{j \neq i}} dN_i \tag{3.5.5}$$

The isobaric temperature derivative of H is the constant-pressure heat capacity (3.3.8), while the isothermal pressure derivative of H is given in § 3.3.3,

$$\left(\frac{\partial H}{\partial P}\right)_{TN} = V(1 - \alpha T) \tag{3.3.35}$$

where α is the volume expansivity (3.3.6). So with (3.3.8), (3.3.35), and the partial molar enthalpy, (3.5.5) becomes

$$dH = C_p dT + V(1 - \alpha T) dP + \sum_i \bar{H}_i dN_i \tag{3.5.6}$$

Similarly, for the entropy we find

$$dS = \frac{C_p}{T} dT - V\alpha \, dP + \sum_i \bar{S}_i dN_i \tag{3.5.7}$$

For changes of state at constant composition, we need C_p together with the volumetric equation of state before we can integrate (3.5.6) and (3.5.7) for ΔH and ΔS. With values for ΔH and ΔS, we can then apply the defining Legendre transforms (3.2.9) for U, (3.2.11) for A, and (3.2.13) for G to obtain changes in the other conceptuals. If the change of state includes a change in composition, then we will also need values for the partial molar enthalpy and entropy. Recall from § 3.4.3 that these partial molar quantities are simply related to the chemical potential.

3.5.2 When T, V, and $\{N\}$ Are Independent

When temperature and volume are the independent variables, the most direct route to the conceptuals is via the internal energy and the entropy. So we consider

$$U = U(T, V, N_1, N_2, \ldots) \tag{3.5.8}$$

for which the total differential is

$$dU = \left(\frac{\partial U}{\partial T}\right)_{VN} dT + \left(\frac{\partial U}{\partial V}\right)_{TN} dV + \sum_i \left(\frac{\partial U}{\partial N_i}\right)_{TVN_{j \neq i}} dN_i \tag{3.5.9}$$

The isothermal volume derivative of U is given in § 3.3.4,

$$\left(\frac{\partial U}{\partial V}\right)_{TN} = T\gamma_v - P \tag{3.3.38}$$

where γ_v is the thermal pressure coefficient (3.3.5). The isometric temperature derivative of U is the constant-volume heat capacity (3.3.7); so using (3.3.7) and (3.3.38) in (3.5.9) gives

$$dU = C_v dT + (T\gamma_v - P)dV + \sum_i \left(\frac{\partial U}{\partial N_i}\right)_{TVN_{j\neq i}} dN_i \tag{3.5.10}$$

The remaining partial derivative can be related to partial molar properties by the procedure developed in Problem 3.26. The final result is

$$dU = C_v dT + (T\gamma_v - P)dV + \sum_i (\overline{H}_i - T\gamma_v \overline{V}_i)\, dN_i \tag{3.5.11}$$

As the second conceptual we consider the entropy,

$$S = S(T, V, N_1, N_2, \ldots) \tag{3.5.12}$$

By a procedure exactly analogous to what we did above for U, we find

$$dS = \frac{C_v}{T} dT + \gamma_v dV + \sum_i (\overline{S}_i - \gamma_v \overline{V}_i)\, dN_i \tag{3.5.13}$$

For changes of state at constant composition, we need C_v and the volumetric equation of state to be able to integrate (3.5.11) and (3.5.13) for ΔU and ΔS. With values for ΔU and ΔS, we can then apply the defining Legendre transforms (3.2.9) for H, (3.2.11) for A, and (3.2.13) for G to obtain changes in the other conceptuals. If the change of state includes a change in composition, then we will also need values for the partial molar volume, enthalpy, and entropy; as shown in § 3.4.3, these partial molar quantities are simply related to the chemical potential.

3.6 GENERALIZED STUFF EQUATIONS

In § 2.4 we presented differential forms of the thermodynamic stuff equations for overall mass, energy, and entropy flows through open systems. Usually, such systems, together with their inlet and outlet streams, will be mixtures of any number of components. Individual components can contribute in different ways to mass, energy, and entropy flows, so here we generalize the stuff equations to show explicitly the contributions from individual components; these generalized forms contain partial molar properties introduced in § 3.4.

Thermodynamic stuff equations are internal constraints on the variables that describe open systems. Therefore, in § 3.6.2 and 3.6.3 we show how those constraints enter determinations of the number of independent quantities needed to analyze open steady-flow systems.

3.6.1 Thermodynamic Stuff Equations in Terms of Components

Consider an open multicomponent system composed of a single homogeneous phase, such as is shown schematically in Figure 3.7. At any instant the system has temperature T, pressure P, and total number of moles N. The system contains C components,

$$N = \sum_{i}^{C} N_i \tag{3.6.1}$$

The temperature outside the system boundary is T_{ext}. Heat Q may cross the boundary, shaft work W_{sh} may act through the boundary, and the boundary itself may be deformed by boundary work W_b. Material may enter the system through any number of feed streams α and leave through any number of discharge streams β.

Material balances. The overall mass balance on the system is written in (2.4.3). The corresponding balance on each component i is therefore

$$dN_i = \sum_{\alpha} dN_{\alpha i} - \sum_{\beta} dN_{\beta i} \tag{3.6.2}$$

Energy balance. The overall energy balance for open systems appears as (2.4.15) in § 2.4. Here we neglect the boundary energy U_b and introduce partial molar quantities for each component i, so (2.4.15) becomes

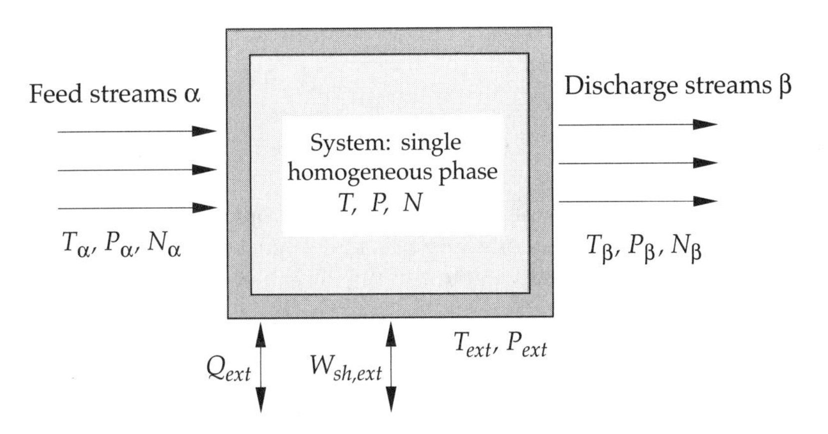

Figure 3.7 Schematic of a single-phase multicomponent system open to exchange of mass and energy with its surroundings

$$d\left(\sum_i N_i \bar{U}_i\right) = \Delta H_{\alpha\beta} + \delta Q + \delta W_b + \delta W_{sh} \tag{3.6.3}$$

Here we have introduced the following shorthand,

$$\Delta H_{\alpha\beta} \equiv \sum_\alpha \sum_i \bar{H}_{\alpha i}\, dN_{\alpha i} - \sum_\beta \sum_i \bar{H}_{\beta i}\, dN_{\beta i} \tag{3.6.4}$$

For processes in which molecular identities are preserved (nonreacting systems), the lhs of (3.6.3) expands to

$$d\left(\sum_i N_i \bar{U}_i\right) = \sum_i \bar{U}_i\, dN_i + \sum_i N_i\, d\bar{U}_i \tag{3.6.5}$$

The second term on the rhs can be replaced with the Gibbs-Duhem equation (3.4.10), so (3.6.5) becomes

$$d\left(\sum_i N_i \bar{U}_i\right) = \sum_i \bar{U}_i\, dN_i + \left(\frac{\partial U}{\partial T}\right)_{PN} dT + \left(\frac{\partial U}{\partial P}\right)_{TN} dP \tag{3.6.6}$$

Then finally, the overall energy balance can be written in terms of components as

$$\sum_i \bar{U}_i\, dN_i + \left(\frac{\partial U}{\partial T}\right)_{PN} dT + \left(\frac{\partial U}{\partial P}\right)_{TN} dP = \Delta H_{\alpha\beta} + \delta Q + \delta W_b + \delta W_{sh} \tag{3.6.7}$$

Entropy balance. The open-system entropy balance appears in (2.4.21). Again, we neglect the boundary term and introduce partial molar entropies for each component, so (2.4.21) becomes

$$d\left(\sum_i N_i \bar{S}_i\right) = \Delta S_{\alpha\beta} + \frac{\delta Q}{T_{ext}} + dS_{gen} \tag{3.6.8}$$

where $\Delta S_{\alpha\beta}$ is defined as in (3.6.4) and dS_{gen} is the entropy created in the system and its boundary. Continuing to limit our attention to nonreacting systems, we expand the lhs and apply the Gibbs-Duhem equation, so the lhs can be written as

$$d\left(\sum_i N_i \bar{S}_i\right) = \sum_i \bar{S}_i\, dN_i + \left(\frac{\partial S}{\partial T}\right)_{PN} dT + \left(\frac{\partial S}{\partial P}\right)_{TN} dP \tag{3.6.9}$$

Therefore, the entropy balance (3.6.8) becomes

$$\sum_i \bar{S}_i dN_i + \left(\frac{\partial S}{\partial T}\right)_{PN} dT + \left(\frac{\partial S}{\partial P}\right)_{TN} dP = \Delta S_{\alpha\beta} + \frac{\delta Q_{ext}}{T_{ext}} + dS_{gen} \tag{3.6.10}$$

3.6.2 Number of Independent Variables for Open Steady-Flow Systems

Many industrial processes take place in open systems in which material enters and leaves the system through process streams and in which energy can cross system boundaries as heat and work. At any instant, a complete identification of the state requires specification of values for such variables as temperatures, pressures, compositions, and flow rates. However, because of the stuff equations in § 3.6.1, not all of these quantities are independent. So we have here the same kinds of questions addressed in § 3.1: How many interactions are available to change the state? How many independent variables must be specified to identify the state of an open steady-flow system? The discussion here extends that in § 3.1 from closed systems to open ones; however, the discussion remains limited to systems composed of a single homogeneous phase with no chemical reactions. The extensions to multiphase systems are given in § 9.1 and to those having chemical reactions in § 10.3.1

As an example of an open system, consider a fixed (control) volume that is open to steady-state mass and energy transfers with its surroundings. Crossing the system boundaries are N_p ports through which one-phase mixtures of C components enter and leave the system. For steady flow situations, we must have at least one inlet and one outlet, so $N_p \geq 2$. The system is in thermal contact with its surroundings and an interaction exists by which shaft work is done, either on or by the system. Note that we do not consider a work mode that could change the size or shape of the control volume.

First we want to determine the number of interactions that are available for manipulating the system state. We assume that each of our one-phase streams obeys (3.1.2); that is, each has $(C + 2)$ interactions with its surroundings. In addition, the control volume has the thermal interaction plus the shaft-work mode. Therefore, the maximum number of orthogonal interactions is given by

$$\mathcal{V}_{max} = N_p(C + 2) + 2 \tag{3.6.11}$$

However, just as in § 3.1.1, the number of orthogonal interactions actually available may be less than this maximum because of external constraints imposed on some interactions. Examples of external constraints include fixed flow rates of some streams, insulated streams, no shaft work, and some components missing from some streams. (For example, the number of constraints is increased by unity for each component missing from each stream.) Let S_{ext} be the total number of external constraints, then the number of available interactions is given by

$$\mathcal{V} = \mathcal{V}_{max} - S_{ext} = N_p(C + 2) + 2 - S_{ext} \tag{3.6.12}$$

Second we want the number of independent variables needed to identify the system state. This number will be less than $\mathcal{V}$ because of internal constraints. For open systems, the internal constraints are a material balance for each component plus an overall energy balance; note that the entropy balance is *not* an internal constraint. Therefore the total number of internal constraints is

$$S = C + 1 \tag{3.6.13}$$

and the number of independent variables is

$$\mathcal{F}_{ex} = \mathcal{V} - S = \mathcal{V} - (C+1) = N_p(C+2) - (C-1) - S_{ext} \tag{3.6.14}$$

The quantities counted by $\mathcal{F}_{ex}$ in (3.6.14) all pertain to streams and energy conduits crossing the system boundary; none are properties of the system itself. This occurs because all flows are steady states. Further note that, unlike for closed systems, the independent quantities needed for open systems may include process variables, such as Q and W. Values of $\mathcal{F}_{ex}$ independent quantities, together with any external constraints and solutions to the material and energy balances, give a complete description of the system. However, if values for some number of variables less than $\mathcal{F}_{ex}$ are known, then the state is not identifiable. Such incomplete descriptions can arise in design situations, and then complete descriptions might be obtained by including additional (nonthermodynamic) feasibility or economic constraints.

Since $\mathcal{V}$, the number of variables available to manipulate the state, is larger than $\mathcal{F}_{ex}$, the number needed to identify the state, simply manipulating variables (such as by changing valve settings) may not set enough variables to provide a complete thermodynamic description of an open system. Instead, additional constraints must be imposed or additional constraint relations must be found to complete the identification of state. Moreover, the values for $\mathcal{V}$ and $\mathcal{F}_{ex}$ depend on your choice of system, so making another choice may simplify an analysis or make an incomplete description complete. This possibility is illustrated in the following example.

3.6.3 Example

How many independent variables must be known to analyze a simple heat exchanger?

We intend to reduce the temperature of a hot nitrogen stream by bringing it into thermal contact with a stream of cooling water. The cooling is done in an insulated, double-tube, countercurrent-flow heat exchanger, as shown schematically in Figure 3.8. We consider two analyses of this one situation.

Analysis 1. First we consider situations in which the heat duty $\dot{Q}$ is to be calculated. Our first problem is then this: how many variables must be known before we can compute $\dot{Q}$? To answer this question, we choose the system to be the water side of the exchanger tube. Therefore, $C = 1$, because the water is pure, and $N_p = 2$, because the water tube has one inlet and one outlet. Hence, the maximum possible number of interactions available for manipulating the system is, from (3.6.11),

$$\mathcal{V}_{max} = N_p(C+2) + 2 = 2(1+2) + 2 = 8 \tag{3.6.15}$$

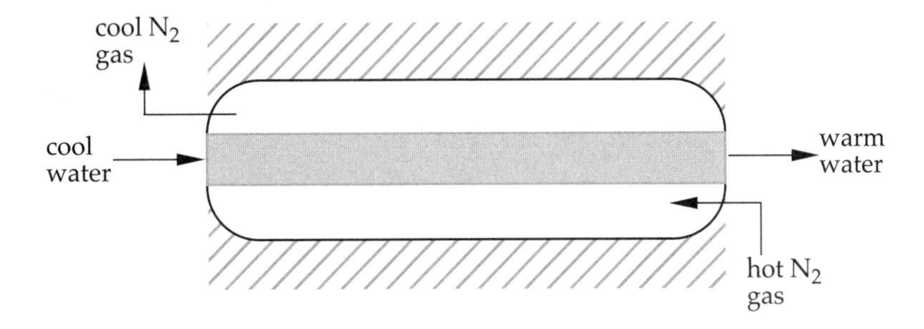

Figure 3.8 Schematic of a steady-flow, insulated, countercurrent, double-tube heat exchanger

The only external constraint on the system (water) is lack of any shaft work, so $S_{ext} = 1$. (Note that this system is not insulated.) Therefore, (3.6.12) gives

$$\mathcal{V} = 8 - 1 = 7 \tag{3.6.16}$$

The seven available interactions are a thermal interaction, a work mode, and a mass flow interaction for the inlet (total of 3), likewise for the outlet (3 more), plus a thermal interaction on the control volume.

To obtain the number of variables needed to identify the state, we apply (3.6.14); hence, we need the number of internal constraints. For water as our system, we have a steady-state material balance and an energy balance. Therefore, (3.6.14) gives

$$\mathcal{F}_{ex} = \mathcal{V} - S = 7 - 2 = 5 \tag{3.6.17}$$

A typical set of the required five variables would be the temperatures and pressures of the inlet and outlet water streams, T_i, T_o, P_i, and P_o, plus the inlet water flow rate $\dot{N}$. With values for these five variables, we can solve the steady-state material balance for the outlet water flow rate (the inlet and outlet mass flow rates are equal here) and we can solve the steady-state energy balance for $\dot{Q}$. In this example the value computed for the heat duty is the actual value for the real process, regardless of reversibility, because the process is workfree. However, in the general case, when heat and work both cross a system boundary, the energy balance gives only their sum. Variations on this problem are also possible; for example, if we knew values for the five variables T_i, T_o, P_i, P_o and $\dot{Q}$, then we could solve the energy balance for the required water flow rate. Or, if we knew T_i, P_i, P_o, $\dot{Q}$, and $\dot{N}$, then we could solve for the outlet water temperature T_o.

Analysis 2. In this second analysis, we consider situations in which the heat duty is unimportant and can be eliminated. In these cases we take the entire exchanger as the system. Now the system involves two substances, so $C = 2$, and it has two inlets plus two outlets, so $N_p = 4$. We also have the following external constraints: no shaft work (1), no heat transfer between system and surroundings (1), and only one component in each of the four streams (4). So the number of interactions available for manipulating this system, given by (3.6.12), is

$$\mathcal{V} = N_p(C+2) + 2 - \mathcal{S}_{ex} = 4(2+2) + 2 - 6 = 12 \qquad (3.6.18)$$

We also have three internal constraints: a material balance on each component plus the energy balance, so (3.6.14) gives

$$\mathcal{F}_{ex} = \mathcal{V} - \mathcal{S} = 12 - 3 = 9 \qquad (3.6.19)$$

A typical set of these nine variables would be the pressures of the four streams, the two inlet flow rates, and the temperatures of three streams. If values for these nine quantities were known, then we could solve the material and energy balances for the gas and water outlet flow rates and for the temperature of the fourth stream. In another version of this problem we might know the temperatures and pressures of all four streams plus one inlet flow rate; then we could obtain the other three flow rates by solving the two component material balances plus the energy balance.

These two analyses illustrate several important points: (a) The number of independent variables $\mathcal{F}_{ex}$ usually depends on what is chosen as the system. (b) The identity of the system also determines the number of dependent variables and the equations used to solve for their values. (c) The quantities counted in (3.6.14) for $\mathcal{F}_{ex}$ can include heat and work effects, which are process variables, not system properties.

3.7 GENERAL EXPRESSIONS FOR HEAT AND WORK

In § 3.2 we combined the first and second laws to eliminate Q and W and thereby obtained forms of the fundamental equation; those forms all contain some conceptual, such as U, S, or G. But as engineers we more often need values for heat and work rather than for changes in conceptuals. Unfortunately we cannot devise a purely theoretical scheme for computing the heat and work requirements for a real process: every real process involves irreversibilities, and the magnitudes of those irreversibilities must either be measured or estimated. Usually such measurements or estimates are made relative to reversible changes, so we need to be able to compute the heat and work that accompany reversible changes. The necessary equations are derived here.

3.7.1 Heat

For the generic, open, nonreacting system represented schematically in Figure 3.7, an expression for Q is obtained by rearranging the entropy balance (3.6.10),

$$\frac{\delta Q_{ext}}{T_{ext}} = \sum_i \bar{S}_i dN_i + \left(\frac{\partial S}{\partial T}\right)_{PN} dT + \left(\frac{\partial S}{\partial P}\right)_{TN} dP - \Delta S_{\alpha\beta} - dS_{gen} \qquad (3.7.1)$$

Here $\Delta S_{\alpha\beta}$ is defined analogously to (3.6.4). In general the entropy generation term is unknown, but if we consider reversible changes, then $dS_{gen} = 0$, and (3.7.1) reduces to

$$\frac{\delta Q_{rev}}{T_{ext}} = \sum_i \bar{S}_i dN_i + \left(\frac{\partial S}{\partial T}\right)_{PN} dT + \left(\frac{\partial S}{\partial P}\right)_{TN} dP - \Delta S_{\alpha\beta} \tag{3.7.2}$$

For this to apply, the external temperature T_{ext} must either equal the system temperature T and all stream temperatures T_α and T_β, or there must be reversible means for transferring heat across any finite temperature difference. For real processes, the amount of heat given by (3.7.2) will bound the actual heat requirements: an upper bound if heat is added to the system ($\delta Q_{rev} > 0$), a lower bound if heat is removed ($\delta Q_{rev} < 0$).

In the special case of workfree processes with negligible kinetic and potential energy changes, the heat can be obtained from the overall energy balance (3.6.7),

$$\delta Q_{wf} = \sum_i \bar{U}_i dN_i + \left(\frac{\partial U}{\partial T}\right)_{PN} dT + \left(\frac{\partial U}{\partial P}\right)_{TN} dP - \Delta H_{\alpha\beta} \tag{3.7.3}$$

In workfree processes, the heat given by (3.7.3) is the actual heat δQ_{ext}, regardless of the reversibility of the process.

Open steady-flow systems. In these cases, no change in accumulation occurs for any component in the system, so $dN_i = 0$, and the material balances (3.6.2) become

$$0 = \sum_\alpha \dot{N}_{\alpha i} - \sum_\beta \dot{N}_{\beta i} \qquad \textit{for each component } i \tag{3.7.4}$$

where the $\dot{N}$ represent molar flow rates. Similarly, (3.7.2) for the reversible heat simplifies to

$$\dot{Q}_{rev} = T_{ext}\left[-\sum_\alpha \sum_i \bar{S}_{\alpha i} \dot{N}_{\alpha i} + \sum_\beta \sum_i \bar{S}_{\beta i} \dot{N}_{\beta i}\right] \tag{3.7.5}$$

Closed systems. For reversible changes of state in closed systems, $dN_i = dN_{\alpha i} = dN_{\beta i} = 0$, and the overall entropy balance (3.6.10) reduces to

$$\frac{\delta Q_{rev}}{T_{ext}} = d\left(\sum_i N_i \bar{S}_i\right) = N ds \qquad \textit{closed system} \tag{3.7.6}$$

Integrating this from an initial state (1) to a final state (2) yields

$$Q_{rev} = T_{ext} N[s(T_2, P_2, \{x\}) - s(T_1, P_1, \{x\})] \qquad \textit{closed system} \tag{3.7.7}$$

If the reversible change is isothermal, then $T_1 = T_2 = T_{ext}$, and (3.7.7) reduces to the first part of the second law for closed systems; cf. (2.3.5).

3.7.2 Work

To obtain a corresponding expression for work in nonreacting systems, we use the rearranged entropy balance (3.7.1) to eliminate δQ_{ext} from the energy balance (3.6.7),

$$\sum_i [\bar{U}_i - T_{ext}\bar{S}_i]\, dN_i + \left(\frac{\partial[U - T_{ext}S]}{\partial T}\right)_{PN} dT + \left(\frac{\partial[U - T_{ext}S]}{\partial P}\right)_{TN} dP \tag{3.7.8}$$

$$= \Delta H_{\alpha\beta} - T_{ext}\,\Delta S_{\alpha\beta} + \delta W_b + \delta W_{sh} - T_{ext}\, dS_{gen}$$

To obtain a computationally more viable form, we consider reversible changes ($dS_{gen} = 0$) and combine the boundary work and shaft work into a total work term,

$$\delta W_{t,\,rev} = \delta W_{b,\,rev} + \delta W_{sh,\,rev} \tag{3.7.9}$$

Then (3.7.8) can be rearranged to read

$$\delta W_{t,\,rev} = \left(\frac{\partial[U - T_{ext}S]}{\partial T}\right)_{PN} dT + \left(\frac{\partial[U - T_{ext}S]}{\partial P}\right)_{TN} dP \tag{3.7.10}$$

$$+ \sum_i [\bar{U}_i - T_{ext}\bar{S}_i]\, dN_i - \Delta H_{\alpha\beta} + T_{ext}\,\Delta S_{\alpha\beta}$$

Note that if $T_{ext} \neq T$, then the work given by (3.7.10) must include the reversible work that accompanies any reversible heat transfer between system and surroundings. For real processes, the amount of work given by (3.7.10) will bound the actual work: an upper bound if work is done by the system ($\delta W_{t,rev} < 0$), a lower bound if work is done on the system ($\delta W_{t,rev} > 0$).

For adiabatic processes, the overall energy balance (3.6.7) simplifies to

$$\delta W_{t,\,ad} = \sum_i \bar{U}_i\, dN_i + \left(\frac{\partial U}{\partial T}\right)_{PN} dT + \left(\frac{\partial U}{\partial P}\right)_{TN} dP - \Delta H_{\alpha\beta} \tag{3.7.11}$$

and the adiabatic work given by (3.7.11) will be the actual work, regardless of reversibility. Note that the rhs of (3.7.11) is the same as the rhs of (3.7.3) for workfree heat.

If, in addition to all the other restrictions we have applied in obtaining (3.7.10), we also consider isothermal processes, then $T = T_\alpha = T_\beta = T_{ext}$ and

$$\bar{U}_i - T_{ext}\bar{S}_i = \bar{U}_i - T\bar{S}_i = \bar{A}_i \tag{3.7.12}$$

while

$$\bar{H}_i - T_{ext}\bar{S}_i = \bar{H}_i - T\bar{S}_i = \bar{G}_i \tag{3.7.13}$$

With these, (3.7.10) simplifies to

$$\delta W_{t,\,rev} = \sum_i \bar{A}_i\, dN_i + \left(\frac{\partial A}{\partial P}\right)_{TN} dP - \Delta G_{\alpha\beta} \tag{3.7.14}$$

where $\Delta G_{\alpha\beta}$ is defined analogously to $\Delta H_{\alpha\beta}$ in (3.6.4).

Open steady-flow systems. For steady flow, each material balance again takes the form (3.7.4), and (3.7.10) reduces to

$$\dot{W}_{t,\,rev} = -\sum_\alpha \sum_i [\bar{H}_{\alpha i} - T_{ext}\bar{S}_{\alpha i}]\,\dot{N}_{\alpha i} + \sum_\beta \sum_i [\bar{H}_{\beta i} - T_{ext}\bar{S}_{\beta i}]\,\dot{N}_{\beta i} \tag{3.7.15}$$

If the process is also isothermal, so $T_\alpha = T_\beta = T_{ext}$, then (3.7.15) simplifies further to

$$\dot{W}_{t,\,rev} = -\sum_\alpha \sum_i \bar{G}_{\alpha i}\,\dot{N}_{\alpha i} + \sum_\beta \sum_i \bar{G}_{\beta i}\,\dot{N}_{\beta i} \tag{3.7.16}$$

$$= -\sum_\alpha G_\alpha\,\dot{N}_\alpha + \sum_\beta G_\beta\,\dot{N}_\beta \tag{3.7.17}$$

For isothermal steady-flow processes, the reversible work is given by the accumulated difference in Gibbs energy between inlets and outlets.

Closed systems. For reversible isothermal changes of state in closed systems, we have $dN_i = dN_{\alpha i} = dN_{\beta i} = 0$, and $T = T_{ext}$. Then, combining (3.7.6) for the reversible heat with the overall energy balance (3.6.3), and ignoring boundary effects, we find

$$\delta W_{t,\,rev} = \sum_i d(\bar{A}_i N_i) = N da \tag{3.7.18}$$

Integrating this from an initial state (1) to a final state (2) yields

$$W_{t,\,rev} = N[a(T, P_2, \{x\}) - a(T, P_1, \{x\})] \tag{3.7.19}$$

For isothermal processes on closed systems, the reversible work is given by the change in Helmholtz energy, as already noted in (3.2.15).

3.7.3 Physical Meaning of the Chemical Potential

In § 3.2.2 we remarked that the chemical potential (3.2.23) is closely related to the reversible isothermal-isobaric work involved in adding a small amount of component i to a mixture. This statement can be proved using expressions developed in § 3.7.2.

Consider a container filled with a one-phase multicomponent mixture of composition $\{x\}$; the container is immersed in a reservoir that imposes its temperature T and pressure P on the mixture. The container is fitted with a single inlet by which more material can be reversibly injected, as shown schematically in Figure 3.9. The process considered here is addition to the container of a small amount of pure component 1. The reversible work associated with this process is given by (3.7.14); for an isobaric injection of material through one inlet with no outlets, (3.7.14) reduces to

$$\delta W_{t,\,rev} = \sum_i \bar{A}_i dN_i - \sum_\alpha \sum_i \bar{G}_{\alpha i} dN_{\alpha i} \tag{3.7.20}$$

Here $\alpha = 1$ because there is only one inlet, dN_{11} is the amount of pure component 1 added, so $dN_1 = dN_{11}$, while all the other mole numbers in the container remain constant; so, $dN_j = 0$ for $j \neq 1$. Therefore, since the small amount added hardly affects the composition, (3.7.20) reduces to

$$\delta W_{t,\,rev} = \bar{A}_1(T, P, \{x\})\, dN_1 - g_{\text{pure 1}}(T, P)\, dN_1 \tag{3.7.21}$$

Here, we have used the fact that a pure component chemical potential is merely the molar Gibbs energy (3.4.6). Now according to (3.7.9), $\delta W_{t,rev}$ accounts for both the boundary work and the shaft work. Separating these two components in (3.7.21) leaves

$$\delta W_{sh,\,rev} = \bar{A}_1(T, P, \{x\})\, dN_1 - g_{\text{pure 1}}(T, P)\, dN_1 - \delta W_{b,\,rev} \tag{3.7.22}$$

For the work to deform the boundary (the boundary must deform to keep the pressure constant), we can write

$$\delta W_{b,\,rev} = -P dV = -P d\left(\sum_i \bar{V}_i N_i\right) \tag{3.7.23}$$

With the help of the isothermal-isobaric Gibbs-Duhem equation, (3.7.23) simplifies to

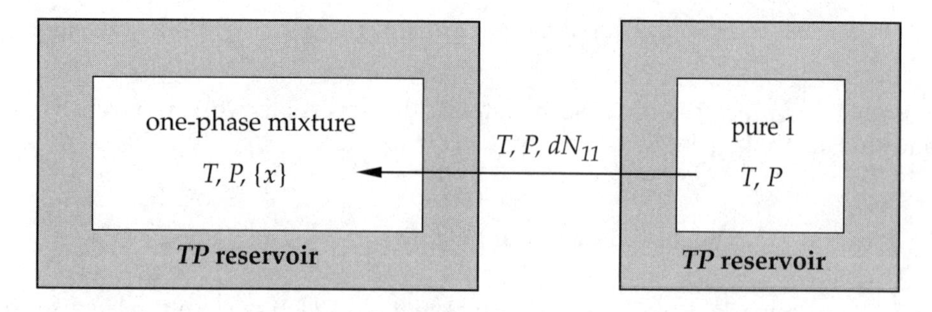

Figure 3.9 Schematic of a one-phase mixture immersed in a TP reservoir. The mixture is open to a single inlet (stream 1) through which a small amount of pure component 1 is added.

$$\delta W_{b,\,rev} = -P \sum_i \overline{V}_i \, dN_i \tag{3.7.24}$$

Because all mole numbers are constant except N_1, only the first term in the sum contributes,

$$\delta W_{b,\,rev} = -P \overline{V}_1 \, dN_1 \tag{3.7.25}$$

Substituting (3.7.25) into (3.7.22) gives

$$\delta W_{sh,\,rev} = [\overline{A}_1(T, P, \{x\}) - g_{\text{pure }1}(T, P) + P\overline{V}_1] \, dN_1 \tag{3.7.26}$$

or

$$\delta W_{sh,\,rev} = [\overline{G}_1(T, P, \{x\}) - g_{\text{pure }1}(T, P)] \, dN_1 \tag{3.7.27}$$

Since only a small amount of component 1 is being added, the composition $\{x\}$ is essentially constant during the process, so

$$\frac{\delta W_{sh,\,rev}}{dN_1} = \overline{G}_1(T, P, \{x\}) - g_{\text{pure }1}(T, P) \tag{3.7.28}$$

The difference between the chemical potential for component 1 in a mixture and that for pure 1 is the reversible work (per mole) that accompanies the transfer of a small amount of 1 from the pure state at T and P to the mixture at the same T and P. This constitutes a physical interpretation of the chemical potential (a conceptual) in terms of reversible work (a measurable).

The result (3.7.28) applies to mixtures containing any number of components. For binary mixtures, we will prove in Chapter 8 that a stable one-phase binary always has

$$\overline{G}_1(T, P, \{x\}) < g_{\text{pure }1}(T, P) \tag{3.7.29}$$

Therefore the work given by (3.7.27) is always negative, so long as the mixture remains a stable single phase; that is, whenever one component is added to a binary mixture at fixed T and P, the system does work on the surroundings. Unfortunately, the work given by (3.7.27) is too small to be useful, and it is usually dissipated.

3.7.4 Minimum Work to Separate a Mixture

A common problem in chemical process design is to develop methods for separating mixtures. Such methods require energy, but the requirements may vary substantially from one method (e.g., distillation) to another (e.g., reverse osmosis). In choosing among alternative methods, it may be useful to know the minimum energy requirements for a particular separation. The minimum requirements are given by reversible changes; here we show that the reversible work required for an isothermal-isobaric separation can be computed from the component chemical potentials.

Consider a vessel containing a one-phase multicomponent mixture of composition $\{x\}$; the vessel is immersed in a *TP* reservoir, as in Figure 3.10. The container is fitted with one outlet for each component. The process is to extract one pure component through each outlet. The reversible work for this process is again given by (3.7.14), which for no inlets becomes

$$\delta W_{t,\,rev} = \sum_i \bar{A}_i dN_i + \sum_\beta \sum_i \bar{G}_{\beta i} dN_{\beta i} \tag{3.7.30}$$

Since each outlet stream carries one pure component i and there is one such stream β for each component, the double sum in (3.7.30) is redundant. Therefore we can write (3.7.30) as

$$\delta W_{t,\,rev} = \sum_i \bar{A}_i dN_i + \sum_i g_{\text{pure } i} dN_{ii} \tag{3.7.31}$$

Here we are removing each component from the mixture, so the change dN_i in the system is related to the flow of component i through its outlet by

$$dN_i = -dN_{ii} \tag{3.7.32}$$

Therefore (3.7.31) can be written as

$$\delta W_{t,\,rev} = \sum_i \bar{A}_i dN_i - \sum_i g_{\text{pure } i} dN_i \tag{3.7.33}$$

As in § 3.7.3, we separate the total work into its boundary and shaft components, and use (3.7.25) for the boundary work. These manipulations give the shaft work as

$$\delta W_{sh,\,rev} = \sum_i [\bar{A}_i + P\bar{V}_i - g_{\text{pure } i}] dN_i \tag{3.7.34}$$

Hence,

$$\delta W_{sh,\,rev} = \sum_i [\bar{G}_i(T, P, \{x\}) - g_{\text{pure } i}(T, P)] dN_i \tag{3.7.35}$$

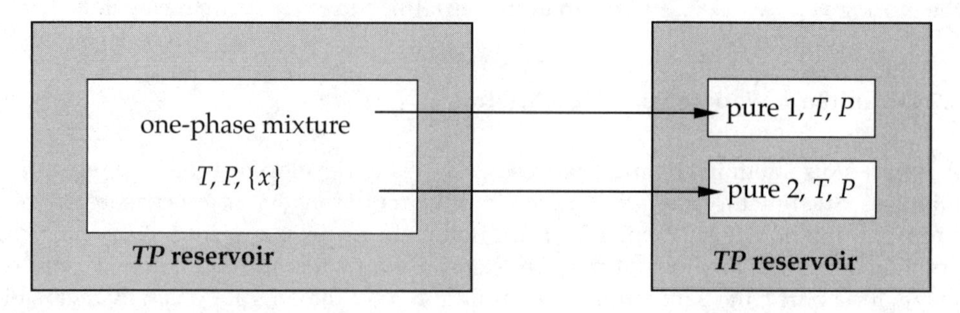

Figure 3.10 Schematic of a one-phase mixture immersed in a *TP* reservoir. The mixture is open to multiple outlets; one pure component can be extracted through each outlet.

If we extract all components in such a way that the composition $\{x\}$ remains constant throughout the separation, then (3.7.35) can be immediately integrated to yield

$$W_{sh,\,rev} = -\sum_i [\bar{G}_i(T, P, \{x\}) - g_{\text{pure }i}(T, P)]\, N_i \tag{3.7.36}$$

where N_i is the total number of moles of component i in the vessel at the start of the separation. Using (3.4.4) for the mixture term, (3.7.36) becomes

$$W_{sh,\,rev} = -\left[G_{mix}(T, P, \{x\}) - \sum_i N_i\, g_{\text{pure }i}(T, P)\right] \tag{3.7.37}$$

Here G_{mix} is the total Gibbs energy of the original mixture. The term in brackets is called the *change of Gibbs energy on mixing*,

$$G^m \equiv G_{mix}(T, P, \{x\}) - \sum_i N_i\, g_{\text{pure }i}(T, P) \tag{3.7.38}$$

Hence, the minimum isothermal-isobaric work needed to separate a mixture into its pure components is given by the negative change of Gibbs energy on mixing. Note that (3.7.37) is not limited to any particular phase: it applies to solids, liquids, and gases. In Chapter 6 we will show how to evaluate the differences in (3.7.37) and (3.7.38) for particular classes of mixtures.

3.8 SUMMARY

In this chapter we have presented fundamental thermodynamic relations among properties—quantities that depend on the system state. But in addition, we need to be able to determine how such properties respond when we change the state. Changes result from interactions—mass and energy crossing the system boundary—and so we need to characterize processes, as well as system states. Those characterizations may involve a description of how a system responds to particular interactions, or it may involve a determination of the interactions required to cause a particular change.

The first important relation we introduced was the fundamental equation, which provides relations among changes in certain thermodynamic properties. The fundamental equation was obtained (§ 3.2) by combining the first and second laws to eliminate the path functions Q and W. In the absence of path functions, we were able to transform the fundamental equation into alternative forms by applying attributes of exact differentials. These alternative forms allow us to choose a convenient set of independent variables to use when performing a thermodynamic analysis.

We then presented important derivatives that explicitly show how particular properties respond to changes in temperature or pressure or mole number (§ 3.3 and 3.4). Some of those many derivatives are measurable and therefore, when the relevant experimental data are available, those derivatives provide means for obtaining

Table 3.3 Selected isomorphisms between the calculus and thermodynamics

Calculus		Thermodynamics
Exact differentials	⇒	Changes in state functions; Maxwell eqs.
Integrating factors	⇒	Definition of S; Gibbs-Helmholtz equations
Partial derivatives	⇒	Response functions; partial molar properties
Legendre transforms	⇒	Definitions of H, A, and G
Implicit function theorem	⇒	Triple product rules
Homogeneous functions	⇒	Integrate fundamental eqs.; Gibbs-Duhem eq.

numerical values for changes in some state functions. More generally, in § 3.5 we cited the experimental data needed to compute changes in any of the conceptuals U, H, A, G, or S. The required information includes thermal data, in the form of heat capacities, and volumetric data, in the form of $PvTx$ equations of state.

Two patterns occur in this chapter, and we draw your attention to them here. One is the degree to which elements in thermodynamics are isomorphic to elements in the calculus. For example, the state functions of thermodynamics are, in the calculus, merely those quantities that form exact differentials. Several such isomorphisms are cited in Table 3.3, suggesting that much of fundamental thermodynamics is merely an application of the calculus. One striking consequence is that although the first and second laws, formulated in Chapter 2, did not explicitly contain anything about mixtures, we were, nevertheless, able to show formally how properties of mixtures may differ from those of pure substances.

A second general pattern occurs in how we use the calculus to formulate the response of a thermodynamic property to a change of state. The pattern can be resolved into the following steps:

(a) Identify the property of interest, call it F.

(b) Determine the appropriate number of other independent properties needed to identify the state, and choose a particular set of those properties (say T, P, and $\{N\}$). Then we might consider the property F to be expressible as

$$F = F(T, P, N_1, N_2, \ldots) \tag{3.8.1}$$

(c) Form the total differential dF, which represents the response of F to a change in the quantities T, P, and $\{N\}$,

$$dF = \left(\frac{\partial F}{\partial T}\right)_{PN} dT + \left(\frac{\partial F}{\partial P}\right)_{TN} dP + \sum_i \bar{F}_i \, dN_i \tag{3.8.2}$$

(d) Relate the partial-derivative coefficients in (3.8.2) to measurables. Here we have made the common (but arbitrary) choice of T and P as independent;

however, if variables other than T and P are chosen, then the partial molar quantities will not appear as simply as they do in (3.8.2).

(e) Integrate dF over the change of state to obtain ΔF. To compute those integrals, we need data or correlations that contain the state dependencies of the measurables introduced into (3.8.2).

We have not yet broached the problems associated with correlating data, so we are not yet ready to perform step (e). However, regardless of the correlation used, the procedure (a)–(e) or its equivalent must be used to obtain values for changes in conceptuals.

Lastly, we recognize that engineers routinely need to know heat and work effects associated with changes of state. Therefore, in § 3.6 and § 3.7 we developed formal expressions that allow us to use state functions to calculate the reversible heat and reversible work. In most cases Q_{rev} and W_{rev} only bound the actual values, but such bounds are often helpful in design and processing situations. To get values of Q and W for real processes we usually estimate the magnitude of the entropy generated and make corrections to Q_{rev} and W_{rev}; such estimates often involve process efficiencies extracted either from experiment or from correlations.

But while we have accomplished much in this chapter, more remains to be done. For example, we have established numerous relations among properties, but we have not addressed the most viable ways for obtaining numerical values for any of them. That task is taken up in Part II of this book. Moreover, throughout Part I we have restricted ourselves to single-phase, homogeneous systems; the problems posed by multiphase systems are tackled in Part III. Nevertheless, you will find that everything done in later chapters builds on the material presented here.

LITERATURE CITED

[1] D. Eisenberg and W. Kauzmann, *The Structure and Properties of Water*, Oxford University Press, New York, 1969, pp. 184–85.

[2] N. B. Vargaftik, *Tables on the Thermophysical Properties of Liquids and Gases*, 2nd ed., Hemisphere Publishing Co. (Wiley), New York, 1975.

[3] J. S. Rowlinson and F. L. Swinton, *Liquids and Liquid Mixtures*, 3rd ed., Butterworth, London, 1982.

[4] E. W. Washburn (ed.), *International Critical Tables of Numerical Data, Physics, Chemistry and Technology*, vol. 3, McGraw-Hill, New York, 1928.

[5] K. S. Pitzer, J. C. Peiper, R. H. Busey, "Thermodynamic Properties of Aqueous Sodium Chloride Solutions," *J. Phys. Chem. Ref. Data*, **13**, 1 (1984).

PROBLEMS

3.1 A liquid mixture of ethanol and water completely fills the cylinder of a piston-cylinder apparatus. The cylinder is closed to mass transfer, but its walls are thermally conducting and the piston can be moved. Determine values for $\mathcal{V}$, $\mathcal{F}_{ex}$, and $\mathcal{F}$, and explain what each of these quantities means. If the cylinder were insulated, which of your values change and which remain the same?

3.2 The ideal-gas equation, $R = PV/NT$, is an example of the generic equation of state written in (3.1.5); it implies that P, V, N, and T are not all independent.

(a) If we choose $F = V$, then what properties should appear in the argument list when the ideal-gas law serves as the basis for the function ψ in (3.1.5)?

(b) If we choose $F = P$, then what properties should be in the argument list?

(c) For a pure ideal gas, how many properties are needed for identifying an extensive equilibrium state; that is, what is the value of $\mathcal{F}_{ex}$?

(d) Evaluate the partial derivatives $(\partial P/\partial N)_{TV}$ and $(\partial P/\partial N)_{Tv}$ for an ideal gas.

3.3 For a fixed amount of a pure gas (not necessarily ideal) at state (P_1, V_1),

(a) Prove that, on a PV diagram, only one reversible isotherm passes through (P_1, V_1); i.e., $\mathcal{V} = 1$.

(b) Prove that there is only one reversible adiabat through (P_1, V_1); i.e., $\mathcal{V} = 1$.

3.4 Starting from the fundamental equation for closed systems, obtain expressions that give each of the following solely in terms of measurables,

$$\left(\frac{\partial H}{\partial T}\right)_{VN}, \quad \left(\frac{\partial H}{\partial T}\right)_{PN}, \quad \left(\frac{\partial H}{\partial P}\right)_{TN}, \quad \left(\frac{\partial H}{\partial V}\right)_{PN}$$

3.5 If a change in the shape of a system can make a difference in its properties, then we must allow for a new interaction: the surface work mode.

(a) Let the surface work be expressed as $\delta W_{sur} = \sigma\, d\mathcal{A}$, where $\mathcal{A}$ is the system surface area (extensive) and σ is the "surface tension" (a measurable). Write this system's fundamental equations in U and G.

(b) If a new function Υ is to be defined whose variables are S, σ, V, and N, what Legendre transform would be used?

(c) Derive at least two relations such that each connects a class I derivative of the surface tension σ to another class I derivative.

3.6 Consider a rubber band that can be elongated in a vacuum. The work of stretching the band is $\delta W_{rev} = -\tau\, dL$, where τ is the tension and L is the length.

(a) For this system, what is the fundamental equation in U?

(b) Obtain two relations between class II derivatives of S and class I derivatives.

(c) If $\tau = kLT$, where k is a constant, show that U depends only on temperature.

(d) The temperature of the rubber band increases during an adiabatic stretching. What does this suggest about the variation of U and S with T?

3.7 (a) Find a relation for $(\partial U/\partial N_i)$ in terms of measurables and accessible partial derivatives, such as the chemical potential for component i. The derivative is to be taken at fixed T, V, and $N_{j\neq i}$.

(b) Find a relation for $\overline{H}_i$ in terms of measurables and accessible partial derivatives, such as the chemical potential for component i.

3.8 Calculate the changes in H and G/T when 1,3 butadiene, in the ideal-gas state, is heated at constant pressure from 300 to 1000 K. The heat capacity for 1,3 butadiene as an ideal gas is given by

$$c_p/R = A + B(T/100) + C(T/100)^2 \qquad \text{(P3.8.1)}$$

where T is in Kelvin, $A = 0.290$, $B = 4.70476$, and $C = -0.15714$.

3.9 One mole of nitrogen undergoes the following three-step cyclic process, which starts and ends at 5 bar and 5 liter:

(a) Reversible expansion at constant isothermal compressibility to 10 liter,

(b) Reversible compression at constant volume expansivity to 5 liter,

(c) Irreversible isometric cooling to 5 bar.

Determine the net work done and the net heat transferred over one cycle. Assume the ideal-gas equation is obeyed with $c_v = 5R/2$.

3.10 (a) Ten moles of liquid water are initially at 20°C and 1 bar. The water is to be compressed isothermally to 500 bar, with the compression done in such a way that the required work is minimized. Use the response functions for water from Table 3.2 to estimate the final density, the amount of work required, and the direction and amount of any heat transferred.

(b) Repeat (a) for ten moles of air, assuming air is an ideal gas with $c_v = 5R/2$.

3.11 (a) Prove (3.3.30): the ratio of compressibilities is the ratio of heat capacities.

(b) Evaluate this ratio for air and for water using data from Table 3.2.

3.12 Consider a pure ideal gas with constant heat capacity c_v. For an arbitrary state (P_1, v_1), prove (a) that the slopes of the reversible adiabat and reversible isotherm through (P_1, v_1) are both negative, and (b) that, as v increases, P along the adiabat decreases faster than does P along the isotherm.

3.13 Heat capacities are functions of state and their response to changes in pressure or volume are related to the equation of state. Prove that

$$\left(\frac{\partial C_p}{\partial P}\right)_{TN} = -T\left(\frac{\partial^2 V}{\partial T^2}\right)_{PN} \quad \text{and} \quad \left(\frac{\partial C_v}{\partial V}\right)_{TN} = T\left(\frac{\partial^2 P}{\partial T^2}\right)_{VN}$$

3.14 Pure water is to be compressed from 1 bar, 20°C. (a) If the compression is done adiabatically to 100 bar, can the final temperature be 30°C? (b) If the compression is done adiabatically to 200 bar, what is the lowest possible final temperature?

3.15 For a substance with constant c_p, show that isobaric cooling and heating processes produce straight lines on a plot of s vs. $\ln T$.

3.16 The sonic velocity w in a fluid is a thermodynamic property related to the adiabatic compressibility by $w = 1/(\rho\kappa_s)^{1/2}$, where ρ is the mass density of the fluid.

(a) Show that w can also be written in terms of the isothermal compressibility

$$w = \frac{1}{\sqrt{\rho\kappa_T(1-(TV\alpha\gamma_v)/c_p)}} \tag{P3.16.1}$$

(b) Use data from Table 3.2 to compare the sonic velocity in air (used in determining Mach numbers for speeds of aircraft) with that in water (used in detecting submarines and other underwater objects).

(c) What are the magnitudes of the absolute and relative errors in the sonic velocity of air if we assume $\kappa_s = \kappa_T$ at 1 bar and 25°C?

3.17 Consider a binary mixture of 1 and 2 having molar volume v. Show that the partial molar volumes can be written in the form of Legendre transforms,

$$\overline{V}_1 = v - x_2\left(\frac{\partial v}{\partial x_2}\right)_{TP} \tag{P3.17.1}$$

and

$$\overline{V}_2 = v - x_1\left(\frac{\partial v}{\partial x_1}\right)_{TP} \tag{P3.17.2}$$

3.18 A mixture of ethanol(1) and water(2) has $x_1 = 0.7$ and a density $\rho = 0.8306\ \text{g/cm}^3$. At these conditions, the partial molar volume of water is $\overline{V}_2 = 15.68\ \text{cm}^3/\text{mole}$.

(a) What is the value of $\overline{V}_1$, the partial molar volume for ethanol?

(b) Estimate the mixture density ρ when x_1 is changed from 0.70 to 0.71 at fixed T and P.

(c) Do you expect $\overline{V}_1$ to increase or decrease when x_1 is increased from 0.70 at fixed T and P? Justify your expectation and clearly cite all assumptions made.

3.19 At 20°C and 1 bar a binary liquid mixture of 1 and 2 has the composition dependence of the partial molar volume of component 1 given by

$$\overline{V}_1 = v_{\text{pure 1}} + Ax_2^2 \tag{P3.19.1}$$

where A is a constant. Find the analogous expression for $\overline{V}_2$.

3.20 Consider a binary mixture of components 1 and 2 at fixed T and P. For such a mixture, show whether or not it is legitimate to represent a partial molar property as a linear function of composition. For example, show whether we may write the partial molar volume as

$$\overline{V}_1 = v_{\text{pure 1}} + Ax_2 \tag{P3.20.1}$$

where A is a constant, independent of composition.

3.21 (a) What information would enable you to integrate (3.5.7) to obtain ΔS?

(b) What information would enable you to integrate (3.5.13) to obtain ΔS?

3.22 For a single-phase substance containing any number of components, show that

$$C_v dT + V(T\gamma_v - P)(\alpha\, dT - \kappa_T\, dP) - \sum_i N_i\, d\overline{U}_i = 0 \tag{P3.22.1}$$

3.23 A stream of air (stream A) initially at 20°C is to be heated to 60°C by bringing it into contact with a second air stream (stream B) in a double-tube heat exchanger. Stream B enters the exchanger at 80°C. Assume the exchanger is well-insulated and operates at constant P; also assume air obeys $Pv = RT$ and has constant c_p.

(a) If the exchanger operates countercurrently and the mass flow rates of the two streams are the same, determine the outlet temperature of stream B.

(b) If the exchanger operates co-currently and the mass flow rates are equal, determine the outlet temperature of stream B.

(c) Schematically sketch, on the same diagram, your results from part (b) in the form of the temperature of each stream versus distance down the exchanger from inlet to outlet. Compare and discuss your results for (a) and (b); in particular, what's the same and what differs in processes (a) and (b)?

3.24 Derive the expression for the shaft work w_{sh} done by a reversible adiabatic compressor, assuming the fluid is an ideal gas with constant heat capacities. Your result should take the form

$$w_{sh} = \frac{\gamma R T_{in}}{(\gamma - 1)}\left[1 - \left(\frac{P_{out}}{P_{in}}\right)^{(\gamma-1)/\gamma}\right] \tag{P3.24.1}$$

where γ is the ratio of heat capacities (a measurable), $\gamma = c_p/c_v$.

3.25 A Joule-Thomson (J-T) expansion occurs whenever a steadily flowing fluid passes through a conduit or device that is well-insulated and that involves no shaft work.

(a) Because of friction, we expect a pressure drop across a J-T expansion; that is, $P_{out} < P_{in}$. Develop a thermodynamic argument to confirm this.

(b) However, show thermodynamically that there is no constraint on the relation between T_{in} and T_{out}: a fluid may be heated or cooled by a J-T expansion.

(c) The J-T coefficient is defined by $(\partial T/\partial P)_h$. Find an expression for this derivative solely in terms of measurables.

(d) Evaluate the J-T coefficient for an ideal gas.

3.26 Complete the derivation of (3.5.11) from (3.5.10) by showing that

$$\left(\frac{\partial U}{\partial N_i}\right)_{TVN_{j\neq i}} = \overline{H}_i - T\gamma_v \overline{V}_i \tag{P3.26.1}$$

3.27 The forms of the fundamental equation contain five conceptuals: U, S, H, A, and G. For these properties we can only obtain changes, such as ΔU, or values relative to a prechosen reference state, such as $U = U_{ref} + \Delta U$. As part of a process design project, you need to construct a table of these five properties for a certain pure substance (like the steam tables). First, you select a reference state: a phase, a temperature T_{ref}, and a pressure P_{ref}.

(a) Of the five conceptuals, how many can be arbitrarily set to zero at the reference state?

(b) Let n represent the number found in part (a); $n \leq 5$. At the reference state, can any n of the five conceptuals be set to zero? Or are there constraints on the identities of the n properties?

3.28 In modern steam tables, the reference state is usually taken to be the saturated liquid at its triple point ($T_{ref} = 273.16$ K, $P_{ref} = 0.611$ kPa); at this state the internal energy and entropy are set to zero: $u_{ref} = s_{ref} = 0$. An excerpt from such a table follows.

	$P = 1$ bar		**$P = 20$ bar**	
T(°C)	h(J/g)	s(J/g K)	h(J/g)	s(J/g K)
100	2676.2	7.3614	. . .	. . .
400	3278.2	8.5435	3247.6	7.1271

(a) For 1 g of steam heated isobarically at 1 bar from 100 to 400°C, compute Δh, Δs, and Δg.

(b) For 1 g of steam compressed isothermally from 1 to 20 bar at 400°C, compute Δh, Δs, and Δg.

(c) If, instead of the saturated liquid used above, the reference state is chosen to be superheated steam at $T_{ref} = 100$°C and $P_{ref} = 10$ kPa, then the above table becomes the one below. For the same processes as in (a), use the following table to compute Δh, Δs, and Δg.

	$P = 1$ bar		**$P = 20$ bar**	
T(°C)	h(J/g)	s(J/g K)	h(J/g)	s(J/g K)
100	–11.3	3.9434	. . .	. . .
400	590.7	5.1255	560.1	3.7091

(d) For the same processes as in (b), use this new table to compute Δh, Δs, and Δg.

(e) Compare your results from (a) with those from (c) and compare your results from (b) with those from (d). Explain any differences.

PART II

SINGLE-PHASE SYSTEMS

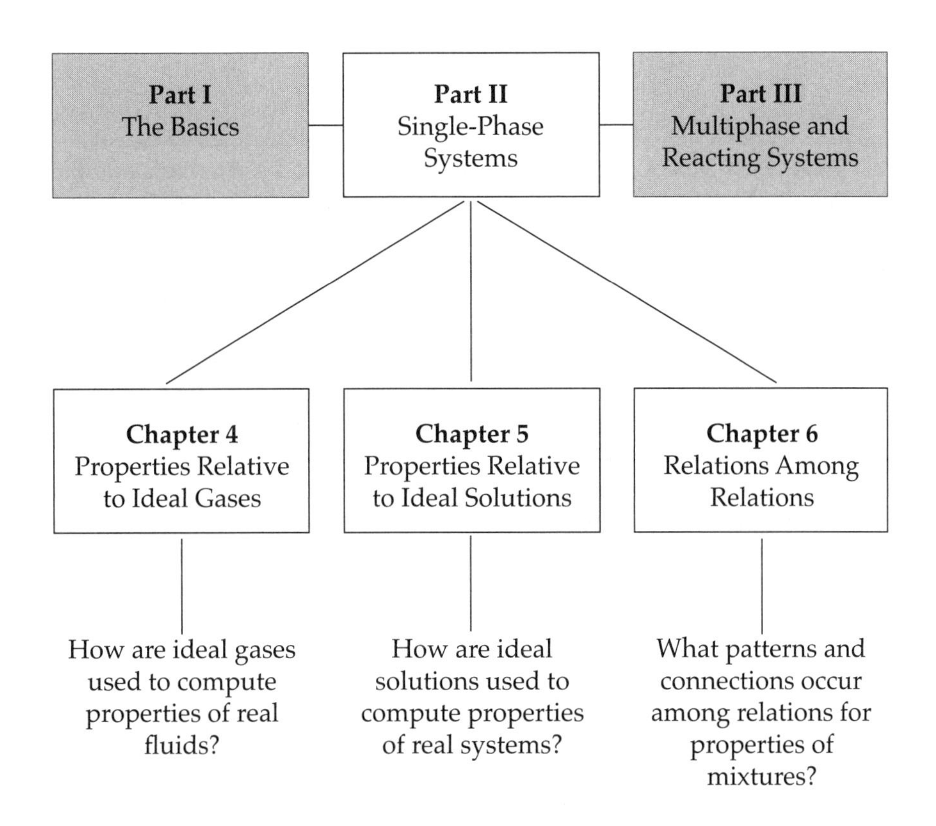

4

PROPERTIES RELATIVE TO IDEAL GASES

In Part I we established formal relations between process variables and system properties and between measurables and conceptuals; we also established relations among various conceptuals. Those relations suggest that properties can serve as the basis for thermodynamics analysis, but Part I does not provide values for any quantities. To get numbers we must do something beyond formal thermodynamics; ultimately, we must rely on experimental data and on models based on that data. In particular, experiments describe systems in terms of measurables, but before we can perform thermodynamic calculations, we need to know how to convert those measurables into conceptuals. In this chapter and the next, we focus on practical strategies for obtaining values for conceptuals.

One strategy, and one much used in thermodynamics, divides a property F into two parts: an ideal contribution, F_{id}, and a deviation or correction term, F_{dev}. This strategy can be realized in at least two ways. In the first, the deviation takes the form of a *difference measure*,

$$F_{dev} = F - F_{id} \tag{4.0.1}$$

While in the second, the deviation takes the form of a *ratio measure*,

$$F'_{dev} = \frac{F}{F_{id}} \tag{4.0.2}$$

In both ways the ideal substance must be well defined and its properties must be easy to compute; beyond that, the choice of ideality is made strictly for convenience. In both approaches the job of computing the property F is reduced to the (hopefully) easier job of computing one of the deviation terms, either F_{dev} or F'_{dev}.

Of the many measurables that exist, the ones whose values can be determined most readily are temperature, pressure, volume, and heat capacities. Of the many substances that exist, the ones whose measurables are most easily related to conceptuals are the substances that obey the ideal-gas equation of state. Therefore, the ideal gas often serves as a computationally convenient ideal substance for the deviation measures defined in (4.0.1) and (4.0.2), even when the magnitudes of the deviations are not small. In this chapter we show how to use the ideal gas as the basis for computing values for conceptual properties.

In § 4.1 we introduce ideal gases and their mixtures, and we derive equations for computing their thermodynamic properties. Then, we use the rest of the chapter to develop expressions for computing deviations from ideal-gas values: the difference measures in § 4.2, the ratio measures in § 4.3.

To compute values for the deviation measures, we need volumetric data for the substance of interest; such data are usually correlated in terms of a model *PvTx* equation of state. In § 4.4 we develop expressions that enable us to use equations of state to compute difference and ratio measures for deviations from the ideal gas. Finally, in § 4.5 we present a few simple models for the volumetric equation of state of real fluids. These few models are enough to introduce some of the problems that arise in attempting to analytically represent the *PvTx* behavior of real substances, and they allow us to compute values for conceptuals, using the expressions from § 4.5. However, more thorough expositions on equations of state must be found elsewhere [1–4].

4.1 IDEAL GASES

An ideal gas has three defining characteristics: (a) Its molecules exert no forces on one another, so there is no intermolecular potential energy. (b) Its atoms and molecules do have motion, so there is molecular kinetic energy and temperature. (c) Its molecules can exchange momentum with the walls of a confining vessel, so the gas has a pressure and volume. The absence of repulsive forces between molecules implies that an ideal gas can be compressed to zero volume without a phase change. The absence of attractive interactions implies that the gas has no driving force for condensation to a liquid or solid phase: an ideal gas remains gaseous at all state conditions. Ignoring intermolecular forces is a drastic assumption, except for supercritical substances at low pressure; however, we will use the ideal gas, not so much as an approximation to real substances, but rather as a basis for obtaining properties of real substances.

4.1.1 Pure Ideal Gases

Historically, the ideal-gas equation of state

$$PV = NRT \qquad \textit{ideal gas} \qquad (4.1.1)$$

was obtained by combining the experimental *PvT* data of low density gases that is codified in the laws of Boyle and Charles. Alternatively, this equation can be derived formally in statistical mechanics, under the assumption that there are no forces acting among the molecules.

For a substance that obeys (4.1.1), we can use (3.3.35) for H and (3.3.38) for U to show that U and H of an ideal gas depend only on temperature,

$$U^{ig} = Nu^{ig}(T) \quad \text{and} \quad H^{ig} = Nh^{ig}(T) \qquad (4.1.2)$$

These results are consistent with our definition of the ideal gas as a substance having no intermolecular forces. Recall the internal energy is the mechanism for storing energy; specifically, U is the combined kinetic and potential energies of the molecules. But if molecules exert no forces on one another, then they can have no molecular potential energy, and energy can be stored only as molecular kinetic energy. Therefore, (4.1.2) is consistent with (2.3.6), which states that molecular kinetic energy is related only to the absolute temperature, not to pressure or volume.

Thermodynamics cannot identify the forms taken by the functions of temperature in (4.1.2), but those functions can be found using either kinetic theory or statistical mechanics. Those functions are determined by the kinds of motions that are allowed to the atoms on a molecule; that is, they are determined by molecular structure. For molecules whose allowed motions (i.e., degrees of freedom) are predominantly external translations and rotations,

$$U^{ig} = \frac{\upsilon}{2} NRT \tag{4.1.3}$$

where υ is the number of degrees of freedom. Spherical molecules, such as argon, have only translational degrees of freedom and $\upsilon = 3$. Rigid diatomics, such as oxygen, have three translational plus two rotational degrees of freedom, so $\upsilon = 5$. But real molecules also have internal degrees of freedom (such as bond vibration, bond bending, and bond rotation), producing internal energies that are more complicated than (4.1.3) and then υ is usually a function of temperature.

With the equation of state (4.1.1) and an expression for the internal energy, such as (4.1.3), we can integrate relations in § 3.3 to obtain expressions for differences in all other thermodynamic properties of a pure ideal gas. The results include

$$c_p^{ig} - c_v^{ig} = R \tag{4.1.4}$$

$$\Delta u^{ig} = u^{ig}(T_2) - u^{ig}(T_1) = \int_{T_1}^{T_2} c_v^{ig}(T)\, dT \tag{4.1.5}$$

$$\Delta h^{ig} = h^{ig}(T_2) - h^{ig}(T_1) = \int_{T_1}^{T_2} c_p^{ig}(T)\, dT \tag{4.1.6}$$

$$\Delta s^{ig} = s^{ig}(T_2, P_2) - s^{ig}(T_1, P_1) = \int_{T_1}^{T_2} \frac{c_p^{ig}(T)}{T}\, dT - R \ln\frac{P_2}{P_1} \tag{4.1.7}$$

$$= \int_{T_1}^{T_2} \frac{c_v^{ig} T}{T}\, dT + R \ln\frac{v_2}{v_1} \tag{4.1.8}$$

Since c_v and c_p are necessarily positive, the ideal-gas enthalpy and internal energy must always increase with increasing temperature. Likewise, the ideal-gas entropy must increase with isobaric or isometric increases in temperature, decrease with isothermal increases in pressure, and increase with isothermal increases in volume. We emphasize that unlike u and h, the entropy s depends on pressure and volume as well as temperature.

The fundamental equations (3.2.12) and (3.2.14) can be used to determine how the ideal-gas Gibbs and Helmholtz energies respond to changes of state. For example, for isothermal changes in pressure,

$$\Delta a^{ig} = \Delta g^{ig} = RT \ln \frac{P_2}{P_1} \qquad \textit{fixed } T \qquad (4.1.9)$$

So in an ideal-gas, G and A *always* increase with isothermal increases in P.

Although no gas is truly ideal, real gases approximate ideal-gas behavior when the gas density is sufficiently small: at low densities there are few collisions or interactions among real molecules. Molecular size correlates with the density at which a gas becomes nearly ideal: the larger the molecules, the lower the density must be. This is because the range and strength of intermolecular forces increase with the number of electrons per molecule. We can make these statements precise by considering the zero density limit of volumetric equations of state. The limit can be expressed in either of two forms, depending on the identity of the independent variables.

If the independent variables are temperature and volume, our equation of state takes the form,

$$P = P(T, v) \qquad (4.1.10)$$

then the ideal-gas limit occurs when the (extensive) volume of the container is made infinitely large (specifically, when the container volume is large compared to the volume of the molecules themselves),

$$\lim_{V \to \infty} (\text{real stuff}) = \text{ideal gas} \qquad \textit{fixed } T \textit{ and } N \qquad (4.1.11)$$

Hence,

$$\lim_{V \to \infty} P = P^{ig} \qquad \textit{fixed } T \textit{ and } N \qquad (4.1.12)$$

All pressure-explicit equations of state should satisfy this limit.

Alternatively, if the independent variables are chosen to be temperature and pressure, then our equation of state takes the form

$$v = v(T, P) \qquad (4.1.13)$$

Now the ideal-gas limit should be expressed in terms of pressure; by inverting (4.1.12) we find that the limit occurs when the pressure is made vanishingly small,

$$\lim_{P \to 0} V = V^{ig} \qquad \textit{fixed } T \textit{ and } N \qquad (4.1.14)$$

All volume-explicit equations of state should satisfy this limit. For gases composed of small rigid molecules such as nitrogen and carbon dioxide at ambient and higher temperatures, properties are generally within 1% of their ideal-gas values for pressures up to roughly 10 bar. So the stuff you are now breathing is essentially an ideal gas.

4.1.2 Mixtures of Ideal Gases

In an ideal-gas mixture the molecules do not exert forces on one another, but molecules of different species are distinguishable; for example, they may have different masses or different structures or both. But because there are no intermolecular forces, each molecule is "unaware" of the presence of other molecules and therefore unaware that other species are present.

Consider an ideal-gas mixture confined to a vessel of volume V at temperature T. For such mixtures, each extensive property F is merely the sum of the corresponding extensive properties of the pure ideal-gas components, with each component at the mixture temperature T and occupying a container of the same volume V [5, 6]:

$$F^{ig}_{mix}(T, V, \{N\}) = \sum_i F^{ig}_{\text{pure } i}(T, V, N_i) \tag{4.1.15}$$

where $V = V_{mix} = V_i$ for each pure i. Here F could be U, H, S, A, or G; it could not, of course, be the volume. The intensive version of (4.1.15) is

$$f^{ig}_{mix}(T, v, \{x\}) = \sum_i x_i f^{ig}_{\text{pure } i}(T, v_i) \tag{4.1.16}$$

where the molar volumes are related by $v_i = Nv/N_i$.

To understand why (4.1.15) is valid, note that in a classical description of matter the values of extensive properties are determined by four attributes: (1) the number and structure of the molecules present, (2) the molecular kinetic energy, (3) the molecular potential energy (i.e., intermolecular forces), and (4) the nature of molecular interactions with the surroundings. These four attributes are identical on the two sides of (4.1.15): both the mixture and the collection of pure gases have the same number and kinds of molecules, they have the same molecular kinetic energies (temperatures), the same molecular potential energies (none), and the same interactions with their surroundings. This last attribute includes not only repulsive interactions between gas molecules and container walls that give rise to a pressure P, but also any spatial constraints imposed on the gas that restrict the molecules to containers of common volumes V. Some form of (4.1.15) is often used as a thermodynamic definition of an ideal-gas mixture, but we prefer to cite (4.1.15) as a consequence of the molecular definition given at the beginning of § 4.1.

From (4.1.15) we can derive expressions for all properties of ideal-gas mixtures; for example, we can immediately determine the pressure. To do so, we use the fundamental equation (3.2.12) to write any mixture or pure-component pressure as

$$P = -\left(\frac{\partial A}{\partial V}\right)_{TN} \tag{4.1.17}$$

Writing (4.1.15) explicitly for the Helmholtz energy A and applying the derivative in (4.1.17) to each term, we obtain

$$\left(\frac{\partial A^{ig}_{mix}(T, V, \{N\})}{\partial V}\right)_{TN} = \sum_i \left(\frac{\partial A^{ig}_{\text{pure i}}(T, V, N_i)}{\partial V}\right)_{TN_i} \tag{4.1.18}$$

Since $V_{\text{pure i}} = V$ for each component i, we merely write V for the extensive volumes of the mixture and all pure components. Using (4.1.17) in (4.1.18) yields the *law of additive pressures* [6]:

$$P^{ig}_{mix}(T, v, \{x\}) = \sum_i P^{ig}_{\text{pure i}}(T, v_{\text{pure i}}) \tag{4.1.19}$$

Note that although the extensive volumes are all the same $V = V_{\text{pure i}}$, the intensive volumes differ, $v < v_{\text{pure i}}$, because each pure gas necessarily contains fewer molecules than the mixture. Equation (4.1.19) states that the pressure of an ideal-gas mixture is the sum of the pure component pressures, when N_i molecules of each pure i are confined to a vessel having the same extensive volume V as that of the mixture vessel and each pure is at the mixture temperature T. Since each pure component is an ideal gas, we can substitute the ideal-gas law (4.1.1) into the rhs of (4.1.19) and find the same equation of state as for pure gases,

$$P^{ig}_{mix} = \sum_i \frac{N_i RT}{V} = \frac{NRT}{V} \tag{4.1.20}$$

4.1.3 Partial Molar Properties of Ideal Gases

To obtain the partial molar properties of ideal-gas mixtures we apply the partial molar derivative (3.4.5) either to the ideal-gas law, to obtain the partial molar volume, or to the general expression (4.1.15), to obtain other properties. The generic expression (4.1.15) yields

$$\bar{F}^{ig}_i = \left(\frac{\partial F^{ig}_{mix}}{\partial N_i}\right)_{TPN_{j\neq i}} = \frac{\partial}{\partial N_i}\left(\sum_k N_k f^{ig}_{\text{pure k}}(T, v_k)\right)_{TPN_{j\neq i}} \tag{4.1.21}$$

Here P is the pressure of the mixture at T and v. But $f_{\text{pure k}}$ is intensive and therefore it does not depend on any N_i; so in the sum, only the term having $i = k$ contributes to the derivative and we find

$$\bar{F}^{ig}_i(T, v, \{x\}) = f^{ig}_{\text{pure i}}(T, v_i) \tag{4.1.22}$$

Note that $v_i \neq v$. Because of (4.1.19), we can also express (4.1.22) in terms of pressure,

$$\bar{F}_i^{ig}(T, P, \{x\}) = f_{\text{pure i}}^{ig}(T, P_{\text{pure i}}^{ig}) \tag{4.1.23}$$

Note that the mixture pressure P is not the same as that of the pure: $P_{\text{pure i}}^{ig} \neq P$.

Recall that F in (4.1.22) can be any of U, H, S, A, or G. For ideal-gas mixtures, the generic result (4.1.22) reduces to either of two forms depending on whether F is a thermal property (i.e., first-law property, U or H) or an entropic property (i.e., second-law property, S, A, or G).

First-law properties. The partial molar volume can be found by applying the partial molar derivative (3.4.5) to the equation of state (4.1.20); the result is

$$\bar{V}_i^{ig}(T, P, \{x\}) = \frac{RT}{P} = v_{\text{pure i}}^{ig}(T, P) \tag{4.1.24}$$

That is, the partial molar volume for component i is the molar volume of pure i at the same T and P as the mixture. We emphasize that (4.1.23) and (4.1.24) are evaluated differently. Specifically, in (4.1.24) the pure molar volume v is to be evaluated at the T and P of the mixture; however, in (4.1.23) the pure component property f is to be evaluated at the mixture T and at the pressure $P_{\text{pure i}} = N_iRT/V$, which is always less than the mixture pressure P. In other words, (4.1.23) does *not* apply to the volume.

To obtain the partial molar internal energy and enthalpy, we use the generic expression (4.1.22) to obtain

$$\bar{U}_i^{ig}(T, v, \{x\}) = u_{\text{pure i}}^{ig}(T, v_i) \tag{4.1.25}$$

$$\bar{H}_i^{ig}(T, v, \{x\}) = h_{\text{pure i}}^{ig}(T, v_i) \tag{4.1.26}$$

But the pure ideal-gas internal energy and enthalpy are independent of pressure and volume, so these reduce to

$$\bar{U}_i^{ig}(T, \{x\}) = u_{\text{pure i}}^{ig}(T) \tag{4.1.27}$$

$$\bar{H}_i^{ig}(T, \{x\}) = h_{\text{pure i}}^{ig}(T) \tag{4.1.28}$$

Second-law properties. Writing (4.1.22) explicitly for the partial molar entropy, we have

$$\bar{S}_i^{ig}(T, v, \{x\}) = s_{\text{pure i}}^{ig}(T, v_i) \tag{4.1.29}$$

This implies that the entropy does not change when ideal gases are mixed at constant T and v. But rather than T and v, usually we want to use T and P as the independent variables. Here P represents the pressure when the mixture has temperature T and molar volume v. Therefore, (4.1.29) can be written as

$$\bar{S}_i^{ig}(T, P, \{x\}) = s_{\text{pure i}}^{ig}(T, P_{\text{pure i}}) \tag{4.1.30}$$

But from (4.1.15) the pure components each occupy a container having the same volume V as the mixture; they are not at the same pressure (the mixture and each pure have different numbers of moles, so their pressures differ, even though their temperatures and extensive volumes are the same). So on the rhs of (4.1.30), we cannot simply replace v_i with P because they refer to different states and because the ideal-gas entropy depends on pressure and volume. Instead, to express the rhs of (4.1.30) in terms of the mixture pressure P, we must correct the pure component entropy on the rhs from the pure component pressure $P_{\text{pure i}}$ to the mixture pressure P.

The correction can be evaluated from the Maxwell relation (3.3.34). For the pure ideal gas it is

$$\left(\frac{\partial S}{\partial P}\right)_{TN} = -\left(\frac{\partial V}{\partial T}\right)_{PN} = -\frac{NR}{P} \qquad \textit{ideal gas} \tag{4.1.31}$$

Separating variables and integrating along the isotherm T from $P_{\text{pure i}}$ to the mixture pressure P, we find

$$S_{\text{pure i}}^{ig}(T, P, N_i) - S_{\text{pure i}}^{ig}(T, P_{\text{pure i}}, N_i) = N_i R \ln\left(\frac{P_{\text{pure i}}}{P}\right) \tag{4.1.32}$$

Since the mixture and pure i are ideal gases at the same T and V, we have

$$\frac{P_{\text{pure i}}}{P} = \frac{N_i RT}{V}\frac{V}{NRT} = x_i \tag{4.1.33}$$

and (4.1.32) can be written as

$$S_{\text{pure i}}^{ig}(T, P_{\text{pure i}}, N_i) = S_{\text{pure i}}^{ig}(T, P, N_i) - N_i R \ln x_i \tag{4.1.34}$$

Substituting the intensive version of (4.1.34) into the rhs of (4.1.30) gives the final result

$$\bar{S}_i^{ig}(T, P, \{x\}) = s_{\text{pure i}}^{ig}(T, P) - R \ln x_i \tag{4.1.35}$$

An analogous procedure yields, for the Gibbs and Helmholtz energies,

$$\bar{A}_i^{ig}(T, P, \{x\}) = a_{\text{pure i}}^{ig}(T, P) + RT \ln x_i \tag{4.1.36}$$

and

$$\bar{G}_i^{ig}(T, P, \{x\}) = g_{\text{pure i}}^{ig}(T, P) + RT \ln x_i \tag{4.1.37}$$

Note that since $x_i < 1$, the last term on the rhs of (4.1.35) is necessarily positive and likewise those on the rhs of (4.1.36) and (4.1.37) are necessarily negative. This means that, for isothermal-isobaric mixing of ideal gases, the entropy increases, while the Gibbs and Helmholtz energies decrease. But note that this behavior differs from that for isothermal-isometric mixing.

4.1.4 Properties of Ideal-Gas Mixtures

To obtain the properties of ideal-gas mixtures we simply accumulate the partial molar properties according to (3.4.4), all at the same T and P,

$$f^{ig} = \sum_i x_i \bar{F}_i^{ig} \tag{3.4.4}$$

Then on substituting (4.1.24) into (3.4.4), we find that the volume of an ideal-gas mixture is the mole-fraction average of the pure molar volumes,

$$v^{ig}(T, P, \{x\}) = \sum_i x_i \bar{V}_i^{ig} = \sum_i x_i v_{\text{pure i}}^{ig}(T, P) \tag{4.1.38}$$

Likewise, substituting (4.1.27) and (4.1.28) into (3.4.4) shows that the internal energy and enthalpy depend only on temperature and composition,

$$u^{ig}(T, \{x\}) = \sum_i x_i u_{\text{pure i}}^{ig}(T) \tag{4.1.39}$$

$$h^{ig}(T, \{x\}) = \sum_i x_i h_{\text{pure i}}^{ig}(T) \tag{4.1.40}$$

To obtain the heat capacities, we apply the definitions (3.3.7) and (3.3.8) to (4.1.39) and (4.1.40), respectively. The results are

$$c_v^{ig}(T, \{x\}) = \sum_i x_i c_{v\text{pure i}}^{ig}(T) \tag{4.1.41}$$

$$c_p^{ig}(T, \{x\}) = \sum_i x_i c_{p\text{pure i}}^{ig}(T) \tag{4.1.42}$$

Further, the difference between the heat capacities for ideal-gas mixtures is the same as for pure ideal gases (4.1.4). In summary, all first-law properties of ideal-gas mixtures are rigorously obtained by mole-fraction averaging pure ideal-gas properties.

For second-law properties, we substitute (4.1.35)–(4.1.37) into (3.4.4) to find

$$s^{ig}(T, P, \{x\}) = \sum_i x_i s_{\text{pure i}}^{ig}(T, P) - R \sum_i x_i \ln x_i \tag{4.1.43}$$

$$g^{ig}(T, P, \{x\}) = \sum_i x_i g^{ig}_{\text{pure i}}(T, P) + RT \sum_i x_i \ln x_i \tag{4.1.44}$$

$$a^{ig}(T, P, \{x\}) = \sum_i x_i a^{ig}_{\text{pure i}}(T, P) + RT \sum_i x_i \ln x_i \tag{4.1.45}$$

The quantity $(-R\,\Sigma x_i \ln x_i)$ is called the *ideal entropy of mixing*; it is always positive, so that the entropy of an ideal-gas mixture is always greater than the mole-fraction average of the pure component entropies, when the mixture and all pures are at the same T and P. But note that the entropy of mixing appears because the pure components are at the same P as the mixture, rather than at the same volume. If the pures had been specified at the same extensive volume as the mixture, then, as implied by (4.1.29), the entropy would not change on mixing. Therefore, an increase of entropy on mixing occurs not only because molecules of different species are distinguishable [5], but also because, for isobaric mixing, the space available to the molecules increases.

4.1.5 Example

When an ideal-gas mixture is separated into its pure components, is less work required for a separation at constant T and P or for one at constant T and V?

One mole of an equimolar mixture of methane and ethane is confined to a vessel at 25°C and 1 bar. The mixture is to be isothermally separated into its pure components.

Isobaric separation. In an isobaric process, the mixture and the pure components are each to be at 1 bar. Under the stated conditions these gases are essentially ideal; hence, by the ideal-gas law, the volumes of the pure gases are each half the volume of the original mixture, as shown in Figure 4.1. The lower bound on the work occurs when the separation is performed reversibly, and the required reversible isothermal-isobaric work was determined in § 3.7.4. The general result was found to be the negative change of Gibbs energy on mixing,

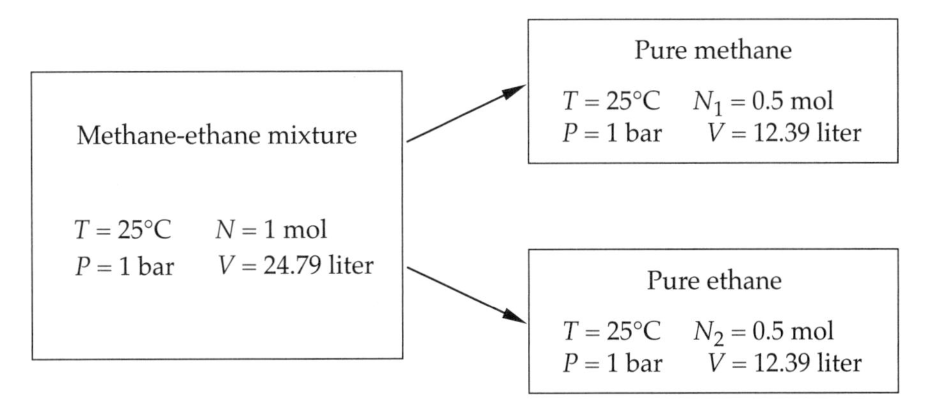

Figure 4.1 Schematic of an isothermal-isobaric process for separating a binary ideal-gas mixture into its pure components

$$W_{sh,\,rev} = -G^{m}(T, P, \{N\}) = -G_{mix}(T, P, \{N\}) + \sum_i N_i\, g_{\text{pure i}}(T, P) \qquad (4.1.46)$$

For ideal gases, we replace G_{mix} in (4.1.46) with the extensive form of (4.1.44), finding

$$W_{sh,\,rev} = -RT\sum_i N_i \ln x_i = 1.72 \text{ kJ} \qquad (4.1.47)$$

The positive value indicates that work must be done on the mixture to achieve an isobaric separation. In a real isothermal-isobaric separation of ideal gases, more than this minimum amount of work would be needed, because a real process would be irreversible. Moreover, when separating real mixtures (whose components have intermolecular forces), the total minimum work would not be given by (4.1.47). However, it could still be determined from G^m using (4.1.46), provided a reliable model were available for the Gibbs energy of the mixture and each pure. Expressions for G^m of real mixtures would be more complicated than the ideal-gas expression (4.1.47) but such expressions could be obtained from model equations of state.

Isometric separation. In the isometric process the mixture and the pure components are each confined to vessels having the same volume, as in Figure 4.2. For this process, the derivation of the expression for the reversible work parallels that given in § 3.7.4 for the isobaric work. We start with the expression for the total, reversible, isothermal work (3.7.14), written for two outlets and no inlets,

$$\delta W_{t,\,rev} = \sum_i \bar{A}_i\, dN_i + \left(\frac{\partial A}{\partial P}\right)_{TN} dP + \sum_i \bar{G}_{\text{pure i}}\, dN_i \qquad (4.1.48)$$

Here, the boundary work is zero because V remains constant. Using the ideal-gas law and the chain rule, the middle term in (4.1.48) becomes $\sum RT\, dN_i = 0$, so we have left

$$W_{sh,\,rev} = -\left[A_{mix}(T, V, \{N\}) - \sum_i N_i\, a_{\text{pure i}}(T, V, N_i)\right] \qquad (4.1.49)$$

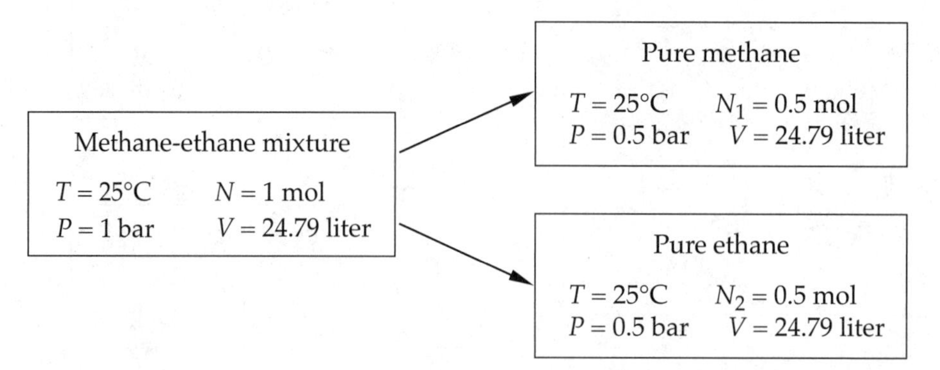

Figure 4.2 Schematic of an isothermal process for separating a binary ideal-gas mixture into its pure components, with each gas held in a vessel of the same volume V

Note that the term in brackets is *not* the change in Helmholtz energy on mixing A^m, because the terms that define A^m must be at a common pressure, but the terms in (E4.1.5) are at a common extensive volume. Substituting the ideal-gas expression (4.1.45) for A_{mix} in (4.1.49), we obtain

$$W_{sh,\,rev} = 0 \tag{4.1.50}$$

This means that it is thermodynamically possible to separate an ideal-gas mixture into its pure components without doing any work, if the process is performed at constant T and V. But in real separations of ideal gases some amount of work would be needed to overcome irreversibilities. The important lesson here is that the minimum work to perform the isometric separation, given by (4.1.50), differs from that for the isobaric separation, given by (4.1.47).

4.1.6 Entropy and Disorder

Entropy is frequently interpreted physically as a measure of the amount of "order" or "disorder" in a system. Specifically, statements are made to the effect that increases in the disorder of a system are reflected by increases in entropy. In this section we explore such claims. Mixing is one process in which substances can be considered to become less ordered, and so, if the conventional wisdom is correct, the mixing of pure substances should be accompanied by entropy increases. To test this, we consider two processes for mixing pure ideal gases: (a) one at fixed T and P, (b) another at fixed T and V.

Isothermal-isobaric mixing. Consider N_1 moles of pure ideal gas 1 and N_2 moles of pure ideal gas 2 initially in separate containers at the same T and P. We mix these two gases in such a way that the mixture remains at the same T and P; note this is the reverse of the process shown in Figure 4.1. We want to determine whether the change in entropy is positive, negative, or zero. The entropy change is given by

$$\Delta S(T, P) = S^{ig}(T, P, N) - \sum_i N_i s^{ig}_{\text{pure i}}(T, P) \tag{4.1.51}$$

From (4.1.43) we have for the mixture

$$S^{ig}(T, P, N) = \sum_i N_i s^{ig}_{\text{pure i}}(T, P) - R \sum_i N_i \ln x_i \tag{4.1.52}$$

So substituting (4.1.52) into (4.1.51) leaves

$$\Delta S(T, P) = -R \sum_i N_i \ln x_i \tag{4.1.53}$$

We consider two cases: (i) If the gases differ (say methane and ethane), then $x_i < 1$ and (4.1.53) shows that S increases, as expected,

$$\Delta S(T, P) > 0 \qquad \textit{mixing different ideal gases} \tag{4.1.54}$$

(ii) However, if the gases are the same (e.g., two samples of methane), then $i = 1$, $x_i = 1$, and (4.1.53) gives

$$\Delta S(T, P) = 0 \qquad \textit{mixing the same ideal gas} \tag{4.1.55}$$

Isothermal-isometric mixing. Now consider N_1 moles of pure ideal gas 1 and N_2 moles of pure ideal gas 2 initially in separate containers at the same T and V. We mix these two gases in such a way that the mixture remains at the same T and V; this is the reverse of the process shown in Figure 4.2. For this situation the entropy change is

$$\Delta S(T, V) = S^{ig}(T, V, N) - \sum_i N_i s^{ig}_{\text{pure i}}(T, V/N_i) \tag{4.1.56}$$

For this process, (4.1.15) gives the mixture entropy,

$$S^{ig}(T, V, N) = \sum_i N_i s^{ig}_{\text{pure i}}(T, V/N_i) \tag{4.1.57}$$

Combining (4.1.57) with (4.1.56) leaves,

$$\Delta S(T, V) = 0 \qquad \textit{mixing different ideal gases} \tag{4.1.58}$$

However, if we mix two samples of the same gas, then (4.1.56) becomes

$$\Delta S(T, V) = \sum_i N_i [s^{ig}_{\text{pure i}}(T, V/N) - s^{ig}_{\text{pure i}}(T, V/N_i)] \tag{4.1.59}$$

The entropy changes because the molar volume of the pure in the final state differs from that of the two pures in their initial states. The response of S to changes in v is given by the Maxwell relation in (3.3.37). For ideal gases it becomes

$$\left(\frac{\partial s}{\partial v}\right)_T = \left(\frac{\partial P}{\partial T}\right)_v = \frac{R}{v} \qquad \textit{ideal gas} \tag{4.1.60}$$

Separating variables and integrating over the volume, we find for each sample i,

$$\Delta s_i = R \int_{V/N_i}^{V/N} \frac{dv}{v} = R \ln \frac{VN_i}{VN} = R \ln \frac{N_i}{N} < 0 \tag{4.1.61}$$

Hence, both terms in the sum in (4.1.59) are negative and therefore,

$$\Delta S(T, V) < 0 \qquad \textit{mixing the same ideal gas} \tag{4.1.62}$$

Table 4.1 Changes in entropy for mixing ideal gases at fixed T and V and at fixed T and P

Gases being mixed	Fixed T & V	Fixed T & P
Samples of the same ideal gas	$\Delta S < 0$	$\Delta S = 0$
Different ideal gases	$\Delta S = 0$	$\Delta S > 0$

Our results, summarized in Table 4.1, imply that entropy does not necessarily measure the amount of "disorder." When ideal gases are mixed (and "disorder" presumably increases), the entropy may increase, decrease, or remain constant, depending on how the mixing is done and on whether we are mixing different gases or samples of the same gas. Note that none of the results in Table 4.1 violate the second law.

4.2 DEVIATIONS FROM IDEAL GASES: DIFFERENCE MEASURES

In § 4.1.4 we found that to compute the thermodynamic properties of ideal-gas mixtures, we need only the mixture composition plus the pure ideal-gas properties at the same state condition as the mixture. In other words, the properties of ideal-gas mixtures are easy to compute. We would like to take advantage of this, even for substances that are not ideal gases. To do so we introduce, for a generic property F, a *residual property* F^{res}, which serves as a difference measure for how our substance deviates from ideal-gas behavior.

By dividing F into an ideal-gas part plus a residual part, we sometimes ease the computational burden incurred when we need to compute the properties F of a real mixture. Of course, this strategy is most successful when our real substance does not differ much from an ideal gas, for then F^{res} is a small portion of the total property F and we may be able to tolerate a sizable error in estimating that small portion. As a result, residual properties have been most useful for nonideal gas mixtures. They are also legitimate entities for liquids and solids, though for condensed phases their magnitudes are large. In traditional practice residual properties were infrequently used for condensed phases; however, recent advances in modeling enable us to evaluate residual properties for dense fluids as well as for gases.

We define two classes of residual properties: isobaric ones and isometric ones. The isobaric residual properties (§ 4.2.1) are the traditional forms and use P as the independent variable. The isometric ones (§ 4.2.2) use v as the independent variable and thereby simplify computations when our equation of state is explicit in the pressure; such equations of state are now commonly used to correlate thermodynamic data for dense fluids. Although isometric property calculations may be more complicated than those for isobaric properties, with the help of computers, this is not really an issue.

4.2.1 Isobaric Residual Properties

These residual properties are defined only for those thermodynamic properties F that can be made extensive:

$$F^{res}(T, P, \{N\}) \equiv F(T, P, \{N\}) - F^{ig}(T, P, \{N\}) \tag{4.2.1}$$

Note that all three terms in (4.2.1) are to be evaluated at the same temperature, pressure, and composition. In general, F^{res} may be positive, negative, or zero. An ideal gas has *all* residual properties equal to zero; if a substance has only some residual properties equal to zero, it is not an ideal gas.

Since the definition (4.2.1) is a linear combination of thermodynamic properties, all relations among extensive properties, such as those in Chapter 3, can be expressed in terms of residual properties. Examples of such relations include the four forms of the fundamental equation and the Maxwell relations. Moreover, using the expressions developed in § 4.1.4 for ideal-gas mixtures, the following intensive forms for residual properties are obtained:

$$v^{res}(T, P\{x\}) = v(T, P\{x\}) - \frac{RT}{P} \tag{4.2.2}$$

$$u^{res}(T, P, \{x\}) = u(T, P, \{x\}) - \sum_i x_i u^{ig}_{\text{pure i}}(T) \tag{4.2.3}$$

$$h^{res}(T, P, \{x\}) = h(T, P, \{x\}) - \sum_i x_i h^{ig}_{\text{pure i}}(T) \tag{4.2.4}$$

$$s^{res}(T, P, \{x\}) = s(T, P, \{x\}) - \sum_i x_i s^{ig}_{\text{pure i}}(T, P) + R\sum_i x_i \ln x_i \tag{4.2.5}$$

$$a^{res}(T, P, \{x\}) = a(T, P, \{x\}) - \sum_i x_i a^{ig}_{\text{pure i}}(T, P) - RT\sum_i x_i \ln x_i \tag{4.2.6}$$

$$g^{res}(T, P, \{x\}) = g(T, P, \{x\}) - \sum_i x_i g^{ig}_{\text{pure i}}(T, P) - RT\sum_i x_i \ln x_i \tag{4.2.7}$$

The residual chemical potential

$$\bar{G}_i^{res}(T, P, \{x\}) = \bar{G}_i(T, P, \{x\}) - \bar{G}_i^{ig}(T, P, \{x\}) \tag{4.2.8}$$

can be found either by applying the partial molar derivative to (4.2.7) or by substituting (4.1.37) for the chemical potential of an ideal gas directly into the definition (4.2.8). Both procedures give the same result,

$$\bar{G}_i^{res}(T, P, \{x\}) = \bar{G}_i(T, P, \{x\}) - g^{ig}_{\text{pure i}}(T, P) - RT \ln x_i \tag{4.2.9}$$

In all these equations (4.2.2)–(4.2.9), the mixture and each pure ideal gas must be at the same temperature, pressure, and composition, except the ideal-gas values for U and H, which only must be at the same temperature and composition because the ideal-gas internal energy and enthalpy are independent of pressure.

4.2.2 Isometric Residual Properties

Instead of using T and P as the independent variables, as we did in § 4.2.1, we could choose T and V. Therefore, we define another set of residual properties in which the real substance and the ideal gas each occupy a container of the same volume V. Extensive isometric residual properties are defined by

$$F^{res}(T, V, \{N\}) \equiv F(T, V, \{N\}) - F^{ig}(T, V, \{N\}) \tag{4.2.10}$$

The corresponding intensive analogs are defined by

$$f^{res}(T, v, \{x\}) \equiv f(T, v, \{x\}) - f^{ig}(T, v, \{x\}) \tag{4.2.11}$$

In the definition (4.2.10), F can be any of the extensive properties U, H, S, A, or G and in (4.2.11) f can be any of their intensive analogs. Of course, the extensive and intensive (isometric) residual volumes are always zero. In addition, (4.2.11) can be used to define a residual pressure,

$$P^{res}(T, v, \{x\}) \equiv P(T, v, \{x\}) - \frac{RT}{v} \tag{4.2.12}$$

For intensive properties, the generic forms for the residual properties are all obtained by combining (4.1.16) with the definition (4.2.11):

$$f^{res}(T, v, \{x\}) = f(T, v, \{x\}) - \sum_i x_i f^{ig}_{\text{pure i}}(T, v_i) \tag{4.2.13}$$

Here, as in (4.1.16), the pure ideal-gas molar volumes are related to the mixture molar volume by $v_i = Nv/N_i$. Equation (4.2.13) applies to both first-law and second-law properties, and we have

$$u^{res}(T, v, \{x\}) = u(T, v, \{x\}) - \sum_i x_i u^{ig}_{\text{pure i}}(T) \tag{4.2.14}$$

$$h^{res}(T, v, \{x\}) = h(T, v, \{x\}) - \sum_i x_i h^{ig}_{\text{pure i}}(T) \tag{4.2.15}$$

$$s^{res}(T, v, \{x\}) = s(T, v, \{x\}) - \sum_i x_i s^{ig}_{\text{pure i}}(T, v_i) \tag{4.2.16}$$

$$a^{res}(T, v, \{x\}) = a(T, v, \{x\}) - \sum_i x_i a^{ig}_{\text{pure i}}(T, v_i) \tag{4.2.17}$$

$$g^{res}(T, v, \{x\}) = g(T, v, \{x\}) - \sum_i x_i g^{ig}_{\text{pure i}}(T, v_i) \tag{4.2.18}$$

Note that here the second-law properties contain no ideal entropy of mixing term, because the mixture and all pure ideal-gas components are at the same temperature and same extensive volume V.

The isometric residual chemical potential can be obtained from that part of (3.2.22) which relates the chemical potential to the Helmholtz energy,

$$\bar{G}_i^{res}(T, v, \{x\}) = \left(\frac{\partial N a^{res}}{\partial N_i}\right)_{TVN_{j \neq i}} \tag{4.2.19}$$

Using (4.2.17) for the residual Helmholtz energy, we find

$$\bar{G}_i^{res}(T, v, \{x\}) = \bar{G}_i(T, v, \{x\}) - a^{ig}_{\text{pure i}}(T, v_i) \tag{4.2.20}$$

In all the equations (4.2.14)–(4.2.20) the mixture and each pure ideal gas must be at the same temperature, composition, and extensive volume.

4.2.3 Relations Between the Two Kinds of Residual Properties

We now relate the two kinds of residual properties introduced in § 4.2.1 and 4.2.2. First write the intensive form of the definition (4.2.1) for isobaric residual properties,

$$f^{res}(T, P, \{x\}) \equiv f(T, P, \{x\}) - f^{ig}(T, P, \{x\}) \tag{4.2.21}$$

and then subtract this from the definition (4.2.11) for $f^{res}(T, v, \{x\})$. The result is

$$f^{res}(T, v, \{x\}) = f^{res}(T, P, \{x\}) + \Delta f^{ig} \tag{4.2.22}$$

where

$$\Delta f^{ig} = f^{ig}(T, P, \{x\}) - f^{ig}(T, v, \{x\}) \tag{4.2.23}$$

We emphasize that in (4.2.23) the independent variables P and v are each properties of the real substance; they are not related by the ideal-gas law, so the value for Δf^{ig} given by (4.2.23) is not necessarily zero.

When f is the volume, $v^{res}(T, v, \{x\}) = 0$, and (4.2.22) reduces to (4.2.2). Likewise, when f is the pressure we have $P^{res}(T, P, \{x\}) = 0$, then (4.2.22) reduces to (4.2.12). Otherwise, for first-law properties, we have

$$u^{res}(T, v, \{x\}) = u^{res}(T, P, \{x\}) \tag{4.2.24}$$

and

$$h^{res}(T, v, \{x\}) = h^{res}(T, P, \{x\}) \tag{4.2.25}$$

because the ideal-gas values of u and h are independent of P and v.

But for second-law properties, the relations are not as simple as (4.2.24) and (4.2.25). To have a representative second-law property, consider the entropy. The Maxwell relation (3.3.34) leads to

$$\Delta s^{ig} = -\int_{P^{ig}}^{P} \left(\frac{\partial v}{\partial T}\right)_{PN}^{ig} d\pi = -R\int_{P^{ig}}^{P} \frac{d\pi}{\pi} \tag{4.2.26}$$

$$= -R \ln \frac{P(T, v, \{x\})}{P^{ig}(T, v, \{x\})} = -R \ln \frac{Pv}{RT} = -R \ln Z \tag{4.2.27}$$

So for the entropy, (4.2.22) combined with (4.2.27) gives

Residual Entropy.

$$s^{res}(T, v, \{x\}) = s^{res}(T, P, \{x\}) - R \ln Z \tag{4.2.28}$$

With u^{res}, h^{res}, and s^{res} determined, we can use the defining Legendre transforms to relate the residual Gibbs energy, Helmholtz energy, and chemical potential. The results are

$$a^{res}(T, v, \{x\}) = a^{res}(T, P, \{x\}) + RT \ln Z \tag{4.2.29}$$

$$g^{res}(T, v, \{x\}) = g^{res}(T, P, \{x\}) + RT \ln Z \tag{4.2.30}$$

$$\overline{G}_i^{res}(T, v, \{x\}) = \overline{G}_i^{res}(T, P, \{x\}) + RT \ln Z \tag{4.2.31}$$

We caution that (4.2.31) cannot be derived in a simple way by applying the partial molar derivative to the difference in residual Gibbs energies given in (4.2.30). The difficulty is that the partial molar derivative imposes a fixed pressure, but when the lhs of (4.2.30), $g^{res}(T, v, \{x\})$, is changed at fixed pressure, the mixture and ideal-gas volumes are no longer the same. Consequently, the isobaric derivative of the lhs of (4.2.30) is not an isometric residual property; in particular, it is not the lhs of (4.2.31).

In all the equations relating second-law residual properties (4.2.28)–(4.2.31), the compressibility factor Z is to be evaluated at the state $(T, P, v, \{x\})$ of the real substance of interest. The state dependence of Z is discussed in the next section.

4.3 DEVIATIONS FROM IDEAL GASES: RATIO MEASURES

Besides difference measures, it is frequently convenient to describe deviations from ideality by using ratio measures. In this section we present the ratio measures commonly employed to measure deviations from ideal-gas behavior: the compressibility factor and the fugacity coefficient.

4.3.1 Compressibility Factor

The compressibility factor Z serves as a ratio measure for how a real-substance volume deviates from that of an ideal gas at the same T and P,

$$Z = \frac{v(T, P, \{x\})}{v^{ig}(T, P, \{x\})} = \frac{PV}{NRT} \tag{4.3.1}$$

Therefore, an ideal gas has $Z = 1$, but the converse is not true: a substance may have $Z \approx 1$ but not be ideal. Nonideal gases may have $Z > 1$, $Z < 1$, or $Z = 1$.

The compressibility factor serves a purpose similar to that of the isobaric residual volume: both measure how the volume of a substance deviates from the ideal-gas volume at the same T and P. The distinction is that one is a difference, while the other is a ratio. But the two are related; the relation is found by combining (4.2.2) with (4.3.1),

$$v^{res} = \frac{RT}{P}(Z-1) \tag{4.3.2}$$

The compressibility factor and residual volume of pure ethane are compared in Figure 4.3 for temperatures above the critical point. Note that neither Z nor v^{res} is constant: both change with T and P. Along isotherms near T_c they decrease with increasing pressure, pass through minima, and increase as P goes to high values. At high temperatures the minima are weaker, until at sufficiently high temperatures, both Z and v^{res} increase monotonically with increasing pressure. This behavior is typical of many gases.

Extrema are usually caused by competing effects. In Figure 4.3 the competition is between repulsive and attractive forces acting among the molecules. Consider the low temperature isotherm for Z in Figure 4.3. When the pressure is low, the molecules are widely separated (on the average), their interactions are infrequent and weak, so the gas is essentially ideal. As the pressure is increased, molecules are pushed together, they begin to attract one another, and the molar volume contracts, causing v^{res} to become negative and Z to decrease below unity. But as the pressure is increased more, electron clouds begin to overlap: repulsive forces become strong enough to counterbalance attractive forces, so v^{res} and Z pass through minima. At still higher pressures, repulsive forces dominate attractive forces and the molar volume becomes greater than the ideal-gas value. Along each low temperature isotherm there is a pressure at which the repulsive and attractive forces balance, making $v^{res} = 0$ and $Z = 1$, although the gas is *not* ideal.

A principal difference between Z and v^{res}, seen in Figure 4.3, is the limiting behavior at $P = 0$. At zero pressure we expect (4.1.14) to be obeyed and the gas to be ideal; the limiting value $Z = 1$ supports that expectation. That is,

$$\lim_{P \to 0} Z = 1 \qquad \textit{fixed } T \tag{4.3.3}$$

Similarly, we might at first expect $v^{res} = 0$ at $P = 0$; however, Figure 4.3 implies that, with temperature fixed, v^{res} approaches a non-zero constant as P approaches zero. Moreover, the value of that constant changes with temperature:

$$\lim_{P \to 0} v^{res} = \lim_{P \to 0} \frac{RT(Z-1)}{P} = \text{constant} \qquad \textit{fixed } T \tag{4.3.4}$$

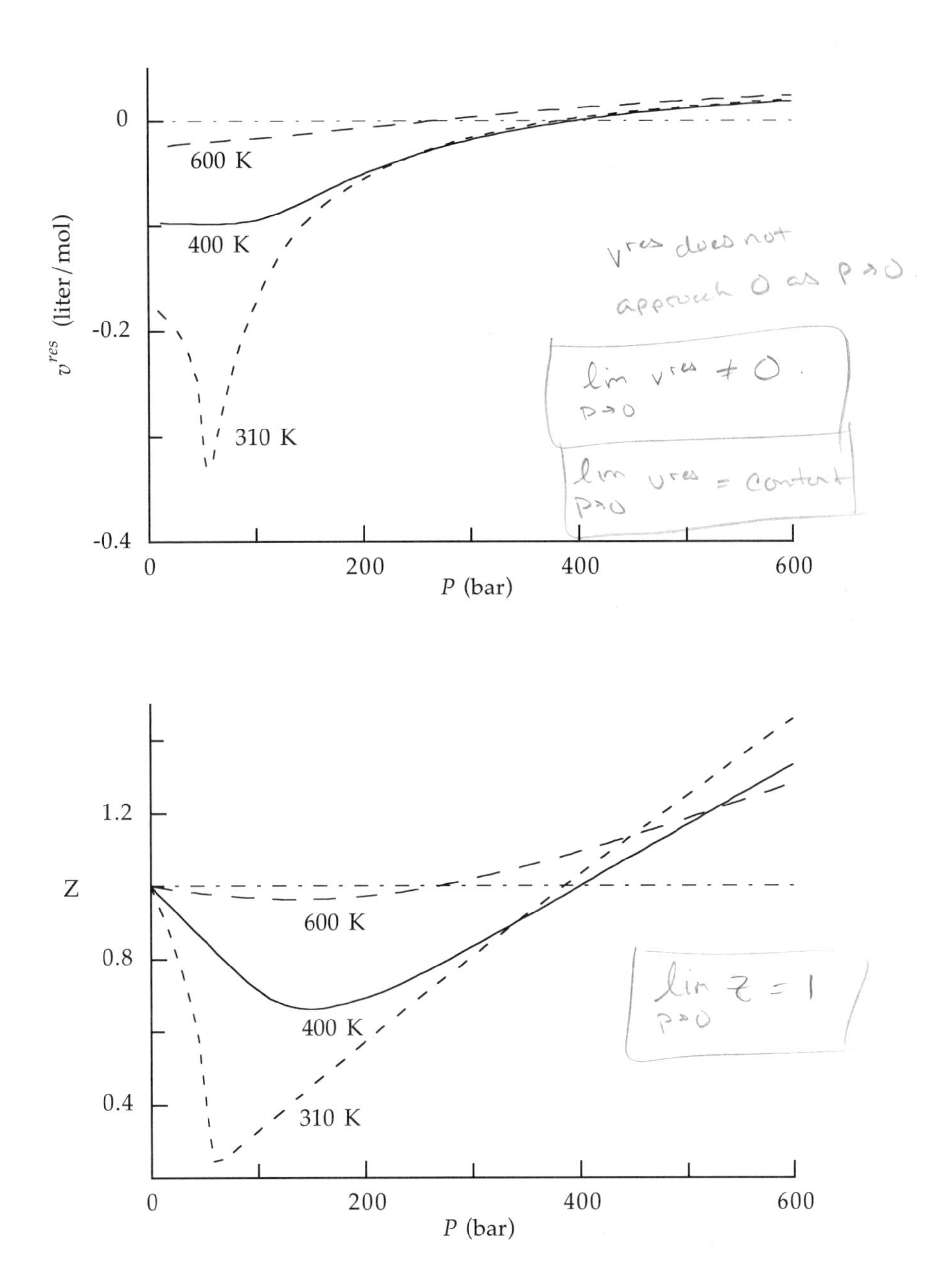

Figure 4.3 Effects of pressure on residual volumes and compressibility factors along three supercritical isotherms for pure ethane. Broken horizontal lines represent values for the ideal gas. The ethane critical point occurs at $T_c = 305.3$ K and $P_c = 48.7$ bar. Note that $Z \rightarrow 1$ as $P \rightarrow 0$, regardless of the temperature; however, the residual volumes do not approach zero in the same limit. Instead, they obey (4.3.4). Further note that each isotherm has $Z = 1$ at some high pressure, although ethane is not an ideal gas at those pressures. Curves calculated from experimental data given in [7].

Later in this chapter we will discover the identity of the constant in (4.3.4), but for now we merely note that it arises because of intermolecular forces. In particular, that limiting value of v^{res} plays a central role in describing nonideal gas behavior at low pressures.

This comparison of Z and v^{res} illustrates the complementary roles that a ratio measure and a difference measure can play in describing the same kind of deviation from ideality. Both describe the same kinds of deviations, but each reveals those deviations in different ways. In a particular situation one measure or the other may prove more illuminating or useful or both.

4.3.2 Fugacity

The fugacity is a property, created by G. N. Lewis, to provide an alternative to the chemical potential [8]. Conceptually, fugacity offers no advantage over the chemical potential, but it does offer computational advantages, particularly for mixtures. The definition of fugacity is motivated by the response of the chemical potential in an ideal gas when the state is changed isothermally. For an ideal-gas mixture, that response is derived from (4.1.44) and found to be

$$d\bar{G}_i^{ig} = dg_{\text{pure i}}^{ig} + RT\,d\ln x_i \qquad \textit{fixed } T \qquad (4.3.5)$$

At fixed T, the pure-fluid term can respond only to changes in P, so we can write

$$d\bar{G}_i^{ig} = v_{\text{pure i}}^{ig}dP + RTd\ln x_i = \frac{RT}{P}dP + RTd\ln x_i \qquad (4.3.6)$$

Hence,

$$d\bar{G}_i^{ig} = RTd\ln(x_iP) \qquad \textit{fixed } T \qquad (4.3.7)$$

Lewis defined the fugacity f_i as an analogy to the ideal-gas expression (4.3.7). The definition contains two parts. For component i in a mixture of any phase, the first part of the definition is

Part 1 of Definition $$d\bar{G}_i \equiv RTd\ln f_i \qquad \textit{fixed } T \qquad (4.3.8)$$

To preserve thermodynamic consistency, we require that the general expression (4.3.8) revert to the special form (4.3.7) if our substance is indeed an ideal gas. Therefore as the second part of the definition, we require that the ideal-gas fugacity obey

Part 2 of Definition $$f_i^{ig} \equiv x_iP \qquad (4.3.9)$$

The two parts, (4.3.8) and (4.3.9), *together* constitute a complete definition of the fugacity. The chemical potential is an intensive, conceptual, state function and it has dimensions of energy/mole; the fugacity, as defined by (4.3.8) and (4.3.9), is also an intensive, conceptual, state function, but it has dimensions of pressure. For a pure component the chemical potential is merely the molar Gibbs energy, so the definition of the fugacity reduces to

$$dg_{\text{pure i}} = RT\, d\ln f_{\text{pure i}} \qquad \textit{fixed } T \tag{4.3.10}$$

$$f^{ig}_{\text{pure i}} = P \tag{4.3.11}$$

We can obtain an algebraic form of (4.3.8) by integrating from a reference state (®), along an isotherm, to the state of interest. The general result is

$$\bar{G}_i(T, P, \{x\}) - \bar{G}_i^{®}(T, P^{®}, \{x^{®}\}) = RT \ln \frac{f_i(T, P, \{x\})}{f_i^{®}(T, P^{®}, \{x^{®}\})} \tag{4.3.12}$$

This algebraic form is *always true,* but it is not computationally useful until we identify the reference state and determine a value for the fugacity in that state. ("Reference state" refers to the lower limit on the integral that produced (4.3.12) from (4.3.8).)

We cite two reasons for introducing fugacity as an alternative to the chemical potential. One is to obtain the algebraic form (4.3.12), which replaces a difference measure with a ratio measure. In many applications, functional forms for ratios are less complicated than the corresponding forms for differences. Such simplifications facilitate numerical calculations. Further, the expressions (4.3.7) and (4.3.8) for the chemical potential become troublesome as $x_i \to 0$; in comparison, the fugacity remains well behaved ($f_i \to 0$ as $x_i \to 0$). A second reason is that the second part of the definition (4.3.9) identifies the ideal gas as the reference state for fugacity, and numerical values for ideal-gas fugacities are readily obtainable.

A principal use of fugacity is in phase equilibrium computations—a use we will develop in Part III of this book—but for now note that f_i is a well-defined quantity, even for systems of a single phase. Incidentally, although Lewis first defined the quantity f_i and chose its name, he did not create the word *fugacity.* The word itself is a nominative form derived from the Latin verb *fugere,* which means *to flee.* According to the *Oxford English Dictionary* the word *fugacity* was used as early as 1666 by Robert Boyle. So the word is old, though it is now rarely used in other than technical discourse.

A response of the fugacity to a change in state is simply found by combining the definition (4.3.8) with the appropriate response of the chemical potential, as given in § 3.4.3. For isothermal changes in pressure, (4.3.8) combined with (3.4.15) gives

$$\left(\frac{\partial \ln f_i}{\partial P}\right)_{Tx} = \frac{\bar{V}_i}{RT} \tag{4.3.13}$$

For isobaric changes in temperature, we choose the ideal gas as the reference state in (4.3.12), divide (4.3.12) by *RT,* and take the temperature derivative of both sides with pressure and composition fixed. On applying the Gibbs-Helmholtz equation (3.4.17), we find

$$\left(\frac{\partial \ln f_i}{\partial T}\right)_{Px} = -\frac{\overline{H}_i^{res}}{RT^2} \tag{4.3.14}$$

In addition, the fugacities of all components in a mixture are not independent; rather, they are related through the Gibbs-Duhem equation. If we use g/RT for the generic function f in the Gibbs-Duhem equation (3.4.10), we find

$$\sum_i x_i \, d\ln f_i = -\frac{h}{RT^2}dT + \frac{v}{RT}dP \tag{4.3.15}$$

In particular, at fixed T and P, a binary mixture must have

$$x_1\left(\frac{\partial \ln f_1}{\partial x_1}\right)_{TP} = -x_2\left(\frac{\partial \ln f_2}{\partial x_1}\right)_{TP} = x_2\left(\frac{\partial \ln f_2}{\partial x_2}\right)_{TP} \qquad \textit{fixed T and P} \tag{4.3.16}$$

So for example, if f_1 increases as x_1 increases, then f_2 *must* decrease.

4.3.3 Fugacity Coefficient

In this subsection we introduce a ratio measure that indicates how the fugacity of a real substance deviates from that of an ideal gas. As the reference state, we choose the ideal-gas mixture at the same temperature, pressure, and composition as our real mixture. Then, on integrating the definition of fugacity (4.3.8) from the ideal-gas state to the real state, we obtain an algebraic form analogous to (4.3.12); that is, we find

$$\overline{G}_i(T, P, \{x\}) - \overline{G}_i^{ig}(T, P, \{x\}) = RT \ln \frac{f_i(T, P, \{x\})}{f_i^{ig}(T, P, \{x\})} \tag{4.3.17}$$

The ratio of fugacities on the rhs is the desired deviation measure; it is called the *fugacity coefficient* φ_i,

$$\varphi_i \equiv \frac{f_i(T, P, \{x\})}{f_i^{ig}(T, P, \{x\})} = \frac{f_i(T, P, \{x\})}{x_i P} \tag{4.3.18}$$

The lhs of (4.3.17) is the isobaric residual chemical potential, so we can write

$$\overline{G}_i^{res}(T, P, \{x\}) = RT \ln \varphi_i(T, P, \{x\}) \qquad \textit{always true} \tag{4.3.19}$$

This relates a ratio measure to a difference measure. Physically, the residual chemical potential (and therefore the fugacity coefficient) measures the reversible isothermal-isobaric work done in extracting a small amount of component i from an ideal-gas mixture and injecting it into the real mixture. This interpretation arises from (6.3.8), which is derived in § 6.3.1. For a pure substance the definition (4.3.18) simplifies to

$$\varphi_{\text{pure i}} \equiv \frac{f_{\text{pure i}}(T, P)}{f^{ig}_{\text{pure i}}(T, P)} = \frac{f_{\text{pure i}}(T, P)}{P} \tag{4.3.20}$$

and (4.3.19) reduces to

$$\overline{G}^{res}_{\text{pure i}}(T, P) = RT \ln \varphi_{\text{pure i}}(T, P) \tag{4.3.21}$$

The fugacity coefficient is always positive; however, it may be greater or less than unity. The ideal gas has $\varphi_i = 1$, but the converse is not true: a substance having $\varphi_i = 1$ is not necessarily an ideal gas. Note that the definition (4.3.18) places no restriction on the kind of phase to which φ_i may be applied; it is a legitimate property of solids, liquids, and gases, though it is most often applied to fluids.

Figure 4.4 compares the two ratio measures, Z and φ, for deviations from ideal-gas behavior for pure ammonia along the subcritical isotherm at 100°C. The figure shows that Z(P) is discontinuous across the vapor-liquid phase transition, while $\varphi(P)$ is not. The discontinuity in Z occurs because the vapor and liquid phases have different molar volumes. In contrast, $\varphi(P)$ appears continuous and smooth, though in fact it is only piecewise continuous. That is, the $\varphi(P)$ curves for vapor and liquid intersect at the saturation point, but they intersect with different slopes. Near the triple point that difference in slopes is marked, but near the critical point the difference is small: the

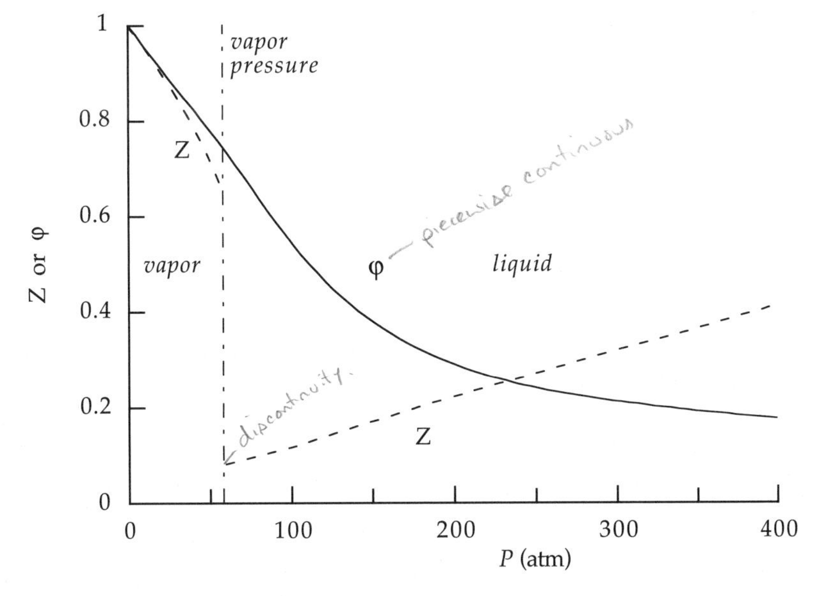

Figure 4.4 Comparison of the ratio measures Z and φ for pure ammonia at 100°C. The ammonia critical point occurs at T_c = 405.6 K and P_c = 111.5 atm. These curves are plots of data tabulated by Walas [9].

curve for $\varphi(P)$ in Figure 4.4 appears to make a smooth transition across the saturation pressure (near 60 atm).

To obtain the response of the fugacity coefficient to a change in state, we combine the relation (4.3.19) with the appropriate response of the residual chemical potential. At fixed composition, the fundamental equation can be written in the form

$$d\overline{G}_i^{res} = -\overline{S}_i^{res} dT + \overline{V}_i^{res} dP \tag{4.3.22}$$

Then the response of φ_i to an isothermal change in pressure is given by

$$\left(\frac{\partial \ln \varphi_i}{\partial P}\right)_{Tx} = \frac{\overline{V}_i^{res}}{RT} \tag{4.3.23}$$

and the response to an isobaric change in T is given by a Gibbs-Helmholtz equation,

$$\left(\frac{\partial \ln \varphi_i}{\partial T}\right)_{Px} = -\frac{\overline{H}_i^{res}}{RT^2} \tag{4.3.24}$$

Note that the temperature derivative of $\ln \varphi_i$ is the same as the temperature derivative of $\ln f_i$; cf. (4.3.14) with (4.3.24). Further, the fugacity coefficients must obey a Gibbs-Duhem equation. Letting g^{res}/RT be the generic function f in the Gibbs-Duhem equation (3.4.10), we find

$$\sum_i x_i \, d\ln \varphi_i = -\frac{h^{res}}{RT^2} dT + \frac{v^{res}}{RT} dP \tag{4.3.25}$$

At fixed T and P this reduces to

$$\sum_i x_i \, d\ln \varphi_i = 0 \qquad \textit{fixed T and P} \tag{4.3.26}$$

Figure 4.5 shows how changes in composition usually affect fugacity coefficients in binary mixtures of nonideal gases. In the figure, note that the values of φ for CO_2 and C_3H_8 at least qualitatively satisfy the Gibbs-Duhem equation (4.3.26); that is, as the mole fraction of CO_2 increases, one fugacity coefficient increases while the other decreases. Note also that for both substances, the slope of each $\varphi_i(x_i)$ is zero in its pure-fluid limit, as required by (4.3.26). Since both these φs are less than unity, the corresponding component residual chemical potentials are negative, by (4.3.19).

Figure 4.6 shows how a gas-phase fugacity coefficient may be affected by increasing pressure: φ increases, passes through a maximum, and decreases at high pressure. Just as for the minimum in Z in Figure 4.3, the maximum in Figure 4.6 can be explained by competition between attractive and repulsive forces among molecules.

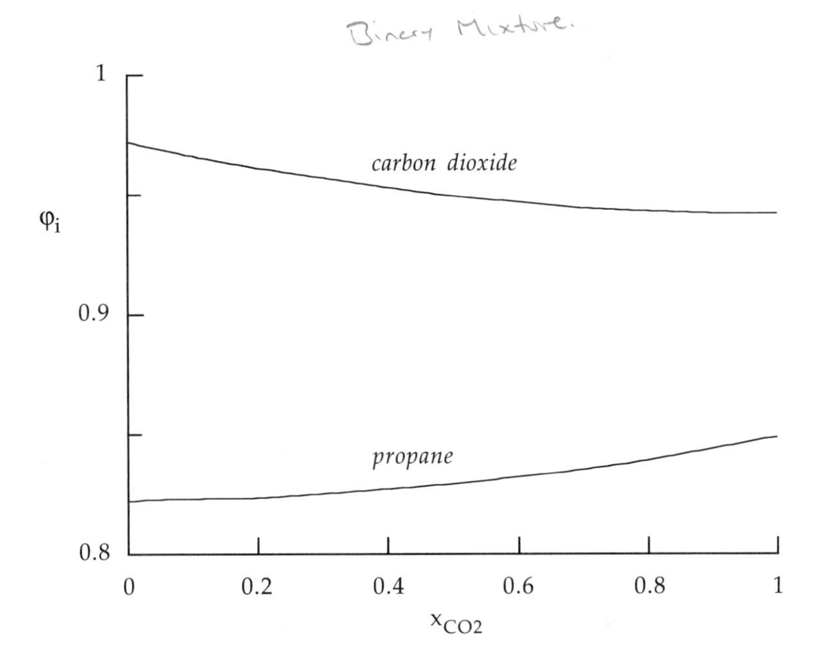

Figure 4.5 Typical composition dependence of fugacity coefficients in gas mixtures. Fugacity coefficients in carbon dioxide + propane mixtures at 100°F, 200 psia. These curves are corrected from results tabulated by Walas [9].

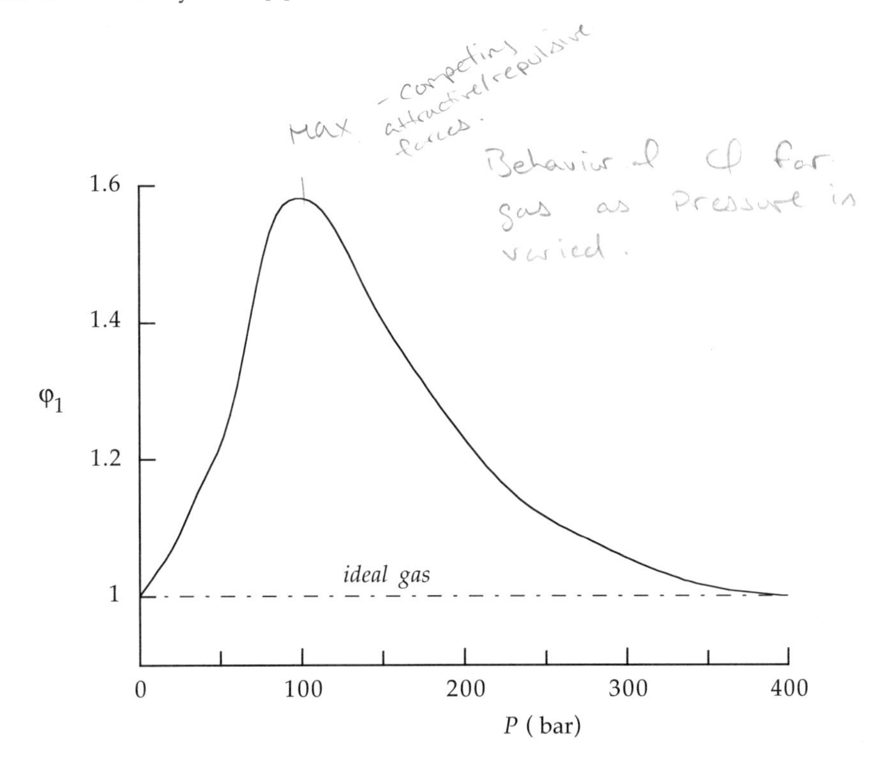

Figure 4.6 Effect of pressure on fugacity coefficient for CO_2 in carbon dioxide(1) + n-butane(2) mixtures containing 85 mole % butane at 171°C. Replotted from a figure in Prausnitz et al. [10].

4.4 CONCEPTUALS FROM MEASURABLES USING EQUATIONS OF STATE

In this section we present explicit expressions that allow us to use the measurables P, v, T, and $\{x\}$ to compute values for u, h, s, a, and g. We can only obtain values for these conceptuals relative to some well-defined reference state, so here we choose the reference to be the ideal gas. As a result, the expressions obtained below provide the residual properties. In addition, from the expression for the residual chemical potential we can readily obtain expressions for the fugacity coefficient. All the relations derived below involve integrals over functions of P, V, T, and $\{x\}$, and to exploit those relations, we need a volumetric equation of state for our substance.

A volumetric equation of state takes one of two forms. A *volume-explicit* equation has the form $v = v(T, P, \{x\})$, while a *pressure-explicit* equation has the form $P = P(T, v, \{x\})$. Therefore, our expressions for conceptuals divide into two classes, depending on whether P (§ 4.4.1) or v (§ 4.4.2) is independent. In a particular problem, calculations are often simplified by using one set of independent variables rather than the other. To choose between the two sets, we follow the steps given in Figure 4.7.

The choice hinges on whether the independent variable (P or v) in our equation of state is appropriate for the conceptual whose value we need to compute. Recall from Chapter 3 that the fundamental equations for u and a have v as the independent variable, while those for h and g have P. Consequently, if we need to compute Δu or Δa, then we prefer to use a pressure-explicit equation of state, $P(T, v, \{x\})$, but if we need to compute Δh or Δg, then we prefer to use a volume-explicit equation, $v(T, P, \{x\})$. Note that if we need Δs, f_i, or φ_i, then little advantage is offered by one kind of equation over the other: both kinds involve about the same computational effort. These possibilities summarize the lhs of the diagram in Figure 4.7.

However, if the independent variable (P or v) in our equation of state differs from the one that is appropriate for a particular conceptual, then a more involved computational procedure must be followed. This procedure appears on the rhs of the figure. For example, if we need $\Delta h(T, P, \{x\})$, but our equation of state uses $(T, v, \{x\})$, then we have an incompatible situation to resolve. It should be resolved not by tampering with the expression given in § 4.4.1 for the residual enthalpy, but instead by computing Δh indirectly via a Legendre transform. Therefore the procedure should be this:

(a) Solve the equation of state for v at the known values of T, P, $\{x\}$.

(b) Compute $u^{res}(T, v, \{x\})$ using the equation given in § 4.4.2.

(c) Form Δu from u^{res} by $\Delta u = \Delta u^{res} + \Delta u^{ig}$.

(d) Obtain Δh using the defining Legendre transform $\Delta h = \Delta u + \Delta(Pv)$.

4.4.1 When T, P, and $\{x\}$ Are Independent

When we have an equation of state in the form $v(T, P, \{x\})$, then pressure is the independent variable (rather than v), and the relevant forms for conceptuals are the isobaric residual properties introduced in § 4.2.1. To evaluate those quantities, we start with the observation that the residual properties are all state functions; this allows us to write

$$f^{res}(T, P, \{x\}) = \int_{f^{res(ig)}}^{f^{res}} df^{res} = \int_{0}^{f^{res}} df^{res} \tag{4.4.1}$$

The lower limit is zero because the residual properties of ideal gases are zero. Our objective is to write the rhs of (4.4.1) as an integral over a measurable, and since our equation of state has P independent, we choose P to be that measurable. Therefore, we write (4.4.1) as

$$f^{res}(T, P, \{x\}) = \int_{0}^{P} \left(\frac{\partial f^{res}}{\partial \pi}\right)_{Tx} d\pi \qquad \textit{fixed } T \textit{ and } \{x\} \tag{4.4.2}$$

where π is the dummy integration variable that corresponds to the pressure. Equation (4.4.2) serves as the starting point for evaluating any isobaric residual property from a

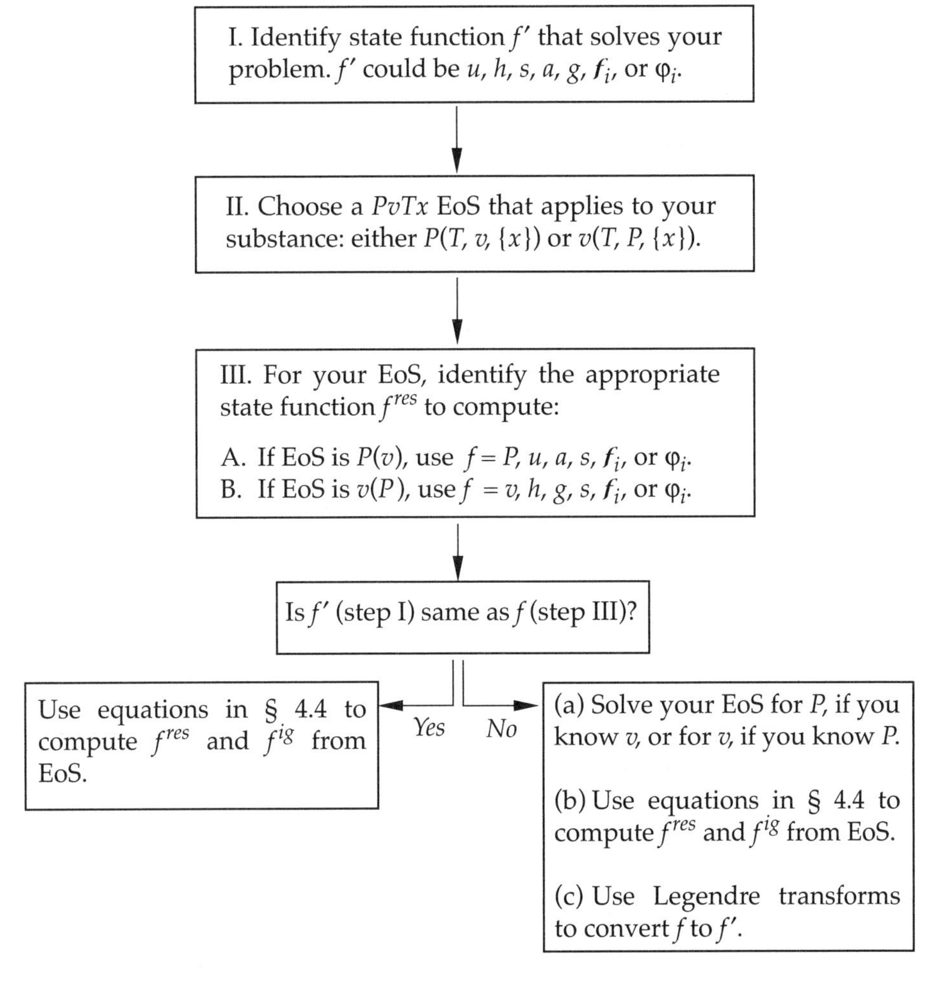

Figure 4.7 Steps involved in identifying the set of independent variables $(T, P, \{x\})$ or $(T, v, \{x\})$ to be used in computing conceptuals from $PvTx$ equations of state (EoS)

volume-explicit equation of state. In practice, we need to evaluate only three properties: the residual volume (v^{res}), one first-law conceptual (u^{res} or h^{res}), and one second-law conceptual (s^{res}, a^{res}, or g^{res}); then the remaining residual properties can be computed from Legendre transforms.

First-law properties. First note that we can obtain the residual volume by evaluating $v(T, P, \{x\})$ directly from the volumetric equation of state, then

$$v^{res}(T, P, \{x\}) = v(T, P, \{x\}) - \frac{RT}{P} \tag{4.2.2}$$

Second, since P is the independent variable, the appropriate first-law conceptual to evaluate is the enthalpy. For the enthalpy, (4.4.2) becomes

$$h^{res}(T, P, \{x\}) = \int_0^P \left(\frac{\partial h^{res}}{\partial \pi}\right)_{Tx} d\pi = \int_0^P \left(\frac{\partial h}{\partial \pi}\right)_{Tx} d\pi \tag{4.4.3}$$

The second equality is valid because the enthalpy of an ideal gas does not depend on pressure. In § 3.3.3 we obtained the derivative in (4.4.3) in terms of measurables; so, on substituting the intensive form of (3.3.35) into (4.4.3), we find

$$h^{res}(T, P, \{x\}) = \int_0^P v(1 - T\alpha)d\pi \qquad \textit{fixed T and \{x\}} \tag{4.4.4}$$

where α is the volume expansivity. This expression allows us to compute values for the isobaric residual enthalpy from an equation of state of the form $v(T, P, \{x\})$. Note that (4.4.4) applies to pure substances as well as to mixtures.

Second-law properties. With P as the independent variable, we could evaluate either g or s as the second-law property; here we choose s. Then (4.4.2) becomes

$$s^{res}(T, P, \{x\}) = \int_0^P \left(\frac{\partial s^{res}}{\partial \pi}\right)_{Tx} d\pi = \int_0^P \left[\left(\frac{\partial s}{\partial \pi}\right)_{Tx} - \left(\frac{\partial s^{ig}}{\partial \pi}\right)_{Tx}\right] d\pi \tag{4.4.5}$$

With the help of the Maxwell relation (3.3.34), we find

$$s^{res}(T, P, \{x\}) = -\int_0^P \left(v\alpha - \frac{R}{\pi}\right) d\pi \qquad \textit{fixed T and \{x\}} \tag{4.4.6}$$

Although each term in the integrand divergences in the $P \to 0$ limit, those divergences cancel, so the integral in (4.4.6) is bounded. Equation (4.4.6) provides a computational route for obtaining the residual entropy from equations of state of the form $v(T, P, \{x\})$; it applies to pure substances as well as mixtures. With v^{res}, h^{res}, and s^{res} now determined, the remaining residual properties, u^{res}, g^{res}, and a^{res}, can be obtained from their defining Legendre transforms.

Residual chemical potential. For the chemical potential of component i in a mixture, (4.4.2) becomes

$$\overline{G}_i^{res}(T, P, \{x\}) = \int_0^P \left(\frac{\partial \overline{G}_i^{res}}{\partial \pi}\right)_{Tx} d\pi = \int_0^P \overline{V}_i^{res} d\pi \tag{4.4.7}$$

Hence,

$$\overline{G}_i^{res}(T, P, \{x\}) = \int_0^P \left[\overline{V}_i - \frac{RT}{\pi}\right] d\pi \qquad \textit{fixed T and \{x\}} \tag{4.4.8}$$

For a pure fluid, (4.4.8) simplifies to

$$\overline{G}^{res}_{\text{pure i}}(T, P) = g^{res}_{\text{pure i}}(T, P) = RT\int_0^P [Z-1]\frac{d\pi}{\pi} \qquad \textit{fixed T} \tag{4.4.9}$$

In both (4.4.8) and (4.4.9), the two terms in the integrand cancel as $P \rightarrow 0$, so no divergence occurs in the ideal-gas limit.

Fugacity coefficient. To obtain computationally useful expressions for fugacity coefficients, we merely need to combine (4.4.8) with (4.3.19), which relates the fugacity coefficient to the residual chemical potential. So for a mixture we have

$$\ln\varphi_i(T, P, \{x\}) = \int_0^P \left[\frac{\pi \overline{V}_i}{RT} - 1\right]\frac{d\pi}{\pi} \qquad \textit{fixed T and \{x\}} \tag{4.4.10}$$

while for a pure substance, combining (4.4.9) with (4.3.19) leaves

$$\ln\varphi_{\text{pure i}}(T, P) = \int_0^P [Z-1]\frac{d\pi}{\pi} \qquad \textit{fixed T} \tag{4.4.11}$$

4.4.2 When T, v, and $\{x\}$ Are Independent

When we have an equation of state in the form $P(T, v, \{x\})$, then the isometric residual properties are easier to compute than are the isobaric ones. However, in applications, we usually need the isobaric residual properties, not the isometric ones. We follow a two-step procedure to obtain values for isobaric residual properties: (1) evaluate the isometric residual properties from the integrals presented in this section, then (2) convert those isometric properties to isobaric ones using the relations given in § 4.2.3.

The procedure for obtaining computationally useful expressions for isometric residual properties parallels that used in § 4.4.1 for the isobaric properties. That is, analogous to (4.4.2), we can obtain each residual property by starting from

$$f^{res}(T, v, \{x\}) = \int_{\infty}^{v} \left(\frac{\partial f^{res}}{\partial \psi}\right)_{Tx} d\psi \qquad \text{fixed } T \text{ and } \{x\} \tag{4.4.12}$$

where ψ is the dummy integration variable that corresponds to the molar volume v and the ideal-gas limit is attained as $v \to \infty$. As in § 4.4.1, of the five conceptuals u, h, s, a, and g, we actually only need to evaluate (4.4.12) for one first-law property and one second-law property, then the others can be obtained from Legendre transforms.

First-law properties. With volume taken to be independent, the appropriate first-law conceptual is the internal energy, and (4.4.12) becomes

$$u^{res}(T, v, \{x\}) = \int_{\infty}^{v} \left(\frac{\partial u^{res}}{\partial \psi}\right)_{Tx} d\psi = \int_{\infty}^{v} \left(\frac{\partial u}{\partial \psi}\right)_{Tx} d\psi \tag{4.4.13}$$

Using the intensive form of (3.3.38) for the volume derivative of u, we find

$$u^{res}(T, v, \{x\}) = \int_{v}^{\infty} [P - T\gamma_v] d\psi \qquad \text{fixed } T \text{ and } \{x\} \tag{4.4.14}$$

where γ_v is the thermal pressure coefficient. This result applies to pure substances and mixtures. Once we have a value for the isometric quantity u^{res} from (4.4.14), we also have the isobaric property u^{res}, by (4.2.24).

Second-law properties. As the second-law property, we again choose s, so (4.4.12) becomes

$$s^{res}(T, v, \{x\}) = \int_{\infty}^{v} \left(\frac{\partial s^{res}}{\partial \psi}\right)_{Tx} d\psi = \int_{\infty}^{v} \left[\left(\frac{\partial s}{\partial \psi}\right)_{Tx} - \left(\frac{\partial s^{ig}}{\partial \psi}\right)_{Tx}\right] d\psi \tag{4.4.15}$$

Using the Maxwell relation (3.3.37), we find

$$s^{res}(T, v, \{x\}) = \int_{v}^{\infty} \left[\frac{R}{\psi} - \gamma_v\right] d\psi \qquad \text{fixed } T \text{ and } \{x\} \tag{4.4.16}$$

which applies to pure substances as well as mixtures. With a value for the isometric s^{res} from (4.4.16), we can obtain the corresponding value of the isobaric property s^{res} by applying (4.2.28). The isobaric v^{res} can be obtained simply from (4.2.2). Then with v^{res}, u^{res}, and s^{res} now determined, the remaining isobaric residual properties, h^{res}, g^{res}, and a^{res}, can be obtained from their defining Legendre transforms.

Residual chemical potential. To obtain the isometric residual chemical potential, we evaluate the general form (4.4.12) by integrating over either the intensive volume v or the extensive volume V; we used the intensive volume above. But the following development is somewhat easier if we use the extensive volume, so we rewrite (4.4.12) as

$$\bar{G}_i^{res}(T, v, \{x\}) = \int_{\infty}^{V} \left(\frac{\partial \bar{G}_i^{res}}{\partial \Psi} \right)_{Tx} d\Psi \tag{4.4.17}$$

where Ψ is the dummy integration variable that corresponds to the extensive volume V. Using (4.2.19), we write

$$\left(\frac{\partial \bar{G}_i^{res}}{\partial V} \right)_{TN} = \left[\frac{\partial}{\partial V} \left(\frac{\partial A^{res}}{\partial N_i} \right)_{TVN_{j \neq i}} \right]_{TN} = \left[\frac{\partial}{\partial N_i} \left(\frac{\partial A^{res}}{\partial V} \right)_{TN} \right]_{TVN_{j \neq i}} \tag{4.4.18}$$

$$= -\left(\frac{\partial P^{res}}{\partial N_i} \right)_{TVN_{j \neq i}} = -\left(\frac{\partial P}{\partial N_i} \right)_{TVN_{j \neq i}} + \left(\frac{\partial P^{ig}}{\partial N_i} \right)_{TVN_{j \neq i}} \tag{4.4.19}$$

At first glance, the rhs of (4.4.19) appears peculiar because we have an intensive property (pressure) responding to changes in the amount of material. But note the variables held fixed on the rhs—not the molar volume v, but the extensive volume V. For example, if at fixed T we add to the amount of gas held in a rigid vessel (so V is fixed), then P changes (usually it increases). So the derivatives on the rhs of (4.4.19) are well-defined; in fact, for the ideal gas we have

$$\left(\frac{\partial P^{ig}}{\partial N_i} \right)_{TVN_{j \neq i}} = \left[\frac{\partial}{\partial N_i} \left(\frac{NRT}{V} \right) \right]_{TVN_{j \neq i}} = \frac{RT}{V} \tag{4.4.20}$$

Then we combine (4.4.17), (4.4.19), and (4.4.20) to obtain

$$\bar{G}_i^{res}(T, v, \{x\}) = \int_{V}^{\infty} \left[\left(\frac{\partial P}{\partial N_i} \right)_{TVN_{j \neq i}} - \frac{RT}{\Psi} \right] d\Psi \qquad \textit{fixed } T \textit{ and } \{x\} \tag{4.4.21}$$

where the integration is over the extensive volume ($V \Leftrightarrow \Psi$). With a value for the isometric residual chemical potential from (4.4.21), we can obtain the corresponding value for the isobaric property by applying (4.2.31). Equation (4.4.21) applies to mixtures and pure fluids, though for pure fluids it simplifies to

$$\bar{G}_{\text{pure }i}^{res}(T, v) = RT \int_{v}^{\infty} [Z - 1] \frac{d\psi}{\psi} + RT(Z - 1) \qquad \textit{fixed } T \tag{4.4.22}$$

Fugacity coefficient. Substituting (4.4.21) into (4.2.31) gives the isobaric residual chemical potential, then substituting that into (4.3.19) leaves

$$\ln \varphi_i(T, P, \{x\}) = \int_{V}^{\infty} \left[\frac{\Psi}{RT} \left(\frac{\partial P}{\partial N_i} \right)_{TVN_{j \neq i}} - 1 \right] \frac{d\Psi}{\Psi} - \ln Z \qquad \textit{fixed } T \textit{ and } \{x\} \tag{4.4.23}$$

For pure fluids this simplifies to

$$\ln\varphi_{\text{pure i}}(T, P) = \int_v^\infty [Z-1]\frac{d\psi}{\psi} + (Z-1) - \ln Z \qquad \textit{fixed } T \qquad (4.4.24)$$

Note that the integrations for mixtures above *must* be done over the extensive volume ($V \Leftrightarrow \Psi$); however, the integrations for pures can be done over either the extensive volume V or the intensive volume ($v \Leftrightarrow \psi$).

4.5 SIMPLE MODELS FOR EQUATIONS OF STATE

The expressions for residual properties derived in § 4.4 all involve integrals over various functions of the measurables P, v, T, and $\{x\}$. Therefore, to actually compute those integrals, and hence to obtain numerical values for residual properties, we must have numerical or analytic forms for volumetric equations of state. In this section we present a few simple but important forms that model the nonideal-gas behavior of real fluids: the hard-sphere fluid (§ 4.5.1), the virial equations (§ 4.5.2–4.5.5), and selections from the van der Waals family of equations (§ 4.5.6–4.5.12). These are not by any means the only analytic forms available for equations of state, but they are enough to allow us to exercise the relations given in the previous section and to obtain qualitative descriptions of fluid behavior. A more thorough discussion of $PvTx$ equations can be found in books by Sandler et al. [3], Sengers et al. [4], and Poling et al. [11].

4.5.1 Hard Spheres

As the density of a fluid is decreased, the effects of forces between molecules weaken, and the fluid behaves more like an ideal gas; that is, the behavior of real fluids may simplify under extreme conditions. Another extreme occurs by making the temperature high, for then many simple fluids behave as if they were composed of hard spheres:

$$\lim_{T \to \text{large}} (\text{real stuff}) = \text{hard-sphere substance} \qquad \textit{fixed } v \qquad (4.5.1)$$

In a hard-sphere fluid each molecule occupies space and the molecules exert forces on one another, but those forces are purely repulsive and act only when spheres collide: the spheres act like billiard balls. So the hard-sphere substance is an extreme model, but under certain conditions it is more realistic than the ideal-gas model.

In a pure hard-sphere fluid, all spheres have the same diameter σ and the compressibility factor Z depends on only the fluid density $\rho = \mathcal{N}/V$, where $\mathcal{N}$ is the number of spheres held in a vessel of volume V. For hard spheres, the density is conventionally cited in terms of the *packing fraction* η, which is the ratio of the volume of the spheres to the volume of their container:

$$\eta = \frac{V_{\text{spheres}}}{V} = \frac{\mathcal{N}4\pi(\sigma/2)^3}{3V} = \frac{\pi N_A \sigma^3}{6v} \qquad (4.5.2)$$

where N_A is Avogadro's number. There is an upper bound on the value of η because a rigid vessel can hold only a finite number of rigid spheres; in particular, voids between closed-packed spheres make $\eta < 1$. The maximum packing fraction occurs when the spheres form a face-centered cubic (fcc) lattice—the structure used by grocers to stack oranges for display. (The hexagonal close-packed structure also gives the same maximum density.) The maximum packing fraction is then

$$\eta_{max} = \frac{\pi\sqrt{2}}{6} = 0.74048\ldots \tag{4.5.3}$$

However, this maximum is for a solid phase, wherein spheres are so closely packed that long-range order is preserved and there is little, if any, net diffusion of spheres. For the pure hard-sphere *fluid*, the upper bound on η is even less; the fluid-solid phase transition occurs at $\eta = 2\eta_{max}/3 = 0.494$ [12]. For $\eta \le 0.494$ the substance is fluid and long-range order is disrupted by molecular motions. Without attractive forces between spheres, no vapor-liquid phase transition occurs and we refer to the material at $\eta < 0.494$ as merely "fluid." The hard-sphere phase diagram is shown in Figure 4.8.

Over the years numerous functional forms have been devised for the hard-sphere compressibility factor $Z(\eta)$. A simple yet accurate expression has been devised by Carnahan and Starling [13]:

$$Z = \frac{1 + \eta + \eta^2 - \eta^3}{(1-\eta)^3} \qquad \eta \le 0.494 \tag{4.5.4}$$

Since $\eta < 1$, the hard-sphere Z is always greater than unity and as $\eta \to 0$ this expression reduces to the ideal-gas value, $Z = 1$.

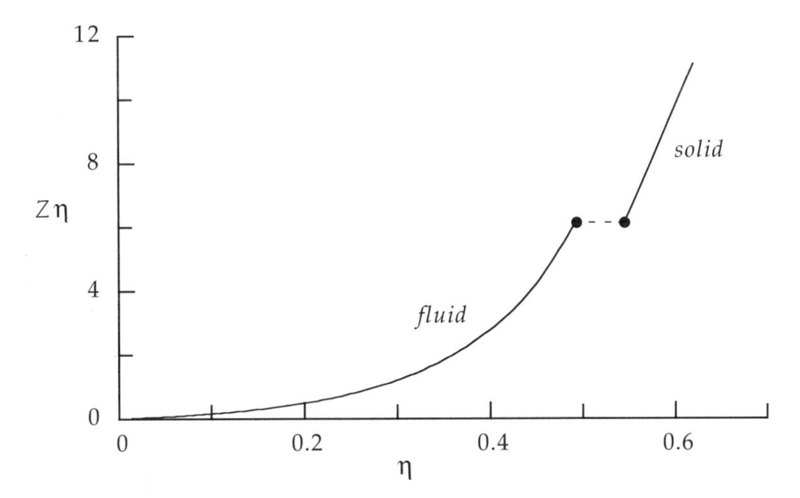

Figure 4.8 Phase diagram for a pure substance composed of hard spheres. The fluid-phase Z was computed from the Carnahan-Starling equation (4.5.4); the solid-phase Z was taken from the computer simulation data of Alder et al. [14]. The broken horizontal line at $Z\eta = 6.124$ connects fluid ($\eta = 0.494$) and solid ($\eta = 0.545$) phases that can coexist in equilibrium, as computed by Hoover and Ree [12].

By combining the Carnahan-Starling equation with the integral forms in § 4.4.2, we can evaluate the residual properties of a pure hard-sphere fluid. The results are [15]

$$v^{res}(T, P) = \frac{\pi N_A \sigma^3}{6\eta}\left(\frac{Z-1}{Z}\right) \tag{4.5.5}$$

$$u^{res}(T, P) = 0 \tag{4.5.6}$$

$$s^{res}(T, P) = R\ln Z + \frac{R\eta(3\eta-4)}{(1-\eta)^2} \tag{4.5.7}$$

Note that since $Z > 1$, v^{res} is always positive, while s^{res} is always negative. With these three residual properties, others can be obtained via Legendre transforms, and we find that h^{res}, a^{res}, g^{res}, and $\ln \varphi_i$ are always positive.

The Carnahan-Starling equation of state (4.5.4) has been extended by Mansoori et al. [16] to binary mixtures of hard spheres having different diameters. Binary mixtures of hard spheres exhibit fluid-solid phase transitions at packing fractions somewhat larger than that for the pure substance; that is, at $\eta > 0.5$. The exact state for the transition depends on composition and on the relative sizes of the spheres. We expect the density of the transition to increase as the size disparity increases; the limited computer simulation data available support this expectation [17]. Certain kinds of hard-sphere mixtures are the simplest substances to exhibit a fluid-fluid phase transition [17], but those phase transitions are more like liquid-liquid than vapor-liquid. Analytic representations of the $Z(\eta)$ for hard-sphere and other hard-body fluids have been critically reviewed by Boublik and Nezbeda [18].

4.5.2 Virial Expansion in Density

Real fluids are neither ideal gases nor are they composed of hard spheres. But if the density is low, a gas might be nearly ideal, or if the temperature is high, a gas might behave somewhat like a fluid of hard spheres. In such cases the ideal-gas or hard-sphere models may serve as references in expansions that approximate real behavior. In this section we consider Taylor expansions (see Appendix A) of the compressibility factor Z about that for the ideal gas. The expansions may be done using either density or pressure as the independent variable; we introduce the density expansion first.

Consider a one-component gas at temperature T and molar density $\rho = 1/v$. At low to moderate densities we write the compressibility factor as an expansion in ρ about the ideal-gas value ($\rho = 0$):

$$Z = Z\big|_{\rho=0} + \rho\left(\frac{\partial Z}{\partial \rho}\right)_T\bigg|_{\rho=0} + \frac{\rho^2}{2!}\left(\frac{\partial^2 Z}{\partial \rho^2}\right)_T\bigg|_{\rho=0} + \ldots \qquad \textit{fixed } T \tag{4.5.8}$$

Note that this expansion is performed along the isotherm T and each derivative is evaluated in the ideal-gas limit. The derivatives are called *virial coefficients*

$$B(T) \equiv \left(\frac{\partial Z}{\partial \rho}\right)_T\bigg|_{\rho = 0} = \text{second virial coefficient} \tag{4.5.9}$$

$$C(T) \equiv \frac{1}{2!}\left(\frac{\partial^2 Z}{\partial \rho^2}\right)_T\bigg|_{\rho = 0} = \text{third virial coefficient} \tag{4.5.10}$$

and similarly for higher-order coefficients. Therefore, one form of the *virial equation* of state is a power series in density

$$Z = 1 + B\rho + C\rho^2 + \ldots \tag{4.5.11}$$

The virial coefficients are measurable state functions; for pure gases, they depend *only* on temperature and are independent of density and pressure.

The second virial coefficient is the limiting slope of an isotherm as the gas-phase density approaches zero; this interpretation is illustrated in Figure 4.9. Different iso-

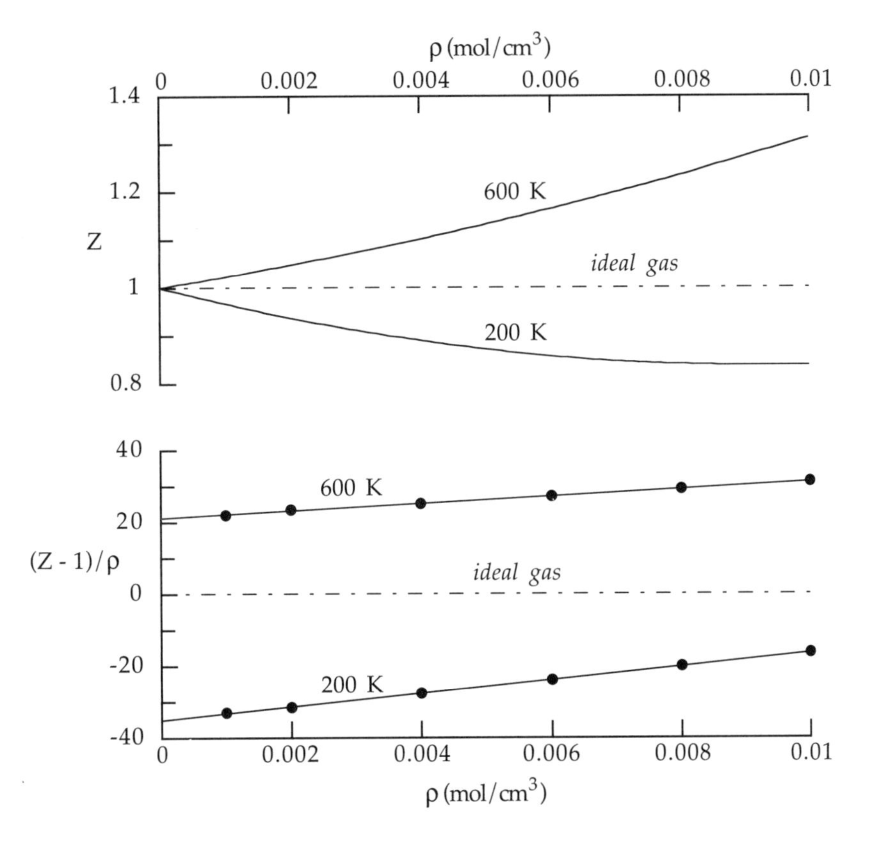

Figure 4.9 Second virial coefficients can be interpreted either as slopes or as intercepts. *Top*: Compressibility factors for pure nitrogen gas at 200 K and 600 K [19]. At each temperature, $B(T)$ is the *slope* of the isotherm as $\rho \to 0$. *Bottom*: The data replotted as $(Z - 1)/\rho$; now $B(T)$ is the *intercept* of an isotherm as $\rho \to 0$. Points are data from [19]; straight lines are least-squares fits. Note that B may be positive or negative.

therms approach zero density in different ways, so B changes with temperature. At low temperatures, attractive forces among molecules dominate repulsive forces so $Z < 1$ and B is negative. At high temperatures, repulsive forces dominate, $Z > 1$, and then B is positive. Consequently, we expect B to be negative at low temperatures and to increase with increasing temperature; this behavior is shown in Figure 4.10 for helium. At very high temperatures B passes through a maximum and then decreases to a finite positive value: some "softness" exists in short-range repulsive forces, which reflects distortion of electron clouds. Note that while B changes sign, C remains positive over most temperatures of interest.

The temperature at which B changes sign is called the *Boyle temperature* T_B; it occurs at roughly two-thirds of the critical temperature, $T_B/T_c \approx 2/3$. The Boyle temperature is used in Figure 4.10 to make the plotted temperature dimensionless. To make B and C dimensionless, we use the Boyle volume which is defined by [20]

$$v_B \equiv T_B \left(\frac{dB}{dT}\right)\bigg|_{T_B} \tag{4.5.12}$$

Then in Figure 4.10 we use

$$B^* = B/v_B \quad \text{and} \quad C^* = C/v_B^2 \tag{4.5.13}$$

Real molecules have impenetrable cores (two molecules cannot occupy the same space at the same time), so the high-temperature limit of $B(T)$ is a finite value characteristic of the size and shape of the core; for nearly spherical molecules, the value will be that for an equivalent hard-sphere fluid. In practice, experimental data rarely

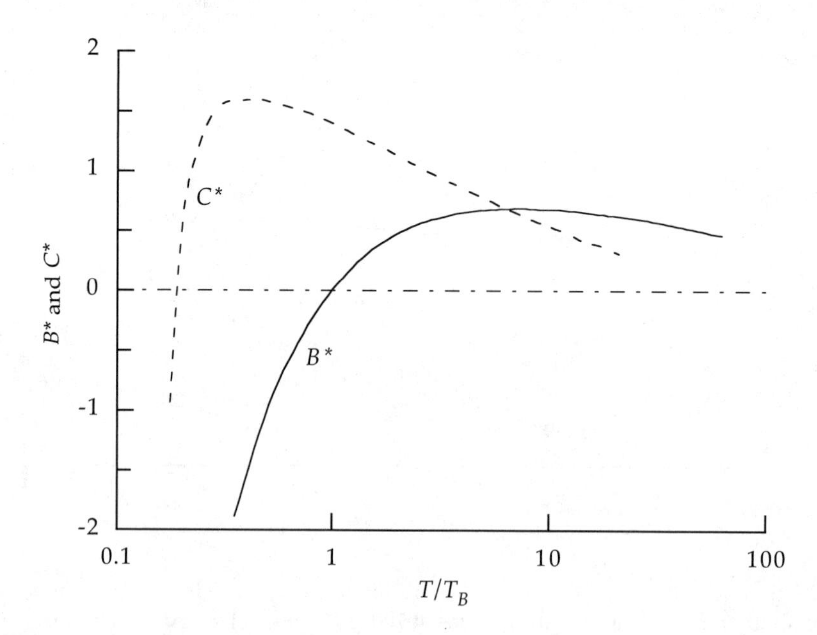

Figure 4.10 Temperature dependence of the second and third virial coefficients for pure helium. Here $B^* = B/v_B$ and $C^* = C/v_B^2$, where v_B is the Boyle volume defined in (4.5.12) and T_B is the Boyle temperature. Data taken from Dymond and Smith [21].

extend to sufficiently high temperatures to determine the limit. However for spherical molecules, we can obtain the limiting value by substituting the hard-sphere equation of state (4.5.4) into the definition of B (4.5.9), performing the differentiation, and taking the low density limit. The result is

$$\lim_{T \to \text{large}} B(T) = B^{hs} = \frac{2\pi\sigma^3}{3} \text{ cm}^3/\text{molecule} \tag{4.5.14}$$

or

$$B^{hs} = \frac{2\pi\sigma^3 N_A}{3} \text{ cm}^3/\text{mol} \tag{4.5.15}$$

Here σ is an effective hard-sphere diameter whose value depends on the kind of gas.

We construe another interpretation of $B(T)$ by rearranging the equation of state (4.5.11) and taking the ideal-gas limit; we find

$$\lim_{\rho \to 0} \frac{Z-1}{\rho} = B(T) \qquad \textit{fixed } T \tag{4.5.16}$$

This limit is illustrated in the bottom panel of Figure 4.9. Further, note that the lhs of (4.5.16) is the same as the limit on the rhs of (4.3.4); hence, the second virial coefficient is the low-density limit of the residual volume,

$$\lim_{\rho \to 0} v^{res} = B(T) \qquad \textit{fixed } T \tag{4.5.17}$$

In a fashion similar to that for $B(T)$, higher-order virial coefficients can be interpreted as limiting derivatives of the slopes of isotherms. For example, the third virial coefficient C is the limiting slope of the slope. Consequently, as we move to higher order, the virial coefficients become progressively more difficult to measure. Moreover, the effects of temperature on the higher order coefficients are more complicated than that on B; for example, Figure 4.10 shows that when T is increased, $C(T)$ quickly increases, passes through a maximum, and slowly decays.

In addition to pure gases, the Taylor expansion (4.5.8) can be applied to gaseous mixtures. The resulting form is still (4.5.11), but the virial coefficients now depend on both temperature and composition. The composition dependence is rigorously obtained from statistical mechanics; here we are interested only in the results. For a mixture containing n components,

$$B(T, x) = \sum_i^n \sum_j^n x_i x_j B_{ij}(T) \tag{4.5.18}$$

$$C(T, \{x\}) = \sum_i^n \sum_j^n \sum_k^n x_i x_j x_k C_{ijk}(T) \tag{4.5.19}$$

The coefficients B_{ij} and C_{ijk} depend only on temperature; that is, *all* the composition dependence of the mixture coefficients appears explicitly on the rhs of these equations. Those coefficients having the same indices, such as B_{22} and C_{111}, are the pure-component coefficients discussed earlier: their values are obtained from PvT measurements on pure gases. However, those coefficients having different indices, such as B_{12}, are properties of the mixtures and their values must be obtained from measurements on mixtures. These coefficients are often called the *unlike interaction coefficients.*

Nature does not know the labels that we have arbitrarily assigned to each component ($i = 1$ or $i = 2$, etc.), so the coefficients must be invariant under permutations of those labels; that is, we have $B_{ij} = B_{ji}$ and similarly $C_{ijk} = C_{ikj} = C_{jik} = C_{jki} = C_{kij} = C_{kji}$. Therefore, although the double sum for B in (4.5.18) contains n^2 terms, the number of unique terms is only $n(n+1)/2$. Likewise, the treble sum for C in (4.5.19) contains n^3 terms, but only $n(n+1)(n+2)/3!$ of them are unique.

Measurements of $B(T)$ have been made for many pure gases and some mixtures; some data also exist for $C(T)$, but few experiments are accurate enough to provide $D(T)$ or higher coefficients. The existing data for $B(T)$ and $C(T)$ up to 2002 have been critically compiled by Dymond et al. [22].

4.5.3 Virial Expansion in Pressure

As an alternative to using density as the independent variable in a virial equation of state, we could use pressure. Then the Taylor expansion takes this form,

$$Z = Z|_{P=0} + P\left(\frac{\partial Z}{\partial P}\right)_T\bigg|_{P=0} + \frac{P^2}{2!}\left(\frac{\partial^2 Z}{\partial P^2}\right)_T\bigg|_{P=0} + \ldots \qquad \text{fixed } T \tag{4.5.20}$$

Again, this expansion is performed along an isotherm T and each derivative is evaluated in the ideal-gas limit. These derivatives are defined to be the *pressure-virial coefficients*

$$B'(T) \equiv \left(\frac{\partial Z}{\partial P}\right)_T\bigg|_{P=0} \tag{4.5.21}$$

$$C'(T) \equiv \frac{1}{2!}\left(\frac{\partial^2 Z}{\partial P^2}\right)_T\bigg|_{P=0} \tag{4.5.22}$$

So this second form of the virial equation of state is a power series in pressure

$$Z = 1 + B'P + C'P^2 + \ldots \tag{4.5.23}$$

We can ascribe the same kinds of mathematical and physical interpretations to the pressure coefficients as we did to the density coefficients, but the two sets of coefficients differ numerically. We can see this merely by considering units: the primed coefficients have dimensions that are powers of P^{-1}, while the unprimed coefficients have dimensions in powers of ρ^{-1}.

But, although the two sets of coefficients differ, they are related. We can find the relations by equating the two expansions (4.5.11) and (4.5.23), keeping all terms; that is, at a particular state condition, the value of Z must be unaffected by whether we represent it by the complete ρ-expansion or by the complete *P*-expansion,

$$Z(T, \rho) = Z(T, P) \tag{4.5.24}$$

Now either we use the ρ-expansion to eliminate *P* from the rhs, or else we use the *P*-expansion to eliminate ρ from the lhs. Then we equate coefficients of terms having the same order; the results are

$$B' = \frac{B}{RT} \tag{4.5.25}$$

$$C' = \frac{C - B^2}{(RT)^2} \tag{4.5.26}$$

So if we have the ρ-coefficients, we may compute the *P*-coefficients, and vice versa: we need measure only one set of coefficients.

4.5.4 Truncated Virial Expansions

As implied by (4.5.24), the ρ-expansion and the *P*-expansion give the same value of Z if all terms in both expansions are used. But in practice we do not have values for all the coefficients; measurements have been done only for $B(T)$ and perhaps $C(T)$. Therefore, we must use truncated versions of the virial equations. However, truncated versions of the ρ-expansion and the *P*-expansion behave differently: they may give different values for Z. We consider four possible equations:

$$Z = 1 + B\rho \tag{4.5.27}$$

$$Z = 1 + \frac{BP}{RT} \tag{4.5.28}$$

$$Z = 1 + B\rho + C\rho^2 \tag{4.5.29}$$

$$Z = 1 + \frac{BP}{RT} + (C - B^2)\left(\frac{P}{RT}\right)^2 \tag{4.5.30}$$

Which of these should we use? On both theoretical and experimental grounds, the density expansion is preferred over the pressure expansion. The theoretical argument is that statistical mechanics provides a rigorous derivation of the density expansion. That derivation shows how the virial equation corrects for deviations from ideal-gas behavior: the second virial coefficient *B* accounts for interactions between pairs of molecules, the third virial coefficient C accounts for interactions among triplets of molecules, etc. In contrast, (4.5.26) shows that C' combines effects from two-body and three-body interactions.

The experimental argument is that, for a finite number of terms, the ρ-expansion is more easily fit to experimental PvT data. This statement can be justified by comparing plots of isotherms of Z vs. P with Z vs. ρ. The pressure plot will contain regions with large slopes, while the density plot will show less drastic variations [20].

But aside from these theoretical and experimental considerations, there are practical ones concerning which truncated expansion is more accurate and which is easier to use [10]. To address the question of accuracy, Figure 4.11 compares the four truncated expansions (4.5.27–30) applied to one isotherm of carbon dioxide. Up to about 50 bar, the first-order equations (4.5.27) and (4.5.28) are equally reliable in reproducing the experimental data; so which to use merely depends on whether we want ρ or P as an independent variable.

Beyond about 50 bar, high accuracy demands the next term in each expansion; otherwise, at high pressures, both first-order equations produce negative Z-values, with the density expansion becoming negative at the lower pressure. With the third virial coefficients included, the figure indicates that up to about 100 bar the two equations are almost equally reliable. Beyond about 120 bar, the density expansion remains reliable, but the pressure expansion qualitatively fails: it misses the minimum in the isotherm and eventually it gives negative values for Z. This behavior is typical; that is, beyond the second term, the density expansion is usually more reliable than the pressure expansion. But we caution that, for a particular gas, the relative accuracy of the

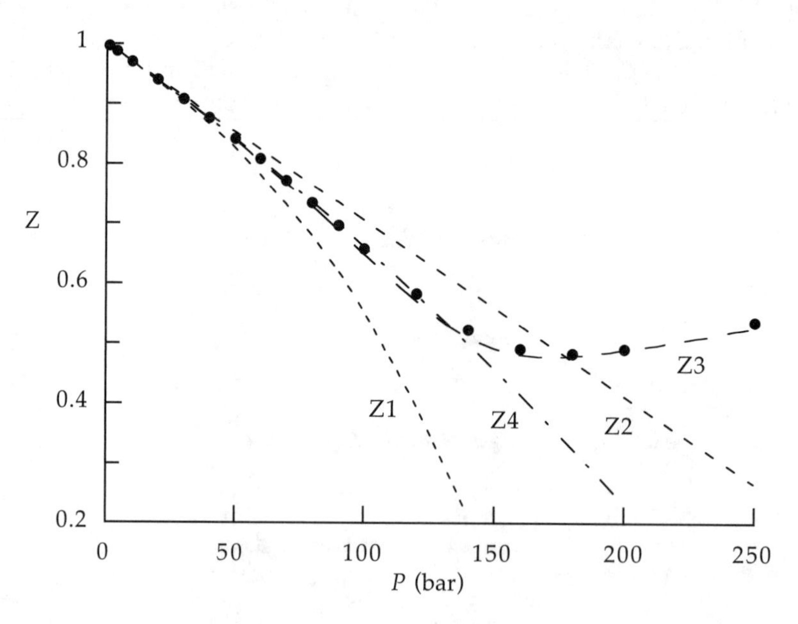

Figure 4.11 Comparison of truncated virial equations applied to carbon dioxide at 350 K. The critical point occurs near $T_c = 304$ K, $P_c = 73.8$ bar. Points are experimental data from Vargaftik [19]. Values of B and C taken from Dymond and Smith [21]: $B = -85.5$ cc/mol and $C = 3500$ (cc/mol)^2. Z1 is first-order density expansion (4.5.27), Z2 is first-order pressure expansion (4.5.28), Z3 is second-order density expansion (4.5.29), Z4 is second-order pressure expansion (4.5.30).

two second-order expansions depends crucially on the values of C and C'—quantities that may be difficult to measure or estimate accurately.

Finally, we emphasize that, even if we had several virial coefficients for a substance, the virial equations still only apply to gases and gas mixtures—both the density expansion and the pressure expansion fail to converge for liquids. Moreover, in practice we can find data or correlations for, at most, B and C, so the expansions should only be used for gases at low to moderate densities.

4.5.5 Example

How do we use the virial equations to compute values for the residual properties of gas mixtures?

Consider gaseous mixtures of methane and sulfur hexafluoride at 60°C, 20 bar; we want to compute u^{res}, h^{res}, and s^{res} across the entire composition range. For these small molecules at this modest pressure, the volumetric behavior is adequately represented by the virial equation in the form

$$Z = 1 + \frac{BP}{RT} \tag{4.5.31}$$

with B given by

$$B(T, x) = \sum_i^n \sum_j^n x_i x_j B_{ij}(T) \tag{4.5.18}$$

The model (4.5.31) is simple enough that it can be written explicitly in both v and P, so we can compare results for the isobaric residual properties with those for the isometric residual properties.

When T, P, and $\{x\}$ are independent. The volume explicit form of the equation of state (4.5.31) is

$$v = \frac{RT}{P} + B \tag{4.5.32}$$

Therefore the residual volume is merely the second virial coefficient,

$$v^{res} = B \tag{4.5.33}$$

To evaluate other isobaric residual properties, we will need the state dependence of the volume expansivity. Applying the definition (3.3.6) of α to the equation of state (4.5.32), we find

$$\alpha = \frac{1}{v}\left(\frac{R}{P} + \frac{dB}{dT}\right) \tag{4.5.34}$$

With the equation of state (4.5.32) and (4.5.34), we can evaluate the integral in (4.4.4) to obtain the residual enthalpy,

$$h^{res}(T, P, \{x\}) = P\left(B - T\frac{dB}{dT}\right) \tag{4.5.35}$$

The residual internal energy is then obtained from a Legendre transform,

$$u^{res}(T, P, \{x\}) = h^{res}(T, P, \{x\}) - Pv^{res} = -PT\frac{dB}{dT} \tag{4.5.36}$$

Finally, we use (4.5.32) and (4.5.34) in (4.4.6) to obtain the isobaric residual entropy,

$$s^{res}(T, P, \{x\}) = -P\frac{dB}{dT} \tag{4.5.37}$$

When T, v, and $\{x\}$ are independent. The model (4.5.31) can be written in a pressure-explicit form as

$$P = \frac{RT}{v - B} \tag{4.5.38}$$

From this we can find the isometric residual pressure,

$$P^{res} = \frac{BRT}{v(v - B)} \tag{4.5.39}$$

To evaluate other isometric residual properties, we will need the thermal pressure coefficient (3.3.5). Applying its definition to the equation of state (4.5.38), we obtain

$$\gamma_v = \frac{R}{v - B} + \frac{RT}{(v - B)^2}\frac{dB}{dT} \tag{4.5.40}$$

Now we can substitute (4.5.38) and (4.5.40) into (4.4.14) to obtain the isometric residual internal energy,

$$u^{res}(T, v, \{x\}) = -PT\frac{dB}{dT} \tag{4.5.41}$$

This is the same as (4.5.36); that is, the isometric and isobaric internal energies are the same. This was proved in (4.2.24). Similarly by (4.2.25), the isometric and isobaric residual enthalpies are also the same,

$$h^{res}(T, v, \{x\}) = P\left(B - T\frac{dB}{dT}\right) \tag{4.5.42}$$

Table 4.2 Values of second virial coefficients for methane(1)-sulfur hexafluoride(2) mixtures. B_{ij} values from [21]. Values of B and dB/dT are for equimolar mixtures.

	313.15 K	333.15 K	353.15 K
B_{11} (cm^3/mol)	–37.9	–31.8	–26.6
B_{12} (cm^3/mol)	–85.	–68.	–57.
B_{22} (cm^3/mol)	–253.	–223.	–192.
B (cm^3/mol)	–115.	–98.	–83.
dB/dT (cm^3/mol K)		0.80	

To obtain the isometric residual entropy, we substitute (4.5.40) into (4.4.16) and find

$$s^{res}(T, v, \{x\}) = -P\frac{dB}{dT} - R\ln Z \tag{4.5.43}$$

So the isobaric and isometric residual entropies differ by $R\ln Z$, as required by (4.2.28).

Sample calculations for the equimolar mixture. To use the above expressions we need values for B and its temperature derivative dB/dT. For CH_4(1)-SF_6(2) mixtures, Dymond and Smith [21] give the experimental values of B_{ij} in Table 4.2. The value of B was then computed from (4.5.18) using $x_1 = x_2 = 0.5$, and the temperature derivative of B was estimated as a central difference,

$$\frac{dB}{dT} \approx \frac{\Delta B}{\Delta T} \tag{4.5.44}$$

With values from Table 4.2, we can compute residual properties for the equimolar mixture at 60°C, 20 bar. The results are in Table 4.3. Plots of the residual properties over the full composition range are presented in Figure 4.12.

Table 4.3 Residual properties for equimolar mixtures of CH_4-SF_6 at 60°C, 20 bar, computed from virial equations

Property	Value	Equation
u^{res}	–534. J/mol	(4.5.36) or (4.5.41)
h^{res}	–730. J/mol	(4.5.35) or (4.5.42)
$s^{res}(T, P, \{x\})$	–1.60 J/mol K	(4.5.37)
$s^{res}(T, v, \{x\})$	–1.00 J/mol K	(4.5.43)

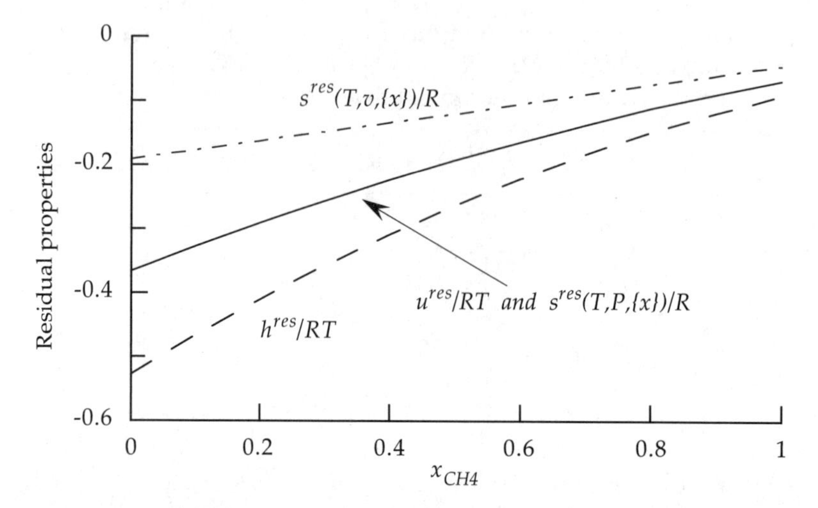

Figure 4.12 Dimensionless residual properties for gaseous CH_4-SF_6 mixtures at 60°C, 20 bar, from the virial equation (4.5.31). It is an artifact of the model that $u^{res}/RT = s^{res}(T, P, \{x\})/R$.

4.5.6 Van der Waals Equation of State

The ideal-gas law, hard-sphere equation of state, and virial equations all have rigorous foundations in statistical mechanics. But they are so simple that none applies over all fluid regions of the phase diagram; in particular, none of these apply to liquids or very dense gases. The development of a widely applicable volumetric equation of state is a formidable theoretical problem, because it must properly account for both short-range repulsive forces and long-range attractive forces among the molecules. We do not consider that problem in any detail here; instead, we must be content to introduce a class of semitheoretical approximations (§ 4.5.8–4.5.9) that are based on the equation originally devised by van der Waals.

Recall that the virial equations originate from Taylor expansions about the ideal gas. Alternatives can be obtained by expanding, not about ideal gases, but about hard spheres. Real fluids approach hard-sphere behavior in the isochoric high-temperature limit (4.5.1), so we use the parameter $\beta = 1/RT$ as the independent variable. Then on truncating the expansion at first-order, we have

$$Z = Z_{hs} + \beta\left(\frac{\partial Z}{\partial \beta}\right)_{\rho}\Bigg|_{\beta=0} \qquad \text{fixed } \rho \qquad (4.5.45)$$

The first term on the rhs accounts for short-range repulsive forces among the molecules, while the second term accounts for long-range attractive forces. We now seek approximate forms for the temperature and density dependence of these two terms.

In the time of van der Waals (1870s) forms for the hard-sphere compressibility factor were unknown, and so he had to contrive an estimate. His approximation can be rationalized as follows. First, consider the definition of the compressibility factor,

$$Z_{hs} = \frac{v(T, P)}{v^{ig}(T, P)} \qquad (4.5.46)$$

Since hard-sphere forces are purely repulsive ($Z_{hs} > 1$), the ideal-gas volume must be *smaller* than the hard-sphere volume to produce the same pressure at the same temperature. So van der Waals wrote

$$Z_{hs} = \frac{v}{v-b} = \frac{1}{1-b\rho} \qquad \textit{van der Waals} \qquad (4.5.47)$$

where $\rho = 1/v$ and v is the molar volume of the hard-sphere fluid. The parameter b is called the *covolume*; it measures the space that cannot be occupied by a molecular center because that space is already occupied by other molecules (the so-called *excluded volume*). Therefore, b attempts to correct the ideal-gas law for the fact that molecules are not points.

The covolume b depends on state condition and on the kinds of molecules. To obtain a value for b, van der Waals devised an argument based in kinetic theory [23]. In practice, the covolume is usually taken to be a constant for a particular substance, with its value obtained by a fit to experimental data. If we do take b to be constant, if the molecules can be approximated as spheres, and if we want the equation of state to reliably reproduce Z at low densities, then the covolume can be taken to be the hard-sphere second virial coefficient,

$$b = B_{hs} = \frac{2\pi}{3}\sigma^3 N_A = \frac{4\eta}{\rho} \qquad (4.5.48)$$

Here σ is the diameter of one sphere, N_A is Avogadro's number, ρ is the molar density, and η is the packing fraction (4.5.2). The derivation of (4.5.48) is straightforward and is left as an exercise. Since the volume of one sphere is $\pi\sigma^3/6$, (4.5.48) indicates that, at low densities, the space excluded by one molecule is not merely the volume of that molecule; rather, it is four times larger.

At this point, the equation of state has the form

$$Z = \frac{1}{1-b\rho} + \beta\left(\frac{\partial Z}{\partial \beta}\right)_\rho\Bigg|_{\beta=0} \qquad (4.5.49)$$

To approximate the second term, we seek qualitative guidance from a simple virial equation,

$$Z = 1 + B\rho \qquad (4.5.50)$$

Taking the isochoric β-derivative, and recalling that B depends only on temperature,

$$\left(\frac{\partial Z}{\partial \beta}\right)_\rho = \rho\left(\frac{dB}{d\beta}\right) \qquad (4.5.51)$$

We contrive a simple expression for the β-dependence of the second virial coefficient by taking $B(T)$ values for a simple gas and plotting them as B vs. β. We find that, over most of the temperature range, B is roughly linear in β with a negative slope. So we approximate the temperature dependence of B as a straight line in β,

$$B \approx -a\beta + c \tag{4.5.52}$$

Hence,

$$\left(\frac{dB}{d\beta}\right) = -a \tag{4.5.53}$$

Combining (4.5.49), (4.5.51), and (4.5.53) yields the van der Waals equation [23, 24]

$$Z = \frac{1}{1-b\rho} - \frac{a\rho}{RT} \tag{4.5.54}$$

In the van der Waals model (4.5.54), the first term makes the compressibility factor larger than the ideal-gas value to account for repulsive forces among the molecules. The second term makes Z smaller, to account for attractive forces. So the two terms compete in their effects on Z; one term or the other may dominate, depending on state condition (T and ρ). In the low-density limit, the van der Waals equation collapses to the ideal-gas law, while in the high temperature limit it approximates the hard-sphere equation of state. Formally, the parameters a and b depend on state condition as well as the kind of molecules, but in practice values for a and b are usually assumed to be constant for a particular substance (see § 4.5.10).

4.5.7 Example

If a fluid has Z = 1, is it necessarily an ideal gas?

Consider the van der Waals equation (4.5.54), which we now write as

$$Z = 1 + \frac{b\rho}{1-b\rho} - \frac{a\rho}{RT} \tag{4.5.55}$$

The issue is whether all residual properties are zero whenever Z = 1. To test this, consider the residual internal energy, which can be obtained by using (4.5.55) in (4.4.14); the result is

$$u_{vdw}^{res} = -a\rho \tag{4.5.56}$$

Then

$$h_{vdw}^{res} = u_{vdw}^{res} + Pv_{vdw}^{res} = -a\rho + RT(Z-1) \tag{4.5.57}$$

When Z = 1, (4.5.57) reduces to (4.5.56),

$$h_{vdw}^{res} = u_{vdw}^{res} = -a\rho \qquad Z = 1 \tag{4.5.58}$$

and (4.5.55) gives

$$\rho = \frac{a - bRT}{ab} \qquad Z = 1 \tag{4.5.59}$$

Combining (4.5.58) and (4.5.59) leaves

$$h_{vdw}^{res} = u_{vdw}^{res} = \frac{bRT - a}{b} \qquad Z = 1 \tag{4.5.60}$$

These residual properties are zero only at the one temperature $T = a/bR$. Hence in general, the fluid is *not* an ideal gas, even though $Z = 1$. Note that in the ideal-gas limit ($\rho \to 0$), (4.5.55) has $Z = 1$, (4.5.56) has $u^{res} = 0$, and (4.5.57) has $h^{res} = 0$, as they should.

4.5.8 Redlich-Kwong Equation of State

The van der Waals equation is historically important because it was the first equation of state to predict the vapor-liquid phase transition. However, although it is qualitatively informative, it is quantitatively unreliable, especially for dense fluids. The principal use of the van der Waals equation has been as a starting point for devising more reliable, and more complex, equations of state. Modified van der Waals equations have been devised by the hundreds, most with only empirical justification. Here we cite two important modifications.

Since van der Waals made approximations in arriving at both terms in his equation, we have two kinds of possible improvements. Historically, more effort has been devoted to the second term—the one that tries to account for attractive forces. At least two corrections can be made to the attractive term.

First, we can improve on the approximation (4.5.52) that the second virial coefficient is linear in β; in fact, B is more nearly linear in $\beta^{3/2}$, so we replace (4.5.52) with

$$B \approx -a'\beta^{3/2} + c \tag{4.5.61}$$

which leads to

$$\left(\frac{dB}{d\beta}\right) = -\frac{3a'}{2}\beta^{1/2} \equiv -aT^{-1/2} \tag{4.5.62}$$

and our equation of state becomes

$$Z = Z_{hs} - \frac{a\rho}{RT^{3/2}} \tag{4.5.63}$$

Second, we expect this improved attractive term to be most reliable at low densities, because it is based on the virial equation. To extend it to higher densities, we could append more terms (terms that roughly correspond to higher-order terms in the virial equation), but those extra terms would introduce more parameters in addition to a and b. Alternatively, we can hope to combine those missing additional terms by resumming their effects into a single factor. There is no unique way to perform this resummation and many forms have been tried. A particularly simple and successful form has proven to be $(1 + b\rho)^{-1}$, perhaps because the desired resummation of omitted terms can be approximated by

$$(1 + b\rho)^{-1} = 1 - b\rho + (b\rho)^2 - (b\rho)^3 + \dots \tag{4.5.64}$$

Our equation then has the form

$$Z = Z_{hs} - \frac{a\rho}{RT^{3/2}(1 + b\rho)} \tag{4.5.65}$$

If we adopt the original van der Waals form for Z_{hs}, then we have the Redlich-Kwong equation of state [25],

$$Z = \frac{1}{1 - b\rho} - \frac{a\rho}{RT^{3/2}(1 + b\rho)} \tag{4.5.66}$$

This development of the Redlich-Kwong equation is not a derivation, but only a systematic rationalization of the modifications. The equation is noteworthy because it provides substantial quantitative improvements over the original van der Waals equation. Nevertheless, Redlich himself remarked that there is no real theoretical justification for the changes made in the attractive term [26]. Modern formulations of the attractive term make the parameter a temperature dependent; examples are the Peng-Robinson [27] and Redlich-Kwong-Soave [28] equations. For other forms, see [3, 4, 11].

4.5.9 Modified Redlich-Kwong Equation of State

We now consider modifications to the repulsive term in the van der Waals equation. Although the van der Waals hard-sphere term is correct at low densities, Figure 4.13 shows that it quickly becomes erroneous as the density is increased: the excluded volume is not constant, but depends on density in some complicated way. Therefore we can improve the equation of state by using the Carnahan-Starling form (4.5.4) for Z_{hs}. Our modified Redlich-Kwong (mRK) equation of state is then [29]

$$Z = \frac{1 + \eta + \eta^2 - \eta^3}{(1 - \eta)^3} - \frac{a\rho}{RT^{3/2}(1 + b\rho)} \tag{4.5.67}$$

which is similar to the DeSantis equation [30]. If we use (4.5.48) to eliminate ρ in favor of the packing fraction η as the density variable, then

$$Z = \frac{1 + \eta + \eta^2 - \eta^3}{(1 - \eta)^3} - \frac{4a\eta}{bRT^{3/2}(1 + 4\eta)} \tag{4.5.68}$$

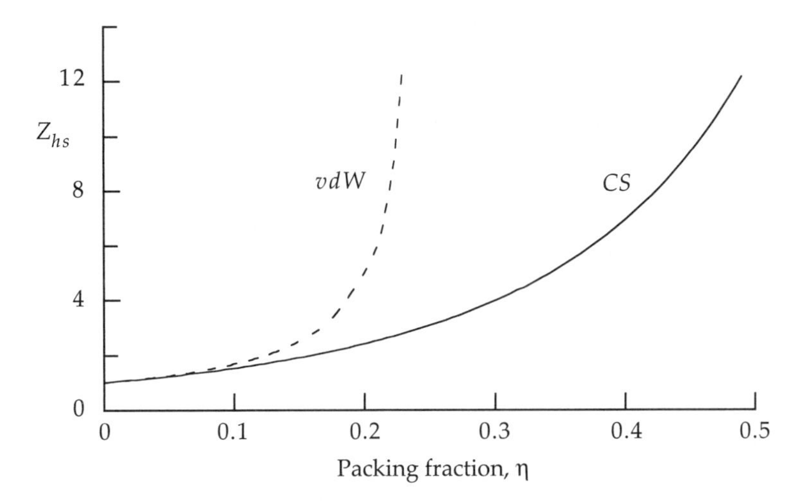

Figure 4.13 At moderate to high densities, the van der Waals (vdW) approximation (4.5.47) to the hard-sphere compressibility factor is in serious error when compared to the essentially exact Carnahan-Starling (CS) expression (4.5.4).

Unlike the van der Waals and Redlich-Kwong equations, which are cubics in density, this mRK equation is fifth-order. It is not unusual that improvements in accuracy are accompanied by increases in algebraic complexity; here the complexity occurs because we have combined a theoretically reliable repulsive term with an empirically proven attractive term.

With expressions from § 4.4.2, we can use the modified Redlich-Kwong equation (4.5.68) to obtain estimates for the residual properties of pure fluids. Those expressions contain the two parameters a and b; the results for the isobaric residual properties are

$$\frac{Pv^{res}(T,P)}{RT} = Z-1 = \frac{8b\rho(8-b\rho)}{(4-b\rho)^3} - \frac{a\rho}{RT^{3/2}(1+b\rho)} \tag{4.5.69}$$

$$\frac{u^{res}(T,P)}{RT} = -\frac{3a}{2bRT^{3/2}}\ln(1+b\rho) \tag{4.5.70}$$

$$\frac{s^{res}(T,P)}{R} = \ln Z + \frac{b\rho(3b\rho-16)}{(4-b\rho)^2} - \frac{a}{2bRT^{3/2}}\ln(1+b\rho) \tag{4.5.71}$$

Other residual properties can be obtained via Legendre transforms. Note that in the zero-density limit, these residual properties all go to zero, as they should. Further, in the hard-sphere limit ($a = 0$) these expressions revert to the Carnahan-Starling expressions (4.5.5)–(4.5.7), as they should.

4.5.10 Parameters in Semitheoretical Models

We have now introduced three semitheoretical equations of state: van der Waals (vdW), Redlich-Kwong (RK), and the modified Redlich-Kwong (mRK). Each contains two parameters, a and b. For a particular pure gas, values for a and b can be obtained by fitting to two or more experimental PvT points. Traditionally, however, values have been obtained by matching the equation of state to the gas-liquid critical point, T_c, P_c, and v_c. At the critical point the critical isotherm passes through a point of inflection, so we have the two conditions

$$\left(\frac{\partial P}{\partial \rho}\right)_T\Bigg|_{T_c} = 0 \quad \text{and} \quad \left(\frac{\partial^2 P}{\partial \rho^2}\right)_T\Bigg|_{T_c} = 0 \tag{4.5.72}$$

These provide two algebraic equations that can be solved simultaneously, yielding expressions for a and b in terms of either T_c and v_c or T_c and P_c. This procedure offers at least two advantages: (1) The critical properties of many pure substances have been measured [11] and if they have not been measured, they can be estimated by group contribution methods [11]. (2) By forcing the equation of state to reproduce the critical point, we ensure that the equation distinguishes the supercritical one-phase region of the phase diagram from the subcritical two-phase region. However, these semitheoretical equations have been found to provide only semiquantitative descriptions of the critical region itself [4].

The resulting expressions for a and b are given in Table 4.4 for each of the three equations of state. Also given in the table are values of the critical compressibility factor Z_c provided by each equation. Those values fall in the range $0.3 < Z_c < 0.4$, but for fluids of small rigid molecules such as argon, oxygen, nitrogen, and methane, the experimental value of $Z_c \approx 0.29$. Judging the three equations on just these values of Z_c, we expect mRK to perform better than RK and, in turn, RK to be better than vdW. Usually, this is so. However, each of these equations predicts that Z_c will have the

Table 4.4 Expressions for parameters a and b in terms of critical properties for three semitheoretical equations of state

	vdW (4.5.54)	**RK (4.5.66)**	**mRK (4.5.67)**
Z_c	3/8	0.333	0.316
		in terms of T_c and v_c	
b/v_c	1/3	0.2599	0.3326
$a/(v_c RT_c)$	9/8	$1.2824\sqrt{T_c}$	$1.4630\sqrt{T_c}$
		in terms of T_c and P_c	
bP_c/RT_c	1/8	0.08664	0.1050
$aP_c/(RT_c)^2$	27/64	$0.4275\sqrt{T_c}$	$0.4619\sqrt{T_c}$

same value for all substances, although in fact, Z_c spans a range of values for different materials, with most substances having $Z_c < 0.3$.

We caution that the expressions in Table 4.4 for $a(T_c, P_c)$ and $b(T_c, P_c)$ are consistent with those for $a(T_c, v_c)$ and $b(T_c, v_c)$ only for the value of Z_c quoted in the table. If, as is likely, a substance has some value of Z_c other than the tabulated one, then values computed for the parameters will differ, depending on whether the T_c-v_c forms or the T_c-P_c forms are used. For example, n-hexane has $Z_c = 0.26$; consequently, for the Redlich-Kwong equation,

$$b(T_c, v_c) = 96.2 \text{ cc/mol} \quad \text{but} \quad b(T_c, P_c) = 123.1 \text{ cc/mol} \tag{4.5.73}$$

Similar discrepancies occur between $a(T_c, v_c)$ and $a(T_c, P_c)$. In general, the parameter values computed from T_c and P_c should be used rather than those from T_c and v_c [31].

4.5.11 Comparisons of Results from vdW, RK, and mRK Equations

We now show predictions of the compressibility factor for pure carbon dioxide along two isotherms, one supercritical and the other subcritical. All results shown here used values of a and b computed from T_c and P_c. Figure 4.14 shows the results for the supercritical isotherm, T = 350 K. Up to about 75 bar, the three equations are all in good agreement with experiment, indicating that, at least at this temperature, all three satisfactorily estimate the second virial coefficient. However, for P > 100 bar, errors in

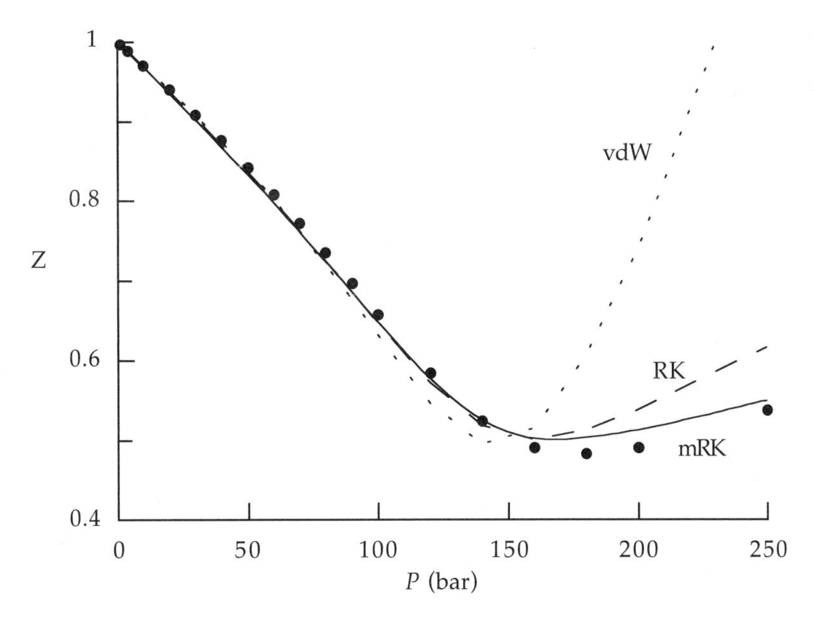

Figure 4.14 Comparison of the van der Waals (vdW), Redlich-Kwong (RK), and modified Redlich-Kwong (mRK) equations for predicting the compressibility factor of carbon dioxide along the supercritical isotherm T = 350 K. For each equation the parameters a and b were computed from expressions in Table 4.4, using T_c = 304.2 K and P_c = 73.82 bar. Points are experimental values taken from Vargaftik [19].

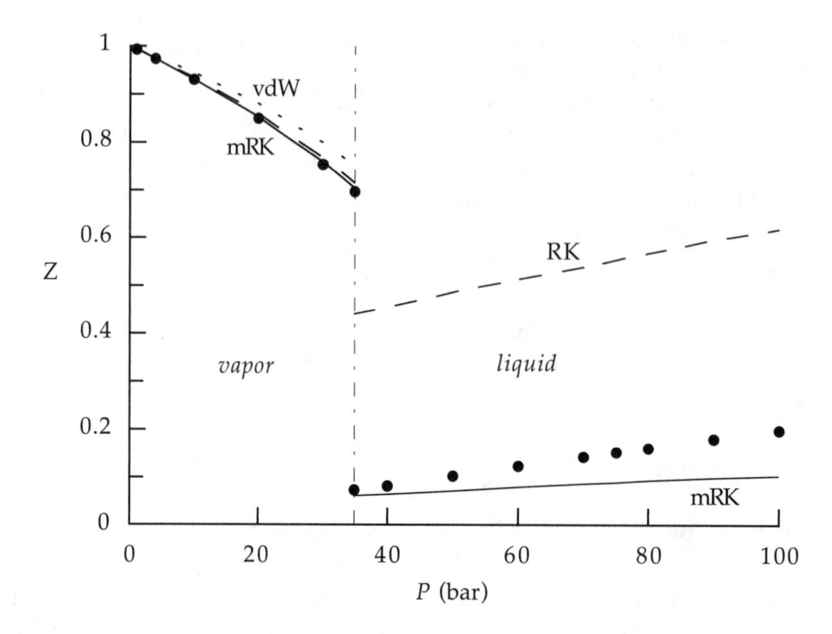

Figure 4.15 Same as Figure 4.14, but for carbon dioxide along the subcritical isotherm T = 273.15 K. Broken vertical line indicates the vapor pressure 34.84 bar at 273.15 K. Points are experimental values taken from Vargaftik [19]. The vdW results for the liquid phase all have Z > 1 and therefore are not shown. These curves were computed by setting T and v to their experimental values and solving each empirical equation for P, and hence Z. The comparisons differ significantly if, instead, experimental values for T and P are specified and the equations are solved by trial for v.

the vdW equation become substantial and for $P > 170$ bar they are intolerable. In contrast, the RK form is qualitatively reliable up to 175 bar, while the mRK form remains quantitatively reliable to 250 bar.

Figure 4.15 shows a similar comparison along the subcritical isotherm 273.15 K. For the vapor phase the three equations are about equally reliable, but for the liquid the three differ substantially. For the liquid, values of Z from the vdW equation exceed unity and therefore do not appear on the plot. Values from RK have the correct slope but are too large by factors of 3 to 4. The mRK equation performs better than the others, but it underestimates Z with errors reaching a factor of 2 at 100 bar.

We emphasize that for these comparisons, the equation-of-state parameters were obtained from T_c and P_c; none of the equations were fit to any data on either isotherm. Considering its simplicity, the mRK equation is a remarkable improvement over the older cubic forms. But it is still only a qualitative guide for the high-pressure liquid: reproducing the behavior of both liquid and dense-gas phases of polyatomic substances is too much to expect of any simple, two-parameter equation of state. But for another perspective on this issue, see Gregorowycz et al. [32].

4.5.12 Mixtures

Lastly we note that extending semitheoretical equations of state to mixtures is not straightforward because we have no theoretical guidance as to how parameters, such as a and b, depend on composition. This situation differs from that for the virial equations, for statistical mechanics tells us exactly how the virial coefficients depend on composition. For semitheoretical equations we are forced to guess the composition dependence of the parameters. Usually such guesses are in the form of *mixing rules,* in which mixture parameters are prescribed as some composition-dependent functions of the pure-component parameters; for example, we might try simple mole fraction averages of the pure parameters:

$$a_{mix} = \sum_i x_i a_{\text{pure i}} \tag{4.5.74}$$

$$b_{mix} = \sum_i x_i b_{\text{pure i}} \tag{4.5.75}$$

But these mixing rules are not particularly reliable, motivating searches for better relations; as usual, improvements in reliability are purchased at the price of increased complexity. For example, one improvement is a set of "van der Waals" mixing rules in which (4.5.75) is retained for b, but a is obtained from

$$a_{mix} = \sum_i \sum_j x_i x_j a_{ij} \tag{4.5.76}$$

where each sum runs over all components. Here $a_{ii} = a_{\text{pure i}}$, but an additional prescription is needed to obtain a_{ij} when $j \neq i$. Other mixing rules are discussed in [11].

It is conventional to estimate values for unlike parameters (such as a_{ij}) by combining the pure-component parameters ($a_{ii} = a_{\text{pure i}}$ and $a_{jj} = a_{\text{pure j}}$); such prescriptions are called *combining rules.* One choice is a simple arithmetic average,

$$a_{ij} = 0.5(a_{ii} + a_{jj}) \tag{4.5.77}$$

But when this is inserted into (4.5.76), we merely recover the simple mixing rule (4.5.74). An alternative is a geometric mean,

$$a_{ij} = \sqrt{a_{ii} a_{jj}} \tag{4.5.78}$$

Substituting this into (4.5.76) yields

$$a_{mix} = \left(\sum_i x_i \sqrt{a_{ii}}\right)^2 \tag{4.5.79}$$

Note that mixing rules depend on composition, but combining rules do not.

These kinds of empirical prescriptions often work reasonably well for properties of gas mixtures at low to moderate pressures. But for gases at high pressure, liquids, and phase-equilibrium calculations, further complications may be needed. One strategy is to introduce an adjustable parameter k_{ij}, so the combining rule (4.5.78) becomes

$$a_{ij} = (1 - k_{ij})\sqrt{a_{ii}\, a_{jj}} \tag{4.5.80}$$

Values of k_{ij} are usually small and positive (between 0.0 and 0.2); they may be assumed constant or they may be allowed to depend on temperature. They are best obtained by fitting experimental data: small variations in the value of k_{ij} can drastically affect the values of some properties, such as liquid-phase fugacity coefficients.

Note that for correlating or predicting properties of mixtures, we invoke approximate models at several different conceptual and computational levels: model equations of state, mixing rules, combining rules, adjustable parameters. Consequently, the possible combinations are numerous and the resulting complications can become subtle. All mixing rules and combining rules are essentially ad hoc and their use can lead to vagaries that are vexing.

4.6 SUMMARY

In this chapter we have developed ways for computing conceptual thermodynamic properties relative to well-defined states provided by the ideal gas. We identified two ways for measuring deviations from ideal-gas behavior: differences and ratios. Relative to the ideal gas, the difference measures are the isobaric and isometric residual properties, while the ratio measures are the compressibility factor and fugacity coefficient. These differences and ratios all apply to the properties of any single homogeneous phase (liquid or gas) composed of any number of components.

We then developed equations for computing the difference and ratio measures from the measurables P, v, T, and x. Data for these measurables are correlated by some volumetric equation of state, usually an analytic equation explicit in pressure $P(v, T, \{x\})$ or explicit in volume $v(P, T, \{x\})$. So the equations we derived for the conceptuals all involve integrals over appropriate functions of the equation of state. Then, in the last section of the chapter, we presented a few simple models for equations of state; these models are sufficient to illustrate the problems that arise both in trying to use simple analytic functions to represent volumetric data and in evaluating the integrals that provide values for conceptual properties.

We emphasize that the difference and ratio measures are means, not ends. That is, in a thermodynamic analysis of a process, the goal is not to determine a value for a residual property itself; instead, the determination of a residual property is an intermediate step in computing how a conceptual responds to a change of state. To determine how a total property F changes from state 1 to state 2, we would write

$$\Delta F_{12} = \Delta F_{12}^{res} + \Delta F_{12}^{ig} \tag{4.6.1}$$

where terms on the rhs are evaluated from a $PvTx$ equation of state and ideal-gas heat capacities. Similarly, when fugacity coefficients occur in problem descriptions, the goal is not to obtain values for fugacity coefficients, but rather to use them to obtain values for fugacities,

$$f_i = x_i\, \varphi_i\, P \tag{4.6.2}$$

In this chapter we have taken a significant and substantial step away from the formal thermodynamics of Part I toward the practical use of thermodynamics contained in Part IV. As we journey from Part I to Part IV, it is important to continually distinguish the approximations from the things that are rigorously correct. In this chapter the difference and ratio measures are all rigorous, because they are merely definitions. In addition, the relations that connect those deviation measures to measurables are also rigorous: no approximations or simplifying assumptions underlie any of the integrals appearing in § 4.4. Approximations occur when we use a model (such as an equation of state) to represent experimental data for measurables. And, of course, when we combine an approximate model with a rigorous integral for a conceptual, the result is an approximate value for the conceptual.

This illustrates an important advantage that accrues in working with conceptual properties: by constructing unambiguous definitions, we can devise computationally viable schemes of analysis without sacrificing thermodynamic rigor. Our computational procedures can therefore be exact, and uncertainties arise only when we implement the procedure; that is, when we choose a model to represent experimental data. Such a strategy limits the possible sources of error to clearly identifiable portions of an analysis.

It then becomes a matter of engineering judgement as to which model should be used in a particular situation. For example, we want to balance computational complexity against numerical reliability, but there are other concerns, such as the availability of experimental data. The proper exercise of engineering judgement is crucial to success in applying thermodynamics to real processes, and therefore it is an issue we will address repeatedly. In fact, if a situation is misjudged, causing an inappropriate model to be used, then even though the computational procedure is exact, the advantages of thermodynamic rigor are lost and the results are unreliable.

LITERATURE CITED

[1] K. C. Chao and R. L. Robinson, Jr. (eds.), "Equations of State in Engineering and Research," *ACS Adv. Chem. Ser.*, **182** (1979).

[2] S. Eliezer and R. A. Ricci (eds.), *High-Pressure Equations of State: Theory and Applications*, North-Holland, Amsterdam, 1991.

[3] S. I. Sandler, H. Orbay, and B.-I. Lee in *Models for Thermodynamic and Phase Equilibria Calculations*, S. I. Sandler (ed.), Marcel Dekker, New York, 1993.

[4] J. V. Sengers, R. F. Kayser, C. J. Peters, and H. J. White, Jr. (eds.), *Equations of State for Fluids and Fluid Mixtures*, Elsevier, Amsterdam, 2000.

[5] J. W. Gibbs, "On the Equilibrium of Heterogeneous Substances," in *The Collected Works of J. Willard Gibbs*, vol. I, Yale University Press, New Haven, 1906; reprinted 1948, 1957; pp. 150f, 166f.

[6] E. A. Guggenheim, *Thermodynamics*, 5th ed., North-Holland, Amsterdam, 1967, p. 174.

[7] R. D. Goodwin, H. M. Roder, G. C. Straty, "Thermophysical Properties of Ethane from 90 to 600K at Pressures to 700 Bar," *NBS Technical Note 684*, 1976.

[8] G. N. Lewis, "The Law of Physico-Chemical Change," *Proc. Am. Acad. Arts Sci.*, **37**, 49 (1901); "Das Gesetz physicko-chemischer Vorgänge," *Z. physik. Chem.* (Leipzig), **38**, 205 (1901).

[9] S. M. Walas, *Phase Equilibria in Chemical Engineering*, Butterworth, Boston, 1985.

[10] J. M. Prausnitz, R. N. Lichtenthaler, and E. G. de Azevedo, *Molecular Thermodynamics of Fluid-Phase Equilibria*, 2nd ed., Prentice-Hall, Englewood Cliffs, NJ, 1986.

[11] B. E. Poling, J. M. Prausnitz, and J. P. O'Connell, *The Properties of Gases and Liquids*, 5th ed., McGraw-Hill, New York, 2001.

[12] W. G. Hoover and F. H. Ree, "Melting Transition and Communal Entropy for Hard Spheres," *J. Chem. Phys.*, **49**, 3609 (1968).

[13] N. F. Carnahan and K. E. Starling, "Equation of State for Nonattracting Rigid Spheres," *J. Chem. Phys.*, **51**, 635 (1969).

[14] B. J. Alder, W. G. Hoover, and D. A. Young, "Studies in Molecular Dynamics. V. High-Density Equation of State and Entropy for Hard Disks and Spheres," *J. Chem. Phys.*, **49**, 3688 (1968).

[15] N. F. Carnahan and K. E. Starling, "Thermodynamic Properties of a Rigid-Sphere Fluid," *J. Chem. Phys.*, **53**, 600 (1970).

[16] G. A. Mansoori, N. F. Carnahan, K. E. Starling, and T. W. Leland, Jr., "Equilibrium Thermodynamic Properties of the Mixture of Hard Spheres," *J. Chem. Phys.*, **54**, 1523 (1971).

[17] T. W. Melnyk and B. L. Sawford, "Equation of State of a Mixture of Hard Spheres with Nonadditive Diameters," *Mol. Phys.*, **29**, 891 (1975).

[18] T. Boublik and I. Nezbeda, "P-V-T Behaviour of Hard Body Fluids. Theory and Experiment," *Collection Czech. Chem. Commun.*, **51**, 2301 (1986).

[19] N. B. Vargaftik, *Tables on the Thermophysical Properties of Liquids and Gases*, 2nd ed., Hemisphere Publishing Co. (Wiley), New York, 1975.

[20] E. A. Mason and T. H. Spurling, *The Virial Equation of State*, Pergamon, Oxford, 1969.

[21] J. H. Dymond and E. B. Smith, *The Virial Coefficients of Pure Gases and Mixtures*, Clarendon Press, Oxford, 1980.

[22] J. H. Dymond, K. N. Marsh, and R. C. Wilhoit, *The Virial Coefficients of Pure Gases and Mixtures*, Springer-Verlag, Heidelberg, 2003.

[23] J. D. van der Waals, *On the Continuity of the Gaseous and Liquid States*, J. S. Rowlinson (ed.), North-Holland, Amsterdam, 1988.

[24] M. M. Abbott, "Thirteen Ways of Looking at the van der Waals Equation," *Chem. Engr. Progr.*, **85**, 25 (1989).

[25] O. Redlich and J. N. S. Kwong, "On the Thermodynamics of Solutions. V. An Equation of State. Fugacities of Gaseous Solutions," *Chem. Rev.*, **44**, 233 (1949).

[26] O. Redlich, *Thermodynamics: Fundamentals, Applications*, Elsevier, Amsterdam, 1976.

[27] D. -Y. Peng and D. B. Robinson, "A New Two-Constant Equation of State," *IEC Fund.*, **15**, 59 (1976).

[28] G. Soave, "Equilibrium Constants from a Modified Redlich-Kwong Equation of State," *Chem. Eng. Sci.*, **27**, 1197 (1972).

[29] D. Henderson, "Practical Calculations of the Equation of State of Fluids and Fluid Mixtures Using Perturbation Theory and Related Theories," *ACS Adv. Chem. Ser.*, **182**, 1 (1979).

[30] R. DeSantis, F. Gironi, and L. Marrelli, "Vapor-Liquid Equilibria from a Hard Sphere Equation of State," *IEC Fund.*, **15**, 183 (1976)

[31] J. M. Smith and H. C. van Ness, *Introduction to Chemical Engineering Thermodynamics*, 4th ed., McGraw-Hill, New York, 1987, p. 83.

[32] J. Gregorowicz, J. P. O'Connell, and C. J. Peters, "Some Characteristics of Pure Fluid Properties that Challenge Equation-of-State Models," *Fluid Phase Equil.*, **116**, 94 (1996).

[33] K. S. Pitzer and R. F. Curl, Jr., "The Volumetric and Thermodynamic Properties of Fluids. III. Empirical Equation for the Second Virial Coefficient," *J. Am. Chem. Soc.*, **79**, 2369 (1957).

PROBLEMS

4.1 Use the fundamental equations with the Maxwell relations (3.3.34) and (3.3.37) to show that, for ideal gases, neither U nor H changes (a) with isothermal changes in pressure nor (b) with isothermal changes in volume.

4.2 Derive (4.1.7) which gives the response of the entropy of an ideal gas when both T and P are changed at constant mass.

4.3 A mixture of ideal gases is to expand adiabatically from 5 bar, 100°C to 20°C. Which mixture would produce the larger amount of work: an equimolar mixture of methane and ethane or an equimolar mixture of ethane and propane? The pure component heat capacities obey [11]

$$c_p^{ig}/R = A + BT + CT^2 + DT^3 + ET^4 \qquad \textit{with } T \textit{ in K} \qquad \text{(P4.3.1)}$$

	A	B(10^3)	C(10^5)	D(10^8)	E(10^{11})
methane	4.568	–8.975	3.631	–3.407	1.091
ethane	4.178	–4.427	5.660	–6.651	2.487
propane	3.847	5.131	6.011	–7.893	3.079

4.4 For a pure ideal gas, sketch a temperature-entropy diagram that contains isobars and isenthalps.

4.5 In a certain plant, a continuous isothermal-isobaric process is needed for extracting pure ethane from ethane-methane mixtures at 1 bar, 300 K. The gases may be assumed to be ideal with heat capacities given in Problem 4.3.

(a) For an equimolar mixture flowing at two mol/s, what would be the minimum rate of work needed to achieve a complete separation into the pure components? How much heat would have to be transferred? Would the heat have to be added or removed from the system?

(b) Repeat part (a) for a mixture composed of 90 mole% methane, flowing at 10 mol/s.

(c) Note that the processes in (a) and (b) both produce one mol/s of pure ethane, yet, even for ideal gases, the two processes require different amounts of work. What do these results suggest about diluting substances in one part of a process if they must be purified later?

4.6 A spherical weather balloon is filled at ground-level (1 bar, 300 K) with 1 m^3 of helium. (a) What would be the diameter of the balloon at an altitude of 4 km, where $T = 260$ K and $P = 0.8$ bar? (b) What would be the diameter in (a) if the balloon were filled with hydrogen rather than helium? (c) What would be the difference in maximum masses that the hydrogen and helium balloons could lift in air to 4 km?

4.7 Determine the difference, if any, between each of the following pairs of derivatives. In each case, your result should be expressed in terms of measurables, including perhaps measurable response functions:

(a) $\left(\frac{\partial S}{\partial P}\right)_{TN}$ and $\left(\frac{\partial S^{res}}{\partial P}\right)_{TN}$

(b) $\left(\frac{\partial P}{\partial T}\right)_{v}$ and $\left(\frac{\partial P}{\partial T}\right)_{v^{res}}$

(c) $\left(\frac{\partial S^{res}}{\partial V}\right)_{TN}$ and $\left(\frac{\partial S^{res}}{\partial V^{res}}\right)_{TN}$

4.8 Determine expressions for the isobaric residual properties u^{res}, h^{res}, s^{res}, and g^{res} for a pure gas that obeys the virial equation $Z = 1 + B\rho + C\rho^2$.

4.9 Use data from steam tables to estimate the values of the fugacity for saturated liquid water and saturated steam, both at the normal boiling point.

4.10 Compute and plot the fugacities $f_1(x_1)$ and $f_2(x_1)$ over the entire composition range for binary mixtures of carbon tetrachloride(1) and sulfur hexafluoride(2) at 271.6 K and 20 bar. Make one plot for each component. On each plot show curves produced from each of the following assumptions: (a) the mixture is an ideal gas, (b) the mixture is nonideal but obeys $Z = 1 + BP/RT$. The expression for the partial molar volume is given in Problem 4.23. Values of the B_{ij} for this mixture are $B_{11} = -112.4$ cc/mol, $B_{22} = -339$ cc/mol, and $B_{12} = -193$ cc/mol.

4.11 (a) For a pure gas that obeys the simple virial equation $Z = 1 + BP/RT$, show that the fugacity coefficient is given by

$$\ln\varphi = \frac{BP}{RT} \tag{P4.11.1}$$

(b) For a binary gas mixture that obeys $Z = 1 + BP/RT$, show that the fugacity coefficient of component 1 is given by

$$\ln\varphi_1 = \frac{P}{RT}(-B + 2x_1 B_{11} + 2x_2 B_{12}) \tag{P4.11.2}$$

where B is the mixture second virial coefficient (4.5.18). The partial molar volume for this situation is given in Problem 4.23.

4.12 Consider a binary mixture of components 1 and 2.

(a) Prove that at fixed T and P, if the fugacity of one component passes through an extremum with mole fraction x_1, then the fugacity of the other component also passes through an extremum at the *same* value of the mole fraction.

(b) For the same conditions as in (a), prove that the two components have opposite extrema; e.g., if one is a maximum, then the other *must* be a minimum.

4.13 Determine a reliable estimate for the maximum work that could be obtained when one mole of pure methane, initially at 25°C, 30 bar, adiabatically expands to twice its original volume. Assume for these conditions that methane obeys the model $Z = 1 + BP/RT$, with

$$\frac{BP_c}{RT_c} = 0.083 - \frac{0.422}{T_R^{1.6}} \tag{P4.13.1}$$

where $T_R = T/T_c$. For the ideal-gas heat capacity, you may assume c_p is independent of temperature, with $c_p = 19R/4$. Methane has $T_c = 190.6$ K and $P_c = 46$ bar.

4.14 (a) Starting from the mixture expression for the fugacity coefficient (4.4.10), derive the pure-fluid expression (4.4.11).

(b) Starting from the mixture expression for the fugacity coefficient (4.4.23), derive the pure-fluid expression (4.4.24).

4.15 At moderate pressures methane obeys $Z = 1 + BP/RT$, where $B = a - b/RT$ and the constants have these values: $a = 0.043$ m^3/kmol and $b = 2.29(10^6)$ MPa m^6/kmol^2. At very low pressures, the methane heat capacity obeys $c_p = \alpha + \beta T$, with $\alpha = 19.87$ kJ/(K kmol) and $\beta = 0.05$ kJ/(K^2 kmol).

(a) Based on this information, obtain an expression for the T and P dependence of c_p that would apply at moderate pressures.

(b) Compute the adiabatic power required to continuously change 1 kmol/s of methane from 290 K, 5 bar to 350 K, 20 bar.

(c) What fraction of your answer in (b) comes from residual contributions?

4.16 Samples of two pure gases, one containing N_1 moles and the other N_2, are initially both at T^o, P^o. The gases are mixed and, by a sequence of heat and work effects, are brought to a final state at T, P. The pure gases and their mixtures obey $Z = 1 + BP/RT$, with $B_{ij} = a_{ij} - b_{ij}/T^2$ for T in K The pure ideal-gas heat capacities can be correlated by $c_p/R = \alpha_i + \beta_i T$ with T in K. The parameters a_{ij}, b_{ij}, α_i, and β_i are all constants, independent of state.

(a) Obtain expressions for the changes in U and S for the process.

(b) You need to perform this process under the following conditions: $N_1 = 1$ kmol, $N_2 = 2$ kmol, $T^o = 300$ K, $P^o = 5$ bar, $T = 400$ K, and $P = 1$ bar. Parameter values: $\alpha_1 = \alpha_2 = 3$, $\beta_1 = 0.01/\text{K}$, $\beta_2 = 0.005/\text{K}$, $a_{11} = 0.2\ \text{m}^3/\text{kmol}$, $a_{22} = 0.1\ \text{m}^3/\text{kmol}$, $b_{11} = 1.25(10^5)\ \text{m}^3\ \text{K}^2/\text{kmol}$, $b_{22} = 1(10^5)\ \text{m}^3\ \text{K}^2/\text{kmol}$. The mixture also has $B_{12} = (B_{11} + B_{22})/2$. To design equipment for performing such a process, you would like to know the minimum energy requirements; that is, you would like to compute the reversible heat and work effects. Show whether this problem is well-posed; that is, show whether enough information is known to enable you to compute Q_{rev} and W_{rev}. If the problem is well-posed, use your results from (a) to compute the Q_{rev} and W_{rev}. If not, what other information would you need?

4.17 For a pure gas that obeys the truncated virial equation, $Z = 1 + BP/RT$, show whether or not the internal energy changes (a) with isothermal changes in pressure and (b) with isothermal changes in volume.

4.18 Pure carbon dioxide is to be compressed, reversibly and isothermally, from 1 bar, 350 K to 200 bar. At 350 K CO_2 has $B = -85.5$ cc/mol and $C = 3500$ (cc/mol)^2. Compute the work required using each of the following equations of state:

(a) ideal-gas law

(b) $Z = 1 + BP/RT$

(c) $Z = 1 + B\rho + C\rho^2$

(d) $Z = 1 + B'P + C'P^2$

4.19 Use each of the following equations of state to estimate the density of an equimolar gaseous mixture of carbon tetrachloride(1) and sulfur hexafluoride(2) at 271.6 K and 75 bar. At 271.6 K the third virial coefficients are $C_{111} = 7620$ (cc/mol)^2, $C_{222} = 18{,}640$ (cc/mol)^2, $C_{112} = 10{,}260$ (cc/mol)^2, and $C_{122} = 14{,}530$ (cc/mol)^2. Values for the second virial coefficients are given in Problem 4.10.

(a) ideal-gas law

(b) $Z = 1 + BP/RT$

(c) $Z = 1 + B\rho + C\rho^2$

(d) $Z = 1 + B'P + C'P^2$

4.20 Use the fact that the critical isotherm passes through a point of inflection at the critical point (4.5.72) to derive all the expressions in Table 4.4 for the parameters a and b in the following equations of state: (i) van der Waals, (ii) Redlich Kwong, and (iii) modified Redlich-Kwong.

4.21 Show that the van der Waals equation of state gives $u^{res} = -a\rho$ for a pure fluid. Here ρ is the molar density while a is the parameter in the equation of state and is assumed to be constant. What is the significance of the sign of u^{res}? What is the ideal-gas limit for the van der Waals expression for u^{res}?

4.22 A stream of pure ethylene is to be cooled from 100°C to 25°C in a single-pass, counter-flow, tube-in-shell heat exchanger. The gas enters the tube at 25 bar and a volumetric flow rate of 3 m^3/min. Cooling water is available at 20°C and can be heated no more than 10 C°. The heat exchanger is well-insulated. Determine the required flow rate of cooling water for the following cases:

(a) Assume ethylene is an ideal gas with

$$c_p^{ig}/R = A + BT + CT^2 \tag{P4.22.1}$$

and $A = 1.424$, $B = 0.0144$, $C = -4.39(10^{-6})$.

(b) Assume ethylene still has the heat capacity (P4.22.1), but now it obeys $Z = 1 + BP/RT$ with B given by the Pitzer correlation [33],

$$\frac{BP_c}{RT_c} = B_o + \omega B_1 \tag{P4.22.2}$$

$$B_o = 0.083 - \frac{0.422}{T_R^{1.6}} \tag{P4.22.3}$$

$$B_1 = 0.139 - \frac{0.172}{T_R^{4.2}} \tag{P4.22.4}$$

where $T_R = T/T_c$, $T_c = 282.4$ K, $P_c = 50.4$ bar, and acentric factor $\omega = 0.085$.

4.23 For a multicomponent gas mixture that obeys $Z = 1 + BP/RT$, show that the partial molar volume of component i is given by

$$\overline{V}_i = \frac{RT}{P} - B + 2\sum_k x_k B_{ik} \tag{P4.23.1}$$

where the sum runs over all components.

4.24 Use the Carnahan-Starling equation

(a) To derive (4.5.6) for the residual internal energy for a pure hard-sphere fluid.

(b) To derive (4.5.7) for the residual entropy for a fluid of pure hard spheres.

4.25 Estimate the volume required of a rigid tank to store one kilogram of gaseous propane at 25 bar, 100°C. Use (a) the Redlich-Kwong equation and (b) the modified Redlich-Kwong equation. Propane has $T_c = 369.9$ K, $P_c = 42.5$ bar, and molecular weight = 44.1.

4.26 Show that the van der Waals covolume b is the same as the hard-sphere second virial coefficient B_{hs}; that is, derive (4.5.48). To do so, rearrange the van der Waals estimate of Z_{hs} (4.5.47) to find

$$b = \lim_{\rho \to 0}\left(\frac{Z-1}{Z\rho}\right) = \lim_{\rho \to 0}\left(\frac{Z-1}{\rho}\right) \tag{P4.26.1}$$

and compare with B_{hs} from the virial equation written for hard spheres.

4.27 For pure substances that obey the Redlich-Kwong equation of state, derive the following expressions for isobaric residual properties:

(a) $$u^{res} = \frac{-3a}{2b\sqrt{T}}\ln(1+b\rho)$$

(b) $$h^{res} = \frac{-3a}{2b\sqrt{T}}\ln(1+b\rho) + \frac{RTb\rho}{1-b\rho} - \frac{a\rho}{\sqrt{T}(1+b\rho)}$$

(c) $$s^{res} = \frac{-a}{2bT\sqrt{T}}\ln(1+b\rho) + R\ln(1-b\rho) + R\ln Z$$

(d) $$\ln\varphi_{pure} = \frac{-a}{bRT\sqrt{T}}\ln(1+b\rho) - \ln(1-b\rho) - \ln Z + \frac{b\rho}{1-b\rho} - \frac{a\rho}{RT\sqrt{T}(1+b\rho)}$$

4.28 (a) Consider a pure fluid of hard spheres that obeys the Carnahan-Starling equation (4.5.4). Show that such a fluid always has positive values for the residual properties h^{res} and g^{res} and a positive value for $\ln\varphi$.

(b) Show that the van der Waals equation of state gives $c_v = c_v(T)$ but it also gives $c_p = c_p(T, P)$.

4.29 The Joule-Thomson (J-T) expansion, introduced in Problem 3.25, is characterized by the J-T coefficient, $\mu = (\partial T/\partial P)_h$.

(a) Evaluate μ for a pure fluid that obeys the Redlich-Kwong equation.

(b) What is the physical significance of states at which $\mu > 0$? Of $\mu < 0$?

(c) The J-T inversion temperature is the temperature at which μ changes sign; i.e., at which $\mu = 0$. Use the Redlich-Kwong equation to obtain an expression for the inversion temperature as a function of molar volume.

4.30 Consider a binary gas that obeys the virial equation $Z = 1 + B\rho$.

(a) Under what conditions, if any, will work have to be done on the gas in order to add a small amount ($x_3 < 10^{-4}$) of a third component at fixed T and P?

(b) Under what conditions, if any, will the fugacity of this dilute component pass through an extremum as P is increased with T and $\{x\}$ fixed?

(c) Repeat (a) and (b) for gases that obey $Z = 1 + B\rho + C\rho^2$.

4.31 This exercise illustrates one approach commonly used in developing thermodynamic models: a reliable functional form for a property of one substance is extended to a class of substances by parameterizing in terms of critical properties, and perhaps improved somewhat by curve fitting. Use the Redlich-Kwong equation of state to obtain an expression for the temperature dependence of the second virial coefficient. Then use the relations $a(T_c, P_c)$ and $b(T_c, P_c)$ from Table 4.4 to replace the equation of state parameters a and b with critical properties. Show that your result can be expressed in reduced form as

$$\frac{BP_c}{RT_c} = 0.08664 - \frac{0.4275}{T_R^{1.5}} \tag{P4.31.1}$$

where $T_R = T/T_c$ is the reduced temperature. Now using this form as a guide, we generalize by writing

$$\frac{BP_c}{RT_c} = \alpha - \frac{\beta}{T_R^{\gamma}} \tag{P4.31.2}$$

We then obtain values of the parameters α, β, and γ by a least squares fit to $B(T)$ data for gases composed of small rigid nonpolar molecules. Using the numerical values in (P4.31.1) as initial guesses in the fit, the result is the expression for B_0 given by the Pitzer correlation in Problem 4.22.

5

PROPERTIES RELATIVE TO IDEAL SOLUTIONS

In Chapter 4 we used differences and ratios to relate the conceptuals of real substances to those of ideal gases. To compute values for those differences and ratios, we use the equations given in § 4.4 together with a volumetric equation of state. Such equations of state are available for many mixtures, particularly gases; however, few of those equations reliably correlate properties of condensed-phase mixtures. Although some equations of state reproduce the behavior of condensed phases of complex substances, those equations are complicated and applying them can require considerable computational skill and resources. This is particularly true when we attempt to apply equations of state to mixtures of liquids.

Therefore we seek ways for computing conceptuals of condensed phases while avoiding the need for volumetric equations of state. One way to proceed is to choose as a basis, not the ideal gas, but some other ideality that is, in some sense, "closer" to condensed phases. By "closer" we mean that changes in composition more strongly affect properties than changes in pressure or density. The basis exploited in this chapter is the *ideal solution*. We still use difference measures and ratio measures, but they will now refer to deviations from an ideal solution, rather than deviations from an ideal gas.

We start the development in § 5.1 by defining ideal solutions and giving expressions for computing their conceptual properties. In § 5.2 we introduce the excess properties, which are the differences that measure deviations from ideal-solution behavior, and in § 5.3 we show that excess properties can be computed from residual properties. In § 5.4 we introduce the activity coefficient, which is the ratio that measures deviations from ideal-solution behavior, and in § 5.5 we show that activity coefficients can be computed from fugacity coefficients. This means that deviations from ideal-solution behavior are formally related to deviations from ideal-gas behavior, but in practice, one kind of deviation may be easier to compute than the other. Traditionally, activity coefficients have been correlated by fitting excess-property models to available experimental data; simple forms for such models are introduced in § 5.6. Those few simple models are enough to allow us to exercise many of the relations presented in this chapter; however, more thorough discussions of models for excess properties and activity coefficients must be found elsewhere [1, 2].

5.1 IDEAL SOLUTIONS

We define an ideal solution to be a mixture in which the molecules of different species are distinguishable (they have different masses or different structures or both); however, unlike the ideal gas, the molecules in an ideal solution exert forces on one another. When those forces are the same for all molecules, independent of species, then a solution is said to be ideal. This insensitivity to differences in molecular interactions does not mean that all properties are independent of composition (even in ideal-gas mixtures, most properties change with composition), but it does mean that when ideal-solution properties change with composition, they do so in regular ways. No real mixture is truly ideal, although many real mixtures are nearly ideal when they contain only molecules that are structurally similar; this includes isotopic mixtures ($H_2O + D_2O$), mixtures of components from a homologous series (methanol + ethanol + propanol), and mixtures of components that have a dominant structural characteristic, such as the aromatic ring in mixtures of benzene + toluene. Note that this definition does not restrict us to a particular phase; that is, gases may form ideal solutions. But the common use of this approach is for condensed phases—liquids and solids.

When intermolecular forces are independent of composition, each fugacity deviates from its ideal-gas value by an amount that is also independent of composition. This means each ideal-solution fugacity coefficient does not depend on composition,

$$\varphi_i^{is}(T,P) = \frac{f_i^{is}}{f_i^{ig}(T,P,\{x\})} = \frac{f_i^{is}}{x_i P} \tag{5.1.1}$$

Since the ideal-gas fugacity is linear in the mole fraction x_i while φ_i^{is} is independent of mole fraction, the ideal-solution fugacity must also be linear in x_i. We write that linearity in this form:

$$f_i^{is} = x_i f_i^o \tag{5.1.2}$$

where the proportionality constant f_i^o is the fugacity of the pure component in some well-defined *standard state.** We denote standard-state properties with a superscript *o*. The standard-state temperature is always taken to be the mixture temperature T, but the standard-state pressure P_i^o need not be the same as that of the mixture. Further, the value of P_i^o may be allowed to change when the mixture P changes, and we may choose different standard-state pressures for different components i.

The linear form (5.1.2) is the simplest expression that can be devised for the composition dependence of a fugacity, and in fact (5.1.2) can be considered to be a thermodynamic definition of *ideal solution*. Even the ideal-gas mixture, for which $\varphi_i^o = 1$, is a special kind of ideal solution; that is, the ideal-gas fugacity takes the form (5.1.2) with

* *Standard state* (o) is a district in the land of *reference states* (®). In contrast to the definition given above for standard state, a reference state (introduced in § 4.3.2) is any well-defined state with respect to which values of conceptuals are computed: a value for a reference-state property amounts to a lower limit on an integral that gives a change in a conceptual. For example, reference states are used in obtaining the values for u, h, and s that appear in steam tables. Reference states may be pure states or mixtures, so their property values may depend on composition. We caution that some authors make other distinctions between standard state and reference state; and some use these two terms synonymously.

$P_i^o = P$ and $f_i^o = P$. But real mixtures (nonideal solutions) have fugacities that are necessarily more complicated functions of composition than (5.1.2). Historically, the expression (5.1.2) was useful because a value for the standard-state fugacity f_i^o can be extracted from experimental data without appealing to a volumetric equation of state. But this advantage is becoming less important as more equations of state are being devised for correlating the $PvTx$ behavior of liquids, as well as dense gases.

On combining (5.1.2) with (5.1.1) we find

$$\varphi_i^{is}(T, P) = \frac{f_i^o(T, P_i^o)}{P} \tag{5.1.3}$$

This shows that, although the ideal-solution fugacity coefficient is independent of composition, it does depend on the choice made for the standard state. Consequently, the ideal-solution fugacity coefficient is not the same as the standard-state fugacity coefficient unless we choose $P_i^o = P$. That is, in general

$$\varphi_i^o(T, P_i^o) = \frac{f_i^o(T, P_i^o)}{P_i^o} \neq \varphi_i^{is}(T, P) \tag{5.1.4}$$

Many choices for the standard state are possible, and in fact, we need not even choose the same standard state for all species in a mixture. But to have an example for use throughout this chapter, we introduce the most common choice: the *Lewis-Randall rule* [3], in which the standard state for each component is taken to be the pure substance in the same phase and at the same T and P as the mixture. With this choice, each standard-state fugacity is given by

$$f_i^o(T, P_i^o) = f_i^o(T, P) = f_{\text{pure i}}(T, P) \tag{5.1.5}$$

and the ideal-solution fugacity in (5.1.2) becomes

$$f_i^{is}(T, P, \{x\}) = x_i f_{\text{pure i}}(T, P) \tag{5.1.6}$$

We refer to such a mixture as a *Lewis-Randall ideal solution.*

5.1.1 Partial Molar Properties of Lewis-Randall Ideal Solutions

To obtain expressions for the partial molar properties of ideal solutions, we first determine the chemical potential. Using the ideal-solution fugacity (5.1.6) in the integrated definition of fugacity (4.3.12) we find

$$\overline{G}_i^{is}(T, P, \{x\}) = g_i^o(T, P) + RT \ln x_i \tag{5.1.7}$$

For a Lewis-Randall ideal solution, g_i^o is the molar Gibbs energy of the pure component at the mixture T and P. The derivatives of $\overline{G}_i$ given in § 3.4.3 provide other properties; for example, the temperature and pressure derivatives of (5.1.7) produce the partial molar entropy and volume, respectively. Further, the Gibbs-Helmholtz equation (3.4.17) applied to (5.1.7) gives the partial molar enthalpy. Then the remaining properties can be found from their defining Legendre transforms. The results fall into two classes: those for first-law properties and those for second-law properties.

The partial molar results for first-law properties are independent of composition: these properties are simply the values of the corresponding first-law properties of the pure component in its standard state:

$$\bar{V}_i^{is}(T, P) = v_i^o(T, P) \tag{5.1.8}$$

$$\bar{U}_i^{is}(T, P) = u_i^o(T, P) \tag{5.1.9}$$

$$\bar{H}_i^{is}(T, P) = h_i^o(T, P) \tag{5.1.10}$$

In contrast, partial molar results for second-law properties depend on composition through entropy of mixing terms:

$$\bar{S}_i^{is}(T, P, \{x\}) = s_i^o(T, P) - R \ln x_i \tag{5.1.11}$$

$$\bar{G}_i^{is}(T, P, \{x\}) = g_i^o(T, P) + RT \ln x_i \tag{5.1.12}$$

$$\bar{A}_i^{is}(T, P, \{x\}) = a_i^o(T, P) + RT \ln x_i \tag{5.1.13}$$

Note that each second-law property diverges in the dilute-solution limit ($x_i \rightarrow 0$). Note also that each expression in (5.1.8)–(5.1.13) has the *same* functional form as the corresponding expression for an ideal-gas mixture (cf. § 4.1.3).

5.1.2 Total Properties of Lewis-Randall Ideal Solutions

With the partial molar properties now known, expressions for the total properties of ideal solutions can be formed from the generic relation between a mixture property and its corresponding component partial molar properties:

$$f(T, P, \{x\}) = \sum_i x_i \bar{F}_i(T, P, \{x\}) \tag{3.4.4}$$

Again, the results divide into expressions for first-law properties,

$$v^{is}(T, P, \{x\}) = \sum_i x_i v_i^o(T, P) \tag{5.1.14}$$

$$u^{is}(T, P, \{x\}) = \sum_i x_i u_i^o(T, P) \tag{5.1.15}$$

$$h^{is}(T, P, \{x\}) = \sum_i x_i h_i^o(T, P) \tag{5.1.16}$$

$$c_p^{is}(T, P, \{x\}) = \sum_i x_i c_{pi}^o(T, P) \tag{5.1.17}$$

and expressions for second-law properties,

$$s^{is}(T, P, \{x\}) = \sum_i x_i s_i^o(T, P) - R \sum_i x_i \ln x_i \tag{5.1.18}$$

$$g^{is}(T, P, \{x\}) = \sum_i x_i g_i^o(T, P) + RT \sum_i x_i \ln x_i \tag{5.1.19}$$

$$a^{is}(T, P, \{x\}) = \sum_i x_i a_i^o(T, P) + RT \sum_i x_i \ln x_i \tag{5.1.20}$$

For the Lewis-Randall ideal solution, all terms in these equations must be at the same temperature, pressure, and phase, even if some of those states are not physically realizable. For example, if P is below the vapor pressure of a pure substance, then that substance cannot exist as a liquid; nevertheless, the properties of a hypothetical liquid at that P might still be useful. Note that these results for ideal solutions are functionally the same as those given in § 4.1.4 for ideal-gas mixtures. This reinforces our comment that an ideal-gas mixture is merely one kind of ideal solution.

5.1.3 Changes of Properties on Mixing

Besides total properties, it is often useful to compare mixture properties to those of the pure components. Such comparisons can be made by defining, for any extensive property F, a change of property on mixing F^m,

$$F^m(T, P, \{N\}) \equiv F(T, P, \{N\}) - \sum_i N_i f_{\text{pure } i}(T, P) \tag{5.1.21}$$

where the mixture and all pures are at the same T and P. A particular instance of (5.1.21) is the change of Gibbs energy on mixing G^m, encountered in § 3.7.4 and § 4.1.5.

To evaluate F^m for the special case of ideal solutions, we merely substitute (5.1.14)–(5.1.20) in turn into (5.1.21). On so doing, we find that each first-law property takes a simple form,

$$v^{is, m} = \sum_i x_i [v_i^o(T, P) - v_{\text{pure } i}(T, P)] \tag{5.1.22}$$

$$u^{is, m} = \sum_i x_i [u_i^o(T, P) - u_{\text{pure } i}(T, P)] \tag{5.1.23}$$

$$h^{is,\,m} = \sum_i x_i\,[h_i^o(T, P) - h_{\text{pure i}}(T, P)] \tag{5.1.24}$$

while each second-law property contains an entropy of mixing term,

$$s^{is,\,m} = \sum_i x_i\,[s_i^o(T, P) - s_{\text{pure i}}(T, P)] - R\sum_i x_i \ln x_i \tag{5.1.25}$$

$$g^{is,\,m} = \sum_i x_i\,[g_i^o(T, P) - g_{\text{pure i}}(T, P)] + RT\sum_i x_i \ln x_i \tag{5.1.26}$$

$$a^{is,\,m} = \sum_i x_i\,[a_i^o(T, P) - a_{\text{pure i}}(T, P)] + RT\sum_i x_i \ln x_i \tag{5.1.27}$$

For the Lewis-Randall ideal solution, these expressions simplify as follows.

(a) Changes in first-law properties (5.1.22)–(5.1.24) all vanish: when an Lewis-Randall ideal solution is formed by mixing pure components at fixed T and P, there are no volumetric or thermal effects associated with the mixing.

(b) Changes in second-law properties (5.1.25)–(5.1.27) are not zero; instead, all reduce to an entropy of mixing term. Consequently, an Lewis-Randall ideal-solution has s^m always positive while g^m and a^m are always negative.

Item (b) means that, at fixed T and P, work must *always* be done on a Lewis-Randall ideal solution to separate it into its pure components. Further note that this work does not depend on phase: the minimum work to separate a liquid ideal solution at T, P, and $\{x\}$ is the same as that to separate an ideal-gas mixture at the same T, P, and $\{x\}$.

5.2 DEVIATIONS FROM IDEAL SOLUTIONS: DIFFERENCE MEASURES

Although no real mixture is truly ideal, we can often use the concept of an ideal solution to reduce the labor needed to compute property values for real mixtures. To do so we introduce, for each property f, an excess property f^E,

$$f^E(T, P, \{x\}) = f(T, P, \{x\}) - f^{is}(T, P, \{x\}) \tag{5.2.1}$$

Here f represents an intensive property value for the real mixture, and all three terms in (5.2.1) are at the same temperature T, pressure P, composition $\{x\}$, and phase. The excess properties provide a convenient way for measuring how a real mixture deviates from an ideal solution. In general, an excess property f^E may be positive, negative, or zero. An ideal solution will have all excess properties equal to zero. Note that the value for f^E depends on the choice of standard state used to define the ideal solution. Further note that the definition (5.2.1) is not restricted to any phase: excess properties may be defined for solids, liquids, and gases, although they are most commonly used for condensed phases.

5.2.1 Excess Properties and Mixing Properties

Excess properties are simply related to the changes of properties on mixing defined in § 5.1.2. Specifically, if we combine the definition of f^E (5.2.1) with the intensive version of the definition of F^m (5.1.21), we obtain

$$f^E(T, P, \{x\}) = f^m(T, P, \{x\}) - f^{is,\, m}(T, P, \{x\}) \tag{5.2.2}$$

That is, the excess properties are the differences between the real and ideal-solution changes of properties on mixing. The result (5.2.2) can be used for any ideal solution defined relative to any standard state; for example, when excess properties are relative to the Lewis-Randall ideal solution, we substitute the ideal-solution expressions (5.1.22)–(5.1.27) into (5.2.2) to obtain the following relations between f^E and f^m. First-law excess properties are identical to the changes on mixing,

$$v^E(T, P, \{x\}) = v^m(T, P, \{x\}) \tag{5.2.3}$$

$$u^E(T, P, \{x\}) = u^m(T, P, \{x\}) \tag{5.2.4}$$

$$h^E(T, P, \{x\}) = h^m(T, P, \{x\}) \tag{5.2.5}$$

while second-law excess properties differ from the changes on mixing by entropy of mixing terms,

$$s^E(T, P, \{x\}) = s^m(T, P, \{x\}) + R\sum_i x_i \ln x_i \tag{5.2.6}$$

$$g^E(T, P, \{x\}) = g^m(T, P, \{x\}) - RT\sum_i x_i \ln x_i \tag{5.2.7}$$

$$a^E(T, P, \{x\}) = a^m(T, P, \{x\}) - RT\sum_i x_i \ln x_i \tag{5.2.8}$$

We emphasize that (5.2.3)–(5.2.8) only apply when we use the Lewis-Randall standard state (5.1.5) for the ideal solution.

Since the definition (5.2.1) is a linear combination of thermodynamic properties, all the usual relations for extensive properties (see Chapter 3) can be expressed in terms of excess properties. Those relations include the Legendre transforms, the four forms of the fundamental equation, the response functions, and the Maxwell relations. Such relations reduce the amount of information needed to compute values for excess properties.

5.2.2 Excess Gibbs Energy and Its Derivatives

If we write the fundamental equation (3.2.28) for the excess Gibbs energy, we have

$$dG^E = -S^E dT + V^E dP + \sum_i \overline{G}_i^E \, dN_i \tag{5.2.9}$$

where $\overline{G}_i^E$ is the excess chemical potential for component i. We can obtain another form of this equation by replacing G^E with (G^E/RT) as the dependent variable,

$$d\left(\frac{G^E}{RT}\right) = \left(\frac{\partial(G^E/RT)}{\partial T}\right)_{PN} dT + \left(\frac{\partial(G^E/RT)}{\partial P}\right)_{TN} dP + \frac{1}{RT}\sum_i \overline{G}_i^E \, dN_i \tag{5.2.10}$$

With the analog of the Gibbs-Helmholtz equation (3.3.17),

$$\left(\frac{\partial(G^E/RT)}{\partial T}\right)_{PN} = -\frac{H^E}{RT^2} \tag{5.2.11}$$

and the analog of the pressure derivative (3.3.32),

$$\left(\frac{\partial(G^E/RT)}{\partial P}\right)_{TN} = \frac{V^E}{RT} \tag{5.2.12}$$

(5.2.10) becomes

$$d\left(\frac{G^E}{RT}\right) = \left(-\frac{H^E}{RT^2}\right) dT + \frac{V^E}{RT} dP + \frac{1}{RT}\sum_i \overline{G}_i^E \, dN_i \tag{5.2.13}$$

This equation has units of moles; it is important because T, P, N_i, V^E, and H^E are all measurable. And although the excess chemical potentials cannot be measured directly, they can be extracted from phase-equilibrium data. (It is instructive to note that while absolute values for conceptuals, such as H, can never be measured, certain kinds of differences in conceptuals, such as H^E and H^m, can be.)

The excess chemical potentials are not independent; rather, they are related through a Gibbs-Duhem equation. In particular, if we let g^E be the function f in the generic Gibbs-Duhem equation (3.4.10), we obtain

$$\sum_i^C x_i \, d\overline{G}_i^E = -s^E dT + v^E dP \tag{5.2.14}$$

And if T and P are fixed, then (5.2.14) reduces to

$$\sum_i^C x_i \, d\overline{G}_i^E = 0 \qquad \textit{fixed T and P} \tag{5.2.15}$$

When excess properties are defined relative to the Lewis-Randall ideal solution, v^E and h^E have simple physical interpretations. For v^E, we combine the ideal-solution form (5.1.14) with the definition (5.2.1) to obtain

$$v^E(T, P, \{x\}) = v(T, P, \{x\}) - \sum_i x_i v_i^o(T, P) \tag{5.2.16}$$

where v^E could be positive, negative, or zero. Let's consider some representative liquid mixtures. For example, mixtures of benzene and toluene are nearly ideal solutions with respect to the Lewis-Randall standard state. At ambient conditions, they have $v^E = 0$: mixing 50 cc of benzene with 50 cc of toluene produces 100 cc of mixture. In contrast, Figure 3.3 shows that ethanol-water mixtures have $v^E < 0$: at ambient conditions, 50 cc of ethanol added to 50 cc of water produces *less* than 100 cc of mixture. Further, Figure 3.3 also shows that mixtures of carbon tetrachloride and benzene have $v^E > 0$: 50 cc of CCl_4 added to 50 cc of C_6H_6 produces *more* than 100 cc of mixture.

In an ideal solution intermolecular forces are the same between all molecules, regardless of species: differences in those forces produce nonzero values for excess properties. In particular, magnitudes and signs of excess properties are determined by imbalances in the strengths of interactions between molecules of the same component (like interactions) as compared to those between molecules of different components (unlike interactions). Figure 5.1 illustrates these points by showing v^E at 25°C for

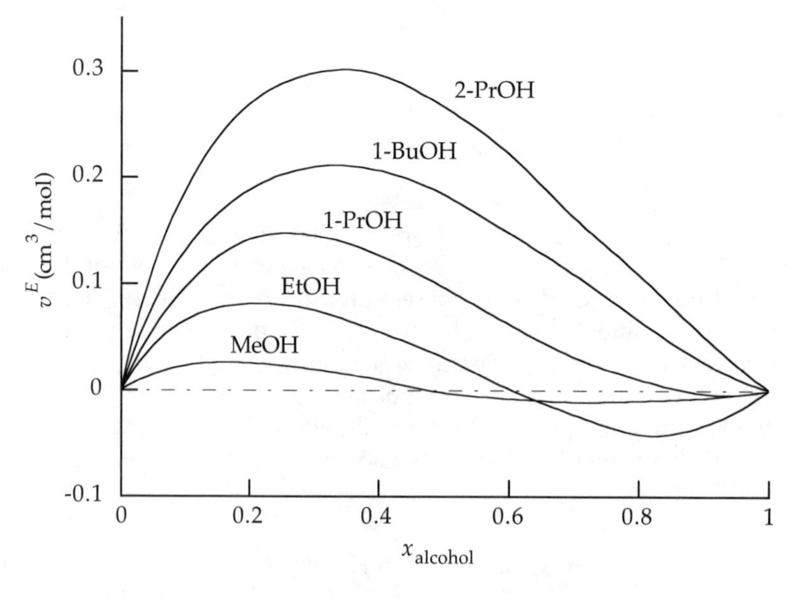

Figure 5.1 Excess volumes (relative to Lewis-Randall ideal solutions) for binary liquid mixtures of benzene plus an alcohol, all at 25°C. MeOH = methanol, EtOH = ethanol, 1-PrOH = n-propanol, 1-BuOH = n-butanol, and 2-PrOH = isopropanol. Adapted from Battino [4].

binary mixtures of benzene plus an alcohol. For mixtures of benzene plus methanol or ethanol, v^E can be either positive or negative, depending on composition. However, as the hydrocarbon chain of the alcohol is made longer, v^E becomes positive over all compositions. Further, v^E is influenced not only by the size of the hydrocarbon chain, but also by the location of the -OH group; consequently, v^E for benzene + 2-propanol is much larger than v^E for benzene + 1-propanol.

For the excess enthalpy, combining (5.1.16) with the definition (5.2.1) leaves

$$h^E(T, P, \{x\}) = h(T, P, \{x\}) - \sum_i x_i h_i^o(T, P) \tag{5.2.17}$$

When pure components are mixed at constant T and P, an energy balance shows that h^E measures the heat effect. In the Lewis-Randall standard state, ideal solutions have no heat effect on mixing, $h^E = h^m = 0$; but for real mixtures, the heat effect may be exothermic ($q = h^E < 0$) or endothermic ($q = h^E > 0$).

Excess properties are usually strong functions of composition; they may also be strong functions of temperature, but for liquids they are usually weak functions of pressure. Figures 5.2 and 5.3 show typical plots of the composition dependence of g^E, h^E, and s^E in sample binary liquid mixtures. The g^E values were obtained from analyses of vapor-liquid equilibrium data, the h^E values are from calorimetric data, and the s^E values were computed from the Legendre transform,

$$s^E = (h^E - g^E)/T \tag{5.2.18}$$

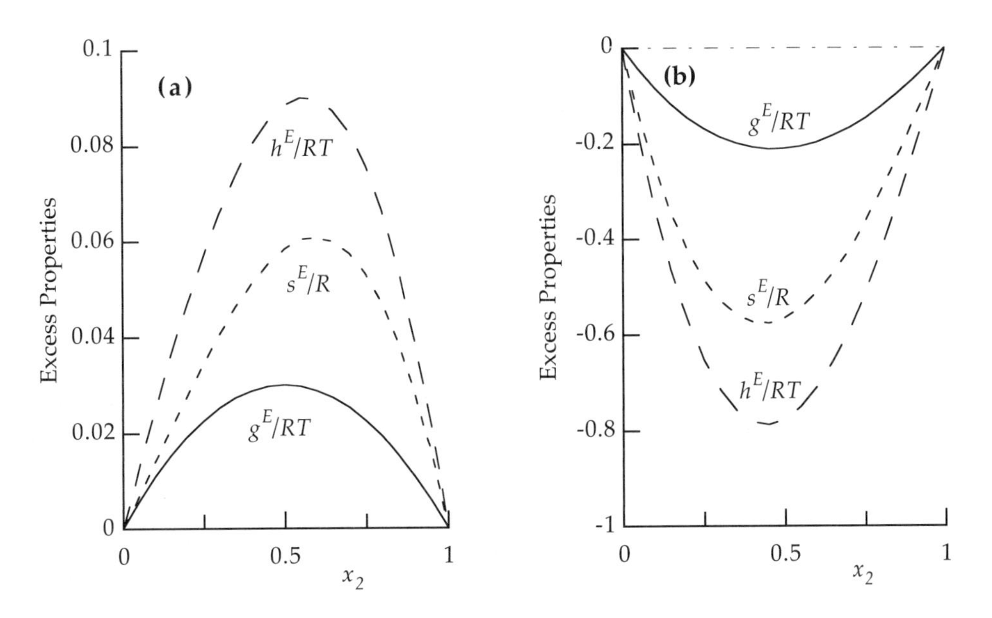

Figure 5.2 Composition dependence of excess properties (relative to Lewis-Randall ideal solutions) in representative binary liquid mixtures. (a) (*left*) n-hexane(1)–cyclohexane(2) at 20°C, (b) (*right*) chloroform(1)–acetone(2) at 25°C. Note different scales on ordinates. Redrawn from plots in [5].

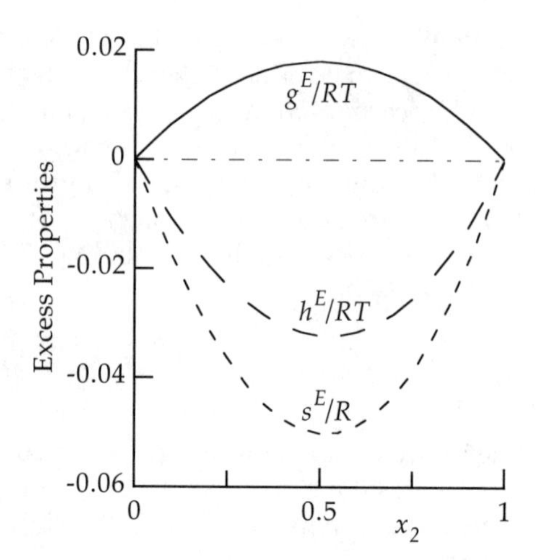

Figure 5.3 Effect of composition on the excess properties in binary liquid mixtures of n-octane(1) and tetraethylmethane(2) at 50°C. Excess properties defined relative to the Lewis-Randall ideal solution. Redrawn from plots in [5].

In Figure 5.2(a) all three excess properties are positive; for binary mixtures, h^E and g^E are often positive, while s^E may be of either sign. In Figure 5.2(b) all three properties are negative; this is less common than (a), but not rare. Note that the magnitudes of the excess properties in Figure 5.2(b) are about an order of magnitude larger than those in Figure 5.2(a). In Figure 5.3, the behavior is more complex: both h^E and s^E are negative, but s^E has the larger magnitude, so by (5.2.18), g^E is positive.

Figure 5.4 shows how temperature affects the excess properties in ethanol-water mixtures. At ambient temperatures, h^E and s^E are negative, with $Ts^E < h^E$, so g^E is positive. As T increases, both h^E and s^E become more positive. Note that at 70°C, h^E may be positive or negative, depending on composition. These changes in excess properties reflect complex and subtle changes in effects of molecular interactions in response to the change in temperature.

An important point to note from Figures 5.2–5.4 is that g^E is a weak function of temperature and is more nearly symmetric in composition than either h^E or s^E. These features are common to many binary mixtures: the nonidealities, as functions of composition, are more uniform in g^E than in either of the separate contributions, h^E and s^E. Furthermore, the relations among g^E, h^E, and s^E lead to patterns of behavior that can be important in applications [6].

Lastly, we emphasize that the definition of the excess properties (5.2.1) is completely general in that it can be used to measure deviations from any kind of ideal solution. In this section we have illustrated that definition using ideal solutions based on the Lewis-Randall standard state (5.1.5). This is a typical application; however, other kinds of ideal solutions, based on other standard states, can be defined and prove useful in special situations. In those cases, the generic definition of the excess property (5.2.1) still applies.

5.3 EXCESS PROPERTIES FROM RESIDUAL PROPERTIES

Traditionally, values for excess properties were obtained either directly from experiment or indirectly, by fitting a small number of measured values to a model. But excess properties can also be obtained from residual properties, which are extracted from *PvTx* measurements. In this section we develop relations that enable us to com-

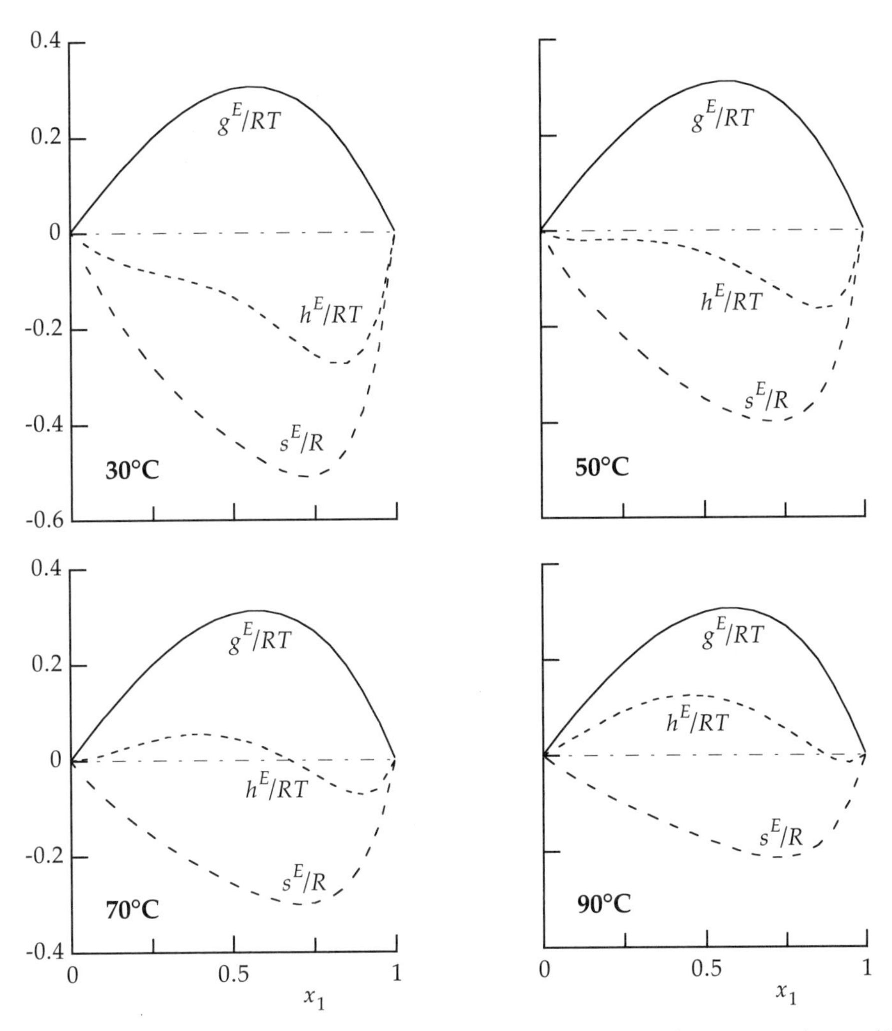

Figure 5.4 Effect of temperature on the excess properties for liquid mixtures of water(1) and ethanol(2). Note that g^E/RT is only weakly affected by these changes in T, while h^E/RT changes sign. Excess properties relative to Lewis-Randall ideal solution. From data tabulated in [7].

pute excess properties from residual properties, and hence from volumetric equations of state. We consider volume-explicit equations of state (§ 5.3.1) first and follow with pressure-explicit equations (§ 5.3.2).

5.3.1 When T, P, and $\{x\}$ Are Independent

When the mixture of interest is described by an equation of state of the form $v(T, P, \{x\})$, then the definition of the excess properties (5.2.1) can be combined with the definition of isobaric residual properties (4.2.1) to yield the intensive form

$$f^E(T, P, \{x\}) = f^{res}(T, P, \{x\}) - f^{is,res}(T, P, \{x\}) \tag{5.3.1}$$

where the second term on the rhs is the residual property of the ideal solution. The general result (5.3.1) enables us to compute excess properties from known values of isobaric residual properties. In addition, (5.3.1) can be used to find equations that enable us to compute excess properties directly from $v(T, P, \{x\})$ equations of state.

To illustrate how (5.3.1) is applied, we use the simple virial equation

$$Z = 1 + \frac{BP}{RT} \tag{5.3.2}$$

to compute excess properties for gaseous mixtures of methane and sulfur hexafluoride at 60°C, 20 bar. In § 4.5.5 we used this same equation of state to compute residual properties for this mixture. The volume-explicit form of this equation is

$$v = \frac{RT}{P} + B \tag{5.3.3}$$

where B is given for mixtures by (4.5.18).

Using expressions from § 5.1.2 for Lewis-Randall ideal-solution properties and those from § 4.1.4 for ideal-gas mixtures, (5.3.1) can be written as

$$f^E(T, P, \{x\}) = f^{res}(T, P, \{x\}) - \sum_i x_i f_i^{o,res}(T, P, \{x\}) \tag{5.3.4}$$

where $f_i^{o,\,res}$ represents the residual property for component i in the Lewis-Randall standard state (5.1.5). Equation (5.3.4) applies to both first-law and second-law properties. For second-law properties, the entropy of mixing terms for the ideal gas and ideal solution are the same, and so they cancel when (4.2.1) and (5.2.1) are combined.

To obtain the excess volume, we substitute the model (5.3.3) into (5.3.4) and obtain

$$v^E(T, P, \{x\}) = \frac{RT}{P} + B - x_1\left(\frac{RT}{P} + B_{11}\right) - x_2\left(\frac{RT}{P} + B_{22}\right) \tag{5.3.5}$$

$$= B - x_1 B_{11} - x_2 B_{22} \tag{5.3.6}$$

Using (4.5.18) for B and simplifying, (5.3.6) becomes

$$v^E(T, P, \{x\}) = x_1 x_2 \delta_{12}(T) \tag{5.3.7}$$

where

$$\delta_{12}(T) \equiv 2B_{12}(T) - B_{11}(T) - B_{22}(T) \tag{5.3.8}$$

Note that δ_{12} quantifies the imbalance of forces acting between molecular pairs of the same component (B_{11} and B_{22}) as compared to pairs of different components (B_{12}); δ_{12} may be positive or negative. Further, an ideal solution has $\delta_{12} = 0$ because all forces are the same ($B_{11} = B_{22} = B_{12}$); however, the mixture would not be an ideal gas unless the forces were not only the same, but also all zero.

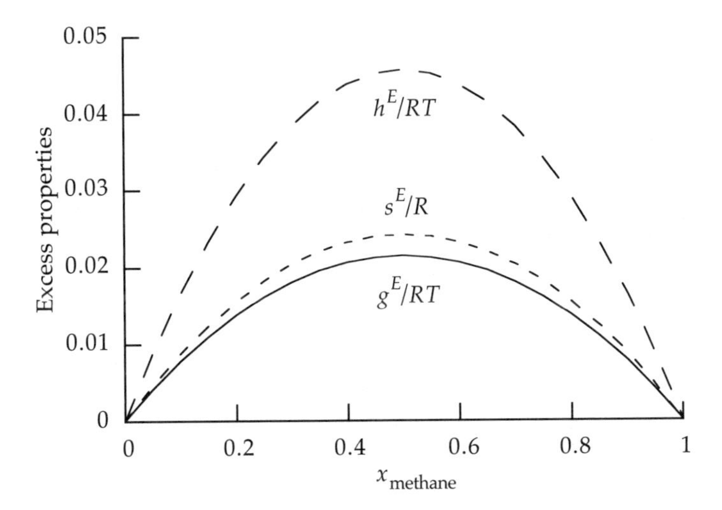

Figure 5.5 Excess properties for gaseous mixtures of methane and sulfur hexafluoride at 60°C and 20 bar; computed from the virial equation (5.3.3) using (5.3.9)–(5.3.11). Excess properties relative to Lewis-Randall ideal solution (5.1.6).

To obtain an expression for the excess enthalpy, we substitute (4.5.35) for the residual enthalpy into (5.3.4) and find

$$h^E(T, P, \{x\}) = x_1 x_2 P \left(\delta_{12} - T \frac{d\delta_{12}}{dT} \right) \tag{5.3.9}$$

Similarly, to obtain the excess entropy we substitute (4.5.37) for s^{res} into (5.3.4),

$$s^E(T, P, \{x\}) = -x_1 x_2 P \frac{d\delta_{12}}{dT} \tag{5.3.10}$$

Then a Legendre transform gives

$$g^E(T, P, \{x\}) = x_1 x_2 P \delta_{12}(T) \tag{5.3.11}$$

To compute excess properties from (5.3.9)–(5.3.11), we need values for δ_{12} and its temperature derivative. For these mixtures, values for B_{ij} are found in Table 4.2, and then (5.3.8) gives

$$\delta_{12} = 119 \text{ cm}^3/\text{mol} \tag{5.3.12}$$

The values in Table 4.2 for the B_{ij} can also be used to estimate the derivative in (5.3.10) as a finite difference; we find

$$\frac{d\delta_{12}}{dT} \approx \frac{\Delta\delta_{12}}{\Delta T} = -0.40 \text{ cm}^3/\text{mol K} \tag{5.3.13}$$

Results from (5.3.9)–(5.3.11) over the entire composition range are shown in Figure 5.5.

5.3.2 When T, V, and $\{x\}$ Are Independent

In this subsection we consider those situations in which our mixture is described by a pressure-explicit equation of state, $P = P(T, v, \{x\})$. Our objective is still to relate excess properties to residual properties and to the equation of state, but with v as an independent variable, we would prefer to express those relations in terms of the isometric residual properties, rather than the isobaric ones. However, the development is not as simple as what we did in the previous section because now we have an inconsistency: the equation of state and the isometric residual properties use $(T, v, \{x\})$ as the independent variables, but the excess properties defined by (5.2.1) use $(T, P, \{x\})$.

For first-law conceptuals (u and h) this inconsistency poses no problem because values of first-law isometric and isobaric residual properties are the same; see (4.2.24) and (4.2.25). However, for second-law conceptuals (g, a, and s) the two kinds of residual properties differ (see § 4.2.3), so we must exercise care when using residual properties to evaluate second-law conceptuals. We need to evaluate only three quantities (v^E, u^E, and s^E) then the remaining three (h^E, g^E, and a^E) can be obtained from Legendre transforms. We also obtain the expression for the excess chemical potential in terms of isometric residual chemical potentials.

To obtain the excess volume at a specified mixture state $(T, P, \{x\})$, we still apply (5.3.4), in which the mixture and all standard states are at the same temperature and pressure. Formally this poses no problem, but for some equations of state we will have to perform trial-and-error calculations to obtain volumes.

From a pressure-explicit equation of state, the internal energy is the appropriate first-law conceptual to evaluate. Since isometric residual internal energies are identical to isobaric ones (4.2.24), we can immediately write (5.3.4) as

$$u^E(T, P, \{x\}) = u^{res}(T, v, \{x\}) - \sum_i x_i u_i^{o,res}(T, v, \{x\}) \tag{5.3.14}$$

To evaluate u^E, we merely substitute (4.4.14) for each term on the rhs of (5.3.14).

As the second-law conceptual, we choose the entropy; combining (4.2.28) for the isometric residual entropy with (5.3.4), we obtain

$$s^E(T, P, \{x\}) = s^{res}(T, v, \{x\}) - \sum_i x_i s_i^{o,res}(T, v_i^o, \{x\}) + R \sum_i x_i \ln \frac{Z}{Z_i^o} \tag{5.3.15}$$

where Z_i^o is the compressibility factor for component i in its standard state. Similarly, the excess chemical potential is given by

$$\overline{G}_i^E(T, P, \{x\}) = \overline{G}_i^{res}(T, v, \{x\}) - \sum_i x_i \overline{G}_i^{o,res}(T, v_i^o, \{x\}) - RT \sum_i x_i \ln \frac{Z}{Z_i^o} \tag{5.3.16}$$

Note that the mixture state of interest is identified by $(T, P, \{x\})$, so the value for the mixture volume v must be obtained by solving the equation of state at $(T, P, \{x\})$. But v_i^o is the molar volume of component i in its standard state at T and P.

To illustrate how these equations are applied, we repeat the calculations in § 5.3.1 to obtain excess properties for gaseous mixtures of methane and sulfur hexafluoride at 60°C, 20 bar. Values for the isometric residual properties of this mixture have already been determined in § 4.5.5. We continue to use the virial equation of state (5.3.2), but now we write it in a pressure-explicit form,

$$P = \frac{RT}{v - B} \tag{5.3.17}$$

with B for mixtures given by (4.5.18). As always, before calculations can be done, we must identify the standard-state for each component. To be consistent with § 5.3.1, we again choose the Lewis-Randall standard state (5.1.5).

The excess volume is still determined by the procedure used in § 5.3.1, leading to the same result (5.3.7). Then to obtain u^{res}, we substitute (4.5.41) for $u^{res}(T, v, \{x\})$ into (5.3.14) and obtain

$$u^E(T, P, \{x\}) = -x_1 x_2 PT \frac{d\delta_{12}}{dT} \tag{5.3.18}$$

Substituting (4.5.43) for $s^{res}(T, v, \{x\})$ into (5.3.15) yields

$$s^E(T, P, \{x\}) = -x_1 x_2 P \frac{d\delta_{12}}{dT} \tag{5.3.19}$$

This is the same result as found in (5.3.10). With v^E, u^E, and s^E known, we can obtain h^E and g^E by Legendre transforms. The results are (5.3.9) for h^E and (5.3.11) for g^E. Since the expressions for the excess properties obtained here are *exactly* those found in § 5.3.1, the numerical results are also the same. In particular, the excess properties for this mixture are still as represented in Figure 5.5.

The approach used here differs from that in § 5.3.1 merely because of the form adopted for the equation of state. For the simple virial equation (5.3.2), we can choose whether we want to use a volume-explicit or a pressure-explicit form. Both forms give the same results for excess properties, and both require about the same computational effort. However for dense fluids, more complicated equations of state must be used; often, they are cubic or higher-order polynomials in the volume. That is, most are pressure-explicit, they cannot be converted into volume-explicit forms, and in such cases, we *must* use the expressions (5.3.14)–(5.3.16) to obtain excess properties.

5.3.3 Compare Values of Excess Properties with Residual Properties

In § 4.5.5 we computed residual properties for gaseous mixtures of methane and sulfur hexafluoride mixtures at 60°C and 20 bar. In § 5.3.1 and 5.3.2 we computed excess properties for this same mixture. We can also compute residual properties for the ideal solution (Lewis-Randall standard state). Comparisons of these three kinds of difference measures are shown in Table 5.1 for equimolar mixtures. We see that the equimolar mixture of methane and sulfur hexafluoride exhibits positive deviations

Table 5.1 Excess properties compared to isobaric residual properties for gaseous equimolar mixtures of methane and sulfur hexafluoride at 60°C and 20 bar. These excess properties are relative to the Lewis-Randall ideal solution.

Property	Excess	Isobaric residual	Ideal-solution residual
Volume (cm^3/mol)	29.7	–97.7	–127.4
Enthalpy (J/mol)	126.	–730.	–856.
Entropy (J/mol K)	0.20	–1.60	–1.80
Gibbs energy (J/mol)	59.4	–197.	–256.

from ideal-solution behavior and negative deviations from ideal-gas behavior. Further, these systems expand slightly on mixing and the mixing is endothermic; however, since the mixture is a gas, these effects are small. This behavior is common; it can usually be attributed to strong attractive forces acting between molecules of the same component, as compared to weaker forces acting between molecules of different components.

5.4 DEVIATIONS FROM IDEAL SOLUTIONS: RATIO MEASURES

In addition to the excess properties, which are difference measures for deviations from ideal-solution behavior, we also find it convenient to have ratio measures. In particular, for phase equilibrium calculations, it proves useful to have ratios that measure how the fugacity of a real mixture deviates from that of an ideal solution. Such ratios are called *activity coefficients*. The activity coefficients can be viewed as special kinds of a more general quantity, called the *activity*; so we first introduce the activity (§ 5.4.1) and then discuss the activity coefficient (§ 5.4.2).

5.4.1 Activity

Consider the algebraic form (4.3.12) that results from an isothermal integration of the first part of the definition of fugacity,

$$\overline{G}_i(T, P, \{x\}) - \overline{G}_i^{®}(T, P_i^{®}, \{x^{®}\}) = RT \ln \frac{f_i(T, P, \{x\})}{f_i^{®}(T, P_i^{®}, \{x^{®}\})} \tag{4.3.12}$$

For the reference state, lets us choose a *pure-component standard state*: the real (or hypothetical) pure substance at the temperature of the mixture and at some convenient pressure P_i^o. This pressure need not be the same as the pressure P of the mixture.* Then (4.3.12) becomes

$$\overline{G}_i(T, P, \{x\}) - \overline{G}_i^o(T, P_i^o) = RT \ln \frac{f_i(T, P, \{x\})}{f_i^o(T, P_i^o)} \tag{5.4.1}$$

and the ratio on the rhs defines the *activity* for component i,

$$\mathfrak{a}_i(T, P, \{x\}; f_i^o) \equiv \frac{f_i(T, P, \{x\})}{f_i^o(T, P_i^o)} \tag{5.4.2}$$

The result (5.4.1) establishes a connection between a difference and a ratio,

$$\overline{G}_i(T, P, \{x\}) - \overline{G}_i^o(T, P_i^o) = RT \ln \mathfrak{a}_i(T, P, \{x\}; f_i^o) \tag{5.4.3}$$

The activity is a dimensionless, conceptual, state function. The notation used in the argument list for $\mathfrak{a}_i$ is intended to emphasize that the numerical value for the activity depends, not only on the state of the mixture (T, P, $\{x\}$), but also on the choice of the standard state. At this point we have not completely identified the standard state; we have said it is the pure substance at T but we have not specified the pressure P_i^o or the phase. This leaves some flexibility in using the activity. For example, we might complete the choice of standard state by identifying it as the real pure liquid i at T and at its vapor pressure $P_i^s(T)$. This is a common choice. However, as an alternative, we might also choose the (hypothetical) pure ideal gas at T and P of the mixture; then the resulting activity would be closely related to the fugacity coefficient. Other choices are also possible, and some are convenient in certain situations.

Numerical values for the activity are always positive, and its value becomes unity only when the mixture fugacity f_i equals the value of the fugacity in the standard state. Since that standard state is a pure state, not an ideal-solution state, the activity does *not* measure deviations from ideal-solution behavior. Nevertheless, the activity proves useful in certain kinds of engineering calculations, which we shall explore in Part IV of this book.

5.4.2 Activity Coefficient

To have a useful ratio that measures how a real fugacity deviates from that in an ideal solution, we return to the definition of the fugacity (4.3.8), and we integrate that definition from an ideal-solution state to the mixture of interest. For the fugacity of i, the

* In the land of pure-component standard states, the Lewis-Randall rule (5.1.5) is but a district. The two differ in their standard-state pressures and phases. The Lewis-Randall standard-state pressure and phase are *always* those of the mixture, but in a generic pure-component standard state, the standard-state pressure and phase need not be the same as those of the mixture. In general, the choice for standard-state is dictated by the availability of a value for the pure-component fugacity: either from a reduction of experimental data, or from a correlation, or from an estimate. We caution that other authors may make other distinctions, and some may make no distinction between the Lewis-Randall rule and the pure-component standard state.

ideal solution is at the temperature and composition of the mixture, but it may be at any convenient pressure P_i^o. In such cases, the integrated definition of the fugacity (4.3.12) takes the form

$$\overline{G}_i(T, P, \{x\}) - \overline{G}_i^{is}(T, P_i^o, \{x\}) = RT \ln \frac{f_i(T, P, \{x\})}{f_i^{is}(T, P_i^o, \{x\})} \tag{5.4.4}$$

and the ratio on the rhs is defined to be the *activity coefficient* γ_i,

$$\gamma_i \equiv \frac{f_i(T, P, \{x\})}{f_i^{is}(T, P_i^o, \{x\})} \tag{5.4.5}$$

We caution that this definition of the activity coefficient* is incomplete because there is no unique ideal solution. Moreover, the pressure P_i^o is chosen for computational convenience; it may or may not be the same as the mixture pressure P. So the value of f_i^{is} in (5.4.5) cannot be determined until we identify our choice for the ideal solution. In the jargon of solution thermodynamics, a value of γ_i is meaningless unless we are also told the standard state to which it refers. We will use the notation

$$\gamma_i = \gamma_i(T, P, \{x\}; f_i^o(T, P_i^o)) \tag{5.4.6}$$

when it is important to emphasize that the value of γ_i depends on the standard-state fugacity f_i^o. Note that the standard-state pressures P_i^o can have different values for different components.

The activity coefficient is a dimensionless, conceptual, state function. Its value is always positive; however, it may be greater or less than unity. The ideal solution has all $\gamma_i = 1$, but the converse is not true: a mixture having all activity coefficients equal to unity may not be an ideal solution. Note that the definition (5.4.5) places no restriction on the kind of phase to which γ_i may be applied: γ_i is a legitimate property of gases, although it is used most often for liquids and solids.

In § 5.1 we observed that every ideal-solution fugacity (5.1.2) is linear in its mole fraction. We now write (5.1.2) in a more explicit form,

$$f_i^{is}(T, P_i^o, \{x\}) = x_i f_i^o(T, P_i^o) \tag{5.4.7}$$

So the definition of the activity coefficient (5.4.5) can be written

$$\gamma_i = \frac{f_i(T, P, \{x\})}{x_i f_i^o(T, P_i^o)} = \frac{a_i(T, P, \{x\}; f_i^o)}{x_i} \tag{5.4.8}$$

* The term "activity coefficient" was apparently first used by Savante Arrhenius in his doctoral dissertation (1884); the modern definition was given by A. A. Noyes and W. C. Bray in a paper published in 1911 [8].

This explicitly relates the activity coefficient to the standard-state fugacity and to the activity a_i defined in (5.4.2). Another important relation can be obtained by combining (5.4.4) and (5.4.5); this produces

$$\overline{G}_i(T, P, \{x\}) - \overline{G}_i^{is}(T, P_i^o, \{x\}) = RT \ln\gamma_i(T, P, \{x\}; f_i^o) \tag{5.4.9}$$

If we let each standard-state pressure P_i^o be that of the mixture ($P_i^o = P$), then the lhs becomes an excess chemical potential,

$$\overline{G}_i^E(T, P, \{x\}) = RT \ln\gamma_i(T, P, \{x\}; f_i^o) \tag{5.4.10}$$

This relates a difference measure to a ratio measure for deviations from ideal-solution behavior.

We emphasize that in writing (5.4.10) we have specified the temperature and pressure of the standard state, but we still have not made a unique choice for the standard state because we have not yet specified its phase. One common choice is the Lewis-Randall standard state, defined in (5.1.5), in which each standard-state fugacity is taken to be that for the pure component in the same phase and at the same temperature T and pressure P as the mixture,

$$f_i^o(T, (P_i^o = P)) = f_{\text{pure i}}(T, P) \qquad \textit{for all } i \tag{5.4.11}$$

Then, substituting this into (5.4.7), we obtain the ideal-solution fugacity, which is that of a Lewis-Randall ideal solution,

$$f_i^{is}(T, P, \{x\}) = x_i f_{\text{pure i}}(T, P) \tag{5.1.6}$$

Note that pure component i may or may not actually exist in the same phase as the mixture at T and P; if it cannot, then the standard state is said to be hypothetical. But whether the standard state is real or hypothetical is immaterial; what is important is that the state is well-defined and that we can assign a value to $f_i^o(T, P_i^o)$.

Using the Lewis-Randall rule (5.4.11) for the standard state fugacity in (5.4.5), the resulting expression for the activity coefficient γ_i approaches unity as the mixture is made more nearly pure in component i:

$$\lim_{x_i \to 1} \gamma_i = \lim_{x_i \to 1} \left[\frac{f_i(T, P, \{x\})}{x_i f_{\text{pure i}}(T, P)} \right] = 1 \tag{5.4.12}$$

If we consider the other extreme, in which the mixture is made infinitely dilute in component i, then (in the Lewis-Randall standard state) the activity coefficient γ_i approaches a finite value, called the *activity coefficient at infinite dilution*,

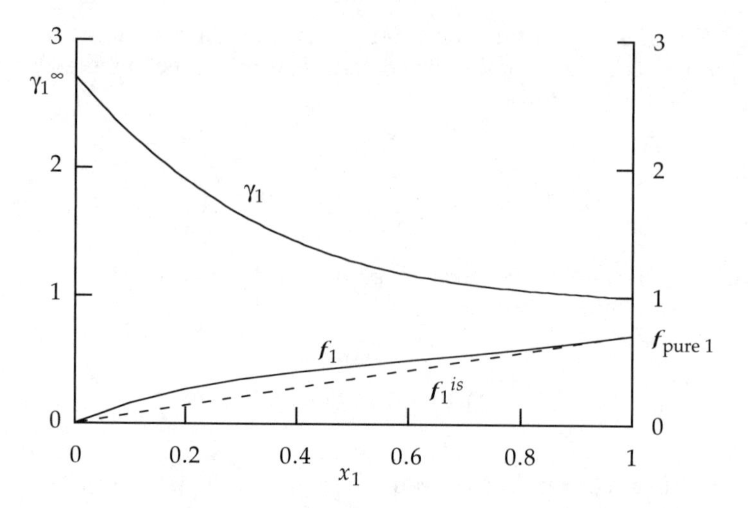

Figure 5.6 Schematic of the composition dependence of the fugacity f_1 and activity coefficient γ_1 in a binary mixture at fixed T and P. This activity coefficient is based on the Lewis-Randall standard state (5.4.11) and therefore satisfies the pure-fluid (5.4.12) and dilute-solution (5.4.13) limits. Note that the fugacity of the ideal-solution (broken line) is linear in the mole fraction and that, in the Lewis-Randall standard state, $f_1^o = f_{\text{pure 1}}$.

$$\lim_{x_i \to 0} \gamma_i(T, P, \{x\}) \equiv \gamma_i^\infty(T, P) \tag{5.4.13}$$

Schematic representations of a component's fugacity and its activity coefficient, relative to the Lewis-Randall standard state, are given in Figure 5.6 for the case in which $\gamma_i > 1$.

If we have the excess chemical potentials for all components in our mixture, then we can combine them via (3.4.4) to obtain the excess Gibbs energy. Further, if we use (5.4.10) to express the chemical potentials in terms of activity coefficients, then we can compute g^E from activity coefficients,

$$g^E(T, P, \{x\}) = \sum_i x_i \overline{G}_i^E(T, P, \{x\}) = RT \sum_i x_i \ln\gamma_i(T, P, \{x\}) \tag{5.4.14}$$

When a mixture has all $\gamma_i > 1$, then $g^E > 0$, and we say the mixture exhibits positive deviations from ideal-solution behavior. Inversely, if a mixture has all $\gamma_i < 1$, then $g^E < 0$, and we say the mixture exhibits negative deviations from ideal-solution behavior. In some mixtures, the intermolecular forces are more complicated, causing some components to have $\gamma_i < 1$ while others have $\gamma_i > 1$.

Activity coefficients can display wide variations in response to changes in composition. For example, consider the three binaries that can be extracted from a ternary mixture of acetone, chloroform, and methanol. Figure 5.7 shows the composition dependence of activity coefficients in those three binary mixtures. Since all these γ_i are in the Lewis-Randall standard state, each γ_i satisfies the pure-component limit given in (5.4.12). But, depending on the kinds of molecules present, γ_i may be greater than unity or less than unity; for example, the acetone-chloroform mixtures have $\gamma_i < 1$, but the mixtures containing the alcohol have $\gamma_i > 1$. Furthermore, the values of the γ_i in

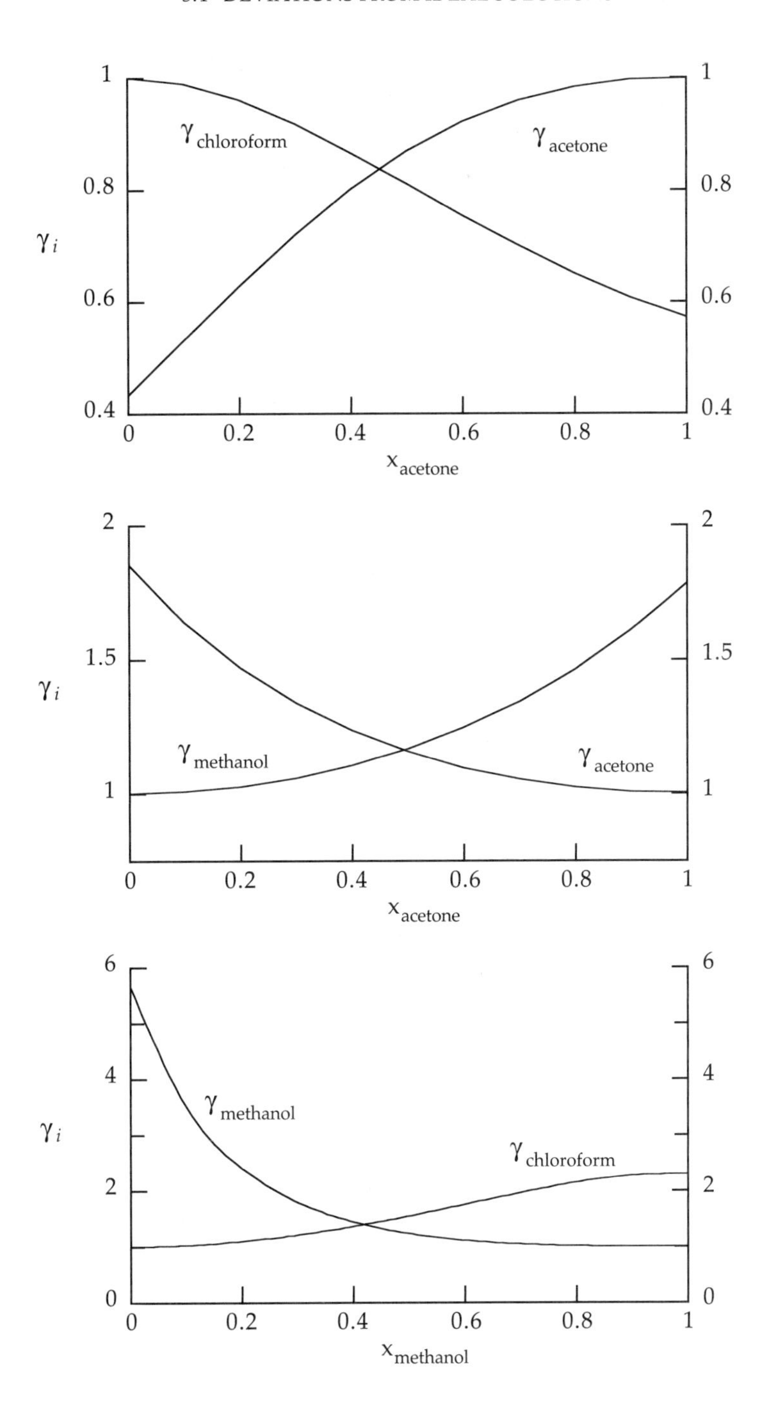

Figure 5.7 Activity coefficients for each component in three binary liquid mixtures, all at 60°C. *Top*: acetone-chloroform, *Middle*: acetone-methanol, *Bottom*: methanol-chloroform. Note the scale change from one ordinate to the next. These γ_i are based on the Lewis-Randall standard state and were computed using the Margules model, with parameters from Table E.2. Note in the top panel that $\gamma_i < 1$, while in the middle and bottom panels $\gamma_i > 1$. After a similar figure in Prausnitz et al. [2] and based on original data in Severns et al. [9].

one binary are not simply related to those in either other binary. This suggests that values of activity coefficients (and hence the kinds of deviations from ideality) are affected by forces acting between molecules of different species.

Substituting (5.4.10) into (5.2.13) gives the fundamental equation for excess properties in terms of activity coefficients,

$$d\left(\frac{G^E}{RT}\right) = \left(-\frac{H^E}{RT^2}\right)dT + \frac{V^E}{RT}dP + \sum_i \ln\gamma_i dN_i \tag{5.4.15}$$

This form can be used to obtain the response of an activity coefficient to a change in T or P. Since the lhs of (5.4.15) is an exact differential, the response of γ_i to an isothermal change in pressure is given by

$$\left(\frac{\partial \ln\gamma_i}{\partial P}\right)_{TN} = \frac{\overline{V}_i^E}{RT} \tag{5.4.16}$$

while the response to an isobaric change in temperature is

$$\left(\frac{\partial \ln\gamma_i}{\partial T}\right)_{PN} = -\frac{\overline{H}_i^E}{RT^2} \tag{5.4.17}$$

Further, in any given mixture the activity coefficients are not independent; rather, they are related through a Gibbs-Duhem equation. We may derive the equation either by letting $f = g^E/RT$ in the generic Gibbs-Duhem equation (3.4.10), or by substituting (5.4.10) into the Gibbs-Duhem equation (5.2.14) for g^E; by either procedure we obtain

$$\sum_i^C x_i d\ln\gamma_i = -\frac{h^E}{RT^2}dT + \frac{v^E}{RT}dP \tag{5.4.18}$$

For liquids h^E can be large, so the first term on the rhs is usually important, unless T is held fixed; however, liquids often have small values for v^E, so the second term is usually negligible. At fixed T and P (5.4.18) reduces to

$$\sum_i^C x_i d\ln\gamma_i = 0 \qquad \textit{fixed T and P} \tag{5.4.19}$$

This form of the Gibbs-Duhem equation can be used to show that, when the pure-component limit is taken, the isothermal-isobaric slope of ($\ln \gamma_i$) vs. x_i is zero,

$$\lim_{x_i \to 1}\left(\frac{\partial \ln\gamma_i}{\partial x_i}\right)_{TP} = 0 \tag{5.4.20}$$

Note that, because of the Gibbs-Duhem equation (5.4.19), if a binary mixture has γ_1 increase (decrease) with x_1, then γ_2 *must* simultaneously decrease (increase); this can be seen in any of the three panels appearing in Figure 5.7.

5.4.3 Example

How are activity coefficients related to the minimum work needed to separate a mixture into its pure components?

In § 3.7.4 we showed that, at fixed T and P, the minimum work needed to completely separate a mixture is given by the negative change in Gibbs energy on mixing,

$$w_{sh,\, rev} = -g^m(T, P, \{x\}) \tag{5.4.21}$$

To obtain an expression for the rhs in terms of activity coefficients, we choose the Lewis-Randall standard state, and then we use (5.2.7) to eliminate g^m in favor of g^E,

$$w_{sh,\, rev} = -g^E(T, P, \{x\}) - RT \sum_i x_i \ln x_i \tag{5.4.22}$$

Now we use (5.4.14) to introduce the activity coefficients; we are left with

$$w_{sh,\, rev} = -RT \sum_i x_i \ln(x_i \gamma_i) \tag{5.4.23}$$

This gives the minimum work needed to achieve an isothermal-isobaric separation. If the mixture were an ideal solution, then all the $\gamma_i = 1$, and we would have

$$w_{sh,\, rev} = -RT \sum_i x_i \ln(x_i) > 0 \qquad \textit{ideal solution} \tag{5.4.24}$$

This also applies to any ideal-gas mixture, which is merely a special kind of ideal solution; therefore, our result is consistent with the ideal-gas expression found in § 4.1.5.

For negative deviations from ideal-solution behavior, all $\gamma_i < 1$, and (5.4.23) gives $w_{sh,\, rev} > 0$. In such cases we must always do work on the mixture to accomplish the separation. Similarly, for small positive deviations, we have $(x_i \gamma_i) < 1$, even though $\gamma_i > 1$, so (5.4.23) still gives $w_{sh,\, rev} > 0$, and work must still be done to cause a separation. However, if the γ_i are positive and large enough, then (5.4.23) may yield $w_{sh,\, rev} < 0$. In these situations, the proposed one-phase mixture is usually unstable and it spontaneously splits into two phases (see Chapter 8). However, the new phases would not be pure components; instead, each phase would be a mixture having a composition that differs from the original proposed mixture. To determine the minimum work to complete the separation, we would need the composition of each phase and then we could apply (5.4.23) to each. The result would be that work would still have to be done on each phase to separate each into the pure components. Note that the physical interpretation of activity coefficients given by (5.4.23) applies only to γ_i in the Lewis-Randall standard state; activity coefficients wrt other standard states will have other interpretations.

5.5 ACTIVITY COEFFICIENTS FROM FUGACITY COEFFICIENTS

In § 5.3 we showed how excess properties, which are difference measures for deviations from *ideal-solution* behavior, can be obtained from residual properties, which are difference measures for deviations from *ideal-gas* behavior. In this section we establish a similar set of equations that relate activity coefficients to fugacity coefficients. As a result, the equations given here, together with those in § 5.3, establish a complete connection between the description of mixtures based on models for $PvTx$ equations of state and the description based on models for g^E and γ_i.

For any one component i in a mixture, the fugacity can be expressed in terms of the fugacity coefficient (4.3.18) or in terms of the activity coefficient (5.4.5). The value for the fugacity must, of course, be the same regardless of how it is obtained, so we can equate (4.3.18) with (5.4.5) and write

$$f_i = \gamma_i f_i^{is} = \varphi_i f_i^{ig} \tag{5.5.1}$$

Hence, the activity coefficient and the fugacity coefficient are related by

$$\gamma_i = \frac{\varphi_i f_i^{ig}}{x_i f_i^o} = \frac{\varphi_i x_i P}{x_i f_i^o} \tag{5.5.2}$$

or more formally,

$$\gamma_i(T, P, \{x\};\ f_i^o(T, P_i^o)) = \frac{\varphi_i(T, P, \{x\})\, P}{f_i^o(T, P_i^o)} \tag{5.5.3}$$

Here f_i^o is the standard-state fugacity at some convenient pressure P_i^o. The standard-state fugacity f_i^o deviates from its ideal-gas value by an amount that is measured by a standard-state fugacity coefficient φ_i^o, so (5.5.3) can also be written as

$$\gamma_i(T, P, \{x\};\ f_i^o(T, P_i^o)) = \frac{\varphi_i(T, P, \{x\})\, P}{\varphi_i^o(T, P_i^o)\, P_i^o} \tag{5.5.4}$$

Three commonly used options are available for dealing with the pressures appearing in (5.5.4); each choice leads to a particular kind of activity coefficient. In what follows, we distinguish among the three using subtle, but vital, differences in notation. In applications, the choice of which to use is based on practical considerations.

5.5.1 Use the Same Pressure for Standard State and Mixture, $P_i^o = P$

If we choose the standard-state pressure P_i^o to be the mixture pressure P, then we have the Lewis-Randall standard state (5.1.5), and (5.5.4) reduces to

$$\gamma_i(T, P, \{x\};\ f_i^o(T, P)) = \frac{\varphi_i(T, P, \{x\})}{\varphi_i^o(T, P)} \tag{5.5.5}$$

So, when γ_i and φ_i are both evaluated at the mixture T and P, the activity coefficient can be interpreted as a ratio measure for how the fugacity coefficient φ_i deviates from the standard-state fugacity coefficient φ_i^o. The result (5.5.5) directly connects ratio measures for deviations from the ideal gas to ratio measures for deviations from an ideal solution. Consequently, it provides a computational means for theories and equation-of-state models based on one kind of ideality (ideal gas) to be used in theories and models based on the other (ideal solution). The activity coefficient (5.5.5) is the one commonly encountered; it is simply related to the excess chemical potential,

$$RT\ \ln\gamma_i(T, P, \{x\};\ f_i^o(T, P)) = \overline{G}_i^E(T, P, \{x\}) \tag{5.4.10}$$

5.5.2 Use $P_i^o \neq P$ and Place Pressure Effect in Fugacity Coefficient

A second possibility occurs when we have, or can readily compute, φ_i not at the mixture pressure P but at some other pressure P_i^o. Then we correct φ_i for the pressure difference. The correction is computed from (4.3.23),

$$\left(\frac{\partial \ln\varphi_i}{\partial P}\right)_{Tx} = \frac{\overline{V}_i^{res}}{RT} = \frac{\overline{V}_i}{RT} - \frac{1}{P} \tag{5.5.6}$$

Separating variables and integrating along the mixture isotherm at fixed composition, we find

$$\varphi_i(T, P, \{x\}) = \varphi_i(T, P_i^o, \{x\})\exp\left[\frac{1}{RT}\int_{P_i^o}^{P} \overline{V}_i^{res}(T, \pi, \{x\})d\pi\right] \tag{5.5.7}$$

Integrating the ideal-gas term in (5.5.6) leaves

$$\varphi_i(T, P, \{x\}) = \varphi_i(T, P_i^o, \{x\})\frac{P_i^o}{P}\exp\left[\frac{1}{RT}\int_{P_i^o}^{P} \overline{V}_i(T, \pi, \{x\})d\pi\right] \tag{5.5.8}$$

Then, on combining (5.5.8) with (5.5.5), we obtain the following relation between the activity coefficient (at P) and the fugacity coefficient (at P_i^o),

$$\gamma_i(T, P, \{x\};\ f_i^o(T, P_i^o)) = \frac{\varphi_i(T, P_i^o, \{x\})}{\varphi_i^o(T, P_i^o)} \exp\left[\frac{1}{RT}\int_{P_i^o}^{P} \overline{V}_i(T, \pi, \{x\})d\pi\right] \tag{5.5.9}$$

The integral on the rhs is to be evaluated at fixed temperature and composition. The exponential term corrects φ_i from the standard pressure to the mixture pressure and is called the *Poynting factor*. Since (5.5.9) has been derived without assumptions from (5.5.5), the two equations are formally equivalent. That is, (5.5.9) offers no formal advantage over (5.5.5), because the values for φ_i, φ_i^o, and $\overline{V}_i$ used in (5.5.9) should be consistent with a particular *PvTx* equation of state. However in some situations, (5.5.9) may be more amenable to reliable approximation than (5.5.5). For example, if the integral in (5.5.9) is over only states of a condensed phase, then we might assume that $\overline{V}_i$ is a constant without seriously affecting the accuracy of the final value computed for γ_i.

The activity coefficient in (5.5.9) is related to chemical potentials by

$$RT\ln\gamma_i(T, P, \{x\};\ f_i^o(T, P_i^o)) = \overline{G}_i(T, P, \{x\}) - \overline{G}_i^{is}(T, P_i^o, \{x\}) \tag{5.4.9}$$

where the ideal solution is at the standard-state pressure. Note that the rhs is a difference between a real and an ideal-solution property, so it is similar to an excess property. But the rhs is not an excess property when $P_i^o \neq P$; cf. (5.2.1).

5.5.3 Use $P_i^o \neq P$ and Place Pressure Effect in Activity Coefficient

A third possibility is to compute the activity coefficient directly at the standard-state pressure P_i^o,

$$\gamma_i(T, P_i^o, \{x\};\ f_i^o(T, P_i^o)) = \frac{\varphi_i(T, P_i^o, \{x\})}{\varphi_i^o(T, P_i^o)} \tag{5.5.10}$$

Then we obtain the activity coefficient at the mixture pressure P by substituting (5.5.10) into (5.5.9),

$$\gamma_i(T, P, \{x\};\ f_i^o(T, P_i^o)) = \gamma_i(T, P_i^o, \{x\};\ f_i^o(T, P_i^o)) \times \exp\left[\frac{1}{RT}\int_{P_i^o}^{P} \overline{V}_i(T, \pi, \{x\})d\pi\right] \tag{5.5.11}$$

where $\overline{V}_i$ is the partial molar volume of component i in the real mixture and the integration is done along the mixture isotherm at constant composition. The form (5.5.11) for γ_i can also be derived starting from (5.4.16), which expresses the pressure derivative of γ_i in terms of the excess partial molar volume.

One advantage offered by (5.5.11) is that it collects all the pressure effects into a single term. As we shall see in § 5.6, many models for g^E contain no pressure dependence; hence, those models provide no pressure dependence for the activity coefficient, and such models are strictly valid only at the standard-state pressure P_i^o. To include pressure in those models, we could use (5.5.11), if we have a reliable estimate for the partial molar volume—say, from a $PvTx$ equation of state.

The activity coefficient (5.5.10) is related to chemical potentials by

$$RT\ln\gamma_i(T, P_i^o, \{x\};\ f_i^o(T, P_i^o)) = \bar{G}_i(T, P_i^o, \{x\}) - \bar{G}_i^{is}(T, P_i^o, \{x\}) \tag{5.5.12}$$

Both terms on the rhs are at the same pressure and so we could identify the rhs as an excess property. However, it is probably better not to do so because we could choose different standard-state pressures for different components $(P_1^o \neq P_2^o)$.

When the standard-state pressure is taken to be the mixture pressure $(P_i^o = P)$, then these distinctions disappear and the three activity coefficients (5.5.5), (5.5.9), and (5.5.11) are the same. But when $P_i^o \neq P$, the numerical values for these three activity coefficients can differ, though the differences are usually not significant at pressures below 10 bar. However, such differences can contribute to the complexity encountered when trying to use a model for activity coefficients as a basis for developing mixing rules for equations of state.

5.6 SIMPLE MODELS FOR NONIDEAL SOLUTIONS

Here we introduce models commonly used to represent the composition dependence of excess properties in liquid mixtures. Just as in § 4.5 for volumetric equations of state, the models considered here are semitheoretical: they may have some limited mathematical or physical basis, but they inevitably contain parameters whose values must be obtained from experimental data. The emphasis here is on the composition dependence of γ_i because, for condensed phases, composition is the most important variable; temperature is next in importance, and pressure is least important.

The strategy for devising models for activity coefficients is based on modeling g^E, rather than modeling the γ_i directly. With a functional form adopted for g^E, the corresponding expressions for the γ_i can be obtained by applying the partial molar derivative in (5.4.10). In addition, if the model parameters are known functions of T and P, then expressions for h^E and v^E can be obtained from (5.2.11) and (5.2.12). This would enable us to obtain the T and P effects on the γ_i from (5.4.16) and (5.4.17).

This indirect approach to modeling activity coefficients is used for at least two reasons: (a) When we model g^E and evaluate the γ_i from (5.4.10), then the γ_i automatically satisfy the Gibbs-Duhem equation (5.4.18). However, if we try to construct independent models for all the γ_i of a mixture, either the proposed equations for the γ_i may fail to satisfy the Gibbs-Duhem equation or else an apparently simple form for one activity coefficient, γ_1, may lead to a complicated form for another, γ_2. (b) For many mixtures, it is easier to develop accurate models for g^E than it is to directly develop accurate models for γ_i. Moreover, when the γ_i are obtained from a g^E model, the resulting expressions for the γ_i are often less complicated than forms devised by a direct modeling procedure. In this section we introduce two classes of models for g^E:

one based on Taylor's expansions (§ 5.6.1–5.6.4) and another with some basis in molecular theory (§ 5.6.5).

5.6.1 Series Representations for G^E

Consider a binary mixture containing components 1 and 2, and let us choose the Lewis-Randall standard state (5.1.5) to define an ideal solution. Our objective is to obtain a functional model to represent the composition dependence of the excess Gibbs energy. If we look back at Figures 5.2–5.5, we see that, for binary liquid mixtures at fixed T, g^E is nearly parabolic in x_1, even when h^E and s^E are not parabolic. This suggests a first approximation to $g^E(x_1)$,

$$\frac{g^E}{RT} = Ax_1x_2 \tag{5.6.1}$$

where the parameter A is dimensionless and independent of composition; it may depend on T and P, but for liquids the pressure dependence is usually ignored. Values for A are usually obtained from fits to experimental data. Because of the Gibbs-Duhem equation (5.2.14), (5.6.1) is the simplest expression we can use to represent the composition dependence of g^E for a binary mixture, provided we choose the Lewis-Randall standard state for both components.

Many ways can be proposed for correcting g^E for deviations from the simple quadratic behavior given in (5.6.1). One simple way is to expand g^E in a power series in the mole fraction of one component; e.g., for a binary at fixed T and P, we can write

$$\frac{g^E}{RTx_1x_2} = A' + B'x_1 + C'x_1^2 + \ldots \tag{5.6.2}$$

Since $x_1 = 1 - x_2$, we could have just as well expanded in x_2, but if we did then the values of the parameters (A', B', C', etc.) would change. That is, the coefficients in (5.6.2) depend on which species is labeled component 1 and which is component 2. This asymmetry in the labels can be reduced (but not eliminated) by using $(x_1 - x_2)$ as the independent variable; then we obtain the Redlich-Kister expansion [10]

$$\frac{g^E}{RTx_1x_2} = A + B(x_1 - x_2) + C(x_1 - x_2)^2 + \ldots \tag{5.6.3}$$

The parameters (A, B, C, etc.) are independent of composition; they do depend on T and P, though the P dependence is usually ignored for liquids. The Redlich-Kister expansion is fully equivalent to (5.6.2), but in (5.6.3) the magnitudes of the parameters (A, B, C, etc.) are unaffected when the component labels are interchanged; however, if the labels are interchanged, the signs of the coefficients on the odd-order terms (B, D, F, etc.) also change. At present, values for these parameters cannot be computed from some more fundamental theory; they can only be obtained from fits to experimental data.

As with any infinite series, the Redlich-Kister expansion can be used for calculations only after it has been truncated. Truncation at low order can account only for small deviations from a quadratic in x_1; for highly nonquadratic behavior, we must use a high-order expansion. However, high-order expansions are troublesome to use, not only because their algebraic forms are complicated, but also because the value for each parameter must be obtained from a fit to experimental data. These complications become problematic when the expansion is applied to mixtures containing more than two components, because ternary and higher-order coefficients appear. Each level of truncation produces a different form for the activity coefficients, but since this is an introductory discussion, we consider only the simple forms that result from truncations after the first and second terms.

5.6.2 Porter Equation

On truncating the Redlich-Kister expansion (5.6.3) after the first term, we are left with the parabolic form in (5.6.1). Traditionally, (5.6.1) has been called the two-suffix Margules equation [11], but this name can be ambiguous and so we prefer to call it *Porter's equation* [12]. Applying (5.4.10) to (5.6.1) shows that the activity coefficients are also quadratic in the mole fractions, as shown in Figure 5.8,

$$\ln\gamma_1 = Ax_2^2 \tag{5.6.4}$$

$$\ln\gamma_2 = Ax_1^2 \tag{5.6.5}$$

These activity coefficients are relative to the Lewis-Randall standard state (5.1.5); hence, they must satisfy the pure-component limit given in (5.4.12). That is, $\gamma_i \to 1$ as

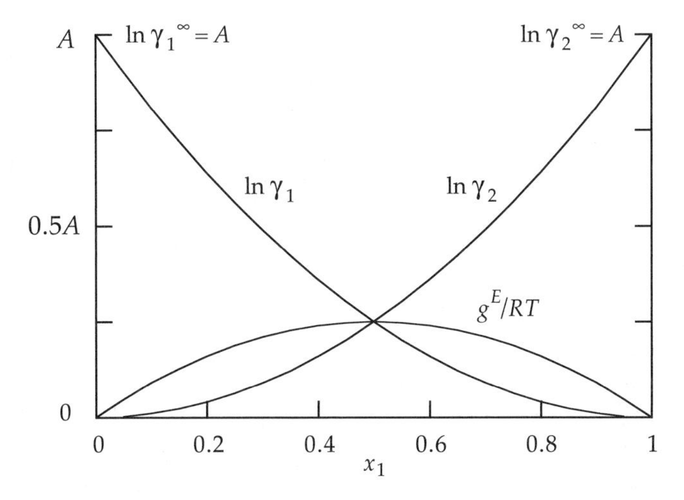

Figure 5.8 Substantial symmetry exists in the composition dependence of the excess Gibbs energy and activity coefficients for binary mixtures that obey the Porter equation (5.6.1).

$x_i \to 1$, where $i = 1$ or 2. In the dilute-solution limit (5.4.13), the activity coefficients in (5.6.4) and (5.6.5) are simply related to the parameter,

$$\ln\gamma_1^\infty = \ln\gamma_2^\infty = A \tag{5.6.6}$$

For some mixtures, values of γ_i^∞ can be extracted from experiment and in those cases we have a convenient means for determining a value for the parameter A. Because the Porter equation (5.6.1) contains only one parameter, a high degree of symmetry exists among the values of g^E, $\ln \gamma_1$, and $\ln \gamma_2$. For binary mixtures, the symmetry appears as in Figure 5.8.

The Porter equation is the simplest expression we can write for nonideal solutions; nevertheless, it can describe both positive ($A > 0$) and negative ($A < 0$) deviations from ideal-solution behavior. Some real mixtures obey the Porter equation fairly well, especially mixtures composed of molecules that are nonpolar and have similar sizes and shapes. Even some mixtures containing polar components may obey the Porter equation over limited ranges of temperatures. Values of A are given in Table E.1 (Appendix E) for some representative mixtures.

If the temperature and pressure dependence of the Porter parameter A is known (from experiment), then we can obtain values for h^E and v^E from (5.2.11) and (5.2.12). The results are

$$v^E = x_1 x_2 RT\left(\frac{\partial A}{\partial P}\right)_T \tag{5.6.7}$$

$$\frac{h^E}{RT^2} = -x_1 x_2\left(\frac{\partial A}{\partial T}\right)_P \tag{5.6.8}$$

Then s^E can be obtained from the Legendre transform for g^E (5.2.18),

$$\frac{s^E}{R} = -x_1 x_2\left[A + T\left(\frac{\partial A}{\partial T}\right)_P\right] \tag{5.6.9}$$

Mixtures that obey (5.6.1) and (5.6.7)–(5.6.9) are variously called simple mixtures, symmetric mixtures, or sometimes regular mixtures (but this last is a misnomer). We follow Rowlinson and Swinton [13] and call them *quadratic mixtures*, because for such mixtures all the excess properties are parabolic in a mole fraction x_1.

We caution that a binary mixture may obey the Porter equation (5.6.1) but still not be a quadratic mixture; that is, g^E may be parabolic in composition but h^E and s^E may not be. An example is the hexane-cyclohexane mixture shown in Figure 5.2. Such behavior occurs because asymmetries in h^E and s^E approximately cancel when they combine via the Legendre transform (5.2.18) to form g^E. Such cancellations are the norm rather than the exception. To say this another way, the Redlich-Kister expansion for g^E (5.6.3) is usually dominated by the first term, which is symmetric in x_1 and x_2. However, in the analogous expansions for h^E and s^E, asymmetric terms are frequently important.

5.6.3 Margules Equation

If we truncate the Redlich-Kister expansion (5.6.3) after the second term, we are left with

$$\frac{g^E}{RTx_1x_2} = A + B(x_1 - x_2) \tag{5.6.10}$$

This is not symmetric in x_1 and x_2; however, by multiplying A by $(x_1 + x_2 = 1)$ and re-collecting terms, we obtain the symmetric form

$$\frac{g^E}{RTx_1x_2} = A_1x_1 + A_2x_2 \tag{5.6.11}$$

where $A_1 \equiv A + B$ and $A_2 \equiv A - B$. Applying the partial molar derivative in (5.4.10) to (5.6.11) produces the corresponding expressions for the activity coefficients,

$$\ln\gamma_1 = x_2^2\,[A_2 + 2x_1(A_1 - A_2)] \tag{5.6.12}$$

?

$$\ln\gamma_2 = x_1^2\,[A_1 + 2x_2(A_2 - A_1)] \tag{5.6.13}$$

Historically, these have been called the 3-suffix Margules equations [11], but we will simply call them the *Margules equations*. The parameters A_1 and A_2 are independent of composition, but they generally depend on T and P. Usually, the effects of P are ignored and the effects of T are obtained experimentally. However, if data are lacking or if the changes in state condition are modest, then A_1 and A_2 are often assumed to be constants. When data are available, we often find that A_1 and A_2 vary as $1/T$.

The Margules expressions for activity coefficients are based on the Lewis-Randall standard state (5.1.5), and therefore they must obey the pure-component limit (5.4.12). In addition, as with Porter's equations, the parameters A_1 and A_2 are simply related to the activity coefficients at infinite dilution. In particular, when we apply the dilute-solution limit (5.4.13) to (5.6.12) and (5.6.13), we obtain

$$\ln\gamma_1^\infty = A_2 \tag{5.6.14}$$

$$\ln\gamma_2^\infty = A_1 \tag{5.6.15}$$

So if we have experimental data for both γ_1^∞ and γ_2^∞, then (5.6.14) and (5.6.15) provide a straightforward way to obtain values for the Margules parameters. If a binary mixture happens to have $\gamma_1^\infty = \gamma_2^\infty$, so that $A_1 = A_2$, then the Margules equations collapse to the Porter equations.

The Margules equations apply to many binary mixtures, including those that display positive deviations from ideality, mixtures that exhibit negative deviations from ideality, mixtures in which one activity coefficient is greater than unity and the other less than unity, and mixtures in which activity coefficients pass through extrema. Values of the Margules parameters for representative mixtures are given in Table E.2. However, the Margules equations can fail when strong specific interactions occur among only some of the constituent molecules. Examples include hydrogen bonding, dimerization in acids, and association in alcohols and aqueous solutions. In such cases, more complicated functional forms are needed to adequately represent the composition dependence of g^E.

5.6.4 Multicomponent Mixtures

The expressions in § 5.6.1–5.6.3 apply only to binary mixtures; however, the Redlich-Kister expansion can be extended to multicomponent solutions. One multicomponent version of the Redlich-Kister expansion is

$$\frac{g^E}{RT} = \frac{1}{2}\sum_{i \neq j}\sum \frac{g_{ij}}{RT} + \frac{1}{3}\sum_{i \neq j \neq k}\sum\sum \frac{g_{ijk}}{RT} + \ldots \tag{5.6.16}$$

where

$$\frac{g_{ij}}{RT} = x_i x_j [A_{ij} + B_{ij}(x_i - x_j) + C_{ij}(x_i - x_j)^2 + \ldots] \tag{5.6.17}$$

and

$$\frac{g_{ijk}}{RT} = x_i x_j x_k K_{ijk} + \ldots \tag{5.6.18}$$

In practice, the ternary and higher-order terms are usually ignored. For example, a ternary mixture might be modeled as

$$\frac{g^E}{RT} = \frac{g_{12}}{RT} + \frac{g_{13}}{RT} + \frac{g_{23}}{RT} \tag{5.6.19}$$

Although (5.6.19) does not explicitly contain high-order parameters that account for multibody interactions among molecules, such interactions are embedded implicitly in those parameters, such as B_{ij} and C_{ij}, that are multiplied by three or more mole fractions. By ignoring any explicit representation of multibody interactions, we obtain a computational advantage: the remaining parameters (A_{ij}, B_{ij}, etc.) have the same values for multicomponent mixtures as they do for binary mixtures of components i and j. Therefore, no data for multicomponent mixtures are needed to evaluate any parameter in (5.6.19). However, this computational advantage may result in a loss of accuracy when applied to some mixtures.

The multicomponent version of Porter's equation is equivalent to (5.6.19),

$$\frac{g^E}{RT} = \sum_{i<j}\sum x_i x_j A_{ij} \tag{5.6.20}$$

and applying (5.4.10) to (5.6.20) gives the activity coefficient of component k as

$$\ln\gamma_k = \sum_{i \neq k} x_i A_{ik} - \sum_{i<j}\sum x_i x_j A_{ij} \tag{5.6.21}$$

For binary mixtures, (5.6.20) reduces to (5.6.1) and (5.6.21) reduces to (5.6.4).

The multicomponent version of the Margules equation is

$$\frac{g^E}{RT} = \sum_{i<j}\sum x_i x_j [A_{ij} + B_{ij}(x_i - x_j)] \tag{5.6.22}$$

where $A_{ij} = A_{ji}$, but $B_{ij} = -B_{ji}$. Applying (5.4.10) to (5.6.22) gives the activity coefficient of component k as

$$\ln\gamma_k = \sum_{i \neq k} x_i [A_{ik} + B_{ik}(x_i - 2x_k)] - \sum_{i<j}\sum x_i x_j [A_{ij} + 2B_{ij}(x_i - x_j)] \tag{5.6.23}$$

?

For binary mixtures, (5.6.22) reduces to (5.6.11) and (5.6.23) reduces to (5.6.12).

5.6.5 Semitheoretical Models for G^E

For mixtures that do not obey the Porter or Margules equations, additional high-order terms must be kept in the Redlich-Kister expansion; hence, more parameters must be evaluated from experimental data. Alternatively, if we want to keep only two parameters, then we must abandon the Redlich-Kister expansion for some more complicated representation of g^E. Many functional forms have been proposed [1, 2], but here we restrict our attention to a useful expression proposed by Wilson in 1964 [14] and now identified as one of the class of "local-composition" models [2]. For binary mixtures Wilson's equation takes the form

$$\frac{g^E}{RT} = -x_1 \ln(x_1 + x_2 \Lambda_{12}) - x_2 \ln(x_2 + x_1 \Lambda_{21}) \tag{5.6.24}$$

where the parameters Λ_{12} and Λ_{21} depend on temperature. Values for these parameters are extracted from experiment.

Applying the partial molar derivative in (5.4.10) to (5.6.24) provides Wilson's expressions for the activity coefficients

$$\ln\gamma_1 = -\ln(x_1 + x_2 \Lambda_{12}) + x_2 \Omega \tag{5.6.25}$$

$$\ln\gamma_2 = -\ln(x_2 + x_1 \Lambda_{21}) - x_1 \Omega \tag{5.6.26}$$

?

where

$$\Omega = \frac{\Lambda_{12}}{x_1 + x_2\Lambda_{12}} - \frac{\Lambda_{21}}{x_2 + x_1\Lambda_{21}} \tag{5.6.27}$$

These activity coefficients are based on the Lewis-Randall standard state (5.1.5), and therefore they must obey the pure-component limits (5.4.12). In addition, the dilute-solution limit (5.4.13) provides relations between the activity coefficients and the parameters,

$$\ln\gamma_1^\infty = 1 - \ln\Lambda_{12} - \Lambda_{21} \tag{5.6.28}$$

$$\ln\gamma_2^\infty = 1 - \ln\Lambda_{21} - \Lambda_{12} \tag{5.6.29}$$

Unlike the corresponding expressions from the Porter and Margules equations, these nonlinear equations must be solved simultaneously by trial to obtain values for the parameters Λ_{12} and Λ_{21} from known values of γ_1^∞ and γ_2^∞. The logarithmic terms in (5.6.28) and (5.6.29) allow the Wilson equations to correlate large values of the γ_i^∞ using small values of the parameters Λ_{ij}. However we caution that, for some mixtures having both γ_1^∞ and γ_2^∞ less than unity, *three* sets of Λ-parameters can be found to satisfy (5.6.28) and (5.6.29) [15].

An ideal solution has $\Lambda_{12} = \Lambda_{21} = 1$, but the converse is not true: mixtures having $\Lambda_{12} = \Lambda_{21} = 1$ at one temperature are not necessarily ideal solutions. Further, (5.6.28) and (5.6.29) require the parameters to be positive. Nevertheless, Wilson's equations apply to both positive and negative deviations from ideal-solution behavior. In particular, Wilson's equations successfully correlate activity coefficients for highly nonideal solutions, including those, such as alcohol-hydrocarbon solutions, that involve hydrogen bonding and chemical association. However, for weakly nonideal solutions, the Wilson equation may offer no improvement over the Margules equation. Moreover, as will be discussed later, the mathematical form of Wilson's equation cannot describe mixtures that undergo liquid-liquid phase splits, despite its ability to correlate large values of infinite-dilution activity coefficients γ_i^∞.

The Redlich-Kister expansion for the excess Gibbs energy g^E provides no guidance about the temperature dependence of its parameters, and so temperature effects can only be obtained from experiment. In contrast, Wilson's equation is based on a theory that estimates the temperature dependence of the parameters,

$$\Lambda_{ij} = \frac{\rho_i}{\rho_j}\exp\left(\frac{-\Delta\lambda_{ij}}{RT}\right) \tag{5.6.30}$$

Here the ρ_i are pure component molar densities and the $\Delta\lambda_{ij}$ are parameters that depend on the identities of species i and j. The $\Delta\lambda_{ij}$ are often assumed to be independent of state condition; alternatively, they may be modeled as simple functions of T. But usually (5.6.30) allows the two temperature-dependent parameters Λ_{12} and Λ_{21} to be replaced by two temperature-independent parameters $\Delta\lambda_{12}$ and $\Delta\lambda_{21}$. Values of ρ_i and the $\Delta\lambda_{ij}$ for selected binary mixtures are given in Table E.3. A more extensive collection of values for the Wilson parameters can be found in the Dechema series [16].

Approximations for h^E and v^E can be obtained by applying (5.2.11) and (5.2.12) to the Wilson expressions (5.6.24) and (5.6.30). The result for the excess volume has never been used. For the excess enthalpy of a binary mixture we find

$$h^E = x_1 x_2[(\alpha_2 - \alpha_1)\Omega RT^2 + \Psi] \tag{5.6.31}$$

where the α_i are pure-component volume expansivities and Ψ is given by

$$\Psi = \frac{\Lambda_{12}\,\Delta\lambda_{12}}{x_1 + x_2\,\Lambda_{12}} + \frac{\Lambda_{21}\,\Delta\lambda_{21}}{x_2 + x_1\,\Lambda_{21}} \tag{5.6.32}$$

This Ψ-term is the dominant contribution to h^E. With g^E and h^E determined, s^E can be obtained from the Legendre transform (5.2.18). But we caution that, just as some mixtures may obey the Porter equation and yet not be quadratic mixtures, so too may some mixtures obey the Wilson equation (5.6.24) for g^E and yet not obey (5.6.31) for h^E. Consequently, while we might obtain values for the parameters $\Delta\lambda_{ij}$ by fitting calorimetric data, the resulting values may or may not reliably predict g^E.

The multicomponent version of Wilson's equation is

$$\frac{g^E}{RT} = -\sum_i x_i \ln\Gamma_i \tag{5.6.33}$$

where

$$\Gamma_i \equiv \sum_j x_j\,\Lambda_{ij} \tag{5.6.34}$$

As before, $\Lambda_{ij} \neq \Lambda_{ji}$, and $\Lambda_{ij} = 1$ when $i = j$. Applying (5.4.10) to (5.6.33) gives the activity coefficient for component k as

$$\ln\gamma_k = 1 - \ln\Gamma_k - \sum_i x_i \frac{\Lambda_{ik}}{\Gamma_i} \tag{5.6.35}$$

Note that the Λ_{ij} in (5.6.34) and the Λ_{ik} in (5.6.35) are all binary parameters; that is, their values are obtained from data for binary mixtures, and their temperature dependence is still usually assumed to be described by (5.6.30). Unlike the multicomponent versions of the Redlich-Kister expansion discussed in § 5.6.4, the theoretical basis for (5.6.33) suggests that high-order multibody parameters are not needed in Wilson's equation; in practice, this appears to be true for many mixtures.

5.7 SUMMARY

In this chapter we developed ways for computing values for conceptuals relative to their values for any well-defined ideal solution. The computational strategy is based on quantities that reveal how a property deviates from its ideal-solution value: the excess properties are difference measures, while the activity coefficient is a ratio measure. In other words, the strategy used in this chapter repeats that used in Chapter 4, with excess properties being analogous to residual properties and activity coefficients being analogous to fugacity coefficients. For example, to determine how a total property F changes from state 1 to state 2, we would use an excess property like this:

$$\Delta F_{12} = \Delta F_{12}^{E} + \Delta F_{12}^{is} \tag{5.7.1}$$

Likewise, to obtain a value for a fugacity, we would use an activity coefficient like this:

$$f_i = x_i \gamma_i f_i^o \tag{5.7.2}$$

Note that (5.7.1) is exactly analogous to (4.6.1) and that (5.7.2) is analogous to (4.6.2). Such analogies are explored more thoroughly in the next chapter; here we point out how the approach developed in Chapter 4 differs from that presented here.

First, we note that the ideal solution is a more general concept than the ideal gas. By an ideal solution we mean one in which the intermolecular forces are all the same, even though the molecules differ; this can be accomplished in many different ways. In contrast, by an ideal gas we mean a substance in which the intermolecular forces are all zero; this can be done in only one way. In other words, in any ideal solution each component fugacity is linear in its mole fraction, $f_i^{is} \propto x_i$, and many choices are available for the (composition-independent) proportionality constant. That constant is called the standard-state fugacity, f_i^o, and it is only when we choose the standard state that we identify a particular ideal solution. For an ideal gas the proportionality constant is the pressure; hence, an ideal-gas mixture is one kind of ideal solution.

It may seem that the residual properties offer additional flexibility because we defined two kinds—isobaric ones and isometric ones—while we introduced only isobaric excess properties. But this difference is mainly one of historical significance. The two kinds of residual properties allow us to perform calculations using both pressure-explicit and volume-explicit equations of state. In contrast, the excess properties were originally applied only to liquids, for which pressure and volume effects are often ignored. We could certainly define isometric excess properties, but in practical applications involving liquids, there seems to be little advantage to doing so. Differences between isometric and isobaric excess properties are discussed by Rowlinson and Swinton [13], but for condensed phases, those differences are usually small.

Since the ideal-solution concept is not restricted to a particular kind of intermolecular force, we have significant flexibility in performing thermodynamic analyses. In many situations, use of one kind of ideal solution may simplify an analysis more than another. For example, calculations are often easier when we use one ideality for nonelectrolyte solutions, another for dilute solutions, another for electrolytes, and yet another for polymeric blends. This degree of flexibility is not obtained by basing all analyses on ideal gases.

The ideal-gas and ideal-solution approaches also differ because they are based on different kinds of experimental data. The residual properties and fugacity coefficients depend on volumetric data: measurements of P, v, T, and $\{x\}$. But the excess properties and activity coefficients depend on density measurements for v^E, calorimetric measurements for h^E, and phase-equilibrium data for g^E and γ_i. Modern modeling tends to rely on volumetric data (equations of state), and a principal feature of this chapter has been to establish how excess properties can be computed from residual properties and how activity coefficients can be computed from fugacity coefficients. But note that such calculations can be performed in either direction; that is, at least in principle, residual properties can be computed from excess properties and fugacity coefficients from activity coefficients. In practice, these latter calculations can be performed only when we know (or can estimate) how parameters in models for the excess properties change with state condition. When this can be done, models for g^E might be used to formulate mixing rules for equations of state [17].

But aside from these practical considerations, another motivation underlies the development of ways for measuring deviations from ideal-solution behavior: the hope that macroscopic quantities can reveal differences in intermolecular forces. Relative differences in intermolecular forces can explain much of the interesting and unusual behavior observed in mixtures—oil and water do not mix because attractive forces between oil and water molecules are much weaker than those acting among just water molecules and among just oil molecules. These kinds of differences can be quantified using values for excess properties extracted from macroscopic experiments. Consequently, excess properties can not only serve as vehicles for computing conceptuals that may be needed in an engineering analysis, but in addition they may also serve to reveal microscopic differences that can explain macroscopic behavior.

LITERATURE CITED

[1] S. I. Sandler, H. Orbey, and B.-I. Lee in *Models for Thermodynamic and Phase Equilibria Calculations*, S. I. Sandler (ed.), Marcel Dekker, New York, 1993.

[2] J. M. Prausnitz, R. N. Lichtenthaler, and E. G. de Azevedo, *Molecular Thermodynamics of Fluid-Phase Equilibria*, 3rd ed., Prentice-Hall, Upper Saddle River, NJ,1999.

[3] G. N. Lewis and M. Randall, *Thermodynamics*, 2nd ed., revised by K. S. Pitzer and L. Brewer, McGraw-Hill, New York, 1961, ch. 18.

[4] R. Battino, "Volume Changes on Mixing for Binary Mixtures of Liquids," *Chem. Rev.*, **71**, 5 (1971).

[5] I. Prigogine and R. Defay, *Chemical Thermodynamics*, D. H. Everett (transl.), Longmans, Green and Co., London, 1954.

[6] M. M. Abbott, M. V. Ariyapadi, N. Balsara, S. Dasgupta, J. S. Furno, P. Futerko, D. P. Gapinski, T. A. Grocela, R. D. Kaminsky, S. G. Karlsruher, E. W. Kiewra, A. S. Narayan, K. K. Nass, J. P. O'Connell, C. J. Parks, D. F. Rogowski, G. S. Roth, M. B. Sarsfield, K. M. Smith, M. Sujanani, J. J. Tee, N. Tzouvaras, "A Field Guide to the Excess Functions," *Chem. Engr. Ed.*, **28**, 18 (1994).

[7] R. C. Pemberton and C. J. Mash, "Thermodynamic Properties of Aqueous Nonelectrolyte Mixtures. II. Vapour Pressures and Excess Gibbs Energies for Water + Ethanol at 303.15 to 363.15 K Determined by an Accurate Static Method," *J. Chem. Thermodynamics*, **10**, 867 (1978).

[8] J. W. Servos, *Physical Chemistry from Ostwald to Pauling*, Princeton University Press, Princeton, NJ, 1990.

[9] W. H. Severns, Jr., A. Sesonske, R. H. Perry, and R. L. Pigford, "Estimation of Ternary Vapor-Liquid Equilibrium," *A. I. Ch. E. J*, **1**, 401 (1955).

[10] O. Redlich and A. T. Kister, "Thermodynamics of Nonelectrolyte Solutions. Algebraic Representation of Thermodynamic Properties and the Classification of Solutions," *Ind. Eng. Chem.*, **40**, 345 (1948).

[11] M. Margules, "Über dir Zusammensetzung der gesättigten Dämpfe von Mischungen," *Sitzunsber Akad. Wiss. Wien.*, **104**, 1243 (1895).

[12] A. W. Porter, "The Vapour-Pressures of Mixtures," *Trans. Faraday Soc.*, **16**, 336 (1921).

[13] J. S. Rowlinson and F. L. Swinton, *Liquids and Liquid Mixtures*, 3rd ed., Butterworth, London, 1982.

[14] G. M. Wilson, "Vapor-Liquid Equilibrium. XI. A New Expression for the Excess Free Energy of Mixing," *J. Am. Chem. Soc.*, **86**, 127 (1964).

[15] K. Miyahara, H. Sadotomo, and K. Kitamura, "Evaluation of the Wilson Parameters by Nomographs," *J. Chem. Eng. Japan*, **3**, 157 (1970).

[16] J. Gmehling and U. Onken, *Vapor-Liquid Equilibrium Data Collection*, Chemistry Data Series (several volumes), DECHEMA, Frankfurt am Main, Germany, 1977.

[17] B. E. Poling, J. M. Prausnitz, and J. P. O'Connell, *The Properties of Gases and Liquids*, 5th ed., McGraw-Hill, New York, 2001.

[18] G. Scatchard and L. B. Ticknor, "Vapor-Liquid Equilibrium. IX. The Methanol-Carbon Tetrachloride-Benzene System," *J. Am. Chem. Soc.*, **74**, 3724 (1952).

PROBLEMS

5.1 Compute the minimum isothermal-isobaric work needed to separate an equimolar mixture of benzene and toluene into its pure components at 80°C and 1 bar.

(a) Assume the mixture is an ideal gas.

(b) Assume the mixture is an ideal solution.

5.2 Consider a multicomponent mixture that obeys $P = RT/(v-b)$ with

$$b = \sum_i x_i b_{\text{pure i}}$$

Show that such a mixture is an ideal solution.

5.3 Amagat's "law" approximates a mixture volume by mole-fraction averaging the pure-component volumes

$$v(T, P, \{x\}) = \sum_i x_i v_{\text{pure i}}(T, P)$$

Show that this leads to the ideal-solution expression: $f_i(T, P, \{x\}) = x_i f_{\text{pure i}}(T, P)$.

5.4 Determine the minimum work needed to remove one mole of solute from each of the following at 1 bar, 25°C: (a) 99 moles of solvent, (b) 999 moles of solvent (1 part per thousand), (c) 1 part per million, (d) 1 part per billion. Is "dilution the solution to pollution" if the solute must ultimately be recovered?

5.5 The behavior of the excess properties for ethanol-water mixtures, shown in Figure 5.4, suggests that modeling excess properties can be difficult.

(a) What is the simplest functional form that could reproduce the figure's response of g^E to changes in T and x_1?

(b) What would be the simplest expression for the excess heat capacity c_p^E that is consistent with the figure?

5.6 Scatchard and Ticknor [18] have reported experimental results for excess properties of methanol-benzene mixtures between 25 and 55°C. For equimolar mixtures their results for g^E can be represented by

$$\frac{g^E}{RT} = 0.25\left(-\frac{423.5}{T} - 2.475\ln T + 17.55\right) \qquad T \text{ in K} \qquad \text{(P5.6.1)}$$

(a) One hundred moles of each pure liquid are added to a double-walled vessel; each pure is initially at 30°C. If the mixing is to be done isothermally, should steam or cooling water be supplied to control the temperature?

(b) If instead of mixing isothermally, the mixing is done adiabatically, by insulating the vessel, estimate the final temperature of the mixture. Pure component heat capacities are $c_p/R \approx 9.94$ for methanol and $c_p/R \approx 16.36$ for benzene.

(c) Determine the value of the isobaric heat capacity for this mixture at 30°C.

5.7 At 50°C a binary liquid mixture has $g^E/RT = 0.5\,x_1x_2$ and $v^E = 4\,x_1x_2$ (cm^3/mol). For the mixture having $x_1 = 0.3$ at 50°C, by how much must the pressure change to cause the activity coefficient γ_1 to increase by 1%?

5.8 If ethyl ether(1) and ethanol(2) were mixed continuously at 2 bar, 310 K, would steam or cooling water be required to maintain the temperature constant at 310 K? Assume these mixtures obey the Margules equation (5.6.11) with

$$A_i = a_i + \frac{b_i}{T(K)} \qquad \text{(P5.8.1)}$$

and $a_1 = 0.1665$, $b_1 = 233.74$, $a_2 = 0.5908$, $b_2 = 197.55$.

5.9 Consider a binary mixture that has $s^E = 0$ and $h^E/RT = 0.6\,x_1x_2$, with the ideal solution relative to the Lewis-Randall standard state. Find the expression for the composition dependence of g^m, the change in Gibbs energy on mixing.

5.10 Derive (5.3.4), which relates excess properties to residual properties. To cover all possibilities you must do the derivation twice: (i) once for a first-law property (u or h) and (ii) again for a second-law property (s, g, or a).

5.11 Consider a binary mixture that obeys the Margules equations. What conditions, if any, must the parameters A_1 and A_2 obey if γ_1 vs. x_1 passes through an extremum at some composition $0 < x_1 < 1$?

5.12 A binary "Flory-Huggins" mixture (often a polymer mixture) has

$$\frac{h^E}{RT} = A\,\varphi_1\,\varphi_2 \tag{P5.12.1}$$

$$\frac{s^E}{R} = -\,x_1 \ln\left(\frac{\varphi_1}{x_1}\right) - x_2 \ln\left(\frac{\varphi_2}{x_2}\right) \tag{P5.12.2}$$

with the ideal solution defined relative to the Lewis-Randall standard state. Here A is a constant, x_i are mole fractions, φ_i are apparent volume fractions,

$$\varphi_i = \frac{x_i v_i}{x_1 v_1 + x_2 v_2} \tag{P5.12.3}$$

and v_i is the molar volume of pure component i. Using $A = 1$, plot the composition dependence of g^E (i.e., g^E vs. x_1) for the cases $v_2/v_1 = 1$, 10, and 100.

5.13 Consider a binary mixture that has

$$\ln\gamma_1 = a x_2^3 + b x_2^2$$

where a and b are constants at fixed T and P. Find the corresponding expression for the composition dependence of γ_2.

5.14 For a certain binary mixture at fixed T and P, Dr. Emmett Brown has proposed that the composition dependence of the component-1 fugacity be represented by

$$f_1 = x_1(2 - x_1) f_{\text{pure }1}$$

Do you find any problem with this proposal?

5.15 Consider a binary mixture at fixed T and P. The composition dependence of the fugacity of component 1 is given by

$$f_1 = x_1(\exp[A x_2^2])\, f_{\text{pure }1}$$

where parameter A is a constant, independent of T and P.

(a) Find the expression for the composition dependence of the fugacity of component 2.

(b) Using $f_{\text{pure }1} = 1$ bar, plot f_1 vs. x_1 for $A = 0$, 0.1, 0.5, 1, 2, and 3.

5.16 Inspired by the simplicity of the Porter equation, Tabitha the Untutored claims that it is easy to contrive models for the composition dependence of activity coefficients.

(a) To illustrate, Tabitha proposes that some binary mixtures must obey

$$\frac{g^E}{RT} = A\,x_1^2 x_2^2$$

where A is constant at fixed T and P. Find the resulting expression for the composition dependence of γ_1, where γ_1 is relative to the pure component standard state. Is there a problem when applying your result to real mixtures?

(b) Undaunted, Tabitha proposes another correlation for binaries,

$$\ln\gamma_1 = A\,x_2^4$$

where A is constant at fixed T and P. Find the corresponding expression for the composition dependence of γ_2. Test whether the expressions for γ_1 and γ_2 satisfy the Gibbs-Duhem equation. Is there any problem with trying to apply your result for γ_2 to real mixtures?

5.17 An article by certain alchemists in an obscure medieval journal reported infinite-dilution activity coefficients for binary mixtures of the rare substances jekyll-hyde(1) and neroburn(2). For 300 K $\le T \le$ 400 K, they gave

$$\ln\gamma_1^\infty = -2.49 + \frac{1500}{T(K)} \quad \text{and} \quad \ln\gamma_2^\infty = -1.47 + \frac{900}{T(K)}$$

Somewhat later, a rival group disputed this and claimed instead that

$$\ln\gamma_1^\infty = 31.08 - 5\ln T(K) \quad \text{and} \quad \ln\gamma_2^\infty = 18.67 - 3\ln T(K)$$

To resolve this discrepancy, you have done calorimetric experiments on these mixtures between 300 and 400 K. Your data can be correlated by

$$c_p^E/R = x_1 x_2 (5x_1 + 3x_2)$$

Which activity coefficients are consistent with the calorimetric data?

5.18 Consider a set of consistent data for activity coefficients of a binary mixture at low pressures. Show that, at constant temperature,

$$\int_0^1 \ln\left(\frac{\gamma_1}{\gamma_2}\right) dx_1 = 0$$

That is, the data should produce a plot of ($\ln\gamma_1 - \ln\gamma_2$) vs. x_1 such that the curve defines two regions of equal area and opposite signs.

5.19 Measurements of the nonidealities for a certain binary organic solution are claimed to be represented by

$$\left(\frac{\partial \ln \gamma_1}{\partial x_2}\right)_{TP} = a \ln T + \frac{b x_2}{T} + \frac{cT}{x_2}$$

where a, b, and c are constants. It is also claimed that $\ln \gamma_1^{\infty} = 1$ at 300 K.

(a) What must be the values of a and c to achieve thermodynamic consistency?

(b) Give expressions for $\ln \gamma_1$, $\ln \gamma_2$, and $\ln \gamma_2^{\infty}$ at 300 K.

(c) Give expressions for g^E and h^E.

5.20 It has been claimed that nonidealities in a certain binary mixture can be described by

$$\ln \gamma_1 = (a x_2^2)/T + bTx_2 + cT^2$$

(a) Find consistent expressions for $\ln \gamma_2$, g^E, and h^E.

(b) A calorimetric experiment on this mixture gave $h^E/RT = 1$ for the equimolar mixture at 300 K. Evaluate γ_1^{∞} and γ_2^{∞} at 325 K.

5.21 For binary mixtures at fixed P, determine the temperature dependence of $\ln \gamma_1$ when (a) g^E/RT is independent of T, (b) g^E is independent of T, (c) h^E is independent of T, (d) $h^E = 0$, (e) $s^E = 0$.

5.22 At 25°C a certain binary mixture has the following values for activity coefficients:

x_1	γ_1	γ_2
0.2	1.12	1.04
0.4	0.94	1.12
0.6	0.92	1.13
0.8	0.97	0.99

Determine whether Porter, Margules, or Wilson equations best represent these data and find the values of the parameters for your choice.

5.23 At 105°C mixtures of ethanol(1) and toluene(2) have activity coefficients at infinite dilution given approximately as $\gamma_1^{\infty} = 5.197$ and $\gamma_2^{\infty} = 4.811$. Compute and plot γ_1 vs. x_1 using (a) the Margules equations and (b) Wilson's equations.

5.24 Write out the complete equation representing g^E/RT for a ternary mixture modeled by the multicomponent Margules equation (5.6.22). Also write out the complete expression for $\ln \gamma_3$.

5.25 Tabitha the Untutored is working with some binary hydrocarbon mixtures that boil above 450 K. She reads in her thermo textbook that "Such substances form athermal mixtures; that is, no change in temperature occurs when the pure components are mixed adiabatically."

(a) What can you say about the signs and values of the excess properties g^E, h^E, and s^E for athermal mixtures?

(b) Reading further, Tabitha finds that, for athermal mixtures, a "good approximation" for g^E is

$$\frac{g^E}{RT} = x_1 \ln\left(\frac{\varphi_1}{x_1}\right) + x_2 \ln\left(\frac{\varphi_2}{x_2}\right) \tag{P5.25.1}$$

where φ_i represents the volume fraction defined in Problem 5.12. Comparing this with (5.4.14), Tabitha quickly concludes that $\gamma_i = \varphi_i/x_i$. But this expression gave poor results when compared to data taken in the company's laboratory. Tabitha argued that the data must be wrong and that the technicians should redo the experiments. Do you agree with her? (Amazingly, it never occurred to her that the authors of the textbook might be wrong!)

5.26 Activity coefficients of water (w) in solutions containing sugar (s) are often correlated by $\ln\gamma_w = \alpha(1 - x_w)^2$, where α is a constant. Write an expression for the composition dependence of g^E and $\ln\gamma_s$, taking into account that the solubility of sugar is limited to $0 < x_s < 0.25$.

5.27 Obtain expressions for the pressure dependence of the fugacity f_1 at fixed T, when the pressure dependence of the partial molar volume is given by each of the following: (a) $\bar{V}_1 = a$, (b) $\bar{V}_1 = bP$, (c) $\bar{V}_1 = c/P$. Here a, b, and c are constants, independent of state.

5.28 Each of the following applies to the fugacity for one component in a binary mixture. In each case, indicate how the quantities γ_1, f_1^o, and $\bar{V}_1$ have been treated if the expression was obtained from (i) Equation (5.5.5), (ii) Equation (5.5.9), and (iii) Equation (5.5.11):

(a) $f_1(T, P, x_1) = x_1 \exp[a + (b + cx_2^2)/T]$

(b) $f_1(T, P, x_1) = x_1 \exp[(a/T + bP)x_2^2 + (c + d/T)P + e + g/T + hT]$

(c) $f_1(T, P, x_1) = x_1 \exp[a + b/T + cx_1^2 x_2 + dx_1 x_2^2 + (eP)/T]$

The parameters a, b, … , h depend on substance but are independent of state.

5.29 Assume mixtures of methanol(1) and water(2) obey Wilson's equations with $\Delta\lambda_{12} = 0.347$ kJ/mol, $\Delta\lambda_{21} = 2.178$ kJ/mol, $\rho_1 = 24.55$ mol/liter, and $\rho_2 = 55.34$ mol/liter. If the temperature of an equimolar mixture is increased from 20°C to 30°C, by how much do both activity coefficients change?

6

RELATIONS AMONG RELATIONS

In previous chapters we have introduced many quantities, and we have developed many relations among those many quantities. We use this chapter to summarize the most important of those relations and to show you that we have consistently used a single approach in developing those relations. We start in § 6.1 by reminding you of the subtle distinctions between system states and constraints on interactions that may be in force when we change a state. Constraints are usually imposed in terms of measurables; for example, constant temperature or constant volume or no heat transfer. But such constraints can have profound effects on conceptuals and, in particular, on our choices for the most useful and economical expressions for relating measurables to conceptuals.

At this point we have developed two principal ways for relating conceptuals to measurables: one based on the ideal gas (Chapter 4) and the other based on the ideal solution (Chapter 5). Both routes use the same strategy—determine deviations from a well-defined ideality—with the deviations computed either as differences or as ratios. Since both routes are based on the same underlying strategy, a certain amount of symmetry pertains to the two; for example, the forms for the difference measures—the residual properties and excess properties—are functionally analogous.

We use § 6.2 to emphasize the symmetries that exist among difference measures and among ratio measures. Difference measures are commonly used to compute thermodynamic properties of single homogeneous phases, while ratio measures are most often used in phase and reaction equilibrium calculations. In § 6.3 we show that similarities among ratio measures extend to their physical interpretations. Then in § 6.4 we collect in one place the five most important ratio measures that are used to compute values for fugacities.

Finally in § 6.5, we illustrate that our two approaches—differences and ratios—are formally equivalent. Consequently, we can, in principle, use differences to compute ratios and vice versa. Whether this can be done in practice depends on the kinds and quantities of experimental data that are available. But in addition, such equivalences can be exploited in thermodynamic modeling, for example, by using g^E models (difference measures) to obtain mixing rules in $PvTx$ equations of state. The resulting $PvTx$ equations would then be used to compute fugacity coefficients (ratio measures).

6.1 EFFECTS OF EXTERNAL CONSTRAINTS ON SYSTEM STATES

In previous chapters we have tried to convince you that if we have a complete equation of state for a one-phase substance, then we can compute values for *all* thermodynamic properties. Up to now, much of our attention has focused on volumetric equations of state, $P(T, v, \{x\})$ or $v(T, P, \{x\})$, because these equations contain only measurables. But those forms are not the only possibilities. For example, our fairy godmother might present us with a complete functional form for the Helmholtz energy

$$A = A(T, V, \{N\}) \tag{6.1.1}$$

or for the Gibbs energy,

$$G = G(T, P, \{N\}) \tag{6.1.2}$$

From either of these we could use relations presented in earlier chapters to obtain all remaining thermodynamic properties.

To determine the number of independent properties required to completely define an equation of state, we use the procedure introduced in § 3.1. There we made a distinction between $\mathcal{V}$, the number of orthogonal interactions available to manipulate a state, and $\mathcal{F}_{ex}$, the number of independent properties needed to identify a state. We also noted that $\mathcal{V}$ is affected by any external constraints imposed on interactions, but that $\mathcal{F}_{ex}$ is not. We elaborate on this distinction here.

Consider two systems, 1 and 2. System 1 is a one-phase mixture of C components, with mole numbers $\{N\}$. This mixture fills a rigid vessel of volume V_1, and the vessel is immersed in a heat bath maintained at temperature T_1. System 2 is another sample of the same mixture, having the same C components and the same mole numbers $\{N\}$. System 2 fills the cylinder of a piston-cylinder apparatus. The cylinder is immersed in a heat bath at T_2. A constant external pressure is imposed on the mixture; at equilibrium the system pressure P_2 balances that external pressure. Therefore, system 2 is at constant pressure, while system 1 is at constant volume.

We adjust the two heat baths so the two temperatures are the same,

$$T_1 = T_2 \tag{6.1.3}$$

and we adjust the external pressure on system 2 so the two volumes are the same,

$$P_2 = P(T_1, V_1, \{N\}) \tag{6.1.4}$$

To identify each state (with $S = 0$), we must by (3.1.7) specify values for $(C + 2)$ properties. We have met this requirement: T_1, V_1, and C mole numbers. Moreover, the two sets of values are identical. Hence, the two states are identical, and consequently, *all* thermodynamic properties are *exactly* the same in the two systems, even though the external constraints differ. (This assumes relations among properties are monotonic; if they are not, we can still adjust T, V, and $\{N\}$ so that the two states are identical.)

But while the two equilibrium states are the same, we may feel that some things about these two situations differ. We bring two things to your attention here. One difference is the identity of the natural variables for describing a state. In system 1 with V

fixed, a natural choice is an expression for the Helmholtz energy (6.1.1), while for system 2 with P fixed, a natural choice would be an expression for the Gibbs energy (6.1.2). "Natural" here means economical in terms of computations needed for an analyis. For example, the entropies are the same in the two situations, but in system 1 the entropy S is economically posed in terms of A, while in system 2 it is better posed in terms of G:

$$S_1 = -\left(\frac{\partial A}{\partial T}\right)_{VN} = -\left(\frac{\partial G}{\partial T}\right)_{PN} = S_2 \tag{6.1.5}$$

Likewise, all chemical potentials are the same,

$$\overline{G}_{i(1)} = \left(\frac{\partial A}{\partial N_i}\right)_{TVN_{j\neq i}} = \left(\frac{\partial G}{\partial N_i}\right)_{TPN_{j\neq i}} = \overline{G}_{i(2)} \qquad \textit{for all } i \tag{6.1.6}$$

This distinction in the choice of appropriate dependent variables will influence our development of the criteria for equilibrium, which appears in Chapter 7.

A second difference is in how the two systems respond to internal fluctuations or to externally imposed disturbances. Such responses are quantified by the thermodynamic response functions and, again, the natural choices for these two systems differ. For example, the first-order response to a change in temperature is given by (6.1.5), but the second-order response is given by a heat capacity: the response for system 1 is given by C_v, while that for system 2 is given by C_p. These two heat capacities differ:

$$C_v = -T\left(\frac{\partial^2 A}{\partial T^2}\right)_{VN} \neq -T\left(\frac{\partial^2 G}{\partial T^2}\right)_{PN} = C_p \tag{6.1.7}$$

This means, for example, that if we increase the temperature of both heat baths by 5°C, the new equilibrium states reached by the two systems will differ. Other inequalities, similar to (6.1.7), occur between other high-order derivatives of A and G, leading to differences between other response functions.

More generally, external constraints affect many aspects of thermodynamic theory and practice. In experiments, certain constraints make particular response functions much easier to measure than others. In statistical mechanics, theoretical descriptions of natural fluctuations are determined by the external constraints imposed on systems. In thermodynamic modeling, external constraints guide us toward those properties that offer the most economical routes to complete descriptions of states. Similarly, in thermodynamic analysis, constraints help us separate dependent variables from independent ones and help us choose those independent variables that are most likely to simplify the analysis. If we merely wanted to develop a thermodynamic description of equilibrium systems, we could ignore external constraints, but since we want to change system states and perform engineering analyses that reveal the consequences of such changes, we must learn to recognize external constraints and account for the limitations they may impose on system behavior and performance. A more complete discussion of relations between thermodynamic properties and external constraints can be found in [1].

6.2 SYMMETRY IN ROUTES TO CONCEPTUALS

We will need values of conceptuals for two classes of problems: (a) calculation of thermodynamic properties for one-phase systems and (b) calculation of multiphase and chemical reaction equilibria. For both kinds of problems, we use the same basic strategy: (i) Compare raw or modeled experimental data with computed properties of an ideal substance to obtain measures for deviations from the ideality, then (ii) exploit the deviation measures to obtain expressions for the required conceptuals in terms of measurables. Calculations of one-phase properties are typically based on differences, while phase and reaction equilibrium calculations typically use ratios. In § 6.2.1 and 6.2.2 we focus on difference measures, while in § 6.2.3 and 6.2.4 we consider ratio measures.

6.2.1 Generalized Difference Measure

In Chapters 4 and 5 we used the same basic strategy for obtaining changes in conceptuals for homogeneous single-phase systems. In both chapters we used a difference f_d to compare a real property value f to that of some ideal substance f^{id}; extending this approach in a completely general way, we define a generalized difference measure by

$$f_d(T, P, \{x\}; T^{id}, P^{id}, \{x^{id}\}) = f(T, P, \{x\}) - f^{id}(T^{id}, P^{id}, \{x^{id}\}) \tag{6.2.1}$$

The ideal substance may be real or hypothetical, so long as its state $(T^{id}, P^{id}, \{x^{id}\})$ is well-defined. To make the difference measure f_d useful, the ideal property value f^{id} must be readily obtained, either from experiment, theory, or correlation. Note that in this most general form, the ideal state $(T^{id}, P^{id}, \{x^{id}\})$ need not be the same as the state of the real substance $(T, P, \{x\})$.

Since the properties f and f^{id} are state functions and the definition (6.2.1) is a linear combination of state functions, the difference f_d is also a state function. This means f_d forms exact differentials, so (6.2.1) can be written as

$$f_d = \int_{f^{id}}^{f} df \tag{6.2.2}$$

In other words, the concept of an ideal substance can be interpreted mathematically as the lower limit of an integration. If the ideal substance is chosen to be the ideal gas at the same state as the real substance ($T^{id} = T$, $P^{id} = P$, $\{x^{id}\} = \{x\}$), then the differences f_d are the residual properties of Chapter 4. Alternatively, if the ideal substance is taken to be the Lewis-Randall ideal solution at $(T, P, \{x\})$, then the differences f_d are the excess properties of Chapter 5. These two possibilities are compared in Table 6.1.

To emphasize that the definition (6.2.1) is completely general and that the ideal substance is at the discretion of the user, we introduce a third class of difference measures, distinct from the residual properties and excess properties. These new differences may be called *generalized changes of properties on mixing* [2-4]; they are defined by the intensive form

Generalized Changes of Properties on Mixing

$$f^M(T, P, \{x\}; T^{id}, P^{id}) = f(T, P, \{x\}) - \sum_i x_i f_{\text{pure i}}(T_i^o, P_i^o) \tag{6.2.3}$$

where the sum runs over all components. In this form, the ideal "substance" is a set of pure components, each in a standard state (T_i^o, P_i^o); these standard states need not be the same as the mixture state (T, P), nor need they be the same for all pures.

The generalized difference measure (6.2.1) provides options for computing changes in conceptuals: it is merely a matter of computational convenience whether we use residual properties, excess properties, or changes of properties on mixing. To illustrate, consider a change from state 1 to state 2. For such a process, we could obtain the change in any conceptual Δf_{12} using residual properties,

$$\Delta f_{12} = \Delta f_{12}^{res} + \Delta f_{12}^{ig} \tag{6.2.4}$$

In such cases, values for the residual properties would be obtained from integrals over appropriate functions of a $PvTx$ equation of state, as discussed in § 4.4.

Or we could obtain Δf_{12} using excess properties,

$$\Delta f_{12} = \Delta f_{12}^E + \Delta f_{12}^{is} \tag{6.2.5}$$

In this approach, values for the excess properties would most likely be obtained from models for g^E, although $PvTx$ equations of state could also be used.

Or we could obtain Δf_{12} using changes of properties on mixing,

$$\Delta f_{12} = \Delta f_{12}^M + \sum_i [x_{i2} f_{\text{pure i}}(T_{i2}^o, P_{i2}^o) - x_{i1} f_{\text{pure i}}(T_{i1}^o, P_{i1}^o)] \tag{6.2.6}$$

In these situations values for the properties f^M would be obtained by integrating appropriate functions of a $PvTx$ equation of state. However, this approach is little used nowadays; instead, when a reliable equation of state is available for a substance, the residual properties are usually used to obtain Δf_{12}. In any case, the strategy based on difference measures is a completely general one that can be implemented in various ways to help reduce the computational burden of an analysis.

Table 6.1 Routes to properties devised in Chapters 4 and 5 use the same strategy: compute deviations from a well-defined ideality

	Deviations from ideality	
Ideality	**Differences**	**Ratios**
Ideal gas	Residual properties	Fugacity coefficient
	$f^{res} = f - f^{ig}$	$\varphi_i = f_i/f_i^{ig}$
Ideal solution	Excess properties	Activity coefficient
	$f^E = f - f^{is}$	$\gamma_i = f_i/f_i^{is}$

6.2.2 Symmetry in Use of Difference Measures

Since the residual properties and excess properties are merely two particular members of the general class of differences defined in (6.2.1), we might expect that the functional forms for relations among excess properties bear similarities to the forms for relations among residual properties. Indeed, many such similarities exist, and in fact the similarities extend beyond functional relations to encompass the entire strategy used in relating conceptuals to experimentally accessible quantities.

That basic strategy is illustrated in Table 6.1. First we define an ideal mixture whose properties we can readily determine. Then for real mixtures we compute deviations from the ideality as either difference measures or ratio measures. In one route the ideality is the ideal-gas mixture, the difference measures are residual properties, and the ratio measure is the fugacity coefficient. In the other route the ideality is the ideal solution, the difference measures are excess properties, and the ratio measure is the activity coefficient.

Figure 6.1 summarizes the strategy we follow to obtain forms for computing property changes of one-phase systems. In route 1A, the required experimental data

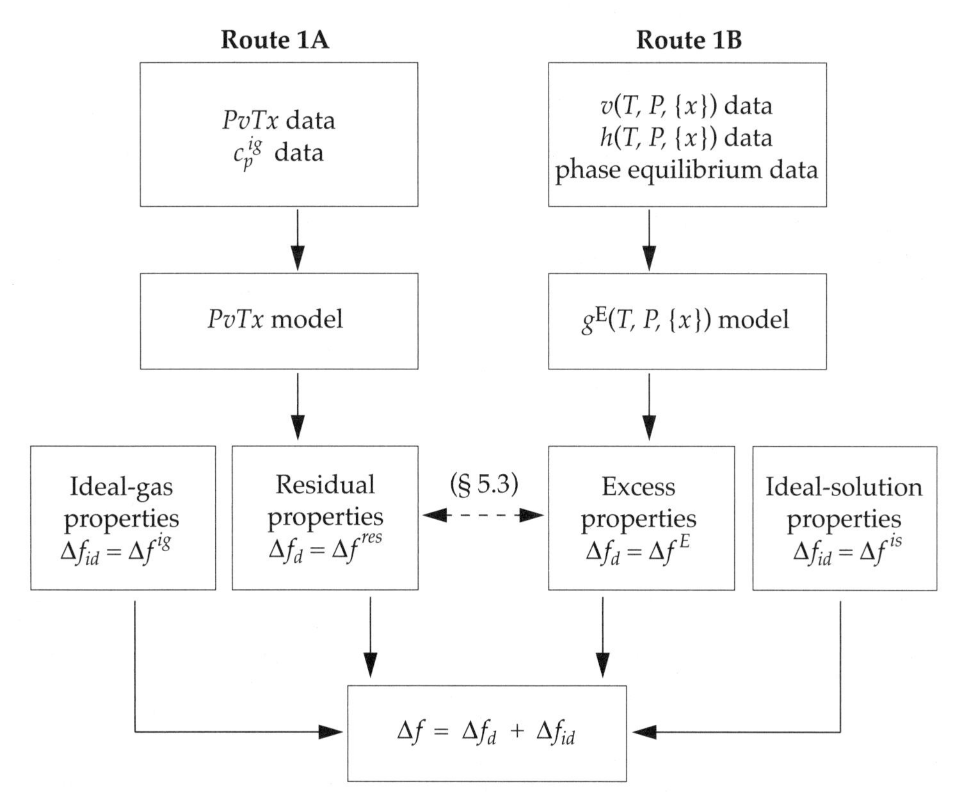

Figure 6.1 To obtain changes in properties of one-phase mixtures, our basic strategy is to compute deviations relative to some ideality. In route 1A (*left*) the ideality is the ideal gas and the deviations are the residual properties. In route 1B (*right*) the ideality is an ideal solution and the deviations are the excess properties. In addition, we could use the relations in § 5.3 to compute residual properties from excess properties and vice versa.

include heat capacity values for the pure-component ideal gases plus volumetric data for the mixture. Those data are correlated as a model $PvTx$ equation of state, and that model together with the relations in § 4.4 provide values for residual properties. Then we combine those residual properties with ideal-gas properties to obtain differences in properties for the substance of interest.

In route 1B, also shown in Figure 6.1, the required experimental data include mixture volumes, enthalpies, and some amount of phase-equilibrium data. From those data, values for excess properties are extracted and fit to a model for g^E. However, before excess properties can be found, we must define the ideal solution; that is, we must choose the standard state for each component. With the excess-property model plus values for ideal-solution properties, we can then compute property differences for the substance of interest.

Traditionally, route 1A was used only for gases and route 1B was used only for liquids. Route 1B is still rarely used for gases because it requires much more experimental data than route 1A; however, when route 1B is applied to liquids and pressure effects can be ignored, then the amount of data required is tolerable. Conversely, route 1A was, in the past, little used for liquids, because older $PvTx$ equations of state were not sufficiently reliable when applied to liquids. Modern volumetric equations of state often overcome this deficiency, so route 1A is now a viable method for liquids, as well as gases. Therefore, the relation between residual properties and excess properties given by

$$f^{res}(T, P, \{x\}) = f^E(T, P, \{x\}) + \sum_i x_i f^{res}_{\text{pure i}}(T, P) \tag{6.2.7}$$

might be used to obtain excess properties if mixture residual properties are known, and conversely. The determination of excess properties from residual properties was developed in § 5.3 and is indicated by the horizontal line in Figure 6.1.

6.2.3 Generalized Ratio Measure

In Chapters 4 and 5 we developed two versions of the same basic strategy for obtaining fugacities: we defined a ratio that compares the real-substance fugacity f_i to that of the substance in some reference state $f_i^{®}$. We generalize this approach by defining a generalized ratio measure, the *generalized activity*,

$$\mathbb{a}_i(T, P, \{x\}; T^{®}, P^{®}, \{x^{®}\}) \equiv \frac{f_i(T, P, \{x\})}{f_i^{®}(T^{®}, P^{®}, \{x^{®}\})} \tag{6.2.8}$$

Just as for the generalized difference (6.2.1), the reference state used in (6.2.8) may be real or hypothetical; it need not be the same as the real state. On taking the logarithm of (6.2.8), we obtain a difference,

$$\ln \mathbb{a}_i(T, P, \{x\}; T^{®}, P^{®}, \{x^{®}\}) = \ln f_i(T, P, \{x\}) - \ln f_i^{®}(T^{®}, P^{®}, \{x^{®}\}) \tag{6.2.9}$$

Moreover, as in (6.2.2), we can interpret the rhs of (6.2.9) as the result of an integration,

$$\ln a_i(T, P, \{x\}; T^{\circledR}, P^{\circledR}, \{x^{\circledR}\}) = \int_{\ln f_i^{\circledR}}^{\ln f_i} d\ln f_i \tag{6.2.10}$$

so we may view the reference state as the lower limit of an integration.

Unfortunately, the general activity a_i is not simply related to a difference in chemical potentials, because the definition of the fugacity (4.3.8) requires that the real and reference states be at the same temperature. Fortunately, we lose almost nothing in computational convenience by taking the reference state to be at the same temperature as the mixture of interest; then, the activity can be written as

$$RT \ln a_i(T, P, \{x\}; T, P^{\circledR}, \{x^{\circledR}\}) = \bar{G}_i(T, P, \{x\}) - \bar{G}_i^{\circledR}(T, P^{\circledR}, \{x^{\circledR}\}) \tag{6.2.11}$$

6.2.4 Symmetry in Use of Ratio Measures

In the above expressions, the reference is chosen by the user; the choice is based on computational convenience. For example, if we choose the reference to be the pure component at the temperature T and pressure P of the mixture, then (6.2.8) becomes the usual activity, and the difference in (6.2.11) becomes the change of chemical potential on mixing,

$$\bar{G}_i^{m} = RT \ln a_i = RT \ln \frac{f_i(T, P, \{x\})}{f_{\text{pure } i}(T, P)} \tag{6.2.12}$$

Alternatively, if the reference is taken to be the ideal gas at the same state as the mixture, then (6.2.8) becomes the fugacity coefficient, and the difference in (6.2.11) is the residual chemical potential. Then, instead of (6.2.12), we would have

$$\bar{G}_i^{res} = RT \ln \varphi_i = RT \ln \frac{f_i(T, P, \{x\})}{f_i^{ig}(T, P, \{x\})} \tag{6.2.13}$$

Further, if the reference is taken to be a Lewis-Randall ideal solution, then the ratio in (6.2.8) is the activity coefficient, while the difference in (6.2.11) becomes the excess chemical potential. Then, instead of (6.2.12) or (6.2.13), we would have

$$\bar{G}_i^{E} = RT \ln \gamma_i = RT \ln \frac{f_i(T, P, \{x\})}{f_i^{is}(T, P, \{x\})} \tag{6.2.14}$$

Table 6.2 The fugacity, fugacity coefficient, activity, and activity coefficient are equivalent representations of the chemical potential; those equivalences extend to their pressure and temperature derivatives.

Property	Pressure effect		Temperature effect	
Chemical potential	$\left[\frac{\partial}{\partial P}\left(\frac{\overline{G}_i}{RT}\right)\right]_{Tx} = \frac{\overline{V}_i}{RT}$	(3.4.15)	$\left[\frac{\partial}{\partial T}\left(\frac{\overline{G}_i}{RT}\right)\right]_{Px} = \frac{-\overline{H}_i}{RT^2}$	(3.4.17)
Fugacity	$\left(\frac{\partial \ln f_i}{\partial P}\right)_{Tx} = \frac{\overline{V}_i}{RT}$	(4.3.13)	$\left(\frac{\partial \ln f_i}{\partial T}\right)_{Px} = \frac{-\overline{H}_i^{res}}{RT^2}$	(4.3.14)
Fugacity coefficient	$\left(\frac{\partial \ln \varphi_i}{\partial P}\right)_{Tx} = \frac{\overline{V}_i^{res}}{RT}$	(4.3.23)	$\left(\frac{\partial \ln \varphi_i}{\partial T}\right)_{Px} = \frac{-\overline{H}_i^{res}}{RT^2}$	(4.3.24)
Activity	$\left(\frac{\partial \ln \mathscr{a}_i}{\partial P}\right)_{Tx} = \frac{\overline{V}_i^{m}}{RT}$	(6.2.15)	$\left(\frac{\partial \ln \mathscr{a}_i}{\partial T}\right)_{Px} = \frac{-\overline{H}_i^{m}}{RT^2}$	(6.2.16)
Activity coefficient	$\left(\frac{\partial \ln \gamma_i}{\partial P}\right)_{Tx} = \frac{\overline{V}_i^{E}}{RT}$	(5.4.16)	$\left(\frac{\partial \ln \gamma_i}{\partial T}\right)_{Px} = \frac{-\overline{H}_i^{E}}{RT^2}$	(5.4.17)

These are the common choices for the reference, but they are not the only possibilities: in special situations other choices may be more useful. For example, for electrolyte solutions it often proves convenient to use, as the reference, a hypothetical mixture at some composition other than the composition of interest.

The relations (6.2.12)–(6.2.14) show that the activity, fugacity coefficient, and activity coefficient are all particular forms of the generalized activity, just as various chemical potentials in (6.2.12)–(6.2.14) are all particular forms of the generalized difference in chemical potentials (6.2.1). In addition, the structural analogies suggested by (6.2.12)–(6.2.14) extend to various derivatives, some of which are summarized in Table 6.2. For example, when pressure changes, each of these quantities responds according to some form of the partial molar volume. Note that $(\overline{G}_i/RT)$ and $(\ln f_i)$ have the same response to changes in pressure. Likewise, when temperature changes, each quantity responds according to some form of the Gibbs-Helmholtz equation, which involves a partial molar enthalpy. Note that $(\ln f_i)$ and $(\ln \varphi_i)$ have the same response to changes in temperature.

The derivatives of φ_i and γ_i in Table 6.2 indicate how the strength of a nonideality responds to changes of state. For example, when a nonideal gas has all $\overline{V}_i^{res} > 0$, then each φ_i increases with isothermal increases in pressure. So if the mixture also has all $\varphi_i > 1$, then the gas becomes more nonideal as pressure increases; this is the common behavior. However, if all $\varphi_i < 1$, then the nonideality weakens with increasing pressure. Similarly, if a nonideal solution has all $\overline{H}_i^E < 0$, then each γ_i increases with isobaric increases in temperature. So, if the mixture is a positive deviant (it has all $\gamma_i > 1$),

Table 6.3 The chemical potential, fugacity, fugacity coefficient, activity, and activity coefficient are all constrained by a form of the Gibbs-Duhem equation.

Property	Identity of f in (3.4.10)	Form of Gibbs-Duhem equation	
Chemical potential	g	$\sum_i x_i\, d\bar{G}_i = -s\,dT + v\,dP$	(3.4.19)
Fugacity	$\dfrac{g}{RT}$	$\sum_i x_i\, d\ln f_i = -\dfrac{h}{RT^2}dT + \dfrac{v}{RT}dP$	(4.3.15)
Fugacity coefficient	$\dfrac{g^{res}}{RT}$	$\sum_i x_i\, d\ln \varphi_i = -\dfrac{h^{res}}{RT^2}dT + \dfrac{v^{res}}{RT}dP$	(4.3.25)
Activity	$\dfrac{g^m}{RT}$	$\sum_i x_i\, d\ln a_i = -\dfrac{h^m}{RT^2}dT + \dfrac{v^m}{RT}dP$	(6.2.17)
Activity coefficient	$\dfrac{g^E}{RT}$	$\sum_i x_i\, d\ln \gamma_i = -\dfrac{h^E}{RT^2}dT + \dfrac{v^E}{RT}dP$	(5.4.18)

then the mixture becomes more nonideal as the temperature is increased. But if the mixture is a negative deviant (all $\gamma_i < 1$), then the nonideality weakens with increasing temperature.

In addition to the similarities among derivatives shown in Table 6.2, each form of the chemical potential is constrained by a Gibbs-Duhem equation, as shown in Table 6.3. For isothermal-isobaric changes in composition, the rhs of each equation in Table 6.3 vanishes; for example, (3.4.19) becomes

$$\sum_i x_i\, d\bar{G}_i = 0 \qquad \textit{fixed T and P} \tag{3.4.20}$$

Further, for changes of state at constant composition, the forms of the Gibbs-Duhem equation in Table 6.3 can be related to derivatives in Table 6.2. For example, for a change in pressure at constant temperature and constant composition, (4.3.13) combines with (4.3.15) to yield

$$\sum_i x_i \left(\frac{\partial \ln f_i}{\partial P}\right)_{Tx} = \sum_i x_i \frac{\bar{V}_i}{RT} = \frac{v}{RT} \tag{6.2.18}$$

Similar relations can be obtained from other quantities appearing in Tables 6.2 and 6.3.

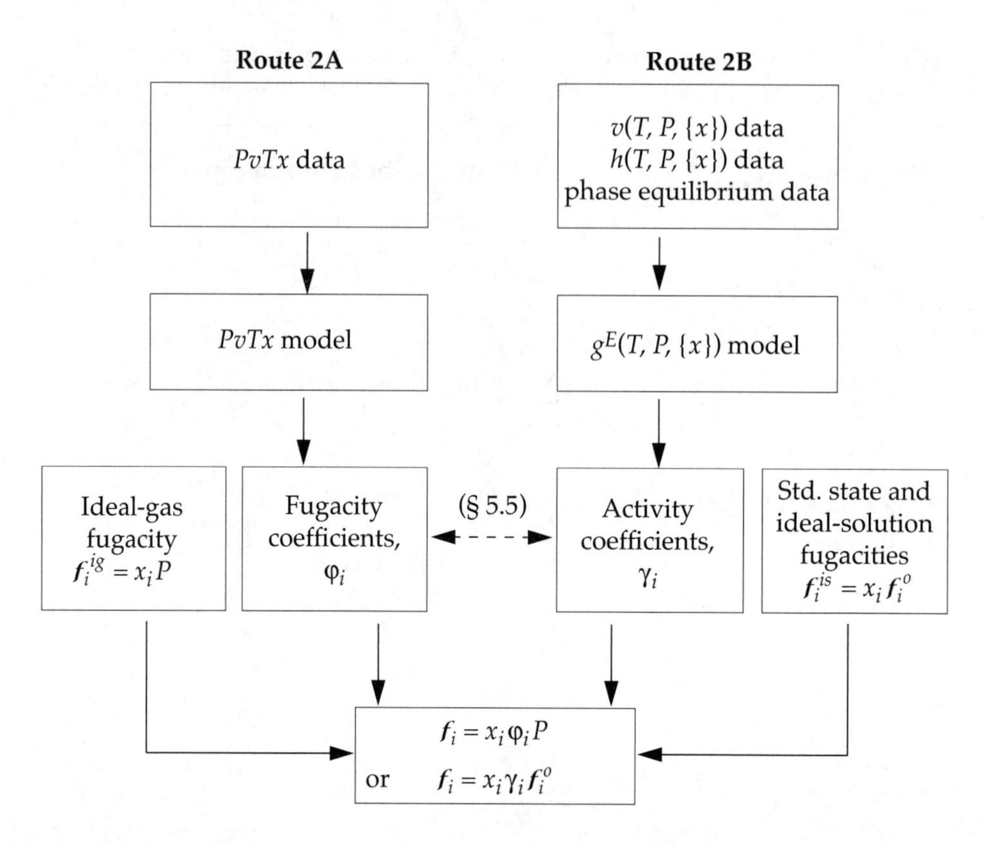

Figure 6.2 Schematic illustration of the strategies used to obtain computational forms for fugacities, which are needed for phase- and reaction-equilibrium calculations. Traditionally, route 2A has been mostly used for gases, while route 2B was confined to condensed phases. However, these uses were dictated, not by thermodynamic limitations, but by limitations of the models used to correlate the data.

To obtain values for fugacities, we must make contact with experimental data, usually through one of the two routes summarized in Figure 6.2. In the figure, route 2A combines experimental *PvTx* data with properties of ideal gases to form the fugacity coefficient. Then the definition of the fugacity coefficient can be used to extract values for the fugacity. Alternatively, fugacities can be obtained by following route 2B in Figure 6.2. Then the required experimental data are mixture volumes, enthalpies, and limited phase-equilibrium data that produce excess properties. Those data, fit to a g^E model and combined with computed properties of ideal solutions, yield activity coefficients. These activity coefficients can then be used to obtain values for fugacities.

Figure 6.2 suggests that route 2B requires considerably more experimental effort than route 2A, because route 2B requires data from volumetric, calorimetric, and phase-equilibrium measurements. But for condensed phases, pressure effects can often be ignored, and then the experimental effort demanded by route 2B may not be excessive. Traditionally route 2A was applied to gases and route 2B was reserved for condensed phases, but now we may be able to use *PvTx* equations of state to determine activity coefficients and fugacities of liquids. Inversely, we may be able to use a g^E model as the basis for devising mixing rules for *PvTx* models, as discussed in § 5.5 and § 6.5. These additional routes are indicated by the horizontal line in Figure 6.2.

6.3 PHYSICAL INTERPRETATIONS OF SELECTED CONCEPTUALS

The similarities in the expressions for the fugacity ratios shown in (6.2.12)–(6.2.14) extend to their physical interpretations; in this section we show that each ratio (hence, each difference in chemical potentials) can be simply interpreted as the reversible work involved in a certain well-defined process: the activity is related to the reversible work involved in adding more of one component to a mixture (§ 6.3.1); the residual Gibbs energy is related to the reversible work involved in changing an ideal gas into a real gas (§ 6.3.2); the excess Gibbs energy is related to the reversible work involved in converting an ideal solution into a real mixture (§ 6.3.3). We also show that the corresponding differences in partial molar entropies can be interpreted as reversible heat effects.

6.3.1 Adding More of One Component to a Mixture

First we consider the reversible addition of a small amount of pure component i to a mixture at fixed temperature and pressure. This process has already been discussed in § 3.7.3; there we showed that, for each mole of substance added, the reversible shaft work is given by

$$w_{sh,\,rev} = \bar{G}_i(T, P, \{x\}) - g_{\text{pure i}}(T, P) \tag{6.3.1}$$

Using (6.2.12) for the rhs, we find

$$w_{sh,\,rev} = RT\ \ln\frac{f_i(T, P, \{x\})}{f_{\text{pure i}}(T, P)} = RT\ \ln \mathbf{a}_i \tag{6.3.2}$$

So, when we choose the pure-substance reference state to be at the same temperature and pressure as the mixture, then the activity of component i is simply related to the reversible isothermal-isobaric work involved in adding a small amount of pure i to the mixture. This provides a physical interpretation for the activity.

For this process, the heat effect is given by the entropy balance (3.6.10), now written for one inlet and no outlets. Since the process is isothermal, the system (T), inlet (T_α), and external boundary (T_{ext}) all have the same temperature, $T = T_\alpha = T_{ext}$, so (3.6.10) becomes

$$q_{rev} = T[\bar{S}_i(T, P, \{x\}) - s_{\text{pure i}}(T, P)] = T\bar{S}_i^{m}(T, P, \{x\}) \tag{6.3.3}$$

For an isothermal-isobaric addition of a small amount of component i to a mixture, the reversible heat effect is given by the change in partial molar entropy on mixing.

In the special case of ideal gases, (6.3.2) reduces to

$$w_{sh,\,rev}^{ig} = RT\ln x_i < 0 \tag{6.3.4}$$

while (6.3.3) reduces to

$$q_{rev}^{ig} = -RT\ln x_i > 0 \tag{6.3.5}$$

Similarly, for Lewis-rule ideal solutions, (6.3.2) again reduces to

$$w_{sh,\,rev}^{is} = RT\ln x_i < 0 \tag{6.3.6}$$

and (6.3.3) reduces to

$$q_{rev}^{is} = -RT\ln x_i > 0 \tag{6.3.7}$$

For both idealities, when material is added, the volume must expand to keep P constant and we must add heat to keep T constant. If we mole-fraction average the work given in (6.3.1), we obtain the change in Gibbs energy on mixing, g^m, which is the reversible isothermal-isobaric work involved in forming a mixture from its pure components; cf. § 3.7.4 in which we consider the reverse process.

The above analysis applies for moving one component from a pure state into a mixture, but we can generalize to moving a component from one mixture to another. In those cases, the reversible shaft work is given by

$$w_{sh,\,rev} = \bar{G}_i(T, P, \{x\}) - \bar{G}_i(T, P, \{x_o\}) = RT\ln\frac{f_i(T, P, \{x\})}{f_i(T, P, \{x_o\})} \tag{6.3.8}$$

If the component fugacity in the original mixture $\{x_o\}$ exceeds that in the target mixture, then the shaft work is negative and the system can be used to do work. However, if the fugacity in the target is larger, then work must be done to force the process to proceed in the desired direction. For this reason, energy must be supplied to concentrate such mixtures as toxic wastes and sewage.

6.3.2 Changing an Ideal Gas into Real Stuff

To obtain a physical interpretation for the residual Gibbs energy, we start with an ideal-gas mixture confined to a closed vessel. As the process, we consider the reversible isothermal-isobaric conversion of the ideal-gas molecules into real ones. Although this process is hypothetical, it is a mathematically well-defined operation in statistical mechanics; the process amounts to a "turning on" of intermolecular forces. We first want to obtain an expression for the work, but since the process involves a change in molecular identities, we must start with the general energy balance (3.6.3). For a system with no inlets and no outlets, (3.6.3) becomes

$$\delta W_{b,\,rev} + \delta W_{sh,\,rev} = d\Big(\sum_i N_i\bar{U}_i\Big) - \delta Q_{rev} \tag{6.3.9}$$

where the sum runs over all components i. Recall W_b is the work involved in deforming the boundary, while W_{sh} is the shaft work (i.e., non-boundary work) associated with the process. Similarly, the general entropy balance (3.6.8) is written for a closed system as

$$\delta Q_{rev} = T_{ext} d\left(\sum_i N_i \bar{S}_i\right) \tag{6.3.10}$$

For an isothermal process, $T = T_{ext}$; then combining (6.3.9) with (6.3.10) to eliminate the heat, and recalling that $A = U - TS$, we find

$$\delta W_{b,\,rev} + \delta W_{sh,\,rev} = d\left(\sum_i N_i \bar{A}_i\right) = d(Na) \tag{6.3.11}$$

where a is the molar Helmholtz energy of the system. The reversible boundary work is given by

$$\delta W_{b,\,rev} = -PdV = -NPdv \tag{6.3.12}$$

We combine (6.3.11) with (6.3.12), recall that the process is isobaric and constant mass, and write

$$\delta W_{sh,\,rev} = Nd(a + Pv) = Ndg \tag{6.3.13}$$

Finally, we integrate (6.3.13) from the ideal-gas state to that of the real stuff,

$$w_{sh,\,rev} = g(T, P, \{x\}) - g^{ig}(T, P, \{x\}) \tag{6.3.14}$$

$$= g^{res}(T, P, x) = RT \sum_i x_i \ln\varphi_i \tag{6.3.15}$$

This shows that the residual Gibbs energy can be interpreted physically as the reversible isothermal-isobaric shaft work involved in "turning on" intermolecular forces, thereby converting ideal-gas molecules into real molecules. In general this work may be positive or negative. For a single component (6.3.15) reduces to

$$w_{sh,\,rev} = RT \ln\varphi_{\text{pure i}} \tag{6.3.16}$$

which is the reversible isothermal-isobaric work involved in transforming one mole of pure ideal gas into a real substance.

If we integrate the differential boundary work in (6.3.12) over the change from ideal gas to real substance, we obtain

$$w_{b,\,rev} = -Pv^{res} \tag{6.3.17}$$

So the residual volume is proportional to the reversible isothermal-isobaric boundary work associated with converting ideal gas into real substance.

The corresponding heat effect associated with the process is obtained from the entropy balance (6.3.10); in particular, for an isothermal constant-mass process, (6.3.10) becomes

$$\delta Q_{rev} = Td\left(\sum_i N_i \bar{S}_i\right) = TNds \tag{6.3.18}$$

And integrating over the process, as we did to obtain (6.3.15), we find

$$q_{rev} = Ts^{res}(T, P, \{x\}) \tag{6.3.19}$$

So we can interpret the residual entropy as proportional to the reversible isothermal-isobaric heat involved in converting ideal-gas molecules into real molecules. With the results from (6.3.15), (6.3.17), and (6.3.19) we can show that process satisfies the first law, $\Delta u = u^{res} = w_b + w_{sh} + q$.

6.3.3 Changing an Ideal Solution into Real Stuff

To obtain a physical interpretation for the excess Gibbs energy, we consider a Lewis-Randall ideal solution confined to a closed vessel, and we determine the reversible isothermal-isobaric work involved in converting the ideal solution into a real mixture. Again this is a hypothetical process: all intermolecular forces are initially the same (but they are nonzero), and the process changes the forces into those of real molecules.

The development of the expression for the reversible work is exactly that already done in § 6.3.2, and the result is functionally the same as (6.3.13),

$$\delta W_{sh,\, rev} = N dg \tag{6.3.20}$$

Integrating this from the ideal-solution state to the real state of interest, we obtain

$$w_{sh,\, rev} = g(T, P, \{x\}) - g^{is}(T, P, \{x\}) \tag{6.3.21}$$

$$= g^E(T, P, x) = RT \sum_i x_i \ln\gamma_i \tag{6.3.22}$$

This provides a physical interpretation for the excess Gibbs energy. Note that the work computed from (6.3.22) may be positive or negative, depending on whether the real mixture exhibits positive or negative deviations from ideality.

The heat associated with the process is also obtained in a manner that parallels that in § 6.3.2; the result is proportional to the excess entropy,

$$q_{rev} = Ts^E(T, P, \{x\}) \tag{6.3.23}$$

The direction of the heat transfer may be into or out of the system. Finally, analogous to (6.3.17), the boundary work for this process is

$$w_{b,\, rev} = -Pv^E \tag{6.3.24}$$

This indicates that the reversible boundary work for the process is proportional to the excess volume. Equations (6.3.22)–(6.3.24) satisfy the first law, $\Delta u = u^E = w_b + w_{sh} + q$.

6.4 FIVE FAMOUS FUGACITY FORMULAE*

In § 6.2.4 we showed the similarities that occur in the fugacity ratios that define the fugacity coefficient, the activity, and the activity coefficient, and in § 6.3 those quantities were given physical interpretations. In this section we summarize certain generalized expressions that relate the fugacity to measurables. Many such relations can be written, but only five forms are in common use.

Fugacity Formula #1. If along an isotherm T, we have a complete Pvx equation of state for our mixture, then we can compute f_i from the definition of the fugacity coefficient (4.3.18). Here we write that definition in the form

FFF #1 $$f_i(T, P, \{x\}) = x_i P \varphi_i(T, P, \{x\}) \tag{6.4.1}$$

If the equation of state is volume-explicit, then

$$\ln \varphi_i(T, P, \{x\}) = \int_0^P \left[\frac{\pi \bar{V}_i}{RT} - 1\right] \frac{d\pi}{\pi} \tag{4.4.10}$$

while if it is pressure-explicit, then

$$\ln \varphi_i(T, P, \{x\}) = \int_V^\infty \left[\frac{\Psi}{RT}\left(\frac{\partial P}{\partial N_i}\right)_{TVN_{j \neq i}} - 1\right] \frac{d\Psi}{\Psi} - \ln Z \tag{4.4.23}$$

Fugacity Formula #2. If we have, from experiment or correlation, the value of a standard-state fugacity f_i^o at the mixture temperature and pressure, so we can use the Lewis-Randall rule (5.1.5), then we recast FFF #1 into an alternative form. First multiply and divide (6.4.1) by the known standard-state fugacity f_i^o,

$$f_i(T, P, \{x\}) = x_i P \frac{\varphi_i(T, P, \{x\})}{f_i^o(T, P)} f_i^o(T, P) \tag{6.4.2}$$

Now replace the denominator with

$$f_i^o(T, P) = P \varphi_i^o(T, P) \tag{6.4.3}$$

and use (5.5.5) to identify the ratio $\varphi_i(T, P, \{x\})/\varphi_i^o(T, P)$ as the activity coefficient; then (6.4.2) becomes

FFF #2 $$f_i(T, P, \{x\}) = x_i \gamma_i(T, P, \{x\}; f_i^o(T, P)) f_i^o(T, P) \tag{6.4.4}$$

This activity coefficient is simply related to the excess chemical potential (5.4.10).

* Professor M. M. Abbott originated this name for the following useful forms for fugacity [5].

Fugacity Formula #3. If we have, from experiment or correlation, the value for a standard-state fugacity f_i^o at the mixture temperature but at some standard-state pressure P_i^o that differs from the mixture pressure P, then we need a Poynting factor to correct the standard-state fugacity from P_i^o to P. The correction is given in Table 6.2:

$$\left(\frac{\partial \ln f_i^o}{\partial P}\right)_{Tx} = \frac{\overline{V}_i^o}{RT} \tag{6.4.5}$$

Separating variables and integrating along the mixture isotherm, we find

$$f_i^o(T, P) = f_i^o(T, P_i^o) \exp\left[\int_{P_i^o}^{P} \frac{\overline{V}_i^o(T, \pi)}{RT} d\pi\right] \tag{6.4.6}$$

where $\overline{V}_i^o$ is the partial molar volume of component i in its standard state. Putting (6.4.6) into FFF #2 leaves

$$\textbf{FFF \#3} \quad f_i(T, P, \{x\}) = x_i \gamma_i(T, P, \{x\}; f_i^o(T, P)) f_i^o(T, P_i^o) \exp\left[\int_{P_i^o}^{P} \frac{\overline{V}_i^o(T, \pi)}{RT} d\pi\right] \tag{6.4.7}$$

Note that the activity coefficient in (6.4.7) is exactly the activity coefficient that appears in FFF #2 of (6.4.4); it is only the expression for the standard-state fugacity in (6.4.6) that has changed in writing (6.4.7).

Fugacity Formula #4. Consider again the situation in which FFF #3 applies: we have a value of a standard-state fugacity f_i^o at the mixture temperature and at some pressure P_i^o other than the mixture pressure P. But rather than divide the pressure effects between two terms, we might want to combine them into a single term. This situation occurs, for example, when neither data nor a model provides a convenient expression for the pressure correction that appears in (6.4.7). In such cases we might choose to keep all the pressure dependence in the activity coefficient.

To accomplish this, we proceed in a manner analogous to that for obtaining FFF #2. First, multiply and divide FFF #1 by the known standard-state fugacity $f_i^o(T, P_i^o)$,

$$f_i(T, P, \{x\}) = x_i P \frac{\varphi_i(T, P, \{x\})}{f_i^o(T, P_i^o)} f_i^o(T, P_i^o) \tag{6.4.8}$$

Then replace the denominator with

$$f_i^o(T, P_i^o) = P^o \varphi_i^o(T, P_i^o) \tag{6.4.9}$$

and use (5.5.4) to identify $(\varphi_i P)/(\varphi_i^o P_i^o)$ as the activity coefficient; so (6.4.8) becomes

FFF #4 $$f_i(T, P, \{x\}) = x_i \gamma_i(T, P, \{x\}; f_i^o(T, P_i^o))\, f_i^o(T, P_i^o) \tag{6.4.10}$$

Note that the activity coefficient in (6.4.10) differs from those in (6.4.4) and (6.4.7) and that it is not simply related to an excess Gibbs energy.

Fugacity Formula #5. In FFF #3 we divided the pressure effects between the activity coefficient and a Poynting factor, while in FFF #4 we placed all the pressure effect in the activity coefficient. Still another possibility is to place all the pressure effect in a Poynting factor. To derive this form, we start with FFF #4 and use (5.5.11) to replace $\gamma_i(T, P, \{x\})$ with $\gamma_i(T, P_i^o, \{x\})$. The result is

FFF #5 $$f_i(T, P, \{x\}) = x_i \gamma_i(T, P_i^o, \{x\}; f_i^o(T, P_i^o))\, f_i^o(T, P_i^o) \tag{6.4.11}$$

$$\times \exp\left[\int_{P_i^o}^{P} \frac{\overline{V}_i(T, \pi)}{RT}\, d\pi\right]$$

Note that in FFF #5 neither the activity coefficient nor the standard-state fugacity depends on the mixture pressure. The Poynting factor in FFF #5 can be computed, provided we can evaluate the partial molar volume for the real substance along the isotherm T from P_i^o to P. In contrast, the Poynting factor appearing in FFF #3 applies to component i in its standard state and involves an integral over the partial molar volume of that standard-state substance.

An alternative derivation of FFF #5 can be performed by starting with FFF #3 and moving the pressure dependence of $\gamma_i(T, P, \{x\})$ into a Poynting factor. That Poynting factor will contain an integral over the partial molar excess volume. Then we would combine that Poynting factor with the one already appearing in FFF #3.

Summary of procedure. To develop the fugacity formulae #2–5 presented above, we follow this procedure:

(a) In every case, we start from FFF #1, which defines the fugacity coefficient.

(b) Then we multiply and divide FFF #1 by a known standard-state fugacity f_i^o:

 (i) for FFF #2, we use $f_i^o = f_i^o(T, P)$,

 (ii) but for FFF #3–5, we use $f_i^o = f_i^o(T, P_i^o)$.

(c) Next we use one of the relations from § 5.5 to identify some ratio as an activity coefficient:

 (i) for FFF #2 and 3 we use $\gamma_i(P; P)$,

 (ii) for FFF #4, $\gamma_i(P; P_i^o)$,

 (iii) and for FFF #5, $\gamma_i(P_i^o; P_i^o)$.

(d) Finally, the appropriate Poynting factor is applied when needed. In FFF #3 the Poynting factor corrects only the standard-state fugacity, but in FFF #5 the Poynting factor corrects the solution fugacity.

Numerical results from FFF. We now show numerical results from each FFF applied to the same mixture: an equimolar gaseous mixture of methane(1) and sulfur hexafluoride(2) at 60°C and 20 bar. We determined residual properties for this mixture in § 4.5.5 and excess properties in § 5.3.1–5.3.3. For the equation of state we use the simple virial equation written in volume-explicit form,

$$v = \frac{RT}{P} + B \tag{4.5.32}$$

with B given by (4.5.18) and values of the B_{ij} given in Table 4.2. We choose the standard state for each component to be the pure gas at 60°C and 10 bar. Then the pure-component fugacity coefficients are given by (P4.11.1) and the Poynting factor PF_1 that appears in FFF #3 is given by

$$PF_1 = \exp\left[\int_{P_i^o}^{P} \frac{\overline{V}_i^o(T, \pi)}{RT} d\pi\right] = \exp\left[\int_{P_i^o}^{P} \frac{v_{\text{pure i}}(T, \pi)}{RT} d\pi\right] \tag{6.4.12}$$

$$= \exp\left[\ln\frac{P}{P_i^o} + \frac{B_{ii}(P - P_i^o)}{RT}\right] \tag{6.4.13}$$

For the model (4.5.32) applied to binary mixtures, the fugacity coefficients are given by (P4.11.2). Similarly, the general expression for the partial molar volume is given in (P4.23.1). Then the Poynting factor that appears in FFF #5 is

$$PF_2 = \exp\left[\int_{P_i^o}^{P} \frac{\overline{V}_i(T, \pi)}{RT} d\pi\right] \tag{6.4.14}$$

$$= \exp\left[\ln\frac{P}{P_i^o} + \frac{(B_{ii} + x_j^2\delta_{12})(P - P_i^o)}{RT}\right] \tag{6.4.15}$$

where $j \neq i$ and δ_{12} is the collection of B_{ij}s given in (5.3.8).

The numerical results leading to the fugacity for methane (component 1) are summarized in Table 6.4. All five fugacity equations provide exactly the same value for the fugacity, as they should. The values of all fugacity coefficients are close to unity, indicating the pure components and the mixture are nearly ideal gases. However, unlike liquids, these gas-phase systems produce large values for the Poynting factors. Further, note that values obtained for activity coefficients change when we change the standard state, even though the final values obtained for the fugacity are unchanged.

Table 6.4 Numerical results from each of the FFF in computations for the fugacity of methane(1) in an equimolar methane-sulfur hexafluoride mixture at 60°C, 20 bar

Property	Eq. used	FFF #1 (6.4.1)	FFF #2 (6.4.4)	FFF #3 (6.4.7)	FFF #4 (6.4.10)	FFF#5 (6.4.11)
x_1	...	0.5	0.5	0.5	0.5	0.5
P/bar	...	20	20	20	20	20
P_1^o /bar	...	...	...	10	10	10
$\varphi_1(P)$	(4.4.10)	0.9985	...	...	0.9985	...
$\varphi_1(P_1^o)$	(4.4.10)	...	...	...	0.9992	0.9992
$\varphi_1^o(P)$	(4.4.11)	...	0.9773	...	...	...
$\varphi_1^o(P_1^o)$	(4.4.11)	...	...	0.9886	0.9886	0.9886
$f_1^o(P)$/bar	(6.4.1)	...	19.546	...	...	...
$f_1^o(P_1^o)$/bar	(6.4.1)	...	...	9.886	9.886	9.886
$\gamma_1(P; P)$	(5.5.5)	...	1.0217	1.0217	...	...
$\gamma_1(P; P_1^o)$	(5.5.9)	...	...	...	2.020	...
$\gamma_1(P_1^o; P_1^o)$	(5.5.10)	...	...	...	...	1.011
PF_1	(6.4.13)	...	...	1.9772	...	...
PF_2	(6.4.15)	...	...	...	1.998	1.998
f_1/bar	(FFF)	9.985	9.985	9.985	9.985	9.985

6.5 MIXING RULES FROM MODELS FOR EXCESS GIBBS ENERGY

We have noted that historically $PvTx$ models and fugacity coefficients were restricted to gas-phase mixtures, while g^E models and activity coefficients were restricted to condensed-phase mixtures. But these restrictions are not thermodynamic; instead, they arose because of limitations in the models themselves and because of computational difficulties that occur in solving sets of nonlinear algebraic equations. But with continuing improvements in models, as well as in the power and availability of digital computers, we can contrive complicated models for nearly any system. In particular, FFF #1 is now being applied to virtually all types of mixtures and phases.

Use of FFF #1 requires a $PvTx$ equation of state for the mixture and each standard state, and we noted in § 4.5.12 that the outstanding problems in applying $PvTx$ equations to mixtures are the choices for mixing rules and combining rules. One approach to this problem is to base mixing rules on models for g^E. The motivation is to combine the composition-dependence in g^E models with the pressure-dependence in $PvTx$

models. This may be attempted in many different ways, and development of viable strategies remains a central problem in thermodynamic modeling. One approach would be to simply rearrange (5.5.5) to read

$$\varphi_i(T, P, \{x\}) = \gamma_i(T, P, \{x\}; f_i^o(T, P))\, \varphi_i^o(T, P) \tag{6.5.1}$$

Formally, this could be used to connect a *PvTx* model for φ_i (lhs) with a model for the activity coefficient (rhs). However, this approach is unattractive because it would require us to deal separately with each component.

A better approach is to start from a particular model for $g^E(T, P, \{x\}; \{A\})$, such as the Porter, Margules, or Wilson models introduced in § 5.6. Here the $\{A\}$ are the model parameters, whose values are usually obtained by fits to phase-equilibrium data. We then select a *PvTx* model; often a cubic is used. In this discussion, we consider the Redlich-Kwong equation (§ 4.5.8). This model contains parameters $\{a, b\}$ that depend on composition via some mixing rules (§ 4.5.12). Our strategy is to find those mixing rules by matching the g^E model to g^E given by the *PvTx* equation.

We can use a *PvTx* model to obtain g^E via the residual Gibbs energy, as described in § 5.3. For the standard state of each component, we choose the pure component at the mixture T and P. Then we can write

$$g_{eos}^E(T, P, \{x\}; \{a, b\}) = g^{res}(T, P, \{x\}; \{a, b\}) - \sum_i x_i\, g_{\text{pure i}}^{res}(T, P; \{a_i, b_i\}) \tag{6.5.2}$$

$$\frac{g_{eos}^E}{RT} = Z - \ln Z + \int_{v(T, P, \{x\})}^{\infty} \left(\frac{Z-1}{v}\right) dv \tag{6.5.3}$$

$$-\sum_i x_i \left[Z_{\text{pure i}} - \ln Z_{\text{pure i}} + \int_{v_i(T, P)}^{\infty} \left(\frac{Z_{\text{pure i}} - 1}{v_{\text{pure i}}}\right) dv_i \right]$$

where $Z = Pv(T, P, \{x\})/RT$ is the mixture compressibility factor at T, P, and $\{x\}$, while $Z_{\text{pure i}} = Pv_{\text{pure i}}(T, P)/RT$ is that for pure i at T and P. The equation-of-state parameters $\{a, b\}$ are obtained by matching, at a single state, the value of g^E from an excess-property model to the value of g^E given by the equation of state via (6.5.3). After the equation-of-state parameters are found, fugacity coefficients are determined from (4.4.10), and fugacities are obtained from FFF #1. Each pairing of a particular equation of state to a particular g^E model produces a unique matching, but the possibilities are many and the resulting expressions for φ_i can be complicated [6, 7].

One of the first implementations of the above procedure was that by Huron and Vidal [8]. They retained the simple mixing rule for the Redlich-Kwong parameter b,

$$b = \sum_i x_i b_i \tag{4.5.75}$$

and they obtained the mixing rule for a under the assumption that, in the limit of infinite pressure, the excess Gibbs and Helmholtz energies are the same. The resulting mixing rule for the Redlich-Kwong parameter a was then found to be

$$a = b\left[\sum_i x_i \frac{a_i}{b_i} - 1.443\sqrt{T}\, g^E_{model}\right] \qquad \textit{Huron-Vidal} \qquad (6.5.4)$$

The factor $1.443\sqrt{T}$ changes when other model equations of state are used. Note that (6.5.4) involves only pure component parameters, a_i and b_i; so, no combining rules are needed. However, these mixing rules do not reproduce the known composition dependence of the second virial coefficient (4.5.18).

Under different assumptions, Wong and Sandler [9] used the Redlich-Kwong equation with the mixing rule (6.5.4) to obtain a quadratic rule,

$$b - \frac{a}{RT\sqrt{T}} = \sum_i \sum_j x_j x_j \left(b - \frac{a}{RT\sqrt{T}}\right)_{ij} \qquad \textit{Wong-Sandler} \qquad (6.5.5)$$

with

$$b = \sum_i \sum_j x_j x_j b_{ij} \qquad (6.5.6)$$

To obtain the unlike parameters, Wong and Sandler chose these combining rules,

$$b_{ij} = (b_{ii} + b_{jj})/2 \qquad (6.5.7)$$

and

$$\left(b - \frac{a}{RT\sqrt{T}}\right)_{ij} = (1 - k_{ij})\,\frac{1}{2}\left[\left(b - \frac{a}{RT\sqrt{T}}\right)_{ii} + \left(b - \frac{a}{RT\sqrt{T}}\right)_{jj}\right] \qquad (6.5.8)$$

The value of the binary interaction parameter k_{ij} must be estimated or found by fitting mixture data. Our brief introduction to this approach has been based on the Redlich-Kwong equation, but the procedure can be implemented with any $PvTx$ equation. More generally, the approach discussed here can provide accurate predictions of fluid properties at high T and P using model g^E parameters fit at low T and P. The procedure is now routinely used in process simulation software.

6.6 SUMMARY

The theme of this chapter is that, while thermodynamic descriptions of mixtures involve a large number of equations, those equations tend to fall into a few repeated patterns. By recognizing the patterns, we not only broaden our understanding, but we also reduce the number of different things that must be mastered.

The first pattern encompasses the difference measures for deviations from some well-defined ideality. In § 6.2.1 we defined a generalized difference, and we showed that this class of generalized differences contains the residual properties, excess properties, and changes of properties on mixing. In principle, any one of these differences can be used to compute the thermodynamic properties of any substance, including substances composed of any number of components and any kind of phase. In practice, the choice of which difference to use is dictated by the available data and by what additional data you can calculate or reliably estimate. Strategies for choosing among computational options will be discussed in Chapter 10. For now the important lesson is to appreciate that different versions of the same pattern provide computational options, and much of an engineer's job is choosing from among the available options.

The second pattern includes the ratio measures for representing chemical potentials. These are all ratios of fugacities, with the general form being the generalized activity defined by (6.2.8). With only a small loss of generality, we choose the real and reference states to be at the same temperature, then the fugacity is related to some difference in chemical potentials, as shown in § 6.2.4. Every form for the fugacity shown in § 6.2.4 involves a reference, and until that reference is identified, those relations carry little meaning and have no computational utility. Fortunately, the choice of reference is at the discretion of the user.

Because the fugacity coefficient, activity, and activity coefficient are each a special case of the generalized activity (6.2.8), each has a similar physical interpretation. As shown in § 6.3, each ratio is simply related to the reversible work involved in moving molecules from a reference substance into the real substance of interest. With this physical interpretation, we can anticipate why the fugacity is intimately involved in calculations for phase equilibria. Consider two phases α and β in contact at the same temperature and pressure. If the phases are out of equilibrium, say with $f_i^{\alpha} > f_i^{\beta}$, then by (6.3.8) work must be supplied to move molecules of component i from β to α; if that work is not supplied, then molecules of component i will naturally diffuse from phase α to phase β. That is, in the absence of temperature and pressure gradients, molecules tend to diffuse from regions of high fugacity to regions of low fugacity; this point will be made mathematically precise in the next chapter.

Values for fugacities are nearly always calculated using one of the five famous fugacity formulae cited in § 6.4. Again, these five formulae represent options that we can exploit in solving all kinds of phase separation and chemical reaction problems. The commonly used procedures for attacking such problems will be developed in Chapter 10, the solution techniques will be described in Chapter 11, and particular examples will be offered in Chapter 12.

LITERATURE CITED

[1] H. Qian and J. J. Hopfield, "Entropy-Enthalpy Compensation: Perturbation and Relaxation in Thermodynamic Systems," *J. Chem. Phys.*, **105**, 9292 (1996).

[2] J. A. Beattie, "The Computation of the Thermodynamic Properties of Real Gases and Mixtures of Real Gases," *Chem. Rev.*, **44**, 141 (1949).

[3] J. A. Beattie and W. H. Stockmayer, *Treatise on Physical Chemistry,* H. S. Taylor and S. Glasstone (eds.), van Nostrand Co., Inc., Princeton, NJ, 1942, ch. 2.

[4] J. A. Beattie, *Thermodynamics and Physics of Matter,* F. D. Rossini (ed.), Princeton University Press, Princeton, NJ, 1955, ch. 3.

[5] M. M. Abbott, personal communication.

[6] B. E. Poling, J. M. Prausnitz, and J. P. O'Connell, *The Properties of Gases and Liquids*, 5th ed., McGraw-Hill, New York, 2001.

[7] S. I. Sandler, H. Orbey, and B.-I. Lee in *Models for Thermodynamic and Phase Equilibria Calculations*, S. I. Sandler (ed.), Marcel Dekker, New York, 1993.

[8] M. J. Huron and J. Vidal, "New Mixing Rules in Simple Equations of State for Representing Vapour-Liquid Equilibria of Strongly Non-Ideal Mixtures," *Fluid Phase Equil.*, **3**, 255 (1979).

[9] D. S. H. Wong and S. I. Sandler, "A Theoretically Correct Mixing Rule for Cubic Equations of State," *A. I. Ch. E. J.*, **38**, 671 (1992).

[10] H. Mansoorian, K. R. Hall, and P. T. Eubank, "Vapor Pressure and PVT Measurements Using the Burnett-Isochoric Method," *Proc. Seventh Symp. Thermophysical Properties*, ASME, New York, 1977, p. 456.

PROBLEMS

6.1 Consider a binary mixture of components A and B. The mixture is initially at (T_1, P_1) and undergoes a change of state to (T_2, P_2). The composition remains fixed during the process. For any extensive property F, show that the change ΔF_{12} for this change of state can be computed either in terms of isobaric residual properties or in terms of excess properties. That is, prove that

$$\Delta F_{12} = \Delta F_{12}^{res} + \Delta F_{12}^{ig} = \Delta F_{12}^{E} + \Delta F_{12}^{is} \tag{P6.1.1}$$

6.2 A certain equimolar binary mixture is at $T = 25°C$ and $P = 10$ bar. At 25°C component 1 has $P_1^s = 1.0$ bar, $f_{pure\,1}(1\text{ bar}) = 0.9$ bar, $f_{pure\,1}(10\text{ bar}) = 2.0$ bar, $f_1(1\text{ bar}, x = 0.5) = 0.35$ bar, $f_1(10\text{ bar}, x = 0.5) = 1.2$ bar. Determine values for each of the following at 25°C and 10 bar:

(a) The fugacity coefficient φ_1 in the equimolar mixture.

(b) The fugacity coefficient $\varphi_{pure\,1}$.

(c) The activity coefficient $\gamma_1(T, P, \{x\}; f_{pure\,1}(T, P))$ in the equimolar mixture.

(d) The activity coefficient $\gamma_1(T, P, \{x\}; f_{pure\,1}(T, P_1^s))$ in the equimolar mixture.

(e) The activity coefficient $\gamma_1(T, P_1^s, \{x\}; f_{pure\,1}(T, P_1^s))$ in the equimolar mixture.

(f) The activity a_1 for the equimolar mixture, with the standard state for a_1 based on the pure-component at the mixture T and P.

6.3 For the process of converting an ideal gas into a real substance, show that the heat and work effects presented in § 6.3.2 and 6.3.3 are consistent with the energy balance on a closed system.

6.4 Consider an isothermal-isobaric process in which a small amount of component i is removed from an ideal-gas mixture at T, P, $\{x\}$ and injected into a real mixture at the same T, P, and $\{x\}$. (a) Show that the required minimum work is given by $RT \ln \varphi_i$. (b) Show that the reversible heat effect is given by $T\bar{S}_i^{res}$.

6.5 Consider a binary mixture of 1 and 2 that obeys the following model equations:

$$\frac{g^E}{RT} = x_1 x_2\left(a_{12} + b_{12}P + \frac{c_{12}}{RT}\right)$$

$$Z = 1 + \frac{P}{RT}\left(\frac{x_1\alpha_1}{T^n} + \frac{x_2\alpha_2}{T^m}\right) \quad \text{and} \quad \ln P_i^s = \beta_i - \frac{\theta_i}{T}$$

where P_i^s is the vapor pressure for pure i. Find the expression for the fugacity of component 1 in terms of the state variables (T, P, x_1) and the constant parameters (a_{12}, b_{12}, c_{12}, α_1, α_2, n, m, β_1, θ_1).

6.6 Consider an isothermal-isobaric process in which a small amount of component i is removed from a Lewis-Randall ideal solution at T, P, $\{x\}$ and injected into a real mixture at the same T, P, and $\{x\}$.

(a) Show that the required minimum work is given by $RT \ln \gamma_i$.

(b) Show that the reversible heat effect is given by $T\bar{S}_i^E$.

6.7 Evaluate the reversible shaft work, the boundary work, and the heat effect when each of the following pure substances is converted isothermally-isobarically from an ideal gas into the real substance.

(a) Pure gaseous methyl chloride at 370 K, 0.2 MPa, with virial coefficients given by Mansoorian et al. [10]: $B = -0.01293 \exp[1110/T]$ liter/mol and $C = 192 \exp[-0.0219\ T]$ (liter/mol)^2, T in K.

(b) Pure n-hexane at $T/T_c = 1.1$ and $v/v_c = 0.9$. If necessary, assume the modified Redlich-Kwong equation (4.5.67) applies.

6.8 Evaluate the reversible work and the heat effect associated with the isothermal-isobaric conversion of an equimolar liquid solution of carbon tetrachloride + chloroform into an ideal gas at 298 K, 0.1 MPa. Assume the real mixture can be modeled by the Porter equation with parameter given in Table E.1.

6.9 Perform calculations to check and confirm the numerical values in Table 6.4.

6.10 Show that when the pressure is low enough, FFF # 3, 4, and 5 are all essentially the same as FFF #2.

6.11 For a pure liquid at 350 K and having a molar volume $v = 0.1$ liter/mol, estimate the pressure at which the Poynting factor deviates by 2% from unity. For the pure vapor pressure at 350 K, use $P^s = 0.1$ MPa.

6.12 At 100°C, estimate the amount by which the pressure must change from 1 bar to

(a) increase the fugacity of water vapor by 5%

(b) increase the fugacity of liquid water by 5%.

6.13 This problem is to illustrate that while the definition of the fugacity is unambiguous, the choice for defining an ideal solution is arbitrary and therefore the definition of the activity coefficient is at our disposal. Consider a binary mixture that obeys Porter's equation

$$\frac{g^E}{RT} = Ax_1x_2$$

where the parameter $A = 0.4$. At the T and P of interest $f_{\text{pure }1} = 5$ bar. Instead of the Lewis-Randall rule, let us define an ideal solution based on the fugacity at the equimolar composition; that is, choose

$$f_1^{is} = x_1 f_1^o(T, P) = 2x_1 f_1(T, P, (x_1 = 0.5)) \tag{P6.13.1}$$

(a) Sketch the fugacity $f_1(x_1)$. On the same plot sketch the ideal-solution fugacities given by the Lewis-Randall rule and that given by (P6.13.1).

(b) Is the standard state used in (P6.13.1) a mixture, a real pure substance, or a hypothetical pure substance?

(c) Define an activity coefficient ζ_1 that measures deviations from the ideal solution defined in (P6.13.1),

$$\zeta_1 = \frac{f_1(T, P, x_1)}{f_1^{is}(T, P, x_1)} \tag{P6.13.2}$$

Show that this activity coefficient must obey the normalization

$$\lim_{x_1 \to 0.5} \zeta_1 = 1$$

(d) Find an expression for ζ_1 solely in terms of x_1, the parameter A, and constants. Plot ζ_1 vs. x_1 and on the same plot show γ_1 (from the Porter equation) in the Lewis-Randall standard state.

(e) For several values of x_1, show that the fugacity given by $f_1 = x_1 \gamma_1 f_{\text{pure}1}$ is numerically the same as the value given by $f_1 = x_1 \zeta_1 f_1^o$.

(f) Define ζ_2 as in (P6.13.2), with $f_2^{is} = 2x_2 f_2(T, P, x_2 = 0.5)$. Find the expression for the composition dependence of ζ_1, and then show that ζ_1 and ζ_2 satisfy the isothermal-isobaric Gibbs-Duhem equation.

6.14 (a) Sketch a plot of the fugacity of a pure substance as it is isothermally compressed from a very low pressure to ten times its vapor pressure. Justify all important features on your plot using appropriate FFF and "always-true" relations.

(b) On the same plot as in (a), add the line for the fugacity if the substance is one component in an equimolar binary mixture.

6.15 (a) Show that the fugacity of a pure liquid at pressure P and a subcritical temperature $T < T_c$ can be written as

$$f_{\text{pure i}}(T, P) = \varphi_i^s(T, P^s)\, P^s(T) \exp\left[\frac{1}{RT}\int_{P^s}^{P} v_{\text{pure i}}\, dP\right] \qquad \text{(P6.15.1)}$$

where P^s is the pure-liquid vapor pressure, φ_i^s is the fugacity coefficient of the pure saturated liquid, and $v_{\text{pure i}}$ is the molar volume of the pure liquid.

(b) Estimate the fugacity of pure liquid water at 100°C and 200 bar. At 100°C the second virial coefficient of pure water is $B = -0.45$ liter/mol.

6.16 An engineer, who works for one of your competitors, reveals that they use the following proprietary expression for the fugacity of component 1 in a certain binary liquid mixture (with P in MPa and T in K):

$$\ln f_1(T, P, x_1) = \ln x_1 + 10 - \frac{3000}{T} - \left(1 + \frac{3000}{T}\right)x_2^2 + \frac{0.1P}{T}$$

(a) Which FFF was probably used to obtain this expression?

(b) Obtain expressions for both activity coefficients, γ_1 and γ_2.

(c) What expressions for g^E/RT and h^E/RT are consistent with this form for f_1?

(d) What expression for $\overline{V}_1(T, P, x_1)$ is consistent with this form for f_1?

6.17 For very dilute mixtures of a gaseous solute(1) in water(2), experimental data show that, over wide temperature ranges, the partial molar volume at infinite dilution can be correlated by

$$\overline{V}_1^{\infty} = \kappa_T RT(1 + \rho_2\{a + b[\exp(c\rho_2) - 1]\})$$

where κ_T is the isothermal compressibility of pure water, ρ_2 is the molar density of pure water, and a, b, c are all constants. Obtain the expression for the infinite-dilution fugacity coefficient and for the fugacity of the solute at low concentrations in water.

6.18 Each of the following gives an expression for the fugacity of one component in a system. For each, what real-substance state (T, P, $\{x\}$, phase) would be appropriate? What models and assumptions were used to obtain the expression (e.g., FFF, standard state, ideal solution, etc.)?

(a) $f_1(T, P, \{x\}) = P$

(b) $f_1(T, P, \{x\}) = x_1 P$

(c) $f_1(T, P, \{x\}) = P \exp\left(\frac{B_{11}P}{RT}\right)$

(d) $f_1(T, P, \{x\}) = x_1 P_1^s(T)$

(e) $f_1(T, P, \{x\}) = x_1 \gamma_1 \varphi_1^s P_1^s \exp\left[\frac{1}{RT}\int_{P_1^s}^{P} \overline{V}_1 dP\right]$

Part III

MULTIPHASE AND REACTING SYSTEMS

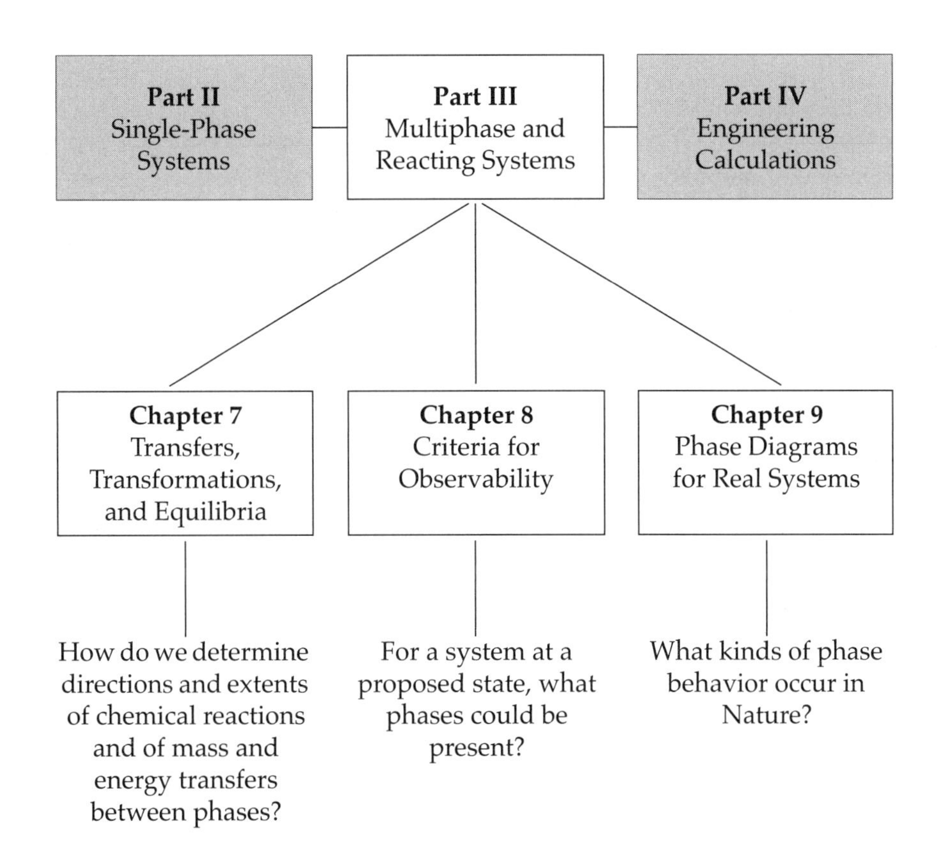

7

TRANSFERS, TRANSFORMATIONS, AND EQUILIBRIA

When two or more homogeneous systems are brought into contact to form a single heterogeneous system, any of several actions may occur before equilibrium is reestablished. The possibilities include mass and energy transfers, chemical reactions, and the appearance or disappearance of phases. In this chapter we provide thermodynamic criteria for determining whether and to what extent such phenomena actually occur. Surprisingly, these criteria invoke no new thermodynamics—we need only combine familiar thermodynamic quantities in new ways and, in some cases, apply to those quantities mathematical operations that we have not used heretofore.

The heterogeneities of most concern to us are those that involve the presence of more than one phase. The analysis of multiphase systems can be important to the design and operation of many industrial processes, especially those in which multiple phases influence chemical reactions, heat transfer, or mixing. For example, phase-equilibrium calculations form the bases for many separation processes, including *stagewise* operations, such as distillation, solvent extraction, crystallization, and supercritical extraction, and *rate-limited* operations, such as membrane separations.

Analysis of multiphase systems is a principal theme of chemistry and chemical engineering; another is analysis of chemical reactions—processes in which chemical bonds are rearranged among species. Rearranging chemical bonds is the most efficient way to store and release energy, it drives many natural processes, and it is used industrially to make substitutes for, and concentrated forms of, natural products.

The chapter divides in two: in early sections we describe the behavior of nonreacting systems, while in later sections we deal with systems in which reactions occur. In § 7.1 we combine the first and second laws to obtain criteria for identifying limitations on the directions of processes and for identifying equilibrium in closed multiphase systems. Then in § 7.2 we develop the analogous relations for heat, work, and material transfers in open systems. With the material in § 7.2 as a basis, we then present in § 7.3 the thermodynamic criteria for equilibrium among phases.

A similar program is used for reacting systems. In § 7.4 we extend the combined first and second laws to closed systems undergoing chemical reactions, then in § 7.5 we show how the combined laws apply to reactions in open systems. In § 7.6 we formulate the thermodynamic criterion for identifying reaction equilibria. By presenting the criteria for both phase and reaction equilibria in the same chapter, we hope to emphasize and exploit similarities that exist between the two. These criteria provide foundations for the engineering calculations described in Part IV of this book.

7.1 THE LAWS FOR CLOSED NONREACTING SYSTEMS

Careful observation teaches us that, left undisturbed, every material system tends to evolve to a unique equilibrium state that is consistent with any imposed constraints. The rates of such evolutions cannot be determined from thermodynamics, but thermodynamics does provide quantitative criteria both for identifying the directions of such evolutions and for identifying equilibrium once it is reached. Those criteria are obtained by combining the first and second laws.

In Chapter 3 we combined the first and second laws to obtain the fundamental equations for closed systems; one example is (3.2.4), which we now write as

$$d(Nu) - Td(Ns) + Pd(Nv) = 0 \tag{7.1.1}$$

But in writing such equations, we assumed that our system is homogeneous—that its values for intensive properties are uniform throughout. Here we want to generalize the development so we can identify equilibrium in heterogeneous systems, especially those in which the heterogeneity results from the presence of more than one part, such as multiple phases. For such systems, the fundamental equation (7.1.1) takes the form

$$\sum_k d(N_k u_k) - \sum_k T_k\, d(N_k s_k) + \sum_k P_k\, d(N_k v_k) = 0 \tag{7.1.2}$$

where each sum runs over all homogeneous parts of the system. The form (7.1.2) allows for the possibility that, during a change of state, different system parts might have different values for some intensive properties, such as temperature and pressure. It allows for material exchange among parts. It also allows rigid or nonconducting walls (or both) to separate different parts, so that even at equilibrium, all parts need not have the same temperature or pressure.

Because of the generality of (7.1.2), a heterogeneous system may not be describable by a single set of intensive system properties. But any resulting ambiguities can be removed by restating the combined first and second laws in a form that contains only extensive system properties plus constant intensive properties of the surroundings. This general form of the combined first and second laws is derived in § 7.1.1. From that general form, we deduce special forms that apply to adiabatic processes (§ 7.1.2) and processes having constant T and V (§ 7.1.4) or constant T and P (§ 7.1.5).

7.1.1 The Combined Laws

Consider a closed system composed of one or more parts. The system has thermal and Pv work-mode interactions with well-defined surroundings. The surroundings have such a large capacity for providing and absorbing energy that such exchanges do not affect its intensive properties. Consequently, the surroundings have a constant tem-

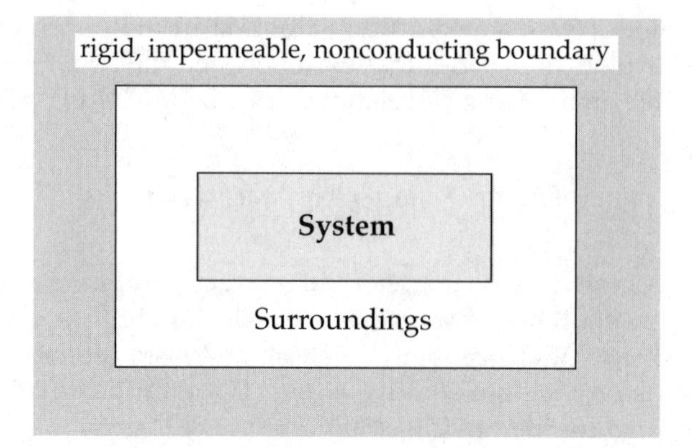

Figure 7.1 Schematic of a system and its surroundings isolated from the universe

perature T_{sur} and pressure P_{sur} throughout any changes of state that occur in the system. Effectively, the system and surroundings are isolated from the rest of the universe, as in Figure 7.1, so that any change of state occurring in the system cannot affect the total internal energy U_t of system plus surroundings,

$$dU_t = dU + dU_{sur} + dU_b = 0 \tag{7.1.3}$$

Subscript b refers to the boundary between system and surroundings. We consider here only those situations in which the boundary is of negligible mass compared to that of the system, so it does not affect the amount of energy being transferred to or from the system. The boundary prevents mass from entering or leaving the system, while allowing energy to enter or leave. (If we need to account for boundary effects, one way to do so would be to include the boundary as another part of the heterogeneous system.) Ignoring boundary effects, (7.1.3) reduces to

$$dU + dU_{sur} = 0 \tag{7.1.4}$$

We now consider the term in (7.1.4) that applies to the surroundings. The first law for the surroundings takes the form

$$dU_{sur} = \delta Q_{sur} + \delta W_{sur} \tag{7.1.5}$$

but note what the signs mean for Q_{sur} and W_{sur} in (7.1.5): δQ_{sur} is positive if heat enters the surroundings from the system. Likewise for δW_{sur}. However, the signs for the system terms Q and W mean the opposite: δQ is positive if heat enters the system from the surroundings. Likewise for δW. So before (7.1.5) can be combined with (7.1.4), the sign conventions for Q_{sur} and W_{sur} must be made consistent with those used for the system terms Q and W. The two sign conventions can be made to agree by setting $\delta Q_{sur} = -\delta Q_{ext}$ and $\delta W_{sur} = -\delta W_{ext}$. (Recall from § 2.3.3 that the subscript *ext* always means external to the system, so $T_{sur} = T_{ext}$.) Therefore we write (7.1.5) as

$$dU_{sur} = -\delta Q_{ext} - \delta W_{ext} \tag{7.1.6}$$

and we combine (7.1.6) with (7.1.4) to obtain

$$dU - \delta Q_{ext} - \delta W_{ext} = 0 \tag{7.1.7}$$

Next we replace the path functions in (7.1.7) with state functions. To eliminate the heat, we appeal to the second law (2.3.8), which for this situation takes the form

$$dS = \frac{\delta Q_{ext}}{T_{sur}} + dS_{gen} \tag{7.1.8}$$

Recall we always have $dS_{gen} \geq 0$. Solving (7.1.8) explicitly for δQ_{ext} and substituting the result into (7.1.7) gives

$$dU - \delta W_{ext} - T_{sur}\, dS = -T_{sur}\, dS_{gen} \leq 0 \tag{7.1.9}$$

Now assume the only work mode is that associated with a volume change, so

$$\delta W_{ext} = -\delta W_{sur} = -(-P_{sur}\, dV_{sur} + \delta W_{lost}) = -P_{sur}\, dV - \delta W_{lost} \tag{7.1.10}$$

where we have used $dV = -dV_{sur}$. Substituting (7.1.10) into (7.1.9), remembering that $\delta W_{lost} \geq 0$, and using N as the total number of moles in the system, we find

$$N du + N P_{sur}\, dv - N T_{sur}\, ds \leq 0 \qquad \textit{closed systems} \tag{7.1.11}$$

Equation (7.1.11) is a general form of the combined first and second laws applied to closed systems; we call it the *combined laws.* Since Nu, Nv, and Ns are extensive properties of the system while T_{sur} and P_{sur} are properties of the surroundings, (7.1.11) applies both to homogeneous systems and to heterogeneous systems. If the system is heterogeneous, but composed of homogeneous parts, then (7.1.11) can be written as a sum over the homogeneous parts, as in (7.1.2).

The equality in (7.1.11) applies only to reversible changes, while the inequality applies for real (i.e., irreversible) processes. The combined laws (7.1.11) differ from the fundamental equation (3.2.4) in that (3.2.4) contains only system properties, while (7.1.11) contains the temperature and pressure of the surroundings. If a change of state occurs with $T_{sur} = T$ and $P_{sur} = P$, then the two equations are identical.

For a finite change of state at constant T_{sur} and P_{sur}, the integrated form of (7.1.11) is

$$N\Delta u + N P_{sur}\, \Delta v - N T_{sur}\, \Delta s \leq 0 \tag{7.1.12}$$

This equation is important because it establishes limits on the kinds of processes that can naturally (spontaneously) occur to change the state of a closed system. If two states satisfy the inequality, then the system can spontaneously evolve from the initial to the final state, but only via some *irreversible* process. If two states fail to satisfy (7.1.12), then the system *cannot* spontaneously evolve from the initial to the proposed

final state. The equality in (7.1.12) pertains to reversible changes, but in practice the equality cannot occur because reversible changes can only proceed differentially.

7.1.2 Adiabatic Processes on Closed Systems

If constraints are applied to the interactions available to our closed system, then (7.1.11) simplifies accordingly. We first consider adiabatic processes in which only work is done on or by a closed system. If we continue to ignore boundary effects, the first law applied to an adiabatic process reduces to

$$N du = \delta W_{ext} \qquad \textit{closed, adiabatic} \qquad (7.1.13)$$

Substituting this into (7.1.11), the combined laws simplify to $NT_{sur}ds \geq 0$, and since T_{sur} is an absolute temperature, we can write

$$N ds = \sum_k d(N_k s_k) \geq 0 \qquad \textit{closed, adiabatic} \qquad (7.1.14)$$

Here the sum runs over all parts of a heterogeneous system. Note that we cannot determine whether the entropy of some parts increases or decreases; it is only the total entropy that is constrained. However, if the system is homogeneous, then the sum contains only a single term, and (7.1.14) still applies.

The result (7.1.14) is merely a restatement of the second law: spontaneous adiabatic changes of state occur only if they either increase the system's total entropy or leave it unaffected. If two states have the same entropy so $S_{final} = S_{initial}$, then the system can evolve along some reversible adiabatic path between the initial and the final states. If $S_{final} > S_{initial}$, then the system spontaneously evolves along some irreversible adiabatic path. But if $S_{final} < S_{initial}$, then the system cannot spontaneously evolve along any adiabat from the initial to the final state. Since all real processes are irreversible, any spontaneous adiabatic process occurring in a closed system must increase the system's total entropy. Such processes might involve heat transfer among parts of the system, even if no heat is exchanged with the surroundings.

When a closed system cannot exchange either heat or work with the surroundings, then the system is said to be *isolated*. For heterogeneous isolated systems, conservation of mass, energy, and volume can be written as

$$dN = \sum_k dN_k = 0 \qquad (7.1.15)$$

$$N du = \sum_k d(N_k u_k) = 0 \qquad (7.1.16)$$

$$N dv = \sum_k d(N_k v_k) = 0 \qquad (7.1.17)$$

When these are substituted into (7.1.11), the combined laws again reduce to

$$N ds = \sum_k d(N_k s_k) \geq 0 \qquad \textit{isolated} \qquad (7.1.18)$$

This shows that thermodynamic constraints on isolated systems are the same as those for adiabatic processes on closed systems.

For an isolated system initially at equilibrium, a spontaneous process away from equilibrium can only be initiated by removing a constraint, thereby allowing system parts to interact. This means that if an isolated system is to undergo a spontaneous change from an equilibrium state, then it must be initially composed of parts whose properties are not all equal. Removing a constraint then allows certain intensive property values to become uniform over parts of the system. For example, some parts of the system may have different temperatures because the parts are separated by insulated walls. By removing the insulation, heat transfer can take place so that, at equilibrium, the temperatures are the same. This suggests that in isolated systems, an increase in entropy is associated with a relaxing of constraints.

Note that (7.1.14) provides the criterion for identifying equilibrium both in isolated systems and in closed, insulated systems. In both cases, a spontaneous change of state can never reduce the system's entropy; so, the equilibrium state has the largest value of entropy that is consistent with the values of the intensive properties used to identify the state.

7.1.3 Example

How does the entropy of an isolated system respond when two system parts, initially at different temperatures, are brought into contact during a workfree process?

Consider a chamber bound by rigid, impermeable nonconducting walls, as in Figure 7.2. The chamber is divided in two by a partition that is also rigid, impermeable, and nonconducting; however, the partition is removable. On one side of the partition we place one mole of a copper alloy having heat capacity $c_{vs} = 3R$. The alloy is initially at temperature T_s. On the other side of the partition we place one mole of an ideal gas whose heat capacity is $c_{vg} = 3R/2$. The gas is initially at temperature T_g.

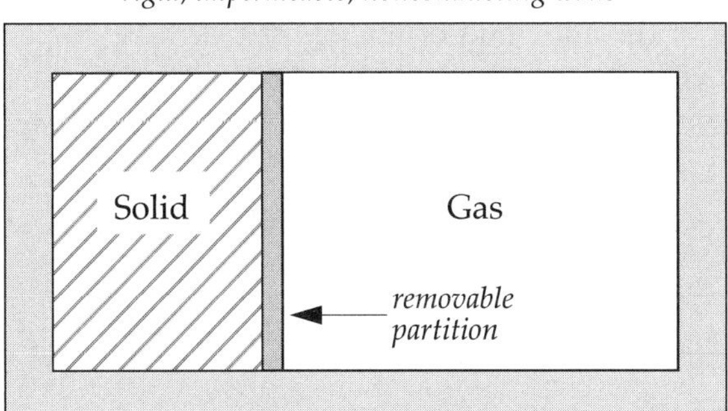

Figure 7.2 Schematic of an isolation chamber divided into two parts. A solid in one part is initially separated by a removable partition from a gas in the other part.

So we have an isolated system divided into two parts; each part is initially in its own equilibrium state, as identified by its intensive properties. To initiate a spontaneous process we relax a constraint: we remove the partition. Our objective is to test (7.1.18); that is, we want to show that no matter what values are used for the initial temperatures of the gas and alloy, the total entropy never decreases. Note that since one part is a solid, no mass transfer occurs between parts: each part is closed.

First we determine the final states of the two parts. After the partition is removed and equilibrium is reestablished, the gas and the alloy are in thermal equilibrium with one another; that is, they have the same final temperature,

$$T_f \equiv T_{sf} = T_{gf} \tag{7.1.19}$$

Applying the first law to the total isolated system, we have

$$\Delta U = \Delta U_g + \Delta U_s = 0 \tag{7.1.20}$$

and since the heat capacities are constant

$$N_g c_{vg}(T_f - T_g) + N_s c_{vs}(T_f - T_s) = 0 \tag{7.1.21}$$

So

$$T_f = \frac{N_g c_{vg} T_g + N_s c_{vs} T_s}{N_g c_{vg} + N_s c_{vs}} \tag{7.1.22}$$

Substituting values for the numbers of moles and heat capacities, we find

$$T_f = \frac{1}{3}T_g + \frac{2}{3}T_s \tag{7.1.23}$$

Now we obtain the change in total entropy. For the entire system we can write

$$\Delta S = \Delta S_g + \Delta S_s \tag{7.1.24}$$

The entropy change of the alloy and of the gas can each be obtained by integrating (3.3.11) over an isometric change on a closed system,

$$\Delta S = N\int_{T_1}^{T_2} \frac{c_v}{T}\, dT = N c_v \ln\frac{T_2}{T_1} \tag{7.1.25}$$

Using (7.1.25) for both the gas and the alloy, (7.1.24) becomes

$$\Delta S = N_g c_{vg} \ln\frac{T_f}{T_g} + N_s c_{vs} \ln\frac{T_f}{T_s} \tag{7.1.26}$$

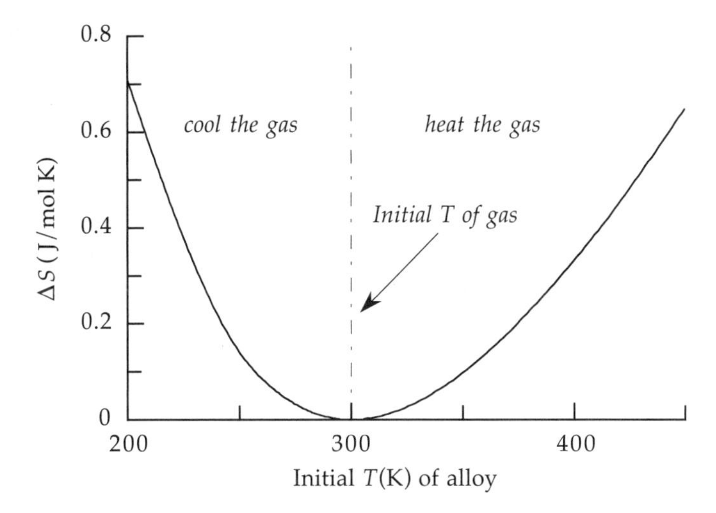

Figure 7.3 Change in total entropy caused by spontaneous workfree heat transfer in the isolated system shown in Figure 7.2. No matter whether the gas is heated or cooled by the solid, the total entropy increases, in agreement with (7.1.18).

Since the mole numbers are the same for each part, we write $N = N_g = N_s$ and substitute (7.1.23) for the final temperature to find

$$\Delta S = 3NR\ln\left[\frac{(1+2(T_s/T_g))^{1/2}(2+T_g/T_s)}{3^{3/2}}\right] \tag{7.1.27}$$

This shows that ΔS is determined solely by the initial temperatures of the gas and the alloy. Further, (7.1.27) shows that ΔS has its minimum value (= 0) when $T_s = T_g$; otherwise, $\Delta S > 0$. Figure 7.3 shows values for ΔS computed from (7.1.27) over a range of temperatures initially assigned to the alloy, with the gas always initially at 300 K. Note that

$$\lim_{T_g \to T_s}\left(\frac{\partial \Delta S}{\partial T_g}\right)_{VN} = 0 \tag{7.1.28}$$

This identifies the minimum in the curve shown in the figure and is consistent with the equality in (7.1.18). Otherwise, the plot shows that the total entropy always increases, in agreement with (7.1.18), no matter whether heat is transferred from the gas to the solid or vice versa.

7.1.4 Isometric Processes with the Same Initial and Final Temperatures

In § 7.1.2 we showed that, for adiabatic processes occurring in closed systems, the combined laws (7.1.11) reduce to a requirement that the system entropy must always increase or remain constant. But if the system can exchange heat with its surroundings, then the entropy may increase, decrease, or remain constant, so for nonadiabatic processes, the entropy no longer serves as an indicator for changes. In this and the next section, we find the appropriate indicators for two processes that will be of particular use in describing phase equilibrium.

Consider the general closed-system situation shown in Figure 7.1, but now let the system boundary be rigid, impermeable, and thermally conducting. Further, let the surroundings be a heat reservoir at a constant temperature T_{sur}. If the system is heterogeneous, then each part is closed to mass transfer, but all parts are in thermal contact with one another. As in § 7.1.2 we want to learn how the system spontaneously responds when its equilibrium is disturbed. We first consider a finite response with N and V fixed, so the finite form of the combined laws (7.1.12) reduces to

$$N\Delta u - T_{sur} N\Delta s \le 0 \qquad \textit{fixed } N \textit{ and } V \qquad (7.1.29)$$

This is almost a change in Helmholtz energy, so we apply the Legendre transform (3.2.11) for A to the entire system and to each part,

$$A = \sum_k N_k a_k = \sum_k N_k (u_k - T_k s_k) \qquad (7.1.30)$$

When the closed parts undergo finite changes of state, this becomes

$$\Delta A = \sum_k [\Delta(N_k u_k) - \Delta(N_k T_k s_k)] \qquad (7.1.31)$$

$$\sum_k \Delta(N_k a_k) = N\Delta u - \sum_k \Delta(N_k T_k s_k) \qquad (7.1.32)$$

Combining (7.1.29) with (7.1.32) to eliminate $N\Delta u$, we have

$$\sum_k \Delta(N_k a_k) + \sum_k \Delta[(T_k - T_{sur}) N_k s_k] \le 0 \qquad (7.1.33)$$

Note that only initial and final values appear for properties of system parts.

Before the process starts and after it ends, all system parts are in thermal equilibrium with one another and with the surroundings; therefore,

$$T_{ki} = T_{kf} = T_{sur} \qquad \textit{for all parts } k \qquad (7.1.34)$$

Here subscript i indicates initial value and subscript f indicates final value. Using (7.1.34) in (7.1.33) leaves

$$\Delta A = \sum_k \Delta(N_k a_k) \le 0 \qquad (7.1.35)$$

During an irreversible process the temperatures T_k are undefined; nevertheless, (7.1.35) still applies, so long as all parts finally reach equilibrium at the temperature

T_{sur} of the surroundings. If the system is homogeneous, then only one term appears in the sum, and (7.1.35) reduces to $\Delta a \leq 0$. So when a fixed *NV* system is in thermal contact with a heat bath and is subjected to a change of state, the system's spontaneous response is confined to processes that either lower the Helmholtz energy or leave it unchanged.

A similar restriction applies to any differential response; for example, consider a fixed *NV* system initially in equilibrium with a surrounding heat bath. If the equilibrium state is differentially disturbed, the response is differential, so throughout the response all $T_k = T_{sur}$. That is, in the differential case, the process is isothermal and (7.1.35) becomes

$$dA \leq 0 \qquad \text{fixed } N, V, T \tag{7.1.36}$$

Furthermore, throughout any reversible change all system temperatures are the same, $T_k = T_{sur} \equiv T$, the equality in the combined first and second laws (7.1.11) applies, and since N and V are fixed, (7.1.11) reduces to

$$dA = dU - TdS = TdS - PdV - TdS = 0 \tag{7.1.37}$$

Therefore, the equality in (7.1.35) applies to reversible changes, while the inequality applies to irreversible processes. If two states have the same values for T, V, and A, so $A_f = A_i$, then the system can evolve along some reversible path between the initial and final states. If the two states have $A_f < A_i$, then the system spontaneously evolves from the initial to the final state along some irreversible path. However if $A_f > A_i$, then the system *cannot* spontaneously evolve from the initial to the proposed final state.

Equation (7.1.35) provides the criterion for identifying equilibrium in *NVT* systems: since a spontaneous change can never increase the Helmholtz energy, the equilibrium state is that state having the smallest value of A that is consistent with the values of those intensive properties used to identify the state. For any *NVT* system the necessary and sufficient condition for equilibrium is that the total Helmholtz energy be a minimum.

7.1.5 Processes with the Same Initial and Final Temperatures and Pressures

By a procedure that is exactly analogous to what we have just done for *NVT* systems, we may also deduce the criteria for equilibrium in *NPT* systems. We again start from the general closed-system situation shown in Figure 7.1, but now consider the special case in which the surroundings contain a heat reservoir at temperature T_{sur} and a mechanical reservoir at pressure P_{sur}. The boundary between system and surroundings is impermeable but flexible and conducting. If the system is heterogeneous, system parts are initially in thermal and mechanical contact with one another.

For a finite response to a disturbance, the combined laws are still

$$N\Delta u + NP_{sur}\Delta v - NT_{sur}\Delta s \leq 0 \tag{7.1.12}$$

As before we introduce a Legendre transform, in this case (3.2.13) which defines the Gibbs energy. In addition, the system is in equilibrium both before and after the response, so analogous to (7.1.34) we have

$$T_{ki} = T_{kf} = T_{sur} \qquad \textit{for all parts k} \qquad (7.1.34)$$

and

$$P_{ki} = P_{kf} = P_{sur} \qquad \textit{for all parts k} \qquad (7.1.38)$$

However, these relations constrain only the initial and final states of the system; during the process, the system may be out of equilibrium, so the temperatures T_k and pressures P_k are undefined. Combining (3.2.13), (7.1.12), (7.1.34) and (7.1.38) leaves

$$\Delta G \leq 0 \qquad \textit{fixed N with (7.1.34) \& (7.1.38)} \qquad (7.1.39)$$

This result limits finite processes that may occur in closed systems that are in thermal contact with a heat bath and in mechanical contact with a constant-pressure reservoir. For a differential disturbance, arguments analogous to those leading to (7.1.36) can be repeated, giving

$$dG \leq 0 \qquad \textit{fixed N, T, P} \qquad (7.1.40)$$

For reversible changes, all system temperatures and pressures satisfy $T_k = T_{sur}$ and $P_k = P_{sur}$ throughout the process and (7.1.40) reduces to the equality. This means that the equality in (7.1.40) applies to reversible changes, while the inequality applies to irreversible processes. Equation (7.1.40) provides the criterion for identifying equilibrium in *NPT* systems: a spontaneous change of state can never increase the Gibbs energy; therefore, the necessary and sufficient condition for equilibrium is that the total Gibbs energy be a minimum.

For other kinds of processes, other criteria apply. For example, any spontaneous isometric-isentropic process must have

$$dU \leq 0 \qquad \textit{fixed N, V, S} \qquad (7.1.41)$$

and the criterion for equilibrium is that the internal energy be a minimum. Similarly, spontaneous isobaric-isentropic processes have

$$dH \leq 0 \qquad \textit{fixed N, P, S} \qquad (7.1.42)$$

and the criterion for equilibrium is that the enthalpy be a minimum. Just as in (7.1.14), (7.1.36), and (7.1.39), the equalities in (7.1.41) and (7.1.42) apply to reversible changes, while the inequalities apply to irreversible processes. However, (7.1.41) and (7.1.42) are generally of less practical use than (7.1.14), (7.1.36), or (7.1.39).

Finally, we caution that U, S, H, A, and G are all state functions, so for specified initial and final states, ΔU, ΔS, ΔH, ΔA, and ΔG are each process independent. However, any criterion for equilibrium is restricted to a particular kind of process. For example, we may certainly contrive a real process that has $\Delta S < 0$ without violating (7.1.14). However, (7.1.14) guarantees us that if a process does have $\Delta S < 0$, then either the process is not adiabatic, or the system is not of constant mass, or both.

7.1.6 Example

How can the combined laws be used to test the feasibility of proposed processes?

Saturated steam, initially at 100°C, is to be completely condensed to liquid. Determine whether the condensation can be done (a) isothermally and isobarically and whether it can be done (b) adiabatically. If either process is possible, determine bounds on the heat and work that would be required. At 100°C, 1.013 bar, saturated steam tables give the values in Table 7.1.

Isothermal-isobaric condensation. To be possible, this process must satisfy the form of the combined laws appearing in (7.1.40), that is,

$$\Delta g = \Delta h - \Delta(Ts) = \Delta h - T\Delta s \leq 0 \tag{7.1.43}$$

Using values from Table 7.1, we find

$$\Delta g = (419.1 - 2676) - 373.15(1.307 - 7.355) = -0.10 \text{ J/g} \tag{7.1.44}$$

Although the sign of the answer is negative, its value is essentially zero within the uncertainties with which properties can be measured. For example, discrepancies of only ± 0.01% in the enthalpies could cause an uncertainty of ± 0.3 J/g in Δg. Nevertheless, we judge that isothermal-isobaric condensation is thermodynamically possible.

Bounds on q and w are obtained by assuming a reversible change; then,

$$w = -P\Delta v = -1.013(1.044 - 1673)(\text{cc bar/g})(1 \text{ J}/10 \text{ cc bar}) = 169 \text{ J/g} \tag{7.1.45}$$

and

$$q = \Delta h = (419 - 2676) = -2257 \text{ J/g} \tag{7.1.46}$$

So to accomplish the proposed condensation, a small amount of work would have to be done on the steam and a large amount of heat would have to be removed. These are the optimal values for q and w; in a real process more work would have to be done and more heat would have to be removed. However in a real condensation, the applied pressure would not have to be much more that the saturation pressure; in that case, the above values for q and w would be close to the actual values.

Adiabatic condensation. The question here is whether we can force condensation by some adiabatic change in the volume. An adiabatic process in a closed system must satisfy the combined laws in the form of (7.1.14); that is, we must have $\Delta S \geq 0$. In Table 7.1 we find the entropy of saturated steam to be s_{vap} = 7.355 J/(g K); so, to achieve an

Table 7.1 Properties of saturated liquid water and saturated steam at 100°C and 1.013 bar

	v(cc/g)	h(J/g)	s(J/g K)
saturated liquid	1.044	419.1	1.307
saturated vapor	1673	2676	7.355

adiabatic condensation, we must find a saturated liquid state that has a higher entropy. But steam tables show that no liquid state has $s_{liq} > 7.355$ J/(g K). This means that without transferring heat, it is not possible to condense all the steam, no matter how we might contrive to manipulate the system volume. However, note that the reverse process is thermodynamically possible; that is, we can flash saturated liquid water by adiabatically increasing the volume.

7.1.7 Selected Processes in Closed Heterogeneous Systems

In § 7.1.4 and 7.1.5 we developed constraints that apply to several kinds of processes:

$$dS \geq 0 \qquad \text{fixed } N, V, U \tag{7.1.18}$$

$$dA \leq 0 \qquad \text{fixed } N, V, T \tag{7.1.36}$$

$$dG \leq 0 \qquad \text{fixed } N, P, T \tag{7.1.40}$$

$$dU \leq 0 \qquad \text{fixed } N, V, S \tag{7.1.41}$$

$$dH \leq 0 \qquad \text{fixed } N, P, S \tag{7.1.42}$$

These constraints apply to both homogeneous and heterogeneous closed systems; in heterogeneous systems, the system parts can exchange both matter and energy with other parts, although they cannot exchange matter with the surroundings.

We consider a special set of heterogeneous systems in which all system parts have the same temperature T and same pressure P. These conditions usually apply when the parts are phases in contact with one another. In these cases, we find that the above constraints all take the *same* form.

First, consider a system at fixed N, V, T. If we substitute the fundamental equation (3.2.27) for dA into the constraint (7.1.36), we have

$$dA = -\sum_k N_k s_k dT - \sum_k P d(N_k v_k) + \sum_k \sum_i \overline{G}_{ki} dN_{ki} \leq 0 \tag{7.1.47}$$

Here $\overline{G}_{ki}$ is the chemical potential for component i in part k of the system. However, any isothermal-isometric process has $dV = \Sigma_k d(N_k v_k) = 0$ and $dT = 0$, so (7.1.47) becomes

$$\sum_k \sum_i \overline{G}_{ki} dN_{ki} \leq 0 \tag{7.1.48}$$

Now consider a system at fixed N, P, T. For these cases, we substitute the fundamental equation (3.2.28) for dG into the constraint (7.1.40) and obtain

$$dG = -\sum_k N_k s_k dT + \sum_k N_k v_k dP + \sum_k \sum_i \overline{G}_{ki} dN_{ki} \leq 0 \tag{7.1.49}$$

But an isothermal-isobaric process has $dT = 0$ and $dP = 0$, so (7.1.49) reduces to (7.1.48).

The form (7.1.48) is also obtained when the fundamental equation (3.2.25) for dU is substituted into the constraint (7.1.41) for fixed N, V, S systems and into the constraint (7.1.18) for fixed N, V, U systems. It is also obtained when the fundamental equation (3.2.26) for dH is substituted into the constraint (7.1.42) for fixed N, P, S systems. As with other constraints derived in § 7.1.4–7.1.5, the equality in (7.1.48) applies to reversible changes, while the inequality applies to irreversible processes.

Although the one form (7.1.48) applies to several kinds of processes, the quantity on the lhs is identified with the Gibbs energy only when T and P are the quantities held fixed. When other quantities are fixed, the lhs takes other names, and for this reason Prigogine and Defay identify the lhs of (7.1.48) as proportional to the *affinity* [1]. However, we reserve this name for the analogous quantity that arises in chemical reaction equilibria (§ 7.4.4).

In addition to its generality, the form (7.1.48) is important because it leads to a computational strategy for analyzing phase-equilibrium situations. In that strategy, a phase-equilibrium problem is treated as a multivariable optimization in which the lhs of (7.1.48) is the quantity to be minimized. An alternative strategy, in which the computational problem is to solve a set of coupled nonlinear algebraic equations, arises from the constraints on open-system processes developed in § 7.2.

7.2 THE LAWS FOR OPEN NONREACTING SYSTEMS

In this section we develop the combined laws for nonreacting systems that are open to mass transfer. Consider a heterogeneous system composed of three parts: bulk phases α and β plus an interface I between them, as in Figure 7.4. Each part contains C components, and the state of each is identified by a temperature, a pressure, and a set of mole numbers. Specifically, phase α has T^{α}, P^{α}, and total moles N^{α}; phase β has T^{β}, P^{β}, and total moles N^{β}; the interface has T^{I}, P^{I}, and total moles N^{I}. The component chemi-

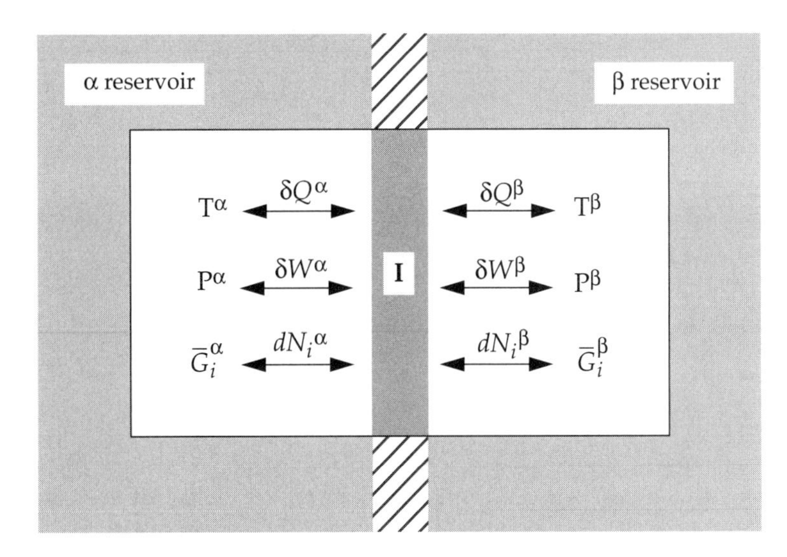

Figure 7.4 A three-part system composed of bulk phases α and β open to material and energy transfers across interface I. The α reservoir maintains constant T and P in the α phase; β reservoir does the same for β phase. (Interface thickness exaggerated for clarity.)

cal potentials are $\overline{G}_i^{\alpha}$ and $\overline{G}_i^{\beta}$. Each phase is immersed in its own constant *T-P* reservoir; we can adjust *T* and *P* in each phase independently by adjusting *T* and *P* of each reservoir.

The interface itself has negligible mass compared to the masses of the phases, and during processes, states of the interface may be undefined or undefinable. We will treat the interface as an open system and interpret each phase as a "port" for the other phase; that is, the open-system energy and entropy balances from § 2.4 will apply. In what follows, we first derive the combined first and second laws (§ 7.2.1). Then we find limits on the directions (§ 7.2.2) and magnitudes (§ 7.2.3) of mass and energy transfers between phases α and β.

7.2.1 Combined Laws

Consider a differential process that transfers material and energy across an interface of negligible mass, as in Figure 7.4. We choose the interface to be the system and write material, energy, and entropy balances for it. Since no accumulation can occur in an interface of negligible mass, those balance equations are merely

$$dN_i^I = 0 \qquad \textit{for } i = 1, 2, \dots, C \tag{7.2.1}$$

$$d(N^I u^I) = 0 \tag{7.2.2}$$

$$d(N^I s^I) = 0 \tag{7.2.3}$$

We have ignored kinetic and potential energy contributions to the energy balance. By identifying the interface as the system, we are able to treat the phases α and β as ports through which material and energy are exchanged with the interface. But we prefer to express changes in terms of properties of the bulk phases, so we will replace interfacial quantities in (7.2.1)–(7.2.3) with quantities pertaining to the phases. For example, the material balance (7.2.1) can be expressed in terms of bulk-phase mole numbers as

$$dN_i^I = -dN_i^{\alpha} - dN_i^{\beta} = 0 \qquad \textit{for } i = 1, 2, \dots, C \tag{7.2.4}$$

The energy balance (7.2.2) represents the open-system form of the first law (2.4.15), which can be written here as

$$d(N^I u^I) = \sum_{k=\alpha}^{\beta} d(h^k N^{k,I}) + \delta Q + \delta W_b + \delta W_{sh} = 0 \tag{7.2.5}$$

This differs from (2.4.15) in that we now allow the intensive states of phases α and β to vary during processes, whereas in (2.4.15) intensive states of feed and discharge streams were assumed to be constant. To obtain the quantities in (7.2.5) in terms of bulk-phase quantities, we employ the following observations.

(a) Since any material entering the $\alpha(\beta)$ side of the interface must necessarily have come from the $\beta(\alpha)$ phase, we have $d(h^k N^{k,I}) = -d(h^k N^k)$, for $k = \alpha$ or β.

(b) Any heat crossing the interface results from net heat transfer between the two phases, so $\delta Q = -\delta Q^{\alpha} - \delta Q^{\beta}$.

(c) Since the interface has negligible mass, no work is involved in deforming the shape of the interface and therefore $\delta W_b = 0$.

(d) But shaft work could be done on or by the interface, so $\delta W_{sh} = -\delta W^{\alpha} - \delta W^{\beta}$; for example, this shaft work could be that which displaces the interface in response to changes in volumes of the phases.

Making all these substitutions, (7.2.5) becomes

$$d(h^{\alpha}N^{\alpha}) + \delta Q^{\alpha} + \delta W^{\alpha} + d(h^{\beta}N^{\beta}) + \delta Q^{\beta} + \delta W^{\beta} = 0 \tag{7.2.6}$$

Now consider the entropy balance (7.2.3) on the interface, which is given by a form of (2.4.21) that is analogous to (7.2.5),

$$d(N^I s^I) = \sum_{k=\alpha}^{\beta} \frac{\delta Q^{k,\,ext}}{T^{k,\,ext}} + \sum_{k=\alpha}^{\beta} d(s^k N^{k,\,I}) + dS^I_{gen} = 0 \tag{7.2.7}$$

where the entropy generated in the interface is $dS^I_{gen} \geq 0$. To replace the interfacial quantities in (7.2.7) with bulk-phase quantities, we use $d(s^k N^{k,I}) = -d(s^k N^k)$ and we use $\delta Q^{k,ext} = -\delta Q^k$, where $k = \alpha$ or β. Then (7.2.7) becomes

$$-\left(\frac{\delta Q^{\alpha}}{T^{\alpha}} + d(s^{\alpha}N^{\alpha}) + \frac{\delta Q^{\beta}}{T^{\beta}} + d(s^{\beta}N^{\beta})\right) + dS^I_{gen} = 0 \tag{7.2.8}$$

We use (7.2.8) to eliminate δQ^{α} from the energy balance (7.2.6); we find

$$\left(\frac{T^{\alpha}}{T^{\beta}} - 1\right)\delta Q^{\beta} - (\delta W^{\alpha} + \delta W^{\beta}) - \sum_{k=\alpha}^{\beta} [d(h^k N^k) - T^{\alpha} d(s^k N^k)] = T^{\alpha} dS^I_{gen} \geq 0 \tag{7.2.9}$$

The terms under the sum in (7.2.9) result from mass transfer across the interface, so those terms are zero when the phases are closed to one another. Then we are left with

$$\left(\frac{T^{\alpha}}{T^{\beta}} - 1\right)\delta Q^{\beta} - (\delta W^{\alpha} + \delta W^{\beta}) \geq 0 \qquad \textit{closed } \alpha \textit{ and } \beta \tag{7.2.10}$$

When the phases are open to one another with T and P fixed in each phase, then each enthalpy term in (7.2.9) can be replaced with

$$d(h^k N^k) = \sum_i \overline{H}^k_i \, dN^k_i \qquad \textit{fixed } T^k \textit{ and } P^k \tag{7.2.11}$$

and likewise, each entropy term can be written as

$$d(s^k N^k) = \sum_i \bar{S}_i^k \, dN_i^k \qquad \text{fixed } T^k \text{ and } P^k \qquad (7.2.12)$$

We use the material balances (7.2.4) to replace each dN_i^β with $(-dN_i^\alpha)$ and use a Legendre transform to introduce the chemical potentials for components in phase α. So (7.2.9) finally becomes

$$\left(\frac{T^\alpha}{T^\beta} - 1\right)\delta Q^\beta - (\delta W^\alpha + \delta W^\beta) - \sum_i [\bar{G}_i^\alpha - (\bar{H}_i^\beta - T^\alpha \bar{S}_i^\beta)] dN_i^\alpha \geq 0 \qquad (7.2.13)$$

Equation (7.2.13) is a form of the combined first and second laws describing processes in which material and energy cross an interface between bulk phases that are each at their own fixed T and P. When only energy can be transferred between the phases, then (7.2.13) reduces to (7.2.10). We now deduce limitations on the directions and magnitudes of transfers by considering special cases of (7.2.10) and (7.2.13); the special cases arise by applying constraints to the interface.

7.2.2 Limits on the Directions of Irreversible Transfers ($dS_{gen} > 0$)

Here we deduce bounds on the directions of irreversible transfers across the interface in Figure 7.4. We consider six processes: workfree constant-mass heat transfer, adiabatic constant-mass work, isobaric constant-mass heat transfer, isothermal constant-mass work, isothermal-isobaric diffusion, and adiabatic-workfree diffusion.

Workfree, constant-mass heat transfer. Let the interface in Figure 7.4 be impermeable, thermal conducting, and fixed in position. When the interface is impermeable, then each phase is closed; when the interface is fixed in position, then the volumes of the two phases are constant: V^α = constant and V^β = constant. We initiate a process by adjusting the reservoirs so the phases have different pressures and temperatures. For this situation, the closed-system form (7.2.10) of the combined laws reduces to

$$\left(\frac{T^\alpha}{T^\beta} - 1\right)\delta Q^\beta \geq 0 \qquad (7.2.14)$$

This constraint applies to heat crossing the interface in either direction, but to have a particular example, say the process transfers heat from phase α to phase β. Then $\delta Q^\beta > 0$, and the inequality in (7.2.14) requires $T^\alpha > T^\beta$; that is, the temperature difference $(T^\alpha - T^\beta)$ drives workfree, constant-mass heat transfer. For such processes, heat *always* flows from regions of high temperature to regions of low temperature.

Adiabatic constant-mass work. Now let the interface be impermeable, thermally nonconducting, and movable. We initiate an adiabatic process by again adjusting the reservoirs so the phases have different temperatures and pressures. Under these conditions, the closed-system form (7.2.10) of the combined laws reduces to

$$-(\delta W^{\alpha} + \delta W^{\beta}) \geq 0 \tag{7.2.15}$$

In this development we have located all irreversibilities in the interface, and since the work terms in (7.2.15) are external to the interface, each is reversible,

$$\delta W = -P dV \tag{7.2.16}$$

Furthermore, the total volume is constant, so $dV^{\beta} = -dV^{\alpha}$; therefore, the work terms in (7.2.15) can be written as

$$-(\delta W^{\alpha} + \delta W^{\beta}) = (P^{\alpha} - P^{\beta}) dV^{\alpha} \tag{7.2.17}$$

Combining (7.2.17) with (7.2.15) leaves

$$(P^{\alpha} - P^{\beta}) dV^{\alpha} \geq 0 \tag{7.2.18}$$

Consider expansion of α against β, so $dV^{\alpha} > 0$. Then, we must have $P^{\alpha} > P^{\beta}$ to make the lhs of (7.2.18) positive. Similarly, when phase α is compressed, then $dV^{\alpha} < 0$ and (7.2.18) requires $P^{\alpha} < P^{\beta}$. That is, for both expansions and compressions of phase α, the pressure difference $(P^{\alpha} - P^{\beta})$ drives an adiabatic, constant-mass change of volume, and the work associated with such volume changes "flows" from regions of high pressure to regions of low pressure. Similar statements apply for other work modes.

Isobaric, constant-mass heat transfer. Let the interface in Figure 7.4 be impermeable, thermal conducting, and movable. We manipulate the reservoirs so the pressures are the same in the two phases; thereafter, the interface moves in response to any (differential) pressure difference so we have $P^{\alpha} = P^{\beta} =$ constant. We initiate a process by adjusting the reservoirs so the temperatures differ in the two phases. Since the interface is impermeable, the closed-system form (7.2.10) of the combined laws applies,

$$\left(\frac{T^{\alpha}}{T^{\beta}} - 1\right) \delta Q^{\beta} - (\delta W^{\alpha} + \delta W^{\beta}) \geq 0 \tag{7.2.10}$$

We again use (7.2.17) for the work terms, obtaining

$$\left(\frac{T^{\alpha}}{T^{\beta}} - 1\right) \delta Q^{\beta} + (P^{\alpha} - P^{\beta}) dV^{\alpha} \geq 0 \tag{7.2.19}$$

But the pressures are balanced, so this reduces to

$$\left(\frac{T^{\alpha}}{T^{\beta}} - 1\right) \delta Q^{\beta} \geq 0 \tag{7.2.20}$$

which is the same as in (7.2.14) for workfree heat transfer. Consequently, a temperature difference drives isobaric heat transfer, and heat *always* flows from regions of high temperature to regions of low temperature.

Isothermal constant-mass work. Let the interface remain impermeable, thermally conducting, and movable, as above. But now adjust the reservoirs so the phases have the same temperature; thereafter, the interface conducts heat in response to any (differential) temperature difference so that $T^\alpha = T^\beta$ = constant. Then further adjust the reservoirs so the phases initially have different pressures; the pressure difference moves the interface, so one phase does work against the other. Under these conditions, the closed-system form (7.2.10) of the combined laws reduces to

$$-(\delta W^\alpha + \delta W^\beta) \ge 0 \tag{7.2.21}$$

We again use (7.2.17) for the work terms, finding

$$(P^\alpha - P^\beta)dV^\alpha \ge 0 \tag{7.2.22}$$

which is the same as (7.2.18). So we find that a pressure difference drives an isothermal volume change, and work always "flows" from regions of high pressure to regions of low pressure.

Isothermal-isobaric single-component diffusion. Now consider both phases to contain samples of the same pure component, and let the interface between them be thermally conducting, movable, and permeable. Adjust the reservoirs so the phases have the same T and P; then we have $T^\alpha = T^\beta$ = constant and $P^\alpha = P^\beta$ = constant. The process is diffusion of the pure component across the interface. Under these restrictions, the combined laws (7.2.13) reduce to

$$-(\delta W^\alpha + \delta W^\beta) - [\overline{G}_1^\alpha - (\overline{H}_1^\beta - T^\alpha \overline{S}_1^\beta)]dN_1^\alpha \ge 0 \tag{7.2.23}$$

Since $T^\alpha = T^\beta$, we can use a Legendre transform ($G = H - TS$) for the β-phase to simplify (7.2.23) to

$$-(\delta W^\alpha + \delta W^\beta) - [\overline{G}_1^\alpha - \overline{G}_1^\beta]dN_1^\alpha \ge 0 \tag{7.2.24}$$

We now use (7.2.17) for the two work terms,

$$(P^\alpha - P^\beta)dV^\alpha - [\overline{G}_1^\alpha - \overline{G}_1^\beta]dN_1^\alpha \ge 0 \tag{7.2.25}$$

but the pressures are the same in the two phases, so we are left with just the term for diffusion of pure-component 1 across the interface,

$$-[\overline{G}_1^\alpha - \overline{G}_1^\beta]dN_1^\alpha \ge 0 \tag{7.2.26}$$

But a pure-substance chemical potential is merely the molar Gibbs energy, so (7.2.26) can also be written as

$$-[g_1^\alpha - g_1^\beta]\, dN_1^\alpha \geq 0 \tag{7.2.27}$$

This constraint applies to diffusion in either direction, but to have an example, assume the pure substance diffuses from the α phase to the β phase. Then $dN_1^\alpha < 0$, and the inequality in (7.2.27) can be satisfied only when $g_1^\alpha > g_1^\beta$; that is, a difference in the molar Gibbs energies drives isothermal-isobaric diffusion of a pure substance. In such cases, the pure substance always diffuses from regions of high g_1 to regions of low g_1. This means that for single-component diffusion occurring at fixed temperature and pressure, the component never diffuses against a gradient of its molar Gibbs energy. However, a pure component may diffuse against its density gradient; for example, pure vapor may condense to liquid.

Equation (7.2.27) may also be expressed in terms of fugacity; to do so, we first integrate the definition of fugacity (4.3.10), at fixed T, from the β-phase to the α-phase. The result is analogous to the algebraic form (4.3.12),

$$g_1^\alpha - g_1^\beta = RT \ln \frac{f_1^\alpha}{f_1^\beta} \tag{7.2.28}$$

Then substituting (7.2.28) into (7.2.27) gives

$$-RT \ln\left(\frac{f_1^\alpha}{f_1^\beta}\right) dN_1^\alpha \geq 0 \tag{7.2.29}$$

Since we have chosen $dN_1^\alpha < 0$, the inequality in (7.2.29) can be satisfied only when $f_1^\alpha > f_1^\beta$; that is, a difference in fugacities is equivalent to a difference in chemical potentials, and for isothermal-isobaric diffusion of one component, the component always diffuses from regions of high fugacity to regions of low fugacity. Equation (7.2.29) illustrates that the fugacity is fully equivalent to the chemical potential: fugacity is more important and informative than might be construed from the common interpretation that fugacity is merely a "corrected" pressure.

Isothermal-isobaric multicomponent diffusion. The constraint (7.2.27) applies to single-component diffusion; now we consider multicomponent diffusion across the interface in Figure 7.4. Using the same analysis as used above for single-component diffusion, the combined laws (7.2.13) reduce to

$$-\sum_i [\overline{G}_i^\alpha - (\overline{H}_i^\beta - T^\alpha \overline{S}_i^\beta)]\, dN_i^\alpha \geq 0 \tag{7.2.30}$$

Since we have $T^\alpha = T^\beta$, we can again use a Legendre transform to obtain

$$-\sum_i [\overline{G}_i^\alpha - \overline{G}_i^\beta]\, dN_i^\alpha \geq 0 \tag{7.2.31}$$

This constraint can also be expressed in terms of fugacity; to do so, we proceed analogously to what was done above to obtain (7.2.29). The result is

$$-RT \sum_i \ln\left(\frac{f_i^\alpha}{f_i^\beta}\right) \geq 0 \tag{7.2.32}$$

For irreversible diffusion, the inequalities in (7.2.31) and (7.2.32) apply; we now identify two ways by which such diffusional processes can satisfy those inequalities.

Uncoupled diffusion. If the diffusion of each component is unaffected by the diffusion of all other components, then the only general way by which (7.2.31) can be satisfied is term-by-term. That is, each term in the sum in (7.2.31) obeys the single-component constraint (7.2.27), so at fixed T and P, each component can only diffuse from regions of high chemical potential (fugacity) to regions of low chemical potential (fugacity). In uncoupled, isothermal-isobaric diffusion, a component never diffuses against its chemical potential gradient, although, if the solution is sufficiently nonideal, some components may diffuse against their concentration gradients.

Coupled diffusion. But in addition, the diffusion of components may be coupled, as alluded to by Prigogine and Defay [1]. In these situations, (7.2.31) need not be satisfied term-by-term; the only thermodynamic constraint is that the entire sum satisfy (7.2.31). As an example, consider binary diffusion in an isothermal workfree process; then (7.2.31) reduces to two terms,

$$-[\overline{G}_1^\alpha - \overline{G}_1^\beta]\, dN_1^\alpha - [\overline{G}_2^\alpha - \overline{G}_2^\beta]\, dN_2^\alpha \geq 0 \tag{7.2.33}$$

This inequality can still be satisfied, even when one term is negative, so long as dN_1^α is coupled to dN_2^α in such a way that the negative term on the lhs is always dominated by the positive term. When $\overline{G}_1^\alpha > \overline{G}_1^\beta$, we expect $dN_1^\alpha < 0$; that is, we expect component 1 to diffuse along its chemical potential gradient from phase α to phase β. But if this occurs with dN_1^α coupled to dN_2^α, then it is possible to also have $dN_2^\alpha < 0$, even when $\overline{G}_2^\alpha < \overline{G}_2^\beta$. In this case component 2 also diffuses from phase α to phase β, but it does so *against its chemical potential gradient*. One way this can occur is when molecules of different substances solvate so strongly that they diffuse together—molecules of one species effectively "carry" those of another species. We will find in § 7.4.4 that this kind of behavior can also occur in systems undergoing coupled chemical reactions when driven by finite reaction rates.

Adiabatic workfree diffusion. Finally, let the interface be thermally nonconducting, fixed in position, and permeable. Now T and P can differ in the two phases, but neither heat nor work can be transferred across the interface. Then the combined laws (7.2.13) reduce to

$$-\sum_i [(\overline{H}_i^\alpha - \overline{H}_i^\beta) - T^\alpha(\bar{S}_i^\alpha - \bar{S}_i^\beta)]dN_i^\alpha \ge 0 \qquad (7.2.34)$$

As in the isothermal case, adiabatic multicomponent diffusion may be coupled or uncoupled, but here we want to emphasize the roles of enthalpy and entropy differences, so consider uncoupled diffusion. In some mixtures diffusion of component i is driven by the enthalpy difference in (7.2.34); then component i diffuses from regions of high partial molar enthalpy to regions having lower values. In other mixtures, diffusion is driven by the entropy difference, and then component i diffuses from regions of low partial molar entropy to regions having higher values. In still other mixtures the enthalpy and entropy terms in (7.2.34) may either compete, so the net diffusion of i is small, or they may reinforce one another, causing large quantities of i to diffuse. In short, (7.2.34) captures a variety of possible behaviors that can be explained thermodynamically by whether enthalpy and entropy effects are competing or compensating. In later chapters, we will find that the relative sizes of enthalpy and entropy effects can be used to interpret other kinds of behaviors in multicomponent mixtures.

7.2.3 Limits on the Magnitudes of Irreversible Transfers ($dS_{\text{gen}} > 0$)

In the previous section we found that, in certain special cases, the directions of energy and mass transfer are limited by gradients in certain intensive properties. In this section we show that, during irreversible transfers of heat and work, not only are there constraints on the directions, but constraints also apply to the magnitudes. To develop the argument, we reconsider irreversible, isothermal, constant-mass work as discussed in § 7.2.2. For such a process, we have already seen that the combined laws reduce to

$$-(\delta W^\alpha + \delta W^\beta) \ge 0 \qquad \textit{fixed } T \textit{ and } N \qquad (7.2.21)$$

To have a particular example, assume phase α does work on phase β, so $\delta W^\alpha < 0$ and $\delta W^\beta > 0$. Then (7.2.21) becomes

$$-(-|\delta W^\alpha|) - \delta W^\beta > 0 \qquad (7.2.35)$$

or

$$|\delta W^\alpha| > \delta W^\beta \qquad (7.2.36)$$

That is, the amount of work done *by* phase α exceeds that done *on* phase β.

What happens to the extra work done by phase α but not applied to phase β? To answer this question, consider the energy balance (7.2.6) written for our constant-mass process,

$$\delta Q^\alpha + \delta Q^\beta + \delta W^\alpha + \delta W^\beta = 0 \qquad (7.2.37)$$

Hence,

$$\delta Q^{\alpha} + \delta Q^{\beta} = -\delta W^{\alpha} - \delta W^{\beta} > 0 \tag{7.2.38}$$

This shows that heat transfer must occur during our proposed process—the process is isothermal, not adiabatic. (For an adiabatic, constant-mass process, the energy balance (7.2.6) requires $(-\delta W^{\alpha} - \delta W^{\beta}) = 0$, regardless of reversibility.)

Further, the heat transfer is constrained by (7.2.38); that is, (7.2.38) requires

$$-\left|\delta Q^{\beta}\right| + \delta Q^{\alpha} > 0 \tag{7.2.39}$$

Hence,

$$\delta Q^{\alpha} > \left|\delta Q^{\beta}\right| \tag{7.2.40}$$

So if $\delta Q^{\alpha} > 0$, then more heat would appear in the α phase than left the β phase, and if $\delta Q^{\alpha} < 0$, then more heat would appear in the β phase than left the α phase. What is the source of the extra heat?

Since the first law (7.2.37) must be obeyed, we can only conclude that the extra work done by phase α, but not accessible to phase β, is converted into the extra heat that appears in the system. That is, part of the work done by phase α is not "usefully" applied to phase β; instead, it is "lost" in overcoming irreversibilities and is dissipated as heat. A general expression for the lost work is given in (2.3.10). But here the process is isothermal, so $T \equiv T^{\alpha} = T^{\beta} = T_{ext}$ and (2.3.10) reduces to

$$\delta W_{lost} = T dS_{gen} \tag{7.2.41}$$

For our example process, in which phase α does work on phase β, we have

$$\delta W_{lost} = -(-\left|\delta W^{\alpha}\right|) - \delta W^{\beta} > 0 \tag{7.2.42}$$

And, simultaneously, for heat transferred from phase β to phase α, we have

$$\delta W_{lost} = -\left|\delta Q^{\beta}\right| + \delta Q^{\alpha} > 0 \tag{7.2.43}$$

This shows that the amount of useful work is limited because some is used to overcome irreversibilities and is, thereby, converted into heat.

In real processes S_{gen} cannot be computed directly, so we usually account for irreversibilities by using an efficiency to correct results that have been calculated assuming reversible changes. For example, isentropic efficiencies are used to correct results computed for reversible adiabatic work generators (e.g., turbines) and consumers (e.g., compressors). Likewise, in stagewise separation processes, stage or overall efficiencies are used to correct results computed for reversible mass transfer based on phase equilibria. Values for such efficiencies are obtained empirically by observing the performance of a real process over actual changes of state and comparing it to the idealized performance. Note that a reversible change will provide at least one property whose value differs from that for the real process. For example, the outlet temperature from a reversible turbine will always be lower than that from a real turbine operated from the same initial temperature and pressure to the same final pressure.

According to (7.2.41), the lost work in an irreversible process is related to entropy generated in the interface, which is the source of the inequality in the combined laws (7.2.13). In complex processes, each term on the lhs of (7.2.13) can contribute to "wasted" energy. However, in real *chemical* processes, the largest contributor is typically the δQ term; in comparison, the δW and dN terms are often smaller.

7.3 CRITERIA FOR PHASE EQUILIBRIUM

In § 7.2 we used the combined first and second laws to obtain limitations on the directions and magnitudes of irreversible transfers of energy and material between fluid phases. Now we use the combined laws to obtain quantitative criteria for identifying thermodynamic equilibrium. In § 1.2.2 we gave a qualitative description of equilibrium: a situation in which no driving forces are present that could change the state. To make this qualitative statement quantitative, we take advantage of the close connection that exists between equilibrium states and reversible changes of state (see § 1.3).

In reversible changes the entropy generation term is zero, the equalities in the combined laws (7.2.10) and (7.2.13) apply, and the system (the interface in Figure 7.4) is in equilibrium with its surroundings (the bulk phases α and β). Consequently, the equations that constrain the driving forces for reversible changes also describe equilibrium situations. We exploit this observation to obtain criteria for thermal, mechanical, and diffusional equilibria. These criteria are equivalent to the extrema found for conceptual property changes in § 7.1; however, the relations developed here are not connected to any process. Rather, they identify the equilibrium state regardless of how it is achieved.

7.3.1 Thermal Equilibrium ($dS_{gen} = 0$)

For constant-mass heat transfer that is either workfree or isobaric, the combined laws (7.2.10) reduce to

$$\left(\frac{T^{\alpha}}{T^{\beta}} - 1\right)\delta Q^{\beta} \geq 0 \qquad (7.2.14)$$

The inequality in (7.2.14) arises because of entropy generation: the inequality applies to irreversible heat transfer, while the equality applies to heat transfer associated with a reversible change of state. In § 1.3, we identified reversible changes as idealized situations attained when all driving forces and their dissipative components are taken to zero. And in § 7.2.2 we found that, for constant-mass heat transfer that is either isobaric or workfree, the driving force is the temperature difference $\Delta \equiv T^{\alpha} - T^{\beta}$. Let $\mathcal{F}$ represent the dissipative components of the driving force Δ; then, according to (1.3.5), a reversible change results when the limits $\Delta \to 0$ and $\mathcal{F} \to 0$ are taken simultaneously,

$$\lim_{\substack{\mathcal{F}\to 0 \\ \Delta \to 0}} \left(\frac{T^{\alpha}}{T^{\beta}} - 1 \right) \delta Q^{\beta} = 0 \qquad \textit{reversible change} \qquad (7.3.1)$$

We also showed in (1.3.2) that when all driving forces are actually (rather than ideally) taken to zero ($\Delta \to 0$), then we obtain an equilibrium state,

$$\lim_{\Delta \to 0} \left(\frac{T^{\alpha}}{T^{\beta}} - 1 \right) \delta Q^{\beta} = 0 \qquad \textit{equilibrium} \qquad (7.3.2)$$

This limit identifies an equilibrium state provided no other internal or external driving forces exist when $\Delta = 0$. In § 1.3 we discussed the subtle distinction between the idealized limit in (7.3.1) and the physically realizable limit in (7.3.2). In § 7.2.2 we found that when $T^{\alpha} > T^{\beta}$ then $\delta Q^{\beta} > 0$, and when $T^{\alpha} < T^{\beta}$ then $\delta Q^{\beta} < 0$; therefore, when the driving force is zero ($\Delta = 0$), then $T^{\alpha} = T^{\beta}$ and we must also have $\delta Q^{\beta} = 0$. Consequently, when the thermal driving force is zero, we have no heat transfer and the system is said to be in *thermal equilibrium*,

$$T^{\alpha} = T^{\beta} \qquad \textit{thermal equilibrium} \qquad (7.3.3)$$

Thermal equilibrium means that *both* terms in (7.3.2) are zero: the thermal driving force is zero ($T^{\alpha} - T^{\beta} = 0$) and the rate of heat transfer is zero ($\delta Q^{\beta} = 0$). Furthermore, neither isobaric nor workfree constant-mass heat transfer can take place ($\delta Q^{\beta} \neq 0$) if the thermal driving force is zero ($T^{\alpha} - T^{\beta} = 0$).

However, it is possible to have a finite driving force ($T^{\alpha} - T^{\beta} \neq 0$) with no apparent transfer of heat ($\delta Q^{\beta} = 0$). Such situations are called *metastable equilibria,* since if such states are perturbed by a small finite disturbance they relax irreversibly to an equilibrium state. For example, when a fluid of nonrigid molecules is allowed to undergo a rapid adiabatic expansion, there is normally a rapid decrease in temperature. However for some molecules, internal molecular modes of bond vibration and rotation relax over much longer time-scales than molecular translational modes. Under a rapid expansion, such fluids can be caught in a metastable state in which all intensive macroscopic properties, including temperature, have stationary values, yet kinetic energies of bond vibration and rotation remain much higher than the translational kinetic energy. Rarefied gases may persist in such metastable states over long periods and relax to true equilibrium states only after sufficient molecular collisions have occurred to properly redistribute the molecular energies among all available modes.

7.3.2 Mechanical Equilibrium ($dS_{gen} = 0$)

For a constant-mass adiabatic expansion and for a constant-mass isothermal expansion of one phase against the other, the combined laws (7.2.10) reduce to

$$(P^{\alpha} - P^{\beta}) dV^{\alpha} \geq 0 \qquad (7.2.22)$$

Again, the inequality occurs because of entropy generation; and therefore, the inequality in (7.2.22) applies to irreversible processes, while the equality applies to reversible changes. In § 7.2.2 we identified the pressure difference ($P^\alpha - P^\beta$) as the driving force for volume changes. Repeating the argument given above for heat transfer, we again establish a correspondence between the limit for reversible changes and the limit for equilibrium states (see § 1.3). So by taking the driving force ($\Delta = P^\alpha - P^\beta$) to zero, we obtain an equilibrium state,

$$\lim_{\Delta \to 0} (P^\alpha - P^\beta)\, dV^\alpha = 0 \tag{7.3.4}$$

This limit identifies an equilibrium state provided no other internal or external driving forces exist when $\Delta = 0$. In § 7.2.2 we found that when $P^\alpha > P^\beta$ then $dV^\alpha > 0$, and when $P^\alpha < P^\beta$ then $dV^\alpha < 0$. Hence when the driving force is zero, $P^\alpha = P^\beta$, then we must also have $dV^\alpha = 0$; this identifies a condition of mechanical equilibrium,

$$P^\alpha = P^\beta \qquad \textit{mechanical equilibrium} \tag{7.3.5}$$

We emphasize that states in mechanical equilibrium have both the driving force ($P^\alpha - P^\beta$) and the volume change (dV^α) equal to zero. Moreover, neither isothermal nor adiabatic constant-mass changes in volume can occur without a mechanical driving force; that is, we cannot have $dV^\alpha \neq 0$ with $P^\alpha - P^\beta = 0$.

However, it is possible to have a finite driving force ($P^\alpha - P^\beta \neq 0$) with no apparent change in volume ($dV^\alpha = 0$). These are metastable states. Mechanical metastabilities can occur in substances, such as certain polymers and metal alloys, that exhibit "memory." When such materials are rapidly deformed, they can retain the deformed shape after the deforming force is removed. However, the material may regain its original shape in response to some disturbance, such as heating.

7.3.3 Single-Component Diffusional Equilibrium ($dS_{gen} = 0$)

For isothermal-isobaric diffusion of pure substance 1, the combined laws (7.2.13) reduce to

$$-[\bar{G}_1^\alpha - \bar{G}_1^\beta]\, dN_1^\alpha \geq 0 \tag{7.2.26}$$

where the inequality applies to irreversible processes, and the equality applies to reversible changes. In § 7.2.2 we found that the difference in chemical potentials serves as the driving force for single-component diffusion. So, analogous to (7.3.2) and (7.3.4), an equilibrium state is obtained by taking the limit as the driving force goes to zero,

$$\lim_{\Delta \to 0} (-[\bar{G}_1^\alpha - \bar{G}_1^\beta])\, dN_1^\alpha = 0 \tag{7.3.6}$$

where $\Delta = \overline{G}_1^{\alpha} - \overline{G}_1^{\beta}$. This limit identifies an equilibrium state provided no other internal or external driving forces exist when $\Delta = 0$. In § 7.2.2 we found that $\Delta < 0$ leads to $dN_1^{\alpha} > 0$, and $\Delta > 0$ causes $dN_1^{\alpha} < 0$. Hence, when the driving force is zero, then $dN_1^{\alpha} = 0$; that is,

$$\overline{G}_1^{\alpha} = \overline{G}_1^{\beta} \qquad \textit{diffusional equilibrium} \qquad (7.3.7)$$

identifies a condition of single-component diffusional equilibrium. Since for pures, the chemical potential is merely the molar Gibbs energy, (7.3.7) can also be expressed as

$$g_1^{\alpha} = g_1^{\beta} \qquad \textit{pure diffusional equilibrium} \qquad (7.3.8)$$

Note that diffusional equilibrium occurs only when both terms in (7.3.6) are zero: $\Delta = 0$ and $dN_1^{\alpha} = 0$. Isothermal-isobaric diffusion cannot occur in the absence of a driving force; that is, we cannot have $dN_1^{\alpha} \neq 0$ with $\Delta = 0$. However, we can observe metastable equilibria in which a finite driving force exists ($\Delta \neq 0$), but apparently no diffusion takes place ($dN_1^{\alpha} = 0$). As an example, such diffusional metastabilities can occur when the pure substance can condense into more than one kind of solid phase. Then, on bringing two forms of the solid into contact at different states, the molar Gibbs energies of the two phases differ, but the rate of diffusion in solids can be so small that the metastability may persist over significantly long times.

The criteria for diffusional equilibrium (7.3.8) can also be expressed in terms of fugacities: at equilibrium the equality in (7.2.29) applies and we have

$$f_1^{\alpha} = f_1^{\beta} \qquad \textit{pure diffusional equilibrium} \qquad (7.3.9)$$

When these fugacities are not equal, then the system is not in diffusional equilibrium, and the difference in fugacities can be interpreted as the driving force for isothermal-isobaric diffusion: material will diffuse from the phase having the higher fugacity to the phase having the lower fugacity.

7.3.4 Multicomponent Diffusional Equilibrium ($dS_{gen} = 0$)

For isothermal-isobaric multicomponent diffusion, the combined laws (7.2.13) reduce to

$$-\sum_i [\overline{G}_i^{\alpha} - \overline{G}_i^{\beta}]\, dN_i^{\alpha} \geq 0 \qquad (7.2.31)$$

where the inequality applies to irreversible processes and the equality applies to reversible changes. The difference in chemical potentials is still the driving force for diffusion, as in the single-component case (see also § 7.2.2), and an equilibrium state is attained by taking all driving forces to zero,

$$-\sum_i \left(\lim_{\Delta_i \to 0} [\overline{G}_i^{\alpha} - \overline{G}_i^{\beta}] \right) dN_i^{\alpha} = 0 \tag{7.3.10}$$

where $\Delta_i = \overline{G}_i^{\alpha} - \overline{G}_i^{\beta}$. This limit identifies an equilibrium state provided no other internal or external driving forces exist when all $\Delta_i = 0$.

In the nonequilibrium situations discussed in § 7.2.2, the mechanism of diffusion may, in some situations, couple some or all of the dN_i^{α}. But the dN_i^{α} in (7.3.10) cannot be coupled when all driving forces are zero. So here we only need consider uncoupled diffusion; then, the dN_i^{α} in (7.3.10) are mutually independent, each term in the sum must be separately zero, and each species must obey the single-component criterion (7.3.7),

$$\overline{G}_i^{\alpha} = \overline{G}_i^{\beta} \qquad \textit{for all i in diffusional equilibrium} \tag{7.3.11}$$

This can also be written in terms of the fugacity of each species,

$$f_i^{\alpha} = f_i^{\beta} \qquad \textit{for all i in diffusional equilibrium} \tag{7.3.12}$$

Since the diffusional equilibrium criterion (7.3.12) applies separately to each term in (7.3.10), we must have, at equilibrium, $dN_i^{\alpha} = 0$ for every component i. This means that diffusional equilibrium requires not only the absence of diffusion of any component i ($dN_i^{\alpha} = 0$) but, in addition, the absence of any driving force for diffusion of any component ($\Delta_i = 0$). We never observe diffusion ($dN_i^{\alpha} \neq 0$) in the absence of a gradient in the chemical potentials ($\Delta_i = 0$); this cannot occur even if the diffusion is coupled, for a zero driving force for component i disrupts any coupling for that component.

However, we may observe metastable states in which the driving force is finite ($\Delta_i \neq 0$), but diffusion is apparently not taking place ($dN_i^{\alpha} = 0$). These diffusional metastabilities occur, for example, in colloidal suspensions, such as foams, surfactant bubbles, and liquid membranes. Systems of these structures can have finite concentration gradients (hence chemical potential gradients); nevertheless, some colloidal structures can persist over macroscopically long times. It then becomes an issue as to whether these life-times are sufficiently long that equilibrium thermodynamics can be applied.

7.3.5 Thermodynamic Equilibrium

Thermodynamic equilibrium encompasses thermal, mechanical, and chemical equilibrium. Chemical equilibrium, in turn, includes both diffusional and reaction equilibrium. In this section we have considered only nonreacting systems, and so, at this point, we have developed only the criteria for thermal, mechanical, and diffusional equilibrium; criteria for reaction equilibrium are given in § 7.6.

Thermodynamic equilibrium occurs when *all* net driving forces are zero ($dS_{gen} = 0$); this includes driving forces between system and surroundings as well as those between different system parts. Since equilibrium criteria apply to different system parts, they can serve as quantitative prescriptions for identifying equilibrium between phases α and β:

thermal equilibrium	$T^\alpha = T^\beta$		(7.3.3)
mechanical equilibrium	$P^\alpha = P^\beta$		(7.3.5)
diffusional equilibrium	$\overline{G}_i^\alpha = \overline{G}_i^\beta$	*for all i*	(7.3.11)

or

diffusional equilibrium	$f_i^\alpha = f_i^\beta$	*for all i*	(7.3.12)

For a system of C components, the sets of equations {(7.3.3), (7.3.5), (7.3.11)} or {(7.3.3), (7.3.5), (7.3.12)} each represent $(C + 2)$ algebraic equations that can be used to identify phase equilibrium situations; of these, the sets of equations containing chemical potentials (7.3.11) and fugacities (7.3.12) each represent C nonlinear equations that usually must be solved by trial.

But even though equilibrium implies the absence of *net* driving forces for change, molecules continually cross the interface in both directions, causing the properties of each phase to fluctuate about stable equilibrium values. Although macroscopic driving forces are in balance when two phases are in equilibrium, the situation is a dynamic one on a microscopic scale.

7.3.6 Example

Can the criteria (7.3.11) for diffusional equilibrium, which require equality of the chemical potentials, be reconciled with the general criterion for isothermal-isobaric equilibrium, namely $dG = 0$ (7.1.40)?

The objective here is to show that the diffusional equilibrium criteria (7.3.11) are a consequence of the more general equilibrium criterion (7.1.40) that applies to any *NPT* system, including systems containing more than one phase.

Consider a multicomponent system contained in a closed vessel and maintained at constant T and P, as represented schematically in Figure 7.5. We seek the conditions under which the system can exist as two phases in equilibrium. Since the external reservoir imposes its temperature and pressure on both phases, no driving forces exist

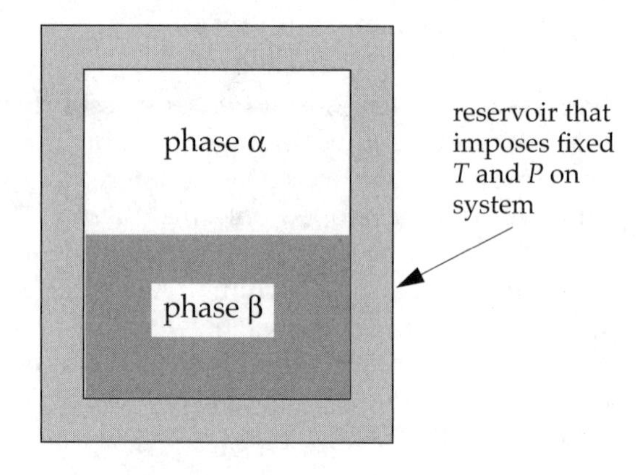

Figure 7.5 Schematic of a two-phase system whose temperature and pressure are held constant by thermal and mechanical interactions with a constant *TP* reservoir. The system cannot exchange mass with the reservoir or the surroundings; however, the two phases can exchange mass with one another.

for net heat transfer, bulk mass transfer, or volume changes. The only possible changes result from diffusional mass transfer across the phase boundary.

Call the phases α and β; they could be any combination of solid, liquid, or gas. Although the interface between two phases is open to mass and energy transfers, the entire system here is closed. Since T and P are fixed for the entire system, the NPT criterion for equilibrium (7.1.40) applies; that is, the system's total Gibbs energy will be a minimum at equilibrium,

$$dG = 0 \qquad \textit{fixed T and P} \tag{7.3.13}$$

The system Gibbs energy is the sum of contributions from each phase

$$G = G^{\alpha} + G^{\beta} \tag{7.3.14}$$

and (3.2.32) can be used to write G for each phase in terms of chemical potentials,

$$G = \sum_i N_i^{\alpha} \bar{G}_i^{\alpha} + \sum_i N_i^{\beta} \bar{G}_i^{\beta} \tag{7.3.15}$$

Now we determine the response of G to diffusion of a differential amount of each component from one phase to the other. Forming the total differential of (7.3.15) and substituting it into (7.3.13) yields

$$dG = \sum_i N_i^{\alpha} d\bar{G}_i^{\alpha} + \sum_i \bar{G}_i^{\alpha} dN_i^{\alpha} + \sum_i N_i^{\beta} d\bar{G}_i^{\beta} + \sum_i \bar{G}_i^{\beta} dN_i^{\beta} = 0 \tag{7.3.16}$$

But at fixed T and P the Gibbs-Duhem equation requires

$$\sum_i N_i^{\alpha} d\bar{G}_i^{\alpha} = 0 \quad \text{and} \quad \sum_i N_i^{\beta} d\bar{G}_i^{\beta} = 0 \tag{7.3.17}$$

So (7.3.16) reduces to

$$\sum_i \bar{G}_i^{\alpha} dN_i^{\alpha} + \sum_i \bar{G}_i^{\beta} dN_i^{\beta} = 0 \tag{7.3.18}$$

The total system is closed, so

$$dN_i^{\beta} = -dN_i^{\alpha} \tag{7.3.19}$$

That is, whatever leaves the α-phase necessarily enters the β-phase, and vice versa. Using (7.3.19) to eliminate the dN_i^{β} from (7.3.18), we find

$$\sum_i (\overline{G}_i^{\alpha} - \overline{G}_i^{\beta})\, dN_i^{\alpha} = 0 \tag{7.3.20}$$

This is (7.3.10), and the argument following (7.3.10) can be repeated leading to (7.3.11), as required. The general criterion for isothermal-isobaric equilibrium (7.1.40) includes the diffusional equilibrium criteria (7.3.11) as a special case. *QED*

7.4 THE LAWS FOR CLOSED REACTING SYSTEMS

It is perhaps surprising that thermodynamics can tell us anything about chemical reactions, for when we encounter a reaction, we naturally think of rates, and we know that thermodynamics cannot be applied to problems posed by reaction rates or mechanisms. However, a chemical reaction is a change, so whenever the initial and final states of a reaction process are well-defined, differences in thermodynamic state functions can be evaluated, just as they can be evaluated for other kinds of processes. In particular, the laws of thermodynamics impose limitations on the directions and magnitudes (extents) of reactions, just as they impose limitations on other processes. For example, thermodynamics can tell us the direction of a proposed reaction; it can tell us what the equilibrium composition of a reaction mixture should be; and it can help us decide how to adjust operating variables to improve the yields of desired products. These kinds of issues can be addressed using equations derived in this and the next section; moreover, these equations are derived without introducing any new thermodynamic fundamentals or assumptions.

In this section we obtain the combined first and second laws for reacting systems. The development parallels that presented in § 7.1 for nonreacting systems. However, the development here is more elaborate than the earlier one because our analysis must account for the fact that, during reactions, chemical species are not conserved. This problem is addressed in § 7.4.1 and examples are offered in § 7.4.2 and 7.4.3; then in § 7.4.4 we derive the combined laws for reactions in closed systems.

7.4.1 Stuff Equations for Material Undergoing Reactions in Closed Systems

In elementary chemistry courses, we are taught that, when analyzing a reacting mixture, we should first write the reaction and balance it. This strategy is appropriate when the system involves a single elementary reaction, such as might represent complete combustion of methane. However, industrial processes typically involve dozens of reactions occurring simultaneously; in those situations, the elementary strategy would require us to write a complete set of *independent* reactions that involve all species present. This can be an overwhelming task, for often we do not even know how many reactions could occur, much less what those reactions might be—we have only a list of reactants and products. Fortunately, this is all we need for a thermodynamic analysis, because changes in state variables are not affected by reaction paths; for a thermodynamic analysis, we only need a systematic procedure for identifying and balancing some set of independent reactions that represent the conversion of reactants into products. Such a systematic procedure can be formulated in several ways [2–4].

We emphasize that the reactions used in the analysis do not have to be the reactions actually occurring—we only need any convenient hypothetical path that connects products to reactants. In fact, we don't even need reactions at all, so long as we can achieve a balance on every element present. Further, "elements" need not be atoms; they can be groups of atoms that may or may not constitute real molecules. Our procedure for identifying and balancing reactions reduces to the stuff equation for material, reformulated to apply to elements. We consider reactions in closed systems here and reactions in open systems in § 7.5.

Consider a closed system containing a total of C chemical species, with N_i moles of species i present at any time. In this section we consider a molecule of each species i to be composed of a_{ki} atoms of element k; the total number of elements present is represented by m_e. Then the total number of atoms b_k for each element k is given by

$$\sum_i^C a_{ki} N_i = b_k \qquad k = 1, 2, \ldots, m_e \tag{7.4.1}$$

These equations provide the fundamental connections between elements and species in a closed system. Since the equations (7.4.1) are linear in the mole numbers, we can write them economically in matrix form,

$$\mathbf{AN} = \mathbf{b} \tag{7.4.2}$$

where $\mathbf{A}$ is the $(m_e \times C)$ array of coefficients a_{ki}, the vector of mole numbers $\mathbf{N}$ is of length C, and the rhs vector $\mathbf{b}$ is of length m_e. (For a review of the jargon and operations of matrix algebra, see Appendix B.) Since each column $\mathbf{A}_i$ of the matrix represents the chemical formula for species i, we refer to $\mathbf{A}$ as the *formula matrix* for the reacting system.

During a reaction in a closed system, it is not species that are conserved, but elements. That is, the N_i in (7.4.2) change, but the b_k in (7.4.2) remain constant. Therefore,

$$\sum_i^C a_{ki}\, dN_i = 0 \tag{7.4.3}$$

or

$$\mathbf{A}(\mathbf{dN}) = \mathbf{0} \tag{7.4.4}$$

where $(\mathbf{dN})$ is a vector of length C. Equation (7.4.4) is a statement of conservation of elements; it is a form of the stuff equation that is useful when the amounts of species change due to reactions.

In a traditional approach to reaction equilibria, we first write a set of $\mathcal{R}$ independent reactions and then balance those reactions. This establishes the *stoichiometry* of the system. However, (7.4.4) represents a useful alternative to balancing reactions because it allows us to impose conservation of elements without explicitly identifying the reactions themselves. Before we can balance any reaction, we must identify all the species that participate in the reaction. In simple situations, the participants may be known, and then we may choose to proceed in the traditional way [2] or by a hybrid method [3, 4]. But many reaction processes are complex; examples include reactions during combustion, in biological processes, and in catalysis. In such cases, it is better to avoid guessing the reactions; instead, we use (7.4.4) and perform the necessary operations on the formula matrix **A**.

One important operation allows us to identify the number of independent reactions. In the traditional approach, we must reduce the proposed reactions to an independent set, but when many reactions occur, finding an independent set can be tedious and prone to error. However, in (7.4.4) the number of independent reactions $\mathcal{R}$ is simply related to the *rank* of the formula matrix **A**; specifically,

$$\mathcal{R} = C - \text{rank}(\mathbf{A}) \tag{7.4.5}$$

The balance equations (7.4.3) express conservation of elements, but they do not tell us how the species mole numbers N_i change as reactions proceed. If we know how the N_i change, then we have a direct way for determining the composition of the mixture. Let N_i^o represent the number of moles of species i initially present and consider a general situation in which $\mathcal{R}$ independent reactions take place. Then at any time during the process, the number of moles of species i is given by the net amounts of i generated and consumed. Those net amounts are the combined results of all reactions, so we write

$$N_i = N_i^o + \Delta N_i = N_i^o + \sum_j^{\mathcal{R}} \Delta N_{ij} \tag{7.4.6}$$

where ΔN_{ij} is the change in N_i caused by reaction j. When N_i is a product of reaction j, then $\Delta N_{ij} > 0$, and when it is a reactant, then $\Delta N_{ij} < 0$.

But during reaction j the changes ΔN_{ij} are not independent; rather, they are coupled through the stoichiometry of the reaction. Consequently, if we designate any one species r as the independent species for reaction j, then we can monitor the changes of all other species in j relative to that of r. This allows us to write (7.4.6) as

$$N_i = N_i^o + \sum_j^{\mathcal{R}} \frac{\Delta N_{ij}}{\Delta N_{rj}} \Delta N_{rj} \tag{7.4.7}$$

For each species i in reaction j, the ratio here is a constant, which we call ν',

$$\nu'_{ij} = \frac{\Delta N_{ij}}{\Delta N_{rj}} = \text{constant} \tag{7.4.8}$$

so (7.4.7) becomes

$$N_i = N_i^o + \sum_j^{\mathcal{R}} \nu'_{ij} \Delta N_{rj} \tag{7.4.9}$$

The relation (7.4.9) allows us to follow the progress of reaction j by monitoring only the one quantity ΔN_{rj}.

But the quantities ν'_{ij} introduce an undesirable asymmetry among the species participating in reaction j; for example, (7.4.8) implies that $\nu'_{rj} = 1$. To avoid treating one species as special, we rescale all the ν'_{ij} parameters by a constant factor, thereby defining

$$\nu_{ij} = \nu'_{ij}\, \nu_{rj} \tag{7.4.10}$$

where the value of the one quantity ν_{rj} can be set arbitrarily to any convenient value. The quantities ν_{ij} are called *stoichiometric coefficients*; their values are constant for a particular reaction. By convention, reactants in j have $\nu_{ij} < 0$, and products have $\nu_{ij} > 0$. Using (7.4.10) in (7.4.9), the number of moles of i present at any time is

$$N_i = N_i^o + \sum_j^{\mathcal{R}} \nu_{ij} \frac{\Delta N_{rj}}{\nu_{rj}} \tag{7.4.11}$$

To follow the progress of reaction j, we monitor one quantity: ΔN_{rj}. But the change in the number of moles of any species i cannot be affected by the identity of the substance chosen to play the role of the independent species r. So we define the ratio on the rhs to be the *extent of reaction j*,

$$\xi_j \equiv \frac{\Delta N_{rj}}{\nu_{rj}} \tag{7.4.12}$$

Note that ξ_j is extensive and has units of moles; also note that there is one extent ξ_j for each independent reaction. The definition (7.4.12) applies for *any* species selected as r, so a particular choice does not have to be made explicitly; moreover, it allows us to use any convenient value for the stoichiometric coefficient ν_{rj}, so that none of the ν_{ij} need necessarily be set to unity.

At the start of reaction j, $\Delta N_{rj} = 0$, and by (7.4.12) the extent is also initially zero: $\xi_j = 0$. If the reaction proceeds in the direction implied by the sign of ν_{ij}, then $\xi_j > 0$. But we might identify r incorrectly; that is, we might designate species r as a reactant when, in fact, it is produced as a product. In such cases $\xi_j < 0$, and the reaction actually proceeds in the "reverse" direction.

Putting (7.4.12) into (7.4.11), we obtain the following general form for the total number of moles of species i present at any time during the $\mathcal{R}$ reactions,

$$N_i = N_i^o + \sum_j^{\mathcal{R}} \nu_{ij}\, \xi_j \tag{7.4.13}$$

For differential changes, the definition (7.4.12) becomes

$$d\xi_j = \frac{dN_{rj}}{\nu_{rj}} \tag{7.4.14}$$

and since the identification of species r is arbitrary (so long as $\nu_{rj} \neq 0$), (7.4.14) must be obeyed by *every* species in j.

To obtain stoichiometric coefficients, we first form the total differential of (7.4.13),

$$dN_i = \sum_j^{\mathcal{R}} \nu_{ij}\, d\xi_j \tag{7.4.15}$$

Then we substitute this into the balance equation (7.4.3),

$$\sum_j^{\mathcal{R}} \left(\sum_i^{C} a_{ki}\, \nu_{ij} \right) d\xi_j = 0 \tag{7.4.16}$$

But the $\mathcal{R}$ reactions are independent and all the $d\xi$s are nonzero, so the quantity in parenthesis must be zero for each reaction j; that is, each reaction must have

$$\mathbf{A}\mathbf{v}_j = 0 \qquad j = 1, 2, \ldots, \mathcal{R} \tag{7.4.17}$$

where $\mathbf{A}$ is the formula matrix in (7.4.2) and $\mathbf{v}_j$ is the vector of stoichiometric coefficients for reaction j, $\mathbf{v}_j^T = (\nu_{1j}\ \nu_{2j}\ \nu_{3j}\ \ldots)$. Equation (7.4.17) represents a balancing of reaction j, and since values are known for the elements in $\mathbf{A}$, (7.4.17) can be solved for the stoichiometric coefficients. However as noted above, the value of one ν_{ij} (that for the reference substance ν_{rj}) can always be chosen arbitrarily, so (7.4.17) does not have a unique solution.

With values known for the stoichiometric coefficients, we can sum (7.4.13) to obtain the total number of moles at any point during the reactions,

$$N = \sum_i^{C} \left(N_i^o + \sum_j^{\mathcal{R}} \nu_{ij}\, \xi_j \right) \neq \text{constant} \tag{7.4.18}$$

$$= N^o + \sum_j^{\mathcal{R}} \sigma_j\, \xi_j \tag{7.4.19}$$

where N^o is the total amount present at the start of all reactions,

$$N^o = \sum_i^{C} N_i^o \tag{7.4.20}$$

and σ_j is the algebraic sum of all stoichiometric coefficients for reaction j,

$$\sigma_j = \sum_i^C \nu_{ij} \tag{7.4.21}$$

Note that σ_j is positive when the total number of moles increases, negative when the total number decreases, or zero when the total number is unchanged by the reaction. With (7.4.13) and (7.4.19), we can form the mole fraction for each species at any time during the reactions,

$$x_i = \frac{N_i}{N} = \frac{N_i^o + \sum_j^{\mathcal{R}} \nu_{ij}\,\xi_j}{N^o + \sum_j^{\mathcal{R}} \sigma_j\,\xi_j} \tag{7.4.22}$$

We have noted that $\xi_j = 0$ at the start of reaction j. Let us consider the normal situation in which ξ_j increases from zero as the reaction proceeds. In such cases, there is an upper bound to ξ_j, based merely on conservation of atoms. That bound occurs when one reactant is first depleted; that is, when the mole fraction of the limiting species x_ℓ first reaches zero. We determine the bound by solving (7.4.22) with $x_\ell = 0$,

$$\xi_j^{ub} = \frac{-N_\ell^o}{\nu_{\ell j}} \tag{7.4.23}$$

The values for this upper bound can be found by computing the rhs of (7.4.23) for each reactant participating in reaction j; the smallest of those values is the upper bound and identifies the limiting reactant. But although (7.4.23) provides a bound on the extent of reaction, that bound is based on material balances; in practice, it is rarely reached. Instead, most reactions reach thermodynamic equilibrium *before* all the initial loading of any reactant is depleted; the equilibrium value obeys $0 < \xi^e < \xi^{ub}$.

The above development should make clear the following points:

(a) For a particular reaction j, the values of the stoichiometric coefficients ν_{ij} are determined only to within an arbitrary multiplicative constant. The value of this constant is set by choosing the value of the coefficient ν_{rj} for one species r; often that value is ± 1, but other choices are sometimes convenient.

(b) The changes in mole numbers ΔN_{ij} for all species i participating in reaction j can be represented by one independent variable ξ_j and those changes are coupled through the vector of stoichiometric coefficients $\mathbf{v}_j$.

(c) The values of the extent of reaction ξ_j depend on the stoichiometric coefficients, so different balances of reaction j set different bounds on the extent.

(d) The extent of reaction ξ_j is extensive, has units, and is *not* restricted to the range [0, 1].

7.4.2 Example

How do we determine the composition of a reacting mixture at any point during a single reaction?

Consider the synthesis of ammonia from nitrogen and hydrogen, with the reactants loaded into the reactor in the ratio $N_2^o/N_1^o = 3/1$, where N_1^o = the initial number of moles of nitrogen and N_2^o = the initial number of moles of hydrogen. We want to obtain the composition of the reaction mixture during the course of the reaction.

In the notation of § 7.4.1, we have total number of species $C = 3$ and total number of elements $m_e = 2$. Let the values of the index over elements be $k = 1$ for nitrogen (N) and $k = 2$ for hydrogen (H). Let the values of the index over species be $i = 1$ for nitrogen (N_2), $i = 2$ for hydrogen (H_2), and $i = 3$ for ammonia (NH_3). Then the number of elements k on each species i is given by a_{ki}; hence, the formula matrix is

$$\mathbf{A} = \overbrace{\begin{bmatrix} 2 & 0 & 1 \\ 0 & 2 & 3 \end{bmatrix}}^{N_2 \;\; H_2 \;\; NH_3} \begin{cases} N \\ H \end{cases} \tag{7.4.24}$$

Note that each column of $\mathbf{A}$ represents the chemical formula for one of the species. The rank of $\mathbf{A}$ is 2 (see Appendix B); therefore, the number of independent reactions is

$$\mathcal{R} = C - \text{rank}(\mathbf{A}) = 3 - 2 = 1 \tag{7.4.25}$$

That is, in this simple example, only one independent reaction occurs. To find the stoichiometric coefficients in that reaction, we solve

$$\mathbf{A}\mathbf{v}_j = 0 \tag{7.4.17}$$

$$\begin{bmatrix} 2 & 0 & 1 \\ 0 & 2 & 3 \end{bmatrix} \begin{bmatrix} \nu_{11} \\ \nu_{21} \\ \nu_{31} \end{bmatrix} = \begin{bmatrix} 0 \\ 0 \end{bmatrix} \tag{7.4.26}$$

This represents two equations in three unknowns. We can pick the value of one stoichiometric coefficient, and since ammonia is the desired product, we set $\nu_{31} = 1$ (recall, products have $\nu > 0$). Then (7.4.26) gives $\nu_{11} = -1/2$ and $\nu_{21} = -3/2$. Consequently, the one independent balanced reaction is

$$-\frac{1}{2}N_2 - \frac{3}{2}H_2 + NH_3 = 0 \tag{7.4.27}$$

or in a traditional form,

$$\frac{1}{2}N_2 + \frac{3}{2}H_2 \rightarrow NH_3 \tag{7.4.28}$$

Choosing a basis of one mole of nitrogen initially loaded into the reactor, we use (7.4.22) to obtain the following expressions for the species mole fractions,

$$x_1 = \frac{1-\xi/2}{4+(-1/2-3/2+1)\xi} = \frac{1-\xi/2}{4-\xi} \tag{7.4.29}$$

$$x_2 = \frac{3(1-\xi/2)}{4-\xi} \quad \text{and} \quad x_3 = \frac{\xi}{4-\xi} \tag{7.4.30}$$

Using (7.4.23) we can find the upper bound on the extent; hence, $\xi^{ub} = 2$ moles and both reactants are completely consumed at the same time. We now use (7.4.29) and (7.4.30) to obtain the composition at any point during the reaction ($0 \le \xi \le 2$). Sample results are given in Table 7.2.

Note that

(a) the stoichiometry of the reaction couples the compositions so that a value for the one variable ξ allows us to determine values for the mole fractions of all reactants and products,

(b) at every value of the extent ξ the mole fractions sum to one,

(c) the total number of moles is not constant during the reaction,

(d) the calculations of the mole fractions did not require us to explicitly write the chemical reactions (7.4.27) and (7.4.28), and

(e) the above procedure is sufficiently systematic so that it can be readily implemented on a computer.

Point (d) is worth emphasizing: the balanced reactions (7.4.27) and (7.4.28) were displayed merely to offer a familiar interpretation for the matrix equation (7.4.17); however, those chemical reactions did not explicitly contribute to the solution of the problem.

Table 7.2 Sample results for composition of reaction mixture during synthesis of ammonia (7.4.27) based on an initial loading of 3 moles of H_2 plus 1 mole of N_2

ξ	$N = 4 - \xi$	x_{N_2}	x_{H_2}	x_{NH_3}
0	4	0.25	0.75	0
0.5	3.5	0.214	0.643	0.143
1	3	0.167	0.5	0.333
1.5	2.5	0.1	0.3	0.6
2	2	0	0	1

7.4.3 Example

How do we determine the composition at any point during a process involving multiple reactions?

Consider the formation of synthesis gas (CO and H_2) by incomplete combustion of methane in oxygen. Let N_1^o be the initial number of moles of methane and let N_2^o be the initial number of moles of oxygen. Assume the feed ratio is $N_2^o/N_1^o = 3/2$. We expect the products will be CO_2, H_2O, CO, and H_2. So we have $C = 6$ species formed from $m_e = 3$ elements (C, H, and O), and the formula matrix looks like this:

$$\mathbf{A} = \overbrace{\begin{bmatrix} 1 & 0 & 1 & 0 & 1 & 0 \\ 4 & 0 & 0 & 2 & 0 & 2 \\ 0 & 2 & 2 & 1 & 1 & 0 \end{bmatrix}}^{CH_4\ O_2\ CO_2\ H_2O\ CO\ H_2} \begin{cases} C \\ H \\ O \end{cases} \tag{7.4.31}$$

The rank of this matrix is 3 (see Appendix B); therefore, the number of independent reactions is $\mathcal{R} = C - \text{rank}(\mathbf{A}) = 6 - 3 = 3$. So we have to find the stoichiometric coefficients for three independent reactions; to do so, we must solve

$$\mathbf{A}\boldsymbol{\nu}_j = 0 \qquad j = 1, 2, 3 \tag{7.4.17}$$

For each reaction, (7.4.17) has this form:

$$\begin{bmatrix} 1 & 0 & 1 & 0 & 1 & 0 \\ 4 & 0 & 0 & 2 & 0 & 2 \\ 0 & 2 & 2 & 1 & 1 & 0 \end{bmatrix} \begin{bmatrix} \nu_{1j} \\ \nu_{2j} \\ \nu_{3j} \\ \nu_{4j} \\ \nu_{5j} \\ \nu_{6j} \end{bmatrix} = \begin{bmatrix} 0 \\ 0 \\ 0 \end{bmatrix} \qquad j = 1, 2, 3 \tag{7.4.32}$$

Equation (7.4.32) represents three equations in six unknowns. To solve these, we must choose, for each reaction, values for any three of the ν_{ij}. This means there are many possible solutions to the three equations in (7.4.32). In general, different values assigned to the three arbitrarily chosen ν_{ij} will produce different balanced reactions, and each trio of reactions will have its own set of extents $\{\xi_j\}$. Nevertheless, every trio of reactions will yield the same mole fractions at any point during the process.

For the first reaction, $j = 1$, we choose $\nu_{11} = -2$, $\nu_{21} = -1$, and $\nu_{31} = 0$; then (7.4.32) gives $\nu_{41} = 0$, $\nu_{51} = 2$, $\nu_{61} = 4$, and the reaction provides the desired products:

$$-2CH_4 - O_2 + 2CO + 4H_2 = 0 \tag{7.4.33}$$

For the second reaction, $j = 2$, we choose $\nu_{12} = -1$, $\nu_{22} = -2$, and $\nu_{62} = 0$. Then (7.4.32) gives $\nu_{32} = 1$, $\nu_{42} = 2$, $\nu_{52} = 0$, and this reaction produces an undesired product (CO_2):

$$-CH_4 - 2O_2 + CO_2 + 2H_2O = 0 \tag{7.4.34}$$

For the third reaction, $j = 3$, we choose $\nu_{13} = 0$, $\nu_{33} = -2$, and $\nu_{63} = 0$. Then (7.4.32) gives $\nu_{23} = 1$, $\nu_{43} = 0$, and $\nu_{53} = 2$, so the third reaction converts the undesired CO_2 to the desired product CO:

$$O_2 - 2CO_2 + 2CO = 0 \tag{7.4.35}$$

If we choose as a basis one mole of CH_4 ($N_1^o = 1$), then $N_2^o = 1.5$ moles, and we can use (7.4.22) to obtain the following expressions for the species mole fractions,

$$x_1 = \frac{1}{N}(1 - 2\xi_1 - \xi_2) \qquad x_4 = \frac{2\xi_2}{N} \tag{7.4.36}$$

$$x_2 = \frac{1}{N}(1.5 - \xi_1 - 2\xi_2 + \xi_3) \qquad x_5 = \frac{1}{N}(2\xi_1 + 2\xi_3) \tag{7.4.37}$$

$$x_3 = \frac{1}{N}(\xi_2 - 2\xi_3) \qquad x_6 = \frac{4\xi_1}{N} \tag{7.4.38}$$

and the total number of moles is given by

$$N = 2.5 + 3\xi_1 + \xi_3 \tag{7.4.39}$$

Sample results for the mole fractions at a few selected values of the extents are given in Table 7.3. In general, the objective would be to maximize the amount of synthesis gas produced (CO and H_2), while minimizing the amounts of other species. But none of the product distributions shown in Table 7.3 are optimal. An optimal distribution

Table 7.3 Sample compositions from synthesis-gas production at selected values of the three extents of reaction

Extents			***N***	**Mole fractions, x_i**					
ξ_1	ξ_2	ξ_3	**mol**	CH_4	O_2	CO_2	H_2O	CO	H_2
0	0	0	2.5	0.4	0.6	0	0	0	0
0	0.5	0	2.5	0.2	0.2	0.2	0.4	0	0
0	1	0.5	3	0	0	0	0.667	0.333	0
0.15	0.3	0.15	3.1	0.129	0.290	0	0.194	0.194	0.194
0.25	0.5	0	3.25	0	0.077	0.154	0.308	0.154	0.308
0.3	0.3	0.15	3.55	0.028	0.211	0	0.169	0.254	0.338
0.5	0	0	4.0	0	0.25	0.	0	0.25	0.5

can be obtained by finding appropriate values for the extents; however, that distribution is not likely to be the equilibrium distribution. To find the equilibrium concentration, we must obtain equilibrium values for the extents ξ_j by applying the criteria for reaction equilibria developed in § 7.6. Ways for applying those criteria are described in Chapter 10.

The examples in § 7.4.2 and 7.4.3 show how material balances are applied in reacting situations to obtain elemental balances, the number of independent reactions, and values for stoichiometric coefficients. In the above examples, we use atoms as the conserved elements, that is, as the reaction invariants. But in some situations the analysis can be simplified by choosing groups of atoms or fragments of molecules as the elemental invariants. Examples of such fragments include a benzene ring and an -OH group. Formally, the procedure is just as we have illustrated above, except that groups (rather than elements) form the rows of the formula matrix. Such an approach can be useful when we know that thermodynamic or kinetic constraints make certain independent reactions unlikely. The net effect is to decrease the number of independent reactions compared to the number provided solely from material-balance considerations. In addition, use of groups in reaction analysis, combined with use of the same groups in phase-equilibrium situations, can simplify calculations in such applications as reactive distillation. The development and use of this method has been described by Pérez Cisneros et al. [3, 4].

7.4.4 Combined Laws for Reactions in Closed Systems ($dS_{gen} > 0$)

With the notation and stuff equations from the previous section, we can now extend the combined first and second laws from unreacting systems (§ 7.1 and 7.2) to reacting systems. To facilitate the presentation, it is useful to introduce a new set of property differences that apply to reacting systems. For any extensive property F in a reacting system, we define a *change in F for each reaction j* by the intensive quantity

$$\Delta F_j \equiv \sum_i^C \nu_{ij} \bar{F}_i \qquad j = 1, 2, \ldots, \mathcal{R} \tag{7.4.40}$$

where $\bar{F}_i$ is the partial molar property, C is the number of species in the mixture, and ν_{ij} is the stoichiometric coefficient for species i in reaction j. Recall that $\nu_{ij} < 0$ if i is a reactant and $\nu_{ij} > 0$ if i is a product. The quantity F could be any of the usual thermodynamic properties, including U, H, S, A, and G. As is often the case, situations in which $F = G$ have special significance, so the (negative) change of Gibbs energy for reaction j is given a special name: it is called the *affinity,* $\mathcal{A}_j$,

$$\mathcal{A}_j \equiv -\Delta G_j = -\sum_i^C \nu_{ij} \bar{G}_i \tag{7.4.41}$$

The affinity is an intensive conceptual property having dimensions of energy/mol.

Consider a closed system containing a total of C chemical species, with N_i moles of species i present at any time. The values of the mole numbers N_i are changing due to $\mathcal{R}$ independent reactions taking place. The value of $\mathcal{R}$ can be obtained using the procedure described in § 7.4.1. Because of the reactions, the state of the system changes and, consequently, the values of properties change. Consider any one such property F,

$$Nf = F = F(T, P, N_1, N_2, \ldots) \tag{7.4.42}$$

The response of F to the change of state can be written as

$$d(Nf) = \sum_i^C N_i\, d\bar{F}_i + \sum_i^C \bar{F}_i dN_i \tag{7.4.43}$$

The first term on the rhs is given by the Gibbs-Duhem equation (3.4.10). Moreover the system is closed, so the change in mole numbers N_i can be caused only by reactions; therefore, we can substitute (7.4.15) for the dN_i in (7.4.43) and obtain

$$d(Nf) = \left(\frac{\partial F}{\partial T}\right)_{PN} dT + \left(\frac{\partial F}{\partial P}\right)_{TN} dP + \sum_j^{\mathcal{R}} \left(\sum_i^C \nu_{ij}\, \bar{F}_i \right) d\xi_j \tag{7.4.44}$$

where ξ_j is the extent of reaction j. Using the definition (7.4.40) of the change of F due to reaction j and holding T and P fixed, (7.4.44) becomes

$$d(Nf) = \sum_j^{\mathcal{R}} \Delta F_j\, d\xi_j \qquad \textit{fixed T and P} \tag{7.4.45}$$

This is a general result for the total differential of any extensive property F responding to $\mathcal{R}$ chemical reactions occurring in a closed system at fixed T and P.

For a closed system in which reactions are occurring, the combined first and second laws should be written as a generalized form of (7.1.11),

$$d(Nu) + P_{sur}\, d(Nv) - T_{sur}\, d(Ns) \le 0 \qquad \textit{closed systems} \tag{7.4.46}$$

This form is appropriate because in reacting systems, the mole numbers N may change, even though the system is closed. We restrict our attention to reactions performed at fixed temperature and pressure: $T = T_{sur}$ and $P = P_{sur}$. Therefore, we can use (7.4.45) for each total differential in (7.4.46), and (7.4.46) becomes

$$\sum_j^{\mathcal{R}} [\Delta U_j + P\Delta V_j - T\Delta S_j]\, d\xi_j \le 0 \qquad \textit{fixed T and P} \tag{7.4.47}$$

Recall the ΔF_j are all intensive. Since T and P are fixed, we can write (7.4.47) as

$$\sum_j^{\mathcal{R}} \Delta G_j \, d\xi_j \leq 0 \qquad \textit{fixed T and P} \qquad (7.4.48)$$

Introducing the affinity via (7.4.41), we find

$$\sum_j^{\mathcal{R}} \mathcal{A}_j v_j \geq 0 \qquad \textit{fixed T and P} \qquad (7.4.49)$$

where v_j is the rate of reaction j, defined by

$$v_j \equiv v(t, \xi_j) = \frac{d\xi_j}{dt} \qquad (7.4.50)$$

At equilibrium all reaction rates are zero; otherwise, they are always finite, although they may be positive or negative.

Equation (7.4.49) is the combined law for closed systems in which chemical reactions are occurring at fixed T and P. For such systems, the combined law imposes a limitation on the direction in which reactions can proceed: they can only proceed in ways that cause the lhs of (7.4.49) to be positive or zero. Recall from § 7.1 that the inequality in the combined law results from entropy generation; so in general, for closed systems at fixed T and P, we expect chemical reactions to be accompanied by generation of entropy in the system. However, we caution that the *total* entropy of the system may increase or decrease because the process is isothermal, not adiabatic. We now interpret the combined law (7.4.49) for single reactions and for multiple reactions.

Single reactions. For one reaction, (7.4.49) reduces to

$$\mathcal{A} v \geq 0 \qquad (7.4.51)$$

If $v = 0$, then no reaction is occurring and the equality in (7.4.51) applies; otherwise, for $v \neq 0$ we have two possibilities: (i) If $v > 0$, then we must have $\mathcal{A} > 0$ for the reaction to proceed in the forward direction, and (ii) if $v < 0$, then we must have $\mathcal{A} < 0$ for the reaction to proceed in the reverse direction. If the inequality in (7.4.51) is violated, then the reaction cannot proceed in the proposed direction at fixed T and P. Note that, for a single reaction, we can never have $\mathcal{A} = 0$ with $v \neq 0$ because reactions occurring at finite rates always generate entropy.

To gain some additional insight into (7.4.51), consider the form of the combined laws in terms of the chemical potentials (7.4.41). For a single reaction, (7.4.51) is

$$\Delta G \, d\xi < 0 \qquad (7.4.52)$$

or

$$\left(\sum_i \nu_i \bar{G}_i\right) d\xi < \left(\sum_k |\nu_k| \bar{G}_k\right) d\xi \qquad \textit{fixed T and P} \qquad (7.4.53)$$

where index i runs over products while k runs over reactants. We distinguish products from reactants in such a way that $d\xi > 0$. Therefore (7.4.53) requires the reaction to proceed from a situation of larger Gibbs energy (reactants) to one having a smaller value (products); this is consistent with the analysis in § 7.1.5.

Uncoupled multiple reactions. Multiple reactions taking place in a closed system at fixed T and P must satisfy the combined law (7.4.49). However, if the reactions are uncoupled, then each term in the sum on the lhs of (7.4.49) is independent of every other term, and therefore each term must be positive, if that reaction proceeds in the proposed direction. This means that each reaction in the system must separately satisfy the single reaction form of the combined law which appears in (7.4.51). Reactions are usually uncoupled when no reactant or product participates in more than one reaction.

Coupled multiple reactions. But multiple reactions may be coupled, often because some reactants or products participate in more than one reaction, though this condition is neither necessary nor sufficient for coupling. When reactions are coupled, not all the terms in the sum in (7.4.49) are independent, and then it is possible for some of the terms to be negative. Nevertheless, the coupled reactions can still proceed, so long as enough positive terms are available to dominate the sum, forcing the combined laws to be obeyed.

An important example of coupling has been cited by Prigogine and Defay [1]: at ambient conditions, the synthesis of urea via the single reaction

$$2NH_3 + CO_2 \rightleftarrows (NH_2)_2CO + H_2O \tag{7.4.54}$$

has an affinity of $\mathcal{A}_1 \approx -46$ kJ/mol. So, when only ammonia and carbon dioxide are present, urea will not be formed by this reaction. However, in the human liver, the reaction (7.4.54) is coupled to oxidation of glucose,

$$C_6H_{12}O_6 + 6O_2 \rightleftarrows 6CO_2 + 6H_2O \tag{7.4.55}$$

which has an affinity of $\mathcal{A}_2 \approx 482$ kJ/mol. Then the combined law (7.4.49) requires

$$-46v_1 + 482v_2 > 0 \tag{7.4.56}$$

So

$$v_1 < \frac{482}{46}v_2 \approx 10v_2 > 0 \tag{7.4.57}$$

This shows that coupling promotes formation of urea in the liver; in fact, a small amount of oxidation "pumps" a significant amount of urea formation, in spite of the fact that, without coupling, urea would not be formed at all. It seems likely that many living organisms use coupling to promote chemical reactions that would not otherwise occur. Unfortunately, the identification of coupled phenomena is not a problem that can be addressed by thermodynamics. In fact, some effects attributed to thermodynamic coupling actually result from faulty analysis that ignores the presence of intermediate species; such intermediates may change the signs (from negative to positive) of the affinities for some reactions in certain reaction sequences [5].

7.5 THE LAWS FOR OPEN REACTING SYSTEMS

In this section we extend the development in § 7.4 to reactions taking place in open systems. First we develop the open-system material-balance equations for reactions (§ 7.5.1) and then we develop the combined laws (§ 7.5.2).

7.5.1 Stuff Equation for Material in a Single Open Phase

Open-system chemical reactions cause changes in many important situations, such as meteorological and biological systems [1]. Early studies of such systems raised questions about the generality of the laws of thermodynamics because workers failed to distinguish open systems from those that are closed. This confusion was largely resolved by Prigogine, whose work on these problems contributed to his Nobel Prize.

Consider an open system having any number of inlets α and any number of outlets β. For such a system, the general stuff equation (1.4.1) can be written in terms of the number of moles of species i,

$$\sum_{\alpha} dN_i^{\alpha} - \sum_{\beta} dN_i^{\beta} + dN_i^{gen} - dN_i^{con} = dN_i^{acc} \tag{7.5.1}$$

Here the superscript *gen* refers to generation of species i, *con* refers to consumption, and *acc* refers to accumulation of i in the system. When chemical reactions are occurring, the difference between the generation and consumption terms reflects the net effect of reactions (*rxn*). Moreover, we already have an expression (7.4.15) for the change in species mole numbers due to reactions; here we write (7.4.15) in the form

$$dN_i^{gen} - dN_i^{con} = dN_i^{rxn} = \sum_{j}^{\mathcal{R}} \nu_{ij}\, d\xi_j \tag{7.5.2}$$

Recall that $\mathcal{R}$ is the number of independent reactions, ξ_j is the extent of reaction j, and each ν_{ij} is a stoichiometric coefficient for species i in reaction j. Combining (7.5.1) and (7.5.2), we have

$$\sum_{\alpha} dN_i^{\alpha} - \sum_{\beta} dN_i^{\beta} + \sum_{j}^{\mathcal{R}} \nu_{ij}\, d\xi_j = dN_i^{acc} \qquad i = 1, 2, \ldots, C \tag{7.5.3}$$

This is the general form of the stuff equation for an open system containing a total of C species, some or all of which are engaged in chemical reactions.

If we want (7.5.3) explicitly in terms of rates of change, we can write

$$\sum_{\alpha} \dot{N}_i^{\alpha} - \sum_{\beta} \dot{N}_i^{\beta} + \sum_{j}^{\mathcal{R}} \nu_{ij}\, \mathbf{v}_j = \dot{N}_i^{acc} \tag{7.5.4}$$

where $\mathbf{v}_j$ is the reaction rate for the jth reaction; see (7.4.50). In the special case of a closed system, there are no inlets or outlets, and (7.5.4) reduces to

$$\sum_{j}^{\mathcal{R}} \nu_{ij}\, \mathbf{v}_j = \dot{N}_i^{acc} \qquad \textit{closed systems} \tag{7.5.5}$$

which merely confirms that the mole numbers of reacting species are not necessarily conserved, even in closed systems.

7.5.2 Combined Laws for Reactions in Open Systems

To obtain the combined first and second laws for open systems with chemical reactions, we proceed just as we did in § 7.2.1 for nonreacting systems. Our situation can still be represented by Figure 7.4, which contains bulk phases α and β separated by an interface I. The temperatures and pressures in the phases are controlled by reservoirs, as in Figure 7.4. We again choose the interface to be the system, and the interface still has negligible mass, so no mass, energy, or entropy accumulate there; that is, (7.2.1)–(7.2.3) still apply. However, we now have $\mathcal{R}$ independent reactions occurring in each bulk phase, although no reactions occur in the interface. The development is exactly that in § 7.2.1, giving the same result

$$\left(\frac{T^{\alpha}}{T^{\beta}} - 1\right)\delta Q^{\beta} - (\delta W^{\alpha} + \delta W^{\beta}) - \sum_{i} \overline{G}_i^{\alpha}\, dN_i^{\alpha} - \sum_{i} [\overline{H}_i^{\beta} - T^{\alpha}\overline{S}_i^{\beta}]\, dN_i^{\beta} \geq 0 \tag{7.2.13}$$

In § 7.2.1, the changes in bulk-phase mole numbers dN^{α} and dN^{β} were caused by diffusion across the interface. But here, those changes may result from diffusion or chemical reaction or both. So for each component in each phase, we write

$$dN_i^k = dN_i^{k,\,dif} + dN_i^{k,\,rxn} \qquad k = \alpha, \beta \tag{7.5.6}$$

Using (7.4.15) for the reaction term, (7.5.6) can be written as

$$dN_i^k = dN_i^{k,\,dif} + \sum_{j}^{\mathcal{R}} \nu_{ij}^k\, d\xi_j^k \qquad k = \alpha, \beta \tag{7.5.7}$$

Further, note that whatever diffuses from one phase, across the interface, must enter the other phase; so we have

$$dN_i^{\alpha,\,dif} = -dN_i^{\beta,\,dif} \tag{7.5.8}$$

Therefore, (7.2.13) becomes

$$\left(\frac{T^\alpha}{T^\beta} - 1\right)\delta Q^\beta - \sum_i [\bar{G}_i^\alpha - (\bar{H}_i^\beta - T^\alpha \bar{S}_i^\beta)]dN_i^{\alpha,\,dif} - (\delta W^\alpha + \delta W^\beta)$$

$$- \sum_j^{\mathcal{R}} \Delta G_j^\alpha \, d\xi_j^\alpha - \sum_j^{\mathcal{R}} (\Delta H_j^\beta - T^\alpha \Delta S_j^\beta)\, d\xi_j^\beta \geq 0 \tag{7.5.9}$$

Introducing the affinity from (7.4.41), we obtain

$$\left(\frac{T^\alpha}{T^\beta} - 1\right)\delta Q^\beta - \sum_i [\bar{G}_i^\alpha - (\bar{H}_i^\beta - T^\alpha \bar{S}_i^\beta)]dN_i^{\alpha,\,dif} - (\delta W^\alpha + \delta W^\beta)$$

$$+ \sum_j^{\mathcal{R}} [\mathcal{A}_j^\alpha \, d\xi_j^\alpha - (\Delta H_j^\beta - T^\alpha \Delta S_j^\beta)\, d\xi_j^\beta\,] \geq 0 \tag{7.5.10}$$

Equation (7.5.10) is the combined first and second laws for open systems undergoing chemical reactions with T and P constant in each phase. It imposes limitations on the combined effects of reactions, material transfers, and energy transfers across an interface between bulk phases α and β.

When the only work mode is Pv work, and when the temperatures and pressures are not only constant, but also the same in the two phases ($T^\alpha = T^\beta$ and $P^\alpha = P^\beta$), then (7.5.10) collapses to

$$-\sum_i [\bar{G}_i^\alpha - \bar{G}_i^\beta]dN_i^{\alpha,\,dif} + \sum_j^{\mathcal{R}} [\mathcal{A}_j^\alpha \, d\xi_j^\alpha + \mathcal{A}_j^\beta \, d\xi_j^\beta] \geq 0 \qquad T^\alpha = T^\beta \;\&\; P^\alpha = P^\beta \tag{7.5.11}$$

For processes occurring at fixed $T^\alpha = T^\beta$ and fixed $P^\alpha = P^\beta$, (7.5.11) imposes limitations on diffusion and reaction taking place in multiphase systems. For finite rates of diffusion and reaction, only the inequality in (7.5.11) applies. Then, as discussed in detail in § 7.2.2 for diffusion and in § 7.4.4 for reactions, there are two general ways by which the inequality can be satisfied: uncoupled situations and coupled ones.

In completely uncoupled situations, every diffusion rate and every reaction rate is independent of all other rates, and then every term in (7.5.11) must be positive. But although the completely uncoupled situation is mathematically possible, it rarely occurs in practice; in most multiphase reacting systems, coupling is present—especially coupling between diffusion and reaction. In such cases, some terms in (7.5.11) can be negative, so long as they are dominated by positive terms so that the inequality is obeyed. Then it is possible for some species to diffuse against their chemical potential gradients or for some reactions to proceed against their affinities. However, these kinds of behavior are often obscured because, in practice, a few terms in (7.5.11) often dominate the sum.

7.6 CRITERIA FOR REACTION EQUILIBRIUM

In this section we use the combined laws from § 7.5 to obtain the criteria for reaction equilibria in both closed and open systems. The development here parallels that presented in § 7.3 for phase equilibrium.

7.6.1 Closed Systems

For $\mathcal{R}$ independent reactions taking place in a closed system at fixed T and P, the combined laws are

$$\sum_j^{\mathcal{R}} \mathcal{A}_j \, v_j \geq 0 \qquad \textit{fixed T and P} \qquad (7.4.49)$$

Just as in the nonreacting situations discussed in § 7.3, the inequality in (7.4.49) applies to irreversible processes and the equality applies to reversible changes. Now we repeat the argument in § 7.3.1 that establishes a correspondence between reversible changes and equilibrium states; the consequence is that the equality in (7.4.49) applies both to reversible changes and to equilibrium states. Therefore, for reaction equilibrium at fixed T and P, we must have

$$\sum_j^{\mathcal{R}} \mathcal{A}_j \, v_j = 0 \qquad \textit{fixed T and P} \qquad (7.6.1)$$

However, this is only necessary but not sufficient for identifying equilibrium. For example, when reactions are coupled it may happen that some terms in (7.6.1) are positive while others are negative, so the sum is zero; nevertheless, reactions are in progress ($v_j \neq 0$) and the system is not at equilibrium.

Equilibrium means that all driving forces are zero. For chemical reaction j, the driving force is the affinity $\mathcal{A}_j$: reaction equilibrium (at fixed T and P) occurs when *each* affinity is zero. When this condition is met, the equality in (7.6.1) is satisfied term-by-term,

$$\mathcal{A}_j \, v_j = 0 \qquad j = 1, 2, \ldots, \mathcal{R} \qquad \textit{fixed T and P} \qquad (7.6.2)$$

However, (7.6.2) is necessary but still not sufficient for reaction equilibrium. The necessary and sufficient conditions are simply

$$\mathcal{A}_j = 0 \qquad j = 1, 2, \ldots, \mathcal{R} \qquad \textit{fixed T and P} \qquad (7.6.3)$$

If no driving force exists for reaction j ($\mathcal{A}_j = 0$), then the reaction is not occurring; so, a consequence of (7.6.3) is that the rates are also zero,

$$v_j = 0 \qquad j = 1, 2, \ldots, \mathcal{R} \qquad \textit{fixed T and P} \qquad (7.6.4)$$

For a system of $\mathcal{R}$ reactions, we obtain the equilibrium composition, not by solving the rate equations (7.6.4), but rather by solving the $\mathcal{R}$ criteria (7.6.3) for equilibrium values of $\mathcal{R}$ extents of reaction, ξ_j^e , $j = 1, 2, \ldots, \mathcal{R}$. Those equilibrium equations are developed and discussed in Chapter 10.

7.6.2 Open Systems

Open-system processes may include chemical reactions, diffusional mass transfer, and energy transfer across system boundaries. All such processes must satisfy the open-system form of the combined laws. When these processes are all complete and equilibrium is established, then it is the equality in (7.5.10) that applies. However, this statement is only a necessary condition; it is not sufficient. The necessary and sufficient conditions for equilibrium are that each term in (7.5.10)—including each term in each sum—must be zero. In other words, the system must simultaneously satisfy the criteria already discussed for thermal equilibrium (7.3.3), mechanical equilibrium (7.3.5), diffusional equilibrium (7.3.11) or (7.3.12), and reaction equilibrium (7.6.3).

7.6.3 Example

Can the criteria (7.6.3) for reaction equilibrium, which require all affinities to be zero, be reconciled with the general criterion for isothermal-isobaric equilibrium, namely $dG = 0$ (7.1.40)?

The objective here is to show that the reaction equilibrium criteria (7.6.3) are a consequence of the more general equilibrium criterion (7.1.40) that applies to any *NPT* system, including reacting systems. Consider a system of C species confined to a closed vessel and maintained at constant T and P by contact with an external heat and work reservoir. The species may undergo $\mathcal{R}$ independent chemical reactions. Since T and P are fixed for the entire system, the *NPT* criterion for equilibrium (7.1.40) applies; that is, when all reactions are complete and equilibrium is reached, the system's total Gibbs energy will be a minimum,

$$dG = 0 \qquad \textit{fixed } T \textit{ and } P \tag{7.6.5}$$

The total Gibbs energy can be obtained from the species chemical potentials,

$$G = \sum_i^C N_i \bar{G}_i \tag{7.6.6}$$

where the sum runs over all species including all products and reactants. Consider a process in which all reactions proceed by a differential amount; the response of G to such a change is, in general,

$$dG = \sum_i^C N_i \, d\bar{G}_i + \sum_i^C \bar{G}_i \, dN_i \tag{7.6.7}$$

But at fixed T and P, the first term on the rhs is zero by the Gibbs-Duhem equation, and the second term can be rewritten using (7.4.15). The result is

$$\sum_{j}^{\mathcal{R}} \mathcal{A}_j \, \nu_j = 0 \qquad \textit{fixed } T \textit{ and } P \qquad (7.6.1)$$

and the arguments following (7.6.1) can be repeated leading to (7.6.3), as required. The general criterion for isothermal-isobaric equilibrium (7.1.40) includes the reaction-equilibrium criteria (7.6.3) as a special case. *QED*

7.7 SUMMARY

In this chapter we formulated the combined first and second laws for closed and open systems, both with and without chemical reactions. We found that each form of the combined laws imposes limitations on the directions and magnitudes of processes; particular forms apply to particular kinds of processes and systems. In addition, the combined laws provide the conditions that must be satisfied when all processes are complete and equilibrium has been established. This means that the material in this chapter can serve as the starting point for any thermodynamic analysis.

In every case, we found that the directions and magnitudes of natural processes arise from entropy generation, which is always positive. This applies not only to processes involving mechanical work, but also to those involving heat transfer, diffusional mass transfer, and chemical reactions. However, we also showed that entropy generation is mandatory only for the overall process. When a process involves two or more coupled mechanisms, then an individual mechanism might proceed in a direction opposite to that followed when the mechanism operates alone. The importance of such coupling is that it can enable certain transfers or transformations to occur as part of a larger process, when otherwise those same transfers or transformations could not occur in isolation.

The constraints imposed by the combined laws all adhere to a single basic pattern,

$$\sum \left[\begin{pmatrix} \text{driving forces} \\ \text{for change} \end{pmatrix} \times \begin{pmatrix} \text{rate of} \\ \text{change} \end{pmatrix} \right] \geq 0 \qquad (7.7.1)$$

During any process a driving force produces a change, and the rate of change is coupled to the driving force in such a way that (7.7.1) is always obeyed. We have encountered the following examples of (7.7.1) applied to individual processes: when mechanical work causes an expansion or contraction of a system, a pressure difference $(P^\alpha - P^\beta)$ drives a volume change (dV); when a thermal interaction exists between two systems, a temperature difference $(T^\alpha - T^\beta)$ drives heat transfer from one system to the other (δQ); when material diffuses between phases α and β, a difference in chemical potentials drives the mass transfer (dN_i); and when chemical reactions occur, the affinity $(\mathcal{A})$ determines the progress of a reaction $(d\xi)$. At equilibrium the equality in (7.7.1) applies because both the driving force and the rate are zero: at equilibrium not only is there no change in the state, but also there is no *tendency* for change. A schematic of (7.7.1) for a single independent process appears in Figure 7.6.

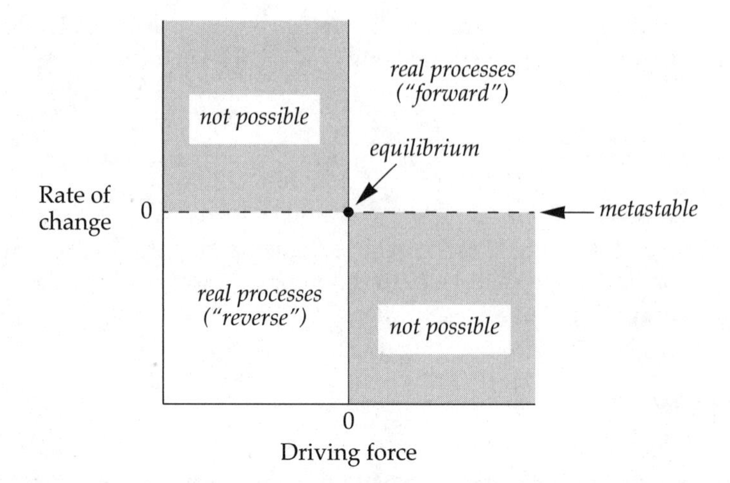

Figure 7.6 Schematic of Equation (7.7.1) for any one independent process, such as work, heat transfer, diffusion, or chemical reaction. Closed circle identifies the equilibrium state in which there is neither a driving force nor a change. Broken horizontal line locates metastable states, in which a nonzero driving force fails to cause a change. Shaded regions cannot be reached by a single process.

For most nonequilibrium situations, the inequality in (7.7.1) applies: a nonzero driving force causes a nonzero rate of change. But nonequilibrium situations can also be found in which the equality in (7.7.1) is satisfied. These occur when a finite driving force is not sufficient to overcome a resistance to change; such situations are metastable and can be catastrophically sensitive to small disturbances. However, we never observe situations in which a finite change is coupled to a zero driving force.

For situations involving mechanical work and heat transfer, we have some intuition and experience to support the notions that driving forces are connected to rates, and that an absence of driving forces implies zero rates. Such familiarity may foster understanding because those driving forces and rates involve common measurables, such as temperature and pressure. However, for phase and reaction equilibrium, the driving forces imposed by Nature appear in terms of conceptuals: the chemical potentials or, equivalently, the fugacities. These conceptual driving forces are only subtly connected to physical reality, as we tried to show in Chapters 4–6.

Nevertheless, the equilibrium criteria presented in this chapter serve as the starting points for performing engineering analyses on situations involving changes of state, phase equilibrium, and reaction equilibrium. However, since these criteria appear in terms of conceptuals (fugacities), we must first restate them in terms of measurables before calculations can be performed. In principle, such restatements are straightforward: we merely select an appropriate form from the five famous fugacity formulae and combine that form with an appropriate model for the required experimental data (a $PVTx$ model or a G^E model). This is easier said than done; to select appropriate models and fugacity formulae, we must exercise considerable engineering judgement. Moreover, the resulting equations must nearly always be solved by trial, and they are best solved on a computer. These are not thermodynamic issues, but they are important practical issues, and they will be discussed in Part IV.

LITERATURE CITED

[1] I. Prigogine and R. Defay, *Chemical Thermodynamics*, D. H. Everett (transl.), Longmans, Green and Co., London, 1954.

[2] W. R. Smith and R. W. Missen, *Chemical Reaction Equilibrium Analysis*, Wiley, New York, 1982; reprinted by Krieger Publishing Co., Malabar, FL, 1991.

[3] E. S. Pérez Cisneros, R. Gani, and M. L. Michelsen, "Reactive Separation Systems. I. Composition of Physical and Chemical Equilibrium," *Chem. Engr. Sci.*, **52**, 527 (1997).

[4] E. S. Pérez-Cisneros, Ph.D. Thesis, Danish Technical University, Lyngby, DK, 1997.

[5] M. Boudart, "Thermodynamic and Kinetic Coupling of Chain and Catalytic Reactions," *J. Phys. Chem.*, **87**, 2786 (1983).

PROBLEMS

7.1 (a) When isometric-isentropic processes are applied to closed systems, show that (7.1.41) applies to the resulting changes of state and show that, when N, V, and S are fixed, equilibrium occurs when the internal energy is minimized. Describe a physical process that has N, V, and S fixed.

(b) When isobaric-isentropic processes are applied to closed systems, show that (7.1.42) describes the resulting changes of state and show that, when N, P, and S are fixed, equilibrium occurs when the enthalpy is minimized. Describe a physical process that has N, P, and S fixed.

7.2 A quantity of pure oxygen is initially at 25°C and 1 bar. The gas is needed at 50°C and 2 bar. Can the required change of state be accomplished by some adiabatic manipulation of the volume? If not, what is the highest pressure that could be attained by an adiabatic process that ends at 50°C? Assume $c_p = 7R/2$. Clearly state any other assumptions made.

7.3 A vessel formed from rigid, thermally conducting walls is immersed in a heat bath at 25°C. The vessel has total volume V and is divided into two compartments, α and β, by a rigid, movable, thermally conducting partition. The partition can slide laterally with little friction; initially the partition is positioned so that one compartment has a volume $V^\alpha = V/5$. The partition is initially held in place by stops. Each compartment is loaded with ten moles of pure nitrogen. A process is initiated by removing the stops, allowing the partition to irreversibly slide to a new equilibrium position. Estimate the amount of entropy generated.

7.4 Consider a system in which the electrostatic work mode is important, $\delta W_e = E\,dq$ where E is the electric field and q is the charge. Show that isothermal transfer of charge across the system boundary occurs from the region of high field to that of low field.

7.5 Tabitha the Untutored has adopted the intuitively seductive position that the driving force for pure-component diffusion is a density difference: material always diffuses from regions of high density to regions of low density. Hence, Tabitha claims that part of the criteria for equilibrium is an absence of density gradients in a system.

(a) Use saturated steam tables to either support or oppose to this claim.

(b) If you do not believe the claim, then what is the driving force for diffusion? Can you use the saturated steam tables to help demonstrate your answer?

7.6 For each of the following phase-equilibrium situations, write a complete set of independent equalities that are always true at equilibrium:

(a) pure carbon dioxide in vapor-solid equilibrium.

(b) a binary mixture of benzene and water in three-phase vapor-liquid-liquid equilibrium.

(c) carbonated water in equilibrium with its vapor ($CO_2 + H_2O$).

(d) a binary mixture of 1 and 2 held in one half of a diffusion cell; the other half contains pure 1 and is separated from the mixture by a semipermeable membrane. The membrane passes 1 but not 2. The cell is immersed in a heat bath.

7.7 Could a difference in strengths of electric fields cause diffusion of a component against its chemical potential gradient? If so, could this occur with neutral molecules or only with charged molecules?

7.8 Consider a liquid solution in equilibrium with its vapor. Show that

$$\Delta\bar{S}_i = (\Delta\bar{H}_i)/T$$

where $\Delta\bar{S}_i = \bar{S}_i^{\ell} - \bar{S}_i^{v}$ is the change in the partial molar entropy for component i on condensation and $\Delta\bar{H}_i$ is the change in partial molar enthalpy of component i on condensation.

7.9 Steam and methane can react to form hydrogen, carbon monoxide, and carbon dioxide. (a) Obtain the stoichiometric coefficients and a set of independent reactions for this system. (b) If a reactor initially contains 4 moles of steam and 2 moles of methane, find the composition of the mixture when 1 mole of steam and 0.1 mole of methane remain.

7.10 Carbon and zinc oxide can react to form Zn, CO, and CO_2. (a) Obtain the stoichiometric coefficients and a set of independent reactions for this system. (b) The reactions can be carried out at 1300 K, 1 bar, with ZnO and C as solids and Zn, CO, and CO_2 in a vapor phase. If the system initially has three moles each of ZnO and C, what will be the vapor-phase composition when one mole each of ZnO and C remain?

7.11 Consider the situation described in the example in § 7.4.3, with an initial loading of 1.5 moles of methane and 1 mole of oxygen. Find values for the extents of reactions that will provide the maximum amounts of CO and H_2.

7.12 Repeat the analysis in the example of § 7.4.2, but for the one arbitrarily selected stoichiometric coefficient, use $\nu_{11} = -1$. Create a table of compositions vs. extent of reaction and compare it with the one in the example. Which quantities change and which remain unaffected by the way the reaction is balanced? Discuss.

7.13 Consider ammonia reacting with propane and propylene to form methane, hydrogen, and hydrogen cyanide.

(a) Obtain the stoichiometric coefficients and a set of independent reactions for this system.

(b) For the reactions carried out in the gas phase at 300 K, 1 bar, determine the affinities for each reaction, as functions of the extents of reaction, when the initial mixture contains 100 moles of ammonia, one mole of propane, and one mole of propylene. Assume the gas is ideal and recall ideal gases have

$$\overline{G}_i = \overline{G}_i^o + RT\ln(x_i P)$$

You may use the following values for the standard-state chemical potentials:

Species	NH_3	HCN	C_3H_6	C_3H_8	CH_4	H_2
$\overline{G}_i^o/R$ (K)	–1,900	15,000	7,500	–2,800	–610	0

7.14 Let a binary mixture of components 1 and 2 form each of the phases α and β in Figure 7.4. The interface between the phases is thermally nonconducting, fixed in position, and permeable to both components. The position of the interface is such that each phase has the same volume. Further, we load the same number of molecules of each component into each phase. We start an adiabatic, workfree process by adjusting the temperatures so that $T^\alpha > T^\beta$.

(a) If component 1 diffuses from phase β to phase α, is the process driven by an enthalpy difference or by an entropy difference?

(b) Is it possible for component 1 to diffuse in one direction across the interface while simultaneously component 2 diffuses in the opposite direction? Explain your answer.

(c) When the process ends and equilibrium is established (and assuming no phase changes occur), will the two phases have the same temperature? The same composition?

8

CRITERIA FOR OBSERVABILITY

During the design and operation of chemical processes, we routinely propose a state for a system by specifying a temperature, pressure, composition, and phase. Then the question is, Can the system be brought to that state? This is a question of observability. In many situations, particularly those involving multicomponent mixtures, the answer is not at all obvious. For example, at certain values for T and P, mixtures of phenol and water can undergo drastic phase changes in response to slight changes in composition: a mixture of phenol in water might be a one-phase vapor, or a one-phase water-rich liquid, or a phenol-rich liquid in equilibrium with a water-rich liquid, or it might be in three-phase vapor-liquid-liquid equilibrium.

In the previous chapter we derived criteria for identifying equilibrium states; for example, in a closed system at fixed T and P, the equilibrium state is the one that minimizes the Gibbs energy. That minimization is equivalent to satisfying the equality of component fugacities. More generally, we derived criteria for thermal, mechanical, and diffusional equilibrium in open systems. But although those criteria can be used to identify equilibrium states, they are not always sufficient to answer the question of observability. Observability requires stability. Thermodynamic states can be stable, metastable, or unstable; a stable equilibrium state is always observable, a metastable state may sometimes be observed, and an unstable state is never observed.

In this chapter we develop the stability criteria for both pure substances and for mixtures. Since we have three kinds of equilibria, we have three kinds of stabilities: thermal stability, mechanical stability, and diffusional stability. If the proposed state of a single phase violates any of these criteria, then the phase might spontaneously split into two or more phases. Therefore, violations of stability criteria contribute to the wealth of phase behavior observed in Nature. In this chapter we introduce some of the phase behavior that results from instabilities, but the subject is an extensive one, so the descriptions of observable phase behavior are continued in the next chapter.

In § 8.1 we derive the thermal and mechanical stability criteria for closed systems, and in § 8.2 we apply those criteria to pure substances. In pure substances only thermal and mechanical instabilities are possible; diffusional instabilities never occur because pure substances cannot exhibit concentration gradients. Then in § 8.3 we derive the diffusional stability criteria for open systems, and in § 8.4 we apply those criteria to fluid mixtures. In general, to answer the question of observability, we have three stability criteria to test, but fortunately, the three are inclusive: if a one-phase mixture is diffusionally stable, then it is also mechanically stable, and if the mixture is mechanically stable, then it is also thermally stable.

Stability criteria are economically posed in terms of conceptuals, such as S, G, or A, but before we can test for stability, we must connect the stability criteria to measurables. The connections can be achieved either via models for volumetric equations of state, say $P(T, v, \{x\})$, or (in cases of mixtures) via models for $g^E(T, P, \{x\})$. Both approaches are viable when all phases are fluid; however, a g^E model should be used for any solid phase. In general, then, we continue to face the ever-present thermodynamic problem of establishing useful relations between conceptuals and measurables.

8.1 PHASE STABILITY IN CLOSED SYSTEMS

The thermodynamics in this book is restricted to a description of well-defined states and to analyses of processes that change the system from one state to another. Thermodynamics deals mainly with equilibrium states, which were discussed in a qualitative way in § 1.2.2 and in a quantitative way in § 7.1. In both § 1.2.2 and § 7.1 we tacitly assumed that the situations under discussion were *stable* equilibrium states. But in general a stable state is only one of several possible kinds of states that are available to systems. In § 8.1.1 we describe the kinds of states that can be legitimately proposed for thermodynamic systems, and we identify those that are observed in practice.

Once we know the states that are available, then we want quantitative criteria that enable us to identify the state actually assumed by the system. Formally, the criteria are contained in § 7.1; for example, if the system is maintained at a constant T and P, then the observed equilibrium state will be the one that satisfies (7.1.40)—the state that minimizes the Gibbs energy. So if at fixed T and P, a system can possibly exist as one phase or as two phases, the observed equilibrium situation will be the one with the lower Gibbs energy. For example, when

$$g_{\text{one phase}} < g_{\text{two phases}} \tag{8.1.1}$$

then the observed equilibrium situation will be a single phase. The criteria (7.1.40) and (8.1.1) are often used to identify phase-equilibrium situations. However, criteria such as (8.1.1) require us to solve the phase-equilibrium problem for the compositions of the two phases. It is often useful to have alternatives to (8.1.1) that involve only the state of the proposed one-phase situation; such forms are derived in § 8.1.2 for closed systems and in § 8.3 for open systems.

8.1.1 Stability of Well-Defined States

By a *well-defined* state we mean a state to which property values can be assigned. The class of well-defined states contains the observable equilibrium states discussed in § 1.2.2; but in addition, the class includes hypothetical states that are not observable but that nevertheless can be identified as points on phase diagrams. Often we need to determine whether a hypothetical state is in fact observable; thermodynamics provides criteria for making such determinations. Sometimes we refer to hypothetical states as *proposed* states, since they often occur in the development of a proposed solution to a thermodynamic analysis.

In the general case (considering not just thermodynamic systems), well-defined states can be divided into two types: *static* and *dynamic*. Static well-defined states are always equilibrium states in which all forces acting on a system are balanced at every instant; however, static states are not accessible to thermodynamic systems, so we do not discuss them further here.

In dynamic situations forces are not balanced at every instant; one special case of dynamic situations is the *steady state*, in which forces are constant but they are not balanced by opposing forces. In addition, dynamic situations may include states at equilibrium. In *dynamic equilibria* forces fluctuate at every instant, but the forces are balanced when they are averaged over finite durations and finite parts of the system. The relevant time and length scales may or may not be sensible or important to an observer. Moreover, these scales can differ substantially for systems in different phases of aggregation; for example, property fluctuations in solids are typically orders of magnitude smaller than those in fluids.

Dynamic states subdivide into various classes, as shown in Figure 8.1. The subdivisions depend on stability characteristics, that is, on how a system spontaneously responds to small perturbations or disturbances. In general the response can take one of three possibilities: a large response, a small bounded response, or no response (that

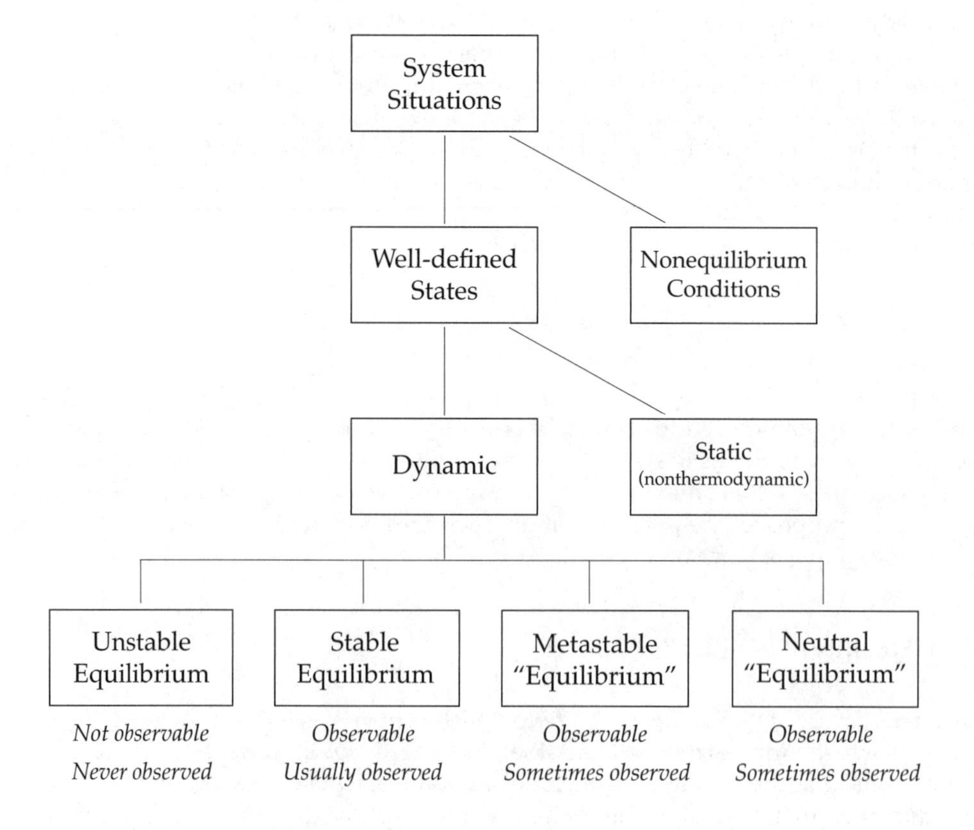

Figure 8.1 The hierarchy of system states

is, no change in the balance of forces). When a system is unaffected by a small disturbance, the state is said to be *neutral*. When a system exhibits a small bounded response to a small disturbance, the state is said to be a *stable equilibrium* state. Systems in stable states often respond not merely in a bounded way to a disturbance, but they may also return to their original unperturbed state.

When a system exhibits a large response to a small disturbance, the state is *unstable*. Unstable states may be proposed for equilibrium and nonequilibrium situations; however, unstable states are not observed in thermodynamic systems [1]. Thermodynamic states are always dynamic situations in which molecular-scale fluctuations are continually disturbing the state. Therefore if a proposed state happens to be unstable, that state will not be observed because spontaneous fluctuations drive the system away from the unstable state and toward some equilibrium state.

These distinctions among states can be illustrated by appealing to a mechanical analogy, as in Figure 8.2. The figure shows a schematic diagram of a ball that rolls on a track; the elevation z of the ball changes with its position x along the track. At any instant the forces acting on the ball are (a) the downward force of gravity and (b) the opposing upward force of the track. (We ignore friction.) Equilibrium occurs when these two forces are balanced.

The gravitational potential energy E_p is given by (2.1.4),

$$E_p = mgz \tag{8.1.2}$$

where m is the mass of the ball, g is the gravitational acceleration, and z is the elevation of the center of the ball relative to some arbitrary datum. Since m and g are constants over modest changes of z, the ordinate plotted in Figure 8.2 is proportional to E_p. This potential energy gives rise to a gravitational force,

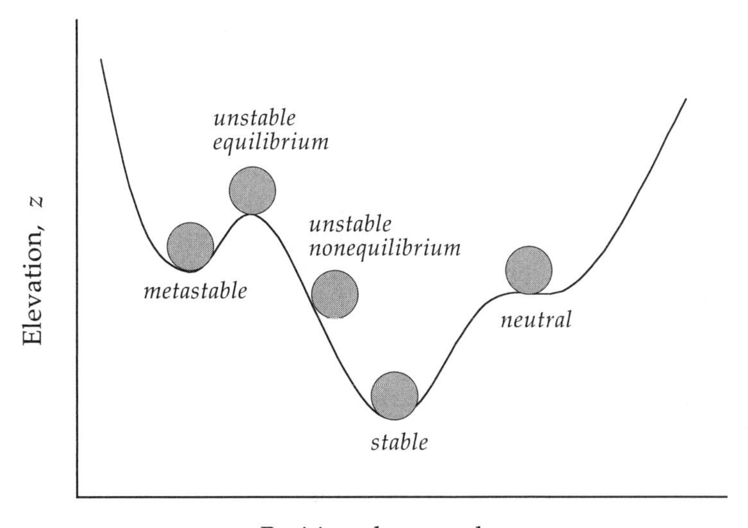

Figure 8.2 Kinds of states that are possible in a mechanical system, such as a roller coaster in a gravitational field

$$F_g = -\left(\frac{dE_p}{dz}\right) = -mg \tag{8.1.3}$$

where the sign indicates an attraction. This is Newton's second law.

Figure 8.2 illustrates that different types of extrema correspond to different responses to small disturbances and therefore to different stability characteristics. *Stable equilibrium* occurs at the global minimum in $z(x)$ or equivalently in $E_p(z)$: if a small disturbance is applied to the ball when it is at the global minimum, the resulting forces return the ball to that minimum. The ball exhibits a small response to a small disturbance. In contrast, *unstable equilibria* occur at maxima in $z(x)$: if a small disturbance is applied to the ball when it is at rest at a maximum, the resulting forces push the ball farther from the maximum. The ball exhibits a large response to a small disturbance. *Neutral equilibria* occur at points of inflection, because at those points a small disturbance has no effect on the balance of forces.

Lastly we mention the troublesome distinction that exists between the global minimum and local minima in Figure 8.2. We emphasize that only the global minimum is identified as the stable equilibrium state. In mechanics, local minima are sometimes called *local equilibrium* states, but in thermodynamics they are usually called *metastable* equilibrium states. Differential criteria cannot distinguish metastable states from stable states: both kinds of equilibria exhibit small bounded responses to differential disturbances. Without knowing the form of the curve, such as in Figure 8.2, a metastable state can be identified only by testing its response to a finite (as opposed to a differential) change of state. The response is monitored by observing an appropriate potential function ψ. If ψ always increases in response to a finite disturbance, then the original state was a stable equilibrium one; but if some finite disturbances cause ψ to decrease, then the original state was metastable. In Figure 8.2, the quantity ψ is the potential energy; in thermodynamic systems the role of ψ is played by the function that identifies equilibrium: U, G, H, A, or $-S$. Recall we found in § 7.1 that the choice from among U, G, H, A, or $-S$ is dictated by the independent properties used to fix the thermodynamic state; for example, if the state of a closed system is set by holding T, P, and N constant, then equilibrium occurs when the Gibbs energy G is minimized.

Metastable equilibrium states are observed in thermodynamic systems; one example is a *superheated liquid*, attained by careful isobaric heating of a pure liquid above its vapor-liquid saturation temperature but without boiling. This metastability can often be disrupted by a small (but finite) mechanical disturbance; the response may be an instantaneous and violent partial flash in which the newly created gas-phase rapidly expands, splashing liquid over a large area. The danger inherent in this sensitive metastability motivates caution when heating liquids over low-temperature flames. Other examples of observable metastabilities include the phenomena known as *antibubbles*, in which a liquid droplet is surrounded by vapor which, in turn, is surrounded by more liquid [2]. In response to external disturbances, antibubbles can undergo violent phase changes.

Still other examples of observable metastabilities include *subcooled phases*, such as subcooled vapors to make liquids, subcooled liquids to make solids, and subcooled solids to make other solids. The lifetimes of such metastable phases can be substantial, because the nucleation of new phases may require particular kinds of fluctuations that occur only rarely. Lastly we mention the huge number of observable conformational metastabilities that can be exhibited by large molecules such as proteins.

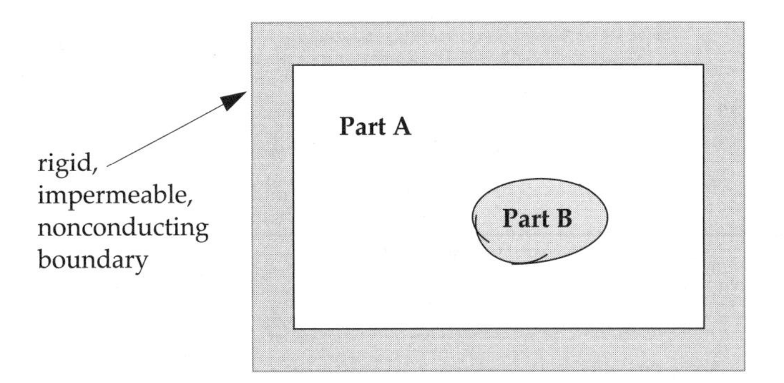

Figure 8.3 Schematic diagram of an isolated system with a local element B distinguished from the rest of the system A. For the stability analysis here, the element B is identified by a particular amount of material that remains constant; however, continual motion and rearrangement of the molecules in B cause the energy U^B and volume V^B to fluctuate.

8.1.2 Criteria for Stability

We have noted that thermodynamic equilibrium is a dynamic situation: because of molecular motions, microscopic regions of a system have intensive properties that continually fluctuate about their equilibrium values. Consequently, localized inhomogeneities in property values occur, at least over some time and length scales. In this section we develop thermodynamic conditions under which such fluctuations would not change the state (disturb the stability) of a pure fluid. For example, if some small fluctuation in temperature or pressure occurs in a localized portion of a system, will the inhomogeneities die away, leaving the overall state undisturbed? Or will such inhomogeneities grow, eventually driving the system to a new state—perhaps, even causing a phase transition?

Consider a pure one-phase fluid at equilibrium and confined to an isolated vessel. To analyze the response to a small fluctuation, imagine dividing the fluid into a large part A and a small part B, as in Figure 8.3. Part B is not necessarily a fixed region of space, but rather a particular collection of molecules whose number is constant over the time-scale of interest and whose average properties are well defined. Now imagine a fluctuation occurring in B, disturbing its energy U^B and volume V^B. Since extensive properties are additive and the total system is isolated, we have

$$U = U^A + U^B = \text{constant} \tag{8.1.4}$$

$$V = V^A + V^B = \text{constant} \tag{8.1.5}$$

and

$$S = S^A + S^B \tag{8.1.6}$$

For small fluctuations, the response of the total entropy can be reliably estimated by a Taylor's expansion,

$$\Delta S = \delta S + \delta^2 S + \ldots \tag{8.1.7}$$

The notation used here is adopted from the variational calculus; it is defined in Appendix G. Since the system is initially at equilibrium, S is a maximum. Such maxima have $\delta S = 0$, and they have

$$\delta^2 S < 0 \tag{8.1.8}$$

This means the surface $S(U, V, N)$ is concave around a stable equilibrium state. If $\delta^2 S = 0$, as it is at critical points, then we would have to consider higher-order variations [3, 4], but this is beyond our present objective. The first-order response to the fluctuation is given by

$$\delta S = \left(\frac{\partial S^A}{\partial U^A}\right)_{V^A} \delta U^A + \left(\frac{\partial S^A}{\partial V^A}\right)_{U^A} \delta V^A \tag{8.1.9}$$

$$+ \left(\frac{\partial S^B}{\partial U^B}\right)_{V^B} \delta U^B + \left(\frac{\partial S^B}{\partial V^B}\right)_{U^B} \delta V^B$$

Since the total internal energy is constant (8.1.4) and so too is the total volume (8.1.5), we can eliminate the variations in B in favor of those in A and write

$$\delta S = \left[\left(\frac{\partial S^A}{\partial U^A}\right)_{V^A} - \left(\frac{\partial S^B}{\partial U^B}\right)_{V^B}\right] \delta U^A + \left[\left(\frac{\partial S^A}{\partial V^A}\right)_{U^A} - \left(\frac{\partial S^B}{\partial V^B}\right)_{U^B}\right] \delta V^A \tag{8.1.10}$$

The variations δU^A and δV^A are arbitrary and independent of one another, so (8.1.10) can satisfy $\delta S = 0$ only if we have, initially,

$$\left(\frac{\partial S^A}{\partial U^A}\right)_{V^A} = \left(\frac{\partial S^B}{\partial U^B}\right)_{V^B} \tag{8.1.11}$$

and

$$\left(\frac{\partial S^A}{\partial V^A}\right)_{U^A} = \left(\frac{\partial S^B}{\partial V^B}\right)_{U^B} \tag{8.1.12}$$

The portions A and B are each representative samples of the same fluid and, at the start of the fluctuation, A and B are in equilibrium; hence, the value of each intensive property in A is initially the same as in B. For example, by the fundamental equation (3.2.4), (8.1.11) merely says that initially $T^A = T^B$, while (8.1.12) states that initially we have $P^A = P^B$.

To apply the stability criterion (8.1.8) we need the second-order variation of S which, from (8.1.9), is found to be

$$\delta^2 S = S^{A}_{uu}(\delta U^{A})^2 + 2S^{A}_{uv}\delta U^{A}\delta V^{A} + S^{A}_{vv}(\delta V^{A})^2 \tag{8.1.13}$$

$$+ S^{B}_{uu}(\delta U^{B})^2 + 2S^{B}_{uv}\delta U^{B}\delta V^{B} + S^{B}_{vv}(\delta V^{B})^2$$

where S_{xy} is a shorthand for the second derivative $(\partial^2 S/\partial X \partial Y)$; see Appendix G. Using the conservation equations (8.1.4) and (8.1.5) together with the fact that all intensive properties are initially the same in parts A and B, (8.1.13) becomes

$$\delta^2 S = [S^{A}_{uu}(\delta U^{A})^2 + 2S^{A}_{uv}\delta U^{A}\delta V^{A} + S^{A}_{vv}(\delta V^{A})^2](N/N^{B}) < 0 \tag{8.1.14}$$

where $N = N^{A} + N^{B}$. Equation (8.1.14) poses the test for stable equilibrium in terms of the response of the A-part of the system. Since we are interested only in the sign of $\delta^2 S$, we delete the A-superscripts and drop the factor N/N^{B}. The criterion for stable equilibrium is then

$$\delta^2 S = S_{uu}(\delta U)^2 + 2S_{uv}\delta U\delta V + S_{vv}(\delta V)^2 < 0 \tag{8.1.15}$$

In the language of linear algebra, the rhs of (8.1.15) is a *quadratic form* (see Appendix B); that is, letting $(\delta U \;\; \delta V)$ be the vector of variations, (8.1.15) can be written as

$$\delta^2 S = \begin{bmatrix} \delta U & \delta V \end{bmatrix} \mathbf{S} \begin{bmatrix} \delta U \\ \delta V \end{bmatrix} < 0 \tag{8.1.16}$$

where **S** is the symmetric matrix

$$\mathbf{S} = \begin{bmatrix} S_{uu} & S_{uv} \\ S_{uv} & S_{vv} \end{bmatrix} \tag{8.1.17}$$

If $\delta^2 S$ is to be negative for all possible variations δU and δV, then the matrix **S** must be negative definite; or equivalently, (–**S**) must be positive definite. The conditions under which **S** is negative definite are given by a theorem from linear algebra: it is necessary and sufficient that the principal minors of **S** satisfy the following inequalities [5]:

$$S_{uu} < 0 \tag{8.1.18}$$

and

$$\begin{vmatrix} S_{uu} & S_{uv} \\ S_{uv} & S_{vv} \end{vmatrix} = S_{uu}S_{vv} - S_{uv}S_{uv} > 0 \tag{8.1.19}$$

When an equilibrium state satisfies (8.1.18) and (8.1.19), then the system is stable to small fluctuations. We now reexpress these criteria in terms of measurables.

8.1.3 Stability Criteria in Terms of Measurables

To evaluate S_{uu}, we first rewrite the fundamental equation (3.2.4) in the form

$$dS = \frac{dU}{T} + \frac{PdV}{T} \tag{8.1.20}$$

Then

$$S_{uu} = \left(\frac{\partial^2 S}{\partial U^2}\right)_V = \frac{\partial}{\partial U}\left(\frac{1}{T}\right)_V = \frac{-1}{T^2}\left(\frac{\partial T}{\partial U}\right)_V \tag{8.1.21}$$

Hence,

$$S_{uu} = \frac{-1}{T^2 C_v} < 0 \tag{8.1.22}$$

Therefore, to satisfy the stability criterion (8.1.18), we must have

$$C_v > 0 \tag{8.1.23}$$

This is the criterion for *thermal stability*: for a system in a well-defined state to be differentially stable, its internal energy must *always* increase in response to any isometric fluctuation that increases the temperature.

To evaluate S_{uv}, we use the fundamental equation (8.1.20) and write

$$S_{uv} = \frac{\partial}{\partial V}\left[\left(\frac{\partial S}{\partial U}\right)_V\right]_U = \frac{\partial}{\partial V}\left(\frac{1}{T}\right)_U = \frac{-1}{T^2}\left(\frac{\partial T}{\partial V}\right)_U \tag{8.1.24}$$

$$= \frac{1}{T^2}\frac{(\partial U/\partial V)_T}{(\partial U/\partial T)_V} = \frac{T\gamma_v - P}{T^2 C_v} \tag{8.1.25}$$

where γ_v is the thermal pressure coefficient defined in (3.3.5).

To evaluate S_{vv}, we again use (8.1.20) to obtain

$$S_{vv} = \left(\frac{\partial^2 S}{\partial V^2}\right)_U = \frac{\partial}{\partial V}\left(\frac{P}{T}\right)_U = \frac{1}{T}\left(\frac{\partial P}{\partial V}\right)_U - \frac{P}{T^2}\left(\frac{\partial T}{\partial V}\right)_U \tag{8.1.26}$$

The second derivative on the rhs is given in (8.1.24). To obtain the first derivative, consider $U = U(P, V)$, and with help from Chapter 3, we eventually find

$$dU = \frac{C_v}{\gamma_v}dP + \left(\frac{C_v}{V\alpha} + T\gamma_v - P\right)dV \tag{8.1.27}$$

Therefore,

$$\left(\frac{\partial P}{\partial V}\right)_U = \frac{-\gamma_v}{C_v}\left(\frac{C_v}{V\alpha} + T\gamma_v - P\right) \tag{8.1.28}$$

where α is the volume expansivity (3.3.6). Combining (8.1.24), (8.1.26), and (8.1.28) yields

$$S_{vv} = \frac{-1}{TV\kappa_T} - \frac{(T\gamma_v - P)^2}{T^2 C_v} \tag{8.1.29}$$

where κ_T is the isothermal compressibility (3.3.25). Finally, putting (8.1.22), (8.1.25), and (8.1.29) into the criterion (8.1.19) gives

$$S_{uu}S_{vv} - S_{uv}S_{uv} = \frac{1}{T^3 V C_v \kappa_T} > 0 \tag{8.1.30}$$

or

$$\kappa_T > 0 \tag{8.1.31}$$

This is the criterion for *mechanical stability*: for a thermally stable system to also be mechanically stable, the system volume must *always* decrease in response to any isothermal fluctuation that increases the pressure.

Note that it appears to be possible for (8.1.30) to be satisfied by having both $C_v < 0$ and $\kappa_T < 0$; however, this is only a mathematical possibility that cannot actually occur. In fact, we expect that the mechanical stability limit will be violated before the thermal limit, because the mechanical limit represents a response of higher-order than the thermal limit [3]; higher-order terms approach zero before lower-order terms. This expectation is confirmed experimentally: whenever an initially stable system is driven into an unstable region of its phase diagram, the mechanical stability limit is always violated before the thermal limit. In other words, a state may be mechanically unstable but remain thermally stable, because κ_T appears only in (8.1.31) and not in (8.1.23). The mechanical stability criterion (8.1.31) is a stronger test than the thermal stability criterion (8.1.23).

With the differential stability criteria (8.1.23) and (8.1.30) plus relations given in Chapter 3, we may identify bounds on other thermodynamic properties. For example, (3.3.31) relates the isometric and isobaric heat capacities,

$$C_p = C_v + \frac{TV\alpha^2}{\kappa_T} \tag{8.1.32}$$

This implies that a differentially stable system must have

$$C_p \geq C_v > 0 \tag{8.1.33}$$

Similarly, (3.3.30) relates the heat capacities and compressibilities, so that (3.3.30) together with (8.1.33) implies

$$\frac{\kappa_T}{\kappa_s} = \frac{C_p}{C_v} > 1 \tag{8.1.34}$$

Hence,

$$\kappa_T \geq \kappa_s > 0 \tag{8.1.35}$$

The equalities in (8.1.33) and (8.1.35) apply only when the expansivity $\alpha = 0$, a condition that rarely occurs.

Lastly, we emphasize that (8.1.23) and (8.1.31) are differential criteria: they provide the conditions under which a system is stable to *small* disturbances (otherwise the Taylor series (8.1.7) does not apply). Unfortunately, those criteria cannot be used to determine whether a proposed state is metastable or stable, because metastable states can also satisfy differential stability criteria. To distinguish metastable states from stable ones, we must observe the system's response to a finite, as opposed to a differential, disturbance.

8.2 PURE SUBSTANCES

We now use the stability criteria from § 8.1.2 to help judge the observability of pure-fluid states and to help describe phase behavior of pure fluids. Issues of observability constitute the theme of this chapter, and so it may be helpful to clarify how an observable state differs from one that is observed. We use *observable* to mean a state that can be realized in a laboratory. To realize an observable state, it is necessary to adjust certain measurables, such as T, P, and $\{x\}$, to particular values; however, such adjustments may not be sufficient to create an observable state. Some observable states can only be observed when measurables are manipulated in certain ways. In general, stable equilibrium states are always observable, but they are not always observed: sometimes a metastable state will be observed instead of a stable state. In contrast, an unstable state is neither observable nor observed (see Figure 8.1).

In the descriptions of pure-fluid phase behavior presented in this section, we rely on the simple yet qualitatively realistic equation of state developed by Redlich and Kwong (4.5.66). That equation is cubic in the volume and can be written in a pressure-explicit form,

$$P = \frac{RT}{v-b} - \frac{a}{\sqrt{T}\, v(v+b)} \tag{8.2.1}$$

The parameters a and b can be related to critical properties, as in Table 4.4.

To start the section, we develop relations by which an equation of state can be used to identify the observability of a proposed pure-fluid state (§ 8.2.1), and we illustrate with an example (§ 8.2.2). Then we qualitatively describe pure-fluid Pv diagrams (§ 8.2.3 and 8.2.4), and follow with quantitative methods for determining vapor pressures (§ 8.2.5) and latent heats of vaporization (§ 8.2.6). We end the section with brief qualitative comments on pure-component phase equilibria involving solids (§ 8.2.7).

8.2.1 Determination of Phase Stability for Pure Fluids

The problem to be considered is this: we have proposed a T and P for a pure one-phase fluid and we want to determine whether that state is observable. To answer this, we apply the following criteria, which are obeyed by any stable pure-fluid state:

(a) It is stable to small disturbances; that is, it satisfies the differential criteria for thermal (8.1.23) and mechanical (8.1.31) stability.

(b) It has a lower Gibbs energy than any other state that can exist at the same T and P.

If (a) is violated, then (b) is also violated, and the proposed single-phase state is unstable: it is not observable. If (a) is obeyed, but (b) is violated, then the proposed single-phase state is metastable: it is observable and it might be observed. If (b) is satisfied, then (a) is also satisfied, and the proposed single-phase fluid is stable and observable. When the proposed state is unstable or metastable, the observed state may be one phase or more; unstable and metastable one-phase states do not always split into two or more phases.

A conventional way to address the criteria (a) and (b) is to employ a volumetric equation of state of the form $P(T, v)$ that applies to all fluid phases of our pure substance. The Redlich-Kwong equation (8.2.1) is an example. Any properly constructed model for a volumetric equation of state should satisfy the thermal stability criterion ($C_v > 0$), and as far as we are aware, all cubic equations of state having constant parameters (a and b) do so. Consequently, thermal stability only needs to be tested when we construct complicated equations of state, such as those that are high-order polynomials in v or that have temperature-dependent parameters. Moreover, as we noted under (8.1.31), the mechanical stability criterion is a stronger test, so we do not consider thermal stability further here.

To test for mechanical stability, we first solve our equation of state for all real roots; these roots correspond to the available volumes at the proposed T and P. If only one real root for v is obtained and (8.1.31) is obeyed, then the proposed single-phase is stable and observable at the given T and P. This solves our problem.

More problematic are those situations in which the equation of state provides multiple roots for v at the given T and P. Which of these are observable? To decide, we first eliminate any v-roots that fail to satisfy the differential criterion for mechanical stability (8.1.31). That criterion can be written in several forms, but it may be more helpful here to state it as a criterion for instabilities,

$$\left(\frac{\partial P}{\partial v}\right)_T > 0 \qquad \textit{unstable} \qquad (8.2.2)$$

Therefore if the isotherm on a Pv diagram has a positive slope at the root v, then that state is unstable and it cannot be observed.

At this point we have eliminated all roots that fail to satisfy the mechanical stability criterion, but we do not yet have a unique root that is stable. To select from among the remaining alternatives, we apply criterion (b), cited at the start of this section. That criterion is a consequence of the equilibrium conditions developed in § 7.1.5: the stable equilibrium state will have a lower value of the Gibbs energy than any other state that might exist at the specified T and P.

We first test single-phase states. Let $v_k(T, P)$ be the desired volume of the proposed state, and let $v_j(T, P)$ be another root. Then the test is whether

$$g_{\text{pure}}(T, P; v_k) < g_{\text{pure}}(T, P; v_j) \tag{8.2.3}$$

If this condition is satisfied for all roots $j \neq k$, then the proposed state $v_k(T, P)$ is the stable state and it is observable.

The criterion (8.2.3) is posed in terms of conceptuals, but to perform the test, it must be connected to measurables. For computations based on a volumetric equation of state, it is usually convenient to relate g to measurables via fugacity coefficients. So we integrate the definition of the fugacity (4.3.10) from pure state $v_j(T, P)$ to pure state $v_k(T, P)$,

$$g_{\text{pure}}(T, P; v_k) - g_{\text{pure}}(T, P; v_j) = RT \ln\left(\frac{f_{\text{pure k}}}{f_{\text{pure j}}}\right) \tag{8.2.4}$$

Then the condition (8.2.3) can be written in terms of the fugacity as

$$f_{\text{pure}}(T, P; v_k) < f_{\text{pure}}(T, P; v_j) \tag{8.2.5}$$

and on substituting FFF # 1 (6.4.1), we obtain (8.2.5) in the form

$$\varphi_{\text{pure}}(T, P; v_k) < \varphi_{\text{pure}}(T, P; v_j) \tag{8.2.6}$$

These pure-component fugacity coefficients can be computed from the known equation of state by evaluating (4.4.24). At the specified T and P, the stable equilibrium state will be that single-phase state whose volume provides the lowest value of the fugacity coefficient.

Finally, we note that two pure volumes might have the same value for the fugacity coefficient, at the same T and P,

$$\varphi_{\text{pure}}(T, P; v_k) = \varphi_{\text{pure}}(T, P; v_j) \tag{8.2.7}$$

When this occurs, the stable state can be a two-phase equilibrium situation. We describe calculations for identifying these situations in § 8.2.5.

8.2.2 Example

How is a cubic equation of state used to test the stability of a proposed state for a pure one-phase fluid?

To make this general question concrete, we repose it this way: Is pure propane a stable one-phase gas at 300 K, 15 bar? To address this question, we adopt the Redlich-Kwong equation of state (8.2.1) and obtain values for the Redlich-Kwong parameters a and b from critical properties. Propane has $T_c = 369.8$ K and $P_c = 42.4$ bar; hence, the expressions in Table 4.4 give

$$a = 0.4275\sqrt{T_c}\,\frac{(RT_c)^2}{P_c} = 18.33(10^7)\ (\text{cc/mol})^2\ (\text{bar K}^{0.5}) \tag{8.2.8}$$

$$b = 0.08664\,\frac{RT_c}{P_c} = 62.82\ \text{cc/mol} \tag{8.2.9}$$

To answer the question, we apply the procedure outlined in § 8.2.1.

Step 1. Determine all real roots at the specified T and P. Since the Redlich-Kwong equation is cubic in v, we can solve it analytically using Cardan's method (Appendix C). That method gives three real roots at 300 K, 15 bar: a gas-phase root, v_{gas} = 1194 cc/mol, plus two others, v_2 = 368.2 cc/mol and v_3 = 100.8 cc/mol.

Step 2. Eliminate any roots having $(\partial P/\partial V)_T > 0$. The Redlich-Kwong equation gives

$$\left(\frac{\partial P}{\partial v}\right)_T = \frac{-RT}{(v-b)^2} + \frac{a(2v+b)}{\sqrt{T}\,v^2(v+b)^2} \tag{8.2.10}$$

At 300 K, 15 bar, 1194 cc/mol, the gas-phase root gives $(\partial P/\partial V)_T = -0.008$ bar mol/cc: the proposed gaseous state is not unstable. For the "middle" root (368.2 cc/mol), (8.2.10) gives $(\partial P/\partial V)_T = 0.07$ bar mol/cc; therefore, this state is unstable and can be eliminated from further consideration. For the liquid root (100.8 cc/mol), (8.2.10) gives $(\partial P/\partial V)_T = -7.0$ bar mol/cc; so this state is also not unstable. We have two volumes to consider further and three possible outcomes: the fluid is a stable single-phase gas, or it is a stable single-phase liquid, or it exists in two-phase vapor-liquid equilibrium (VLE).

Step 3. Of the two remaining one-phase states, at the given T and P, which has the smaller value of φ? That is, we apply (8.2.6), which offers the following possibilities:

(a) If $\varphi_{gas} < \varphi_{liq}$, then the gas is the stable phase.
(b) If $\varphi_{gas} > \varphi_{liq}$, then the liquid is the stable phase.
(c) If $\varphi_{gas} = \varphi_{liq}$, then two-phase VLE is the stable situation.

The Redlich-Kwong equation is pressure-explicit, so we compute the fugacity coefficient from (4.4.24). We find the Redlich-Kwong expression for φ to be

$$\ln\varphi = \frac{b}{v-b} - \ln\left(\frac{v-b}{v}\right) - \ln Z - \frac{\beta b}{v+b} - \beta\ln\left(\frac{v+b}{v}\right) \tag{8.2.11}$$

where β is a dimensionless group,

$$\beta \equiv \frac{a}{bRT\sqrt{T}} \tag{8.2.12}$$

At 300 K, 15 bar, (8.2.11) gives $\varphi_{gas} = 0.78$ and $\varphi_{liq} = 0.65$. Therefore at the given T and P, the stable state is liquid. The gas phase is metastable, and the two phases cannot exist in equilibrium with one another.

8.2.3 Stable One-Phase States of Pure Fluids

The discussions in § 8.2.1 and 8.2.2 imply that a Pv diagram can be divided into two regions: one in which each isotherm can have only one value for the volume v at every pressure P, and a second region in which an isotherm may have multiple values for v at some pressures P. These two regions are separated by the critical isotherm; for the fluid in Figure 8.4, the critical isotherm occurs at $T_c = 304.2$ K.

Every state "above" the critical isotherm ($T > T_c$) is a stable single-phase fluid, because it has no alternatives: each (T, P)-point has only one volume available. Therefore each state satisfies the mechanical stability criterion and all supercritical isotherms have negative slopes on Pv diagrams, as in Figure 8.4. Gases and vapors have high molar volumes, while liquids have smaller volumes. But along supercritical isotherms (such as $T = 350$ K in Figure 8.4), there is no clear distinction between gas and liquid. If a supercritical fluid can be condensed by decreasing T at fixed P or by increasing P at fixed T, then we call it a *vapor*. But if a supercritical fluid can only be condensed by changing both T and P, then we call those substances *fluids* (though they could also be called *gases*). When the distinction among liquid, gas, vapor, etc. is unimportant, we will also use *fluid* as a generic term to mean any non-solid phase.

Along any isotherm below T_c (a subcritical isotherm), multiple volumes occur for pressures $P < P_c$. When calculated from analytic equations of state, each subcritical isotherm has one or more regions of positive slope and two or more regions of negative slope (such as along $T = 250$ K in Figure 8.4). At the smallest molar volumes the slope is negative $[(\partial P/\partial v)_T < 0]$ and the fluid is one-phase liquid. Similarly, at the larg-

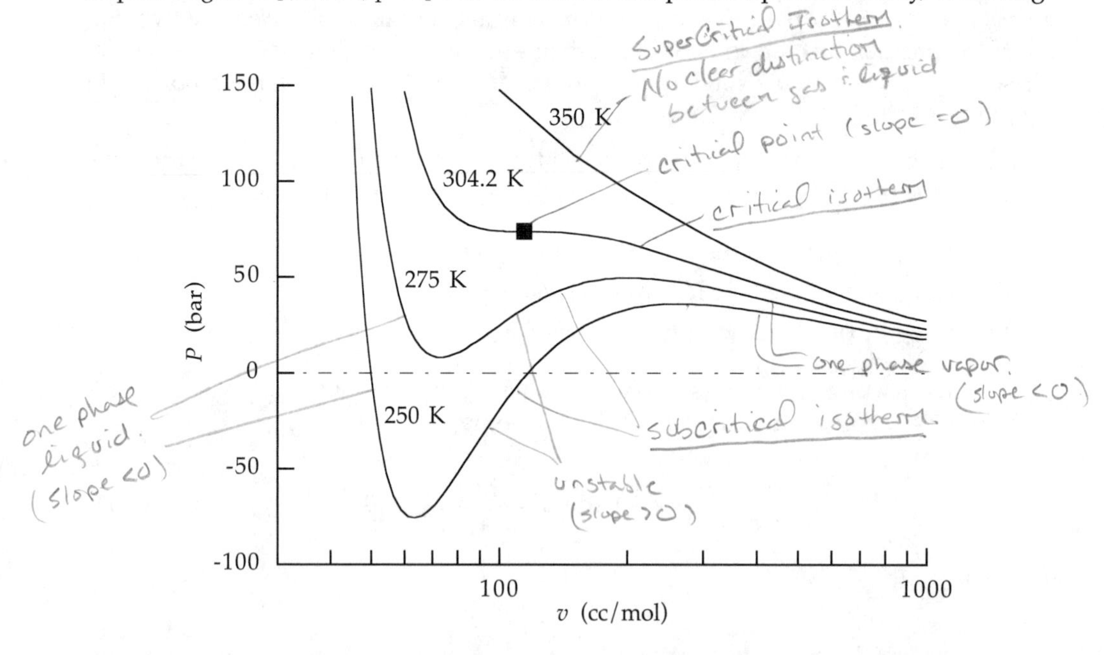

Figure 8.4 Four isotherms of a pure fluid computed from the Redlich-Kwong equation of state. Parameters a and b were computed from T_c and P_c using the relations in Table 4.4. The critical point (filled square) was taken to be $T_c = 304.2$ K and $P_c = 73.8$ bar, which is that for carbon dioxide. However, with these values the Redlich-Kwong equation gives $v_c = 114$ cc/mol, which is not a good approximation to the experimental value of 94 cc/mol for CO_2. Note that the two isotherms below T_c contain metastable and unstable states.

est molar volumes the slope is also negative and the fluid is one-phase vapor. But over some range of volumes the slope is positive $[(\partial P/\partial v)_T > 0]$ and by (8.2.2) the fluid is unstable. Such states are not observable, and the fluid will spontaneously relax to some other situation that is stable. The stable situation may be one phase or two.

The critical isotherm (T_c = 304.2 K in Figure 8.4) separates those fluids that are always one phase from those that can split in two. The critical isotherm does not contain any unstable points (points having positive slopes), but it does pass through one point of zero slope. This is a point of inflection and identifies the *critical point*; any pure-fluid critical point has

$$\left(\frac{\partial P}{\partial v}\right)_T = 0 \qquad \textit{pure critical point} \qquad (8.2.13)$$

and

$$\left(\frac{\partial^2 P}{\partial v^2}\right)_T = 0 \qquad \textit{pure critical point} \qquad (8.2.14)$$

Note that a critical fluid is a one-phase substance, not two. To locate a critical point, we use our particular equation of state to solve the one-phase equations (8.2.13) and (8.2.14); we do not solve any phase-equilibrium equations to find T_c, P_c, and v_c.

Since the isotherm has zero slope at the critical point, the isothermal compressibility at the critical point obeys

$$\kappa_T = \infty \qquad (8.2.15)$$

This suggests that near the critical point a fluid displays unusual behavior. The behavior is unusual because natural fluctuations are not completely suppressed, as they are when κ_T is bounded and positive, but neither are fluctuations able to grow so as to force a phase change, as they can when κ_T is negative. Such fluctuations cause the observable phenomenon known as *critical opalescence*; moreover, critical fluctuations are independent of molecular constitution, so that near their critical points all fluids have certain traits in common. Descriptions of critical phenomena are beyond the scope of this book; see instead [6].

The division of a Pv diagram into supercritical and subcritical regions helps relate the diagram to the stability criteria derived in § 8.1. In addition, that division corresponds to certain mathematical descriptions of critical and stability phenomena. Recall, supercritical isotherms provide only one real root for v from an analytic equation of state, while subcritical isotherms provide more than one real root; such a change in the number of real roots is called a *bifurcation* of an algebraic equation. The existence of critical points and the (mathematical) possibilities of unstable states are reflected in bifurcations of the algebraic equations of state that attempt to describe the phenomena [7]: the critical point can be called a bifurcation point. But although cubic equations of state, such as the Redlich-Kwong, exhibit bifurcations, they do not reliably describe the behavior of fluids in the critical region: such classical models fail to reproduce the correct scaling laws [6].

The jargon associated with bifurcations will be used in the following ways. When we say an equation has not bifurcated, we merely mean that an equation provides only one real root for the quantity of interest. When we say an equation has bifurcated, we mean the equation provides more than one root. An example of the context would be this: All states along supercritical isotherms are stable single phases because the equation of state does not bifurcate at supercritical states. For pure fluids, this usage probably provides only limited benefits. But when we come to mixtures, the jargon is convenient because mixtures have other equations, in addition to equations of state, that can bifurcate, and bifurcations of equations for different properties lead to different kinds of stability behavior, as we shall see in § 8.4.2.

8.2.4 Metastable and Two-Phase States of Pure Fluids

Recall from § 8.1 that differential stability criteria, such as (8.2.2), cannot distinguish stable states from metastable states. In fact Figure 8.4 contains states on the 250 K isotherm that have negative slope ($\kappa_T > 0$), but which are metastable and so are not normally observed. In Figure 8.4, one metastable region includes a range of volumes over which the 250 K isotherm has negative pressures. In engineering practice, negative pressures are rarely observed; nevertheless, they are not necessarily artifacts of the equation of state.

A negative pressure implies that a substance is under tension rather than compression (i.e., a pull rather than a push). Negative pressures are not possible in ideal gases because without intermolecular forces there is no resistance to a tension. Even in most real gases, the collective effects of intermolecular forces are so weak that a negative pressure could be achieved only with difficulty, if at all. However, liquids are another matter. In liquids, molecules exert attractive forces on one another, so liquids can resist tension and sustain negative pressures. Negative pressures are commonly used by Nature to move water from roots, through narrow xylem vessels, to leaves of trees and other plants [8]: so long as the fluid remains a continuous phase, transpiration literally pulls water up from plant roots.

For a pure substance, such as in Figure 8.5, metastable states on an isotherm lie between stable states and unstable states. At one end of the metastable range, metastable states are separated from unstable states by a curve called the *spinodal*. For a pure substance, the spinodal is the locus of points at which the differential stability criterion (8.2.2) is first violated, that is, the points at which

$$\left(\frac{\partial P}{\partial v}\right)_T = 0 \qquad \textit{pure spinodal} \tag{8.2.16}$$

Since this condition is also satisfied by the critical point, a pure-fluid critical point must lie on the spinodal. For a pure substance that obeys the Redlich-Kwong equation of state, the spinodal temperatures and volumes are related by

$$T_{sp} = \left(\frac{a(2v+b)(v-b)^2}{Rv^2(v+b)^2}\right)^{2/3} \tag{8.2.17}$$

By substituting (8.2.17) into the Redlich-Kwong equation (8.2.1), we can relate the pressure to the volume along the spinodal; a plot appears in Figure 8.5. Note that, at the critical point, the spinodal intersects the critical isotherm at its point of inflection.

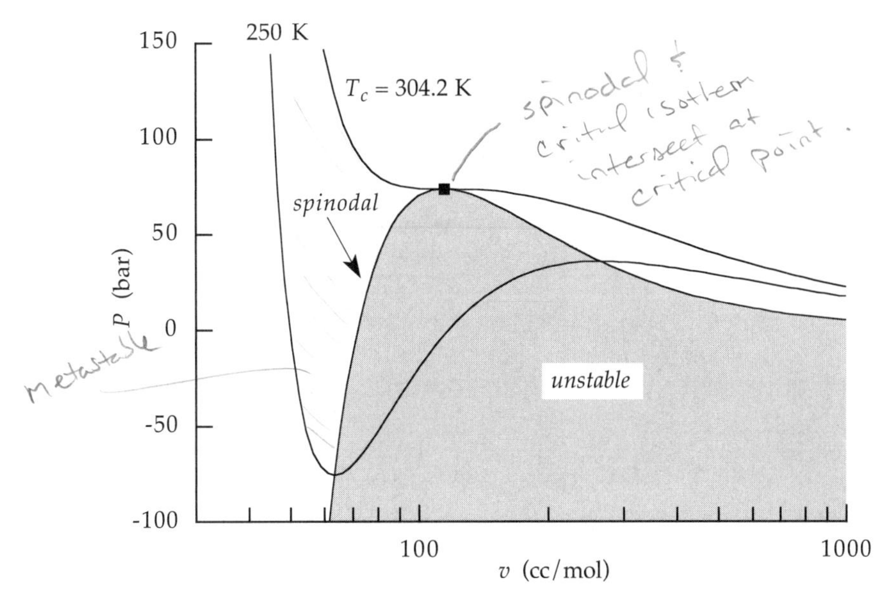

Figure 8.5 Two isotherms taken from Figure 8.4 together with the spinodal, all computed from the Redlich-Kwong equation of state. The critical point is marked with a filled square. States under the spinodal (shaded) are unstable and cannot exist as single phases; states above the spinodal may be stable or metastable.

Along any pure-fluid, subcritical isotherm, the spinodal separates unstable states from metastable states. At the other end of an isotherm's metastable range, metastable states are separated from stable states by the points at which vapor-liquid, phase-equilibrium criteria are satisfied. Those criteria were stated in § 7.3.5: the two-phase situation must exhibit thermal equilibrium, mechanical equilibrium, and diffusional equilibrium. Since we are on an isotherm, the temperatures in the two phases must be the same, and the thermal equilibrium criterion is satisfied.

We use superscript v to indicate the vapor phase and use ℓ to indicate the liquid. Then mechanical equilibrium will occur when there is no net driving force tending to change the volume of either phase; this occurs when

$$P^v = P^\ell \equiv P^s \tag{8.2.18}$$

where P^s is the *saturation pressure* common to both phases; P^s is usually called the *vapor pressure*. For a pure fluid, the vapor pressure depends only on temperature.

For a pure fluid, diffusional equilibrium will occur when there is no net driving force for diffusion of material from one phase to the other. This occurs when

$$f^v = f^\ell \tag{8.2.19}$$

For a pure substance, the fugacity depends on temperature, pressure, and phase. The locus of saturated liquid and saturated vapor states that satisfy both (8.2.18) and (8.2.19) forms the vapor-liquid *saturation curve* (also called the *vapor pressure curve*). Pure-fluid vapor pressures P^s increase with increasing T. But along the liquid branch

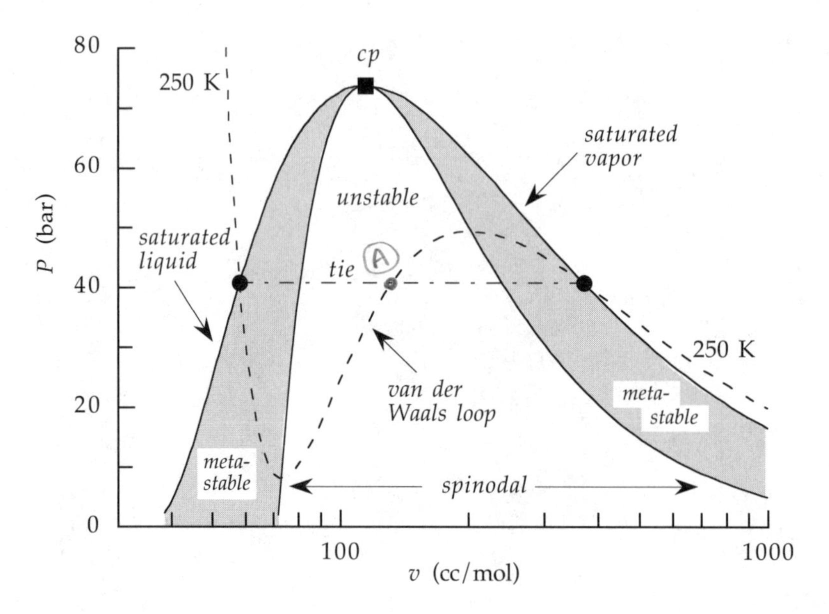

Figure 8.6 Vapor-liquid equilibrium curve for the substance of Figure 8.4, computed from the Redlich-Kwong equation of state. The critical point is marked with a filled square. Also shown is the 250 K isotherm taken from Figure 8.4. At this temperature the Redlich-Kwong equation gives $P^s = 40.8$ bar; the saturated vapor and liquid volumes occur at the filled circles. The van der Waals loop is that part of the 250 K isotherm between the saturated phases.

of the vapor-pressure curve, P^s increases with increasing molar volume v, while along the vapor branch, P^s decreases with increasing v. This is shown in Figure 8.6. Hence at some v, the vapor-pressure curve passes through a maximum: that maximum coincides with the spinodal at the critical point. Subcritical isotherms, such as that at 250 K in the figure, cut the saturation curve at two points, one for the saturated liquid, the other for saturated vapor. Those two phases have the same pressure (the vapor pressure P^s), so they can be connected by a horizontal *tie line*, which "ties" together the two phases that are in equilibrium. Cubic equations of state approximate the tie line by a "van der Waals loop" between the two saturated volumes. Isotherms computed from more complicated equations of state may exhibit more complicated behavior.

The behavior of metastable and unstable fluids is determined by the external constraints imposed on the system (see § 6.1). For example, the behavior at fixed T and P differs from that at fixed T and v, where v is the overall molar volume.

(a) Fluids at proposed states (fixed T and v) under the spinodal will *always* spontaneously split into a saturated vapor phase in equilibrium with a saturated liquid phase; the final pressure will be the vapor pressure $P^s(T)$. Fluids at proposed states (fixed T and v) between the spinodal and the saturation curve are metastable; those metastable one-phase fluids may be observed or the fluid may split into two phases at $P^s(T)$.

(b) In contrast, fluids at fixed T and P will only split into two phases if P is the vapor pressure for T. Otherwise, unstable fluids at fixed T and P *always* relax to the stable one-phase fluid having the lowest molar Gibbs energy. Further, metastable fluids at fixed T and P may be observed, or those fluids may also relax to the stable one-phase condition.

8.2.5 Vapor Pressures from Equations of State

Before we can solve the equilibrium criteria (8.2.18) and (8.2.19) to obtain the saturation curve, we must replace the fugacities with measurables. One way to proceed is to use famous fugacity formula #1 (6.4.1), which connects fugacities to measurables via fugacity coefficients φ. Since both vapor and liquid are pure phases and the pressures in each phase are the same, (8.2.19) combined with (6.4.1) reduces to

$$\varphi^{v}_{\text{pure}}(T) = \varphi^{\ell}_{\text{pure}}(T) \tag{8.2.20}$$

We will find in § 9.1 that, for a pure substance in two-phase equilibrium, only one property is needed to specify the intensive state; in (8.2.20) we have used temperature. However, even after we set a value for the subcritical temperature, (8.2.20) remains implicit in three unknowns: the vapor pressure P^s plus the molar volumes of the liquid and vapor phases, v^v and v^ℓ. To close the problem we need another equation, typically, a PvT equation of state that relates P^s to both saturated volumes at the specified T. Therefore, we must choose an equation of state that is sufficiently complicated that it bifurcates and provides multiple roots for the volume over some range of states. Such equations of state are explicit in the pressure [$P = P(T, v)$], and then we would compute φ from

$$\ln\varphi_{\text{pure}}(T, P) = \int_v^\infty [Z-1]\frac{d\psi}{\psi} + (Z-1) - \ln Z \tag{4.4.24}$$

We apply (4.4.24) to each phase; for each, the integration in (4.4.24) is to be performed along the same subcritical isotherm. When we apply (4.4.24) to the liquid, the lower integration limit is v^ℓ, and when we apply it to the vapor, the lower limit is v^v. Note that the value of the compressibility factor Z in the liquid phase differs from that in the vapor ($Z^\ell \neq Z^v$) because the molar volumes of the two phases differ. Under the integral, the value of Z is not constant, but changes with molar volume: $Z = Z(T, \psi)$. However, outside the integral, the other two values of Z are fixed at the saturation conditions: $Z = Z^\ell(T, v^\ell)$ and $Z = Z^v(T, v^v)$.

Using (4.4.24) for both sides of (8.2.20) and simplifying algebraically, we obtain the following important result,

$$P^s(T) = \frac{1}{v^v - v^\ell}\int_{v^\ell}^{v^v} P(T, v)\, dv \qquad \textit{fixed } T < T_c \tag{8.2.21}$$

In passing from (4.4.24) to (8.2.21) we have changed the dummy integration variable from ψ to v once that dummy variable can be clearly distinguished from the integration limits. The functional form for the integrand $P(T, v)$ is provided by the pressure-explicit equation of state. The integration in (8.2.21) is on the chosen subcritical isotherm T, along the van der Waals loop, from the saturated liquid volume v^ℓ to the saturated vapor volume v^v.

Equation (8.2.21) says that at any subcritical temperature, the vapor pressure is given by a mean-value theorem (Appendix A): P^s is the mean of the pressures along

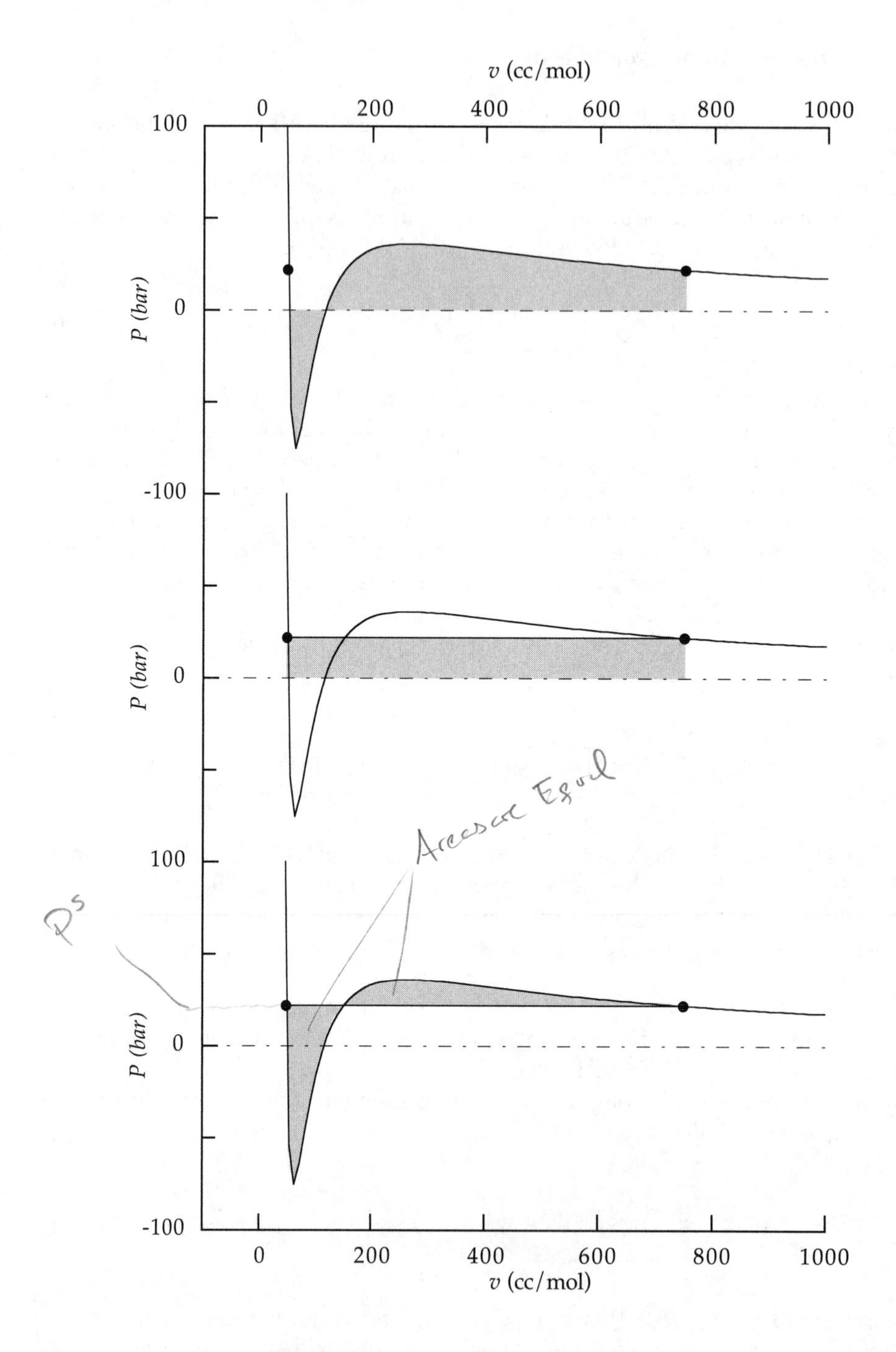

Figure 8.7 Three interpretations of the integrations (8.2.21) and (8.2.22) that determine the vapor-liquid saturation pressure for a pure substance. In each panel the solid curve is the 250 K isotherm from Figure 8.5. The filled circles locate the saturated volumes at $P^s = 21.9$ bar. *Top*: The shaded area is that given by the integral in (8.2.21). *Middle*: The shaded region is the rectangular area $P^s(v^v - v^\ell)$. According to (8.2.21), the shaded regions in the top two panels have the same area. *Bottom*: The shaded area is that provided by the equal-area construction (8.2.22); in this panel, the positive and negative areas cancel one another.

the van der Waals loop. Hence, the area under the van der Waals loop (top panel in Figure 8.7) is the same as the area of a rectangle of width $(v^v - v^\ell)$ and height P^s (middle panel in Figure 8.7).

An alternative form of (8.2.21) can be attributed to Clerk Maxwell [9]. Cross multiply the denominator in (8.2.21) from the rhs to the lhs and then subtract the rhs from the lhs. We obtain

$$\int_{v^\ell}^{v^v} [P(T, v) - P^s(T)]\, dv = 0 \qquad \textit{fixed } T < T_c \qquad (8.2.22)$$

This form is called Maxwell's *equal area construction* and is illustrated in the bottom panel of Figure 8.7. The form (8.2.22) states that the van der Waals loop and the tie line bound two areas whose magnitudes cancel when combined algebraically.

8.2.6 Latent Heats of Vaporization from Equations of State

A PvT equation of state not only provides the saturation pressure and volumes of a pure substance in vapor-liquid equilibrium, it can also provide the latent heat associated with the phase change,

$$\Delta h_{vap} = h^v - h^\ell \qquad (8.2.23)$$

By adding and subtracting the ideal-gas enthalpy, (8.2.23) can be expressed in terms of the residual enthalpies,

$$\Delta h_{vap} = h^{res,\, v} - h^{res,\, \ell} \qquad (8.2.24)$$

We now use the Legendre transform (3.2.9) to relate h^{res} to u^{res} and v^{res}, use (4.4.14) for u^{res}, (4.2.2) for v^{res}, and with the help of the mean-value form for the vapor pressure (8.2.21), we find

$$\Delta h_{vap}(T) = T \int_{v^\ell}^{v^v} \gamma_v(T, v)\, dv \qquad \textit{fixed } T < T_c \qquad (8.2.25)$$

where γ_v is the thermal pressure coefficient (3.3.5) and the integration is along the subcritical isotherm T around the van der Waals loop.

We can also show that the latent heat is simply related to the slope of the vapor pressure curve $P^s(T)$. Let us differentiate the vapor pressure in (8.2.21) wrt temperature; we recognize that the integration limits v^ℓ and v^v must change with temperature, so we apply the Leibniz rule for differentiating such integrals (Appendix A). The result is

$$\frac{dP^s}{dT} = \frac{1}{v^v - v^\ell} \int_{v^\ell}^{v^v} \gamma_v(T, v)\, dv \qquad \textit{fixed } T < T_c \qquad (8.2.26)$$

This is another mean-value theorem; on a PT diagram, the slope of the vapor-pressure curve is the mean of the values of the thermal pressure coefficient along the van der Waals loop. Combining (8.2.25) and (8.2.26) gives *Clapeyron's equation*

$$\frac{dP^s}{dT} = \frac{\Delta h_{vap}}{T\Delta v} \tag{8.2.27}$$

where $\Delta v = v^v - v^\ell$. Clapeyron's equation is always true; moreover, a form analogous to (8.2.27) also applies to pure-component liquid-solid and solid-vapor equilibria. Consequently, on a *PT* diagram the slope of the melting curve is proportional to the latent heat of melting and the slope of the sublimation curve is proportional to the latent heat of sublimation.

For vapor-liquid equilibria, Clapeyron's equation simplifies. Multiply and divide the rhs by P/RT, and then Clapeyron's equation (8.2.27) can be written as

$$\frac{dP^s}{dT} = \frac{P}{RT^2}\frac{\Delta h_{vap}}{\Delta Z} \tag{8.2.28}$$

where ΔZ is the difference in compressibility factors of the two phases, $\Delta Z = Z^v - Z^\ell$. Both Δh_{vap} and ΔZ vary with temperature; however, their ratio $(\Delta h_{vap}/\Delta Z)$ is roughly constant, as shown in Figure 8.8. Therefore we assume $(\Delta h_{vap}/\Delta Z)$ is constant, separate variables in (8.2.28), and integrate along the saturation curve. The result is

$$\ln P^s = \frac{-\Delta h_{vap}}{R\Delta Z}\left(\frac{1}{T}\right) + A \tag{8.2.29}$$

where A is an integration constant. Near the triple point, $\Delta Z \approx Z^v \approx 1$, and (8.2.29) becomes an integrated form of the *Clausius-Clapeyron equation,*

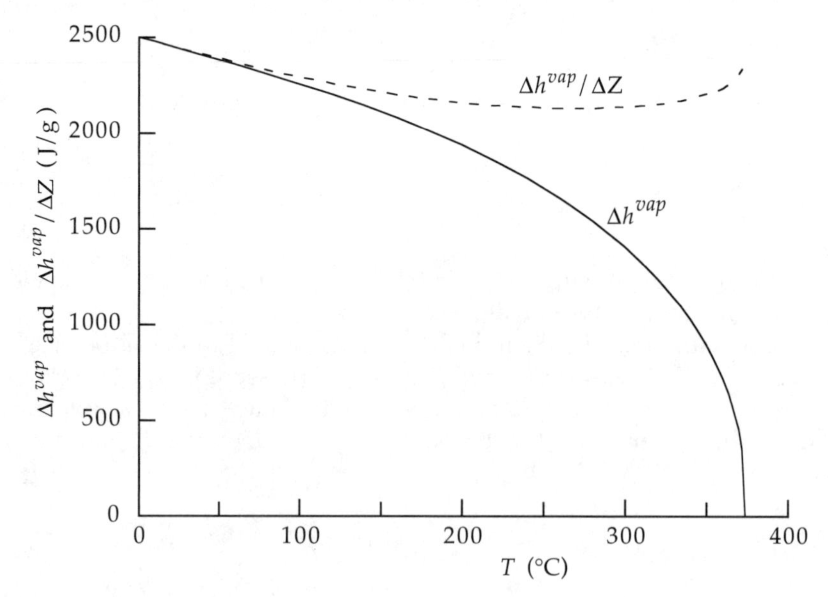

Figure 8.8 Latent heats of vaporization Δh_{vap} for pure water. The heat of vaporization of a pure substance is not constant; rather, it varies from a maximum at the triple point (0.01°C for water) to zero at the critical point (374.15°C for water). In contrast, the ratio $\Delta h_{vap}/\Delta Z$ is roughly constant over the entire range of saturation temperatures.

$$\ln P^s = \frac{-B}{T} + A \qquad (8.2.30)$$

where B is another positive constant. Away from the triple point, (8.2.29) is a much better approximation to the vapor-pressure curve than is (8.2.30). Note that, fortuitously, the Clausius-Clapeyron equation is also obtained from (8.2.29) by assuming, incorrectly, that $\Delta Z \approx 1$ and $\Delta h_{vap} \approx$ constant.

Equation (8.2.29) suggests that a straight line will be obtained when logarithms of pure-component vapor pressures are plotted against reciprocal absolute temperatures; further, the slope of that line provides an estimate for the latent heat of vaporization. This is tested in Figure 8.9 using vapor-pressure data of water; the degree of linearity is striking and is typical of most pure substances. Any deviation from a straight line is often taken into account by including additional terms in (8.2.30). For example, at low temperatures a commonly used alternative is *Antoine's equation* [10],

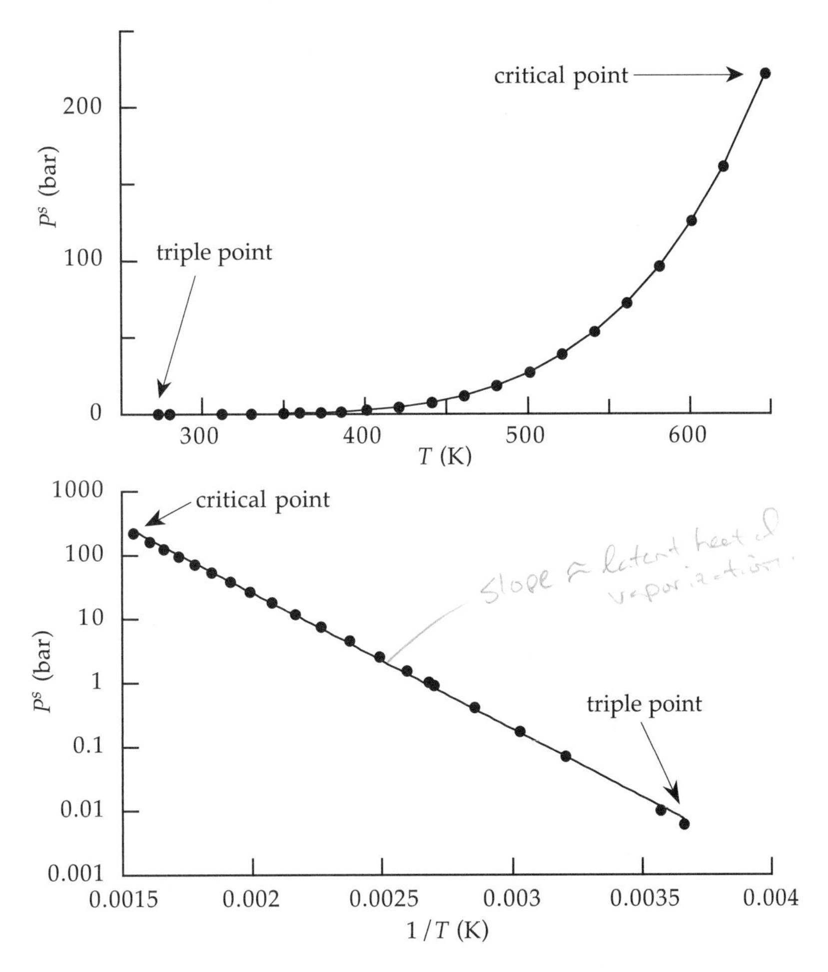

Figure 8.9 Vapor pressures of pure water from triple point to critical point. In both panels the points are from steam tables. *Top*: curve is a simple interpolation through the points. *Bottom*: line is a least-squares fit to the Clausius-Clapeyron form (8.2.30).

$$\ln P^s = \frac{-B}{T - C} + A \tag{8.2.31}$$

Values for the parameters *A*, *B*, and *C* are obtained by fitting to experimental vapor-pressure data; usually, they are all positive, as in Appendix D. But over a wide range of temperatures, a better correlation is the *Wagner equation* [11],

$$\ln\left(\frac{P^s}{P_c}\right) = \frac{a\tau + b\tau^2 + c\tau^{2.5} + d\tau^5}{1 - \tau} \tag{8.2.32}$$

Here, $\tau = 1 - T/T_c$, with T_c the critical temperature and P_c the critical pressure. Values for the parameters *a-d* are obtained by fitting vapor-pressure data $P^s(T)$. If the available data approach the critical point, then a reliable estimate to P_c can be obtained by making P_c an additional parameter in the fit [12].

8.2.7 Pure-Component Phase Equilibria Involving Solids

Properties of solids differ from those of fluids because in solids the motions of molecules are highly restricted. The molecules may be confined to periodic arrays, producing crystalline structures such as the face-centered cubic (fcc) and body-centered cubic (bcc), or they may be periodic only in certain directions, producing layered or amorphous structures such as graphite. Besides equilibrium structures, many solids can exist for prolonged periods in metastable structures; examples include glasses.

Solid-fluid equilibria include coexistence of solids with liquids and coexistence of solids with vapors. On a pure-component *Pv* diagram, such as the one shown in Figure 8.10, the melting lines mark the transition from states of one-phase solid to those of one-phase liquid. The melting lines are a pair of essentially straight, nearly vertical lines, separated by a region of metastable and unstable states, analogous to those appearing under the vapor-pressure curve in Figure 8.6. The melting lines are nearly vertical because $\Delta h / \Delta Z$ is large. In addition, the sublimation curves denote the transition from one-phase solid directly to one-phase vapor. Again, the sublimation curves appear in two branches, separated by a region of metastable and unstable states.

The melting lines, sublimation curves, and branches of the vapor-pressure curve all terminate at the horizontal broken line in Figure 8.10. That line, which is both an isobar and an isotherm, contains the *triple point*: an equilibrium situation in which three phases coexist simultaneously. A triple point occurs at one pressure and one temperature, but at three different molar volumes—one for each phase; hence, the triple point is marked by three filled circles on Figure 8.10. For liquid water in contact with water vapor and the normal phase of ice, the triple point occurs at 0.01°C and 0.0061 bar.

The criteria for equilibria involving solid phases are exactly those given in § 7.3.5 for any phase-equilibrium situation: phases in equilibrium have the same temperatures, pressures, and fugacities. Moreover, pure-component solid-fluid equilibria obey the Clapeyron equation (8.2.27). This means the latent heat of melting is proportional to the slope of the melting curve on a *PT* diagram and the latent heat of sublimation is proportional to the slope of the sublimation curve. In the case of solid-gas equilibria, the Clausius-Clapeyron equation (8.2.30) often provides a reliable relation between temperature and sublimation pressures, analogous to that for vapor-liquid equilibria.

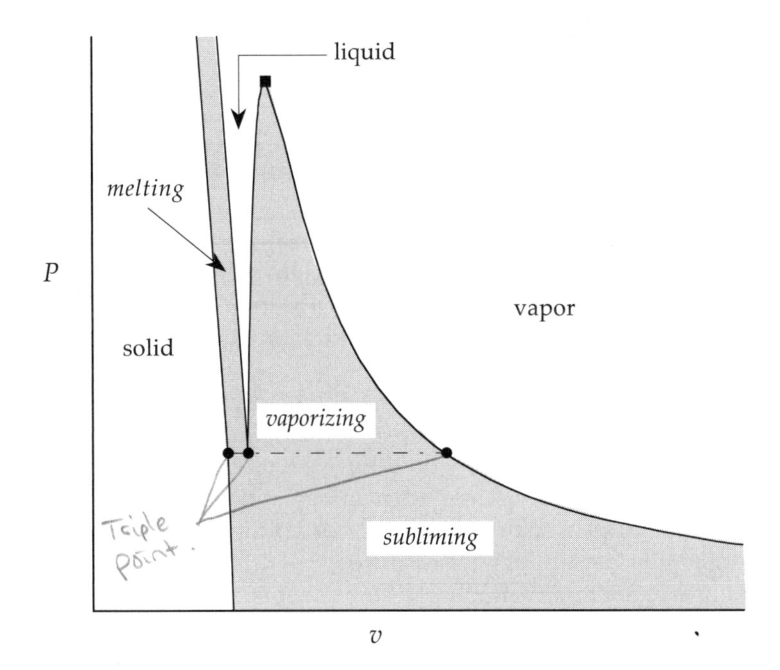

Figure 8.10 Schematic Pv diagram for a pure substance with the solid phase included. Shaded regions are metastable and unstable states. Vapor-liquid critical point (filled square) occurs at the maximum in the vapor-pressure curve. Filled circles are the triple-point volumes at which solid, liquid, and vapor all coexist in three-phase equilibrium.

In principle the stability of pure solid phases can be judged using the thermal and mechanical stability criteria derived in § 8.1.2, but those criteria are not useful for solids when they are implemented via volumetric equations of state. To use an equation of state to test for solid-phase stability, the equation would have to extend an isotherm from a fluid phase into a solid region of the phase diagram. But any analytically continuous, differentiable function that provides such an extension also predicts a solid-fluid critical point—a point that does not actually exist.

To test whether a proposed state will involve one or more solid phases, we usually use the criterion (7.1.40) which states that the equilibrium situation is the one that minimizes the Gibbs energy at the specified T and P. To perform such a calculation we need a model for the solid-phase Gibbs energy, and those models, in turn, require experimental data for the solid phase. The solid-phase data most often used are thermal data, such as heat capacities and latent heats for phase transitions.

Besides solid-fluid equilibria, some pure materials can exist in more than one stable solid structure, giving rise to solid-solid equilibria. Examples include equilibria between the fcc and bcc forms of iron, equilibria between rhombic and monoclinic sulfur, and equilibria among the many different phases of ice. Such solid-solid phase transitions are accompanied by a volume change and a latent heat, and these two quantities are related through the Clapeyron equation (8.2.27). When a pure material can undergo solid-solid phase transitions, then the substance usually exhibits multiple triple points. Besides the usual solid-vapor-liquid point, the pure substance might also exist in solid-solid-liquid or solid-solid-solid equilibria. Several such triple points occur in water, caused by equilibria involving various forms of ice [13].

8.3 PHASE STABILITY IN OPEN SYSTEMS

The thermal and mechanical stability criteria (8.1.23) and (8.1.31) apply both to pure fluids and to mixtures; however, for homogeneous mixtures, those criteria are not sufficient to identify stable systems because, in addition to energy and volume fluctuations, mixtures have concentration fluctuations. These fluctuations occur in localized regions of a system when material spontaneously aggregates and redisperses. If such fluctuations are not to disturb a system's stability, then the mixture must satisfy a set of conditions known as the *material* or *diffusional* stability criteria. These criteria are derived in a manner similar to that given in § 8.1.2 for (8.1.23) and (8.1.31), so we only sketch the procedure here.

For the derivation of stability criteria in § 8.1.2 we divided the fluid of interest into regions A and B; both regions were of constant mass, but their volumes and energies could fluctuate. Those same criteria would have been obtained if we had considered A and B to be regions of fixed volume, with energy and mass fluctuations, or by considering A and B to be open, so that their energies, volumes, and masses could all fluctuate. We employ this last strategy for mixtures.

Consider a one-phase binary mixture of components 1 and 2 confined to an isolated vessel, and imagine dividing the fluid into parts A and B. But unlike the pure case, region B is open to A, so that a fluctuation occurring in part B disturbs not only its internal energy U^B and volume V^B, but also the mole numbers N_1^B and N_2^B. Consequently, the concentration in B fluctuates by transfers of material to and from part A. In addition to the constraints on U, V, and S given by (8.1.4)–(8.1.6), the total amounts of each component are conserved,

$$N_1^A + N_1^B = \text{constant} \tag{8.3.1}$$

$$N_2^A + N_2^B = \text{constant} \tag{8.3.2}$$

As in § 8.1.2, stable equilibrium occurs when the total entropy is a maximum; hence,

$$\delta^2 S < 0 \tag{8.1.8}$$

and the response of the total entropy takes a form analogous to (8.1.10),

$$\delta S = \left[\left(\frac{\partial S^A}{\partial U^A}\right)_{NV^A} - \left(\frac{\partial S^B}{\partial U^B}\right)_{NV^B}\right]\delta U^A + \left[\left(\frac{\partial S^A}{\partial V^A}\right)_{NU^A} - \left(\frac{\partial S^B}{\partial V^B}\right)_{NU^B}\right]\delta V^A \tag{8.3.3}$$

$$+ \left[\left(\frac{\partial S^A}{\partial N_1^A}\right)_{N_2^A U^A V^A} - \left(\frac{\partial S^B}{\partial N_1^B}\right)_{N_2^B U^B V^B}\right]\delta N_1^A$$

$$+ \left[\left(\frac{\partial S^A}{\partial N_2^A}\right)_{N_1^A U^A V^A} - \left(\frac{\partial S^B}{\partial N_2^B}\right)_{N_1^B U^B V^B}\right]\delta N_2^A$$

At equilibrium we must have $\delta S = 0$, which requires each bracketed term in (8.3.3) to be zero. Just as in (8.1.11) and (8.1.12), the first term implies $T^A = T^B$ and the second that $P^A = P^B$. In addition, the last two terms imply that the chemical potentials for component 1 are the same in parts A and B; likewise for those of component 2.

Continuing the derivation in parallel to the steps from (8.1.13) to (8.1.15), we find the stability criterion, analogous to (8.1.15), to be

$$\delta^2 S = \begin{bmatrix} \delta U & \delta V & \delta N_1 & \delta N_2 \end{bmatrix} \mathbf{S} \begin{bmatrix} \delta U \\ \delta V \\ \delta N_1 \\ \delta N_2 \end{bmatrix} < 0 \tag{8.3.4}$$

where $\mathbf{S}$ is the symmetric matrix

$$\mathbf{S} = \begin{bmatrix} S_{uu} & S_{uv} & S_{un} & S_{um} \\ S_{uv} & S_{vv} & S_{vn} & S_{vm} \\ S_{un} & S_{vn} & S_{nn} & S_{nm} \\ S_{um} & S_{vm} & S_{nm} & S_{mm} \end{bmatrix} \tag{8.3.5}$$

Again, S_{xy} is a shorthand for the second derivatives; for example,

$$S_{un} = \left[\frac{\partial}{\partial U}\left(\frac{\partial S}{\partial N_1}\right)_{UVN_2}\right]_{VN} = \left[\frac{\partial}{\partial N_1}\left(\frac{\partial S}{\partial U}\right)_{VN}\right]_{UVN_2} \tag{8.3.6}$$

and

$$S_{um} = \left[\frac{\partial}{\partial U}\left(\frac{\partial S}{\partial N_2}\right)_{UVN_1}\right]_{VN} = \left[\frac{\partial}{\partial N_2}\left(\frac{\partial S}{\partial U}\right)_{VN}\right]_{UVN_1} \tag{8.3.7}$$

As in (8.1.16), if the inequality in (8.3.4) is to be obeyed for all possible variations δU, δV, δN_1, and δN_2, then $\mathbf{S}$ must be negative definite; that is, the principal minors of $\mathbf{S}$ must satisfy the following four inequalities:

$$S_{uu} < 0 \tag{8.1.18}$$

$$\begin{vmatrix} S_{uu} & S_{uv} \\ S_{uv} & S_{vv} \end{vmatrix} > 0 \tag{8.1.19}$$

$$\begin{vmatrix} S_{uu} & S_{uv} & S_{un} \\ S_{uv} & S_{vv} & S_{vn} \\ S_{un} & S_{vn} & S_{nn} \end{vmatrix} < 0 \tag{8.3.8}$$

$$|\mathbf{S}| > 0 \tag{8.3.9}$$

The first two are the conditions for thermal and mechanical stability derived in § 8.1.2: the constraints on the isometric heat capacity (8.1.23) and isothermal compressibility (8.1.31) apply to mixtures as well as pure fluids.

To pose the condition (8.3.8) in terms of measurables, we need to evaluate the six unique derivatives that appear in the determinant. Three have already been determined in § 8.1.2: S_{uu} is given by (8.1.22), S_{uv} by (8.1.25), and S_{vv} by (8.1.29). For the other three derivatives, we find

$$S_{un} = \frac{\overline{H}_1 - T\gamma_v \overline{V}_1}{T^2 C_v} \tag{8.3.10}$$

$$S_{vn} = \frac{\overline{V}_1}{TV\kappa_T} - \frac{(T\gamma_v - P)(\overline{H}_1 - T\gamma_v \overline{V}_1)}{T^2 C_v} \tag{8.3.11}$$

$$S_{nn} = \frac{-\overline{V}_1^2}{TV\kappa_T} - \frac{(\overline{H}_1 - T\gamma_v \overline{V}_1)^2}{T^2 C_v} - \frac{1}{T}\left(\frac{\partial \overline{G}_1}{\partial N_1}\right)_{TPN_2} \tag{8.3.12}$$

Substituting the six elements (8.1.22), (8.1.25), (8.1.29), (8.3.10), (8.3.11), and (8.3.12) into the matrix (8.3.8) and evaluating the determinant, we find, after some lengthy algebra,

$$\frac{1}{T^2 C_v T^2 V \kappa_T}\left(\frac{\partial \overline{G}_1}{\partial N_1}\right)_{TPN_2} > 0 \tag{8.3.13}$$

This is the criterion for material or *diffusional stability*: for a binary mixture to be differentially stable, the mixture must have $C_v > 0$, $\kappa_T > 0$, and (at fixed T and P) the chemical potential of component 1 must *always* increase in response to any increase in N_1. This means that if an isothermal-isobaric plot of the chemical potential (or fugacity) passes through an extremum with x_1, then the mixture is unstable for some x_1-values. The result (8.3.13) confirms (3.7.29) in which we claimed that the chemical potential of a pure component is always greater than its value in any mixture at the same T and P.

The fourth inequality (8.3.9) does not provide any new constraint, but merely gives the analog of (8.3.13) for component 2. In other words, because the labeling of components is arbitrary, an expression like (8.3.13) must be obeyed by each component in the mixture. This can also be deduced in a different way: for a binary, if component 1 obeys (8.3.13), then the Gibbs-Duhem equation demands that component 2 obey the analogous constraint.

Since a mixture must have $C_v > 0$ and $\kappa_T > 0$ for thermal and mechanical stability, many authors simplify (8.3.13) to

$$\overline{G}_{11} \equiv \left(\frac{\partial \overline{G}_1}{\partial N_1}\right)_{TPN_2} > 0 \tag{8.3.14}$$

For states at which the equation of state provides only one real root for v, then $\kappa_T > 0$ and the simplification (8.3.14) is legitimate. But when the equation of state has bifurcated, producing multiple roots for v, then we must exercise care when using (8.3.14) in place of (8.3.13). Some of those volume roots will have $\kappa_T < 0$ and therefore will be mechanically unstable, even if those roots also have $\overline{G}_{11} > 0$, so they satisfy (8.3.14). Consequently, those fluids are diffusionally unstable because (8.3.13) is violated. For cubic equations of state, it is the "middle" root for v that has $\overline{G}_{11} > 0$, but $\kappa_T < 0$, as illustrated in Figure 8.11. Equations of state that are higher-order polynomials in v will have additional roots that behave as in Figure 8.11. So when we test for the observability of proposed states and we do not know where that state lies on a phase diagram, we should apply the complete stability criterion (8.3.13), rather than the abbreviated form (8.3.14).

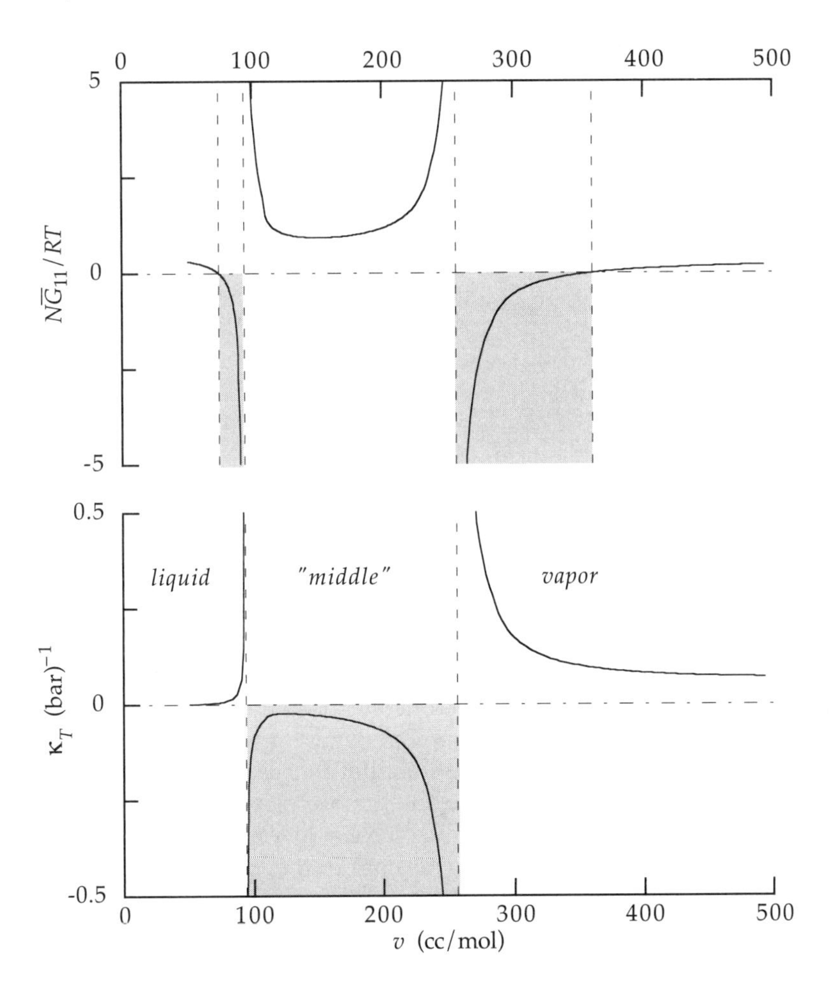

Figure 8.11 An unstable fluid may be misjudged to be stable if the criterion used is (8.3.14), rather than the complete criterion (8.3.13). These plots show how $\overline{G}_{11}$ and κ_T change along a line of fixed $T = 220$ K and fixed $x_1 = 0.75$ for a mixture of methane(1) and propane(2), as computed from the Redlich-Kwong equation. Shaded regions indicate unstable fluids. Here the fluid having v given by the "middle" root of the cubic is diffusionally unstable because $\kappa_T < 0$, even though $\overline{G}_{11} > 0$.

The criterion (8.3.13) implies that if a mixture is mechanically unstable ($\kappa_T < 0$), then it is also diffusionally unstable, just as (8.1.30) implies that if a fluid is thermally unstable ($C_v < 0$), then it is also mechanically unstable. But a fluid may be diffusionally unstable while remaining mechanically and thermally stable. In fact, whenever a stable mixture is driven into an unstable region of its phase diagram, the diffusional stability limit is always violated before the mechanical or thermal limits are violated, because higher-order terms approach zero before lower-order terms [3]. This can be seen in Figure 8.11. This means that the diffusional stability criterion (8.3.13) is a stronger test for thermodynamic stability than the mechanical criterion and (as noted in § 8.1.2) the mechanical criterion, in turn, is a stronger test than the thermal criterion.

Note that the arrangement of the independent variables in (8.3.4) is arbitrary, so if we change the order, we obtain other forms for the stability criteria. However, these other forms are not additional constraints; they are merely other versions of the constraints already found. For example, if we change the order so that (8.3.4) reads

$$\delta^2 S = \begin{bmatrix} \delta N_1 & \delta N_2 & \delta V & \delta U \end{bmatrix} \mathbf{S}' \begin{bmatrix} \delta N_1 \\ \delta N_2 \\ \delta V \\ \delta U \end{bmatrix} < 0 \tag{8.3.15}$$

then the first inequality, analogous to (8.1.18), becomes

$$S_{nn} < 0 \tag{8.3.16}$$

with S_{nn} still given by (8.3.12). This inequality is obviously obeyed by a system in a stable equilibrium state, because (8.3.12) is merely a linear combination of the thermal, mechanical, and diffusional criteria already derived. The lesson is that (8.3.16) does not convey any information not already contained in the conditions (8.1.23), (8.1.31), and (8.3.13). Beegle et al. provide an extensive list of possible forms for the differential stability criteria involving various orderings of the independent variables, including several choices for the independent variables themselves [4].

The above procedure can be repeated to obtain the stability criteria for multicomponent mixtures. For a mixture of C components, the criterion is still (8.3.4) in which $\mathbf{S}$ is the $(C + 2)^2$ matrix of second derivatives analogous to (8.3.5). The fluid is stable to small disturbances when $\mathbf{S}$ is negative definite; that is, when odd-order principal minors of $\mathbf{S}$ are negative and simultaneously those of even order are positive. The reduction of those minors to economical forms is a tedious exercise that can often be alleviated by posing the criteria in terms of G or A rather than S.

8.4 FLUID MIXTURES

In this section we describe the common stability behavior displayed by binary mixtures (§ 8.4.1), including a scheme for classifying that behavior (§ 8.4.2). Then we show how models can be used to test for the observability of one-phase binary mixtures; first we consider $PvTx$ models (§ 8.4.3 and 8.4.4) and then models for the excess Gibbs energy (§ 8.4.5).

8.4.1 Stability of Binary Mixtures

Ultimately, we want to develop a computational procedure for determining the observability of a state proposed for a binary fluid. The motivation is that we want to avoid trying to solve phase-equilibrium problems that do not exist. Therefore we first test for observability, and if multiphase situations are observable, then we solve for phase compositions, if they are required. In this section we consider situations in which the proposed state is identified by specifying values for T, P, and x_1. Such a state could be in any one of three observable conditions: (a) a stable single phase, (b) a stable multiphase equilibrium, or (c) a metastable single phase. Some metastable phases can only relax to a stable single phase, but other metastable phases can split into multiple phases. Multiphase equilibria in binaries are predominantly two-phase situations, so we will restrict our attention to those possibilities here; however, three and four-phase binaries are also possible.

To connect mixture stability to mixture state, we show in Figure 8.12 a Pv diagram for equimolar mixtures of methane and propane. This diagram was calculated using the Redlich-Kwong equation (8.2.1) together with the simple mixing rules given in § 8.4.4. The diagram is typical of many binary mixtures, especially those whose vapor-liquid critical lines are continuous curves between the pure component critical points. However, we caution that not all binary mixture Pv diagrams appear as in Figure 8.12;

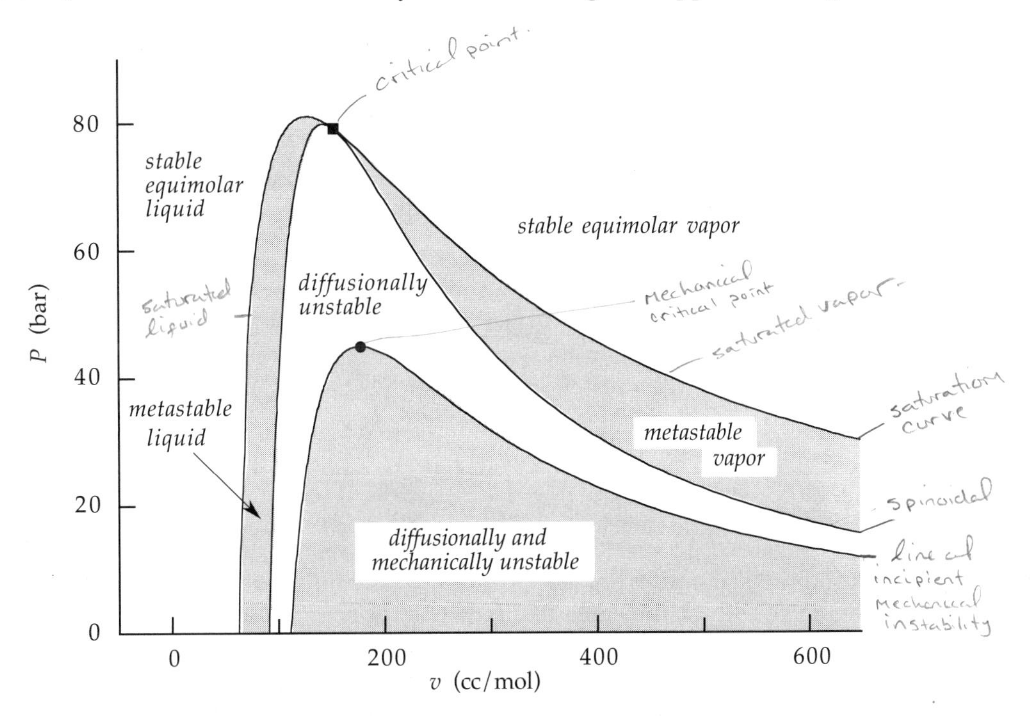

Figure 8.12 Pressure-volume diagram for equimolar mixtures of methane + propane, computed from the Redlich-Kwong equation of state. Filled square is the critical point; filled circle is the mechanical critical point. The two branches of the saturation curve separate stable states from metastable states. The spinodal separates metastable states from unstable states and the line of incipient mechanical instability separates diffusionally unstable states from states that are both diffusionally and mechanically unstable. Since every point on this diagram represents an equimolar mixture, no tie lines can be drawn.

there is no single such diagram that is typical of all mixtures at all compositions. Many of the possible diagrams are described in Chapter 9.

In Figure 8.12 the outer envelope is the locus of saturated equimolar liquid states and saturated equimolar vapor states. However, note that Figure 8.12 is not a phase-equilibrium diagram: in Figure 8.12 every point on the two-phase line represents an equimolar mixture, but phases in vapor-liquid equilibrium generally do not have the same composition. Consequently, Figure 8.12 contains no tie lines across the two-phase region. Outside the saturation envelope, the mixtures are stable one-phase fluids. Underneath that envelope, the mixtures may be metastable one-phase fluids or they may be unstable to one phase (that is, they may exist as two-phases).

The middle envelope is the *spinodal*: the set of states that separate metastable states from unstable states. Recall from § 8.3 that one-phase mixtures become diffusionally unstable before becoming mechanically unstable. Therefore, the mixture spinodal is the locus of points at which the diffusional stability criterion (8.3.14) is first violated; that is, it is the locus of points having

$$\bar{G}_{11} \equiv \left(\frac{\partial \bar{G}_1}{\partial N_1}\right)_{TPN_2} = \frac{x_2}{N}\left(\frac{\partial \bar{G}_1}{\partial x_1}\right)_{TP} = 0 \qquad \textit{mixture spinodal} \qquad (8.4.1)$$

Between the spinodal and the saturation envelope, mixtures may exist as metastable one-phase systems or as stable two-phase systems. The spinodal cannot cross the saturation envelope, but the spinodal becomes tangent to the saturation envelope at the critical point.

For binary mixtures it is conventional to express the conditions for the critical point in terms of the change in Gibbs energy on mixing (3.7.38):

$$\left(\frac{\partial^2 g^m}{\partial x_1^2}\right)_{TP} = 0 \qquad \textit{binary critical point} \qquad (8.4.2)$$

and

$$\left(\frac{\partial^3 g^m}{\partial x_1^3}\right)_{TP} = 0 \qquad \textit{binary critical point} \qquad (8.4.3)$$

These conditions identify both vapor-liquid and liquid-liquid critical points. For vapor-liquid equilibria, they are satisfied when the spinodal coincides with the vapor-liquid saturation curve. However, that point need not occur either at the maximum in the saturation envelope or at the maximum in the spinodal; see Figure 8.12. Along a spinodal the one-phase metastable system is balanced on the brink of an instability; at a critical point that balance coincides with a two-phase situation and the resulting fluctuations cause critical opalescence, just as they do at pure-fluid critical points.

The inner envelope in Figure 8.12 is the line of *incipient mechanical instability*: the line separating states that are only diffusionally unstable from states that are both diffusionally and mechanically unstable. The line of incipient mechanical instability is the locus of points at which (8.1.31) is first violated; that is, the points at which

$$\left(\frac{\partial P}{\partial v}\right)_T = 0 \tag{8.4.4}$$

The maximum in the line of incipient mechanical instability is called the *mechanical critical point*. Note that the true critical point and the mechanical critical point occur at roughly the same molar volume. This often occurs for mixtures whose spinodals (at fixed composition) pass through maxima with v.

Figure 8.12 shows that if a mixture is mechanically unstable, then it is also diffusionally unstable, because the line of incipient mechanical instability lies under the spinodal, or equivalently because κ_T appears in both stability criteria (8.1.30) and (8.3.13). Moreover, a one-phase mixture may be diffusionally unstable but remain mechanically stable, because the spinodal lies above the line of incipient mechanical instability, or equivalently because the mechanical criterion (8.1.30) can be satisfied while the diffusional criterion (8.3.13) is violated. Further, Figure 8.12 contains states at which no differential stability criteria are violated, but at which one-phase mixtures are metastable rather than stable. This means that a violation of any *differential* stability criteria (thermal, mechanical, or diffusional) is only sufficient, but not necessary, for a phase separation to occur.

Phase stability can be described in terms of the Gibbs energy by appealing to the equilibrium criterion (7.1.40): at fixed T and P, the system Gibbs energy must be a minimum. Therefore, if a mixture is a stable single phase, then it must have a lower Gibbs energy than the combined values of the pures; that is, the change of Gibbs energy on mixing must be negative,

$$g^m(x_1) < 0 \qquad \textit{fixed } T \textit{ and } P \tag{8.4.5}$$

But this is only necessary for one-phase stability; it is not sufficient. This means if a mixture violates (8.4.5), then it is definitely not stable; however, a mixture can obey (8.4.5) but still split into two phases. An additional requirement is (8.3.14), which can be expressed in terms of g^m as

$$\bar{G}_{11} \equiv \left(\frac{\partial \bar{G}_1}{\partial N_1}\right)_{TPN_2} = \frac{x_2^2}{N}\left(\frac{\partial^2 g^m}{\partial x_1^2}\right)_{TP} > 0 \qquad \textit{not unstable} \tag{8.4.6}$$

A stable one-phase mixture satisfies (8.4.6), but the converse is not true: a mixture obeying (8.4.6) might be stable or metastable. However, if a mixture violates (8.4.6), then the mixture is definitely unstable and not observable.

When phase splits occur at fixed T and P, the compositions of the new phases generally differ from one another and they differ from that of the original one-phase mixture. Those compositions are computed by solving the equality of fugacities (7.3.12), using appropriate models for each phase. Such calculations will be the focus of our attention in Chapter 10. Later in this section, we develop a procedure for determining whether a proposed binary mixture can exist as a stable single phase, without solving the phase equilibrium problem.

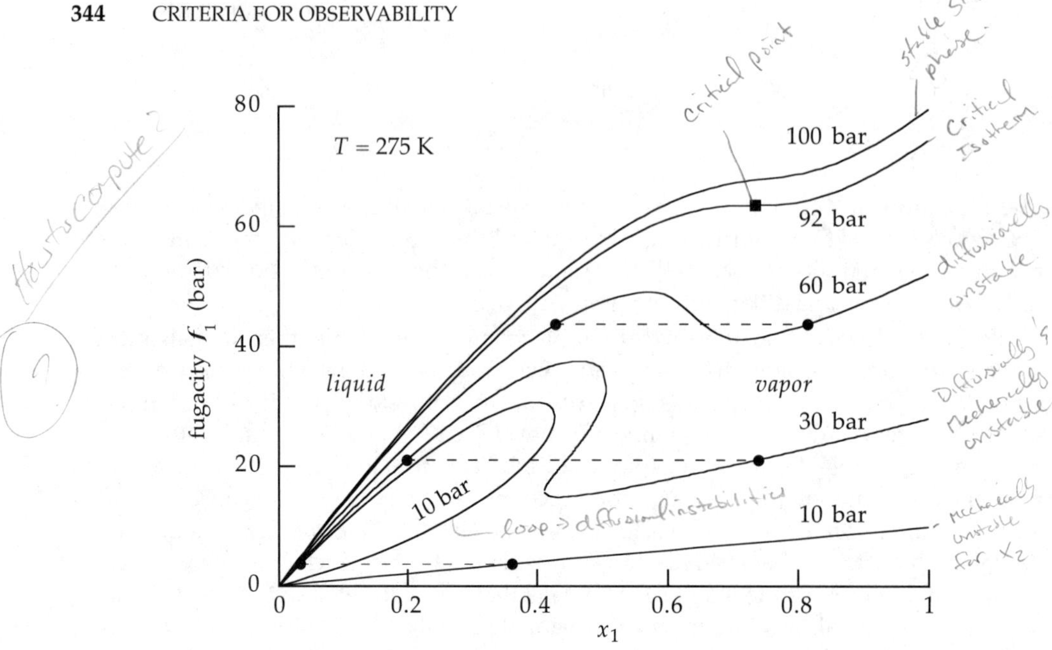

Figure 8.13 When computed from a cubic equation of state, violations of stability criteria affect the form of an isothermal plot of fugacity for a binary mixture. At 100 bar and 275 K, this mixture is a stable single-phase at all compositions, and $f_1(x_1)$ increases monotonically. At 92 bar the fugacity passes through a point of inflection: the critical point for this isotherm (filled square). At 60 bar the mixture violates the diffusional stability criterion for some x_1, while the equation of state provides only one real root for v. The fugacity curve forms a loop, but $f_1(x_1)$ remains single-valued at every x_1. At 30 bar the mixture violates both the diffusional and mechanical stability criteria over some range of x_1. The equation of state provides three real roots for $v(T, P, x)$; hence, the fugacity is now multivalued over some range of x_1. At 10 bar, the mechanical instabilities extend to pure 2, so the three branches of f_1 emanate from the origin. Broken horizontal lines are tie lines connecting phases in equilibrium (dots). Curves computed from Redlich-Kwong equation.

8.4.2 Classes of Stability Behavior in Binary Mixtures

When mixture states are computed from a volumetric equation of state, then instabilities can be related to bifurcations in an algebraic equation, just as we found for pure fluids in § 8.2. Inversely, if no bifurcations occur, then the mixture remains a stable single phase over all compositions, and the fugacity $f_1(x_1)$ is a smooth monotonically increasing curve, as shown for 100 bar in Figure 8.13. Analogous behavior is observed for $g^m(x_1)$: the stability requirement (8.4.6) on the second derivative of g^m defines a simple convex curve for $g^m(x_1)$, like that shown on the left in Figure 8.14.

However if, over some range of compositions, the mixture splits into two phases, then the single-phase equilibrium curve for $g^m(x_1)$ will not be convex over all x_1. Similarly, the monotonicity of $f_1(x_1)$ will be disrupted either by oscillations or by branching. These possibilities appear in Figure 8.13: oscillations occur in the $f_1(x_1)$ curves for 30 and 60, while at 10 bar, the $f_1(x_1)$ curve has divided into two distinct branches. These phenomena are caused by bifurcations in either the equation of state or the fugacity equation or both. Here we use those possibilities to identify four classes of instabilities that can lead to vapor-liquid phase separations in binary mixtures.

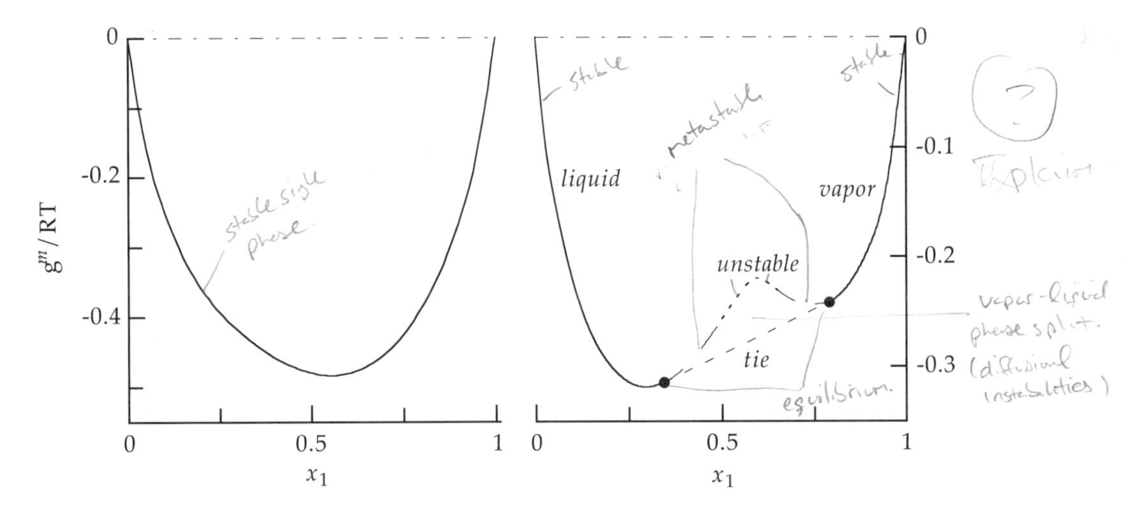

Figure 8.14 Isothermal-isobaric plots of change of Gibbs energy on mixing for binary mixtures. *Left*: Neither the equation of state nor the fugacity equation bifurcate, so the mixtures remain stable single phases at all compositions. *Right*: Class I stability behavior: the fugacity equation bifurcates, but the equation of state does not. This produces a region in g^m that is concave and a vapor-liquid phase split. Filled circles are phases in equilibrium; solid lines stable; long dashes metastable; short dashes unstable. All curves computed from the Redlich-Kwong equation.

Class I: Only the fugacity equation bifurcates. In these situations the equation of state does not bifurcate for either pure or for the mixture, so there is only one real root for the volume at each x_1 and the mechanical stability criterion cannot be violated. A plot of $g^m(x_1)$ provides a smooth continuous curve spanning all x_1; however, the curve will be concave over some x_1 (as on the right in Figure 8.14). The concave region in $g^m(x_1)$ is caused by diffusional instabilities; that is, (8.4.6) is violated over some range of x_1. The corresponding isothermal-isobaric plot of the fugacity $f_1(x_1)$ passes through a loop, but f_1 remains single-valued at each x_1; see the curve for 60 bar in Figure 8.13. This is analogous to the van der Waals loop on a pure-substance Pv diagram. Class I behavior is also exhibited by mixtures in liquid-liquid and gas-gas equilibria; that is, liquid-liquid and gas-gas phase splits are driven only by diffusional instabilities.

Class II: Both mixture equations bifurcate but the pure equations do not. In these systems the mixture fugacity equation and the mixture equation of state both bifurcate. When the equation of state bifurcates, multiple roots occur for v, so $g^m(x_1)$ appears in distinct branches. Each branch corresponds to one root for v, but since bifurcations do not occur in either pure-fluid equation of state, neither branch spans all x_1. In class II mixtures, instabilities may be caused by violations of the diffusional criterion (8.4.6) or by violations of both the diffusional and mechanical stability criterion (8.3.13). A sample plot of $g^m(x_1)$ is shown in Figure 8.15. In some mixtures the metastable regions of g^m extend to positive values, violating (8.4.5). The fugacity remains a single continuous curve that spans all x_1, but because the fugacity equation bifurcates, there is some range of x_1 over which the fugacity is multivalued, like the curve at 30 bar in Figure 8.13. On mixture PT diagrams, class II behavior occurs at states below the mechanical critical line and at pressures below the spinodal of pure 1 but above the spinodal of pure 2 (component 1 is more volatile) [14].

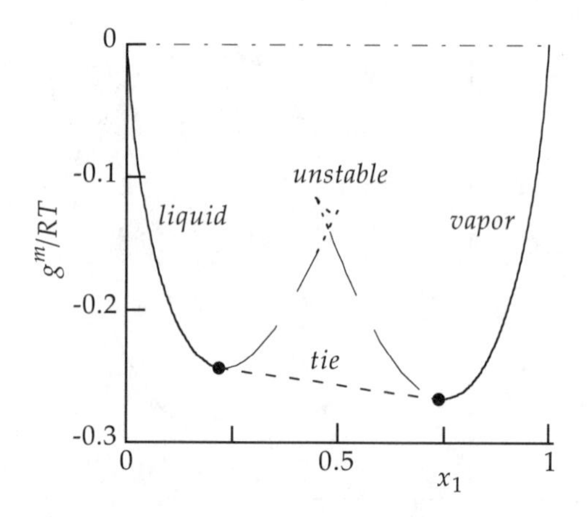

Figure 8.15 Change of Gibbs energy on mixing for class II stability behavior at constant T and P. Both the mixture equation of state and the fugacity equation bifurcate, producing distinct branches in g^m and a vapor-liquid phase separation. However, no branch spans all x_1. Filled circles are phases in equilibrium; long dashes metastable; short dashes unstable. Curves computed using Redlich-Kwong equation.

Class III: Both mixture equations and one pure equation bifurcate. This behavior differs from class II in that now one branch of $g^m(x_1)$ spans all x_1. This happens when the equation of state for one pure bifurcates in addition to the bifurcations that occur in both the mixture equation of state and the mixture fugacity equation. We distinguish two subclasses: in class IIIA mixtures the pure-2 equation bifurcates, while in class IIIB mixtures the pure-1 equation bifurcates. Since $f_1 = 0$ when $x_1 = 0$, the three branches of f_1 in class IIIA mixtures must all emanate from the origin, like the curve for 10 bar in Figure 8.13. In class IIIB mixtures, the pure-1 fugacities will generally have different values, as in Figure 8.16; the smallest identifies the stable pure phase.

Consider those branches of g^m and f_1 that extend over all x_1. In both class IIIA and class IIIB, those branches will contain (at least) some region that is stable; the remaining portion (if any) will be metastable, but not unstable. In class IIIA the stable phase corresponds to parts of the curve near $x_1 = 1$, while in class IIIB it will occupy parts near $x_1 = 0$. The unstable phase will be confined to its own branch, as in Figure 8.16.

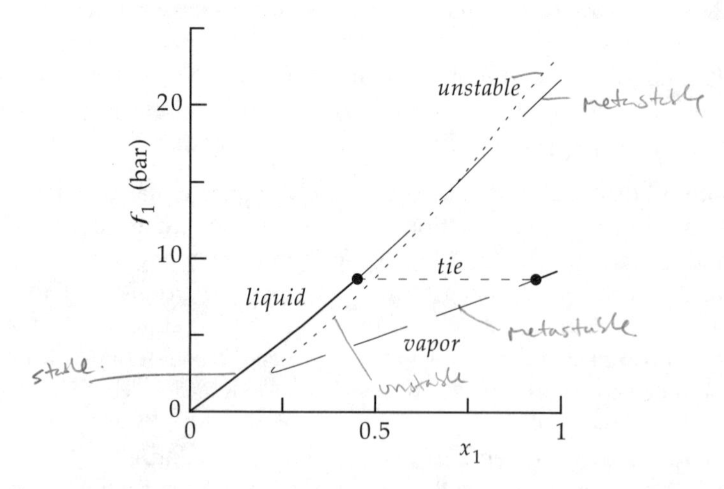

Figure 8.16 Isothermal-isobaric plot of fugacity f_1 for a binary mixture exhibiting class IIIB stability behavior. The pure component-1 equation of state bifurcates, producing three branches in f_1; however, since the pure component-2 equation of state does not bifurcate, only one branch spans all x_1. Filled circles mark phases in equilibrium. Solid lines stable; long dashes metastable; short dashes unstable. All curves computed from the Redlich-Kwong equation of state.

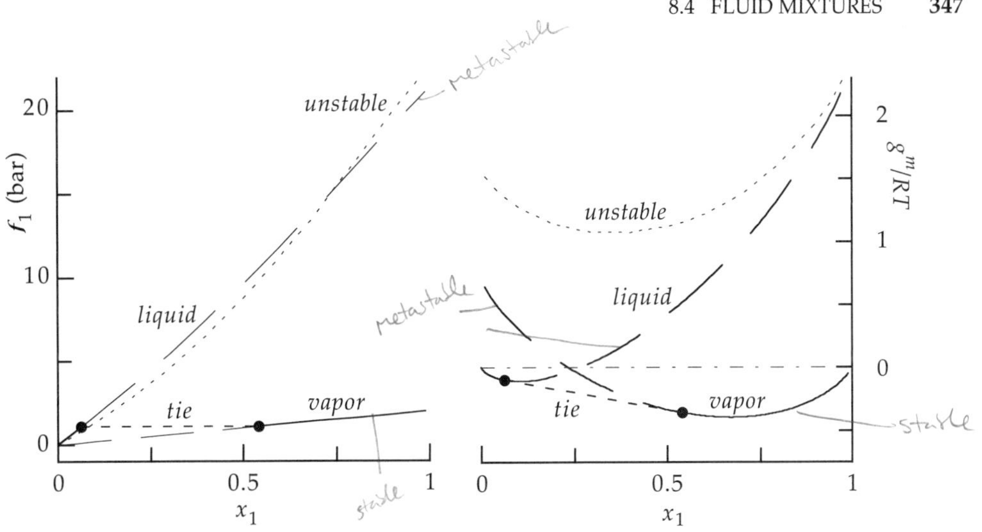

Figure 8.17 Isothermal-isobaric plots of fugacity and change of Gibbs energy on mixing for binary mixtures exhibiting class IV stability behavior. The equation of state bifurcates, but the fugacity equation does not. *Left*: The fugacity appears in three branches that span all x_1; one branch contains all unstable states. *Right*: Each branch of g^m also spans the entire composition range. In both panels, filled circles are phases in equilibrium; solid lines stable, long dashes metastable, short dashes unstable. Computed from Redlich-Kwong equation.

Class IV: Only the equation of state bifurcates. In these cases the fugacity equation does not bifurcate, so no differential diffusional stability criteria are violated. Nevertheless, metastabilities may occur and those metastabilities can lead to phase changes. In these mixtures both pure-component equations of state bifurcate, so g^m and f_1 each divide into three distinct branches, with each branch spanning the entire range of compositions. Typical curves are shown in Figure 8.17. Unstable phases are confined to one branch; however, portions of some branches may have $g^m > 0$, violating the one-phase requirement (8.4.5). On mixture *PT* diagrams, class IV behavior occurs at pressures below those of the spinodals of both pure vapors and at temperatures less than those of the spinodals of both pure liquids [14]. The existence of class IV behavior illustrates that differential stability criteria are only necessary, but not sufficient, to identify stable one-phase mixtures.

8.4.3 Determining Stability Using Fugacities from Equations of State

It is traditional to base determinations of phase stability on the change in Gibbs energy of mixing g^m. But computations of phase equilibria are now more often done via volumetric equations of state, so it may prove more useful to base stability determinations on fugacities. We develop the necessary relations here and illustrate their application with an example in the following section. We limit the presentation to stability of binary mixtures.

The one-phase stability criteria are posed in terms of g^m in (8.4.5) and (8.4.6), but before we use those criteria to test for stability, it will prove more convenient to repose them in terms of the fugacity. We can rewrite (8.4.5) and (8.4.6) in terms of fugacities by combining the definition of g^m (3.7.38) with the integrated definition of the fugacity in (4.3.12). Then (8.4.5) requires that stable phases have

$$\sum x_i \ln \frac{f_i}{f_{\text{pure i}}} < 0 \tag{8.4.7}$$

and (8.4.6) requires that stable and metastable phases have

$$\left(\frac{\partial \ln f_i}{\partial x_i}\right)_{TP} > 0 \qquad \textit{not unstable} \tag{8.4.8}$$

Since mole fractions and fugacities are always positive, (8.4.7) suggests that stable one-phase mixtures have

$$f_i(T, P, \{x\}) < f_{\text{pure i}}(T, P) \qquad i = 1, 2 \tag{8.4.9}$$

Nevertheless, it is mathematically possible for some components to violate (8.4.9) while the mixture still might obey (8.4.7). But with the help of (8.4.8) we can show that, in fact, both components of a stable, one-phase binary must satisfy (8.4.9). The proof is given in Appendix F. Consequently, if a single-phase binary mixture has $f_i > f_{\text{pure i}}$, then that phase cannot be stable.

However, (8.4.9) is only necessary, not sufficient. So if we find a mixture that obeys (8.4.9) we cannot say whether it is stable, metastable, or unstable. This is illustrated in Figure 8.18. Therefore (8.4.9) is useful, but it is not complete. For example, assume we are at the state α in Figure 8.18. We need to know whether or not that state is a stable single-phase mixture. The state satisfies (8.4.9), but that is not enough to determine stability. Note on the figure that at this T, P, and f_1, the stable mixture might be one-phase α, one-phase β, one-phase γ, or some two-phase combination of the three.

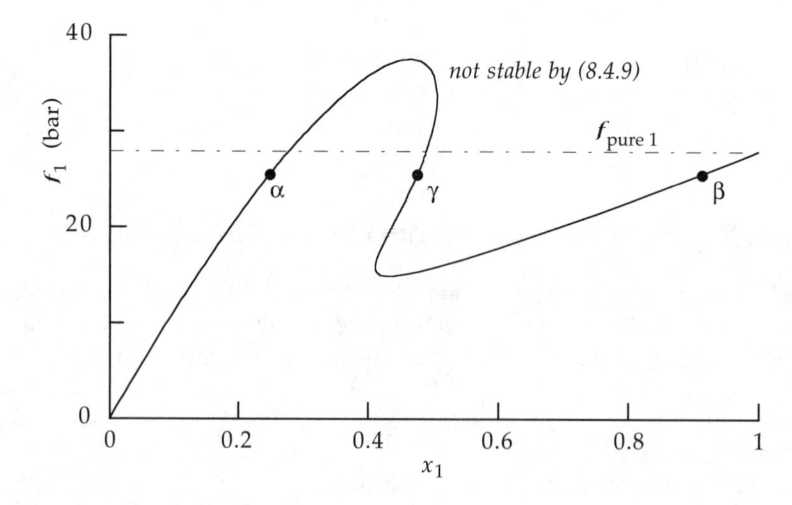

Figure 8.18 Isothermal-isobaric plot of fugacity for component 1 in a binary mixture. Portions of the curve above the broken horizontal line are not stable because of (8.4.9). The three mixtures α, β, and γ have the same value for the fugacity, but only one of the three forms a stable single phase; e.g., mixture at γ violates (8.4.8) and so it is unstable even though it satisfies (8.4.9).

To decide among these possibilities we need a stability criterion for mixtures at fixed T, P, and fugacity f_1. Equivalently, we can develop the criterion in terms of T, P, and the chemical potential $\overline{G}_1$, then convert it to fugacities at the end. Imagine a one-phase binary mixture surrounded by a reservoir that imposes its temperature, pressure, and chemical potential $\overline{G}_1$ on the system. The latter is accomplished by a semi-permeable membrane that separates the system from the reservoir. The membrane allows molecules of component 1 to pass, but it blocks passage of molecules of component 2. When diffusional equilibrium is established, the value of the chemical potential $\overline{G}_1$ is the same in the system and in the reservoir. The extensive state of the system is identified by giving values for the fixed quantities T, P, $\overline{G}_1$, and N_2.

These independent variables motivate us to define a new thermodynamic quantity Ψ using this Legendre transform:

$$\Psi \equiv G - N_1\overline{G}_1 \tag{8.4.10}$$

The quantity Ψ is an extensive conceptual having dimensions of energy. Forming the total differential and using (3.2.28) for dG, we obtain for the binary,

$$d\Psi = -SdT + VdP + \overline{G}_2 dN_2 - N_1 d\overline{G}_1 \tag{8.4.11}$$

For equilibrium at fixed T, P, $\overline{G}_1$, and N_2 (8.4.11) reduces to

$$d\Psi = 0 \tag{8.4.12}$$

and in fact Ψ must be a minimum at equilibrium. This means if two states have the same values for T, P, $\overline{G}_1$, and N_2, the stable equilibrium state will be that having the lower value of Ψ. The two states could differ, for example, in their compositions.

Fortunately, the quantity Ψ is a familiar property. To discover its identify, recall that we can use (3.2.32) to write the Gibbs energy of any binary as

$$G = N_1\overline{G}_1 + N_2\overline{G}_2 \tag{8.4.13}$$

Substituting this into (8.4.10) leaves

$$\Psi = N_2\overline{G}_2 \tag{8.4.14}$$

Dividing by N_2 gives the intensive version

$$\psi = \overline{G}_2 \tag{8.4.15}$$

This means at fixed T, P, $\overline{G}_1$, and N_2, the equilibrium state of the binary mixture is that which minimizes the chemical potential of component 2. Since the chemical potential is conceptually equivalent to the fugacity, we can also say that the equilibrium state is that which minimizes the fugacity of component 2.

Therefore, of the alternatives α, β, and γ in Figure 8.18, the stable one-phase mixture is that which has the lowest value for f_2. We would compute f_2 from an appropriate equation of state. If two of those states had the same value of f_2, then a two-phase equilibrium situation could occur. The condition (8.4.9) together with minimization of f_2 give us sufficient tools for determining the stability of states proposed for binary mixtures. Note we can make such judgements *without* solving the phase-equilibrium problem. We illustrate with an example.

8.4.4 Example

How do we use a volumetric equation of state to determine whether a proposed state of a binary mixture is a stable single phase?

During a process design we need to formulate a mixture of methane(1) and propane(2) that has $x_1 = 0.25$ at 275 K and 30 bar. Can this mixture exist as a stable single phase?

To address this issue, we use the Redlich-Kwong equation of state (8.2.1) with the simple mixing rules from § 4.5.12,

$$a = \sum_i \sum_j x_i x_j a_{ij} \tag{8.4.16}$$

$$b = \sum_i x_i b_{\text{pure i}} \tag{8.4.17}$$

Values for the a_{ij} and b_i can be obtained from pure-component critical properties using these relations from Table 4.4

$$a_{ij} = 0.4275 \frac{R^2 T_{cij}^{2.5}}{P_{cij}} \tag{8.4.18}$$

$$b_i = 0.08664 \frac{RT_{cij}}{P_{cij}} \tag{8.4.19}$$

together with these empirical combining rules:

$$T_{cij} = \sqrt{T_{cii} T_{cjj}} \tag{8.4.20}$$

$$v_{cij} = [0.5(v_{cii}^{1/3} + v_{cjj}^{1/3})]^3 \tag{8.4.21}$$

$$Z_{cij} = 0.5(Z_{cii} + Z_{cjj}) \tag{8.4.22}$$

Table 8.1 Values of Redlich-Kwong parameters for methane(1)-propane(2) mixtures; computed using (8.4.18)–(8.4.23)

ij	T_{cij} (K)	P_{cij} (bar)	v_{cij} (cc/mol)	Z_{cij}	a_{ij} (cc/mol)^2 bar K$^{0.5}$	b_i cc/mol
11	190.6	46.	99.	0.2874	3.222 (10^7)	29.85
22	369.8	42.4	203.	0.2800	18.33 (10^7)	62.82
12	265.5	43.2	145.	0.2837	7.850 (10^7)	

and

$$P_{cij} = \frac{Z_{cij} R T_{cij}}{v_{cij}} \tag{8.4.23}$$

The subscripts *cii* and *cjj* indicate pure-component critical properties. Note that all quantities in (8.4.18)–(8.4.23) are invariant under exchange of labels i and j; for example, $a_{21} = a_{12}$. Resulting values for these parameters are given in Table 8.1.

The fugacity is obtained from the Redlich-Kwong equation by evaluating (4.4.23) for the fugacity coefficient and then applying FFF#1. The result from (4.4.23) is

$$\ln\varphi_1 = \frac{b_1}{v-b} - \ln\left(\frac{v-b}{v}\right) - \ln Z - \frac{\beta b_1}{v+b} - \Omega_{11}\ln\left(\frac{v+b}{v}\right) \tag{8.4.24}$$

Here v is the mixture molar volume, while β and Ω_{11} are dimensionless groups:

$$\beta \equiv \frac{a}{bRT\sqrt{T}} \tag{8.4.25}$$

$$\Omega_{11} \equiv \beta\left(\frac{2\sigma_{11}}{a} - \frac{b_1}{b}\right) \tag{8.4.26}$$

with

$$\sigma_{11} \equiv x_1 a_{11} + x_2 a_{12} \tag{8.4.27}$$

The expression for φ_2 is functionally the same as (8.4.24), but with subscripts 1 and 2 interchanged. To evaluate pure-component fugacities, (8.4.24) still applies, but in a simplified form because a pure substance has $b_1 = b$ and $a_{11} = a = \sigma_{11}$, so $\Omega_{11} = \beta$; the result appears in (8.2.11). We caution that in evaluating φ_{pure} from (8.4.24), the pure-component T and P must be the same T and P as the mixture. Often a T-P pair will produce multiple pure states (volumes) that satisfy the analytic equation of state, even if a single state is found for the mixture. Of those multiple solutions, only the stable equilibrium state is the appropriate state to be used in the following calculations. The stable pure state can be identified by the procedure illustrated in § 8.2.2. To determine the stability of the proposed mixture, we proceed as follows.

Step 1. Determine whether the equation of state bifurcates when applied to each pure substance at the proposed mixture T and P. The mixture temperature (275 K) is above the critical temperature of pure methane (190.6 K), so pure methane is a single-phase fluid and the equation of state cannot bifurcate. For pure propane we solve the Redlich-Kwong equation (8.2.1) for v at 275 K and 30 bar. We find a single real root (v = 90.5 cc/mol), so pure propane is a single-phase liquid and, again, the equation of state does not bifurcate. Since the equation of state does not bifurcate for either pure substance, the mixture fugacity f_1 forms a single continuous curve that spans all x_1: the mixtures exhibit either class I or class II stability behavior.

Step 2. Evaluate the fugacity for pure 1 at the mixture T and P. Applying (8.2.11) to pure methane at 275 K and 30 bar, we find $\varphi_{\text{pure 1}} = 0.931$. Then FFF#1 gives

$$f_{\text{pure 1}} = \varphi_{\text{pure 1}} P = 0.931 \times 30 = 27.9 \text{ bar} \tag{8.4.28}$$

Step3. Evaluate the fugacity for component 1 in the mixture at the given T, P, x_1. At $x_1 = 0.25$ the mixing rules (8.4.16) and (8.4.17) give these values for the mixture parameters: $a = 13.45(10^7)(\text{cc/mol})^2 \text{ bar K}^{0.5}$ and $b = 54.58$ cc/mol. With these, the Redlich-Kwong equation gives a single real root for the mixture volume ($v = 88.9$ cc/mol); the stability behavior is class I. Then (8.4.24) gives $\varphi_1 = 3.396$ and FFF#1 gives

$$f_1 = x_1 \varphi_1 P = 0.25 \times 3.396 \times 30 = 25.47 \text{ bar} \tag{8.4.29}$$

Step 4. Check whether $f_1 > f_{\text{pure 1}}$; if so, the proposed mixture state is not stable. The values in (8.4.28) and (8.4.29) do not obey this inequality; that is, (8.4.9) is satisfied. Unfortunately, this is not sufficient for us to draw any conclusion about the stability of the proposed mixture. But for mixtures in which (8.4.9) is violated, this test would identify the proposed mixture as not stable and our problem would be solved.

Step 5. Determine whether the mixture fugacity equation has bifurcated at the same value of f_1 and the given T and P. This can be done graphically or analytically by solving (8.4.24) using a trial-and-error procedure. For pedagogical reasons we use the graphical approach here. First, we use (8.4.24) to compute f_1 over the entire range of x_1, then we plot the results. The plot appears in Figure 8.18; on that plot, point α represents our proposed mixture. The plot indicates that two other mixtures have the same values for T, P, and f_1: mixture γ at $x_1 = 0.477$ and mixture β at $x_1 = 0.911$.

Step 6. Determine the value of f_2 for all roots at the specified T, P, and f_1. We apply the Redlich-Kwong equation together with (8.4.24) to find the values in Table 8.2. We

Table 8.2 Values of fugacity f_2 for mixtures of methane and propane having $T = 275$ K, $P = 30$ bar, and $f_1 = 25.47$ bar

Root	x_1	v (cc/mol)	f_2 (bar)
α	0.25	88.9	4.54
γ	0.477	184.	5.48
β	0.911	686.	1.78

emphasize that the three roots in Table 8.2 are caused by bifurcations in the fugacity equation (8.4.24), not by bifurcations in the Redlich-Kwong equation of state (8.2.1).

Step 7. Identify the root having the lowest value of f_2 as the stable one-phase mixture at the proposed T, P, and f_1. From Table 8.2 we see that the stable one-phase mixture is root β. Therefore root α, which is our proposed mixture, is not a stable one-phase mixture. Further, Figure 8.18 shows that root α satisfies the requirement on the derivative (8.4.8), so the proposed mixture is not unstable. Hence, it must be metastable: it might be observed, but more likely it will split into two phases. To find the compositions of those phases, we would solve the phase-equilibrium problem. Other procedures for identifying stable one-phase mixtures include the tangent-plane method which originates with Gibbs [15] and has been fully developed by Michelsen, especially for multicomponent mixtures [16].

8.4.5 Determining Stability Using Models for Excess Gibbs Energy

We have shown how models for volumetric equations of state can be used with stability criteria to predict vapor-liquid phase separations. However, not all phase equilibria are conveniently described by volumetric equations of state; for example, liquid-liquid, solid-solid, and solid-fluid equilibria are usually correlated using models for the excess Gibbs energy g^E. When solid phases are present, one motivation for not using a PvT equation is to avoid the introduction of spurious fluid-solid critical points, as discussed in § 8.2.5. A second motivation is that properties of liquids and solids are little affected by moderate changes in pressure, so PvT equations can be unnecessarily complicated when applied to condensed phases. In contrast, g^E-models often do not contain pressure or density; instead, they attempt to account only for the effects of temperature and composition. Such models are thereby limited to descriptions of phase separations that are driven by diffusional instabilities, and the stability behavior must be of class I (see § 8.4.2). In this section we show how a g^E-model can describe liquid-liquid and solid-solid equilibria.

To pose the diffusional stability criterion (8.4.6) in terms of $g^E(x)$, we rearrange (5.2.7) to express g^m in terms of g^E,

$$g^m(T, P, \{x\}) = g^E(T, P, \{x\}) + RT\sum_i x_i \ln x_i \tag{8.4.30}$$

Applying (8.4.6) to (8.4.30), the diffusional stability criterion for a binary is obtained in terms of g^E as

$$\left(\frac{\partial^2 g^E}{\partial x_1^2}\right)_{TP} + \frac{RT}{x_1 x_2} > 0 \qquad \textit{binary, not unstable} \tag{8.4.31}$$

To illustrate, we use Porter's equation, which is the simplest possible model of g^E for binary mixtures (see § 5.6.2),

$$\frac{g^E}{RT} = A(T)\, x_1 x_2 \tag{8.4.32}$$

Note that the parameter A is dimensionless and depends only on temperature. Although simple, Porter's equation can reproduce states that violate the diffusional stability criterion, thereby giving rise to liquid-liquid or solid-solid equilibria. Whether or not such violations occur depends on the parameter A. To identify the stability bound on A, we substitute Porter's equation (8.4.32) into (8.4.31), and find

$$A(T) < 2 \qquad \textit{stable binary} \qquad (8.4.33)$$

At a given temperature, if $A < 2$ then the binary is a stable one-phase mixture at all compositions. However if at some other temperature, $A > 2$, then over some range of x_1 the mixture is either metastable or unstable and a phase split can occur.

When a split does occur, the compositions of the two phases, call them α and β, are obtained by solving the equilibrium conditions on the fugacities,

$$f_i^{\alpha} = f_i^{\beta} \qquad i = 1, 2 \qquad (8.4.34)$$

For Porter's equation, this becomes

$$x_i^{\alpha} \exp[A(x_j^{\alpha})^2] = x_i^{\beta} \exp[A(x_j^{\beta})^2] \qquad i = 1, 2;\ j = 1, 2;\ i \neq j \qquad (8.4.35)$$

Equation (8.4.35) represents two nonlinear algebraic equations that must be solved by trial. Examples of such roots, which represent the compositions of the two phases in equilibrium, are shown in Figure 8.19. Because of the symmetry in Porter's equation (8.4.32), the equilibrium curve in Figure 8.19 is symmetric about the equimolar composition. For example, the equilibrium compositions obtained from Porter's equations satisfy $x_1^{\beta} = 1 - x_1^{\alpha}$. (This relation is model-dependent and rarely occurs in practice.)

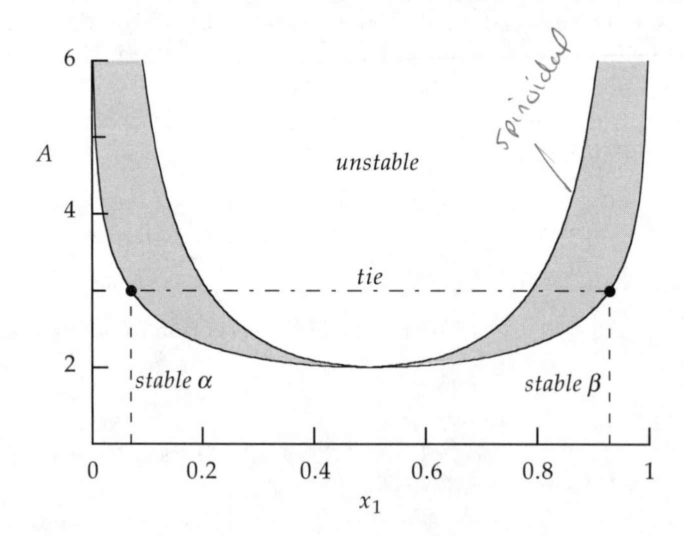

Figure 8.19 Stability of binary mixtures as given by the Porter equation (8.4.32) over a range of values for the parameter A. For $A < 2$, mixtures are stable in all proportions. For $A > 2$, mixtures can be stable, unstable, or metastable, depending on composition. Shaded regions are metastable. Curve separating stable from metastable states is the two-phase equilibrium curve, obtained by solving (8.4.35). A sample solution to (8.4.35) is shown for $A = 3$; filled circles give compositions of phases in equilibrium.

Porter's equation also provides an estimate for the spinodal, which separates unstable states from metastable ones. In terms of g^E, the spinodal of a binary occurs when the diffusional stability criterion is first violated, that is, when

$$\left(\frac{\partial^2 g^E}{\partial x_1^2}\right)_{TP} + \frac{RT}{x_1 x_2} = 0 \tag{8.4.36}$$

Substituting Porter's equation (8.4.32) into (8.4.36) gives the composition of the spinodal at a specified temperature,

$$x_1 = \frac{1}{2}\left(1 \pm \sqrt{1 - \frac{2}{A(T)}}\right) \tag{8.4.37}$$

The two roots of (8.4.37) represent the compositions of each phase on the two branches of the spinodal. In Figure 8.19, the spinodal is the curve that separates unstable states from metastable ones (shaded).

Since pressure and density are often unimportant to descriptions of liquids and solids, binary liquid-liquid and solid-solid phase diagrams are often limited to plots of temperature vs. composition. Figure 8.20 shows such a *Txx* diagram computed from the Porter equation with the temperature dependence of *A* given by

$$A = 2 + 0.02[50 - T] \qquad T \text{ in } ^\circ\text{C} \tag{8.4.38}$$

For $T > 50$°C, $A < 2$ and the mixture is a single stable phase at all compositions. However, for $T < 50$°C, the diffusional stability criterion is violated and the mixture can

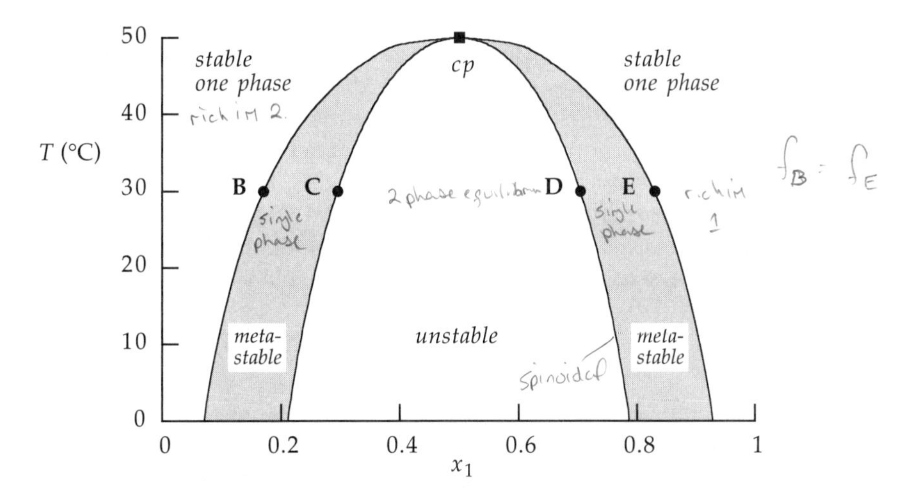

Figure 8.20 *Txx* diagram for liquid-liquid or solid-solid equilibria in binary mixtures that obey the Porter equation (8.4.32) with parameter *A* given by (8.4.38). Filled square is the critical point; filled circles lie on the isotherm at 30°C. The inner envelope, with labels C and D, is the spinodal and satisfies (8.4.37). The outer envelope is the equilibrium curve, which satisfies the equilibrium conditions (8.4.35).

split into two phases, over some range of compositions. At T = 50°C, A = 2 and the mixture exhibits a critical point, analogous to a gas-liquid critical point.

As a particular example, consider the isotherm at 30°C. In the figure this isotherm is marked by the letters B and E on the phase equilibrium curve, where the equality of fugacities (8.4.34) is satisfied, and it is marked with C and D on the spinodal, where (8.4.36) is obeyed. Therefore at 30°C,

if $x_1 < x_1^B$	then the mixture is a stable single phase, rich in species 2,
if $x_1^B < x_1 < x_1^C$	then the mixture may be a metastable single phase,
if $x_1^C < x_1 < x_1^D$	then the mixture must be in two-phase equilibrium,
if $x_1^D < x_1 < x_1^E$	then the mixture may be a metastable single phase, and
if $x_1^E < x_1$	then the mixture is a stable single phase, rich in species 1.

Analogous to the descriptions of vapor-liquid equilibria presented in § 8.4.2, the stability of condensed phases can be described in terms of the change of Gibbs energy on mixing $g^m(x)$ and its second mole-fraction derivative. Figure 8.21 shows $g^m(x)$ and its second derivative along two isotherms for the binary mixture of Figure 8.20. Along the isotherm at 60°C, Figure 8.21 shows that the second derivative of g^m is positive at all compositions, so the mixture remains a stable single phase. This is consistent with the diagram in Figure 8.20, which shows that no phase split occurs for $T > 50$°C.

However at 30°C, Figure 8.21 shows that the second derivative becomes negative at the points labeled C and D, and therefore at 30°C the mixture separates into two phases. The compositions of the two phases are given by the points B and E, obtained by solving the phase equilibrium conditions (8.4.35); those equilibrium points are connected by a tie line. The four points B–E correspond to the points having the same labels on the *Txx* diagram in Figure 8.20. Along the isothermal segments BC and DE the mixture can exist as a single metastable phase, or it can separate into two phases. But along the segment CD the diffusional stability criterion is violated and the mixture always splits into two phases.

The liquid-liquid or solid-solid equilibrium situation in Figure 8.21 is analogous to the vapor-liquid equilibrium situation in right panel of Figure 8.14; in each case the phase separation is driven by diffusional instabilities. However, most correlations for $g^E(x)$ do not allow for the possibility of mechanical instabilities because they do not involve the mixture pressure or density. Therefore such correlations produce curves for g^m that are always continuous through the unstable region: the stability behavior is class I.

8.5 SUMMARY

In this chapter the central issue has been the observability of a proposed state: if we need a mixture at a particular T, P, $\{x\}$, and phase, Can that phase actually exist at the specified T, P, and $\{x\}$? If the proposed state is unstable, then it is neither observable nor observed; if it is metastable, it is observable and sometimes observed; and if it is stable, it is observable and usually observed. To distinguish among these possibilities, we have brought to bear two general tests: (i) differential stability criteria, which dis-

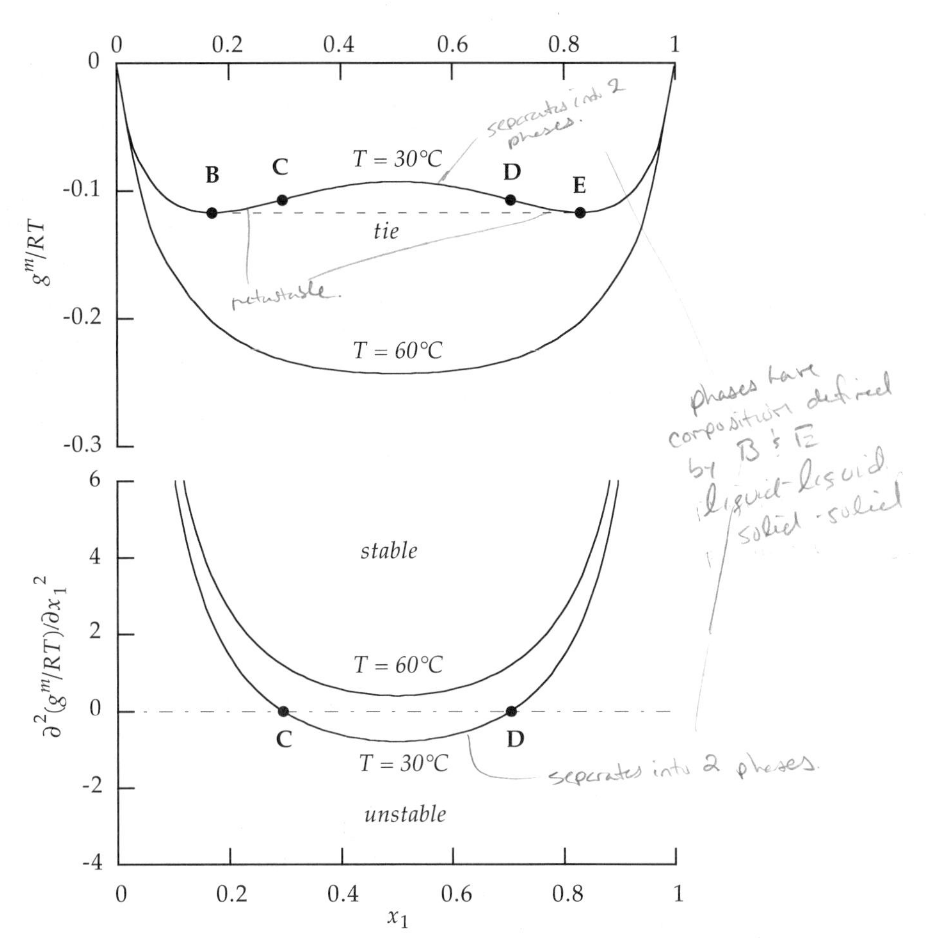

Figure 8.21 $g^m(x)$ and its second mole-fraction derivative computed from Porter's equation for the binary mixtures in Figure 8.20. At 60°C the diffusional stability criterion is satisfied at all compositions and the mixture is a stable single phase. However at 30°C, states between C and D violate the diffusional stability criterion and the mixture splits into two phases: C and D lie on the spinodal. Filled circles at 30°C correspond to states of the same labels in Figure 8.20.

tinguish unstable states from the others, and (ii) equilibrium criteria, which distinguish stable states from the others.

The differential stability criteria were derived by finding conditions that maximize the total entropy in an isolated system. Those conditions constrain how the system responds to thermal, mechanical, and diffusional fluctuations. In the derivations, those constraints are conveniently posed as *stability criteria*; they show us that a stable substance must always obey the thermal criterion (8.1.23), the mechanical criterion (8.1.31), and the diffusional criterion (8.3.14). But the converses of those statements are not always true; for example, a mechanically stable fluid always has $\kappa_T > 0$, but a fluid having $\kappa_T > 0$ is not necessarily stable—it might be metastable. Therefore, in using these differential criteria (as opposed to merely deriving them), many ambiguities can be avoided if we repose each constraint in the form of an *instability criterion*; such criteria identify those thermodynamic states at which a pure substance or mixture is differentially unstable.

The first instability criterion is that a thermally unstable substance always has

$$C_v < 0 \qquad \textit{thermally unstable} \tag{8.5.1}$$

The second is that a mechanically unstable substance always has

$$C_v\,\kappa_T < 0 \qquad \textit{mechanically unstable} \tag{8.5.2}$$

These first two criteria (8.5.1) and (8.5.2) apply to pure substances and to mixtures. The third criterion is that a diffusionally unstable mixture always has

$$\frac{\overline{G}_{11}}{C_v\,\kappa_T} < 0 \qquad \textit{diffusionally unstable} \tag{8.5.3}$$

Not only are these statements always true, but their converses are also always true. For example, a mechanically unstable substance has $C_v \kappa_T < 0$, and conversely, a substance that has $C_v \kappa_T < 0$ is always mechanically unstable.

The forms (8.5.1)–(8.5.3) show that these differential criteria are inclusive: a mixture that is diffusionally stable is also mechanically stable, and a mechanically stable substance is also thermally stable. Inversely, a thermally unstable fluid is also mechanically unstable, and a mechanically unstable mixture is also diffusionally unstable. In addition, use of the diffusional instability criterion (8.5.3), may remind us that a binary mixture can be diffusionally unstable because $\kappa_T < 0$ even when $\overline{G}_{11} > 0$.

However, the full instability criteria (8.5.1)–(8.5.3) still cannot distinguish stable states from metastable states; but then, no differential test can make this distinction. To distinguish stable states from metastable states, we must apply an appropriate equilibrium criteria. For example, if T and P have been specified for a proposed state, then the stable state is the one that minimizes the Gibbs energy. Using this as a basis, we showed how to identify the stable state for pure fluids and for binary mixtures.

A second theme of this chapter is that phase transitions decouple from unstable states. Unstable fluids may or may not split into two phases, depending on where the state lies on the phase diagram and on what external constraints are imposed. If T and v are fixed, then unstable pure fluids will undergo phase splits. But if T and P are fixed, then an unstable pure fluid will not necessarily separate into two phases: it may relax to another one-phase situation. In addition, unstable binary fluids at fixed T and P above the mechanical critical line always split into two phases, but below the mechanical critical line they do not necessarily split. Moreover, phase separations do not necessarily originate from unstable states; metastable fluids may also separate into two phases. These comments mean that, at fixed $(T, P, \{x\})$, differential stability criteria alone may not be enough to help us decide whether a phase split will occur.

Although methods for identifying phase splits generally involve more that just differential stability criteria, they do not require us to solve the phase-equilibrium problem for the compositions of any new phases. Such methods are particularly useful when we only need to know whether or not a one-phase fluid can separate. Even when we need to compute equilibrium compositions, it is wise to precede the calculations with a determination as to whether a phase separation can actually be observed. In such cases, the phase stability tests presented in this chapter can serve as informative preliminaries to solving phase-equilibrium problems.

LITERATURE CITED

[1] A. Münster, *Classical Thermodynamics*, Wiley-Interscience, London, 1970.

[2] C. L. Strong and J. E. Connett, "Curious Bubbles in Which a Gas Encloses a Liquid Instead of the Other Way Around," *Sci. Am.*, **230**(4), 116 (1974).

[3] M. Modell and R. C. Reid, *Thermodynamics and Its Applications*, 2nd ed., Prentice-Hall, Englewood Cliffs, NJ, 1983.

[4] B. L. Beegle, M. Modell, and R. C. Reid, "Thermodynamic Stability Criterion for Pure Substances and Mixtures," *A. I. Ch. E. J.*, **20**, 1200 (1974); **21**, 826 (1975).

[5] G. E. Shilov, *An Introduction to the Theory of Linear Spaces*, R. A. Silverman (transl.), Prentice-Hall, Englewood Cliffs, NJ, 1961, pp. 130–132.

[6] J. V. Sengers, R. F. Kayser, C. J. Peters, and H. J. White, Jr. (eds.), *Equations of State for Fluids and Fluid Mixtures*, Elsevier, Amsterdam, 2000.

[7] G. Iooss and D. D. Joseph, *Elementary Stability and Bifurcation Theory*, Springer-Verlag, New York, 1980.

[8] S. Vogel, *Life's Devices*, Princeton University Press, Princeton, NJ, 1988, p. 92–95.

[9] J. C. Maxwell, "On the Dynamical Evidence of the Molecular Constitution of Bodies," reprinted in *The Scientific Papers of James Clerk Maxwell*, W. D. Niven (ed.), vol. II, Dover, New York, 1965, p. 418.

[10] C. Antoine, *C. R. Acad. Sci. (Paris)*, "Tensions des Vapeurs: Nouvelle Relation Entre les Tensions et les Températures," **107**, 681 (1888).

[11] W. Wagner, "New Vapour Pressure Measurements for Argon and Nitrogen and a New Method for Establishing Rational Vapour Pressure Equations," *Cryogenics*, **13**, 470 (1973).

[12] D. Ambrose and N. B. Ghiassee, "Vapour Pressures, Critical Temperatures, and Critical Pressures of Benzyl Alcohol, Octan-2-ol, and 2-Ethylhexan-1-ol," *J. Chem. Thermodynamics*, **22**, 307 (1990).

[13] D. Eisenberg and W. Kauzmann, *The Structure and Properties of Water*, Oxford University Press, New York, 1969, p. 80.

[14] S. S. Kulkarni, *Bifurcations in the Equation of State and Stability of Binary Mixtures*, MS Thesis, Clemson University, Clemson, SC, 1996.

[15] J. W. Gibbs, "A Method of Geometrical Representation of the Thermodynamic Properties of Substances by Means of Surfaces," *Trans. Connecticut Acad.*, II, 382 (1873); reprinted in *The Collected Works of J. Willard Gibbs*, vol. I., Yale University Press, New Haven, 1957.

[16] M. L. Michelsen, "The Isothermal Flash Problem. I. Stability Analysis," *Fluid Phase Equil.*, **9**, 1 (1982); "II. Phase-Split Calculation," *Fluid Phase Equil.*, **9**, 21 (1982).

[17] N. B. Vargaftik, *Tables on the Thermophysical Properties of Liquids and Gases*, 2nd ed., Halsted Press (Wiley), New York, 1975.

PROBLEMS

8.1 Consider a gas that obeys the simple virial equation $Z = 1 + BP/RT$. Determine whether this substance can become mechanically unstable. Is your conclusion affected by whether the gas is pure or a mixture?

8.2 (a) Consider a pure fluid composed of spherical molecules. At low densities this fluid is essentially an ideal gas with internal energy $u = 3RT/2$. Determine whether this fluid can become thermally unstable and mechanically unstable.
(b) Consider the same substance as in (a), but now at a higher density where it obeys the Redlich-Kwong equation of state (8.2.1). Determine whether the fluid can now become thermally unstable.

8.3 Start with the equality of fugacities (7.3.12) for vapor-liquid equilibrium and perform the steps cited in § 8.2.5 to derive (8.2.21) for pure-component vapor pressures. Continue the derivation to obtain the equal-area form (8.2.22).

8.4 Use the Redlich-Kwong equation (8.2.1) along with (8.2.21) to estimate the vapor-pressure curve $P^s(T)$ for pure carbon dioxide. Then use your results to test the Clausius-Clapeyron equation by preparing a plot analogous to that in the bottom of Figure 8.9. Include on your plot the following experimental values of the vapor pressure (from Vargaftik [17]):

T (K)	220	235	250	265	280	295	304.2
P (bar)	6.0	10.75	17.9	27.9	41.6	59.8	73.8

8.5 Use the Antoine's equation in Appendix D to estimate the latent heat of vaporization for toluene at its normal boiling point, 110.63°C, and its normal melting point, –95°C. Compare your estimates with the experimental values, which are near 364 J/gm and 453 J/gm, respectively [17].

8.6 Use Figure 8.9 to estimate the latent heat of vaporization for pure water. Compare your value with that at the normal boiling point, as given by steam tables.

8.7 Starting with the definition of the latent heat of vaporization in (8.2.23), perform the steps cited in § 8.2.6 to derive (8.2.25), which allows us to compute the latent heat from a volumetric equation of state.

8.8 Tabitha the Untutored claims that a simple quadratic form such as

$$P/RT = A + B/v + C/v^2$$

should be sufficient to reproduce vapor-liquid equilibrium data for pure fluids. Here A, B, and C are empirical parameters that depend only on temperature. Values of A, B, and C may be positive or negative. Tabitha points out that at fixed P and fixed $T < T_c$, such an equation could yield two roots for the volume: one could be that for saturated liquid, while the other could be for saturated vapor. Do you agree with the claim that such a form is sufficient? Justify your position.

8.9 The stability test for a pure substance, as illustrated in § 8.2.2, applies when the proposed state is at a fixed T and P. But when the state is identified by fixing T and v, then the procedure in § 8.2.2 must be modified. To help develop a basis for a new procedure, perform the following.

(a) For a pure, stable, one-phase substance, prove that an isothermal plot of the Helmholtz energy vs. molar volume $a(v)$ is a convex curve with negative slope; i.e., prove that, for all v,

$$\left(\frac{\partial a}{\partial v}\right)_T < 0 \quad \text{and} \quad \left(\frac{\partial^2 a}{\partial v^2}\right)_T > 0$$

(b) For a pure substance, sketch a subcritical isotherm on an a-v diagram and show the vapor-liquid tie line. Also sketch a supercritical isotherm.

8.10 Write a computer program that uses a cubic equation of state for determining the stability of a pure fluid at a proposed state (T, P). Use the Redlich-Kwong equation of state and check your program by repeating the calculations outlined in § 8.2.2. Then use your program to determine the stability of the following states; if any of the following are not stable, find the stable state at the specified T and P.

	Species	**Phase**	**T (K)**	**P (bar)**
(a)	propane	gas	298	1
(b)	propane	gas	298	12
(c)	propane	liquid	350	25
(d)	propane	liquid	350	32
(e)	n-butane	gas	298	1
(f)	n-butane	gas	298	4
(g)	n-butane	liquid	350	11
(h)	n-butane	liquid	350	12

8.11 Consider N moles of a pure substance in a closed system at a proposed state (T, v) that is unstable. With T and v fixed, an unstable pure substance always separates into two phases, α and β. The final pressure would be the saturation pressure $P^s(T)$. Let v^α and v^β be the molar volumes of the equilibrium phases.

(a) Use a material balance to derive the *Lever Rule*, which gives the relative amounts in the two phases,

$$\frac{N^\alpha}{N^\beta} = \frac{v - v^\beta}{v^\alpha - v} \tag{P8.11.1}$$

(b) Let the equilibrium phases be vapor (α) and liquid (β). Sketch a subcritical isotherm on a Pv diagram for a pure fluid and draw the tie line at the vapor pressure P^s. For a particular value of the overall volume v, show on your plot how the tie line is related to the numerator and denominator in (P8.11.1).

8.12 A pure one-phase substance completely fills a closed rigid vessel at fixed temperature. Maynard Malaprop claims that it is sometimes possible to reduce the system pressure by isothermally adding more material. That is, for an extensive volume V, he claims that there are states at which

$$\left(\frac{\partial P}{\partial N}\right)_{TV} < 0 \tag{P8.12.1}$$

(a) To illustrate his claim, Maynard uses the van der Waals equation,

$$P = \frac{NRT}{V - Nb} - \frac{aN^2}{V^2}$$

He says that if a van der Waals fluid is at a state such that

$$v_R T_R \left(\frac{v_R}{v_R - 1/3}\right)^2 < \frac{9}{4}$$

then the inequality in (P8.12.1) is satisfied and the pressure will decrease with increasing N. Confirm this. (Here $T_R = T/T_c$ and $v_R = v/v_c$.)

(b) In spite of the result in (a), you may remain skeptical; after all, a mathematically correct result is not necessarily sound thermodynamically, is it? Maynard scoffs at this: surely you don't believe that thermodynamics can violate mathematics? Construct a thermodynamically rigorous argument that proves or disproves Maynard's claim about the inequality in (P8.12.1).

(c) Now consider a mixture. The question is whether we can identify any constraint on the sign of the response of the pressure to an increase in the mole number of one species; that is,

$$\left(\frac{\partial P}{\partial N_i}\right)_{TVN_{j\neq i}} \overset{?}{\gtrless} 0$$

To do so, use a triple product rule to relate this derivative to measurables. For a mixture, is there some constraint which demands that the pressure must always increase or decrease when N_i is increased?

8.13 Use the Redlich-Kwong equation (8.2.1) along with (8.2.25) to estimate the latent heat of vaporization for pure isobutane at 20°C. Compare your estimate with the experimental value of 336 J/gm [17].

8.14 (a) Use the definition of a derivative to derive the Leibniz rule for differentiating integrals (see Appendix A).

(b) Starting from (8.2.21) for vapor pressure, derive Clapeyron's equation (8.2.27).

(c) From (8.2.27), derive the Clausius-Clapeyron equation (8.2.30).

8.15 Consider binary liquid mixtures of benzene and toluene at 20°C and 1 bar. Show whether, at any composition, such mixtures can exhibit diffusional instabilities; if so, they would split into two liquid phases.

8.16 Consider a binary gas mixture that obeys the virial equation $Z = 1 + BP/RT$, where the mixture B is given by (4.5.18). Show whether or not this mixture can be diffusionally unstable.

8.17 Following the procedure outlined in § 8.4.4, use the Redlich-Kwong equation (8.2.1) to compute the fugacity $f_1(x_1)$ for the following mixtures. Prepare plots of your results and identify the regions over which one-phase mixtures are definitely stable and definitely not stable. Will phase splits occur from those situations that are not stable? Let the first named component be 1.

(a) carbon dioxide and n-butane at 260K and 10 bar

(b) carbon dioxide and n-butane at 300K and 8.5 bar

(c) methane and propane at 165K and 1 bar

(d) methane and propane at 278K and 10 bar

8.18 At 30°C binary liquid mixtures of methanol(1) and heptane(2) roughly obey Porter's equation and have $\gamma_1^\infty \approx \gamma_2^\infty \approx 11.0$. Determine whether, at 30°C, these mixtures exhibit liquid-liquid phase splits over some range of compositions.

8.19 Consider a binary liquid mixture that obeys Porter's equation, $g^E/RT = Ax_1x_2$, where the dimensionless parameter A depends on temperature.

(a) Derive the diffusional stability criterion (8.4.33).

(b) Derive the expression (8.4.37) for the spinodal.

(c) Assume the temperature dependence of A is given by (8.4.38). Compute $g^m(x)/RT$ and its second composition derivative at 20°C, 40°C, and 55°C. Plot your results as in Figure 8.21. At each temperature, indicate whether a phase split occurs; if a split does occur, label the regions of stable, metastable, and unstable phases on your plot. (If a split occurs, you do not have to compute the compositions of the two phases.)

8.20 Use the Redlich-Kwong equation and the mixing rules given in § 8.4.4 to compute the spinodal and line of incipient mechanical instability for equimolar mixtures of carbon dioxide and n-butane. Plot your curves on a Pv diagram. (You do not have to compute the saturation curves, since methods for doing so are not presented until Chapter 10.)

8.21 Consider a binary mixture that obeys the van Laar equation

$$\frac{g^E}{RT} = \frac{x_1x_2}{Ax_1 + Bx_2}$$

where A and B are constants. Find the expression for the liquid-liquid critical temperature, if there is one.

8.22 Consider a binary mixture for which one activity coefficient obeys

$$RT\ln\gamma_1 = A(x_1 - 1)^2$$

where A is a constant. Find an expression for the liquid-liquid critical temperature in terms of A, if such a critical point exists.

8.23 For a certain binary liquid mixture the excess volume and excess enthalpy obey $v^E = ATx_1x_2$ and $h^E = Bx_1x_2$, where A and B are independent of T and P.

(a) Find the consistent expression for g^E in terms of T, P, x_1, and x_2.

(b) The mixture has a liquid-liquid critical point at 330 K and 1 bar. It also has $v^E = -1$ cm^3/mol, $h^E/RT = 0.2$ for the equimolar mixture at 330 K. Estimate the liquid-liquid critical temperature at 100 bar.

8.24 A certain binary liquid mixture exists in two-phase liquid-liquid equilibrium. What should be the expression for g^E if the mole fractions of the two phases are independent of temperature?

8.25 Sketch an isothermal-isobaric plot of the change of Gibbs energy on mixing g^m vs. mole fraction x_1 for a binary mixture in three-phase vapor-liquid-liquid equilibrium. Include the tie lines on your plot and indicate the compositions of the three phases.

8.26 (a) Derive the thermal stability criterion for a binary mixture that undergoes only fluctuations in U at fixed N_1, N_2, and V.

(b) Derive the mechanical stability criterion for a binary mixture that undergoes only volume fluctuations at fixed T, N_1, and N_2.

(c) Derive the diffusional stability criterion for a binary mixture that undergoes only fluctuations in N_1 at fixed T, P, and N_2.

8.27 For a binary mixture that splits into two liquid phases, prepare plots of $\ln f_1$ vs. x_1 along three isotherms: one below, one above, and one at the liquid-liquid critical temperature.

8.28 For the classes of binary-mixture stability behavior discussed in § 8.4.2 make a table that tells whether the equation of state and the fugacity equation bifurcate. Your table should contain five rows, one for each class (I, II, IIIA, IIIB, IV), and it should have four columns, one for each equation (pure-1 equation of state, pure-2 equation of state, mixture equation of state, and mixture fugacity equation).

8.29 Write a computer program that determines the stability of a one-phase binary mixture at a proposed T, P, and x_1. Use the Redlich-Kwong equation of state with the simple mixing rules given in § 8.4.4. Test your program by applying it to the situation described in § 8.4.4.

8.30 Use your computer program from Problem 8.29 to determine the stability of the following proposed states for mixtures of methane(1) and propane(2). If the proposed state is not stable, is the stable situation one phase or two?

	T (K)	P (bar)	x_1	Phase
(a)	165	1	0.3	liquid
(b)	216	34	0.82	liquid
(c)	278	10	0.15	liquid
(d)	278	10	0.35	liquid
(e)	278	10	0.4	liquid
(f)	278	20	0.167	vapor
(g)	278	20	0.2	vapor
(h)	278	33.6	0.47	vapor
(i)	278	50	0.3	liquid
(j)	278	50	0.4	liquid
(k)	278	50	0.6	vapor
(l)	300	15	0.2	liquid
(m)	300	50	0.28	liquid

8.31 Derive the following stability criteria, given in terms of the Helmholtz energy A, for a binary mixture at fixed T and V.

$$A_{11} < 0 \quad \text{and} \quad A_{11}A_{22} - A_{12}A_{21} > 0$$

where

$$A_{ij} = \left(\frac{\partial \bar{G}_i}{\partial N_j}\right)_{TVN_{k \neq j}}$$

To do so, use a system like that in Figure 8.3, but now consider the small region B to be of fixed volume and temperature. However, region B is open to the larger region, so the mole numbers (N_1 and N_2) fluctuate in both regions.

9

PHASE DIAGRAMS FOR REAL SYSTEMS

With many million pure substances now known, an essentially infinite number of mixtures can be formed, resulting in a diversity of phase behavior that is overwhelming. Consider just two components: not only can binary mixtures exhibit solid-gas, liquid-solid, and liquid-gas equilibria, but they might also exist in liquid-liquid, solid-solid, gas-gas, gas-liquid-liquid, solid-liquid-gas, solid-solid-gas, solid-liquid-liquid, solid-solid-liquid, and solid-solid-solid equilibria. That's a dozen different kinds of phase equilibrium situations—just for binary mixtures. For multicomponent mixtures the possibilities seem endless.

In this chapter we describe the kinds of phase behavior that are commonly observed in pure fluids, binary mixtures, and some ternary mixtures. The descriptions typically take the form of phase diagrams, and we show how studies of phase behavior can be made systematic by identifying classes of diagrams. Since we are interested in describing what is actually seen, the mixture diagrams presented in this chapter are plotted in terms of measurables: usually temperature, pressure, composition, or a subset of those. Calculations of phase equilibria necessarily involves conceptuals, and such calculations are discussed in Chapter 10. Here we only describe phenomena.

We start in § 9.1 by giving prescriptions for determining the number of properties needed to identify the thermodynamic state in multicomponent mixtures. Those prescriptions include Duhem's theorem and the Gibbs phase rule as special cases. The required number of properties determines the dimensionality of the state diagram needed to represent phase behavior. Then in § 9.2 we summarize some features of pure-component diagrams that have not been discussed in earlier chapters.

Sections 9.3–9.5 present the common phase behavior of binary mixtures: § 9.3 describes vapor-liquid, liquid-liquid, and vapor-liquid-liquid equilibria at low pressures; § 9.4 considers solid-fluid equilibria; and § 9.5 discusses common high-pressure fluid-phase equilibria. Then § 9.6 briefly describes the basic vapor-liquid and liquid-liquid equilibria that can occur in ternary mixtures. This chapter describes many apparently different phase behaviors, and so we try to show when those differences are more apparent than real. The organization is intended to bring out underlying similarities, thereby reducing the number of different things to be learned.

9.1 THERMODYNAMIC STATE FOR MULTIPHASE SYSTEMS

In § 3.1 we discussed the thermodynamic state for closed systems composed of a single homogeneous phase; we now extend that discussion to heterogeneous systems, especially, systems containing more than one phase. The fundamental questions addressed in § 3.1 are revisited here: How many interactions are available for manipulating the state (§ 9.1.1)? How many property values are needed to identify the state (§ 9.1.2)? Even when we specify the correct number of properties for identifying the state, is there still a possibility of encountering computational difficulties (§ 9.1.3)?

9.1.1 Number of Interactions to Change a State

Consider a system composed of C components in a single homogeneous phase. The system can interact with its surroundings through the thermal interaction, a PV work mode, and the exchange of any of the components. For such a system, we found in § 3.1.1 that the number of interactions available for changing the state is given by

$$\mathcal{V} = C + 2 - S_{ext} \tag{9.1.1}$$

Here S_{ext} is the number of any external constraints that block interactions. If other work modes, such as electrical or surface work, are present, then the rhs of (9.1.1) increases accordingly. Note that the number of interactions applied to a system is independent of the condition of material within the system. For example, instead of being homogeneous, the system might consist of two phases, such as vapor and liquid. Nevertheless, we still interact with such a system by exchanging, at most, any of C components, heat, and PV work. Therefore, (9.1.1) also applies to heterogeneous systems composed of $\mathcal{P}$ homogeneous phases. Just as in § 3.1.1, (9.1.1) applies to changes in both intensive and extensive states. Further, just as in § 3.1.1, if we block all mass-transfer interactions (so $S_{ext} = C$), then (9.1.1) reduces to Duhem's theorem for multiphase systems,

$$\mathcal{V} = 2 \tag{9.1.2}$$

9.1.2 Number of Properties to Identify an Equilibrium State

For a single homogeneous phase containing C components, we found in § 3.1.2 that the number of properties needed to identify the extensive state is given by

$$\mathcal{F}_{ex} = \mathcal{V}_{max} - S \qquad \textit{one phase} \tag{9.1.3}$$

where

$$\mathcal{V}_{max} = \mathcal{V}\big|_{S_{ext}=0} = C + 2 \tag{9.1.4}$$

Again this assumes only the thermal interaction and a single work mode are present. For a heterogeneous system containing $\mathcal{P}$ homogeneous phases, (9.1.4) applies to each, so

$$\mathcal{V}_{max} = \mathcal{P}(C+2) \tag{9.1.5}$$

But at equilibrium we also have internal constraints imposed by Nature. For example, if the $\mathcal{P}$ homogeneous phases are all open to one another through $(\mathcal{P}-1)$ different interfaces, then each interface imposes the $(C+2)$ phase-equilibrium constraints given in § 7.3.5. For the one interface between phases α and β in equilibrium, these constraints are

$$T^{\alpha} = T^{\beta} \tag{9.1.6}$$

$$P^{\alpha} = P^{\beta} \tag{9.1.7}$$

$$\overline{G}_i^{\alpha} = \overline{G}_i^{\beta} \qquad i = 1, 2, \ldots, C \tag{9.1.8}$$

Therefore instead of (9.1.3), we have

$$\mathcal{F}_{ex} = \mathcal{P}(C+2) - (\mathcal{P}-1)(C+2) - S \tag{9.1.9}$$

or

$$\mathcal{F}_{ex} = C + 2 - S \qquad \textit{any number of phases} \tag{9.1.10}$$

Here S counts any *additional* internal constraints besides the phase-equilibrium constraints in (9.1.6)–(9.1.8). Examples include constraints imposed by critical points (certain stability relations must be obeyed) and azeotropes (certain relations must exist among T, P, and the compositions of the phases). The number given by (9.1.10) can be much less than the total number of variables given by (9.1.5). For example, a four-component system in three-phase equilibrium has $\mathcal{V}_{max} = 18$, but only $\mathcal{F}_{ex} = 6$ of those are needed to identify the extensive state (with $S = 0$). Values for the other twelve would be computed by solving stuff equations together with the phase-equilibrium equations (9.1.6)–(9.1.8); those calculations may or may not be easily performed.

To determine the number of properties needed for identifying the intensive equilibrium state, we remove the total amount of material as a possible variable; hence,

$$\mathcal{F}' = \mathcal{F}_{ex} - 1 \tag{9.1.11}$$

Then for $\mathcal{P}$ phases, using (9.1.10) in (9.1.11) leaves

$$\mathcal{F}' = C + 1 - S \qquad \textit{any number of phases} \tag{9.1.12}$$

Counted in $\mathcal{F}'$ are the relative sizes of the phases. For example, for ethanol and water in vapor-liquid equilibrium, we have $C = 2$, $\mathcal{P} = 2$, and $S = 0$, so (9.1.11) gives $\mathcal{F}' = 3$: we need values for three independent intensive properties to identify the intensive state. The three could be T, P, and N_v/N, where N_v/N is the fraction of material in the vapor phase. Another legitimate set is T, ρ, and z_E, where z_E represents the overall mole fraction of ethanol in the system.

Often we ignore the relative sizes of the phases when describing the intensive state; doing so removes $(\mathcal{P}-1)$ variables from the number in (9.1.12), leaving the *generalized phase rule*,

$$\mathcal{F} = C + 2 - \mathcal{P} - S \tag{9.1.13}$$

The phase rule gives the number of properties needed for identifying the intensive state of closed systems. However, the form (9.1.13) applies *only* to those situations that conform to our assumptions:

(a) we have only one work mode,

(b) we have ignored the relative sizes of phases, and

(c) we have no chemical reactions.

Systems with reactions are discussed in § 10.3.1. When no other internal constraints apply, then $S = 0$, and the general rule (9.1.13) reduces to the *Gibbs phase rule*,

$$\mathcal{F} = C + 2 - \mathcal{P} \tag{9.1.14}$$

On subtracting (9.1.13) from (9.1.12), we find that $\mathcal{F}$ differs from $\mathcal{F}'$ by the $(\mathcal{P}-1)$ ratios that represent the relative amounts in the phases,

$$\mathcal{F}' - \mathcal{F} = \mathcal{P} - 1 \tag{9.1.15}$$

For one-phase nonreacting systems $(\mathcal{P} = 1)$, (9.1.15) gives $\mathcal{F}' = \mathcal{F}$; otherwise, the relative amounts contribute to the number of properties counted by $\mathcal{F}'$, but they do not contribute to the number counted by $\mathcal{F}$. This difference between $\mathcal{F}$ and $\mathcal{F}'$ allows us to distinguish between two kinds of phase diagrams. On an $\mathcal{F}'$ diagram, the relative amounts in the phases must be known to locate a multiphase state (a point); an example of such a plot is a pure substance Pv diagram. However, on an $\mathcal{F}$ diagram, the relative amounts do not help us locate a multiphase state; an example is any PT diagram. If a mixture diagram has composition plotted, then it is an $\mathcal{F}'$-diagram.

Therefore, one important use of $\mathcal{F}'$ is in constructing and interpreting phase diagrams. When we intend to represent the behavior of a system on a phase diagram, $\mathcal{F}'$ (not $\mathcal{F}$) gives the dimensionality of the space needed for the plot. For example, to represent the states of pure water with no constraints ($C = 1$ and $S = 0$), (9.1.12) gives $\mathcal{F}' = 2$; that is, all intensive states of pure water can be represented on a two-dimensional surface, such as a plot of P vs. T or one of P vs. v. Note that the value given by (9.1.12) for $\mathcal{F}'$ is independent of the number of phases present; for example, if the water is in vapor-liquid equilibrium, (9.1.12) still gives $\mathcal{F}' = 2$ because the relative amounts in the two phases can change at fixed pressure. However, we caution that states identified by $\mathcal{F}'$ variables may not be unique; see § 9.1.3.

A principle use of $\mathcal{F}$ occurs when analyzing constrained equilibria: the value of $\mathcal{F}$ gives the dimensionality of the object that represents a constrained equilibrium on an $\mathcal{F}$-diagram. For example, if we have pure water constrained to states in vapor-liquid

Table 9.1 Kinds of geometric objects appearing on $\mathcal{F}$ diagrams when constraints apply to phase-equilibrium situations; values of $\mathcal{F}$ from (9.1.13)[a]

$\mathcal{C}$	$\mathcal{P}$	Constraints (S)	Example	$\mathcal{F}$	Object
1	1	none (0)	L, S, V	2	surface
	1	mechanical stability[b] (2)	VL critical point	0	point
	2	none (0)	SL, SV, VL	1	line
	3	none (0)	SLV triple point	0	point
2	1	none (0)	L, S, V	3	volume
	1	diffusional stability[c] (2)	VL critical points	1	line
	1	diff. stab. (2); Tx extremum (1)	critical azeotrope	0	point
	2	none (0)	SL, LL, LV	2	surface
	2	Tx extremum (1)	azeotrope	1	line
	2	diffusional stability (2)	VLL crit. end pt.	0	point
	3	none (0)	SVL, VLL	1	line
	4	none (0)	SLLV, SSLV	0	point

[a] This is a modified version of a table originally devised by de Loos [1].
[b] The two mechanical stability constraints are (8.2.13) and (8.2.14).
[c] The two diffusional stability constraints are (8.4.2) and (8.4.3).

equilibrium, then $P = 2$ and the Gibbs phase rule (9.1.14) gives $\mathcal{F} = 1$: states of two-phase equilibria appear as lines on a one-component $\mathcal{F}$-diagram (such as a PT diagram). Table 9.1 gives examples of the kinds of geometric objects that appear on $\mathcal{F}$-diagrams when constraints are imposed on pure components and on binary mixtures.

Another principal use of $\mathcal{F}$ and $\mathcal{F}'$ is in testing whether equilibrium problems are well posed. To use $\mathcal{F}$ and $\mathcal{F}'$ properly for this purpose, we must first decide whether we have an $\mathcal{F}$-problem or an $\mathcal{F}'$-problem. For an $\mathcal{F}$-problem, one of the phase rules, (9.1.13) or (9.1.14), tells us the number of property values we must know to have a well-posed problem. But for an $\mathcal{F}'$-problem, the required number is given by (9.1.12). Versions of (9.1.12) and (9.1.13) for reacting systems are developed in § 10.3.1.

9.1.3 Indifferent States

Situations can arise in which we have apparently specified values for enough properties, and yet the state is still not uniquely identified. We follow Prigogine and Defay [2] and call these *indifferent* states. The existence of these situations can frustrate some trial-and-error procedures for solving phase-equilibrium problems.

One kind of indifference occurs when we have specified too few property values to solve a problem; for example, we give an $\mathcal{F}$-specification when we actually need an $\mathcal{F}'$-specification (recall, $\mathcal{F} < \mathcal{F}'$). An example occurs when we specify T and P for a one-component vapor-liquid equilibrium system, but we need to determine the fraction of material in the vapor phase. This is an indifferent situation because, at the specified T and P, our system can be at any of an infinite number of points along the tie line between liquid and vapor.

A second kind of indifference occurs when we specify values for the correct number of property values, but those properties are not all independent, or if they are independent initially, they become coupled (via internal constraints) during a calculation. Examples include azeotropes and critical points that could be encountered during vapor-liquid equilibrium calculations, because at azeotropes and critical points, T, P, and $\{x\}$ are not mutually independent. In such situations, the number of properties required by an $\mathcal{F}'$-specification is not wrong, but the particular properties chosen to satisfy the requirement are no longer independent. The possibility of computational algorithms entering indifferent situations can lead to frustration or erroneous interpretations of results; this problem will be discussed further (but not resolved) when we present computational algorithms in Chapter 11. Here are some examples.

Example 1. For a binary mixture in vapor-liquid equilibrium with no other constraints, we have $C = 2$, $\mathcal{P} = 2$, $S = 0$, so $\mathcal{F}' = 3$. Therefore, specifying values for T, P, and z_1 provides an $\mathcal{F}'$-specification. Knowing the overall mole fraction z_1 allows us to compute the relative amounts in the two phases. Hence, the state is not indifferent.

Example 2. For a binary mixture in vapor-liquid equilibrium at a homogeneous azeotrope, we have $C = 2$, $\mathcal{P} = 2$, $S = 1$, so $\mathcal{F}' = 2$. Specifying values for T and P creates an indifferent situation because T and P are coupled through the azeotropic condition. For the same reason, specifying T and z_1 is not appropriate (at an azeotrope $z_1 = x_1 = y_1$). But specifying values for T and an overall system density ρ does provide a unique $\mathcal{F}'$-specification and avoids an indifferent situation.

Example 3. For a binary mixture in vapor-liquid-liquid equilibrium, we have $C = 2$, $\mathcal{P} = 3$, $S = 0$, so $\mathcal{F}' = 3$. But setting values for T, P, and z_1 creates an indifferent situation because T and P are coupled through the three-phase equilibrium criteria. However, specifying values for T and the ratios of amount of vapor to the amounts in each liquid phase does provide a unique $\mathcal{F}'$-specification and avoids an indifferent situation.

9.2 PURE SUBSTANCES

Pure substance phase diagrams may be created using any combination of independent properties. First we consider diagrams containing only measurables (§ 9.2.1), and then diagrams containing one conceptual (§ 9.2.2).

9.2.1 Diagrams Containing Only Measurables

For a pure substance existing as a single phase with no internal constraints, (9.1.12) gives $\mathcal{F}' = 2$, indicating that intensives states can be represented on phase diagrams of two dimensions. Those diagrams may be $\mathcal{F}$-diagrams, such as the PT diagram on the

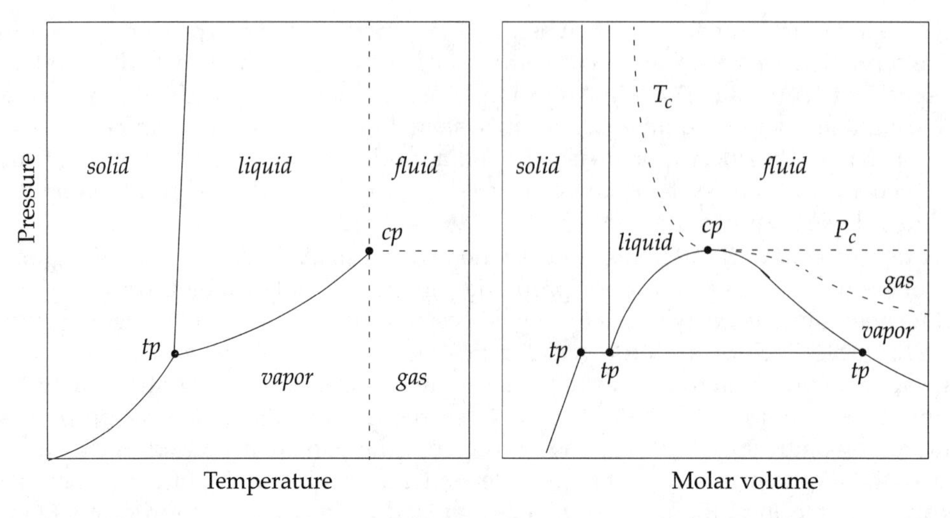

Figure 9.1 Schematic phase diagrams for a pure substance. (*left*) An $\mathcal{F}$-diagram, which cannot show the relative amounts in each phase when two phases are present. (*right*) An $\mathcal{F}'$-diagram, which can show relative amounts. cp = critical point and tp = triple point.

left in Figure 9.1, or they may be $\mathcal{F}'$-diagrams, such as the Pv diagram on the right in Figure 9.1. As listed in Table 9.1, one-phase situations appear as areas on the pure-component diagrams in Figure 9.1; two-phase equilibrium situations appear as lines; three-phase situations (triple points) occur as points. On the $\mathcal{F}$-diagram, the triple point is a single point because all three phases have the same T and P; however, on the $\mathcal{F}'$-diagram, it appears as three points because each phase has its own molar volume.

On the PT diagram in Figure 9.1, non-solid areas divide into four distinct regions. One-phase *vapor* states lie below the vapor-pressure curve at temperatures $T < T_c$, while one-phase *gas* states have $T > T_c$ and $P < P_c$. This means that a vapor can be condensed either by an isothermal compression or by an isobaric cooling, but a gas can be condensed only by some process that involves cooling. In a similar manner, one-phase *liquid* states lie above the vapor-pressure curve at temperatures $T < T_c$, while one-phase *fluid* states have $T > T_c$ and $P > P_c$. Unfortunately, these distinctions are not universally used: some authors do not distinguish vapor from gas or gas from fluid.

Note in Figure 9.1 that multiphase situations on $\mathcal{F}$-diagrams form objects of different dimensionality from the same situations on $\mathcal{F}'$-diagrams. This occurs because one variable plotted on $\mathcal{F}'$-diagrams takes different values for each phase in equilibrium; in Figure 9.1 that variable is the molar volume. Phases in equilibrium have the same T and P, but their molar volumes differ. For example, on the PT diagram in Figure 9.1, two-phase situations are lines, but on the Pv diagram, two-phase situations span areas. When an $\mathcal{F}$-specification is made, the molar volumes of equilibrium phases are fixed, regardless of the quantities present. However, even if we keep T and P fixed, we might change the distribution of material between the two phases, thereby changing the $\mathcal{F}'$-specification. The distribution of material (i.e., the relative amounts in the two phases) can be computed by solving a material balance, that is, by applying a lever rule.

In § 8.2.6 we found that the slope of any two-phase line on a pure-component PT diagram obeys the Clapeyron equation,

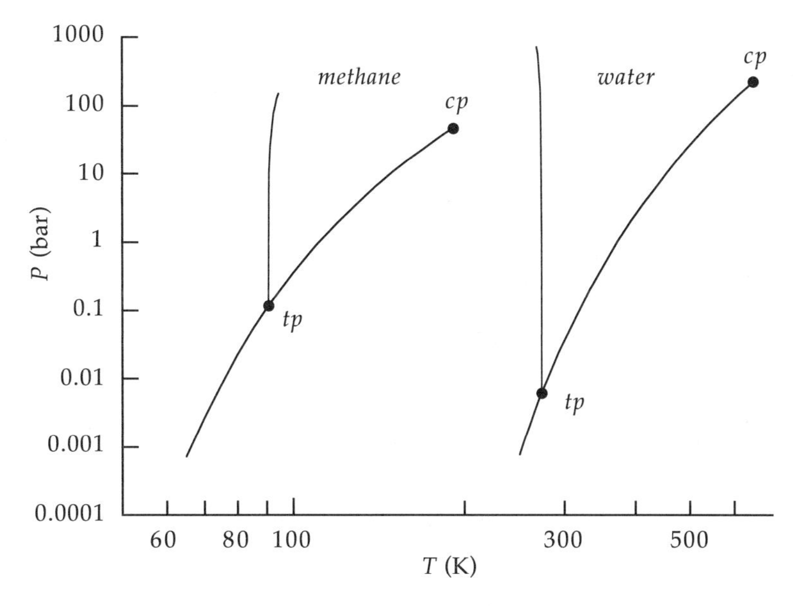

Figure 9.2 *PT* diagrams for two pure substances: one that expands on melting (methane) and one that contracts on melting (water). Data for methane taken from Tester [3] and for water from Eisenberg and Kauzmann [4].

$$\left(\frac{\partial P}{\partial T}\right)_\sigma = \frac{\Delta h}{T \Delta v} \qquad \textit{pure, always true} \qquad (9.2.1)$$

where Δh is the enthalpy change (or "latent heat") across the phase transition and Δv is the corresponding change in the molar volumes. The subscript σ in (9.2.1) reminds us that the derivative is evaluated along a two-phase saturation line. According to the Clapeyron equation (9.2.1), the slope of the vapor-pressure curve is related to the latent heat of vaporization Δh_{vap}, the slope of the fusion curve is related to the latent heat of melting Δh_m, and the slope of the sublimation curve is related to the latent heat of sublimation Δh_{sub}. For phase changes from solid to liquid, from solid to vapor, and from liquid to vapor, the latent heats Δh are always positive.

But while the vapor-pressure and sublimation curves always have positive slopes, the slope of the melting curve may be positive or negative. Most pure materials expand on melting, so $v_{liq} > v_{sol}$, $\Delta v > 0$, and therefore $(\partial P/\partial T)_\sigma > 0$. The corresponding *PT* diagram is like that shown on the left in Figure 9.2. However, a few materials, including water, contract on melting (ice floats on water); these have $v_{liq} < v_{sol}$, then $\Delta v < 0$, and therefore the Clapeyron equation gives $(\partial P/\partial T)_\sigma < 0$. The resulting *PT* diagram is like that shown on the right in Figure 9.2. In both cases, the melting curves are essentially vertical.

9.2.2 Diagrams Containing a Conceptual

Pure substance $\mathcal{F}'$ diagrams can be constructed using conceptuals as well as measurables. The ones commonly encountered are pressure-enthalpy and enthalpy-entropy

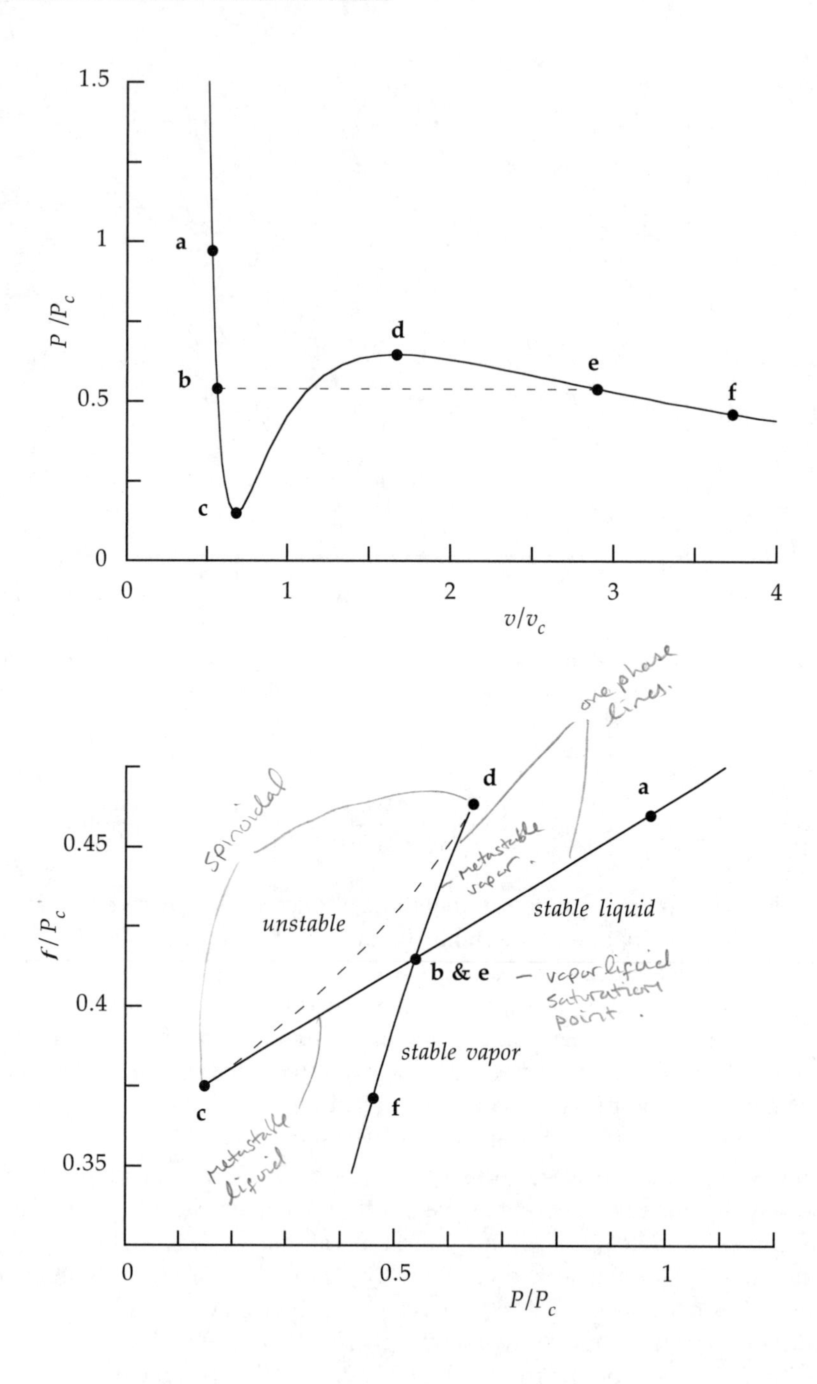

Figure 9.3 Along subcritical isotherms for pure fluids, the fugacity passes through stable, metastable, and unstable regions just as does the pressure. Here we have plotted the subcritical isotherm $T/T_c = 0.863$ for a van der Waals fluid. Each point (a–f) on the fugacity plot corresponds to the point of the same label on the Pv diagram. Points b and e have the same fugacity and pressure ($P^s/P_c = 0.539$) and therefore locate the vapor-liquid equilibrium state. Points c and d are on the spinodal. Line segment bc locates metastable liquid states; segment de locates metastable vapor states; segment cd locates unstable states.

diagrams, which are often used to analyze the performance of machines that interconvert heat and work. For pure fluids, such diagrams present the same kinds of features already discussed: single phases occupy areas and two-phase situations span areas bounded by two-phase saturation curves. When first confronting such diagrams, you should orient yourself using what you already know: (a) vapor-liquid tie lines are both isobars and isotherms, (b) the vapor-liquid saturation curves intersect at the critical point, (c) at fixed T and P, saturated vapor always has larger values for enthalpy and entropy than saturated liquid.

In Chapter 8 we discussed the mechanical stability of a pure fluid in terms of the behavior of a subcritical isotherm on a Pv diagram. A sample isotherm is shown at the top of Figure 9.3, computed using the van der Waals equation of state. Also in Chapter 8 we showed that pure-fluid vapor-liquid equilibrium states are found by solving the equilibrium conditions (9.1.8). The equality of chemical potentials in (9.1.8) can also be expressed as an equality of fugacities; in the case of pure-fluid vapor-liquid equilibria,

$$f^{v}(T, P) = f^{\ell}(T, P) \tag{9.2.2}$$

The bottom of Figure 9.3 shows the fugacity computed from the van der Waals equation along the same isotherm shown in the top of the figure. In the fP plot, note that curves for each one-phase fluid must have positive slopes because

$$\left(\frac{\partial g}{\partial P}\right)_T = RT\left(\frac{\partial \ln f}{\partial P}\right)_T = \frac{RT}{f}\left(\frac{\partial f}{\partial P}\right)_T = v > 0 \tag{9.2.3}$$

The intersection of those two one-phase lines satisfies (9.2.2) and therefore identifies the vapor-liquid saturation point. The lines for one-phase liquids terminate at the spinodal—they become unstable—and the unstable portion of the van der Waals loop is represented by the broken line on the fP plot.

9.3 BINARY MIXTURES OF FLUIDS AT LOW PRESSURES

We now describe the phase behavior exhibited by binary mixtures at modest pressures. The kinds of behavior observed in Nature include vapor-liquid equilibria (VLE, § 9.3.1–9.3.3), azeotropes (§ 9.3.4), critical points (§ 9.3.5), liquid-liquid equilibria (LLE, § 9.3.6), and vapor-liquid-liquid equilibria (VLLE, § 9.3.7). When solid-fluid equilibria occur (§ 9.4), many (but not all) of the resulting phase diagrams are analogous to their counterparts in fluid-fluid equilibria; for example, many liquid-solid diagrams are analogous to vapor-liquid diagrams.

9.3.1 Isothermal *Pxy* Diagrams for Binary VLE

Consider a mixture of components 1 and 2 in vapor-liquid equilibrium in a closed vessel at temperature T and pressure P. Let the composition of the liquid phase be represented by the mole fraction x_1 and that of the vapor by mole fraction y_1. The properties of first importance are the four measurables T, P, x_1, and y_1. In the absence

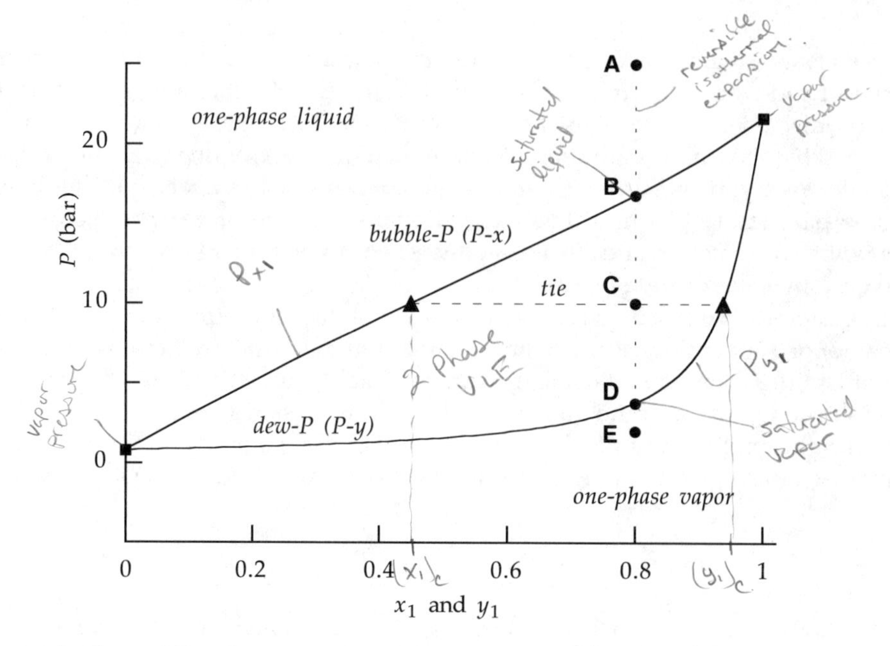

Figure 9.4 Isothermal *Pxy* diagram at 330 K computed from the Redlich-Kwong equation of state. Component 1 is an alkane; 2 is an aromatic. Broken vertical line represents a reversible isothermal expansion from one-phase liquid at A to one-phase vapor at E. Broken horizontal line is the vapor-liquid tie line at 10 bar. Filled squares mark pure-component vapor pressures at 330 K. Note that component 1 is more volatile than component 2.

of additional constraints, (9.1.12) gives $\mathcal{F}' = 3$; that is, three-dimensional plots give a complete representation of binary VLE. But the Gibbs phase rule (9.1.14) gives $\mathcal{F} = 2$: only two properties are needed to create an $\mathcal{F}$-diagram for a binary. So rather than create 3D plots, it is conventional to fix T (or P) and plot P (or T), x_1, and y_1, with both sets of mole fractions plotted on the abscissa. We describe isothermal *Pxy* plots here.

A typical isothermal *Pxy* diagram is shown in Figure 9.4. This figure was calculated from the Redlich-Kwong equation of state using critical properties for an aromatic and a short-chain alkane. Component 1 is the alkane and is the more volatile component; that is, at fixed $T < T_c$, pure 1 has a higher vapor pressure than pure 2. The entire diagram in Figure 9.4 is at one temperature: 330 K.

The curves in Figure 9.4 are saturated one-phase lines. The upper curve (nearly straight) is Px_1; it gives compositions of liquids in equilibrium with vapors. The lower curve is Py_1; it gives compositions of equilibrium vapors. Therefore, the diagram divides into three regions. At high pressures (above Px_1), these mixtures are single-phase liquids; they have $\mathcal{F} = 3$, so values must be given for T, P, and x_1 to identify a state. Similarly, at low pressures (below Py_1), these mixtures are single-phase vapors; they also have $\mathcal{F} = 3$, so values must be given for T, P, and y_1 to identify a state. The third region lies between the two-phase curves; these are two-phase VLE states. Those states require $\mathcal{F}' = 3$ if the relative amounts in the two phases are needed; otherwise, they only require $\mathcal{F} = 2$ to identify the intensive state without relative amounts.

Consider the one-phase liquid state A, which is at 330 K, 25 bar, and overall composition $z_1 = 0.8$. From point A a reversible isothermal expansion will trace the vertical path through points B, C, D, and E. During the expansion along AB, the mixture remains one-phase liquid, but the pressure decreases as the volume expands. At point

B the pressure has reached 16.8 bar and the mixture is a saturated liquid; any further expansion will cause a bubble of vapor to form. Hence the line through B is called the *bubble-P curve*; it relates pressures to the compositions of saturated liquid mixtures.

If we continue the expansion from point B, the mixture progresses through a sequence of two-phase states BE. In this two-phase expansion, the pressure continues to decrease, and the vapor phase grows at the expense of the liquid. During the expansion, the compositions of the vapor and liquid change, but the overall composition remains constant at $z_1 = 0.8$.

If we stop the expansion momentarily at 10 bar (point C), we can use the diagram to obtain the compositions and relative amounts in the two phases. Construct the horizontal through C, then x_1 and y_1 are given by the intersections of the horizontal (a *tie line)* with the two-phase curves. These intersections are marked with triangles; we find $x_1 = 0.449$ and $y_1 = 0.937$. On an isothermal *Pxy* diagram, tie lines are horizontal because at equilibrium both phases have the same pressure. The fraction of the system in the vapor phase can be determined by material balance (a lever rule):

$$\frac{N^v}{N^v + N^\ell} = \frac{z_1 - x_1}{y_1 - x_1} = 0.72 \tag{9.3.1}$$

Now continue the expansion from point C to D. At D the pressure is only 3.77 bar, and as the last drop of liquid disappears the mixture becomes all vapor. Therefore the line through D is called the *dew-P curve*, it relates pressures to the compositions of saturated vapor mixtures. In the region below D, the mixture is one-phase vapor.

If we change the temperature from 330 K, as in Figure 9.5, then the two-phase region must shift to other pressures. For example, if we increase the temperature, then

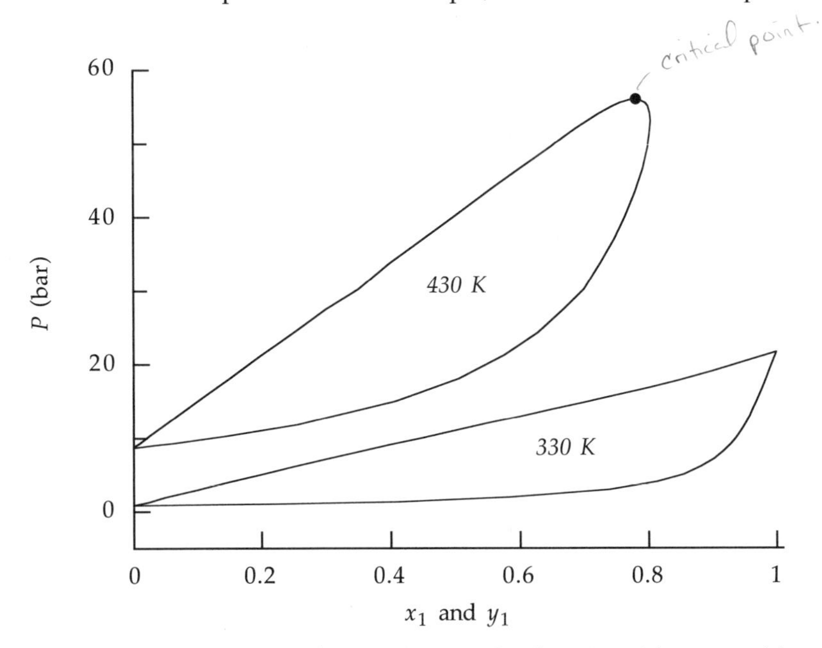

Figure 9.5 Effect of temperature on the *Pxy* diagram for the alkane(1)-aromatic(2) mixture of Figure 9.4. The 330 K-isotherm is subcritical, but the 430 K-isotherm has a critical point (dot) at 56.1 bar and $z_1 = 0.781$. Computed from Redlich-Kwong equation.

to keep both liquid and vapor phases, we must also increase the pressure. Figure 9.5 shows how the two-phase region moves when T is changed from 330 K to 430 K. At 430 K the bubble-P and dew-P curves no longer span the entire composition range because a mixture critical point occurs. That critical point has P = 56.1 bar and overall composition z_1 = 0.781 at 430 K. Note in Figure 9.5 that the mixture critical point occurs at the maximum pressure on the two-phase envelope.

In Figure 9.5 we see that, except near the critical point, the slope of the bubble curve has the same sign as the slope of the dew curve. We now prove that this is usually the case. First, use Table 6.3 to write the Gibbs-Duhem equation for the fugacities in each phase:

$$RT\sum_i x_i\, d\ln f_i = -\frac{h^\ell}{T}dT + v^\ell dP \tag{9.3.2}$$

and

$$RT\sum_i y_i\, d\ln f_i = -\frac{h^v}{T}dT + v^v dP \tag{9.3.3}$$

Since the temperatures, pressures, and fugacities are the same in the two phases, we can subtract (9.3.2) from (9.3.3) to find

$$RT\sum_i (y_i - x_i)d\ln f_i = -\frac{\Delta h}{T}dT + \Delta v\, dP \tag{9.3.4}$$

where Δ indicates differences between properties of vapor and liquid. The mole fractions must sum to unity in each phase, so for a binary,

$$y_1 - x_1 = -(y_2 - x_2) \tag{9.3.5}$$

Using this and considering fixed T, (9.3.4) reduces to

$$RT(y_1 - x_1)(d\ln f_1 - d\ln f_2) = \Delta v\, dP \qquad \textit{fixed } T \tag{9.3.6}$$

For the slope of the isothermal Px curve, this becomes

$$RT(y_1 - x_1)\left[\left(\frac{\partial \ln f_1}{\partial x_1}\right)_{T\sigma} - \left(\frac{\partial \ln f_2}{\partial x_1}\right)_{T\sigma}\right] = \Delta v\left(\frac{\partial P}{\partial x_1}\right)_{T\sigma} \tag{9.3.7}$$

The subscript σ reminds us that the derivatives are to be evaluated along the vapor-liquid saturation curve. Here, we seek an expression for the derivative on the rhs.

At fixed T, we can write the total differential of the fugacity as

$$d\ln f_i = \left(\frac{\partial \ln f_i}{\partial P}\right)_{Tx} dP + \left(\frac{\partial \ln f_i}{\partial x_i}\right)_{TP} dx_i \tag{9.3.8}$$

Introducing the partial molar volume from Table 6.2 and rearranging, we have

$$\left(\frac{\partial \ln f_i}{\partial x_i}\right)_{T\sigma} = \frac{\overline{V}_i}{RT}\left(\frac{\partial P}{\partial x_i}\right)_{T\sigma} + \left(\frac{\partial \ln f_i}{\partial x_i}\right)_{TP} \tag{9.3.9}$$

Substituting (9.3.9) into (9.3.7) for each component, we find

$$RT(y_1 - x_1)\left[\left(\frac{\partial \ln f_1}{\partial x_1}\right)_{TP} - \left(\frac{\partial \ln f_2}{\partial x_1}\right)_{TP}\right] = [\Delta v + (y_1 - x_1)(\overline{V}_2^{\ell} - \overline{V}_1^{\ell})]\left(\frac{\partial P}{\partial x_1}\right)_{T\sigma} \tag{9.3.10}$$

With the help of the Gibbs-Duhem equation on the lhs, (9.3.10) simplifies to

$$\frac{RT(y_1 - x_1)}{x_2}\left(\frac{\partial \ln f_1}{\partial x_1}\right)_{TP} = [\Delta v + (y_1 - x_1)(\overline{V}_2^{\ell} - \overline{V}_1^{\ell})]\left(\frac{\partial P}{\partial x_1}\right)_{T\sigma} \tag{9.3.11}$$

But

$$\Delta v = v^v - v^{\ell} = y_1\overline{V}_1^v + y_2\overline{V}_2^v - x_1\overline{V}_1^{\ell} - x_2\overline{V}_2^{\ell} \tag{9.3.12}$$

So (9.3.11) finally can be written as

$$\left(\frac{\partial P}{\partial x_1}\right)_{T\sigma} = \frac{RT(y_1 - x_1)}{x_2(y_1\Delta\overline{V}_1 + y_2\Delta\overline{V}_2)}\left(\frac{\partial \ln f_1}{\partial x_1}\right)_{TP} \tag{9.3.13}$$

By an analogous procedure, starting from (9.3.6), we obtain the expression for the slope of an isothermal Py saturation curve,

$$\left(\frac{\partial P}{\partial y_1}\right)_{T\sigma} = \frac{RT(y_1 - x_1)}{y_2(x_1\Delta\overline{V}_1 + x_2\Delta\overline{V}_2)}\left(\frac{\partial \ln f_1}{\partial y_1}\right)_{TP} \tag{9.3.14}$$

where $\Delta\overline{V}_i = \overline{V}_i^v - \overline{V}_i^{\ell}$. To satisfy the diffusional stability criterion (8.3.14), the mole-fraction derivatives of the fugacities in (9.3.13) and (9.3.14) must always be positive. Further, away from critical points, the differences in partial molar volumes are positive, so the lhs of both (9.3.13) and (9.3.14) have the same sign: each has the same sign as $(y_1 - x_1)$. However, near a mixture critical point for $y_1 > x_1$, it is possible for the partial molar volume of component 2 to become negative in the vapor. Then the denominator of (9.3.14) is negative while that of (9.3.13) is positive; hence, the slopes of the Px and Py curves may differ very near a mixture critical point. This can be seen near the critical point shown in Figure 9.5 at 430 K.

9.3.2 Isobaric *Txy* Diagrams for Binary VLE

Consider the same binary VLE mixtures as in § 9.3.1, but now let us hold pressure fixed and plot T against x_1 and y_1. A typical plot appears in Figure 9.6. The locus of temperatures and saturated liquid compositions defines the *bubble-T curve*, while that of temperatures and saturated vapor compositions defines the *dew-T curve*. At subcritical pressures, both curves span the entire composition range from the boiling point of pure 1 to that of pure 2. The *Txy* diagram divides into three regions: a one-phase vapor region at high temperatures, a one-phase liquid region at low temperatures, and a two-phase region at intermediate temperatures.

In Figure 9.6 the broken vertical line represents a reversible isobaric cooling from one-phase vapor at point A through the two-phase region to one-phase liquid at E. When the one-phase vapor reaches point B, it is saturated and any further cooling condenses a drop of liquid. Continued cooling from B to D increases the amount of liquid at the expense of the vapor, until at D the last bubble of vapor disappears.

At any point C on the line BD we may construct a horizontal tie line and obtain the compositions of the liquid and vapor phases. On *Txy* diagrams, tie lines are horizontal because at equilibrium both phases are at the same T. In the figure, point C lies on the 415 K tie line and the equilibrium compositions are $x_1 = 0.418$ and $y_1 = 0.774$. The cooling process in Figure 9.6 is done at a constant overall mole fraction $z_1 = 0.6$, so we can apply a lever rule to obtain the fraction of material that is vapor at point C:

$$\frac{N^v}{N^v + N^\ell} = \frac{z_1 - x_1}{y_1 - x_1} = 0.51 \tag{9.3.15}$$

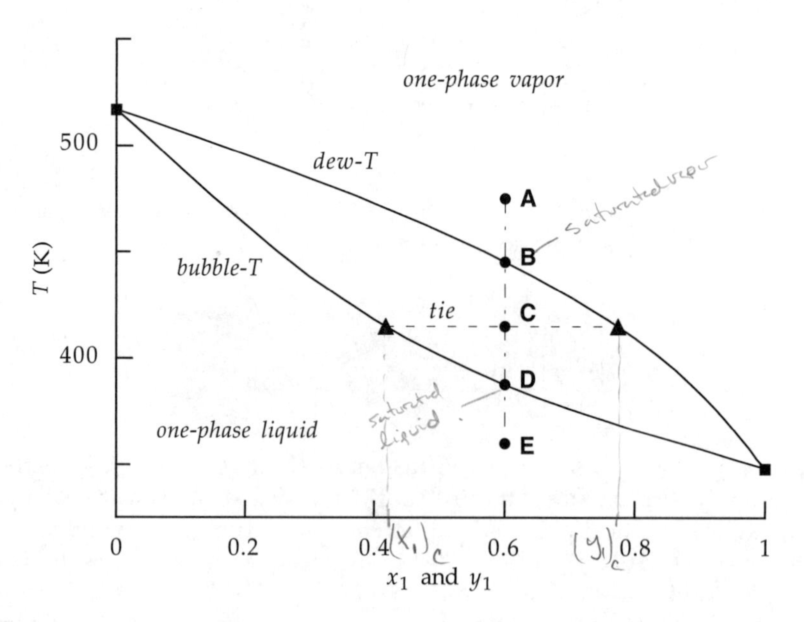

Figure 9.6 Isobaric *Txy* diagram at 30 bar for the same alkane(1)-aromatic(2) mixture shown in Figure 9.4. The broken vertical line represents a reversible isobaric cooling from one-phase vapor at A to one-phase liquid at E. The broken horizontal line is the vapor-liquid tie line at 415 K. Filled squares are pure-component boiling points at 30 bar.

Expressions for the slopes of isobaric Tx and Ty curves can be derived using a procedure analogous to that used in § 9.3.1 for Px and Py curves. The expression for Tx is

$$\left(\frac{\partial T}{\partial x_1}\right)_{P\sigma} = \frac{-RT^2(y_1 - x_1)}{x_2(y_1\Delta\bar{H}_1 + y_2\Delta\bar{H}_2)}\left(\frac{\partial \ln f_1}{\partial x_1}\right)_{TP} \tag{9.3.16}$$

where $\Delta\bar{H}_i$ is the difference in partial molar enthalpies between vapor and liquid phases. Further, we can use a triple product rule [see (9.3.18) below] to relate slopes of isobaric saturation curves [such as (9.3.13)] to slopes of isothermal ones [such as (9.3.16)]; the result is that most binary mixtures in VLE have

$$\left(\frac{\partial P}{\partial x_1}\right)_{T\sigma} \propto -\left(\frac{\partial T}{\partial x_1}\right)_{P\sigma} \quad \text{and} \quad \left(\frac{\partial P}{\partial y_1}\right)_{T\sigma} \propto -\left(\frac{\partial T}{\partial y_1}\right)_{P\sigma} \tag{9.3.17}$$

Again, we use the subscript σ to emphasize that we are considering only changes of state that preserve two-phase equilibria. Exceptions to (9.3.17) usually occur only near mixture critical points. Otherwise, particularly at low pressures, if the isothermal Px curve increases (decreases) with x_1, then the corresponding isobaric Tx curve usually decreases (increases). This relation between slopes can be seen by comparing the Pxy diagram in Figure 9.4 with the Txy diagram in Figure 9.6.

If we increase the pressure from 30 bar, as in Figure 9.7, then the two-phase region in Figure 9.6 shifts to higher temperatures, just as the pure-component boiling points

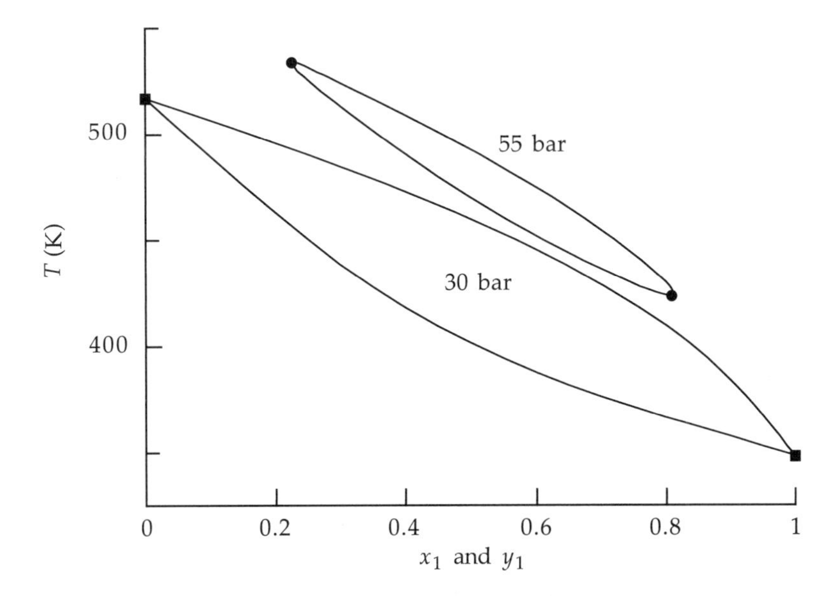

Figure 9.7 Effect of pressure on the Txy diagram for the alkane(1)-aromatic(2) mixture of Figure 9.6. At 30 bar all these mixtures are subcritical, but at 55 bar two critical points occur (dots): one at 423.55 K with z_1 = 0.8075 and another at 533.60 K with z_1 = 0.2255. Filled squares mark pure-component boiling points at 30 bar.

must increase. Figure 9.7 shows how the two-phase region moves when the pressure is changed from 30 bar to 55 bar. At 55 bar the bubble-T and dew-T curves no longer span the entire composition range: this isobar lies above the critical pressures of both pure components, so neither pure exhibits VLE. However, over a mid-range of compositions, the mixtures can exist in two phases. At 55 bar these mixtures have two critical points, one at the lowest T, the other at the highest T. In § 9.3.5 we show that critical points *always* occur at such extrema.

9.3.3 *PT* Diagrams for Binary VLE

The *Pxy* and *Txy* diagrams shown in the previous sections are $\mathcal{F}'$ diagrams: they allow us to determine the relative distribution of material between two phases in equilibrium, such as in (9.3.1) and (9.3.15). Now we describe a typical $\mathcal{F}$-diagram, which is obtained by plotting two-phase pressures against the corresponding temperatures. An example appears in Figure 9.8, which was computed from the Redlich-Kwong equation for the same mixture shown in Figures 9.4–9.7.

In Figure 9.8, the solid curves are the pure-component vapor pressure curves; each ends at its pure critical point (closed circles). For these systems, the locus of mixture critical points (dashed line) connects the pure-component critical points. This behavior is common to many binary mixtures, but it is not universal. The figure shows representative lines of constant composition at $x_1 = y_1 = 0.25$ and $x_1 = y_1 = 0.75$. For this mixture every line of constant composition passes through a mixture critical point, but not all mixtures behave this way. For some binaries, certain mixtures of constant composition have no critical points; for others, some constant-composition mixtures

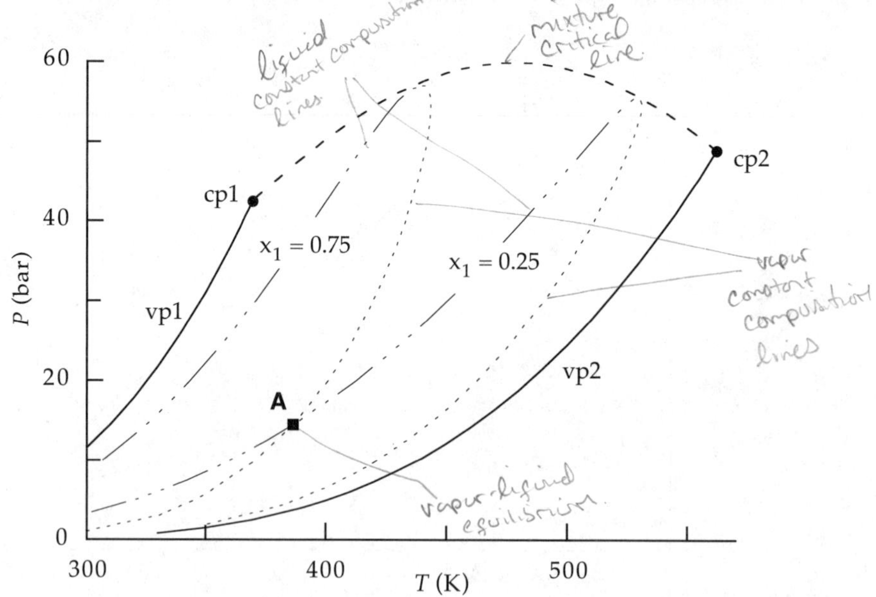

Figure 9.8 Pressure-temperature diagram for the alkane(1)-aromatic(2) mixture in Figures 9.4-9.7. Solid lines are pure vapor-pressure curves, ending at pure critical points (filled circles). Dashed line is the mixture critical line. Dash-dot lines are liquid constant-composition lines; small dashed lines are vapor constant-composition lines. Filled square at A is a vapor-liquid equilibrium point; it occurs at 14.5 bar, 386.7 K, $x_1 = 0.25$, $y_1 = 0.75$.

have multiple critical points. On *PT* diagrams, two-phase VLE points occur at intersections of x_1 and y_1 curves; an example is shown in Figure 9.8. For mixtures having continuous critical lines, such intersections occur at $T < T_{c1}$ for all x_1 and y_1 lines; however, for $T > T_{c2}$, no such intersections occur. This is consistent with Figure 9.7.

On *PT* diagrams the lines of constant composition are related to the saturation curves that appear on *Pxy* and *Txy* diagrams; specifically, these three sets of saturation curves are related through a triple product rule,

$$\left(\frac{\partial P}{\partial T}\right)_{x\sigma} = -\frac{\left(\dfrac{\partial P}{\partial x_1}\right)_{T\sigma}}{\left(\dfrac{\partial T}{\partial x_1}\right)_{P\sigma}} \tag{9.3.18}$$

The numerator in (9.3.18) is the slope of an isothermal *Px* curve (9.3.13) and the denominator is the slope of an isobaric *Tx* curve (9.3.16). Combining those expressions for numerator and denominator, we obtain

$$\left(\frac{\partial P}{\partial T}\right)_{x\sigma} = \frac{(y_1 \Delta \overline{H}_1 + y_2 \Delta \overline{H}_2)}{T(y_1 \Delta \overline{V}_1 + y_2 \Delta \overline{V}_2)} \tag{9.3.19}$$

This result is the analog of the Clapeyron equation (8.2.27) extended from pure substances to binary mixtures. It gives the slope of a saturation line of constant composition plotted on a *PT* diagram. The differences in partial molar enthalpies and volumes in (9.3.19) are usually positive, so the slope given by (9.3.19) is usually positive (see Figure 9.8). However, negative values of those slopes are observed for some mixtures at states near mixture critical points; these are usually caused by negative partial molar volumes of the heavier component.

9.3.4 Extrema on *Pxy* and *Txy* Diagrams: Azeotropes

The phase diagrams shown in Figures 9.4–9.7 all have *T* and *P* monotonic in the compositions of both phases. Consequently, at any fixed pressure the mixture boiling points are bounded by the pure-component boiling points, and at any fixed temperature the mixture pressures are bounded by the pure-component vapor pressures. But binary mixtures can have *T* and *P* pass through extrema with composition. Consider the slope of an isothermal *Px* curve for a binary mixture in VLE,

$$\left(\frac{\partial P}{\partial x_1}\right)_{T\sigma} = \frac{RT(y_1 - x_1)}{x_2(y_1 \Delta \overline{V}_1 + y_2 \Delta \overline{V}_2)} \left(\frac{\partial \ln f_1}{\partial x_1}\right)_{TP} \tag{9.3.13}$$

This derivative may become zero, hence extrema can occur, in either of two ways:

(a) If the compositions of the two phases become equal ($y_1 = x_1$), but the composition derivatives of the fugacity remain positive, so the phases remain stable, then the extrema are called *homogeneous* azeotropes.

(b) If the compositions become equal, and if in addition the composition derivatives of the fugacity also vanish, then the resulting extrema are mixture critical points. These are discussed in § 9.3.5.

A mixture VLE state that satisfies the criterion (a) cited above is called a homogeneous azeotrope, a word formed from *a* (without) + *zeo* (Greek for boil) + *tropos* (Greek for turning or changing). That is, a homogeneous azeotrope boils without changing its composition ($y_1 = x_1$), and according to (9.3.13) and (9.3.16), when $y_1 = x_1$, extrema occur on both the binary *Pxy* diagram and the *Txy* diagram. The extrema may be maxima or minima, but maxima in *Pxy* are more common; see Figure 9.9. Note that the extremum in the P-x_1 curve coincides with that in the P-y_1 curve. Further, (9.3.17)

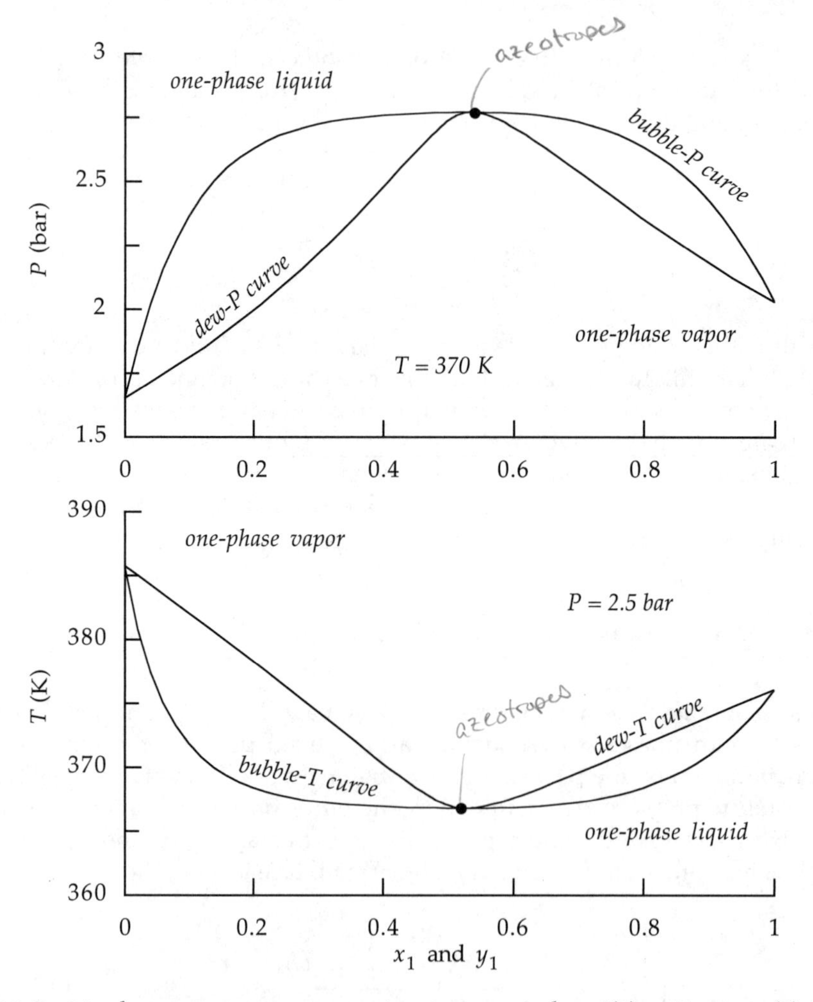

Figure 9.9 Positive homogeneous azeotropes in mixtures of ethanol(1) and benzene(2). The *Pxy* diagram is at 370 K, and the *Txy* diagram is at 2.5 bar. Filled circles locate the azeotropes. These diagrams were calculated using the Margules equations for activity coefficients, with parameters taken from Appendix E. (Computations discussed in Chapter 11.)

implies that when a maximum (minimum) occurs on an isothermal Pxy diagram, then a minimum (maximum) occurs on the corresponding isobaric Txy diagram.

Mixtures that have a maximum in the Px curve exhibit positive deviations from ideal-solution behavior; that is, the activity coefficients are greater than unity. Such mixtures are called positive deviants and their azeotropes are called *positive azeotropes*. Since such mixtures have minima in their Tx curves, those same azeotropes are also called *minimum boiling-point* azeotropes. Positive deviants usually occur when attractive intermolecular forces between molecules of the same species are stronger than those between molecules of different species.

A few binary mixtures exhibit *negative azeotropes*; these are caused by negative deviations from ideality as reflected in activity coefficients that are less than unity. An example is the azeotrope formed by mixtures of acetone and chloroform, shown in Figure 9.10. Now both the Px and Py curves pass through minima at the same composition, while the corresponding Tx and Ty curves pass through maxima. Hence nega-

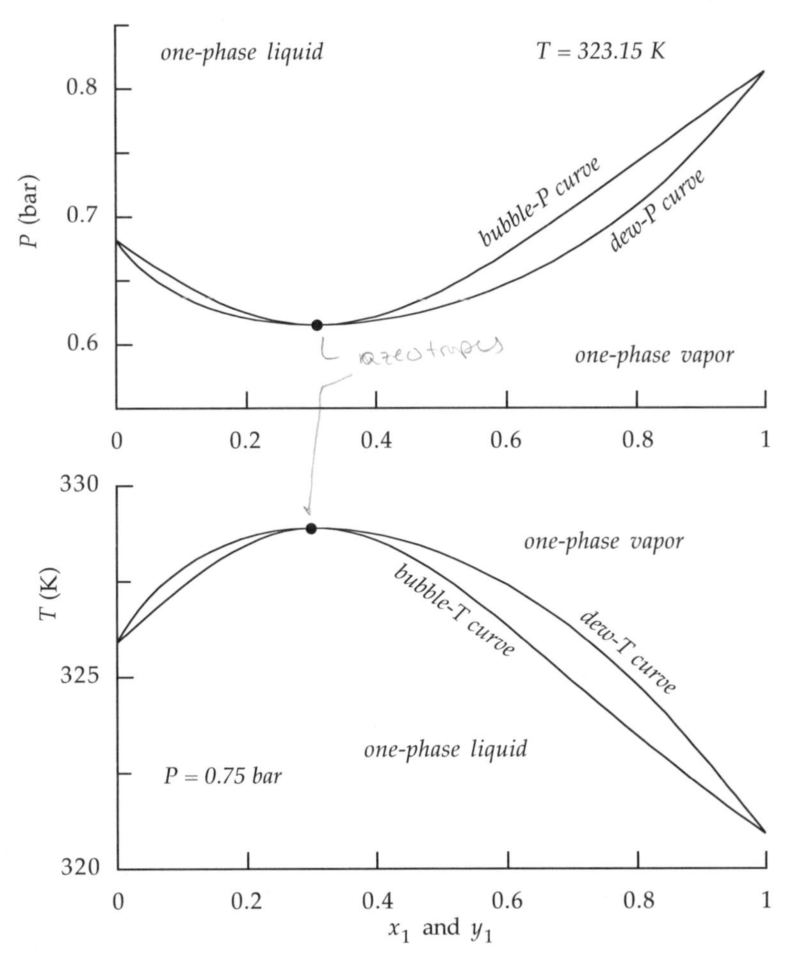

Figure 9.10 Negative homogeneous azeotropes (dots) in mixtures of acetone(1) and chloroform(2). The Pxy diagram is at 50°C; the Txy diagram is at 0.75 bar. Computed using the Margules model for activity coefficients, with parameter values from Walas [5]. (Computations discussed in Chapter 11.)

tive azeotropes are also called *maximum boiling-point* azeotropes. Negative deviants usually occur when molecules of different species attract one another more strongly than molecules of the same species. In addition, negative deviants may also occur when large disparities exist in the sizes or conformational structures of molecules, such as may be found in many polymer solutions.

When azeotropes form, mixtures at nearby states may not be uniquely identified by an $\mathcal{F}$-specification; e.g., for a binary in VLE, the Gibbs phase rule (9.1.14) gives

$$\mathcal{F} = C + 2 - \mathcal{P} = 2 + 2 - 2 = 2 \tag{9.3.20}$$

So we expect that setting values for T and P would allow us to determine the compositions of vapor and liquid phases. However, near an azeotrope, the saturation curves pass through extrema, causing some tie lines to separate into two branches: one on either side of the azeotrope with both branches having the same T and P. Therefore, the phase-equilibrium equations have two solutions for the compositions of both vapor and liquid. The Gibbs phase rule cannot account for nonmonotonicity of properties. To distinguish between the two pairs of roots, we need to set an additional property; typically, the overall composition z_1. That is, we need an $\mathcal{F}'$ specification to uniquely identify states near azeotropes.

Azeotropes shift to other pressures and compositions in response to changes in temperature (see Figure 9.11), but we can devise a simple relation between the azeotropic T and P. We start from (9.3.4), which applies to any binary VLE situation,

$$RT\sum_i (y_i - x_i) d\ln f_i = -\frac{\Delta h}{T} dT + \Delta v\, dP \tag{9.3.4}$$

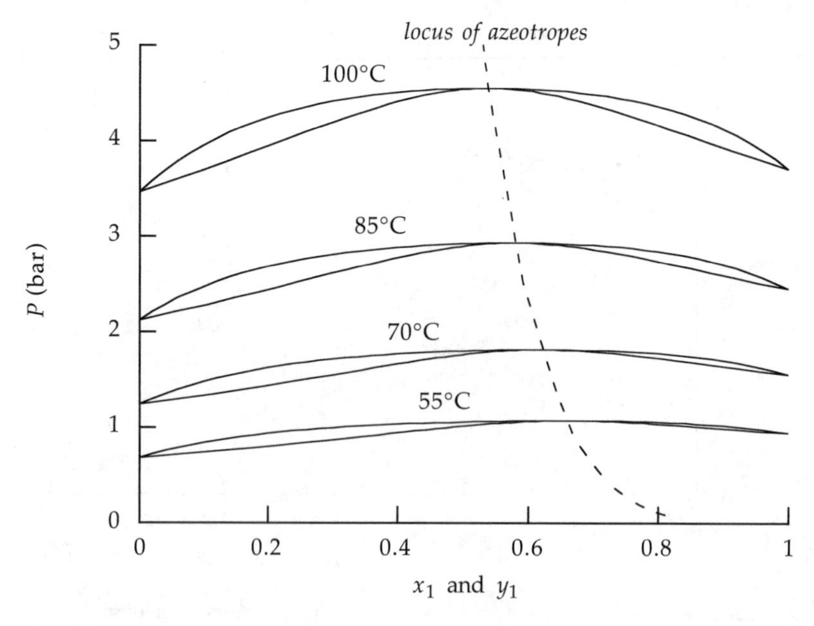

Figure 9.11 Effect of temperature on positive azeotropes formed in mixtures of methyl acetate(1) and methanol(2). Calculated using the Wilson model for activity coefficients, with parameters from Appendix E.

But at a homogeneous azeotrope $x_i = y_i$, so the lhs of (9.3.4) vanishes, and we have

$$\frac{dP^{Az}}{dT} = \frac{\Delta h}{T\Delta v} \tag{9.3.21}$$

This is a remarkable result, for it is merely the Clapeyron equation (8.2.27) extended from pure substances to azeotropic mixtures. The derivative on the lhs represents the slope of the locus of azeotropes on a PT diagram. We can use (9.3.21) as a basis for correlating azeotropic temperatures and pressures, just as we used it in § 8.2.6 for correlating pure-component vapor pressures. We obtain the same generalized form of the Clausius-Clapeyron equation,

$$\ln P^{Az} = A - \frac{B}{T} \tag{9.3.22}$$

Here P^{Az} is the azeotropic pressure at absolute temperature T, while A and B are parameters whose values are obtained by fits to azeotropic data. This correlation works well for both positive and negative azeotropes, as shown in Figure 9.12. The principal drawback to (9.3.22) is that the azeotropic compositions remain implicit; to find those compositions, we must solve the phase-equilibrium problem.

Since vapor and liquid have the same composition at a homogeneous azeotrope, the occurrence of an azeotrope prevents a separation by simple distillation. Once an azeotrope forms on a stage or plate of a distillation column, no further separation occurs and the azeotropic mixture becomes one of the product streams. For example, simple distillation cannot be used to extract pure-grain alcohol from ethanol-water mixtures because an azeotrope forms at atmospheric pressure.

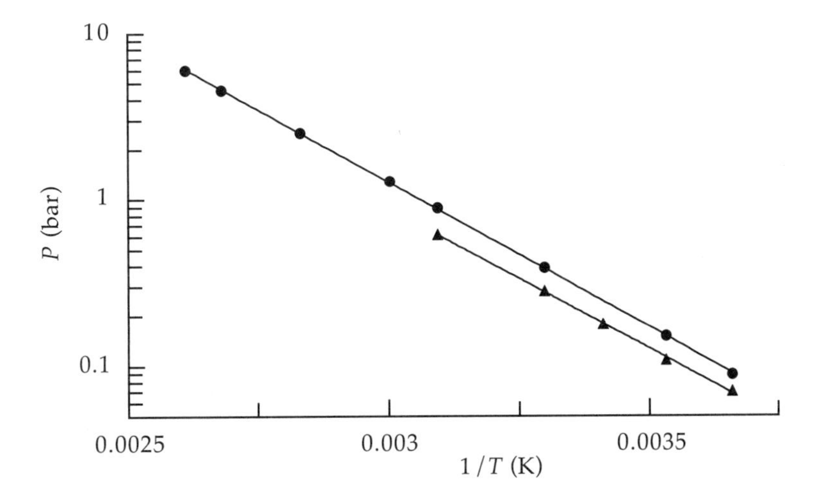

Figure 9.12 Tests of the Clausius-Clapeyron equation (9.3.22) for correlating azeotropic pressures and temperatures. Circles are positive azeotropes formed in mixtures of methyl acetate and methanol (see Figure 9.11). Triangles are negative azeotropes that occur in mixtures of acetone and chloroform (see Figure 9.10). Lines are least-squares fits.

However, we have seen that azeotropic compositions change in response to changes in temperature and pressure (Figure 9.11), and such changes can sometimes be exploited to avoid azeotropes in distillation columns; for example, we might work around azeotropes by using two distillation columns operating at different pressures. Alternatively, some azeotropic binaries can be separated by performing the distillation in the presence of a third component; this is called *extractive* distillation. But these strategies have limited flexibility, and often azeotropic mixtures must be separated by some method other than distillation; alternatives include solvent extraction, membrane separation, and combinations of different processes.

Homogeneous azeotropes occur in a great many binary mixtures, and tables of azeotropic temperatures, pressures, and compositions can be found in the compilations by Horsley [6] and by Gmehling et al. [7]. Such azeotropes occur when one vapor phase is in equilibrium with one liquid phase. In addition, extrema in isothermal *Pxy* and isobaric *Txy* diagrams occur in some three-phase VLLE situations for binary mixtures. These are called *heterogeneous* azeotropes. But at heterogeneous azeotropes the composition of the vapor is rarely the same as that of either liquid; these situations are discussed in § 9.3.7.

9.3.5 Extrema on *Pxy* and *Txy* Diagrams: Mixture Critical Points

In (8.4.2) and (8.4.3) the conditions for a binary critical point are given in terms of the change of Gibbs energy on mixing. Those conditions can also be expressed in terms of the fugacity of either component,

$$\left(\frac{\partial \ln f_1}{\partial x_1}\right)_{TP} = 0 \qquad \textit{mixture critical point} \qquad (9.3.23)$$

and

$$\left(\frac{\partial^2 \ln f_1}{\partial x_1^2}\right)_{TP} = 0 \qquad \textit{mixture critical point} \qquad (9.3.24)$$

These mean that an isothermal-isobaric plot of f_1 vs. x_1 passes through a point of inflection at the critical point, as was illustrated in Figure 8.13. Points that satisfy only (9.3.23) locate the spinodal, and when the spinodal coincides with the vapor-liquid saturation curve, then both (9.3.23) and (9.3.24) are satisfied and a vapor-liquid critical point occurs.

If we substitute (9.3.23) into (9.3.13), we find that critical points occur at extrema on isothermal *Px* and *Py* plots. Likewise if we substitute (9.3.23) into (9.3.16), we find that critical points occur at extrema on isobaric *Tx* and *Ty* plots. However, the converses of these statements are not true: extrema on such plots are not necessarily critical points; we have already seen that they could be azeotropes. Further, those extrema mean that the numerator and denominator in the triple product rule (9.3.18) are both zero; however, that ratio need not be zero, so on *PT* diagrams, critical points rarely occur at extrema of constant-composition lines.

9.3.6 Binary Liquid-Liquid Equilibria

In § 8.4.5 we described the stability conditions that, when violated, can cause a one-phase liquid mixture to separate into two liquid phases. We also showed in Figure 8.20 an isobaric, liquid-liquid, *Txx* diagram on which one-phase states divide into stable, metastable, and unstable states. Liquid-liquid separations occur in nonideal mixtures that have strong positive deviations from ideal-solution behavior; in such mixtures the activity coefficients become much greater than unity. This occurs when attractive forces between molecules of the same species are stronger than those between molecules of different species. Liquid-liquid separations have never been observed in mixtures that are negative deviants over the entire composition range.

The *Txx* diagram shown in Figure 8.20 is typical of most binary liquid-liquid systems: the two-phase curve passes through a maximum in temperature. The maximum is called a consolute point (also known as a critical mixing point or a critical solution point), and since T is a maximum, the mixture is said to have an *upper critical solution temperature* (UCST). A particular example is phenol and water, shown in Figure 9.13. At $T > T_c$, molecular motions are sufficient to counteract the intermolecular forces that cause separation.

A few binaries have lower critical solution temperatures (LCST), in which the mixture is a one-phase liquid at low temperatures, but splits into two liquid phases at high temperatures. Solutions forming LCSTs include mixtures of a light hydrocarbon and a substance composed of small polar molecules (such as carbon dioxide or ethyl ether), mixtures of a short-chain hydrocarbon and a long-chain hydrocarbon, mixtures of water with a glycol ether or an organic base or a surfactant, and mixtures of a polymer with a hydrocarbon. An example is presented on the right in Figure 9.13.

In many mixtures having LCSTs, relatively strong attractive forces act between molecules of different species as well as between molecules of the same species; often such forces are caused by hydrogen bonding. At low temperatures $T < T_c$ attractions between unlike molecules dominate and prevent a liquid-liquid split. But the strength

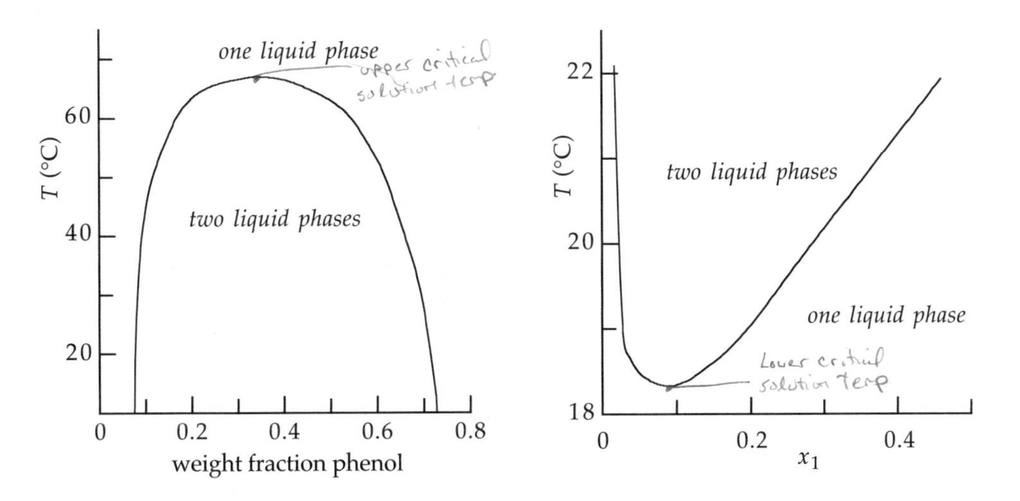

Figure 9.13 *Left*: Mixtures of phenol (C_6H_6O) and water have a UCST near 67°C and 0.35 weight fraction phenol [8–11]. *Right*: Mixtures of triethylamine(1) ($C_6H_{15}N$) and water(2) have an LCST near 18.3°C and $x_1 \approx 0.095$ [12].

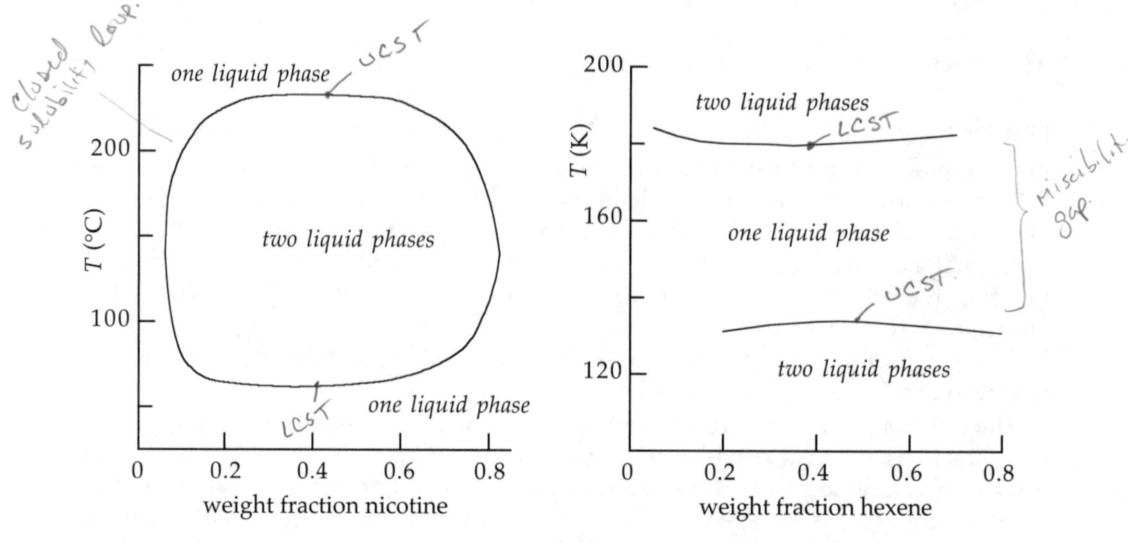

Figure 9.14 Examples of binary mixtures that have both a UCST and an LCST. *Left*: Mixtures of nicotine ($C_{10}H_{14}N_2$) and water have a closed solubility loop, with UCST = 233°C and LCST = 61.5°C [13]. *Right*: Mixtures of 1-hexene (C_6H_{12}) and methane have a miscibility gap, with UCST = 133.8 K and LCST = 179.6 K [14]. Pure hexene solidifies at 133.3 K, so the UCST occurs just above the melting curve of the mixtures.

of forces such as hydrogen bonding decrease rapidly as temperature increases, and if the attractions between unlike molecules are weakened more than those between like molecules, then a phase separation can occur. For mixtures composed of components of very different molecular sizes, the entropy increase on mixing, which prevents a phase split at low temperatures, is diminished, and a phase split can occur, if energetic effects are large enough.

A few binaries have both a UCST and an LCST, and these divide into two classes. Those having UCST > LCST are said to exhibit a *closed solubility loop*; an example is nicotine and water, shown in Figure 9.14. Others have UCST < LCST and are said to exhibit a *miscibility gap* (also shown in Figure 9.14); examples include mixtures of methane with 1-hexene and of benzene with polyisobutene. (Some mixtures of sulfur with a hydrocarbon (such as sulfur + benzene) also have miscibility gaps, but in these mixtures the gap probably occurs because the molecular structure of sulfur changes with temperature [15].) A closed loop would be observed for more binaries except that some other phase transition intervenes as T is changed. For example, increasing T may cause vaporization before a UCST can appear; this happens in mixtures of water with 3-ethyl-4-methyl pyridine. Similarly, decreasing T may cause freezing before an LCST can occur; this happens in mixtures of water and methyl(ethyl)ketone wherein solidification prevents formation of an LCST at 1 atm. [5]. Over 6000 critical solution points have been tabulated in a book by Francis [16].

9.3.7 Vapor-Liquid-Liquid Equilibria in Binary Mixtures

$\mathcal{F}'$-phase diagrams for binary VLLE situations combine VLE diagrams from § 9.3.2 with LLE diagrams from § 9.3.6. This is illustrated in Figure 9.15. At the high pressure P_1 of Figure 9.15, three-phase VLLE does not occur. Instead, the binary may exist in any of four conditions: (i) a single-phase vapor at very high T, (ii) two-phase VLE at

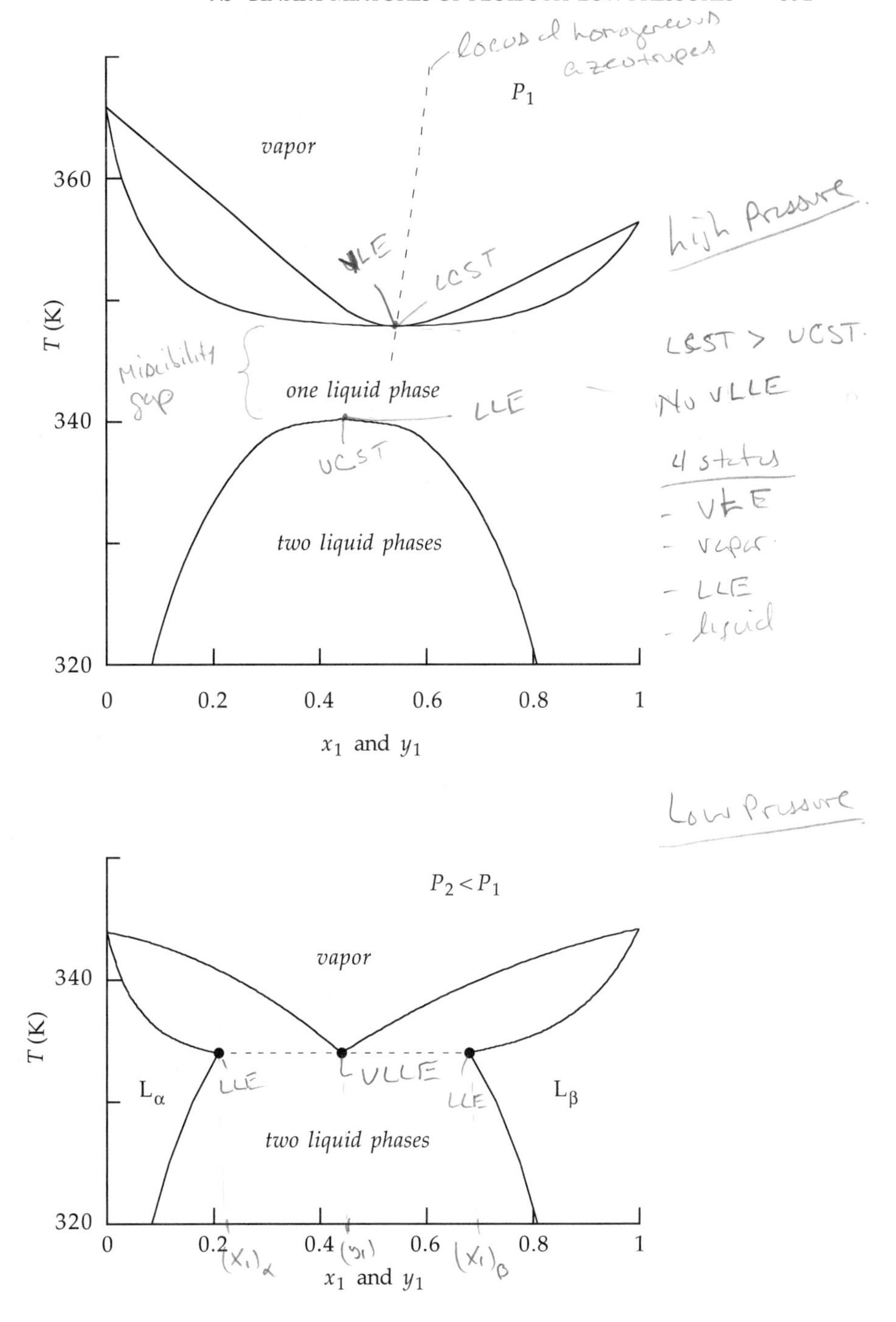

Figure 9.15 Effect of pressure on *Txy* diagram for a binary mixture that exhibits vapor-liquid-liquid equilibrium via a heterogeneous azeotrope. At high pressures (*top*) the VLE and LLE regions are separated by a one-phase liquid region, and no VLLE occurs. Broken line at top is locus of homogeneous azeotropes. But at low pressures (*bottom*) the VLE and LLE regions intersect along an isotherm (broken horizontal line) at which the three phases coexist. Filled circles give compositions of the three phases in equilibrium; center circle gives composition of the vapor. Note that the change in pressure has little effect on the LLE envelope.

high temperatures, (iii) a single liquid phase at moderate T, or (iv) two-phase LLE at low temperatures. Note that the mixture forms a positive azeotrope; such azeotropes are common in mixtures that undergo liquid-liquid phase splits. (Very few binaries are known to exhibit both negative azeotropes and LLE, but when they do, the composition of the azeotrope is well away from the compositions of the liquid immiscibility; an example is H_2O + HCl, in which the negative azeotrope occurs at low HCl concentrations while LLE occurs at high HCl concentrations.)

If we decrease the pressure in Figure 9.15, then the VLE region moves to lower temperatures. The broken line in the top of Figure 9.15 shows how the azeotropic temperature and composition respond to changes in P. The LLE region may also move to higher T, but such movement is usually slight because liquids are little affected by moderate pressure changes. Nevertheless, the movement of the VLE region reduces the area of the miscibility gap that lies between the LLE and VLE regions.

If the pressure is decreased to P_2 in Figure 9.15, then the VLE and LLE regions intersect along an isotherm. That is the temperature of the three-phase VLLE situation, and the isotherm joins the compositions of the three equilibrium phases. For this situation, the Gibbs phase rule (9.1.14) gives

$$\mathcal{F} = C + 2 - \mathcal{P} = 2 + 2 - 3 = 1 \qquad (9.3.25)$$

This means that at any given pressure, such as P_2 in Figure 9.15, three-phase VLLE occurs at only one temperature. That temperature identifies a *heterogeneous* azeotrope: at the pressure P_2 the azeotropic temperature is the lowest at which vapor can exist in equilibrium with liquid. At this T, boiling of the two-phase liquid will produce a vapor of fixed composition, regardless of the overall composition of the system.

In the bottom panel of Figure 9.15, the region labeled L_α marks one-phase liquid that is rich in component 2; similarly, region L_β is occupied by one-phase liquid rich in 1. Along the horizontal three-phase isotherm, the points give the compositions of the three phases: the left point gives x_1 for liquid α, the right point gives x_1 for liquid β, and the center point gives y_1 for the vapor. At temperatures above the three-phase line but below the pure boiling points, the mixture can be in one of five situations: (i) one-phase liquid α, (ii) VLE between liquid α and vapor, (iii) one-phase vapor, (iv) VLE between liquid β and vapor, or (v) one-phase liquid β. At temperatures below the three-phase line, the mixture can be in one of three situations: (i) one-phase liquid α, (ii) two-phase LLE, or (iii) one-phase liquid β.

In Figure 9.16 we show the isothermal Pxy diagram that corresponds to the isobaric Txy diagram in Figure 9.15. Because of the triple product rule (9.3.18), the Pxy diagram is qualitatively inverted compared to the isobaric Txy diagram. At the one T in Figure 9.16, the VLLE situation occurs at one pressure. At pressures above the three-phase pressure, the mixture can exist in one of three situations: (i) one-phase liquid α, (ii) two-phase LLE, or (iii) one-phase liquid β. At pressures below that of VLLE but greater than the pure vapor pressures, the mixture can exist in one of five situations: (i) one-phase liquid α, (ii) VLE between vapor and liquid α, (iii) one-phase vapor, (iv) VLE between vapor and liquid β, or (v) one-phase liquid β.

The behavior shown in Figures 9.15 and 9.16 is common to many mixtures of immiscible liquids, but several variations are also possible. Here are some of the alternatives: (i) Most immiscible liquids have UCSTs, as in Figure 9.15, but mixtures with

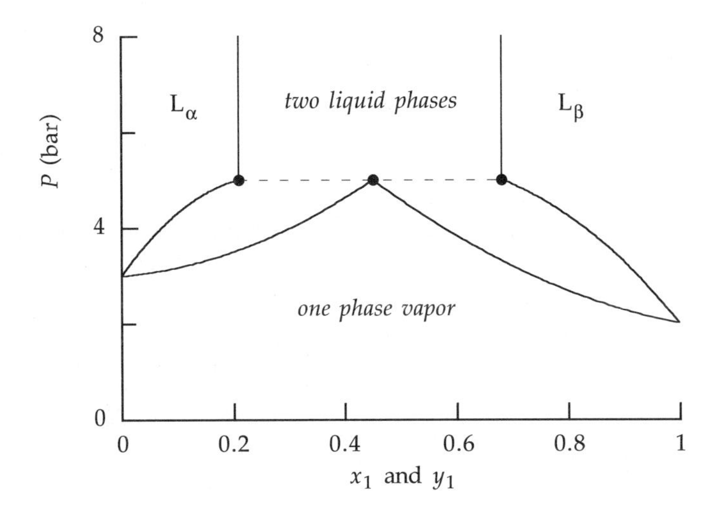

Figure 9.16 Schematic isothermal *Pxy* diagram for the binary shown in Figure 9.15. Broken horizontal line is the three-phase tie line for VLLE (heterogeneous azeotrope). At this temperature, LLE occurs at high pressures and VLE occurs at low pressures.

LCSTs can also exhibit VLLE; examples include water + 2-butanol and water + 2-butanone. In such cases, VLLE prevents formation of a closed solubility loop. (ii) Many immiscible liquids form homogeneous azeotropes at high pressures, as in Figure 9.15, but some do not. Those without azeotropes include CO_2 + long-chain alkanes, such as n-octane and n-decane. (iii) Often VLLE occurs at heterogeneous azeotropes, as in Figure 9.15, and then the vapor-phase composition lies between the compositions of the two liquid phases. However, VLLE also occurs in some mixtures in which the vapor-phase composition does not lie between the compositions of the liquid phases. Examples of the latter include ammonia + toluene and water + phenol.

Can binary mixtures coexist as three *liquid* phases? Such behavior is not prevented by the phase rule, but very unusual intermolecular forces would be required. As far as we are aware, binary LLLE occurs only when the components react to form either a physical or a chemical compound, so the mixture effectively becomes a ternary. An example is n-butyl chloral hydrate in water [17, 18].

9.4 BINARY MIXTURES CONTAINING SOLIDS

Simple fluid-solid phase behavior is analogous to the liquid-vapor behavior discussed in § 9.3. Since properties of solid phases are largely unaffected by changes in pressure, we need consider only temperature-composition diagrams. Then simple solid-phase equilibria can be described using the *Txy* diagrams from § 9.3.2, relabeling vapor regions as liquid regions and relabeling liquid regions as solid regions.

But not all solid-phase equilibria are simple; complications may occur due to (i) partial immiscibility of solid phases, (ii) the presence of more than one crystalline structure (polymorphism) of a solid phase, and (iii) reaction of pure compounds to form solid intermolecular compounds. When any of these occur, the phase diagrams are complex; but fortunately, those diagrams can be generally understood as superpositions of simpler diagrams.

9.4.1 Completely Miscible Mixtures

Solid components are rarely miscible at all concentrations. Those that are completely miscible have basic structural units (atoms, molecules, or ions) that are similar. Examples include binary alloys (gold + silver) in which the atomic diameters of the two metals are nearly the same, mixtures of similar salts (NaCl + AgCl), and mixtures of large organic molecules having similar structures, such as β-methylnaphthalene + β-chloronaphthalene. In such cases the temperature-composition diagrams are similar to the *Txy* diagram in Figure 9.6. The two-phase lines change continuously and monotonically from the melting point of one pure solid to that of the other. The two-phase liquid-solid region separates one-phase liquid mixtures, at high temperatures, from one-phase solid mixtures, at low temperatures. The upper two-phase curve, the *liquidus*, is analogous to the dew-T curve and gives compositions of saturated liquids. The lower curve, the *solidus*, is analogous to the bubble-T curve and gives compositions of saturated solids. Tie lines are isotherms and are therefore horizontal, just like those in Figure 9.6.

Extrema can occur on *Txx* diagrams that describe liquid-solid equilibria for completely miscible solids; these extrema are analogous to homogeneous azeotropes and are called *solutropes*. Such liquid-solid *Txx* diagrams are analogous to the vapor-liquid *Txy* diagrams shown in Figures 9.9 and 9.10. For example, mixtures of d-carvoxime and 1-carvoxime have a *maximum melting-point solutrope*, while binary mixtures of p-dichlorobenzene and p-chloroiodobenzene have a *minimum melting-point solutrope*.

In addition, miscible liquid-solid systems can display phase behavior more complex than vapor-liquid systems. For example, mixtures of carbon tetrachloride and cyclohexanone form a compound from one molecule of each pure; this compound (x_1 = 0.5) melts at –39.6°C. Below this temperature, the compound exhibits two minimum melting temperatures; so the melting curve for this binary has three extrema, two minima and a maximum, and all three lie below the melting points of the pure components. Compound formation in a solid phase can also cause constant-composition melting without an extremum in temperature. This occurs in mixtures of bromine and iodine. At 40°C the compound IBr melts at constant composition, although this temperature lies between the melting points of pure iodine and pure bromine. Phase diagrams for these kinds of solid systems can be found in the book by Walas [5].

9.4.2 Immiscible Solids with Miscible Liquids

Solid solutions are usually composed of dissimilar components, with the result that most solid mixtures exhibit partial or (essentially) complete immiscibility. Partial immiscibility is often observed in metallic alloys [19], while complete immiscibility is common in solid mixtures of organic substances. We first consider partially immiscible components. In most partially immiscible solids, as temperature increases the concentration range for solid-solid equilibria decreases. If the melting point is sufficiently high, then the solid-solid equilibrium curve will close at a maximum temperature and the SSE diagram will be analogous to an LLE diagram with a UCST; see Figure 9.17(a).

More often, however, an increase in temperature produces melting before the SSE situation disappears. Then the resulting phase diagram is a fusion of the LSE curves and the SSE curves, as shown in Figure 9.17(b). The isotherm at which the LSE and SSE curves join, T^* in Figure 9.17(b), is a three-phase line representing a liquid mixture in equilibrium with two solid-phase mixtures. The minimum in the liquidus at T^* is analogous to a heterogeneous azeotrope and is called a *eutectic*; T^* is called the *eutectic temperature*. As an example, eutectics occur in mixtures of lead + tin, commonly used as solder to join wires and metal pipes. If the volume of a eutectic is allowed to expand, forming a vapor phase, then we can have two components in four-phase VLLS equilibrium. At such a point the phase rule (9.1.14) gives $\mathcal{F} = 0$, and the state is called a *quadruple* point, in analogy to the triple point of pure-component equilibria.

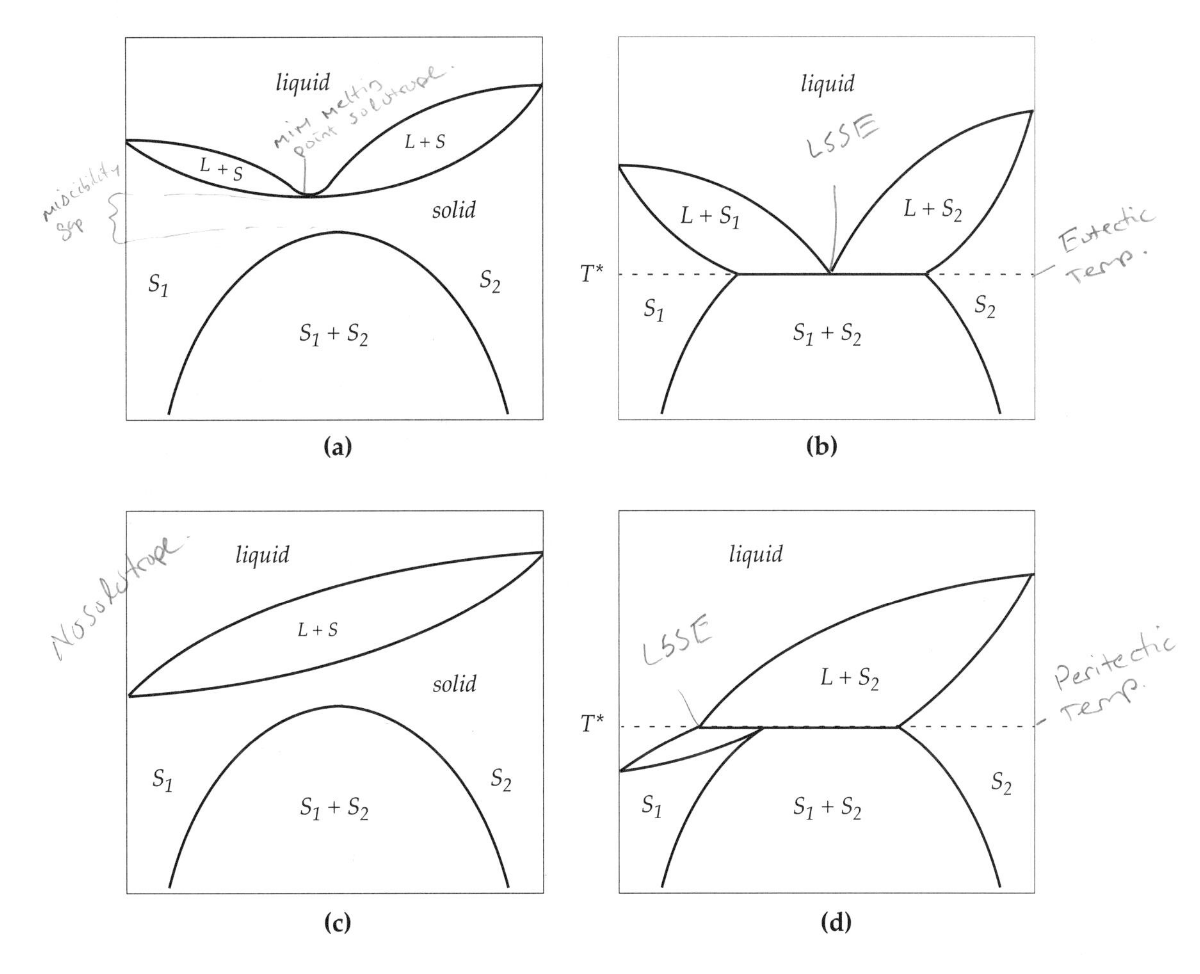

Figure 9.17 Typical *Txx* diagrams for binary LSE situations in which two solid phases occur. (a) A binary alloy in which a miscibility gap separates the region of two-phase SSE from that of two-phase LSE. The LSE situation exhibits a minimum melting-point solutrope. (b) Another alloy in which no miscibility gap occurs; instead, the LSE and SSE regions join at the eutectic temperature T^*, which represents a three-phase LSSE condition. At T^* the composition of the liquid phase lies between those of the two solid phases. (c) A binary alloy in which a miscibility gap exists, but without a solutrope. (d) An alloy in which the LSE and SSE regions join at the peritectic temperature T^*, where three-phase LSSE occurs. At T^* the mole fraction of the liquid phase is less than those of the two solid phases. Note that panel (b) can be formed by combining the LSE and SSE regions in panel (a), and panel (d) can be formed by combining the LSE and SSE regions in panel (c).

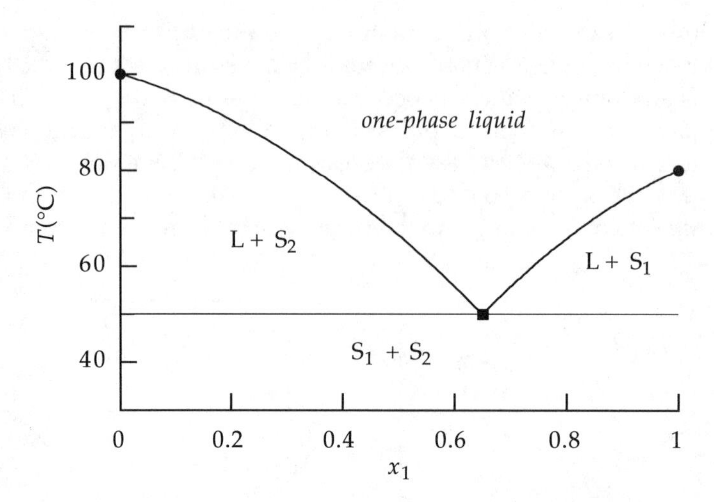

Figure 9.18 Schematic *Txx* diagram for a binary mixture that is completely immiscible in the solid phase. Filled circles are pure solid melting points; filled square is the eutectic. Note that the melting point of the eutectic is lower than that of both pure solids.

Eutectics occur when a discontinuity in the liquidus corresponds to a minimum temperature. In addition, three-phase LSSE can occur at liquidus discontinuities that are not minima in temperature; these are called *peritectics*. An example is shown in Figure 9.17(d), which can be interpreted as a superposition of the LSE and SSE regions of the diagram in Figure 9.17(c). A few binaries are known to exhibit both eutectics and peritectics; examples include mixtures of methylcyclopentane and cyclohexane, which have a eutectic at –144.7°C and a peritectic at –100°C.

Finally, we consider substances that are almost completely immiscible as solids. In these cases the mixtures often contain an organic compound with water; the resulting phase diagrams are relatively simple. Usually adding a pure component lowers the melting point, causing a eutectic on the liquidus, as in Figure 9.18; but now the eutectic liquid is in equilibrium with essentially pure solids. These eutectic solids are usually agglomerates of fine crystals, though still separate phases. Since eutectic liquids freeze at temperatures below those of either pure component, they are routinely used when it is necessary to maintain fluid phases at subfreezing temperatures. An example is ethylene glycol + water, used as a coolant for internal combustion engines.

9.4.3 Experimental Determination of Liquidus and Solidus

Consider a one-phase substance being cooled or heated isobarically at a constant rate; assuming the heat capacity is constant, a plot of temperature versus time will be linear. But when a phase transition is encountered, the heat effect changes. This change occurs for two reasons: one is the latent heat that accompanies the transition and the other is a change in the heat capacity that occurs as the old phase is transformed into the new. These changes cause a discontinuity in the slope of the temperature-time plot, thereby providing a signature for the onset of a phase change.

This procedure has been a traditional experimental method for locating the solidus and liquidus on LSE phase diagrams. To avoid effects of metastabilities, the solidus is

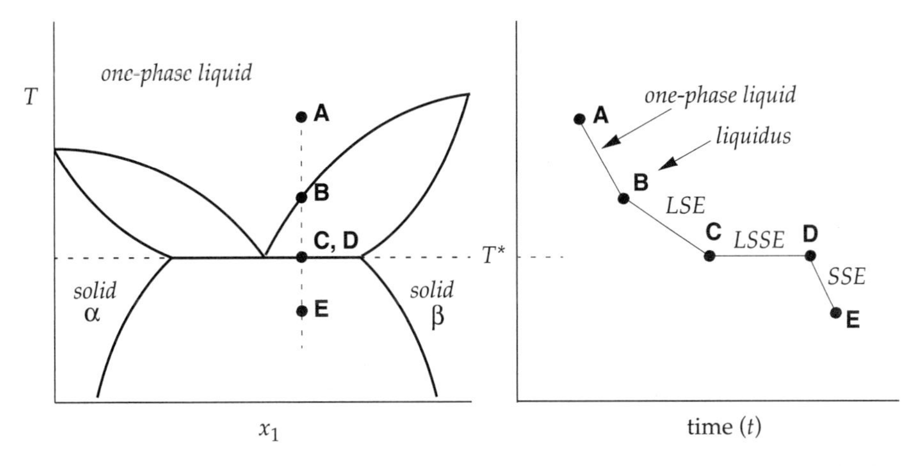

Figure 9.19 Schematic *Txx* diagram and the temperature-time diagram by which cooling curves are used to locate point B on the liquidus and to locate the eutectic temperature T^*

usually located by heating, while the liquidus is usually located by cooling. Figure 9.19 shows a schematic cooling curve for a binary alloy, such as a mixture of lead and tin. The process is performed on a mixture of known constant composition, and it is started at a high temperature T_A in the one-phase liquid region of the phase diagram. By cooling at a constant rate, the temperature falls from A to B on the vertical line on the *Txx* diagram; this path corresponds to the linear segment AB on the temperature-time (*Tt*) diagram. At point B solidification begins to occur, forming a solid phase β in equilibrium with the liquid. As cooling continues, the system passes through a sequence of two-phase equilibrium states in which the amount of solid β increases, while the amount of liquid decreases. (These changes in amounts can be determined from a lever rule.) This process corresponds to the linear path BC on the *Tt* diagram in Figure 9.19.

At the eutectic temperature T^*, the solid phase α begins to form and we can observe all three phases of liquid-solid-solid equilibrium. At this point, continued cooling removes latent heat from the remaining liquid, so more solid phases α and β form, but no further change in temperature can occur; as a result, the path CD on the *Tt* diagram is horizontal. When all liquid has disappeared, removing more heat cools the system below T^* and pushes the mixture into the two-phase solid-solid region of the phase diagram. The slope of each branch of the cooling curve is related to the heat capacities and latent heats for the corresponding phase change.

The cooling curve in Figure 9.19 is an idealized schematic in which the discontinuities in slope occur at sharply defined temperatures. In reality the transitions from one straight line to the next on the *Tt* diagram are rounded and may appear continuous. Such indistinct discontinuities may be caused by too rapid cooling, supercooling, or incomplete mixing, which produces local nonuniformities in the sample.

Supercooling can also cause the spontaneous appearance of new phases, accompanied by large latent heat effects and a temporary increase in temperature. Since liquids can be so easily supercooled, cooling curves often fail to locate the solidus reliably; cooling curves typically yield values of the solidus temperature that are too low. Therefore, once the liquid phase has completely solidified, the solidus is often determined by reversing the process and heating the solid back to the solidus [20]. To locate the entire liquidus and solidus, this procedure is repeated for many overall compositions. More examples of cooling curves are available in the book by Adamson [21].

9.4.4 Immiscible Solids and Immiscible Liquids

Many nonmetallic substances are almost completely immiscible in both the solid and the liquid phases. For such mixtures the *Txx* diagram is composed of an LLE diagram superimposed on an LSE diagram. Almost any mixture having a UCST will give such a diagram; an example is phenol and water, shown in Figure 9.20. On that diagram, at high temperatures, $T >$ UCST, the mixture is one-phase liquid. At $42.5 > T > 1.3$°C the system can exist in an LLE situation or in an LSE situation; in the latter, a phenol-rich liquid is in equilibrium with pure phenol solid. At 1.3°C the system is in three-phase equilibrium involving a water-rich liquid, a phenol-rich liquid, and pure phenol solid. At –1.3°C a eutectic forms in which the two pure solids are each in equilibrium with a nearly pure-water liquid phase.

The lesson here is that, while the phase behavior of just two components can involve complications among several solid, liquid, and gaseous phases, the resulting phase diagrams can be understood as superpositions of simple diagrams discussed in this chapter. For example, the phenol-water diagram in Figure 9.20 combines its LLE diagram from the left of Figure 9.13 with its LSE diagram, which is like Figure 9.18.

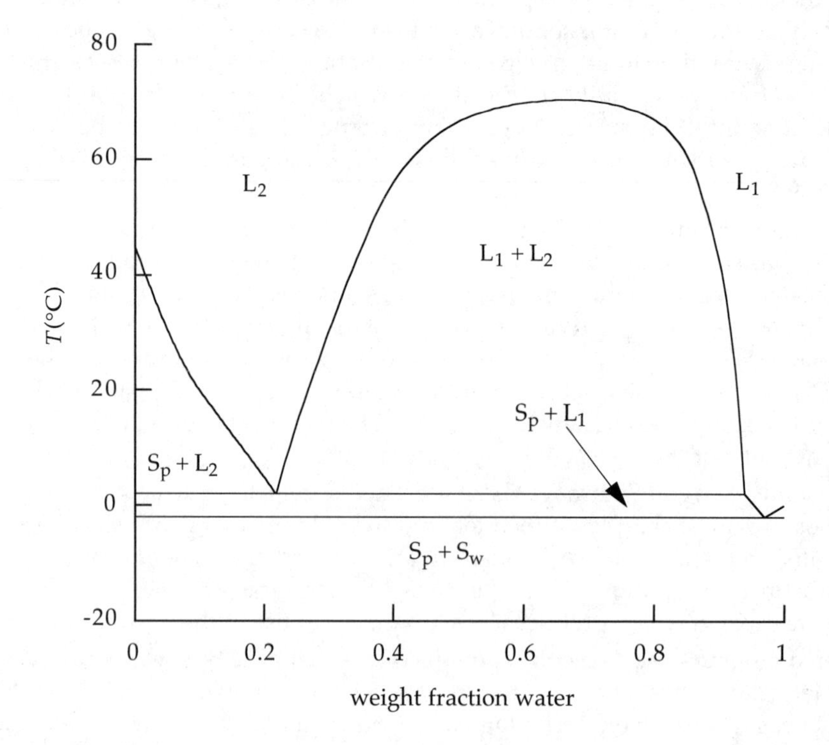

Figure 9.20 *Txx* diagram for phenol-water mixtures at 1 bar. These mixtures are partially immiscible as liquids and completely immiscible as solids. S_p = pure phenol solid; S_w = pure ice. Note the eutectic for nearly pure-water mixtures. Data from [8–11].

9.5 BINARY MIXTURES OF FLUIDS AT HIGH PRESSURES

In previous sections we discussed phase diagrams for binary mixtures at moderate pressures. When mixtures contain only condensed phases, their phase diagrams are little affected by increases in pressure. However, phase diagrams involving vapors can show substantial changes with increasing pressure. For example, mixture critical lines may not be continuous curves connecting the pure-component critical points; liquid phases may split in two, causing three-phase vapor-liquid-liquid equilibria; and high-pressure gas mixtures may themselves split, causing gas-gas equilibria. Many different kinds and combinations of fluid-phase behavior have been observed in binary mixtures, but in this section we can summarize only the commonly observed behavior. Our presentation summarizes more detailed descriptions of high-pressure phase equilibria given by Rowlinson and Swinton [22] and by Schneider [23].

9.5.1 Classification of Phase Diagrams

Binary mixtures can exhibit so many different kinds of fluid-phase behavior [23, 24] that we need a way to organize and classify them. The classification presented here is based on a scheme first suggested by Scott and van Konynenburg [25, 26]. In their original work Scott and van Konynenburg used the van der Waals equation to identify different forms for mixture critical lines. Since that work, more complicated PvT equations of state have been used to identify additional shapes for critical lines; however, many of those have not been observed experimentally.

Identification of classes can be simplified if we work from an $\mathcal{F}$-diagram, rather than an $\mathcal{F}'$-diagram, so we base the classification on features of PT diagrams. For many binary mixtures, the gas-liquid critical line on a PT diagram is a continuous curve between the pure-component critical points. This occurs for many binaries, but not for all. So, as in Figure 9.21, the Scott-van Konynenburg scheme first divides all

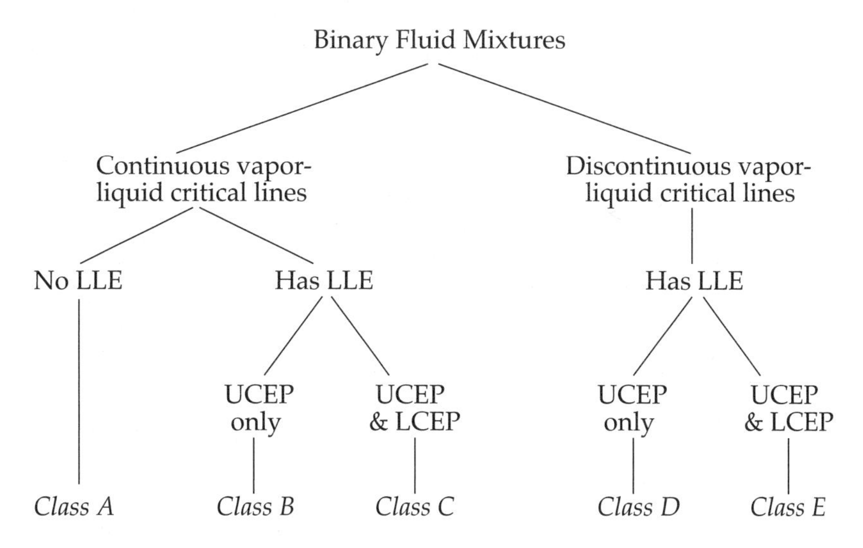

Figure 9.21 Hierarchy of binary fluid mixtures in the Scott-van Konynenburg classification, based on high-pressure fluid-phase behavior

binaries into two groups: those whose gas-liquid critical lines are continuous from one pure-component critical point to the other, and those whose critical lines are discontinuous. Then the mixtures in each group are divided into classes, depending on whether liquid-liquid phase separations occur and, if so, on features of the three-phase VLLE loci. This classification hierarchy is summarized in Figure 9.21.

A three-phase line will terminate when two of the three phases become identical; such states mark the intersection of the three-phase line with a critical line. Consider the three-phase vapor-liquid-liquid situation. If the two liquid phases become identical, then the VLLE line has intersected a locus of liquid-liquid critical points. Similarly, if the vapor phase becomes identical to one of the liquid phases, then the VLLE line has intersected a gas-liquid critical line. These intersections are called *upper critical end points* (UCEP) if they occur at a maximum temperature on the VLLE locus; they are called *lower critical end points* (LCEP) if they occur at a minimum temperature. The number and kinds of critical end points help distinguish the classes in the Scott-van Konynenburg scheme.

Representative *PT* diagrams for the five classes are sketched in Figure 9.22; that figure has been constructed with the following horizontal and vertical symmetries. The top row in the figure contains all mixtures that have continuous critical lines between the pure-component critical points; mixtures in the bottom row do not have their pure critical points connected by a mixture critical line. The second column in Figure 9.22 contains those mixtures in which the VLLE line ends only at a UCEP, while the third column is composed of those mixtures that have both an LCEP and a UCEP. We caution that each class further divides into subclasses, so the schematic *PT* diagrams in Figure 9.22 only represent some members of each class. We now briefly summarize the broad features of the five principal classes; more details can be found elsewhere [22–26].

9.5.2 Binaries with Continuous Vapor-Liquid Critical Lines

Mixtures in this first group all have gas-liquid critical points that form a continuous line between the critical points of the pure components. The three classes in this group can be distinguished by the number of critical end points (CEPs).

Class A. These binaries never exhibit LLE and therefore have no critical end points, although many form azeotropes. However, most mixtures in class A would exhibit LLE with UCEPs, except that solidification occurs at temperatures above that at which a liquid-liquid split would occur. Although the mixture critical line is continuous, the mixture critical T and P may not be bounded by the pure-component critical points. In a few class A mixtures phase splits occur at temperatures above the critical points of both pures. This could hardly be called liquid-liquid equilibrium; instead, such an immiscibility is called *gas-gas equilibrium* (GGE). Three kinds of gas-gas equilibria have been identified; that in class A is called GGE of the *third kind*.

Class B. These binaries have continuous gas-liquid critical lines and undergo liquid-liquid phase splits at low temperatures. The LLE curves have only UCSTs and the three-phase VLLE line terminates at a UCEP by intersecting the locus of UCSTs (see Figure 9.22). From the UCEP the UCST locus extends to higher pressures, but it does not intersect the gas-liquid critical line. The slope of the UCST locus may be positive or negative.

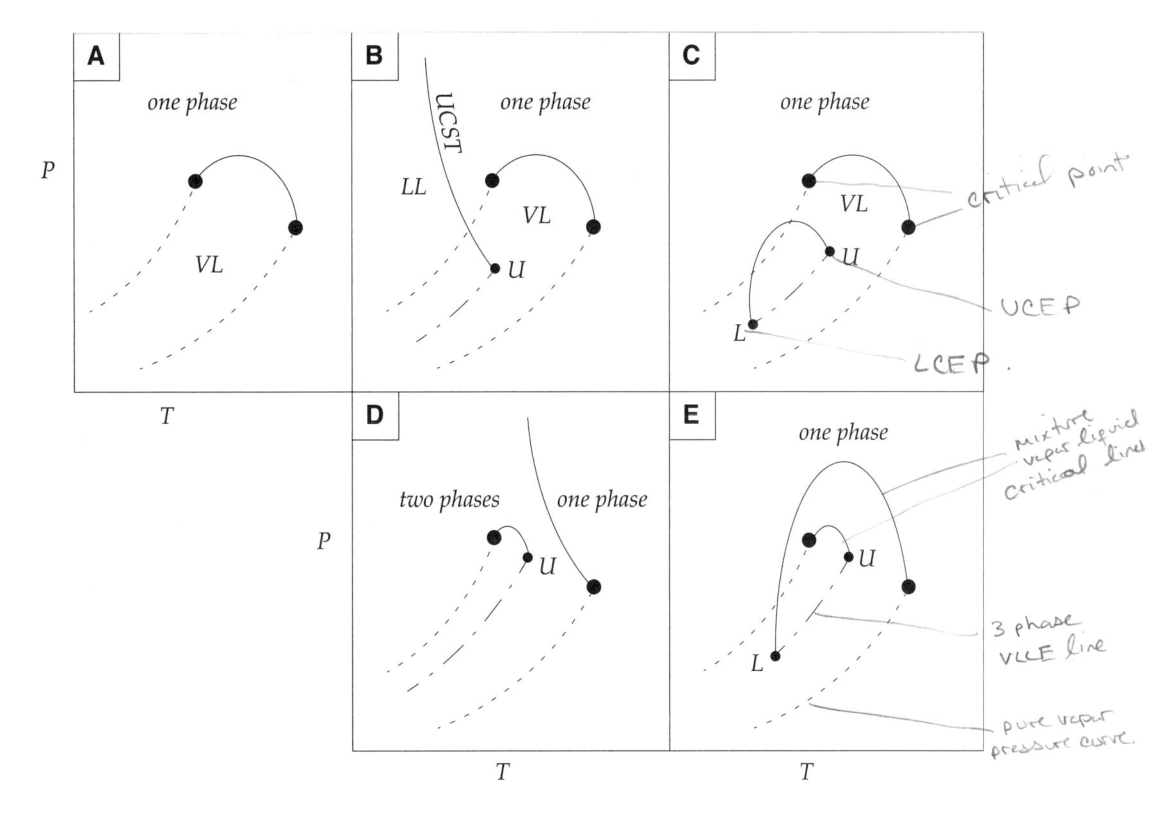

Figure 9.22 Schematic *PT* diagrams for the five major classes of binary fluid mixtures. Large dots are pure vapor-liquid critical points; dashed lines are pure vapor-pressure curves. Solid lines starting from the pure, high-pressure critical point are mixture vapor-liquid critical lines; other solid lines are mixture liquid-liquid critical lines. Small dots are upper (U) and lower (L) critical end points; dash-dot lines are three-phase VLLE lines. Diagrams shown here are representative of the classes, but they do not exhaust the possibilities.

Class C. These binaries have both a UCST and an LCST at temperatures removed from the critical temperatures of the pure components. The locus of UCSTs intersects the VLLE line at a UCEP, while that for LCSTs intersects the VLLE line at an LCEP. The loci of UCSTs and LCSTs may or may not form a continuous line of fluid-fluid critical points. Few binaries have both UCSTs and LCSTs, so few fall into class C.

9.5.3 Binaries with Discontinuous Vapor-Liquid Critical Lines

When the mixture critical line is discontinuous, it is divided into two branches by a three-phase VLLE line, as shown in Figure 9.22. The high temperature end of the VLLE line is a UCEP that connects one branch of the critical line to the critical point of the more volatile component. This branch of the critical line is usually short. Based on the behavior of the critical branch emanating from the critical point of the less volatile component, we divide these mixtures into two classes.

Class D. From the critical point of the less volatile component, the critical branch in these mixtures traces a path to high pressures without terminating at a fluid-phase critical end point; however, at sufficiently high pressures solidification may occur. In most class D mixtures the VLLE line lies between the pure-component vapor pressure curves. However, in some mixtures the VLLE line lies above the vapor-pressure curve of the more volatile component, causing heterogeneous azeotropes at low pressures. A few class-D mixtures exhibit gas-gas equilibrium of the *second kind*, in which the critical line passes through a temperature minimum, while others exhibit gas-gas equilibrium of the *first kind*, in which the critical line has a maximum in temperature. Mixtures in class D are the most highly nonideal and display the richest phase behavior of any class in the Scott-van Konynenburg scheme.

Class E. In these systems one branch of the critical line originates at the critical point of the less volatile component and circles back to low temperatures and pressures, terminating at an LCEP on the three-phase VLLE curve. At low temperatures the range of liquid-liquid immiscibility ends at an LCST. So if we start at the critical point of the less volatile component, we can experimentally trace the mixture critical loci in a continuous fashion from vapor-liquid critical states through liquid-liquid critical states. Some mixtures in this class have a second region of liquid-liquid immiscibility with another UCST at still lower temperatures: a miscibility gap.

9.5.4 Relations Among Classes

The Scott-van Konynenberg classification of binary mixtures helps us organize observed behavior. For example, when we examine an oil reservoir, we know that the number of phases observed will depend on the substances present, as well as on the temperature, pressure, and composition. As material is removed from the reservoir, the state changes, causing the phase behavior to change. Such changes may be, at least qualitatively, anticipated and understandable in terms of the classification of binaries, even though reservoir fluids are not binary mixtures.

But besides organizing observations, the classification scheme can also help us devise strategies for manipulating phase behavior to achieve engineering goals. Phase behavior reflects differences in intermolecular forces, and we have at least one crude means for changing intermolecular forces—change a component. For example, recent research has shown how variations in critical lines and shifts among classes can be achieved by mixing alkanes of different chain lengths [24]. Similar shifts can also be achieved by mixing polymers that have different molecular weights and different molecular weight distributions.

Such shifts in phase behavior are exploited in many applications. For example, in "miscible flooding," carbon dioxide, propane, or butane may be injected into reservoirs to recover underground oil. These light components tend to solubilize liquid oil into a fluid phase that more readily flows to a recovery well; however, the phase diagrams for such mixtures are complicated by the presence of water. Likewise, deasphalting and dewaxing processes in refineries introduce light components into heavy crudes to precipitate heavy components while solubilizing lighter ones.

We now offer some specific examples of how manipulating intermolecular forces can cause a change from one class to another. Recall that at one extreme we have mixtures in class A, which are most nearly ideal, and at the other extreme we have those in class D, which are most strongly nonideal. To change a mixture systematically from ideal to highly nonideal, we must change the degree of disparity in intermolecular forces. This can be done in many ways; two are illustrated in Figure 9.23. In the top part of Figure 9.23 we consider mixtures containing methane plus a second alkane. These mixtures become more nonideal as the length of the second component is increased; their nonideality also increases when branching or unsaturated bonds are added to the molecules of one component. For example, methane and propane are miscible in all proportions and their mixtures are class A, but when we reach methane + n-hexane, the nonideality is strong enough to cause liquid-liquid immiscibility and we have a class E mixture. As we move to mixtures of methane + 1-hexene, we remain in class E, but the mixtures are immiscible at low and high temperatures, with a miscibility gap at intermediate temperatures. Finally, when we reach methane + methylcyclopentane we have arrived in class D, and these mixtures exhibit GGE.

In another kind of progression, we can exploit not only size and shape effects, but also differences in molecular polarity. An example appears in the lower part of Figure 9.23, wherein the intermediate class is B rather than E. This path is common to mixtures of polar + nonpolar molecules or strongly polar + weakly polar molecules. Again the strategy is to increase the disparity in intermolecular forces as we move from class A through B to D [22, 27–29].

A third progression carries us from class A directly to class C; nearly all known class C mixtures contain water as one component. To have class C behavior, a mixture apparently must have strong unlike-molecule (solvation) interactions, as well as strong like-molecule (association) interactions. These effects often occur as a result of hydrogen bonding. A particular example of the progression from A to C is provided by aqueous-alcohol solutions, in which we systematically change the size and shape of the alcohol. For example, liquid mixtures of water plus methanol, or ethanol, or a propanol are all miscible and are class A mixtures; however, some butanol + water mixtures are only partially miscible and are class C mixtures. In fact, water + dimethylethanol is a member of class A, while aqueous solutions of n-butanol and of both methylpropanols are members of class C.

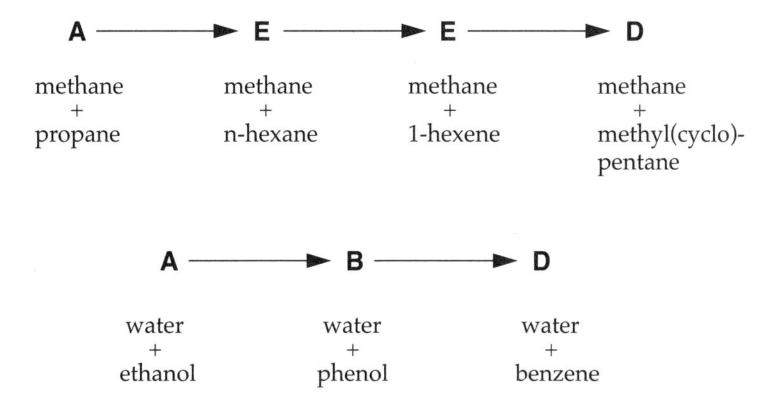

Figure 9.23 Two paths by which we can use systematic replacements of one component to move among classes of binary mixtures, thereby changing phase behavior. Mixtures of methane and n-hexane exhibit LLE with a UCST and are in class E; mixtures of methane and 1-hexene are also in class E, but they exhibit both a UCST and an LCST, as shown in Figure 9.14.

These kinds of changes in composition can serve as a basis for improving process designs. For example, because of the differences in phase behavior, liquid-liquid phase separations (extraction) are more energy efficient than vapor-liquid phase separations (distillation). So, if we have a VLE situation that can be converted to LLE by a small change in composition, then we may be able to improve the economics and efficiency of a particular separation. Subtle changes in composition can sometimes produce large changes in phase behavior.

9.5.5 Do All Mixtures Have Vapor-Liquid Critical Points?

Mixtures in classes D and E have discontinuous vapor-liquid critical lines, which suggests that some mixtures may not have critical points at all compositions. To test this, we used the Redlich-Kwong equation of state to compute the critical lines for a mixture in class D. The equations to be solved are the conditions for a binary critical point (9.3.23) and (9.3.24). The mixing rules used for the Redlich-Kwong parameters are given in § 8.4.4, along with the Redlich-Kwong expression for the fugacity coefficient.

The conditions (9.3.23) and (9.3.24) represent two algebraic equations that can be solved for two of the three unknowns (T_c, P_c, and x_{mc}, where x_m is mole fraction of methane). The equations are nonlinear in all three unknowns and must be solved simultaneously by trial. We found the critical lines by setting a value for T_c, then solving (9.3.23) and (9.3.24) for P_c and x_{mc}. The calculations were performed using pure-component critical properties characteristic of methane (T_c = 190.6 K, P_c = 46 bar) and ammonia (T_c = 405.6 K, P_c = 112.8 bar). Mixtures of methane and ammonia are known to be members of class D; however, we caution that the critical lines provided by the Redlich-Kwong equation with our simple mixing rules are only semiqualitative. But since our intent is only to show qualitative behavior, this simple model is adequate.

The computed critical line is shown on the Tx diagram in Figure 9.24. As expected, the critical line has two branches: a short branch extending from the pure methane

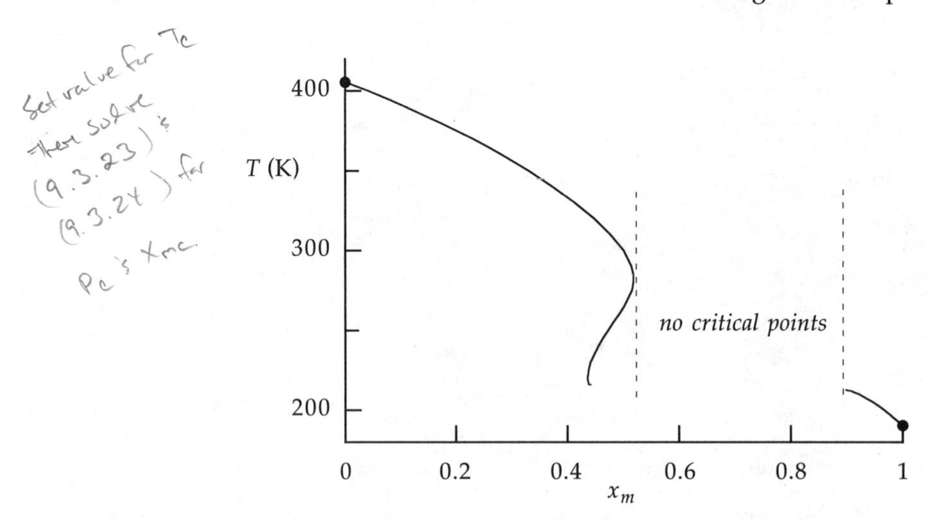

Figure 9.24 Vapor-liquid critical lines computed from the Redlich-Kwong equation of state for binary mixtures of a methane-like component and an ammonia-like component. Filled circles are pure-component critical points. Mixtures having compositions ($0.43 < x_m < 0.52$) have two critical points, but mixtures having ($0.52 < x_m < 0.89$) have none.

critical point to about 212 K with $x_m = 0.9$, and a long branch extending from the pure ammonia critical point to about 216 K with $x_m = 0.435$. There is a small range in temperature (212.5 K < T < 215.5 K) and a large range in composition ($0.52 < x_m < 0.89$) over which this model mixture displays no critical points: for these mixtures at temperatures T < 215 K, vapor-liquid (fluid-fluid?) equilibrium can be observed at pressures to hundreds—even thousands—of bars. Note that there is a small range of compositions ($0.43 < x_m < 0.52$) over which the mixtures exhibit more than one critical point.

9.6 TERNARY MIXTURES

In this section we briefly describe some of the phase behavior that has been observed in ternary mixtures. When three components are present, mixtures can exhibit a wealth of phase behavior, including equilibria among solid, gas, and multiple liquid phases. We have space here only to show the most common diagrams (§ 9.6.1) observed for simple liquid-liquid (§ 9.6.2) and vapor-liquid (§ 9.6.3) equilibria. More extensive descriptions can be found elsewhere [5, 17].

9.6.1 Phase Equilibria on Triangular Diagrams

Phase equilibria for ternary mixtures are conventionally represented on equilateral triangular diagrams. Such diagrams provide a convenient way to present basic material balance relations; these are reviewed in Appendix H. Triangular diagrams are $\mathcal{F}'$ diagrams, and for $C = 3$ components, (9.1.12) gives

$$\mathcal{F}' = C + 1 - S = 4 - S \tag{9.6.1}$$

where S is the number of internal constraints, excluding phase-equilibrium relations. Typically, we plot isothermal-isobaric triangular diagrams, so (9.6.1) reduces to $\mathcal{F}' = 2$: we need two overall system mole fractions to locate the state on a triangular diagram.

To determine the number of properties needed to identify the state, we apply the phase rule (9.1.13); for ternaries at fixed T and P, it becomes

$$\mathcal{F} = C + 2 - \mathcal{P} - S = 3 - \mathcal{P} \tag{9.6.2}$$

So a one-phase ternary has $\mathcal{F} = 2$; these states span areas on a triangular diagram. At fixed T and P, a two-phase ternary has $\mathcal{F} = 1$, which defines a line. Two-phase lines appear in pairs, each giving the composition of one phase. Areas between two-phase lines are traversed by tie lines, and, if the overall mole fractions are known, the relative amounts in the two phases can be found by lever rules.

At fixed T and P, a three-phase ternary has $\mathcal{F} = 0$, which defines a point. On a triangular diagram, a three-phase situation produces three points, each giving the composition of one of the phases. The three points can be connected to form a triangle, and the relative amounts in the three phases can be found by a tie-triangle rule (see Appendix H).

9.6.2 Liquid-Liquid Equilibria

Triangular diagrams are commonly used to depict liquid-liquid equilibria, and in ternary mixtures many different kinds of diagrams can occur. Figure 9.25 shows schematics of six common kinds of isothermal-isobaric diagrams, with the diagrams (a)–(f) arranged according to the number of two-phase regions. This same arrangement is obtained if we use, as the organizing principle, the number of binaries that undergo liquid-liquid phase splits.

Many ternaries display the simple behavior appearing in Figure 9.25(a), in which only one binary undergoes LLE and the third component is completely miscible in both phases. The resulting triangular diagram contains one single-phase region and one two-phase region, and the two-phase boundary must contain a consolute point. A

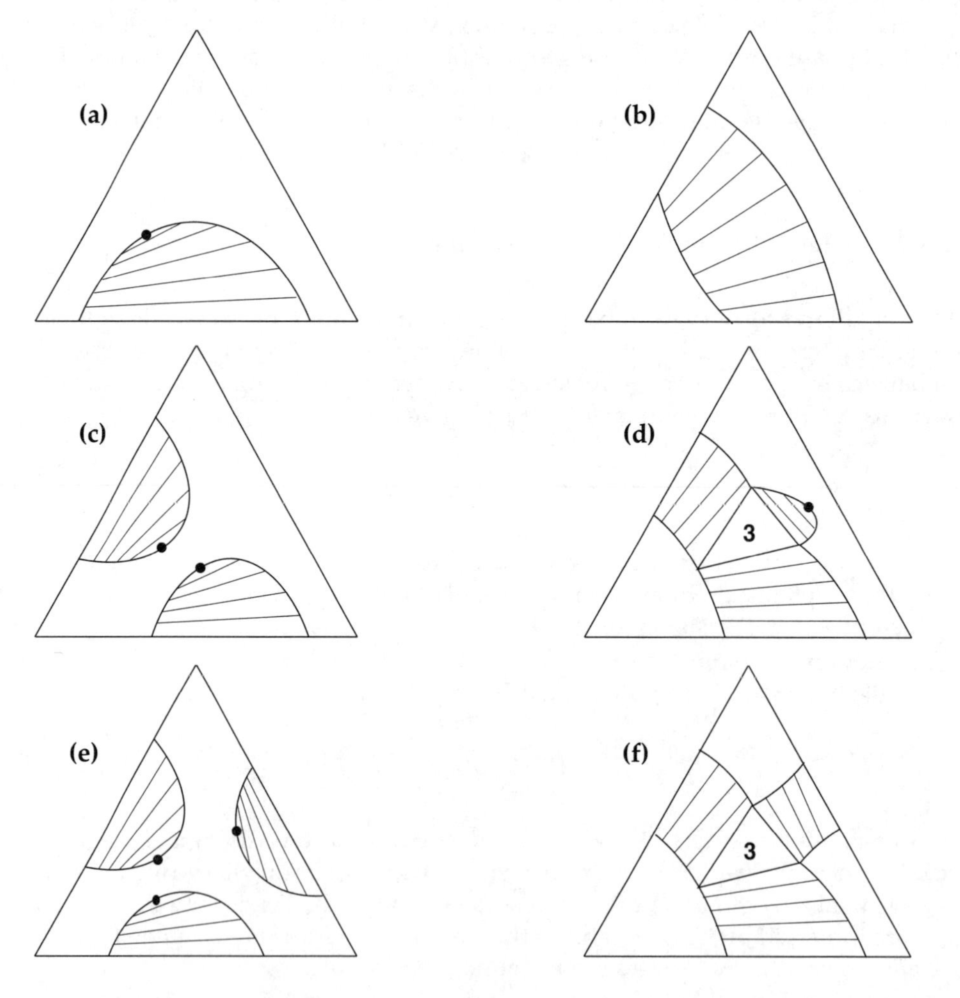

Figure 9.25 Six common types of isothermal-isobaric triangular diagrams for ternary mixtures that exhibit liquid-liquid equilibria. Filled circles locate consolute points. Numeral 3 inside a triangle identifies three-phase LLLE; the compositions of the three phases are given by the vertices of the triangles. These six diagrams are arranged by the number of two-phase regions: (a) and (b) each have one, (c) has two, and (d)-(f) each have three. Adapted from Walas [5].

consolute point is not necessarily an extremum on an isothermal-isobaric two-phase line; instead, the consolute point corresponds to a tie line of zero length.

When two of the binaries can exhibit LLE, the resulting ternary diagram usually takes one of the forms shown in (b), (c), or (d) of Figure 9.25. Of these, the simplest appears in Figure 9.25(b). This diagram contains two single-phase regions separated by one two-phase region that extends from one immiscible binary to the other. Consequently, no consolute point occurs for any ternary mixture.

In Figure 9.25(c) the two immiscible binaries give rise to two different two-phase regions separated by one single-phase region. Each two-phase boundary contains a consolute point, and at states between the two-phase regions, the ternary is completely miscible. This behavior is common; however, in some cases, mixtures at states between the two-phase regions are not miscible, but instead split into three phases. This possibility appears in Figure 9.25(d). The compositions of the three phases form the vertices of a triangle, and each side of the triangle is bounded by a two-phase region. Neither the vertices nor the sides of the three-phase triangle contain consolute points; but since only two of the binaries exhibit LLE, the third two-phase region must end at a consolute point.

If all three binaries can exhibit LLE, the resulting ternary diagram usually appears as in either (e) or (f) of Figure 9.25. In (e) a one-phase region separates the three two-phase regions, and each two-phase boundary contains a consolute point. Alternatively, if a three-phase region separates the two-phase regions, we obtain a diagram like that in (f). Now none of the two-phase boundaries contain consolute points.

Solutropes. In § H.1 of Appendix H, we remark on the invariance in composition that occurs when a line on a triangular diagram lies parallel to an edge of the triangle. This invariance takes on special significance when the parallel line is a tie line across a two-phase region. When this occurs, the two phases in LLE have the same composition in one component, and the mixture is called a *solutrope*. The component of common composition is the one represented by the vertex that lies opposite the tie line. An example is the solutrope formed by mixtures of benzene, pyridine, and water, shown in Figure 9.26. In those mixtures, the solutropic benzene-rich phase has the same composition in pyridine as the solutropic water-rich phase.

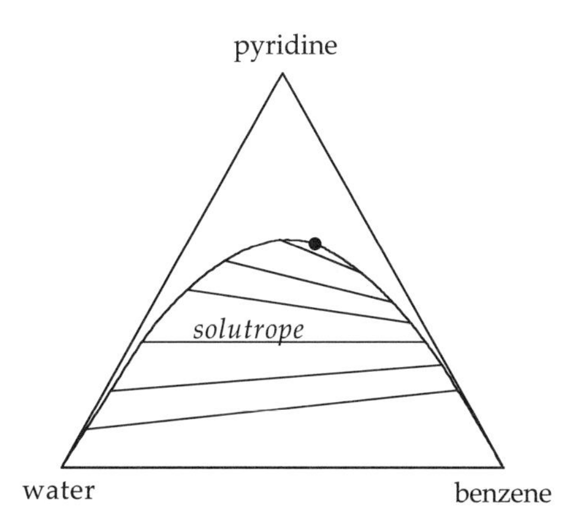

Figure 9.26 Formation of a solutrope in mixtures of water, benzene (C_6H_6), and pyridine (C_5H_5N). Filled circle marks consolute point. Adapted from Francis [17].

Solutropes are not uncommon; they may occur in any of the classes of ternary mixtures shown in Figure 9.25. Their practical significance arises from their ability to inhibit separations by liquid extraction, because transfers of components between phases are often hindered when a mole fraction becomes the same in both phases. Such inhibitions may be compounded if the densities of the phases also become equal, as they may near solutropes. Since liquid extractions exploit density differences, no separation occurs in an extraction process when the densities of the two phases become equal, even if their compositions differ.

Changes in Temperature. In general, the diagrams shown in Figure 9.25 apply to mixtures of different components, but in some cases, they may apply to one mixture at different values of T or P or both. Such changes in the diagram of one ternary can be explained by changes in the miscibility of its component binaries. Further, not only might changes of state shift a ternary among the classes shown in Figure 9.25, but such changes may also cause more subtle shifts within a class. As an example, consider mixtures of water, phenol, and triethylamine at 10°C and at 75°C. First, consider the phase behavior of the three binaries: (a) at 10°C water and phenol are partially miscible (see Figure 9.13), but they are completely miscible at 75°C. (b) At 10°C water and triethylamine are completely miscible but they are partially miscible at 75°C. (c) Phenol and triethylamine are completely miscible at both temperatures. Based on the behaviors of these binaries, we would expect the ternary mixture to have a triangular diagram like that in Figure 9.25(b); that is, we expect the ternary to have two one-phase regions separated by a two-phase region without a consolute point. But, in fact, the diagrams for both temperatures appear as in Figure 9.27. Both diagrams have one single-phase region and one two-phase region with a consolute point; further, the two-phase regions differ in subtle but significant ways. For example, at 10°C the consolute mixture contains very little phenol, but at 75°C it contains very little triethylamine. Note that at 10°C pure phenol solidifies, producing a small region of liquid-solid equilibrium in the ternary, but this does not affect our analysis of the LLE.

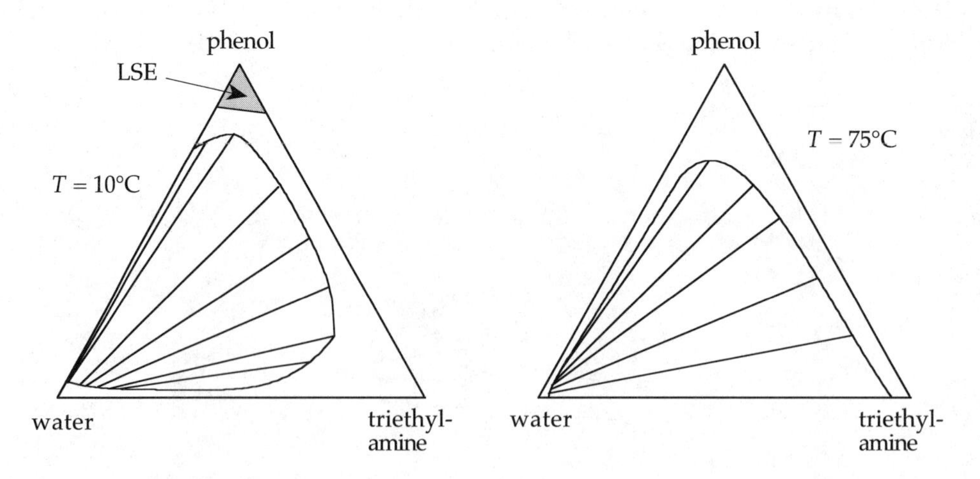

Figure 9.27 Effect of temperature on LLE in mixtures of phenol, water, and triethylamine. At 10°C a consolute point occurs in mixtures lean in phenol, while at 75°C a consolute point occurs in mixtures lean in triethylamine. At 10°C pure phenol solidifies. Compositions plotted here as weight fractions. Adapted from Walas [5].

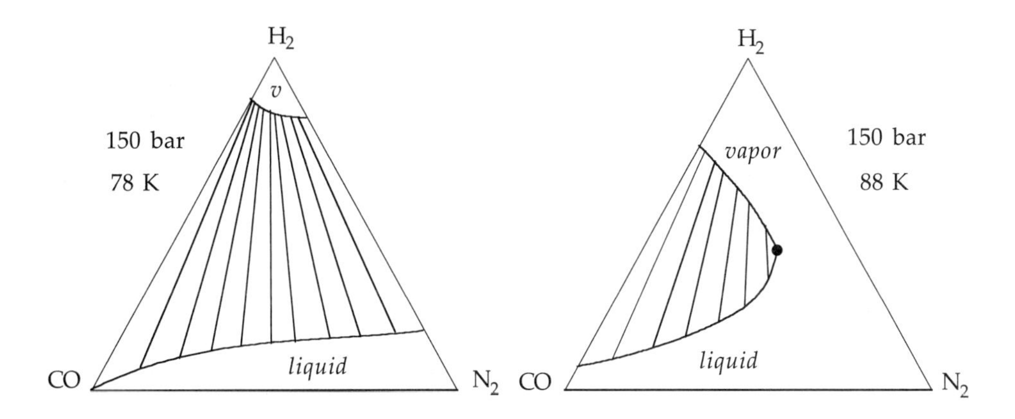

Figure 9.28 Effect of a temperature change on vapor-liquid equilibria in ternary mixtures of carbon monoxide, nitrogen, and hydrogen at 150 bar. Mixtures rich in hydrogen are vapors; those lean in hydrogen are liquids. At 78 K (*left*) a vapor-liquid phase split occurs for most of these mixtures; however, at 88 K (*right*) VLE is confined to mixtures lean in nitrogen, and a vapor-liquid critical point appears (filled circle). Data from Walas [5].

9.6.3 Vapor-Liquid Equilibria

To show how vapor-liquid equilibria usually appear on triangular diagrams, we consider the effects of temperature on mixtures of CO, N_2, and H_2 at 150 bar. This pressure is above the critical pressures of all three pure components. First, we consider mixtures at 78 K, as shown on the lhs of Figure 9.28. Pure hydrogen is supercritical at this state, but 78 K is below the critical temperatures of both N_2 and CO. Under these conditions, binary mixtures of N_2 with CO are single-phase liquids at all compositions, but binary mixtures of H_2 with N_2 and of H_2 with CO exhibit VLE. In all cases, the vapor phase is rich in hydrogen. For ternary mixtures, the two-phase region spans most CO and N_2 compositions. Across the two-phase region tie lines have positive slopes when the liquid contains a large fraction of CO and negative slopes when the liquid contains a small fraction of CO.

When the temperature is raised to 88 K, then the H_2-N_2 binary is a single-phase fluid at all concentrations, while the H_2-CO binary can still exhibit VLE; see the rhs of Figure 9.28. At this higher temperature, ternary mixtures undergo vapor-liquid phase splits over a smaller region of the triangular diagram than they do at 78 K, though the vapor phase is still rich in hydrogen. Since only one binary exhibits VLE, a critical point must occur along the boundary of the two-phase region; such a critical point represents a tie line of zero length, and it separates saturated ternary vapors from saturated ternary liquids. The critical point does not necessarily occur at an extremum on a two-phase line plotted on a triangular diagram at fixed T and P. Figure 9.28 illustrates that, on triangular diagrams, the shapes of two-phase regions can change drastically in response to relatively minor changes of state.

Lastly, we mention that ternary fluid mixtures can exhibit homogeneous azeotropes. These are vapor-liquid equilibrium states at which both phases have the same composition and the pressure is an absolute maximum or minimum with respect to composition at fixed temperature [6].

9.7 SUMMARY

Even though we have limited this chapter to one-, two-, and three-component substances, we have seen a rich variety of phase-equilibrium behavior. We have tried to show that much of that behavior can be organized to take advantage of similarities and analogies. These similarities and analogies form basic patterns that promote understanding; our understanding deepens when we can recognize the limitations and exceptions to such organizing principles. Here is a summary of the more important patterns.

First, recall the close analogies that exist between pure-component critical points and those of binary-mixtures. Unlike simple phase changes, which represent transitions between stable and metastable behavior, all critical points represent transitions between stable and unstable behavior. For pure components, the transition is driven by mechanical instabilities, and at vapor-liquid critical points pure fluids have

$$\left(\frac{\partial P}{\partial v}\right)_T = 0 \tag{8.2.13}$$

with

$$\left(\frac{\partial^2 P}{\partial v^2}\right)_T = 0 \tag{8.2.14}$$

These equations imply that, at a critical point on a Pv diagram, the critical isotherm passes through a point of inflection.

Similarly, for binary mixtures, transitions at critical points are driven by diffusional instabilities; so, at both vapor-liquid and liquid-liquid critical points, binary mixtures have

$$\left(\frac{\partial \ln f_1}{\partial x_1}\right)_{TP} = 0 \tag{9.3.23}$$

with

$$\left(\frac{\partial^2 \ln f_1}{\partial x_1^2}\right)_{TP} = 0 \tag{9.3.24}$$

These equations imply that, at a critical point on a fugacity-composition diagram, an isothermal-isobaric plot of the fugacity passes through a point of inflection. The analogies among critical points are even stronger than implied by the structures of the pairs of equations (8.2.13)–(8.2.14) and (9.3.23)–(9.3.24): *all* critical points display certain universal features that are independent of the kinds of components present and independent of the kind of criticality. See Sengers et al. [30].

Second, recall the many analogies that occur in the phase behavior of binary mixtures. For example, many features that occur on binary vapor-liquid phase diagrams have counterparts on liquid-solid diagrams. Some of those equivalent features are listed in Table 9.2. Furthermore, such equivalences include not only the kinds of behavior but may also extend to the general shapes of two-phase lines. That is, many

Table 9.2 Examples of equivalent features that can appear on isobaric Txy and Txx diagrams for vapor-liquid and liquid-solid equilibria of binary mixtures

VLE		LSE
dew-point curve	⇔	liquidus
bubble-point curve	⇔	solidus
homogeneous azeotrope	⇔	solutrope
heterogeneous azeotrope	⇔	eutectic

Txy diagrams for vapor-liquid equilibria are equivalent to Txx diagrams for liquid-solid equilibria; we merely relabel the lines and regions. When two structures are the same but have different labels, we say the structures are *isomorphisms*; an example is shown in Figure 9.29. However, we caution that not all liquid-solid diagrams are isomorphic to vapor-liquid diagrams; liquid-solid systems can also display phase behavior, such as peritectics, that rarely or never occurs in vapor-liquid systems.

Third, recall those fundamental features that are used to organize high-pressure vapor-liquid behavior of binary mixtures. The relevant features are the vapor-liquid and liquid-liquid critical lines on PT diagrams. Binaries are first divided into two groups: mixtures in group 1 have continuous vapor-liquid critical lines between the pure-component critical points, while those in group 2 do not. Then these two groups are each divided into classes, based on their liquid-liquid critical lines. If a mixture does not exhibit LLE, then it is in class A. Otherwise, group-1 (or group-2) mixtures that exhibit only UCSTs are placed in class B (or class D). Similarly, group-1 (or group-2) mixtures that exhibit both UCSTs and LCSTs are placed in class C (or class E). These

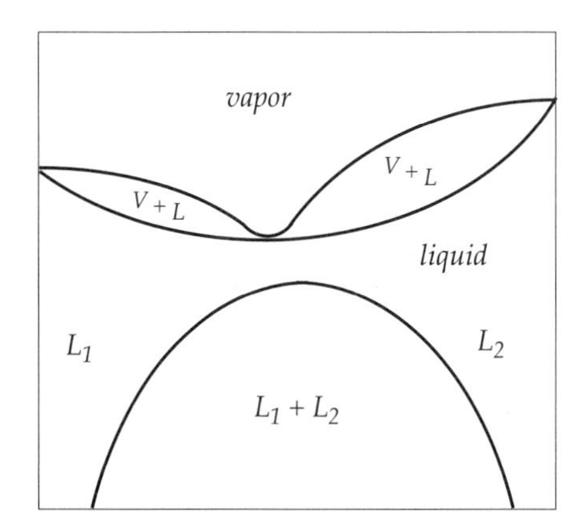

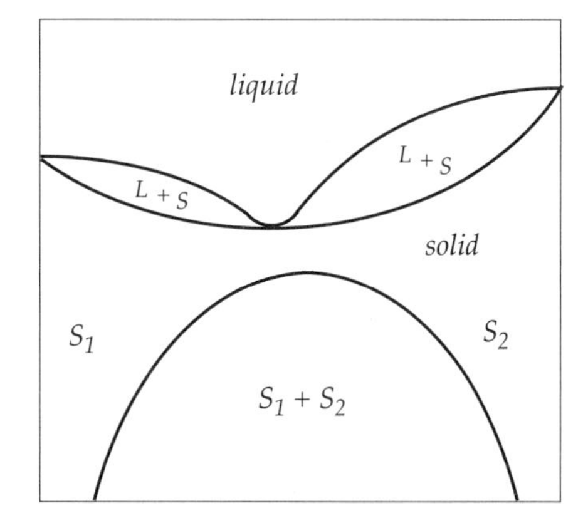

Figure 9.29 One of the many isomorphisms that exist between vapor-liquid and liquid-solid phase diagrams for binary mixtures. (*left*) An isobaric Txy diagram with a minimum boiling-point azeotrope and a miscibility gap above an LLE situation; (*right*) an isobaric Txx diagram with a minimum melting-point solutrope and a miscibility gap above an SSE situation.

five classes can then be divided into subclasses, based on more detailed behavior, but these fundamental patterns provide a solid foundation for more detailed study.

Lastly, recall that complicated phase diagrams are usually superpositions of simple diagrams. For example, diagrams showing vapor, liquid, and solid phases are usually combinations of vapor-liquid and liquid-solid diagrams. Diagrams showing vapor-liquid critical end points are combinations of vapor-liquid and liquid-liquid diagrams. And liquid-solid diagrams containing both a eutectic and a peritectic can usually be resolved into a liquid-solid eutectic diagram plus a liquid-solid peritectic diagram. In analyzing such diagrams, your strategy should be to divide and conquer.

It is a sobering fact that this chapter introduces only the more common and least complex kinds of multiphase behavior: the chapter fails to do full justice to Nature's diversity. Yet, the features described here should provide a structure by which you can effectively analyze and use phase diagrams.

LITERATURE CITED

[1] T. W. de Loos, "Understanding Phase Diagrams," in *Supercritical Fluids: Fundamentals for Applications*, E. Kiran and J. M. H. Levelt Sengers (eds.), NATO ASI Series E, vol. 273, Kluwer Academic Publishers, Dordrecht, 1994, p. 65.

[2] I. Prigogine and R. Defay, *Chemical Thermodynamics*, D. H. Everett (transl.), Longmans, Green and Co., London, 1954.

[3] H. E. Tester, "Methane," in *Thermodynamic Functions of Gases*, vol. 3, F. Din (ed.), Butterworth, London, 1961.

[4] D. Eisenberg and W. Kauzmann, *The Structure and Properties of Water*, Oxford University Press, Oxford, 1969.

[5] S. M. Walas, *Phase Equilibria in Chemical Engineering*, Butterworth, Boston, 1985.

[6] L. H. Horsley, *Azeotropic Data*, Adv. Chem. Series (American Chemical Society), **6** (1952); **35** (1962); **116** (1973).

[7] J. Gmehling, J. Menke, J. Krafczyk, and K. Fischer, *Azeotropic Data*, three volumes, Wiley, New York, 2004.

[8] A. Findlay, *The Phase Rule and Its Applications*, 8th ed., Longmans, Green and Co., London, 1938; reprinted by Dover Publications, New York.

[9] A. N. Campbell and A. J. R. Campbell, "Concentrations, Total and Partial Vapor Pressures, Surface Tensions and Viscosities, in the Systems Phenol-Water and Phenol-Water-4%Succinic Acid," *J. Am. Chem. Soc.*, **59**, 2481 (1937).

[10] A. E. Hill and W. M. Malisoff, "The Mutual Solubility of Liquids. III. The Mutual Solubility of Phenol and Water," *J. Am. Chem. Soc.*, **48**, 918 (1926).

[11] F. H. Rhodes and A. L. Markley, "The Freezing-Point Diagram of the System Phenol-Water," *J. Phys. Chem.*, **25**, 527 (1921).

[12] F. Kohler and O. K. Rice, "Coexistence Curve of the Triethylamine-Water System," *J. Chem. Phys.*, 26, 1614 (1957).

[13] A. N. Campbell, E. M. Kartzmark, and W. E. Falconer, "The System Nicotine-Methylethylketone-Water," *Can. J. Chem.*, **36**, 1475 (1958).

[14] A. J. Davenport, J. S. Rowlinson, and G. Saville, "Solutions of Three Hydrocarbons in Liquid Methane," *Trans. Faraday Soc.*, **62**, 322 (1966).

[15] R. L. Scott, "Phase Equilibriums in Solutions of Liquid Sulfur. I. Theory," *J. Phys. Chem.*, **69**, 261 (1965).

[16] A. W. Francis, *Critical Solution Temperatures*, Adv. Chem. Series (American Chemical Society), **31** (1961).

[17] A. W. Francis, *Liquid-Liquid Equilibriums*, Wiley-Interscience, New York, 1963.

[18] E. F. Zhuravlev, "Nonvariant Equilibrium of Three Liquid Phases in Two-Component Condensed Systems," *Uchenye Zapiski Molotov Univ.*, **9** No. 4, 113 (1955); cited in *Chem Abstr.* **53**, 54645 (1959).

[19] M. Hansen, *Constitution of Binary Alloys*, 2nd ed., McGraw-Hill, New York, 1958; supplements published thereafter.

[20] W. Hume-Rothery, J. W. Christian, and W. B. Pearson, *Metallurgical Equilibrium Diagrams*, The Institute of Physics, London, 1952.

[21] A. A. Adamson, *Understanding Physical Chemistry*, 2nd ed., W. A. Benjamin, Inc., Menlo Park, CA, 1969.

[22] J. S. Rowlinson and F. L. Swinton, *Liquids and Liquid Mixtures*, 3rd ed., Butterworth, London, 1982.

[23] G. M. Schneider, "Physico-Chemical Properties and Phase Equilibria of Pure Fluids and Fluid Mixtures at High Pressures," in *Supercritical Fluids: Fundamentals for Applications*, E. Kiran and J. M. H. Levelt Sengers (eds.), NATO ASI Series E, vol. 273, Kluwer Academic Publishers, Dordrecht, 1994, p. 91.

[24] C. J. Peters, "Multiphase Equilibria in Near-Critical Solvents," in *Supercritical Fluids: Fundamentals for Applications*, E. Kiran and J. M. H. Levelt Sengers (eds.), NATO ASI Series E, vol. 273, Kluwer Academic Publishers, Dordrecht, 1994, p. 117.

[25] R. L. Scott and P. H. van Konynenburg, "Static Properties of Solutions. 2. Van der Waals and Related Models for Hydrocarbon Mixtures," *Disc. Faraday Soc.*, **49**, 87 (1970).

[26] P. H. van Konynenburg and R. L. Scott, "Critical Lines and Phase Equilibriums in Binary van der Waals Mixtures," *Phil. Trans.*, **A298**, 495 (1980).

[27] G. M. Schneider, "High-Pressure Phase Diagrams and Critical Properties of Fluid Mixtures," in *Chemical Thermodynamics*, A Specialist Periodical Report, vol. 2, The Chemical Society, London, 1978, p. 105.

[28] G. M. Schneider, "Phase Equilibria in Fluid Mixtures at High Pressures," *Adv. Chem. Phys.*, **17**, 1 (1970).

[29] M. St. Marie, J. Schwedock, and K. E. Gubbins, "Binary Fluid Phase Diagrams," unpublished report (1978).

[30] J. V. Sengers, R. F. Kayser, C. J. Peters, and H. J. White, Jr. (eds.), *Equations of State for Fluids and Fluid Mixtures*, Elsevier, Amsterdam, 2000.

PROBLEMS

9.1 Table 9.1 presents several phase-equilibrium situations and gives the value of the variable $\mathcal{F}$. For each entry in the table, give the corresponding value of $\mathcal{F}'$.

9.2 (a) On a pure-component Pv diagram, the triple point(s) are points, even though (9.1.12) gives $\mathcal{F}' = 2$, which suggests they should define a surface. Explain.

(b) On an isothermal Pxy diagram for a binary mixture, what would be the object that represents three-phase VLLE situations? Is your answer consistent with the value of $\mathcal{F}'$ given by (9.1.12)? If not, explain.

9.3 (a) Use the Pxy diagram in Figure 9.4 to estimate the number of moles of liquid present at 10 bar, 330 K when $z_1 = 0.75$ and $N = 10$.

(b) Use the Pxy diagram in Figure 9.4 to estimate the number of moles of liquid present at 2 bar, 330 K when $z_1 = 0.75$ and $N = 10$.

(c) Using the Pxy diagrams in Figure 9.5, describe the number and relative amounts of the phases that appear when a mixture, initially having $z_1 = 0.79$ at 430 K and 60 bar, is isothermally expanded to 30 bar.

9.4 Each entry in the following table represents a proposed problem concerning a two-phase, VLE situation. For each problem, answer these four questions:

(a) Is it an $\mathcal{F}$ problem or an $\mathcal{F}'$ problem?

(b) Is it well-posed, underspecified, or overspecified?

(c) If it is not well-posed, what reasonable changes will make it well-posed?

(d) What always-true equations start a solution to the well-posed problem?

Case[a]	C	Knowns	Constraints	To find
1	2	T, y_1	...	P, x_1
2	4	T, P, x_1, y_1, y_2	...	x_2, x_3, y_3
3	4	$\{x\}, y_1$	...	T, P, y_2, y_3, y_4
4	3	T, x_1, x_3	...	P, y_2, y_3
5	3	T, x_1	...	P, y_1, V, L
6	3	T, P, z_1, V	isothermal*	$z_2, \{x\}, \{y\}, L$
7	3	$\{z\}, V, L, x_1$	adiabatic*	$x_2, \{y\}, T_{in}, P_{in}$

[a] Notation: $\{x\}$ = set of liquid mole fractions, $\{y\}$ = set of vapor mole fractions, $\{z\}$ = set of feed mole fractions; V, L = fraction in vapor, in liquid; * = steady flow; subscripts 1, 2, 3, 4 = components; subscript in = inlet.

9.5 What equation, relating temperature and pure-component vapor pressures, would be consistent with the assumption that the difference in heat capacities between vapor and liquid is constant over a range of temperatures up to the boiling point?

9.6 For pure substance 1, which appears in Figure 9.18, draw a semiquantitative plot of the molar Gibbs energy as a function of temperature from 0 to 100°C.

9.7 Each entry in the following table represents a proposed problem concerning a multiphase equilibrium situation. For each entry, answer these questions:

(a) Is it an $\mathcal{F}$ problem or an $\mathcal{F}'$ problem?

(b) Is it well-posed, underspecified, or overspecified?

(c) If it is not well-posed, what reasonable changes will make it well-posed?

(d) What always-true equations start a solution for the well-posed problem?

Case[a]	C	$\mathcal{P}$	Knowns	Constraints	To find
1	4	2(LS)	T, P, x_1, w_1, w_2	...	x_2, x_3, w_3
2	3	4(VLLS)	V, T_{in}, P_{in}	adiabatic*	$\{z\}, L^{\alpha}, L^{\beta}, S, P_o, \{y\}$
3	2	2(VS)	T, P	solid phase pure	y_1
4	3	3(VLL)	$\{z\}, P, y_1, V$	...	$y_2, \{x^{\alpha}\}, \{x^{\beta}\}, L^{\alpha}, T$
5	3	2(VL)	$\{z\}, T_o, P_o, P_{in}$	adiabatic, reversible*	$V, \{x\}, \{y\}, T_{in}$

[a] Notation same as in Problem 9.4, with following additions: S = fraction solid; $\{w\}$ = set of solid mole fractions; superscripts α, β = liquid phases; subscripts in = inlet, o = outlet.

9.8 Using only the following information, estimate the triple-point temperature and pressure of pure ethane: $T_m = 89.9$ K, $P_c = 48.8$ bar, $T_c = 305.4$ K, and

$$\ln\left(\frac{P^s}{P_c}\right) = \frac{x(A + Bx^{0.5} + Cx^2 + Dx^5)}{1 - x} \quad \text{(P9.8.1)}$$

where P^s is the vapor pressure, $x = (1 - T/T_c)$, and the parameter values are as follows: $A = -6.34307$, $B = 1.01630$, $C = -1.19116$, and $D = -2.03539$.

9.9 Draw an approximate, but quantitative, PT diagram for pure silicon based only on the following information:

$T_m = 1700$ K	$\Delta h_m/\mathrm{R} = 5600$ K	$v^s = 12$ cc/mol
$T_{boil} = 2700$ K	$\Delta h_{vap}/\mathrm{R} = 36{,}400$ K	$v^{\ell}/v^s = 0.9$ at T_m

9.10 If the vapor pressure of a pure substance obeys $\ln P^s = A + B/T$, what other information would you need to obtain a consistent estimate for the temperature dependence of the sublimation pressure, P^{sub}?

9.11 (a) For the (overall) equimolar mixture in Figure 9.17(c), draw the constant heat-removal cooling curve, analogous to that appearing on the right in Figure 9.19.

(b) For the (overall) equimolar mixture in Figure 9.17(d), draw the constant heat-removal cooling curve.

(c) For the (overall) equimolar mixture in Figure 9.18, draw the constant heat-removal cooling curve.

9.12 Fluid mixtures that exhibit vapor-liquid and liquid-liquid equilibria can be subjected to isothermal expansions performed at constant rates of work production. The results would be pressure versus time plots with lines behaving like those on the temperature versus time plot shown in Figure 9.19. For such expansions, explain why the pressure-time lines would change slope.

9.13 (a) Using the plots in Figures 9.11 and 9.12 as a guide, develop an expression in which the enthalpy of vaporization Δh_{vap} for azeotropes in methyl acetate-methanol mixtures is related to the enthalpy of vaporization for azeotropes in acetone-chloroform mixtures.

(b) Compare values of Δh_{vap} for the azeotropes to those of the pure components in Figure 9.12. Conclusion?

9.14 Draw a Txx phase diagram consistent with the following information; in each region of your diagram, indicate what phases would be present. Pure substance A melts at 500 K while pure B melts at 800 K. Compound AB_2, which behaves as a pure component, melts at 600 K. The solid form of AB_2 can dissolve substance B up to $x_B = 0.8$ for $T < 700$ K. Heating this solid at 700 K yields only a liquid and pure solid B. When the liquid having $x_B = 0.10$ is allowed to cool, pure solid A separates first. Then, at 400 K, no further change in temperature occurs until all liquid has disappeared, leaving a system containing solid A and solid AB_2.

9.15 (a) One mole of chloroform(1) is added to one mole of carbon tetrachloride(2), completely filling a closed vessel. The mixture is then brought to 50°C at 0.55 bar. Under these conditions, the system is in vapor-liquid equilibrium; the compositions of the phases are $x_1 = 0.428$ and $y_1 = 0.556$. Determine the total number of moles in each phase.

(b) At the situation in (a) an additional 0.2 moles of chloroform are added. The system is then returned to equilibrium at 50°C and 0.55 bar. How many phases are present in this new situation? If your answer is two, what are the compositions of the two phases?

9.16 (a) Derive (9.3.16), which gives the slope of an isobaric bubble-T curve for a binary mixture in VLE.

(b) Derive the expression, analogous to (9.3.16), that gives the slope of an isobaric dew-T curve.

(c) Under what conditions are the slopes in both (a) and (b) positive? negative? of opposite signs?

9.17 It is observed experimentally that two miscible subcritical components such as water(1) and n-propanol(2) will change from one-phase liquid to two phases (LL or VL) and even to three phases (VLL) when ethylene(3) is added at pressures above 2 MPa and temperatures above the pure ethylene critical point T_c. Sketch a triangular diagram for $P > 2$ MPa and $T > T_c$. Include on your diagram the boundaries of the various regions and the expected orientations of the two-phase tie lines.

9.18 Using an energy balance for a closed system, derive a relation for the slope of the temperature-time diagram on the right of Figure 9.19, assuming removal of heat at a constant rate. Assume the heat capacities of the phases are constant, but include the latent heats for liquid-solid and solid-solid phase transitions. Does your relation give a slope of zero at the three-phase line?

9.19 Consider the following isothermal-isobaric diagram for a ternary mixture that exhibits liquid-liquid immiscibility.

(a) On the diagram, label any one, two, and three-phase regions that appear.

(b) Of the three binary mixtures that can be formed from these three components, which are completely miscible at all compositions and which exhibit LLE?

(c) Show typical tie-lines for the ternary two-phase regions and if a consolute point occurs, circle it.

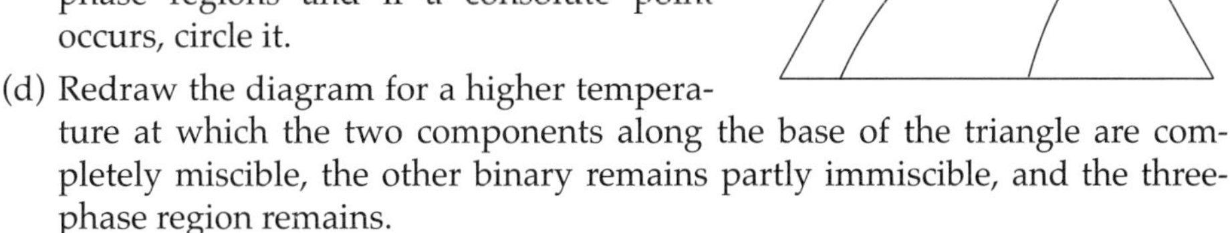

(d) Redraw the diagram for a higher temperature at which the two components along the base of the triangle are completely miscible, the other binary remains partly immiscible, and the three-phase region remains.

9.20 Five moles of benzene(1), five moles of acetonitrile(2), and five moles of water(3) are confined to a closed vessel at 1.0133 bar, 333 K. The mixture is observed to be in three-phase equilibrium: a water-rich liquid (α), an organic-rich liquid (β), and a vapor. Analyses of samples drawn from each phase give the following mole fractions:

i	**Species**	x_i^α	x_i^β	y_i
1	C_6H_6	0.0026	0.4786	0.4784
3	H_2O	0.9212	0.0674	0.2397

(a) Determine the total number of moles in each phase.

(b) Two more moles of water are added to the vessel, but the mixture remains three phases with $P = 1.0133$ bar, $T = 333$ K. What are the new compositions of the phases? Now how many moles are in each phase?

9.21 Using the following information for a binary mixture, draw a quantitative *Txx* diagram at ambient pressure:

(a) Pure component 1 melts at 285 K and boils at 350 K.

(b) Pure component 2 melts at 300 K but no boiling point is seen.

(c) Pure component 1 has a crystalline phase change from α to β at 250 K.

(d) The solubility of component 2 in solid component 1 is negligible at all *T*.

(e) The solubilities of component 1 in phases α, β, and liquid (ℓ) at various temperatures are given by the following data:

	240 K	250 K	260 K	270 K	280 K	290 K
x_1^{α}	0.33	0.32	0.31	...	...	...
x_1^{β}	...	0.0	0.1	0.2	0.15	0.1
x_1^{ℓ}	...	...	...	0.3	0.2	0.15

(f) Constant-rate cooling curves from the vapor to a solid have breaks (b) in slope and horizontals (h) at the following temperatures:

x_1	*T*(K)	*T*(K)	*T*(K)	*T*(K)	*T*(K)
0.25	285 (b)	270 (h)	263 (h)	255 (b)	...
0.5	340 (b)	322 (b)	280 (b)	270 (h)	263 (h)
0.75	350 (b)	325 (b)	285 (b)	270 (h)	263 (h)

Part IV

ENGINEERING CALCULATIONS

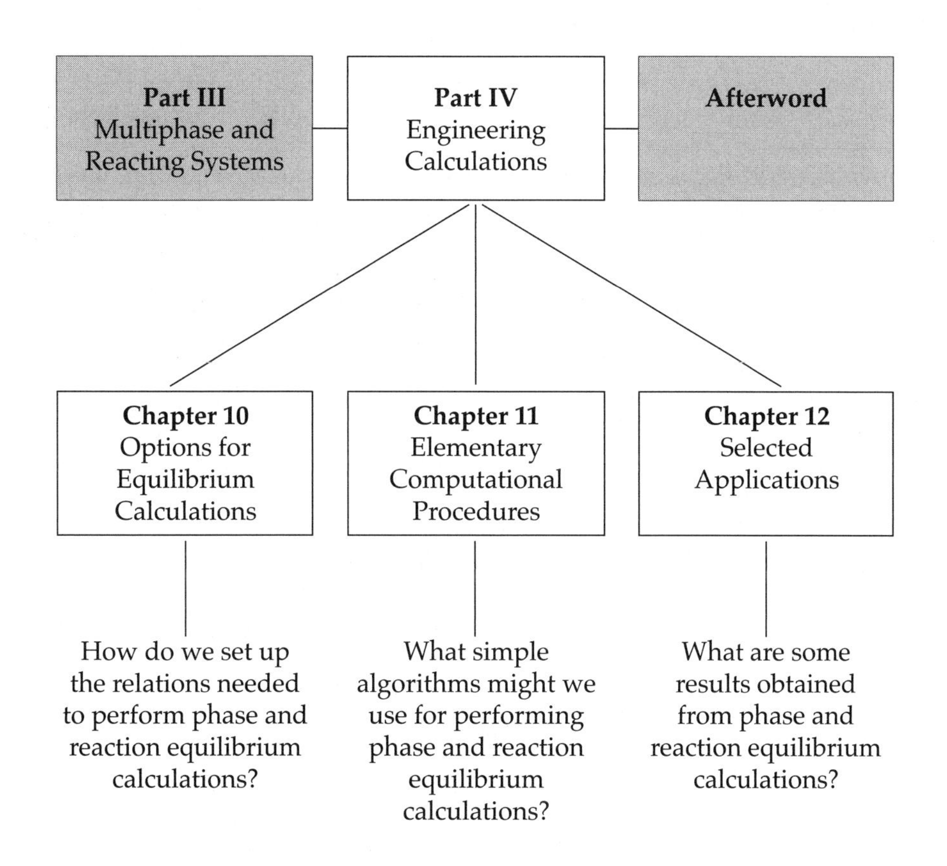

10

OPTIONS FOR EQUILIBRIUM CALCULATIONS

In previous parts of this book we have developed rigorous, generalized, thermodynamic descriptions of phenomena. With this chapter we begin to convert those descriptions into specific forms that can be applied to phase and reaction equilibrium calculations. Such calculations always requires us to make decisions—to select from among alternative computational strategies. For example, a common decision to be faced is this: Which of the five famous fugacity formulae should I use? If, in a given situation, our models are all reliable and their parameters are all known, and if we can solve all appropriate thermodynamic relations, then our choices are relatively simple: our decisions are dictated by the process, the substances involved, and their states. Unfortunately, most situations are not so simple: we usually have limited information about the process, all necessary properties of the substances may not be known, some models may be of limited reliability, and rigorous computational routes may be inaccessible. Such constraints complicate the selection process, forcing us to balance thermodynamic rigor, model reliability, and computational simplicity.

In § 10.1 we present the basic thermodynamic relations that are used to start phase-equilibrium calculations: we discuss vapor-liquid, liquid-liquid, and liquid-solid calculations. We have seen that the most interesting phase behavior occurs in nonideal solutions, but when we describe nonidealities using an ideal solution as a basis, we must select an appropriate standard state. Common options for standard states are discussed in § 10.2; they include pure-component standard states and dilute-solution standard states.

In § 10.3 we introduce two fundamental approaches to reaction equilibrium calculations: the traditional stoichiometric method and a nonstoichiometric method, which is useful when many reactions are occurring and when products and reactants are known but the reactions are unknown. In setting up both kinds of calculations, we must again confront issues related to standard states, and we must select appropriate computational forms. These issues are discussed in § 10.4 for reacting systems.

Finally, in § 10.5 we offer advice to help you select from among the available options. We are not able to cover all possible options in this book, but the basics presented here should help organize your thinking when you come to apply thermodynamics to both traditional and contemporary problems.

10.1 BASIC PHASE-EQUILIBRIUM RELATIONS

Consider any two homogeneous phases α and β in equilibrium with one another at temperature T and pressure P. Each phase contains any number of nonreacting components C that can freely cross the phase boundary. No other constraints apply, so the intensive state can be identified by specifying values for $\mathcal{F}$ properties, with the number $\mathcal{F}$ given by the phase rule (9.1.13),

$$\mathcal{F} = C + 2 - \mathcal{P} = C + 2 - 2 = C \tag{10.1.1}$$

For example, if in addition to T (or P), we know values for the $(C - 1)$ mole fractions in one phase, then the state is specified, and we should be able to compute values for P (or T) plus the $(C -1)$ mole fractions in the other phase. The computation requires us to solve the C phase-equilibrium conditions

$$f_i^{\alpha} = f_i^{\beta} \qquad i = 1, 2, \ldots, C \tag{7.3.12}$$

These equilibrium conditions are *always true* in that they apply to every phase equilibrium situation, and therefore, they serve as the starting point for solving *every* phase-equilibrium problem. But to solve the equations in (7.3.12), the fugacities must be replaced with expressions containing measurables and the measurables must be replaced with numbers. Consequently, we are faced with two kinds of decisions: (a) Which of the five famous fugacity formulae (FFF) from § 6.4 will we use for the fugacities? (b) Which models (ideal gas, vapor-pressure correlations, Redlich-Kwong-Soave equation of state, Wilson equations, etc.) will we use as the basis for introducing numbers into the problem? The first decision leads to one of three common strategies for solving phase-equilibrium problems: the phi-phi method, the gamma-phi method, and the gamma-gamma method.

10.1.1 The Phi-Phi Method for VLE

If we choose to use FFF #1 for all components in both phases, then we have selected the phi-phi method, and the equilibrium conditions (7.3.12) become

$$x_i^{\alpha} \varphi_i^{\alpha} P = x_i^{\beta} \varphi_i^{\beta} P \qquad i = 1, 2, \ldots, C \tag{10.1.2}$$

Since the pressures are the same in the two phases, this reduces to

$$x_i^{\alpha} \varphi_i^{\alpha} = x_i^{\beta} \varphi_i^{\beta} \qquad i = 1, 2, \ldots, C \tag{10.1.3}$$

The fugacity coefficients are to be obtained from a model for the $PvTx$ equation of state. The volume explicit form, $v(P, T, \{x\})$, should *not* be used for multiphase systems; instead, a pressure explicit model, $P(T, v, \{x\})$, should be chosen. Then the fugacity coefficients would be computed from the integral in (4.4.23).

In the past, the phi-phi method was little used either because volumetric equations of state did not reliably reproduce the behavior of liquids and dense fluids or because the computational difficulties were too great. But in recent years, significant improvements have been made in equations of state and in computers, so that now the phi-phi method is often the method of choice for vapor-liquid equilibria and for high-pressure liquid-liquid and gas-gas equilibria.

When the same equation of state is used to represent two-phase equilibria, the fugacity must have the same value at two different mole fractions x_1 at fixed T and P. This is illustrated in Figure 10.1 for a binary mixture that can exhibit VLE over a range of states. The figure shows the fugacity of one component at five pressures along the 275 K isotherm. At this temperature, the mixture critical pressure occurs near 92 bar, while the pressure of the mechanical critical point occurs near 45 bar. At supercritical states along the isobar at 100 bar, the fugacity is single-valued in x_1, the fugacity satisfies the diffusional stability criterion (8.4.8) at every x-value, and no phase separation occurs. But at 60 bar, which lies below the critical point but above the mechanical critical point, the fugacity passes through a loop, the diffusional stability criterion is violated over a range of x_1, and VLE occurs.

If the pressure is reduced to 30 bar, below the mechanical critical point, the loop in the fugacity becomes more pronounced. Here f_1 violates the diffusional stability criterion ($\partial f_1/\partial x_1 > 0$ for stability) over only a small range of x_1; but the mechanical stability criterion ($\kappa_T > 0$) is also violated at states between the extrema in f_1. Finally, at the lowest pressure (10 bar), the loop in f_1 has completely closed, dividing f_1 into two parts: a vapor part that is linear and obeys the ideal-gas law, and a fluid part that includes stable and metastable liquid states at small x_1 values plus mechanically unstable fluid states at higher x_1 values. The broken horizontal lines in Figure 10.1 are vapor-liquid tie lines, computed by solving the phi-phi equations (10.1.3) simultaneously for both components.

The behavior of the fugacity shown in Figure 10.1 is representative, but it is not the only way that volumetric equations of state can produce changes in $f_1(x_1)$ with changes in state. For example, at 10 bar, but at temperatures well below 275 K, the liquid branch of f_1 extends over all x_1, and it is the vapor branch that can form a closed loop. At still other states, both the liquid and the vapor branches extend over all x_1, and no loop (open or closed) occurs at all. These possibilities are discussed briefly in § 8.4.2 and in more detail elsewhere [1]. The lesson here is that even simple cubic equations of state can provide relatively complicated forms for the fugacity, forms sufficiently complicated to satisfy the phi-phi equations (10.1.3) for phase equilibria.

Conceptually, the simplest method for solving phase-equilibrium problems is the phi-phi method, but computationally it is usually more complicated than other methods. The conceptual simplifications arise in part because no decisions need to be made about reference states: the reference state is the ideal gas and the choice of the ideal-gas reference is implicit in choosing to work with fugacity coefficients. Usually, the same pressure-explicit equation of state is used for all components in all phases, for this produces consistency in the results and helps in organizing the calculations. (The same calculations are to be done for all components in all phases, and therefore computer programs can be structured in obvious modular forms.) However, this need not be done; different equations of states can be used for different phases.

Offsetting these advantages in the phi-phi method are certain disadvantages that must be seriously considered. Instead of decisions about reference states, we are faced with decisions about mixing rules: how should we represent the composition dependence of parameters appearing in our equation of state (such as a and b in a cubic)? We have briefly mentioned this problem in § 4.5.12, but the problem is thorny and beset with pitfalls. The computations themselves tend to be complicated on two counts: (i) to obtain values for the φ_i we must perform nontrivial evaluations of the derivatives and integrals that appear in (4.4.23), and (ii) to actually solve the phi-phi equations (10.1.3) we must perform a trial-and-error search for the roots to a coupled set of non-linear algebraic equations. Such searches involve several complications; for example, at each step in the search, the molar volumes for each phase must be recomputed from

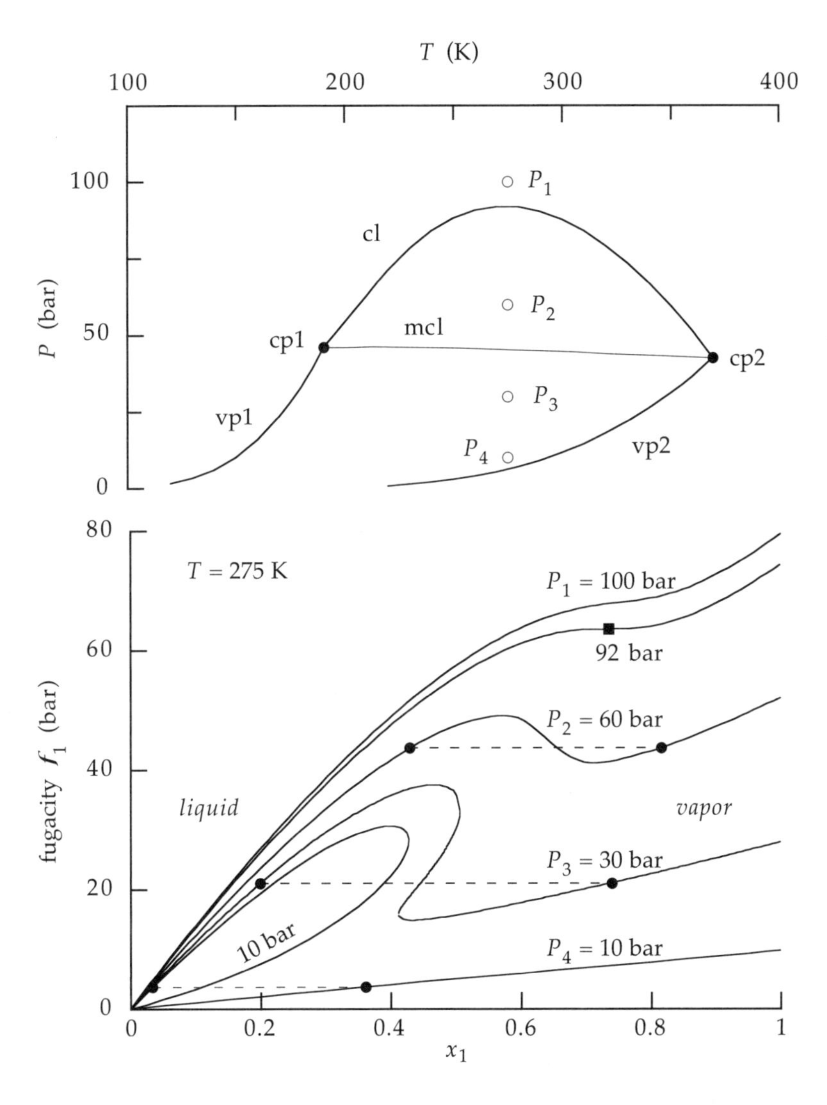

Figure 10.1 When computed from an analytic equation of state using FFF #1, the fugacity vs. composition curve may change significantly with state condition. *Top*: *PT* diagram for a binary mixture. Filled circles are pure critical points; vp1 and vp2 are pure vapor-pressure curves; cl = critical line; mcl = mechanical critical line. *Bottom*: Corresponding fugacity of the more volatile component at 275 K. Broken lines are vapor-liquid tie lines. Isobars at bottom correspond to open circles at top. Bottom same as Figure 8.13. Computed from Redlich-Kwong equation.

the equation of state. This means we actually solve for $C + 2$ variables, rather than just the C required variables. For cubics this can be done analytically, but more complicated models require additional trial-and-error searches. Complications in equations of state and mixing rules can improve accuracy and reliability, but those advantages are purchased at the expense of increased complications in algorithms that are used to solve the equations.

10.1.2 The Gamma-Phi Method for VLE

In this method we choose FFF #1 for the components in the vapor (α), but we use one of FFF #2–5 for the components in the liquid (β). Therefore, we have four general versions of the gamma-phi method; however, at low pressures FFF #2–5 are all equivalent, and we have this one general form:

$$x_i^\alpha \varphi_i^\alpha(T, P, \{x^\alpha\})P = x_i^\beta \gamma_i^\beta(T, \{x^\beta\}) f_i^o(T) \qquad i = 1, 2, \ldots, C \qquad (10.1.4)$$

That is, at low pressures we ignore the pressure dependence of all activity coefficients and all standard-state fugacities. In the β phase, values for the activity coefficients depend on the choice made for the standard-state fugacity; for example, if the Lewis-Randall standard state is chosen for all components (5.1.5), then the γ_i would be obtained from a model for the excess Gibbs energy. Common choices for the standard state are discussed in § 10.2. In the α phase, values for the fugacity coefficients are obtained from a volumetric equation of state; now, either pressure-explicit or volume-explicit models may be chosen. Fortunately at low pressures, either the ideal-gas law or a virial equation may be sufficiently accurate.

The gamma-phi method is illustrated graphically in Figure 10.2 for low-pressure vapor-liquid equilibria in mixtures of ethanol + benzene. These mixtures exhibit azeotropes, as shown in Figure 9.9. Here, we set the temperature to 370 K and the pressure to 2.5 bar. Then vapor-phase fugacities were computed from FFF #1 using the ideal-gas equation of state, while the liquid-phase fugacities were computed from FFF #2 using the Margules equations for activity coefficients, (5.6.12) and (5.6.13). At the specified T and P, phase equilibrium occurs at those mole fractions at which the gamma-phi equations (10.1.4) are satisfied simultaneously by *both* components. Although mixtures of ethanol + benzene form azeotropes, an azeotrope is not observed at 370 K and 2.5 bar; instead, there are two sets of (x_1, y_1) values that satisfy the gamma-phi equations. So in this situation, the gamma-phi equations have two roots, representing two tie lines for VLE.

At high pressures, the rhs of (10.1.4) should be replaced by one of FFF #3–5. Which of those to choose depends on what other data are available for the component; general considerations have been discussed in § 6.4. When FFF # 3 or 5 is chosen, a Poynting factor must be evaluated, often using a volumetric equation of state for the liquid; that equation of state need not be the same as the one used for obtaining the fugacity coefficients for the vapor (phase α).

The gamma-phi method has served as the traditional way for solving low-pressure vapor-liquid equilibrium problems. The elimination of pressure (and hence volume) from the representation of one phase considerably simplifies the calculations relative to the phi-phi method, though the calculations generally still involve trial-and-error searches. The disadvantages of the gamma-phi method include the three different decisions that must be made for modeling the situation: (a) an equation-of-state model must be chosen for the vapor phase, (b) a model must be selected for the fugacity in the standard-state, and (c) a model must be chosen for the composition dependence of the activity coefficients. At high pressures an additional disadvantage occurs because inconsistencies may appear in the results computed from the different models used for the two phases; such inconsistencies are most prominent near critical points.

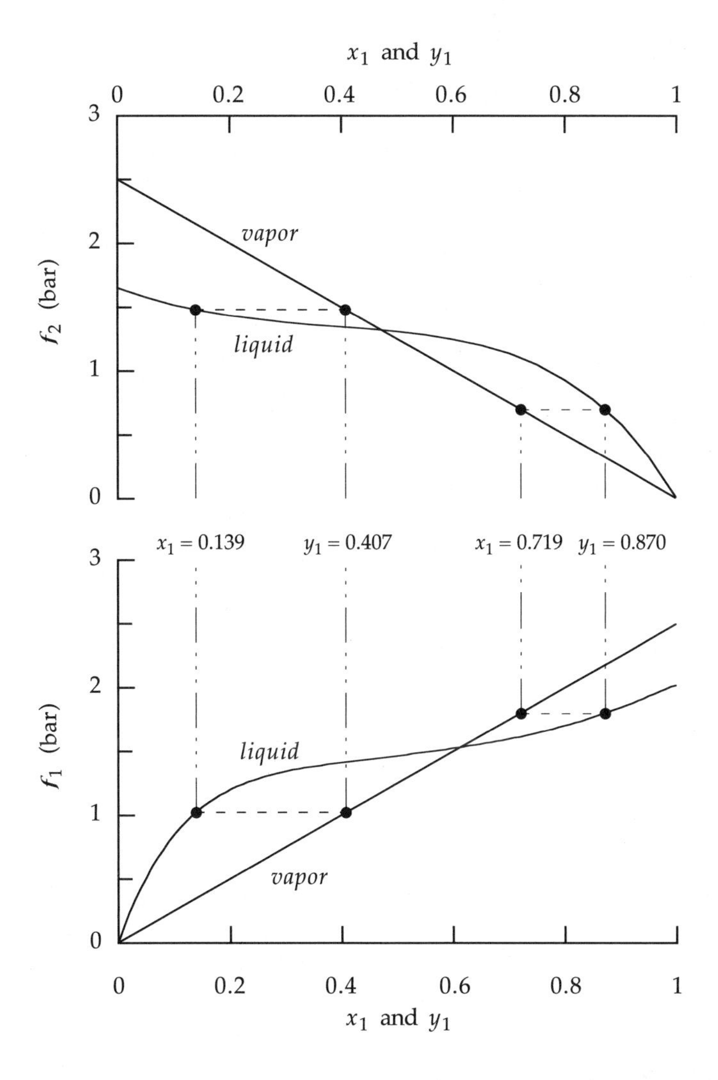

Figure 10.2 Low-pressure vapor-liquid equilibrium computed from the gamma-phi method for binary mixtures of ethanol(1) + benzene(2) at 370 K and 2.5 bar. *Top*: Fugacities of component 2 in each phase; *Bottom*: Fugacities of component 1 in each phase. The vapor-phase fugacities (solid straight lines) were computed from FFF #1 using the ideal-gas law; the liquid-phase fugacities (solid curves) were computed from FFF #2 using the Margules model with parameters taken from Appendix E. Broken horizontal lines are tie lines for VLE. An azeotrope forms near this temperature and pressure, so two pairs of tie lines occur. Note that the fugacities in each phase obey the Gibbs-Duhem equation.

10.1.3 The Gamma-Gamma Method for LLE

If we choose to use one of FFF #2–5 for all components in both phases, then we have the gamma-gamma method. Invariably, the same FFF is used for both phases, so there are four common versions of this method; but in principle, we could use any of the four fugacity formulae 2–5 for each phase, so that for two-phase equilibrium, we have ten possible versions of the method. At low pressures, the FFF #2–5 are equivalent, and the gamma-gamma method simplifies to this one form

$$x_i^\alpha \gamma_i^\alpha(T, \{x^\alpha\}) f_i^{o\alpha}(T) = x_i^\beta \gamma_i^\beta(T, \{x^\beta\}) f_i^{o\beta}(T) \qquad i = 1, 2, \ldots, C \qquad (10.1.5)$$

Often the same standard state is chosen for a component in both phases, so that (10.1.5) reduces further to

$$x_i^\alpha \gamma_i^\alpha(T, \{x^\alpha\};\ f_i^o(T)) = x_i^\beta \gamma_i^\beta(T, \{x^\beta\};\ f_i^o(T)) \qquad i = 1, 2, \ldots, C \qquad (10.1.6)$$

Now the standard-state fugacity no longer enters explicitly into the calculation; however, the notation in (10.1.6) reminds us that the choice of standard state still affects the values for the γs. Common choices for standard states are discussed in §10.2.

The gamma-gamma method is illustrated graphically in Figure 10.3 for a binary mixture that exhibits liquid-liquid equilibrium. For this example, the fugacities were computed from the low-pressure form of FFF #2 using Porter's equations for the activity coefficients, (5.6.4) and (5.6.5). The value of the Porter parameter ($A = 2.4$) corresponds to the temperature $T = 30°C$ that occurs on the Txx diagram in Figure 8.20. Note that, since the standard-state fugacities do not appear in (10.1.6), the ordinates in Figure 10.3 have been normalized by those values; so in the Lewis-Randall standard state, the plotted ratios f_i/f_i^o go to unity in the pure-component limits. At the fixed T and P of the figure, phase equilibrium occurs at the values of x_1 at which the gamma-gamma equations (10.1.6) are satisfied simultaneously by each component. The symmetry in f_1 and f_2, evident in Figure 10.3, results from the simplicity of the Porter equations and does not occur for most binary mixtures.

The gamma-gamma method should be used only for computing equilibria among condensed phases, though this can include not only liquid-liquid but also liquid-solid and solid-solid equilibria. At low pressures, the forms (10.1.5) and (10.1.6) are simple to set up, but finding solutions may be problematic because we are seeking multiple roots from a single model for activity coefficients. At high pressures, the full forms for FFF #2–5 must be used, and then we have the same computational difficulties as already discussed in § 10.1.2 and § 6.4.

10.1.4 The Gamma-Gamma Method for LSE

Liquid-solid equilibria are attacked with the gamma-gamma method in the same general way as liquid-liquid systems; however, the two applications differ in how the standard-state fugacities are treated. We still start from the equality of fugacities,

$$x_i^\ell \gamma_i^\ell f_i^{o\ell}(T) = x_i^s \gamma_i^s f_i^{os}(T) \qquad (10.1.7)$$

For the solid-phase standard state, we use pure solid i at the system T and P; for the liquid-phase standard state, we use the pure liquid at T and P. If the system temperature happens to be on the melting line for pure i, so $T = T_m$ at P, then the standard-state fugacities are equal, $f_i^{o\ell} = f_i^{os}$, and they cancel from (10.1.7). In this case, (10.1.7) simplifies to $x_i^\ell \gamma_i^\ell = x_i^s \gamma_i^s$.

But if $T < T_m(P)$, then the standard-state fugacities do not cancel, and the standard state for the liquid phase is the *subcooled* liquid at T; nevertheless, that subcooled state is well-defined and can be legitimately used as a standard state. The standard-state fugacities can be handled easily, if we rearrange (10.1.7) so that the ratio $f_i^{os}/f_i^{o\ell}$ appears rather than the individual terms. If we have models for the activity coeffi-

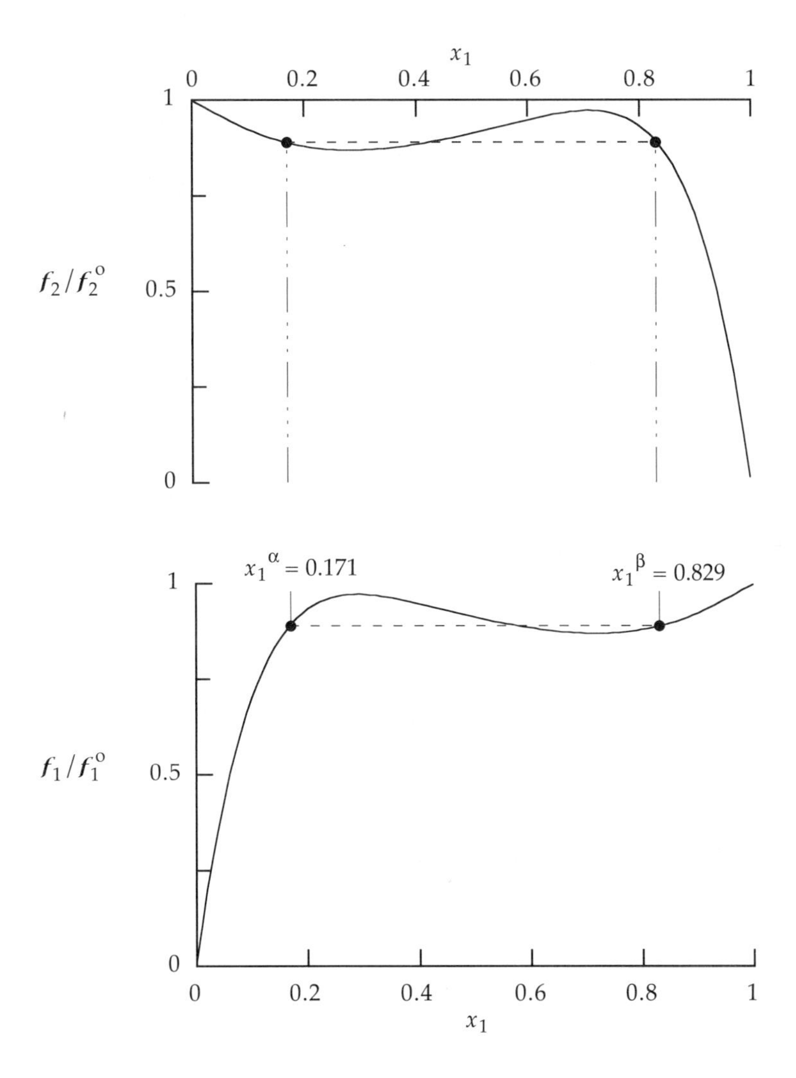

Figure 10.3 Liquid-liquid equilibrium computed from the gamma-gamma method for a binary mixture at fixed T and P. *Top*: Fugacity of component 2. *Bottom*: Fugacity of component 1. The fugacities were computed from the low-pressure form of FFF #2 using Porter's equations with A = 2.4. Broken horizontal lines are tie lines.

cients and if we know the solid-phase mole fraction x_i^s, then that ratio is sufficient to allow us to solve (10.1.7) for the liquid-phase mole fraction x_i^ℓ,

$$x_i^\ell = x_i^s \left(\frac{\gamma_i^s}{\gamma_i^\ell}\right)\left(\frac{f_i^{os}}{f_i^{o\ell}}\right) \tag{10.1.8}$$

To evaluate the ratio of standard-state fugacities, we first integrate the definition of the fugacity from the subcooled liquid at T to pure solid i at T,

$$\ln\left(\frac{f_i^{os}(T)}{f_i^{o\ell}(T)}\right) = \frac{-\Delta g_i(T)}{RT} \tag{10.1.9}$$

where Δg_i is the difference between the pure solid and pure subcooled-liquid molar Gibbs energies at T. To correct the rhs of (10.1.9) from the melting temperature T_m of pure i to the system temperature T, we perform a double integration of the Gibbs-Helmholtz equation, as in § 3.3.2. This gives

$$\ln\left(\frac{f_i^{os}(T)}{f_i^{o\ell}(T)}\right) = \frac{-\Delta h_i^m(T_m)}{R}\left(\frac{1}{T} - \frac{1}{T_m}\right) + \int_{T_m}^{T} \frac{\Delta c_p^o(\tau)}{R}\left(\frac{1}{\tau} - \frac{1}{T}\right) d\tau \tag{10.1.10}$$

where Δh_i^m is the latent heat of melting for pure component i and $\Delta c_p = c_p^\ell - c_p^s$ is the difference in pure-i heat capacities across the melting line.

Equation (10.1.10) is rigorous, but in using it, simplifying assumptions are usually made. One approximation is to take c_p^s from thermal experiments but estimate c_p^ℓ by extrapolating liquid values measured at states above the pure-i melting point T_m. An alternative is to approximate Δc_p from the latent heat via

$$\Delta c_p^o = T_m\left(\frac{\partial \Delta s_i^m}{\partial T}\right)_P \approx T_m\, \Delta h_i^m\left(\frac{\partial(1/T)}{\partial T}\right)_P = \frac{-\Delta h_i^m}{T_m} \tag{10.1.11}$$

A third approximation is to neglect the second term on the rhs of (10.1.10). In any case, when $T < T_m$ then $f_i^{os} < f_i^{o\ell}$, so we expect the rhs of (10.1.10) to be negative. Typical applications of this approach are to determine the freezing point depression when a pure liquid is contaminated by a solid impurity or to estimate eutectic temperatures, such as when salt is spread on highways to prevent icing.

10.2 CHOICES FOR STANDARD STATES IN GAMMA METHODS

The gamma-phi and gamma-gamma methods involve activity coefficients, which measure how component fugacities in a real mixture deviate from those in an ideal solution. Many kinds of ideal solutions are available, and yet all have one common attribute: *every* ideal solution has each fugacity linear in its mole fraction,

$$f_i^{is}(T, P_i^o, \{x\}) = x_i f_i^o(T, P_i^o) \tag{5.4.7}$$

and the proportionality constant is called the *standard-state fugacity.* Therefore, before we can obtain values for activity coefficients, we must choose a particular ideal solution; hence, we must choose a standard-state for each component. The standard state *must* be at the same temperature as the real mixture, but the standard-state pressure P_i^o and its phase can both be chosen for computational convenience. The generic relation (5.4.7) represents any of an infinite number of straight lines on an isothermal plot of f_i vs. x_i; hence, there are an infinite number of ideal solutions.

But while we can pick any straight line and use it to represent an ideal-solution fugacity, in practice we always choose a line that intersects or lies tangent to the curve for the real fugacity at the standard-state pressure. This means that we choose $f_i^{is}(T, P_i^o, \{x^o\}) = f_i(T, P_i^o, \{x^o\})$ at some composition $\{x^o\}$; then, at that composition, the activity coefficient must be unity. At other mole fractions, the fugacity of the ideal solution is given by the equation for the straight line,

$$f_i^{is}(T, P_i^o, \{x\}) = f_i(T, P_i^o, \{x^o\}) + \Delta x_i \left(\frac{\partial f_i^{is}}{\partial x_i}\right)_{TP} \tag{10.2.1}$$

Here $\Delta x_i = x_i - x_i^o$ and the derivative is constant with composition because it represents the slope of the ideal-solution straight line. For binary mixtures, this derivative is well defined; but for multicomponent mixtures, it is not, because many ways exist to vary one mole fraction while constraining others. The resolution of this ambiguity provides alternative standard states for multicomponent mixtures, as we shall see.

When we choose a standard state, we are merely identifying a particular ideal solution on which to base an activity coefficient. The standard state may be real or hypothetical, so long as it is well-defined and so long as a value for its fugacity can be obtained. Ultimately, the choice of standard state is made for computational convenience; normally, this means either that reliable models for γ_i exist, or else that the value of γ_i is close to unity over the states of interest. When neither of these conditions pertain, we should consider changing the standard state. In many situations the appropriate choice is one of the possibilities discussed in § 10.2.1–10.2.3; however, when the mole fraction is not a convenient measure of composition, such as occurs for mixtures of electrolytes or of polymers, then other standard states may be preferred.

10.2.1 Fugacities Based on Pure-Component Standard States

Often we know or can easily compute values of the fugacity for the pure component at the mixture T and at some convenient pressure P_i^o; then it is natural to base the definition of the ideal solution on this known pure-component fugacity. To do so, we choose $x_i^o = 1$; then the slope of the ideal-solution straight line is

$$\left(\frac{\partial f_i^{is}}{\partial x_i}\right)_{TP} = f_i^{is}(T, P_i^o, x_i = 1) = f_{\text{pure i}}(T, P_i^o) \tag{10.2.2}$$

and the generic expression (10.2.1) for the ideal solution becomes

$$f_i^{is}(T, P_i^o, x_i) = x_i f_{\text{pure i}}(T, P_i^o) \tag{10.2.3}$$

The choice

$$f_i^o(T, P_i^o) = f_{\text{pure i}}(T, P_i^o) \tag{10.2.4}$$

is called a *pure-component standard state*; it was introduced in § 5.4.1. Note we have made no restriction as to phase: we may use the ideal-solution expression (10.2.3) to model liquid mixtures, gas mixtures, and solid mixtures. Gas-phase ideal solutions differ from ideal-gas mixtures.

In a generic pure-component standard state, the activity coefficient is expressed as

$$\gamma_i(T, P, \{x\};\ f_{\text{pure i}}(T, P_i^o)) = \frac{f_i(T, P, x_i)}{x_i\, f_{\text{pure i}}(T, P_i^o)} \tag{10.2.5}$$

which is an example of FFF #4. For the special case in which $P_i^o = P$, then (10.2.4) becomes the Lewis-Randall rule (5.1.5), (10.2.3) becomes (5.1.6) for the Lewis-Randall ideal solution, and (10.2.5) becomes

$$\gamma_i(T, P, \{x\};\ f_{\text{pure i}}(T, P)) = \frac{f_i(T, P, x_i)}{x_i\, f_{\text{pure i}}(T, P)} \tag{10.2.6}$$

which is a particular example of FFF #2. In this case the pure-component limit is unity, as given in (5.4.12), and the infinite-dilution limit (5.4.13) defines the infinite-dilution activity coefficient. Activity coefficients greater than unity indicate positive deviations from the Lewis-Randall ideal solution, while values less than unity indicate negative deviations. Most binary mixtures are positive deviates.

If a value for a pure-component fugacity cannot be obtained at the mixture pressure P, then one may be available at the pure vapor pressure $P_i^s(T)$. In such cases we use a Poynting factor to correct the known fugacity to the system pressure,

$$f_{\text{pure i}}(T, P) = f_{\text{pure i}}(T, P_i^s) \exp\left[\frac{1}{RT}\int_{P_i^s}^{P} v_{\text{pure i}}(T, \pi)\, d\pi\right] \tag{10.2.7}$$

Using FFF #1 for the fugacity at P_i^s, and substituting (10.2.7) into (10.2.6), we find

$$\gamma_i(T, P, \{x\};\ f_i^o(T, P)) = \frac{f_i(T, P, x_i)}{x_i\, \varphi_i^s(T)\, P_i^s(T)\, \exp\left[\frac{1}{RT}\int_{P_i^s}^{P} v_{\text{pure i}}(T, \pi)\, d\pi\right]} \tag{10.2.8}$$

where the fugacity coefficient φ_i^s is for the pure saturated liquid or vapor; it is more readily evaluated for the saturated vapor. The molar volume is for the pure condensed phase. Equation (10.2.8) is a particular example of FFF #3.

For pure condensed phases, the Poynting factor in (10.2.8) is straightforward to compute, but unless $P \gg P_i^s$, it is usually small enough to neglect. For example, for liquid water at 25°C and 10 bar, the error introduced by neglecting the Poynting factor is less than 1%. Therefore, for condensed phases at low pressures, we usually approximate the pure-component standard-state fugacity (10.2.7) by

$$f_{\text{pure i}}(T, P) \approx f_{\text{pure i}}(T, P_i^s) = \varphi_i^s(T)\, P_i^s(T) \tag{10.2.9}$$

When we cannot obtain a pure-component fugacity at the mixture pressure P, but we can get a value at another pressure P_i^o, then we may elect to evaluate the activity coefficient at P_i^o, rather than at P. Now, instead of (10.2.6), we use an activity coefficient defined by

$$\gamma_i(T, P_i^o, \{x\}; f_{\text{pure i}}(T, P_i^o)) = \frac{f_i(T, P_i^o, x_i)}{x_i\, f_{\text{pure i}}(T, P_i^o)} \tag{10.2.10}$$

and then we use FFF #5 to evaluate the fugacity at the mixture pressure P. Note that the activity coefficient in (10.2.10) does not depend on the system pressure P; this is the basis for those models, such as Wilson equations, which contain no pressure dependence. Furthermore, it is common to use (10.2.10) whenever the pressure is low, and then there is no distinction among FFF #2–FFF #5.

10.2.2 Example

What approximations might be considered when reducing the equality of fugacities to computational forms for solving phase-equilibrium problems?

Consider low-pressure, multicomponent, vapor-liquid equilibrium (VLE). The relevant measurables are T, P, the liquid-phase mole fractions $\{x\}$, and the vapor-phase mole fractions $\{y\}$; see Figure 10.4. A typical problem is that we know T and $\{x\}$ and we need to compute P and $\{y\}$. Note this is an $\mathcal{F}$-problem (see § 9.1).

Every phase-equilibrium problem is solved by starting from the phase-equilibrium conditions (7.3.12), which express the equality of fugacities,

$$f_i^\ell = f_i^v \qquad i = 1, 2, \ldots, C \tag{10.2.11}$$

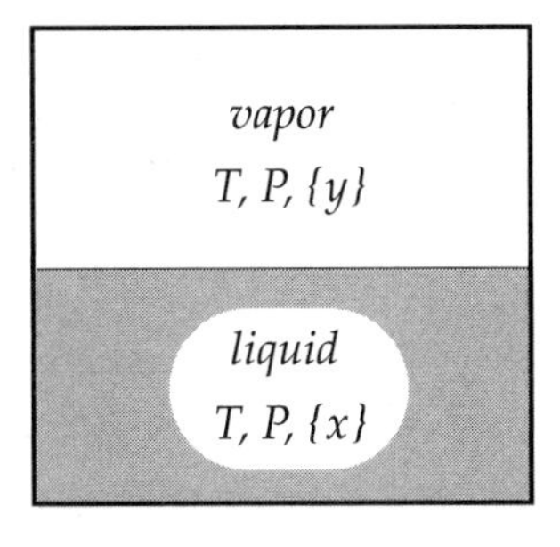

Figure 10.4 Schematic of a closed system containing a mixture of any number of components in a vapor-liquid equilibrium situation

Note we *always* start with an equation that is *always true*. But before we can perform computations, (10.2.11) must be reduced to a form involving measurables. Our purpose here is to illustrate the reduction procedure and to consider options for making approximations.

For this low-pressure VLE situation, we choose the gamma-phi method (10.1.4), using fugacity coefficients for the vapor phase and activity coefficients for the liquid phase. In particular, we choose FFF #1 for the vapor and since we are considering only low pressures, any one of FFF #2–5 for the liquid; at low pressures, FFF #2–5 are all the same. Then, the gamma-phi equations take the form

$$x_i \gamma_i f_i^o = y_i \varphi_i P \qquad i = 1, 2, \ldots, C \qquad (10.2.12)$$

For the liquid-phase ideality we choose the Lewis-Randall ideal solution (5.1.6), and since P is low, we approximate f_i^o using (10.2.9). Then (10.2.12) becomes

$$x_i \gamma_i \varphi_i^s P_i^s = y_i \varphi_i P \qquad i = 1, 2, \ldots, C \qquad (10.2.13)$$

where the fugacity coefficients φ_i^s and vapor pressures P_i^s depend only on the mixture T because they are for pure saturated components. Values for the vapor pressures would be computed from an appropriate model, such as the Antoine correlation in Appendix D. Recall that in the pure limit, $\gamma_i = 1$. In addition, if we retain VLE in the pure limit (so $P \to P_i^s$ at fixed T), then we maintain full consistency only if $\varphi_i \to \varphi_i^s$ when $x_i \to 1$ and $y_i \to 1$.

In general, the fugacity coefficients appearing in (10.2.13) would be computed from a volumetric equation of state using (4.4.10) or (4.4.23). However, if the pressure is sufficiently low over the entire composition range, then the vapor mixture may be an ideal gas, so all $\varphi_i = \varphi_i^s = 1$, or it may obey Amagat's "law", so all $\varphi_i / \varphi_i^s \approx 1$. In either case, (10.2.13) reduces to

$$x_i \gamma_i P_i^s = y_i P \qquad i = 1, 2, \ldots, C \qquad (10.2.14)$$

This is a computationally convenient form for solving low-pressure VLE problems. The only remaining conceptuals are the γ_i, which we assume depend only on T and $\{x\}$; their values are usually obtained from an appropriate model, such as the Margules or Wilson equations in Chapter 5. In general, (10.2.14) constitutes C nonlinear algebraic equations that can be solved for C unknowns from T, P, $\{x\}$, and $\{y\}$.

In the very special case that the system is composed only of molecules having similar intermolecular forces, so that the liquid is indeed a Lewis-Randall ideal solution, then all the $\gamma_i = 1$, and (10.2.14) simplifies to

$$x_i P_i^s = y_i P \qquad i = 1, 2, \ldots, C \qquad (10.2.15)$$

This is *Raoult's law,* which applies *only* for ideal gases in equilibrium with ideal solutions. It assumes that every component has vapor and liquid-phase fugacities that are linear in the mole fractions. If T is known, then the C equations in (10.2.15) can be solved analytically for C unknowns. For example in our problem here, we know T and $\{x\}$, we are to compute P and $\{y\}$, and the strategy is to eliminate the unknown mole fractions $\{y\}$ by summing (10.2.15) over all components; this yields the pressure,

$$P = \sum_{i}^{C} x_i P_i^s \tag{10.2.16}$$

With P determined, we can solve Raoult's law (10.2.15) for each y_i. A similar strategy would be used if we knew T and $\{y\}$ and had to compute P and $\{x\}$. However, Raoult's law is nonlinear in T (through the vapor pressures), so if T is unknown, then (10.2.15) must be solved by trial.

Note that we start from a rigorous expression (10.2.12) that applies to *any* VLE situation. Then we systematically introduced three levels of approximation:

(a) If the system pressure is low enough that its effects on liquid-phase fugacities can be ignored, then we have (10.2.13), which applies to nonideal gases in equilibrium with nonideal liquid solutions.

(b) If in addition to (a), the pressure is so low that the fugacity coefficients are independent of composition, then we have (10.2.14): the situation is like an ideal-gas mixture in equilibrium with a nonideal liquid solution.

(c) If in addition to (a) and (b), the molecules are all so similar that the liquid is essentially a Lewis-Randall ideal solution, then Raoult's law (10.2.15) applies.

To decide which of these can be used in a particular situation, we must know how liquid and gas phase properties are affected (i) by common operating variables, such as temperature and pressure, and (ii) by differences among the molecules that determine the nonidealities. In short, we must exercise engineering judgement.

10.2.3 Fugacities Based on the Solute-Free, Henry's Law, Ideal Solution

Although pure-component standard states are the ones most commonly used, situations arise in which a pure-liquid fugacity is unknown or difficult to determine. These situations occur, for example, when the mixture temperature T is above the critical temperature of the pure component (the gas solubility problem) and when T is below the pure-component melting temperature (the solid solubility problem). In such cases, we seek alternatives to the pure-component standard state. One way is to exploit any data available for mixtures that contain only small amounts of the component; however, we emphasize that this approach does not require the real mixture to be dilute in that component. We are merely seeking an alternative to pure-component data to use as a basis for defining an ideal solution.

Therefore we define a new kind of ideal solution by choosing $x_i^o = 0$ in the generic form (10.2.1). Since $f_i = 0$ when $x_i^o = 0$, (10.2.1) reduces to

$$f_i^{is}(T, P_i^o, \{x\}) = x_i \left(\frac{\partial f_i^{is}}{\partial x_i}\right)_{TP} \tag{10.2.17}$$

To get a value for the slope of the ideal-solution straight line, we use the slope of the real fugacity curve in the dilute-solution limit,

$$\left(\frac{\partial f_i^{is}}{\partial x_i}\right)_{TP} = \lim_{x_i \to 0} \left(\frac{\partial f_i}{\partial x_i}\right)_{TP} \tag{10.2.18}$$

For binary mixtures, the derivative on the rhs is well-defined, but for multicomponent mixtures it is not. For example, consider a ternary and let $i = 1$. Then there are an infinite number of binary mixtures that can be formed from components 2 and 3 when we set $x_1 = 0$. Each of those binaries may produce a different value for the limiting derivative in (10.2.18).

One way to remove this ambiguity is to identify a set of components as *solutes*; we use $\{s\}$ to indicate that set. The component of interest is one of the solutes, $i \in \{s\}$. The remaining components are identified as *solvents*; we use $\{sf\}$ to indicate that the set of solvents is solute-free. We can define solute-free mole fractions for the solvents by

$$x_j^{sf} = \frac{x_j}{\sum_{\{sf\}} x_k} \qquad j \in \{sf\} \tag{10.2.19}$$

where the sum runs only over solvents.

We choose to evaluate the derivative on the rhs in (10.2.18) by taking all solute mole fractions to zero while holding fixed the solute-free mole fractions of the solvents; the resulting derivative is called the *solute-free Henry's constant*,

$$H_{is}(T, P, \{x^{sf}\}) \equiv \lim_{\{x^s\} \to 0} \left(\frac{\partial f_i}{\partial x_i}\right)_{TP\{x^{sf}\}} = \lim_{\{x^s\} \to 0} \left(\frac{f_i}{x_i}\right)_{TP\{x^{sf}\}} \tag{10.2.20}$$

Henry's constants are intensive, measurable properties having dimensions of pressure; (10.2.20) indicates that the solute-free Henry's constant depends on temperature, pressure, and the solute-free mole fractions, but it does not depend on the solute mole fractions. With (10.2.20), the fugacity of a solute i in a Henry's law ideal solution is, as required, linear in the mole fraction of i,

$$f_i^{is}(T, P, \{x\}) = x_i H_{is}(T, P, \{x^{sf}\}) \qquad i \in \{s\} \tag{10.2.21}$$

For binary mixtures with $x_i < 0.03$, this often reliably estimates the fugacity for component i; that is, near $x_i = 0$, the real fugacity becomes linear in the mole fraction.

When the solute-free mixture is a condensed phase at T and P, we might find a value for H_{is} at the saturation pressure of the solvent mixture P_{sf}^s, rather than at the mixture pressure P. When this happens we can apply a Poynting factor to correct the known Henry's constant,

$$H_{is}(T, P) = H_{is}(T, P_{sf}^s) \exp\left[\frac{1}{RT} \int_{P_{sf}^s}^{P} \overline{V}_{is}^{\infty}(T, \pi) d\pi\right] \tag{10.2.22}$$

Here the partial molar volume is evaluated in the infinite-dilution limit with solvent-free mole fractions held fixed,

$$\overline{V}_{is}^{\infty} = \lim_{\{x^s\} \to 0} \overline{V}_i(T, P, \{x^{sf}\}) \tag{10.2.23}$$

For solid solubility problems, the Poynting factor in (10.2.22) can be ignored, but it must be included when analyzing gas solubility problems at high pressures.

For binary mixtures we can use a simple plot, as in Figure 10.5, to compare the Henry's law ideal solution to the Lewis-Randall ideality. The plot shows the real fugacity for component 1, as well as the Lewis-Randall and Henry's law straight lines. The Lewis-Randall fugacity coincides with the real value at $x_1 = 0$ and at $x_1 = 1$, but the Henry's law fugacity coincides only at $x_1 = 0$. Also, since x_1 lies on [0, 1], the intercept of the Henry's law curve at $x_1 = 1$ is the Henry's constant at the given T and P.

Since the value of an activity coefficient depends on the standard state, an activity coefficient based on (10.2.21) will differ numerically from one that is based on a pure-component standard state. To emphasize that difference, we make a notational distinction between the two: we use γ for an activity coefficient in a pure-component standard state and use γ^* for an activity coefficient in the solute-free infinite-dilution standard state. Then for γ^*, the generic definition of the activity coefficient (5.4.5) gives

$$\gamma_i^* = \frac{f_i(T, P, \{x\})}{x_i H_{is}(T, P, \{x^{sf}\})} \tag{10.2.24}$$

The normalization occurs in the dilute-solution limit, taken with T, P, and solute-free mole fractions held fixed,

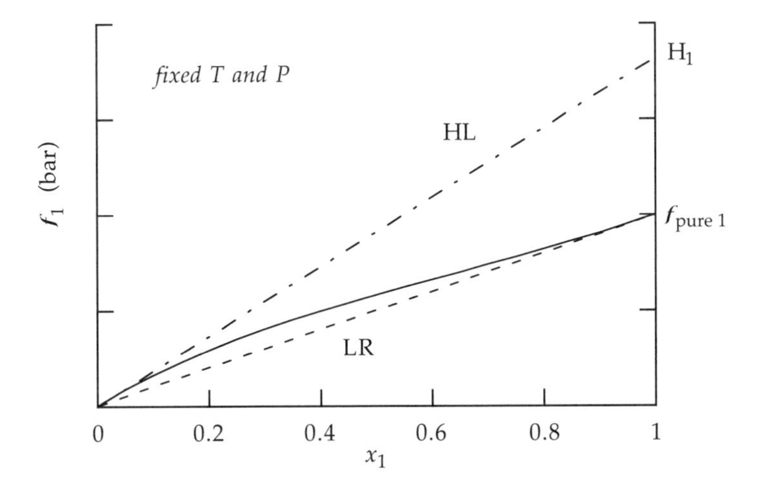

Figure 10.5 Fugacity for one component in a binary mixture at fixed T and P (solid line). Upper broken line is the fugacity for the Henry's law (HL) ideal solution; lower broken line is that for the Lewis-Randall (LR) ideal solution. Both idealities are based on standard states that are pure fluids (intercepts at $x_1 = 1$): LR uses the fugacity for real pure 1, while HL uses the fugacity (H_1) of a hypothetical pure fluid. Scale on ordinate arbitrary.

$$\lim_{\{x^s\}\to 0} \gamma_i^* = 1 \tag{10.2.25}$$

Otherwise we usually have $\gamma_i^* < 1$, as suggested by Figure 10.5; in contrast, in the Lewis-Randall standard state we usually have $\gamma_i > 1$, also suggested by Figure 10.5. This means that deviations from Lewis-Randall ideal-solution behaviors differ qualitatively from deviations from Henry's law ideal-solution behaviors.

Figure 10.5 shows that the Henry's constant is not only the slope of a real fugacity curve at infinite dilution, but it is also the intercept of the ideal-solution line at $x_1 = 1$. It is the intercept that is a fugacity, not the slope, and therefore the standard state here is *not* a mixture at infinite dilution, in fact, it is not a mixture at all. Instead, the standard state is a hypothetical *pure* substance whose fugacity equals the Henry's constant H_{1s}. This point is important, but subtle; it can lead to confusion, because, although the standard state is a pure substance, the identity of that pure substance changes when we change the solute-free mole fractions. Since standard-state properties change with the composition of the real mixture, we might conclude that the standard state is not pure—but this is not so. When the $\{x^{sf}\}$ change, the value of the standard-state fugacity H_{1s} changes, not because the composition of the standard state changes (it is always pure), but because the *identity* of every standard-state molecule changes.

10.2.4 Fugacities Based on the Reference-Solvent, Henry's Law, Ideal Solution

Another way to remove the ambiguity that occurs in the limiting derivative of the ideal-solution expression (10.2.18) is to declare one component r to be a *reference solvent*. We then evaluate the derivative in (10.2.18) by taking $x_r \to 1$ while letting all other mole fractions go to zero,

$$\left(\frac{\partial f_i^{is}}{\partial x_i}\right)_{TP} = \lim_{x_r\to 1}\left(\frac{\partial f_i}{\partial x_i}\right)_{TP} \tag{10.2.26}$$

This defines a reference-solvent Henry's law constant,

$$H_{ir}(T,P) \equiv \lim_{x_r\to 1}\left(\frac{\partial f_i}{\partial x_i}\right)_{TP} = \lim_{x_r\to 1}\left(\frac{f_i}{x_i}\right)_{TP} \tag{10.2.27}$$

and the expression for the ideal-solution fugacity becomes

$$f_i^{is}(T,P,\{x\}) = x_i\, H_{ir}(T,P) \tag{10.2.28}$$

Note that the reference-solvent Henry's constant does not depend on composition. The form (10.2.28) states that when a mixture is nearly pure in the reference solvent r, then the fugacity of any other component i is linear in its mole fraction. The reference-solvent version of Henry's law applies to real mixtures when the composition is dominated by one component (the reference solvent). Both the reference-solvent and the solute-free forms of Henry's law are expected to apply when a mixture contains only a very small amount of the solute *i*.

For binary mixtures the reference-solvent Henry's constant is exactly the same as the solute-free Henry's constant

$$H_{ir}(T, P) = H_{is}(T, P) \qquad \textit{binary} \tag{10.2.29}$$

because in a binary, we must have $x_1 = 0$ when $x_2 = 1$. But for multicomponent mixtures, the two kinds of Henry's constants generally differ.

Often we can find values for H_{1r}, not at the mixture pressure P, but at the saturation pressure of the reference solvent, $P_r^s(T)$. Then we can use a Poynting factor to correct those values to the pressure of interest,

$$H_{ir}(T, P) = H_{ir}(T, P_r^s) \exp\left[\frac{1}{RT}\int_{P_r^s}^{P} \overline{V}_{ir}^{\infty}(T, \pi)d\pi\right] \tag{10.2.30}$$

where

$$\overline{V}_{ir}^{\infty} = \lim_{x_r \to 1} \overline{V}_i(T, P, \{x\}) \tag{10.2.31}$$

To measure deviations from a reference-solvent, Henry's law ideal solution, we introduce another activity coefficient defined by

$$\gamma_i^+ = \frac{f_i(T, P, \{x\})}{x_i H_{ir}(T, P)} \tag{10.2.32}$$

The normalization is obtained in the pure reference-solvent limit,

$$\lim_{x_r \to 1} \gamma_i^+ = 1 \tag{10.2.33}$$

Otherwise, values of γ_i^+ may be greater than one or less than one, but values less than one are more common. For a binary, the plot in Figure 10.5 applies to both a reference-solvent ideal solution and a solute-free ideal solution. This activity coefficient is particularly useful when we can choose the reference-solvent vapor pressure to be the standard-state pressure P_i^o, for then H_{ir} is a function only of T, and we may place all pressure dependence either in an activity coefficient, as in FFF #4, or in a Poynting factor, as in FFF #5.

10.2.5 Relations Among Activity Coefficients

We have now introduced three kinds of standard states for activity coefficients: one based on pure-components (§ 10.2.1), a second based on the solute-free Henry's law (§ 10.2.3), and a third based on the reference-solvent Henry's law (§ 10.2.4). The prin-

Table 10.1 Comparison of activity coefficients based on three choices for the standard state[a]

Standard state	Std state fugacity	Activity coefficient	Normalization
Pure component i at T, P_i^s	$f_i^o = f_{\text{pure i}}$	$\gamma_i = \dfrac{f_i}{x_i f_{\text{pure i}}}$	$\lim\limits_{x_i \to 1} \gamma_i = 1$
Solute-free solvent at T, P_{sf}^s, $\{x^{sf}\}$	$f_i^o = H_{is}$	$\gamma_i^* = \dfrac{f_i}{x_i H_{is}}$	$\lim\limits_{\{x^s\} \to 0} \gamma_i^* = 1$
Pure reference solvent at T, P_r^s	$f_i^o = H_{ir}$	$\gamma_i^+ = \dfrac{f_i}{x_i H_{ir}}$	$\lim\limits_{x_r \to 1} \gamma_i^+ = 1$

[a] Standard-state pressures quoted here are those typically used for condensed phases; they apply when FFF #5 is used to obtain fugacities from activity coefficients.

cipal features of these three are compared in Table 10.1. Although these activity coefficients differ, they are related because they merely represent three separate routes to the same quantity: the fugacity. To develop relations between any two constructs (such as activity coefficients) for the fugacity, we always start by equating the two expressions; that is, we start by writing

$$f_i = f_i \tag{10.2.34}$$

This identity merely states that *the fugacity is always the fugacity, no matter how you calculate it.*

Relate γ_i^* to γ_i. If we extract the fugacity from (10.2.6) and use it on the lhs of (10.2.34), and if we extract the fugacity from (10.2.24) and use it on the rhs, we obtain

$$x_i \gamma_i f_{\text{pure i}} = x_i \gamma_i^* H_{is} \tag{10.2.35}$$

Therefore, the two activity coefficients are simply proportional,

$$\gamma_i^* = \gamma_i \frac{f_{\text{pure i}}}{H_{is}} \tag{10.2.36}$$

where all quantities are at the same T and P. At fixed T, P, and $\{x^{sf}\}$, the ratio in (10.2.36) is independent of composition, and then, as illustrated in Figure 10.6, a plot of $\ln \gamma_i^*$ vs. x_i is merely the same curve as $\ln \gamma_i$ vs. x_i, but displaced vertically by a constant amount,

$$\ln \gamma_i^* = \ln \gamma_i + \ln\left(\frac{f_{\text{pure i}}}{H_{is}}\right) \tag{10.2.37}$$

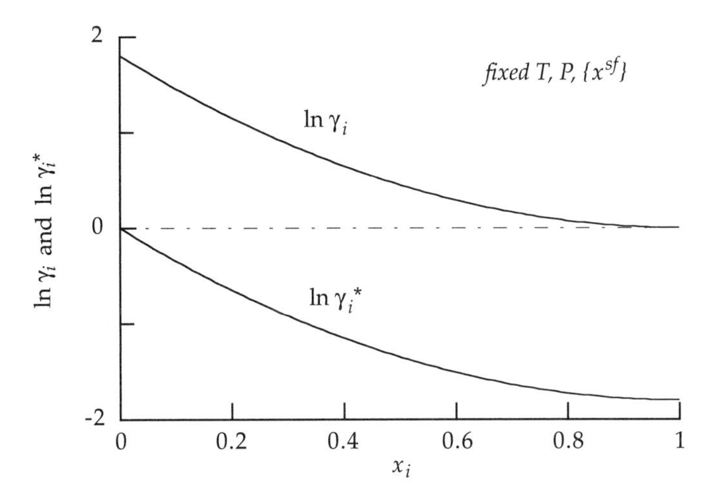

Figure 10.6 When T, P, and the solute-free mole fractions are fixed, then the solute-free Henry's law activity coefficient, γ_i^*, is simply proportional to the Lewis-Randall activity coefficient, γ_i; that is, ln γ_i^* vs. x_i is the same curve as ln γ_i vs. x_i, except the two are vertically displaced from one another by the constant amount given in (10.2.37).

Taking the pure-solute limit of (10.2.36) and appealing to the normalization of γ_i, we find

$$\lim_{x_i \to 1} \gamma_i^* = \frac{f_{\text{pure i}}}{H_{is}} \tag{10.2.38}$$

For a binary, the ratio on the rhs of (10.2.38) is that of the two intercepts appearing at x_i = 1 in Figure 10.5. Similarly, taking the dilute-solute limit of (10.2.36) and applying the normalization of γ_i^*, we obtain

$$\lim_{x_i \to 0} \gamma_i^* = \frac{f_{\text{pure i}}}{H_{is}} \left(\lim_{x_i \to 0} \gamma_i \right) = 1 \tag{10.2.39}$$

The limit on the rhs can be defined to be an activity coefficient at infinite dilution γ_i^∞,

$$\gamma_i^\infty \equiv \lim_{x_i \to 0} \gamma_i \qquad \textit{fixed } T, P, \{x^{sf}\} \tag{10.2.40}$$

So (10.2.39) becomes

$$\gamma_i^\infty = \frac{H_{is}}{f_{\text{pure i}}} \tag{10.2.41}$$

This establishes a simple relation between two standard-state fugacities: that in the Lewis-Randall standard state ($f_{\text{pure i}}$) and that in the solute-free dilute-solution standard state (H_{is}). This relation may be useful in obtaining values for H_{is} because infinite-dilution activity coefficients are measurables; for example, they can be extracted from gas chromatographic experiments [2].

If we use (10.2.39) to eliminate H_{is} from (10.2.36), we obtain

$$\gamma_i^* = \frac{\gamma_i}{\gamma_i^\infty} \tag{10.2.42}$$

So γ_i^* can be interpreted as a ratio measure for how the Lewis-Randall activity coefficient γ_i deviates from its value at infinite dilution.

Relate γ_i^+ to γ_i. Proceeding in a fashion similar to that above, we find that these two activity coefficients are also simply proportional,

$$\gamma_i^+ = \gamma_i \frac{f_{\text{pure i}}}{H_{ir}} \tag{10.2.43}$$

Here the ratio on the rhs is *always* independent of composition. The pure-solute limit applied to (10.2.43) yields

$$\lim_{x_i \to 1} \gamma_i^+ = \frac{f_{\text{pure i}}}{H_{ir}} \tag{10.2.44}$$

while the pure reference-solvent limit gives

$$\lim_{x_i \to 1} \gamma_i^+ = \frac{f_{\text{pure i}}}{H_{ir}} \left(\lim_{x_r \to 1} \gamma_i \right) = 1 \tag{10.2.45}$$

Define the limit on the rhs as the infinite-dilution activity coefficient in the reference-solvent standard state,

$$\gamma_{ir}^\infty \equiv \lim_{x_r \to 1} \gamma_i \tag{10.2.46}$$

This limit differs from the infinite-dilution limit in (10.2.40) because here the mole fractions for all components $i \neq r$ go to zero as $x_r \to 1$, while in (10.2.40) all solute-free mole fractions $\{x^{sf}\}$ are held fixed as $x_i \to 0$. Using (10.2.46) in (10.2.45) leaves

$$\gamma_{ir}^\infty = \frac{H_{ir}}{f_{\text{pure i}}} \tag{10.2.47}$$

And if we use this to eliminate H_{ir} from (10.2.43), we find

$$\gamma_i^+ = \frac{\gamma_i}{\gamma_{ir}^\infty} \tag{10.2.48}$$

So γ_i^+ can be interpreted as a ratio measure for how the activity coefficient γ_i deviates from its value in the limit of pure reference solvent.

10.2.6 Example

Do the three kinds of activity coefficients produce the same value for a fugacity?

We need to determine the value of the fugacity f_1 for a ternary liquid mixture at 25°C and 1 bar with compositions $x_1 = 0.2$, $x_2 = 0.5$, $x_3 = 0.3$. We choose component 1 to be a solute and take components 2 and 3 to be solvents. We have found data for the following infinite-dilution activity coefficients:

$$\text{Using } r = \text{component 2} \qquad \gamma_{1r}^{\infty} = 2.226 \tag{10.2.49}$$

$$\text{Using } r = \text{component 3} \qquad \gamma_{1r}^{\infty} = 3.320 \tag{10.2.50}$$

$$\text{Using fixed } \{x^{sf}\} \qquad \gamma_{1s}^{\infty} = 1.819 \tag{10.2.51}$$

We assume the mixture can be modeled by the ternary form of Porter's equation (5.6.20),

$$g^E/(RT) = x_1x_2A_{12} + x_1x_3A_{13} + x_2x_3A_{23} \tag{10.2.52}$$

The corresponding expression for the activity coefficient in the Lewis-Randall standard state is found from (5.6.21) to be

$$\ln\gamma_1 = x_2A_{12} + x_3A_{13} - x_1x_2A_{12} - x_1x_3A_{13} - x_2x_3A_{23} \tag{10.2.53}$$

Evaluate Porter parameters. Applying the limit in (10.2.46) to (10.2.53), we obtain

$$r = \text{component 2} \qquad \ln\gamma_{1r}^{\infty} = \ln(2.226) = A_{12} = 0.800 \tag{10.2.54}$$

Likewise,

$$r = \text{component 3} \qquad \ln\gamma_{1r}^{\infty} = \ln(3.320) = A_{13} = 1.200 \tag{10.2.55}$$

To get A_{23} from the remaining infinite-dilution activity coefficient, we first rewrite (10.2.53) in terms of solute-free mole fractions. A ternary must have

$$x_2 + x_3 = 1 - x_1 \tag{10.2.56}$$

So we divide (10.2.53) by $(x_2 + x_3)$ and apply the definition of solute-free mole fractions (10.2.19). This gives

$$\frac{\ln\gamma_1}{1-x_1} = (1-x_1)(x_2^{sf}A_{12} + x_3^{sf}A_{13} - x_2^{sf}x_3^{sf}A_{23}) \tag{10.2.57}$$

Taking the limit $x_1 \to 0$ with the solute-free mole fractions held fixed, we obtain

$$\ln\gamma_{1s}^{\infty} = x_2^{sf}A_{12} + x_3^{sf}A_{13} - x_2^{sf}x_3^{sf}A_{23} \tag{10.2.58}$$

For $x_2 = 0.5$ and $x_3 = 0.3$, the values of the solute-free mole fractions are

$$x_2^{sf} = \frac{0.5}{0.5+0.3} = 0.625 \quad \text{and} \quad x_3^{sf} = \frac{0.3}{0.5+0.3} = 0.375 \tag{10.2.59}$$

Using these with the values of A_{12} and A_{13} found above, (10.2.58) yields

$$A_{23} = 1.50 \tag{10.2.60}$$

This completes the determination of values for the three Porter parameters.

Lewis-Randall standard state. If we have a value for the pure 1 fugacity at 25°C and 1 bar, then we can use the activity coefficient in the Lewis-Randall standard state to evaluate f_1. A value for $f_{\text{pure }1}$ might be obtained from a correlation, an estimate, or a reduction of experimental data. For this situation we find $f_{\text{pure }1} = 1.4$ bar. Then

$$f_1 = x_1\gamma_1 f_{\text{pure }1} \tag{10.2.61}$$

Using our values of A_{12}, A_{13}, and A_{23}, Porter's equation (10.2.53) gives

$$\gamma_1 = 1.467 \tag{10.2.62}$$

Then (10.2.61) becomes

$$f_1 = (0.2)(1.467)(1.4) = 0.411 \text{ bar} \tag{10.2.63}$$

Solute-free dilute-solution standard state. In this case we do not have a value for $f_{\text{pure }1}$, so we cannot apply (10.2.61). But we are able to find or estimate a value for the solute-free Henry's constant at our solute-free mole fractions (10.2.59). The value is found to be $H_{1s} = 2.547$ bar. Then we obtain the fugacity from

$$f_1 = x_1\gamma_1^{*}H_{1s} \tag{10.2.64}$$

We obtain the value for the activity coefficient by using (10.2.51) and (10.2.62) in (10.2.42),

$$\gamma_1^* = \frac{\gamma_1}{\gamma_1^\infty} = \frac{1.467}{1.819} = 0.8065 \tag{10.2.65}$$

Note that the value is less than unity. Now (10.2.64) becomes

$$f_1 = (0.2)(0.8065)(2.547) = 0.411 \text{ bar} \tag{10.2.66}$$

This is the same value found in (10.2.63) using the Lewis-Randall standard state.

Reference-solvent dilute-solution standard state. In this case we do not have values for $f_{\text{pure }1}$ or for H_{1s}, so we cannot apply (10.2.61) or (10.2.64). However, we are able to find or estimate a value for the reference-solvent Henry's constant, with component 2 as the reference solvent. The value is found to be $H_{1r} = 4.648$ bar. Then we obtain the fugacity from

$$f_1 = x_1 \overset{+}{\gamma}_1 H_{1r} \tag{10.2.67}$$

To obtain the value for the activity coefficient, we use (10.2.49) and (10.2.62) in (10.2.48),

$$\overset{+}{\gamma}_i = \frac{\gamma_i}{\gamma_{ir}^\infty} = \frac{1.467}{3.320} = 0.4419 \tag{10.2.68}$$

Again the value is less than unity. Now (10.2.67) becomes

$$f_1 = (0.2)(0.4419)(4.648) = 0.411 \text{ bar} \tag{10.2.69}$$

This is the same value found in (10.2.63) using the Lewis-Randall standard state and found in (10.2.66) using the solute-free dilute-solution standard state.

In general we can say that the reference-solvent dilute-solution standard state is easier to use than the solute-free dilute-solution standard state (except, of course, when γ_i^* can be assumed to be unity). This is because H_{1r} is completely independent of composition, while H_{1s} depends on the solute-free mole fractions. But more generally, the lesson is that the three kinds of activity coefficients are simply proportional; they are all embedded with the same information, so they all give the same value for a fugacity. We use the particular standard state that allows us to take advantage of available data and that simplifies calculations.

10.3 BASIC REACTION-EQUILIBRIUM RELATIONS

We now do for reaction-equilibrium problems what we have done in § 10.1 for phase-equilibrium problems: we show how fundamental thermodynamic relations are used to develop computational strategies. We start by discussing the number of independent properties required to identify states in reacting systems (§ 10.3.1); then we present a stoichiometric approach to solving reaction equilibrium problems (§ 10.3.2–10.3.5) and follow that with a nonstoichiometric approach (§ 10.3.6–10.3.7).

10.3.1 Thermodynamic State for Reacting Systems

Extending the development presented in Chapters 3 and 9, we discuss the number of interactions available to change a state, the number of independent properties needed to identify a final equilibrium state, and the number needed to establish a well-posed reaction-equilibrium problem.

Number of interactions to change a state. Consider a system containing C chemical species distributed among $\mathcal{P}$ phases. The system can interact with its surroundings through the thermal interaction, a PV work mode, and a mass transfer interaction for each component. For such a situation we found in (9.1.1) that the number of interactions available for changing the state is given by

$$\mathcal{V} = C + 2 - S_{ext} \tag{9.1.1}$$

where S_{ext} counts any external constraints on interactions. If chemical reactions are occurring in the system, those reactions do not affect any interactions, so the number of interactions available is still given by (9.1.1). Further, if we block all C mass-transfer interactions, we still recover Duhem's theorem,

$$\mathcal{V} = 2 \tag{9.1.2}$$

So Duhem's theorem applies to any closed system of any number of components, any number of phases, and regardless of whether or not chemical reactions are taking place.

Number of properties to identify final equilibrium states. For a closed nonreacting system containing C components, we found in § 9.1.2 that the number of properties needed to identify the extensive state is given by

$$\mathcal{F}_{ex} = C + 2 - S \tag{9.1.10}$$

where S counts any additional internal constraints beyond those for phase equilibrium. Equation (9.1.10) applies to a system containing any number of phases. When reactions are occurring, we have for each reaction j, a new extensive property—the extent of reaction ξ_j. But for each reaction we also have a new internal constraint, the criterion for reaction equilibrium,

$$\mathcal{A}_j = 0 \qquad j = 1, 2, \ldots, \mathcal{R} \tag{7.6.3}$$

Consequently, the number of new properties is the same as the number of new constraints, so the total number of properties needed to satisfy (9.1.10) is unchanged. The number given by (9.1.10) is that required to identify the final *extensive* state after all reactions have ceased. To get the number needed to identify the final *intensive* state, we do not count the total amount of material, so (9.1.10) becomes

$$\mathcal{F}' = C + 1 - S \tag{9.1.12}$$

Recall that if more than one phase is present, the number counted by $\mathcal{F}'$ includes the relative amounts in the phases.

We may also ask for an $\mathcal{F}$-specification of the final equilibrium state. In this case we ignore the relative amounts in the phases and we ignore the extents of reaction. However, the $\mathcal{R}$ reaction-equilibrium constraints (7.6.3) still apply, so the generalized phase rule (9.1.13) becomes

$$\mathcal{F} = C + 2 - \mathcal{P} - \mathcal{R} - S \tag{10.3.1}$$

Here S counts any additional internal constraints besides those for phase and reaction equilibria. When $S = 0$ (10.3.1) reduces to the traditional form of the Gibbs phase rule extended to reacting systems.

To illustrate, consider the gas-phase synthesis of ammonia,

$$N_2 + 3H_2 \rightarrow 2NH_3 \tag{10.3.2}$$

We have $C = 3$ species, $\mathcal{P} = 1$ phase, $S = 0$, and $\mathcal{R} = 1$ reaction, so (10.3.1) gives the number of variables needed for an $\mathcal{F}$-specification as

$$\mathcal{F} = 3 + 2 - 1 - 1 - 0 = 3 \tag{10.3.3}$$

A legitimate set of these three properties would be T, P, and the equilibrium mole fraction of ammonia y_3. With these three, we can obtain values for the other equilibrium mole fractions by solving material balance and reaction-equilibrium expressions. This illustrates a typical use of the traditional phase rule (10.3.1): the $\mathcal{F}$-specification tells us the number of property values needed to identify the final equilibrium state after all reactions are complete.

Number of properties to identify initial states. But for reacting systems, it is not an $\mathcal{F}$-specification that we usually need. Instead, we usually need to know how many properties of the *initial* state are required, so that we can compute the final state. For this purpose, an $\mathcal{F}$-specification is insufficient because it identifies only a class of *indifferent* states (see § 9.1.3 and [3, 4]). This situation is entirely analogous to certain problems that arise in phase-equilibrium calculations. For example, an $\mathcal{F}$-specification of the isothermal flash problem is indifferent because many different feeds produce the same final T, P, $\{x\}$, and $\{y\}$ in the flash chamber (they differ in the relative amounts in the two phases). So to close the isothermal flash problem, we need an $\mathcal{F}'$-specification, not an $\mathcal{F}$-specification.

This same comment applies to reacting systems: to close a reaction equilibrium problem we need an $\mathcal{F}'$-specification, not an $\mathcal{F}$-specification. For example, reconsider the ammonia synthesis (10.3.2), which has $\mathcal{F} = 3$; so, we might try to close the problem by setting values for T, P, and the initial mole fraction of nitrogen y_1^o. But these three values are not sufficient to solve for the equilibrium extent; hence, we cannot get the equilibrium mole fractions. Instead of y_1^o, what we need are the relative amounts of the components initially present; for example, N_1^o/N_2^o. But you are now thinking, wait: if we know y_1^o, then we also know the initial mole fraction of hydrogen, y_2^o, because the initial mole fractions sum to one. But in making this statement you have tacitly assumed that no ammonia is present initially; i.e., you have set $y_3^o = 0$. So you have really set values for *four* properties, not three, and in fact, (9.1.12) gives

$$\mathcal{F}' = C + 1 - S = 3 + 1 - 0 = 4 \tag{10.3.4}$$

So to determine whether a reaction-equilibrium problem is well-posed in terms of the initial state, the traditional version of the phase rule (10.3.1) does not help. Instead, we use (9.1.10) if we want the extensive state, or we use the $\mathcal{F}'$-specification (9.1.12) if we want the intensive state. Finally, note that the forms for (9.1.10) and (9.1.12) are unaffected either by the number of phases present or by the number of reactions occurring.

10.3.2 Stoichiometric Development

In this subsection we begin to develop equations that are commonly used to solve reaction-equilibrium problems. Consider a one-phase system containing C species and recall from § 7.1.7 that when we fix any one of the pairs (T and P), (T and v), (s and P), or (s and v), then the criterion for equilibrium is always

$$\sum_i^C \bar{G}_i \, dN_i = 0 \tag{10.3.5}$$

For example, if T and P are fixed, then (10.3.5) is a consequence of G being a minimum at equilibrium; similarly, if T and v are fixed, then (10.3.5) is a consequence of A being a minimum at equilibrium. But when chemical reactions occur, the N_i in (10.3.5) are not independent; rather, they can only change in ways that conserve the total number of atoms b_k for each element k in the system,

$$\sum_i^C a_{ki} N_i = b_k = \text{constant} \tag{7.4.1}$$

Here each a_{ki} is the number of atoms of element k contained by one molecule of species i. Since the a_{ki} and b_k are constants, the changes dN_i are coupled via

$$\sum_i^C a_{ki} \, dN_i = 0 \tag{10.3.6}$$

This means that in reaction equilibrium problems, we are to solve the equilibrium condition (10.3.5) subject to the constraints imposed by conservation of atoms (10.3.6).

Such problems can be attacked by either a *stoichiometric* development (discussed here) or a *nonstoichiometric* development (discussed in § 10.3.6).

In the stoichiometric approach, the equilibrium condition (10.3.5) is imposed via a known *equilibrium constant* K_j, while the relations among the dN_i are found explicitly by determining the stoichiometric coefficients ν_{ij} for each reaction j. Then the coupled dN_i are replaced by one independent extent ξ_j for each reaction. For a system undergoing $\mathcal{R}$ independent reactions, the combination of the Ks and ξs provides $\mathcal{R}$ algebraic equations that can be solved for the equilibrium values of $\mathcal{R}$ extents of reaction ξ_j^e; from those, the equilibrium mole fractions can be obtained.

Since the determinations of stoichiometric coefficients ν_{ij} and extents of reaction ξ_j have already been discussed and illustrated in § 7.4, we need only introduce the equilibrium constant to complete the description of the stoichiometric approach to reaction equilibrium problems. The full implementation of the stoichiometric approach is described in § 10.4.3, after we have reviewed common choices for standard states.

Consider C species in a closed vessel undergoing $\mathcal{R}$ chemical reactions at fixed T and P. In § 7.6.1 we found that these reactions are finished and equilibrium is reached when the affinity $\mathcal{A}_j$ for each reaction j comes to zero,

$$\mathcal{A}_j = -\sum_i^C \nu_{ij}\, \overline{G}_i(T, P, \{x\}) = 0 \qquad j = 1, 2, \ldots, \mathcal{R} \tag{10.3.7}$$

Here ν_{ij} is the stoichiometric coefficient for species i in reaction j and $\overline{G}_i$ is the chemical potential for component i. To translate (10.3.7) into a computational form, we choose to use fugacities rather than chemical potentials, for then we can exploit the five famous fugacity formulae presented in § 6.4. Recall the fugacity is defined in terms of the chemical potential by

$$d\overline{G}_i = RT\, d\ln f_i \tag{4.3.8}$$

Integrating this definition from a convenient reference state (®) to the final equilibrium state (T, P, $\{x\}$) gives (4.3.12), which we now write as

$$\overline{G}_i(T, P, \{x\}) - \overline{G}_i^{®}(T, P_i^{®}, \{x^{®}\}) = RT \ln \frac{f_i(T, P, \{x\})}{f_i^{®}(T, P_i^{®}, \{x^{®}\})} \tag{10.3.8}$$

Since the definition of the fugacity (4.3.8) imposes a fixed temperature, the integration leading to (10.3.8) *must* be done at the system temperature T. Moreover, we invariably choose the reference state for species i to be pure i ($x_i = 1$); then the reference state becomes a standard state (® $\to$ o), and (10.3.8) can be written in terms of the activity (6.2.8),

$$\overline{G}_i(T, P, \{x\}) = \overline{G}_i^o(T, P_i^o) + RT \ln \mathfrak{a}_i(T, P, \{x\};\, f_i^o(T, P_i^o)) \tag{10.3.9}$$

However, note that the standard state for species i is not completely specified until we choose a pressure P_i^o and a phase. Common choices for standard states will be presented in § 10.4.1; for now, we continue by substituting (10.3.9) into (10.3.7),

$$\mathcal{A}_j = -\sum_i^C \nu_{ij} \overline{G}_i^o - RT \sum_i^C \nu_{ij} \ln a_i = 0 \tag{10.3.10}$$

The first sum is the change in standard-state Gibbs energies for reaction j,

$$\Delta g_j^o(T, \{P^o\}) \equiv \sum_i^C \nu_{ij} \overline{G}_i^o(T, P_i^o) \tag{10.3.11}$$

The notation $\{P^o\}$ represents the set of standard-state pressures for all reactants and products in reaction j; this means that we may choose different standard-state pressures for different species i. Combining (10.3.11) with (10.3.10) gives

$$\ln \prod_i^C a_i^{\nu_{ij}} = \frac{-\Delta g_j^o(T, \{P^o\})}{RT} \tag{10.3.12}$$

We define the product of activities to be the *equilibrium constant* K for reaction j,

$$K_j \equiv \prod_i^C a_i^{\nu_{ij}} \tag{10.3.13}$$

So (10.3.12) becomes

$$\ln K_j(T, \{P^o\}) = \frac{-\Delta g_j^o(T, \{P^o\})}{RT} \tag{10.3.14}$$

In general, a change in Gibbs energy Δg_j^o can be positive or negative, and therefore a reaction may have $K_j > 1$ or $K_j < 1$. Since products have $\nu_{ij} > 0$ while reactants have $\nu_{ij} < 0$, and since, to first order, each component's activity is proportional to the mole fraction ($a_i \propto x_i$), we can say the following: when $K_j \gg 1$, the final mixture will contain a high proportion of products, but inversely, when $K_j \ll 1$, the final mixture will contain a high proportion of reactants.

We emphasize that the standard-state change in Gibbs energy Δg_j^o and the equilibrium constant K_j depend *only* on the equilibrium temperature T and the prechosen standard-state pressures $\{P^o\}$. That is, even though individual activities depend on the full state (T, P, $\{x\}$) of the equilibrium mixture, the product of activities in (10.3.13) is independent of P and $\{x\}$. For many common reactions, values of K are tabulated in handbooks at particular temperatures; then we correct those values to our temperature, as in § 10.3.5. For other reactions we must obtain values of K by computing the standard-state change in Gibbs energy; this strategy is discussed in § 10.4.2.

10.3.3 Example

How do we evaluate and interpret the equilibrium constant for a single reaction?

Consider synthesis of ammonia from nitrogen and hydrogen carried out in the gas phase at 1 bar, 25°C. The stoichiometry for this reaction was determined in § 7.4.2. At 25°C and a standard pressure $P_i^o = 1$ bar, the JANAF tables [5] give the standard change in Gibbs energy to form one mole of ideal-gas NH_3 from its elements; that value is $\Delta g^o = -16.45$ kJ/(mol ammonia formed). With this we can use (10.3.14) to obtain the value for K,

$$\ln K = \frac{-\Delta g^o}{RT} = \frac{-(-16.45\ \text{kJ/mol})1000\ \text{J/kJ}}{(8.314)\text{J/(mol K)}298.15\ \text{K}} = 6.64 \qquad (10.3.15)$$

Therefore

$$K = 765 \qquad (10.3.16)$$

The equilibrium mole fractions are related to K through the activities, so we appeal to (10.3.13),

$$K = a_1^{\nu_1}\, a_2^{\nu_2}\, a_3^{\nu_3} = 765 \qquad (10.3.17)$$

where the ν_i are stoichiometric coefficients and we use 1 = nitrogen, 2 = hydrogen, and 3 = ammonia. In § 7.4.2 we chose $\nu_3 = 1$ and then found $\nu_1 = -1/2$ and $\nu_2 = -3/2$. Further, the activities can be expressed in terms of fugacities, so (10.3.17) becomes

$$K = \frac{f_3/f_3^o}{(f_1/f_1^o)^{1/2}(f_2/f_2^o)^{3/2}} = 765 \qquad (10.3.18)$$

The fugacities are related to the equilibrium mole fractions via the equilibrium extent of reaction ξ^e. To proceed further, we must choose standard states (to get the f_i^os) and we must choose one of the FFF (to get the f_is). All these quantities are estimated or obtained from models; they cannot contain any unknowns except mole fractions or the extent of reaction. Once these decisions are made, (10.3.18) becomes one equation that can be solved for the equilibrium value of the one extent ξ^e. These steps are performed in § 10.4.3, after we discuss options for standard states.

Note that since $K \gg 1$, we expect this reaction to favor formation of product, leaving an equilibrium mixture that is predominantly ammonia. While this is a valid thermodynamic conclusion, it is incomplete because, in fact, at ambient conditions this ammonia-synthesis reaction proceeds slowly. To be industrially viable, the reaction must be carried out at elevated temperatures, where the equilibrium constant is actually smaller than it is at 25°C; compensation is achieved by increasing the reaction pressure and using a catalyst. The controlling factor is a meager reaction rate, but thermodynamics cannot address rates: in analyzing any reaction-equilibrium situation, thermodynamics can only bound what will be observed at the completion of a reaction. In the ammonia synthesis, as in most practical situations, a satisfactory engineering solution requires us to combine thermodynamic analysis with other factors and manipulate operating conditions in a creative way.

10.3.4 Response of Equilibrium Constants to Changes in Temperature

We have noted that the equilibrium constant K_j for reaction j depends only on the system temperature T and the standard state. Often, we need to determine how the equilibrium constant changes with temperature. For example, during a reactor design we routinely want to know whether product yield can be improved by an increase or decrease in operating temperature. Furthermore, many tables (discussed at the end of § 10.4.2) give values for equilibrium constants only at selected temperatures; then we must correct those values to the temperature of our situation.

To address such questions, we first form the temperature derivative of (10.3.14) and invoke the Gibbs-Helmholtz equation (3.3.17),

$$\left(\frac{\partial \ln K_j}{\partial T}\right)_{PN} = \left(\frac{\partial(-\Delta g_j^o/RT)}{\partial T}\right)_{PN} = \frac{\Delta h_j^o(T, \{P^o\})}{RT^2} \tag{10.3.19}$$

Here Δh_j^o is the heat of reaction for reaction j carried out with all species in their standard states. If the heat of reaction is positive, then the reaction is *endothermic* and K increases with increasing T; inversely, if the heat of reaction is negative, then the reaction is *exothermic* and K decreases with increasing T. Therefore, for endothermic reactions we tend to increase the equilibrium fraction of product by increasing T, while for exothermic reactions we tend to increase the equilibrium fraction of product by decreasing T. However, we caution that such simple thermodynamic rules must be tempered by other considerations, such as kinetic constraints. For example, most simple reactions are exothermic, so the equilibrium product yield increases with decreasing T, while the rate of reaction usually decreases. In such cases, the choice of operating temperature must balance the maximum theoretical yield against competing kinetic effects. For some reactions, values of the standard heat of reaction have been measured and tabulated, but in many cases we must compute the standard heat of reaction from standard heats of formation (see § 10.4.2).

Integrating (10.3.19) allows us to use a value for an equilibrium constant at one temperature T_1 to compute its value at another temperature T_2. The result is the integrated form of the Gibbs-Helmholtz equation given in § 3.3.2,

$$\ln\left(\frac{K_j(T_2, \{P^o\})}{K_j(T_1, \{P^o\})}\right) = \frac{-\Delta h_j^o(T_1, \{P^o\})}{R}\left(\frac{1}{T_2} - \frac{1}{T_1}\right) \tag{10.3.20}$$

$$- \frac{1}{RT_2}\int_{T_1}^{T_2} \Delta c_{pj}^o(T)\, dT + \int_{T_1}^{T_2} \frac{\Delta c_{pj}^o(T)}{RT}\, dT$$

Values for Δc^o_{pj} are usually computed from tabulated values for the standard heat capacities of reactant and products,

$$\Delta c^o_{pj}(T) = \sum_i^C \nu_{ij}\, c^o_{pi}(T) \tag{10.3.21}$$

The result (10.3.20) is rigorous, but simplifying assumptions are often made. For example, when temperature changes are moderate, the last two integrals in (10.3.20) are often ignored. This is equivalent to assuming that the heat of reaction is constant, independent of T. Then (10.3.20) simplifies to

$$\ln\left(\frac{K_j(T_2, \{P^o\})}{K_j(T_1, \{P^o\})}\right) = \frac{-\Delta h^o_j(T_a, \{P^o\})}{R}\left(\frac{1}{T_2} - \frac{1}{T_1}\right) \tag{10.3.22}$$

where T_a is some "average" temperature between T_1 and T_2. Equation (10.3.22) implies that plots of $\ln K$ vs. $1/T$ will give straight lines; such lines have positive slopes for exothermic reactions and negative slopes for endothermic reactions. This approximation is tested for a particular reaction in § 10.3.5.

A second approximation is to keep all terms in (10.3.20), but assume the Δc^o_{pj} are constants, independent of T; then the integrals in (10.3.20) can be immediately evaluated. Because of the opposite signs on the two integrals, this may be a reliable assumption for some reactions, even when the heat capacities of the individual species change over the temperature range of interest.

10.3.5 Example

How do we determine the response of an equilibrium constant when temperature changes?

Consider formation of hydrogen sulfide via the gas-phase reaction at 1 atm.,

$$H_2 + \frac{1}{2}S_2 \rightarrow H_2S \tag{10.3.23}$$

At 1 atm. and 300 K, all species are ideal gases, and Bett et al. [6] give the following values for standard-state changes of ideal-gas properties at 300 K and P^o_i = 1 atm.: $\Delta g^o = -73$ kJ/mol, $\Delta h^o = -84.7$ kJ/mol, and $\Delta c^o_p = -11$ J/(mol K). First, we find the value of the equilibrium constant at 300 K. Then we determine its value at 700 K; this is done in two ways: (a) assuming the heat of reaction is independent of T and (b) assuming the heat of reaction changes with T, but that Δc^o_p is constant.

Value of equilibrium constant at 300 K. We substitute the value for the standard change in Gibbs energy into (10.3.14) and find

$$\ln K = \frac{-\Delta g^o}{RT} = \frac{-(-73 \text{ kJ/mol})1000 \text{ J/kJ}}{8.314 \text{ J/(mol K) } 300 \text{ K}} = 29.3 \tag{10.3.24}$$

$$K = 5.14(10^{12}) \qquad \textit{at 300 K} \tag{10.3.25}$$

This large value suggests that the reaction goes to completion at 300 K.

Estimate K at 700 K assuming a constant heat of reaction. Since the reaction is exothermic, we expect the value of K at 700 K will be smaller than its value at 300 K. Here we estimate K at 700 K, assuming Δh^o is constant; that is, we use (10.3.22).

$$\ln\left(\frac{K(T_2)}{K(300 \text{ K})}\right) = 10{,}200 \text{ K}\left(\frac{1}{T_2} - \frac{1}{300 \text{ K}}\right) \tag{10.3.26}$$

This represents a straight line when $\ln K$ is plotted against $1/T_2$. Setting $T_2 = 700$ K, we find

$$K = 1.92(10^4) \qquad \textit{at 700 K} \tag{10.3.27}$$

As expected, the value decreases with increasing T, but it remains large.

Estimate K at 700 K allowing the heat of reaction to change with *T*. We now include the effects of T on the heat of reaction. Those effects are contained in (10.3.20) as the integrals over the heat capacity difference. Since we have only the one value for Δc_p^o, we can only assume it is constant; then, (10.3.20) gives

$$\ln\left(\frac{K(T_2)}{K(T_1)}\right) = \frac{-\Delta h^o(T_1)}{R}\left(\frac{1}{T_2} - \frac{1}{T_1}\right) - \frac{\Delta c_{pj}^o}{RT_2}(T_2 - T_1) + \frac{\Delta c_{pj}^o}{R}\ln\left(\frac{T_2}{T_1}\right) \tag{10.3.28}$$

Using $\Delta c_p^o = -11$ J/(mol K), $T_1 = 300$ K, and $T_2 = 700$ K, this gives

$$\ln\left(\frac{K(700 \text{ K})}{K(300 \text{ K})}\right) = -19.406 + 0.7560 - 1.12 = -19.77 \tag{10.3.29}$$

Then

$$K = 1.33(10^4) \qquad \textit{at 700 K} \tag{10.3.30}$$

Assuming this is a better estimate than the value found in (10.3.27), the two values of K differ by 44%, but both values are so large that this difference may be of little importance in practice. Since the effort required to evaluate (10.3.28) is little more than that required to evaluate (10.3.22), we might as well use (10.3.28) when heat capacity data are available. But when such data are not available for all reactants and products, then we may be forced to use (10.3.22).

10.3.6 Nonstoichiometric Development

In the typical reaction-equilibrium problem for a single reaction, we are to determine the equilibrium composition at the end of the reaction. The problem is solved when we find a set of mole fractions $\{x\}$ that minimize G at fixed T and P, or that minimize A at fixed T and v, etc. That is, in general we have a minimization problem of this form,

$$\underset{\{x\}}{\text{Min}} \sum_i^C N_i \bar{G}_i \tag{10.3.31}$$

subject to the constraints that the mole numbers N_i can vary only in ways that conserve the total number of atoms b_k for each element k. This conservation constraint (7.4.1) can be stated in this way:

$$\sum_i^C a_{ki} N_i - b_k = 0 \qquad k = 1, 2, \ldots, m_e \tag{10.3.32}$$

Here a_{ki} is the number of atoms of element k on one molecule of species i, and m_e is the total number of elements present.

We cannot solve (10.3.31) merely by forming the total differential wrt the mole numbers and setting that differential to zero, because the dN_i are not independent; instead, they are related through (10.3.32). In the stoichiometric development in § 10.3.2, the constraint (10.3.32) was included in the problem through stoichiometric coefficients and an extent of reaction ξ. Here we impose the constraint in a different way; namely, we allow the N_i in the equilibrium condition (10.3.31) to vary independently and enforce the constraints (10.3.32) via Lagrange multipliers (see Appendix I).

Therefore for each element k, we scale the constraint (10.3.32) by a constant factor called a Lagrange multiplier λ_k,

$$\lambda_k \left(\sum_i^C a_{ki} N_i - b_k \right) = 0 \qquad k = 1, 2, \ldots, m_e \tag{10.3.33}$$

Since the lhs of (10.3.33) is still zero, we can sum it over all elements and add the sum to the quantity to be minimized (10.3.31); that is, we merely add zero to (10.3.31). So the minimization problem is now written as

$$\underset{\{x\}, \{\lambda\}}{\text{Min}} \left[\sum_i^C N_i \bar{G}_i + \sum_k^{m_e} \lambda_k \left(\sum_i^C a_{ki} N_i - b_k \right) \right] \tag{10.3.34}$$

Note that the multipliers λ_k must have appropriate units to preserve dimensional consistency in (10.3.34). Our problem is to minimize the quantity in brackets over all possible variations in the set of mole fractions and the set of Lagrange multipliers. The advantage to (10.3.34) over the original problem (10.3.31) is that all xs and all λs can vary independently. Therefore, we can form the total differential and set it to zero,

$$\sum_i^C \left(\bar{G}_i + \sum_k^{m_e} \lambda_k a_{ki} \right) dN_i + \sum_k^{m_e} \left(\sum_i^C a_{ki} N_i - b_k \right) d\lambda_k = 0 \tag{10.3.35}$$

With all the dN_i and $d\lambda_k$ independent, this can only be zero if, in general, every coefficient is separately zero; therefore, we must have

$$\bar{G}_i + \sum_k^{m_e} \lambda_k a_{ki} = 0 \qquad i = 1, 2, \ldots, C \tag{10.3.36}$$

and

$$\sum_i^C a_{ki} N_i - b_k = 0 \qquad k = 1, 2, \ldots, m_e \tag{10.3.37}$$

Note that the equations (10.3.37) are merely the constraints (10.3.32).

To start toward a computational form for (10.3.36), recall that the chemical potential can be expressed in terms of the activity $\mathbf{a}_i$. So using (10.3.9) in (10.3.36), we obtain

$$\bar{G}_i^o + RT \ln \mathbf{a}_i + \sum_k^{m_e} \lambda_k a_{ki} = 0 \qquad i = 1, 2, \ldots, C \tag{10.3.38}$$

But we can make no further progress until we choose standard states for all species i; the options are discussed in § 10.4.1. Nevertheless, the results (10.3.37) and (10.3.38) represent $(C + m_e)$ coupled algebraic equations that can be solved for C unknown equilibrium mole numbers and m_e unknown Lagrange multipliers. The final form will be developed in § 10.4.5.

10.3.7 Example

How do we use the nonstoichiometric method to set up equations for computing the equilibrium composition at the completion of a single reaction?

Let us reconsider the ammonia synthesis already studied in § 7.4.2 and § 10.3.3. The conditions are the same as in those examples: a gas-phase reaction at 1 bar, 25°C, with three moles of H_2 fed to the reactor for each mole of N_2 fed. At 25°C and $P_i^o = 1$ bar, the standard change in Gibbs energy is $\Delta g^o = -16.45$ kJ/(mol ammonia formed) [5]. Let subscripts 1 = nitrogen, 2 = hydrogen, and 3 = ammonia, and choose a basis of four moles of feed.

The initial numbers of atoms of nitrogen and hydrogen are

$$b_1 = \sum_i a_{1i} N_1^o = 0 + 2(1) + 0 = 2 \tag{10.3.39}$$

$$b_2 = \sum_i a_{2i} N_2^o = 2(3) + 0 + 0 = 6 \tag{10.3.40}$$

Then the atom balances (10.3.37) are

$$2N_1 + N_3 - 2 = 0 \tag{10.3.41}$$

$$2N_2 + 3N_3 - 6 = 0 \tag{10.3.42}$$

Let N be the total number of moles present at any time,

$$N = N_1 + N_2 + N_3 \neq \text{constant} \tag{10.3.43}$$

Then the balances (10.3.41) and (10.3.42) can be expressed in terms of mole fractions,

$$2x_1 + x_3 - 2/N = 0 \tag{10.3.44}$$

$$2x_2 + 3x_3 - 6/N = 0 \tag{10.3.45}$$

The particular forms for (10.3.38) are

$$\overline{G}_1^o + RT \ln \mathfrak{a}_1 + 2\lambda_1 = 0 \tag{10.3.46}$$

$$\overline{G}_2^o + RT \ln \mathfrak{a}_2 + 2\lambda_2 = 0 \tag{10.3.47}$$

$$\overline{G}_3^o + RT \ln \mathfrak{a}_3 + \lambda_1 + 3\lambda_2 = 0 \tag{10.3.48}$$

As in § 10.3.3, the activities can be written in terms of fugacities,

$$\overline{G}_1^o + RT \ln(f_1/f_1^o) + 2\lambda_1 = 0 \tag{10.3.49}$$

$$\overline{G}_2^o + RT \ln(f_2/f_2^o) + 2\lambda_2 = 0 \tag{10.3.50}$$

$$\overline{G}_3^o + RT \ln(f_3/f_3^o) + \lambda_1 + 3\lambda_2 = 0 \tag{10.3.51}$$

Recall the fugacities f_i depend on the unknown mole fractions, but the standard state fugacities f_i^o are constants whose values are obtained via judicious choices for standard states (§ 10.4.1). Note that the standard states must be applied consistently to both fugacities and chemical potentials. Once those choices have been made, the six equations (10.3.43)–(10.3.45) and (10.3.49)–(10.3.51) can be solved for the three equilibrium mole fractions, the total number of moles (relative to the selected basis), and the two Lagrange multipliers. The calculation will be finished in § 10.4.6.

10.4 PRELIMINARIES TO REACTION-EQUILIBRIUM CALCULATIONS

In this section we discuss standard states commonly chosen for reacting systems (§ 10.4.1), then we show how values for standard-state properties can be determined from properties of formation (§ 10.4.2). Lastly we develop computational forms used in applying the stoichiometric (§ 10.4.3) and nonstoichiometric (§ 10.4.5) approaches.

10.4.1 Common Choices for Standard States

Before a reaction-equilibrium calculation can be performed, we must select an appropriate standard state for each species. Moreover, we must clearly distinguish quantities, such as fugacities and activities, that depend on the final equilibrium state (T, P, $\{x\}$), from those quantities, such as equilibrium constants, that depend only on the equilibrium temperature T, the standard-state pressures $\{P^o\}$, and the phase. Typically, the standard-state pressure and phase are chosen according to whether the real substance is gas, liquid, or solid at the equilibrium conditions. Those three possibilities are discussed, in turn, here, and each discussion culminates with a particular expression for the activity. Those expressions can be used either in the stoichiometric development, via (10.3.14), or in the nonstoichiometric development, via (10.3.38). We emphasize that when we use the stoichiometric approach, the standard states used for the fugacities *must* be consistent with those associated with the equilibrium constant.

Standard states for gases. When species i is a gas at the equilibrium conditions, the standard state is usually taken to be the pure ideal gas at the equilibrium temperature T and $P_i^o = 1$ bar. (Caution: in older literature, the standard pressure was usually taken as 1 atm = 1.0133 bar.) Then, the standard-state fugacity becomes

$$f_i^o(T, P_i^o) = 1 \text{ bar} \tag{10.4.1}$$

If we choose FFF #1 for gas-phase fugacities,

$$f_i(T, P, \{x\}) = x_i \, \varphi_i(T, P, \{x\})P \tag{10.4.2}$$

then the activity for species i takes the form

$$\mathfrak{a}_i(T, P, \{x\}) = \frac{f_i(T, P, \{x\})}{f_i^o(T, P_i^o)} = \frac{x_i \, \varphi_i(T, P, \{x\})P}{P_i^o} \tag{10.4.3}$$

The activity is dimensionless, and therefore, since the standard-state pressure P_i^o was specified in bars, the value for the system pressure P used in (10.4.3) *must* also be in bars. Values for the fugacity coefficients φ_i appearing in (10.4.3) would be, as usual, computed from a volumetric equation of state. The expression (10.4.3) for the activity can be used both in (10.3.14) of the stoichiometric development and in (10.3.38) of the nonstoichiometric development.

Standard states for liquids. For a liquid species i, the standard state is usually taken to be the pure liquid at the equilibrium temperature T and at the pure vapor pressure $P_i^o = P_i^s(T)$. Then the standard-state fugacity becomes

$$f_i^o(T, P_i^o) = f_{\text{pure i}}(T, P_i^s) = \varphi_i^s(T)\, P_i^s(T) \tag{10.4.4}$$

where φ_i^s is the fugacity coefficient for the pure saturated vapor at T.

If we obtain the fugacity of the real liquid species i from FFF #3, then

$$f_i(T, P, \{x\}) = x_i\, \gamma_i(T, P, \{x\})\, \varphi_i^s(T)\, P_i^s(T) \exp\left[\frac{1}{RT}\int_{P_i^s}^{P} v_{\text{pure i}}(T, \pi)\, d\pi\right] \tag{10.4.5}$$

and the corresponding expression for the activity is

$$a_i(T, P, \{x\}) = x_i\, \gamma_i(T, P, \{x\}) \exp\left[\frac{1}{RT}\int_{P_i^s}^{P} v_{\text{pure i}}(T, \pi)\, d\pi\right] \tag{10.4.6}$$

This has the advantage of making the Poynting factor easy to compute, for it involves a simple integral over the pure molar volume of the liquid; however, (10.4.6) has the disadvantage of requiring us to know the value of the activity coefficient at the equilibrium pressure.

Alternatively, but with the same choice of standard state, we might obtain the fugacity of liquid species i from FFF #5; then, instead of (10.4.6) we would have

$$a_i(T, P, \{x\}) = x_i\, \gamma_i(T, P_i^s, \{x\}) \exp\left[\frac{1}{RT}\int_{P_i^s}^{P} \overline{V}_i(T, \pi, \{x\})\, d\pi\right] \tag{10.4.7}$$

The advantage to this choice is that the activity coefficient is now to be evaluated at the standard-state pressure (P_i^s), but the disadvantage is that the Poynting factor requires an integration over the partial molar volume for i in the reaction mixture.

If the equilibrium temperature T is above the critical temperature of pure i, then the vapor pressure P_i^s does not exist and we seek alternatives to the above choice for the standard state. Other possibilities include a standard state based on one of the versions of Henry's law, discussed in § 10.2. For example, for species i we might choose the standard state to be the (hypothetical) pure i whose fugacity equals the solute-free Henry's constant at T and any convenient pressure P_i^o. Then the standard-state fugacity would be

$$f_i^o(T, P_i^o) = H_{is}(T, P_i^o, \{x^{sf}\}) \tag{10.4.8}$$

where H_{is} is the solute-free Henry's constant and $\{x^{sf}\}$ represents the set of solute-free mole fractions (species i would be one of the solutes). The standard-state pressure P_i^o would be any pressure at which a value could be obtained for the Henry's constant. The fugacity for the real species i would usually be obtained from FFF #3,

$$f_i(T, P, \{x\}) = x_i \gamma_i^*(T, P, \{x\}) H_{is}(T, P_i^o, \{x^{sf}\}) \times \exp\left[\frac{1}{RT}\int_{P_i^o}^{P} \overline{V}_i^{\infty}(T, \pi, \{x^{sf}\})d\pi\right] \tag{10.4.9}$$

and the activity would be given by

$$a_i(T, P, \{x\}) = x_i \gamma_i^*(T, P, \{x\}) \exp\left[\frac{1}{RT}\int_{P_i^o}^{P} \overline{V}_i^{\infty}(T, \pi, \{x^{sf}\})d\pi\right] \tag{10.4.10}$$

The activity coefficient in (10.4.10) is to be evaluated at the system pressure, while the Poynting factor involves the partial molar volume at infinite dilution. Usually, we use (10.4.10) for some species and either (10.4.6) or (10.4.7) for the others. An expression exactly analogous to (10.4.10) can also be developed using the reference-solvent Henry's constant.

Standard states for solids. For a solid species i, the standard state is usually chosen to be the pure solid at the equilibrium temperature T and 1 bar. Then the standard-state fugacity is

$$f_i^o(T, P_i^o) = f_{\text{pure } i}(T, P{=}1) = 1 \text{ bar} \tag{10.4.11}$$

In writing this, we have ignored the Poynting factor that could be used to correct the fugacity from the saturation pressure at T to $P = 1$ bar. We now choose FFF #5 for the real solid, giving

$$f_i(T, P, \{x\}) = x_i \gamma_i(T, P_i^o, \{x\})(1 \text{ bar})\exp\left[\frac{1}{RT}\int_{1}^{P} \overline{V}_i(T, \pi)d\pi\right] \tag{10.4.12}$$

and the corresponding expression for the activity is

$$a_i(T, P, \{x\}) = x_i \gamma_i(T, P_i^o, \{x\})\exp\left[\frac{1}{RT}\int_{1}^{P} \overline{V}_i(T, \pi)d\pi\right] \tag{10.4.13}$$

The upper limit on the integral must be in bars. If all species were solids, then (10.4.13) could be used for all, but this rarely happens, because most industrial-scale reactions are carried out in fluid phases with few, if any, solid species present.

10.4.2 Standard-State Properties from Properties of Formation

In both the stoichiometric and nonstoichiometric approaches to reaction-equilibrium calculations, we need values for the standard change in the Gibbs energy Δg^o. In the stoichiometric development, Δg^o is used in (10.3.14) to obtain values for the equilibrium constant K; in the nonstoichiometric development, Δg^o is used to obtain values for the standard-state chemical potentials that appear in (10.3.38). Since g is a state function, values for Δg^o can be measured or computed along *any* convenient process path that starts with the desired reactants in their standard states and ends with products in their standard states. Of those many possibilities, the most convenient is to determine Δg^o by combining the Gibbs energies of formation for each species. That procedure is developed here. However, values for molecular properties of formation are often available only at a particular temperature T^o, so we must be able to correct those values to the reaction temperature T. Such corrections for Δg^o require values for the standard heat of reaction Δh^o and, perhaps, values for the standard isobaric heat capacities Δc_p^o.

Let F be any extensive thermodynamic property, and let $f = F/N$ be its intensive analog. Then for any reaction j, Δf_j^o represents the difference between the value of F for stoichiometric amounts of reactants and that for stoichiometric amounts of products, all in their standard states,

$$\Delta f_j^o(T^o, \{P^o\}) \equiv \sum_i^C \nu_{ij} \, \bar{F}_i^o(T^o, P_i^o) = \sum_i^C \nu_{ij} \, f_i^o(T^o, P_i^o) \tag{10.4.14}$$

Here f_i^o is the pure-component value of the intensive quantity f at the standard-state temperature and pressure. A particular example of (10.4.14) appears in (10.3.11). For reaction equilibria, we are interested in situations for which $f = g$, h, and c_p. We can consider many ways for obtaining values for the quantities f_i^o, so long as sufficient data are actually available. But here we consider one way: that in which the f_i^o are computed from properties of formation.

Properties of formation. Let a_{ki} be the number of atoms of type k contained on one molecule of species i. Then the *property of formation*, Δf_{if}^o, is defined as the change in f_i^o that occurs when the molecule is created from its constituent atoms,

$$\Delta f_{if}^o \equiv f_i^o(T^o, P_i^o) - \sum_k a_{ki} \, f_k^o(T^o, P_k^o) \tag{10.4.15}$$

Here, f_k^o is the molar value for f of atom k in the standard state. Putting (10.4.15) into (10.4.14) gives

$$\Delta f_j^o = \sum_i^C \nu_{ij} \, \Delta f_{if}^o + \sum_i^C \nu_{ij} \sum_k a_{ki} \, f_k^o(T^o, P_k^o) \tag{10.4.16}$$

Recall that index i runs over species, j runs over reactions, and k runs over atoms on a molecule of species i. The second term on the rhs can be rearranged to

$$\sum_i^C \nu_{ij} \sum_k a_{ki} f_k^o = \sum_k f_k^o \sum_i^C \nu_{ij} a_{ki} = \sum_k f_k^o \mathbf{A} \boldsymbol{\nu}_j = 0 \tag{10.4.17}$$

where $\mathbf{A}$ is the formula matrix and $\boldsymbol{\nu}_j$ is the vector of stoichiometric coefficients for reaction j. The quantity $\mathbf{A}\boldsymbol{\nu}_j$ vanishes because it merely expresses conservation of atoms for a reaction; see (7.4.17). Then (10.4.16) reduces to

$$\Delta f_j^o(T^o, \{P^o\}) = \sum_i^C \nu_{ij} \Delta f_{if}^o \tag{10.4.18}$$

This states that the standard-state change of F for reaction j is simply given by the sum of the properties of formation for each molecule i, with each weighted by its stoichiometric coefficient. Recall that $\nu_{ij} > 0$ for products, but $\nu_{ij} < 0$ for reactants. Properties of formation are zero when the molecule is an element (e.g., H_2, O_2, N_2, etc.). Otherwise, values for properties of formation depend, not only on the standard temperature and pressure, but also on the phase. We first consider corrections for changes in temperature and then for changes in phase.

Temperature corrections. To correct Δg_{if}^o from the standard temperature T^o to the operating temperature T, we use the integrated form of the Gibbs-Helmholtz equation from § 3.3.2,

$$\frac{\Delta g_{if}^o(T, P_i^o)}{RT} = \frac{\Delta g_{if}^o(T^o, P_i^o)}{RT^o} + \frac{\Delta h_{if}^o(T^o, P_i^o)}{R}\left(\frac{1}{T} - \frac{1}{T^o}\right) \tag{10.4.19}$$

$$+ \frac{1}{RT}\int_{T^o}^{T} \Delta c_{pi}^o(\tau)\, d\tau - \int_{T^o}^{T} \frac{\Delta c_{pi}^o(\tau)}{R\tau}\, d\tau$$

Usually, the standard heat capacities are represented empirically by simple polynomials in temperature.

Phase corrections. Occasionally we have a value of a formation property for a substance in one phase, but need the value for another phase. When the phases are vapor and liquid and the pressure is low, the following approximations are common.

Let T^o and P_i^o be the standard-state temperature and pressure, and let P_i^s be the pure-i vapor pressure at T^o. Then the difference between the vapor and liquid enthalpies of formation at P_i^s gives the latent heat of vaporization, Δh_{vap},

$$\Delta h_{ifv}^o(T^o, P_i^s) - \Delta h_{if\ell}^o(T^o, P_i^s) = \Delta h_{i,vap}^o(T^o, P_i^s) \tag{10.4.20}$$

When the standard-state pressure P_i^o and the pure vapor pressure P_i^s are both low, then the liquid enthalpies are essentially the same at both pressures; likewise for the vapor enthalpies. Then we have this approximation

$$\Delta h^o_{i,\,vap}(T^o, P^s_i) \approx \Delta h^o_{ifv}(T^o, P^o_i) - \Delta h^o_{if\ell}(T^o, P^o_i) \tag{10.4.21}$$

So if we know the latent heat of vaporization for pure i at T^o and if we also know the standard heat of formation for one phase, then we can use (10.4.21) to estimate the standard heat of formation for the other phase.

To obtain an approximation for the Gibbs energy of formation, we start from the phase-equilibrium relation, which applies at T^o and P^s_i,

$$\Delta g^o_{if\ell}(T^o, P^s_i) = \Delta g^o_{ifv}(T^o, P^s_i) \tag{10.4.22}$$

When P^s_i is low, the vapor is an ideal gas and we can write

$$\left(\frac{\partial\,(\Delta g^o_{ifv})}{\partial P}\right)_{TN} = v = \frac{RT}{P} \tag{10.4.23}$$

Integrating from P^o_i to P^s_i, we obtain

$$\Delta g^o_{ifv}(T^o, P^s_i) = \Delta g^o_{ifv}(T^o, P^o_i) + RT\ln\left(\frac{P^s_i}{P^o_i}\right) \tag{10.4.24}$$

Substituting this into (10.4.22) and neglecting the effect of pressure on the liquid, we have

$$\Delta g^o_{if\ell}(T^o, P^o_i) \approx \Delta g^o_{ifv}(T^o, P^o_i) + RT\ln\left(\frac{P^s_i}{P^o_i}\right) \tag{10.4.25}$$

Therefore, once we know the value of Δg^o_{if} for one phase, then we can use (10.4.25) (at low pressures) to estimate its value for the other phase. If species i is a solid phase in the reacting mixture, we can evaluate the difference between properties of formation of liquids and solids using the procedure presented in § 10.1.4.

Literature sources. For ideal-gas standard states, properties of formation can be extracted from spectroscopic experiments via statistical mechanics [7]. In these cases, the final values obtained for the formation properties are usually accurate. However, for condensed-phase standard states, properties of formation are usually obtained by combining values from other kinds of reactions; for example, for organics, properties of formation may be obtained by combining property changes during combustion reactions. In these cases, the values for properties of formation may be less accurate than those obtained for ideal gases.

Values for properties of formation can be found in the following standard references. Values of Δg^o_f, Δh^o_f, and c^o_p at 25°C are tabulated in two publications from the U.S. National Bureau of Standards (now NIST, the National Institute for Standards and Technology). The original publication was *NBS Circular 500* [8], which is updated in a later series under the title *NBS Technical Note 270* [9]. A second set of tables has its origins in a publication by the American Petroleum Institute, *API 44* [10]. The newer versions are published by the Thermodynamics Research Center at Texas A & M University and contain Δg_f^o and Δh_f^o at 25°C, along with other related quantities [11].

The tables originally published by Stull et al. [12] have been updated as later editions of the JANAF (Joint Army-Navy-Air Force) tables [5]. These references provide values for Δg_f^o, Δh_f^o, and c_p^o as functions of temperature; similar tables are also available in Landolt-Börnstein [13]. Comparisons of these tables, in terms of content and notation, can be found in Bett et al. [6] and Poling et al. [14]. Poling et al. also discuss how errors in Δg_f^o values affect results from equilibrium calculations.

10.4.3 Computational Forms for Stoichiometric Approach

The typical reaction-equilibrium problem is to determine the equilibrium composition when species are allowed to react under specified conditions. In previous sections we have addressed various aspects of such problems; now we summarize the computational forms usually used in solving them. Common forms used for the stoichiometric method are presented here; those for the nonstoichiometric method are presented in § 10.4.5.

In the stoichiometric development we need values for the stoichiometric coefficients, which we obtain by solving

$$\mathbf{A}\boldsymbol{\nu}_j = 0 \qquad j = 1, 2, \ldots, \mathcal{R} \qquad (7.4.17)$$

The system is composed of C total species and m_e total elements. In (7.4.17) $\boldsymbol{\nu}_j$ is the vector of stoichiometric coefficients for reaction j and $\mathbf{A}$ is the (m_e x C) matrix of a_{ki} coefficients, with a_{ki} the number of atoms of element k on one molecule of species i. With values for the stoichiometric coefficients, we can form expressions for the species mole fractions in terms of an extent ξ_j for each of the $\mathcal{R}$ reactions,

$$x_i = \frac{N_i^o + \sum_j^{\mathcal{R}} \nu_{ij}\, \xi_j}{N^o + \sum_j^{\mathcal{R}} \sigma_j\, \xi_j} \qquad (7.4.22)$$

Here N_i^o is the number of moles of species i initially present, N^o is the total number of moles initially present, and $\sigma_j = \Sigma_i\, \nu_{ij}$ is the algebraic sum of stoichiometric coefficients for reaction j.

The equilibrium mole fractions can be obtained by using the equilibrium values for the extents ξ_j in (7.4.22). The equilibrium values of the extents are those that provide the correct value for the equilibrium constant at the reaction T and P,

$$K_j \equiv \prod_i^C a_i^{\nu_{ij}} \qquad (10.3.13)$$

Values of K_j for use in (10.3.13) are typically obtained from the standard Gibbs energies of formation (§ 10.4.2). Relations between the K_j and the ξ_j^e are determined by the choice of standard state and the corresponding expressions for the activities. We present a few representative forms here to show how expressions from previous sections are combined; but the following, while common, are not the only possibilities.

For gases we would use the standard state from § 10.4.1, with the activities given by (10.4.3). Then if all species were gases, the expression for the equilibrium constant (10.3.13) would become

$$K_j = \prod_i^C (x_i \varphi_i)^{\nu_{ij}} P^{\nu_{ij}} = P^{\sigma_j} \prod_i^C (x_i \varphi_i)^{\nu_{ij}} \tag{10.4.26}$$

Recall that the standard state pressure here is 1 bar for all species, so the reaction pressure P in (10.4.26) must also be in bars. Also recall that K_j does not depend on P; therefore, since σ_j can be positive, negative, or zero, the product on the rhs may increase, decrease, or remain unchanged when P differs from P^o.

For liquids we often use the pure saturated liquid for the standard state, as described in § 10.4.1; if we then choose FFF #3, so the activities are given by (10.4.6), and if all species are liquid, then (10.3.13) for the equilibrium constant takes the form

$$K_j = \prod_i^C (x_i \gamma_i)^{\nu_{ij}} \exp\left[\frac{\nu_{ij}}{RT}\int_{P_i^s}^{P} v_{\text{pure i}}(T, \pi)d\pi\right] \tag{10.4.27}$$

where γ_i is evaluated at P. If $P \approx P_i^s$, then the Poynting factor can be ignored; otherwise, unless $P \gg P_i^s$, we assume the liquid-phase molar volume is constant with P so the integral in the Poynting factor can be easily evaluated. If we prefer to use FFF #5, then the activities are given by (10.4.7), and instead of (10.4.27), and we obtain

$$K_j = \prod_i^C (x_i \gamma_i)^{\nu_{ij}} \exp\left[\frac{\nu_{ij}}{RT}\int_{P_i^s}^{P} \bar{V}_i(T, \pi, \{x\})d\pi\right] \tag{10.4.28}$$

where γ_i is evaluated at P_i^s.

The expressions for the equilibrium constants in (10.4.26)–(10.4.28) apply only when the same standard state has been chosen for all species. When different standard states are used for different species, K is still given by (10.3.13), with the appropriate expression for each activity taken from § 10.4.1. In any event, (10.3.26)–(10.3.28) illustrate how the compositions occur in the expressions for K. On substituting (7.4.22) for those mole fractions, we obtain $\mathcal{R}$ algebraic equations that can be solved for the equilibrium values of $\mathcal{R}$ extents of reaction. Then with those values for the ξ_j^e, the equilibrium mole fractions are obtained from (7.4.22). This procedure is illustrated for a single reaction in the following example. More elaborate reaction-equilibrium problems are discussed in Chapter 11.

10.4.4 Example

How do we use the stoichiometric method to complete the computation started in § 10.3.3 for the equilibrium composition from an ammonia synthesis?

The ammonia synthesis is carried out in the gas phase at 25°C, 1 bar, using an initial feed containing three moles of hydrogen for each mole of nitrogen. The stoichiometry for the reaction was determined in § 7.4.2. In § 10.3.3 we found the value of the equilibrium constant to be

$$K = 765 \tag{10.3.16}$$

Since all species are gases, we use (10.4.26) to relate K to mole fractions. Further, at 1 bar the mixture is an ideal gas, so (10.4.26) simplifies to

$$K = P^{\sigma} \prod_{i}^{C} (x_i \varphi_i)^{\nu_i} = \prod_{i}^{C} (x_i)^{\nu_i} = \frac{x_3}{x_1^{0.5} x_2^{1.5}} \tag{10.4.29}$$

where 1 = nitrogen, 2 = hydrogen, and 3 = ammonia. In § 7.4.2 we found expressions for these mole fractions in terms of the one extent of reaction:

$$x_1 = \frac{1 - \xi/2}{4 - \xi}, \quad x_2 = \frac{3(1 - \xi/2)}{4 - \xi}, \quad \text{and} \quad x_3 = \frac{\xi}{4 - \xi} \tag{10.4.30}$$

These apply at every point during the reaction, but putting them into (10.4.29) restricts us to the final mixture when equilibrium is reached:

$$K = \frac{1}{3^{1.5}} \left(\frac{\xi^e}{4 - \xi^e} \right) \frac{(4 - \xi^e)^2}{(1 - \xi^e/2)^2} = 765 \tag{10.4.31}$$

This is a quadratic for the equilibrium value of the extent ξ^e; solving analytically, we find $\xi^e = 1.937$ and 2.063. The smaller root is correct; this choice can be justified in two ways:

(a) When ξ increases from zero as the reaction proceeds, the equilibrium value is the one encountered first; that is, it is the smallest positive value.

(b) Using (7.4.23), we can compute the upper bound on ξ, based on the limiting reactant. Since the reactants are fed in their stoichiometric ratio, both reactants give the same upper bound; for example, using hydrogen, (7.4.23) gives $\xi^{ub} = -N_1^o/\nu_1 = -3/(-3/2) = 2$. Therefore, any roots $\xi^e > 2$ are meaningless.

Using $\xi^e = 1.937$ in (10.4.30), the equilibrium mole fractions are found to be

$$x_1 = 0.0153, \quad x_2 = 0.0460, \quad \text{and} \quad x_3 = 0.9387$$

As expected, the equilibrium mixture contains a significant fraction of product (ammonia); this is consistent with $K > 1$.

10.4.5 Computational Form for Nonstoichiometric Approach

In this subsection we complete the development of the nonstoichiometric method to obtain a form suitable for computations. For one reaction occurring among C species we have already obtained

$$\overline{G}_i^o + RT \ln \mathbf{a}_i + \sum_k^{m_e} \lambda_k a_{ki} = 0 \qquad i = 1, 2, \ldots, C \tag{10.3.38}$$

subject to the constraints (10.3.37), which we now write as

$$\sum_i^C a_{ki} x_i - \frac{b_k}{N} = 0 \qquad k = 1, 2, \ldots, m_e \tag{10.4.32}$$

One molecule of species i contains a_{ki} atoms of element k, and the mixture contains a total of m_e elements. The quantities λ_k are the unknown Lagrange multipliers whose values will be determined. In (10.4.32), the b_k are the total number of atoms of element k, and the mole numbers sum to a total number of moles N,

$$N = \sum_i^C N_i \neq \text{constant} \tag{10.4.33}$$

We need a way to obtain values for the standard-state chemical potential appearing in (10.3.38). Each standard state is a pure species, so the chemical potential reduces to the pure molar Gibbs energy, and the pure molar property g_i^o is simply related to the Gibbs energy of formation by (10.4.15). So we rewrite (10.4.15),

$$\overline{G}_i^o = g_i^o = \Delta g_{if}^o + \sum_k^{m_e} a_{ki}\, g_k^o \tag{10.4.34}$$

Substituting (10.4.34) into (10.3.38) leaves

$$\frac{\Delta g_{if}^o}{RT} + \ln \mathbf{a}_i + \sum_k^{m_e} a_{ki}\, \lambda_k^* = 0 \qquad i = 1, 2, \ldots, C \tag{10.4.35}$$

where

$$\lambda_k^* \equiv \frac{\lambda_k + g_k^o}{RT} \tag{10.4.36}$$

is a dimensionless constant. The result (10.4.35) together with the constraints (10.4.32) and the material balance (10.4.33) close our problem: the a_{ki}, b_k, and initial feed ratios are known, expressions for the activities can be obtained from the possibilities cited in § 10.4.1, and values for the properties of formation Δg^o_{if} can be found in standard tables (see the end of § 10.4.2). Therefore (10.4.32), (10.4.33), and (10.4.35) constitute $(C + m_e + 1)$ algebraic equations that can be solved simultaneously for the C equilibrium mole fractions, the m_e Lagrange multipliers, and the equilibrium value of N. The procedure is illustrated for a single reaction in an example below.

In this nonstoichiometric method, part of the solution is the set of values for the Lagrange multipliers λ_k. In most situations these multipliers have little physical significance; they merely serve to ensure conservation of atoms, so their values are a necessary but nonphysical by-product of the calculation. When the number of elements m_e is less than the number of species C, the C equations (10.4.35) could be combined to eliminate the m_e multipliers λ, so their values would not obtained explicitly. However, if such a combination is done, the result is equivalent to the stoichiometric expression for the equilibrium constant, and the computational advantages of the nonstoichiometric method are lost.

Note that in the nonstoichiometric approach, we do not obtain values for stoichiometric coefficients and we have no parameters, such as the extent ξ, that track the progress of individual reactions. Moreover, the computational forms (10.4.32)–(10.4.35) contain no quantities that are specific to a particular reaction (e.g., no subscripts j appear). So although the nonstoichiometric equations (10.4.32)–(10.4.35) were derived with one reaction in mind, they actually apply to situations involving *any* number of reactions. In fact, we can use the nonstoichiometric method without knowing how many reactions are occurring or even what those reactions might be: we only need a complete identification of all reactants and products. This constitutes a principal advantage of the nonstoichiometric development.

10.4.6 Example

How do we use the nonstoichiometric method to complete the calculation of the equilibrium composition for the problem started in § 10.3.7?

The reaction is ammonia synthesis by a gas-phase reaction at 1 bar, 25°C. The feed contains three moles of hydrogen for each mole of nitrogen. At 25°C and 1 bar, the standard Gibbs energy of formation for ideal-gas ammonia is $\Delta g^o_f = -16.45$ kJ/(mol ammonia formed) [5]. Let subscripts 1 = nitrogen, 2 = hydrogen, and 3 = ammonia.

The equations of constraint for conservation of atoms were found in § 10.3.7 to be

constraints

$$N = N_1 + N_2 + N_3 \neq \text{constant} \tag{10.4.37}$$

$$2x_1 + x_3 - 2/N = 0 \tag{10.4.38}$$

$$2x_2 + 3x_3 - 6/N = 0 \tag{10.4.39}$$

It remains for us to develop a computational form from the presentation in § 10.4.5. The mixtures are gases, so we choose the standard state for each component to be the pure ideal gas at 1 bar and 25°C because this is the state at which we can find values for the standard-state change in Gibbs energy. Then

$$\boldsymbol{a}_i = \frac{x_i \varphi_i P}{P_i^o} \tag{10.4.3}$$

But at the operating pressure $P = 1$ bar, the gases are in fact ideal, so $\boldsymbol{a}_i = x_i P/P_i^o = x_i$, and (10.4.35) gives for each species,

$$\frac{\Delta g_{1f}^o}{RT} + \ln x_1 + 2\lambda_1^* = 0 \tag{10.4.40}$$

$$\frac{\Delta g_{2f}^o}{RT} + \ln x_2 + 2\lambda_2^* = 0 \tag{10.4.41}$$

$$\frac{\Delta g_{3f}^o}{RT} + \ln x_3 + \lambda_1^* + 3\lambda_2^* = 0 \tag{10.4.42}$$

The dimensionless multipliers λ^* are defined by (10.4.36). For elements, properties of formation are zero, and we have the value for Δg_{3f}^o, so (10.4.40)–(10.4.42) reduce to

$$\ln x_1 + 2\lambda_1^* = 0 \tag{10.4.43}$$

$$\ln x_2 + 2\lambda_2^* = 0 \tag{10.4.44}$$

$$\ln x_3 + \lambda_1^* + 3\lambda_2^* = 6.636 \tag{10.4.45}$$

The equations (10.4.37)–(10.4.39) and (10.4.43)–(10.4.45) constitute six equations that can be solved for the three equilibrium mole fractions, the total number of moles, and the two Lagrange multipliers. In general, such equations, which result from the nonstoichiometric development, are nonlinear and must be solved by trial. Here the results are found to be

$$x_1 = 0.0154, \quad x_2 = 0.0462, \quad \text{and} \quad x_3 = 0.9384 \tag{10.4.46}$$

$$N = 2.064 \text{ moles/mol nitrogen fed} \tag{10.4.47}$$

$$\lambda_1^* = 2.087 \quad \text{and} \quad \lambda_2^* = 1.537 \tag{10.4.48}$$

The mole fractions found here are the same as those found by the stoichiometric method in § 10.4.4. We can make further contact with § 10.4.4 by noting that the equilibrium value of N in (10.4.47) is simply related to the extent of reaction found in § 10.4.4; the relation is given by (7.4.12), which for one reaction becomes

$$\xi^e = \frac{N - N^o}{\sigma} \tag{10.4.49}$$

From § 10.4.4 we have $\sigma = 1 - 1/2 - 3/2 = -1$, and with a feed of $N^o = 4$, we use (10.4.49) to find $\xi^e = (2.064 - 4)/(-1) = 1.936$, which is the same as found in § 10.4.4.

Finally note that we could combine the three equations (10.4.43)–(10.4.45) to eliminate the two Lagrange multipliers. But doing so produces the stoichiometric equation (10.4.31) that relates the equilibrium constant to the mole fractions. In other words, the stoichiometric and nonstoichiometric developments are merely two different formulations of the same equations, though in particular applications one approach or the other may be easier to use.

10.5 CHOOSING AN APPROPRIATE FORM IN APPLICATIONS

At this point we have developed several alternatives for setting up computations. We must select from these alternatives before calculations can be undertaken; we now discuss the issues that should be considered in making the selection. Careful selection helps us reach a reliable result from an economical investment of computational resources. That is, our decisions constitute the classic optimization problem that weighs convenience against reliability. Such judgements are at the heart of engineering practice, for if there were no such decisions to be made, then engineers could be largely replaced by computing machines.

10.5.1 General Considerations

Here we can only begin to address general elements that should be included in reaching a correct decision. More specific elements involve considerations of the behaviors and limitations of particular models that might apply to the problem. We divide the general considerations into three parts.

Assess the risks. What use is to be made of the calculated property? Do we need an exact value (accuracy) or are we only trying to avoid major blunders (reliability)? What is the impact if the computed property is in error by 1%, 10%, or 100%? How accurate and reliable are the data that will be used as inputs to the calculation? Remember, no property is ever measured or computed *exactly*. This means we must understand the problem well enough to be able to determine the desired accuracy.

Make reasonable assumptions. By considering the state of the system and the nature of its components, we can introduce sensible approximations that may vastly simplify the analysis but do little harm to the accuracy of the calculation. A consideration of the state would include identification of the phases present (solid, liquid, gas), estimates for temperature and pressure, and rough estimates for the composition. (For example, is a mixture dominated by one component or are any components present in very small amounts?). In general, it is helpful to locate known state points on phase diagrams, or at least to find where a mixture temperature lies relative to the pure-component melting and critical temperatures. By nature of the components we mean the kinds of intermolecular forces, such as simple van der Waals interactions, hydrogen bonding, polarity, dimer formation, conformational structures, and coulombic interactions in electrolytes.

Attention should also be given to *sensitivity*: how will computational results be affected by changes in input values, such as state conditions, model parameters, and model approximations? For example, *what-if* scenarios can be studied either by examining derivatives of outputs wrt inputs or by solving the full problem under changes in the inputs. In highly sensitive situations, small changes in inputs propagate into large changes in output, and then even "reasonable" assumptions can lead to disaster. But when the sensitivity is low, so large changes in inputs have little effect on outputs, then a simple approach may be sufficient to achieve the desired accuracy.

Identify the resources available. What computational methods can be applied and what parameters and data are needed to implement a particular method? Critical properties? Heat capacities? Vapor pressures? Parameters for a $PvTx$ equation of state? Parameters in models for excess properties? When available data are sparse (the usual situation) or unreliable or conflicting, then set upper and lower bounds on the property and do a sensitivity analysis (which input data have the largest impact on the calculated property?). Considerations should also be given to the resources needed to set up the calculation (pencil and paper, calculator commands, computer software, original computer codes) and the hardware needed to carry them out (brain, fingers, calculator, PC, workstation).

10.5.2 Rules of Thumb for Selecting from the FFF

All phase and reaction equilibrium computations require expressions for the fugacities of all components. The possible expressions are presented in § 6.4 as the five famous fugacity formulae (FFF). Here are some general guidelines for choosing from those possibilities.

FFF #1 should be used whenever the volumetric behavior of the substance is reliably correlated by a $PvTx$ equation of state: FFF #1 is *always* used for gases, but it should *never* be used for solids. FFF #1 can be used for liquids, if a reliable equation of state is available for the liquid phase. However, if all that is wanted is a quick estimate of liquid properties, FFF #2–5 are generally faster to implement than FFF #1.

At the lowest pressures, FFF #2–5 are equivalent, and their use only requires values for the standard-state fugacities plus models for activity coefficients. For condensed phases at low pressures, we typically use either FFF #2 or #3 and ignore the pressure effects on activity coefficients. When pressure effects are important, then we also need volumetric data for computing Poynting factors (or their equivalents). For condensed phases at high pressures, we typically use either FFF #4 or 5; generally we prefer FFF #5 because making rough estimates of the composition dependence of partial molar volumes is usually more reliable than roughly estimating the pressure dependence of activity coefficients. When we have gases dissolved in liquids, we usually use FFF #4.

When using FFF #5, we choose a pure-component standard-state for all condensed components i (solid and liquid) whose pure critical temperatures T_{ci} are not much beyond the system temperature T; say, $T < 1.2T_{ci}$. We choose one of the Henry's law standard states for any component i that has $x_i < 0.01$ and $T > 1.5T_{ci}$. Exceptions to this might include liquid-solid equilibria or situations in which a $PvTx$ equation of state has been directly fit near the conditions of interest.

10.5.3 Selecting an Appropriate Phase-Equilibrium Relation

The choices here are the phi-phi, gamma-phi, and gamma-gamma methods. The phi-phi method can be used whenever an accurate $PvTx$ equation of state is available for all phases. This includes various fluid-fluid situations, particularly high-pressure vapor-liquid, liquid-liquid, and gas-gas equilibria. The method is also being extended to low-pressure vapor-liquid and liquid-liquid equilibria, as better equations of state are developed. The gamma-phi method is the traditional method for low-pressure vapor-liquid and vapor-solid equilibria. The trend seems to be toward using gamma-phi for simple situations, such as when the liquid phase can be assumed to be nearly an ideal solution. The gamma-gamma method is used for equilibria among condensed phases, including liquid-liquid, liquid-solid, and solid-solid equilibria.

10.5.4 Selecting an Appropriate Reaction-Equilibrium Relation

Here the choices are between the stoichiometric and the nonstoichiometric developments. The stoichiometric approach is best suited for small numbers of reactions, because it involves several distinct steps: determination of values for stoichiometric coefficients, construction of expressions for each mole fraction in terms of an extent for each reaction, determination of values for all equilibrium constants. These activities must be accomplished before the final equations can be written. Further, those equations tend to be application-specific, so many (but not all) forms of the stoichiometric approach are less amenable to a general computer implementation.

In contrast, the nonstoichiometric approach is generally better for complicated situations involving many reactions, including those many situations, such as combustion and biological processes, in which all reactions cannot be explicitly identified. This method involves many fewer preparatory steps: no stoichiometric coefficients need be computed, no reactions are identified or balanced, and no equilibrium constants are evaluated. The principal price paid for this convenience is the larger number of equations to be solved. But, even though the nonstoichiometric method produces more equations than the stoichiometric method, the nonstoichiometric development is more systematic and its general form can often be more readily implemented on a computer.

LITERATURE CITED

[1] S. Kulkarni, *Bifurcations in the Equation of State and the Stability of Binary Mixtures*, MS Thesis, Clemson University, 1996.

[2] C. A. Eckert and S. R. Sherman, "Measurement and Prediction of Limiting Activity Coefficients," *Fluid Phase Equil.*, **116**, 333 (1996).

[3] I. Prigogine and R. Defay, *Chemical Thermodynamics*, D. H. Everett (transl.), Longmans, Green, and Co., London, 1954.

[4] R. A. Heidemann, "Non Uniqueness in Phase and Reaction Equilibrium Computations," *Chem. Engr. Sci.*, **33**, 1517 (1978).

[5] M. W. Chase, *NIST JANAF Thermochemical Tables*, 4th ed., American Chemical Society, Washington, D.C., 1998; see also, *J. Phys. Chem. Ref. Data*, **14**, Supplement #1 (1985).

[6] K. E. Bett, J. S. Rowlinson, and G. Saville, *Thermodynamics for Chemical Engineers*, MIT Press, Cambridge, MA, 1975.

[7] G. N. Lewis, M. Randall, K. S. Pitzer, and L. Brewer, *Thermodynamics*, 2nd ed., McGraw-Hill, New York, 1961, ch. 27.

[8] F. C. Rossini, D. D. Wagman, W. H. Evans, S. Levine, and I. Jaffe, *Selected Values of Chemical Thermodynamic Properties*, National Bureau of Standards Circular 500, Washington, 1952.

[9] D. D. Wagman, W. H. Evans, V. B. Parker, I. Halow, S. M. Bailey, and R. H. Schuum, *Selected Values of Chemical Thermodynamic Properties*, National Bureau of Standards Technical Note 270, Part 3 (1968), Part 4 (1969), Part 5 (1971), *et seq*.

[10] F. D. Rossini, K. S. Pitzer, R. L. Arnett, R. M. Brown, and G. C. Pimental, *Selected Values of Physical and Thermodynamic Properties of Hydro Carbons and Related Compounds*, American Petroleum Institute, Research Project 44, Carnegie Press, Pittsburgh, 1953.

[11] *Selected Values of Properties of Chemical Compounds*, Thermodynamics Research Center, Texas A & M University, College Station, 1971–2000.

[12] D. R. Stull, E. F. Westrum, Jr., and G. C. Sinke, *The Chemical Thermodynamics of Organic Compounds*, Wiley, New York, 1969.

[13] Landolt-Börstein, *Tabellen*, vol. 2, Part 4, *Kalorische Zustandsgrössen*, Springer-Verlag, Berlin, 1961.

[14] B. E. Poling, J. M. Prausnitz, and J. P. O'Connell, *The Properties of Gases and Liquids*, 5th ed., McGraw-Hill, New York, 2001.

[15] T. J. V. Findlay and J. L. Copp, "Thermodynamics of Binary Systems Containing Amines. V. Alcohols and Pyridine," *Trans. Faraday Soc.*, **65**, 1463 (1969).

[16] S. M. Walas, *Phase Equilibria in Chemical Engineering*, Butterworth, Boston, 1985.

[17] C. M. McDonald and C. A. Floudas, "Global Optimization and Analysis for the Gibbs Free Energy Function Using the UNIFAC, Wilson, and ASOG Equations," *IEC Res.*, **34**, 1674 (1995).

PROBLEMS

10.1 For each of the following two-phase equilibrium situations, assume you know T and the composition of one phase. Indicate whether you should use φ-φ, γ-φ, or γ-γ to solve for the pressure and the composition of the other phase.

(a) solid naphthalene in contact with air

(b) carbon dioxide vapor in contact with champagne

(c) LSE for melting of a zinc-copper alloy

(d) benzene, toluene, and water in LLE

(e) n-propanol and isopropanol in VLE.

10.2 In Figure 10.2 we show the fugacities computed for ethanol(1) and benzene(2) in VLE at 370 K and 2.5 bar. According to the models used, an azeotrope occurs near 367 K and 2.5 bar. For this azeotropic state, make qualitative sketches of the fugacities $f_1^\ell(x_1), f_1^v(y_1), f_2^\ell(x_1), f_2^v(y_1)$ analogous to those shown in Figure 10.2. Mark the azeotrope on your plots. You do not have to do any calculations.

10.3 Since neither an ideal gas nor an ideal solution can violate the diffusional stability criteria derived in Chapter 8, Tabitha the Untutored maintains that Raoult's law must be nonsense. Do you agree? If so, then how do you explain that ideal gases can exist in VLE with ideal liquid solutions? If you do not agree, then how do you explain phase separation without instabilities?

10.4 A binary liquid mixture is composed of components 1 and 2. Over the range of compositions $0 < x_1 \leq 0.05$ component 1 obeys Henry's law. Obtain the expression for the composition dependence of the fugacity of component 2 over this same range of compositions.

10.5 Still trying to debunk Raoult's law, Tabitha the Untutored has noted that Raoult's law (10.2.15) gives a straight line for isothermal plots of VLE Px_1 data. Therefore, she says, Raoult's law asserts this: whenever we find a straight line for low-pressure VLE Px_1 data, then we have found an ideal liquid solution. Tabitha says this is nonsense because we can find nonideal mixtures that give such straight lines. For example, here are some data from Findlay and Copp [15] for mixtures of pyridine(1) and ethanol(2) at 75°C:

x_1	0	0.2	0.4	0.6	0.8	1.
P (mm Hg)	201.2	299.3	393.6	484.4	573.7	665.3
y_1	0	0.530	0.690	0.832	0.930	1.

(a) Plot these data as a Pxy diagram and fit a straight line to the Px_1 points.

(b) Tabitha says that a pyridine molecule is not much like an ethanol molecule, so they cannot form ideal liquid solutions (Lewis-Randall standard state), yet the Px_1 curve is essentially straight. Hence, Raoult's law is nonsense. Do you agree or disagree? Justify your answer.

10.6 Consider binary mixtures of benzene(1) and toluene(2) in a closed system at vapor-liquid equilibrium. Expressions for pure-component vapor pressures are given in Appendix D.

(a) Compute and construct a Px_1y_1 diagram at 100°C.

(b) Compute and construct a Tx_1y_1 diagram at 1 bar.

(c) Estimate the latent heat of vaporization for pure benzene.

10.7 Tabitha the Untutored claims the fugacity is the fugacity, no matter what, so it is legitimate to use FFF #1 for one component in a binary liquid mixture and use FFF #2 for the other component. Do you agree? If so, justify. If not, what fundamental thermodynamic relations would be violated?

10.8 Determine the dew pressure and liquid composition for an equimolar vapor mixture of ether(1) and acetone(2) at 45°C. At this T the pure vapor pressures are $P_1^s = 1.453$ bar and $P_2^s = 0.68$ bar. The liquid obeys $g^E/RT = 0.712\, x_1 x_2$.

10.9 Consider binary mixtures of benzene(1) and toluene(2) in VLE, with the pure-component vapor pressures given in Appendix D.

(a) What is the numerical value of $\mathcal{F}$ (see § 9.1) and what is its significance? Parts (b)–(e) are examples of $\mathcal{F}$-problems.

(b) Determine P and y_1 when $T = 110$°C and $x_1 = 0.4$.

(c) Determine P and x_1 when $T = 95$°C and $y_1 = 0.65$.

(d) Determine T and y_1 when $P = 1.2$ bar and $x_1 = 0.3$.

(e) Determine T and x_1 when $P = 1.5$ bar and $y_1 = 0.7$.

(f) What is the numerical value of $\mathcal{F}'$ (see § 9.1) and what is its significance? Solve the following $\mathcal{F}'$-problem: Find x_1, y_1, and V (the fraction of total moles in the vapor phase) when $T = 100$°C, $P = 1$ bar, and the overall mole fraction for benzene is $z_1 = 0.4$.

10.10 At 50°C a binary liquid mixture of 1 and 2 obeys $g^E/RT = 1.5\, x_1 x_2$. At 50°C the pure-component vapor pressures are 0.8 bar for component 1 and 0.933 bar for component 2. Determine whether this mixture exhibits an azeotrope at 50°C. Clearly state any assumptions made.

10.11 (a) Derive an expression that relates the activity coefficient γ_1^* to the reversible isothermal-isobaric work needed to extract a small amount of pure 1 from a Lewis-Randall ideal solution and inject it into a real mixture at T, P, $\{x\}$.

(b) Derive the expression that relates the activity coefficient $\gamma_1^\dagger$ to the reversible isothermal-isobaric work needed to extract a small amount of pure 1 from a reference-solvent, Henry's law ideal solution and inject it into a real mixture at T, P, $\{x\}$.

10.12 Find an expression for the amount of reversible isothermal-isobaric work needed to inject a small amount of pure 1 from a vacuum into a real binary mixture that is essentially pure component 2. That is, find the expression for the work, and then take the infinite-dilution limit. You may assume the mixture obeys Henry's law.

10.13 Assume air is a binary mixture of nitrogen and oxygen. Use the following models to estimate the fugacities of each component at 25°C and 40 bar. (a) Ideal gas. (b) Lewis-Randall ideal solution with components obeying the simple virial equation $Z = 1 + BP/RT$ and values of the second virial coefficients provided by the Pitzer correlation in Problem 4.22.

10.14 Nitrogen(1) and carbon tetrachloride(2) are in vapor-liquid equilibrium at 25°C and 1 bar. The liquid has $x_1 = 5(10^{-4})$; a correlation for vapor pressures of pure CCl_4 is given in Appendix D. Estimate the liquid composition when the system pressure is increased isothermally to 2 bar while vapor-liquid equilibrium is maintained.

10.15 For certain ternary mixtures of components 1, 2, and 3, the molar volume obeys

$$v = x_1 v_1 + x_2 v_2 + x_3 v_3 + A_{12} x_1 x_2 + A_{13} x_1 x_3 + A_{23} x_2 x_3 + B x_1 x_2 x_3$$

where $v_1 = 18$, $v_2 = 10$, $v_3 = 12$, $A_{12} = 10$, $A_{13} = -20$, $A_{23} = 15$, and $B = -20$, all in cc/mole.

(a) Determine the expression for the composition dependence of the partial molar volume of component 1.

(b) Let component 2 be a reference solvent, $r = 2$. Obtain the value of $\overline{V}_1$ in the pure reference-solvent limit. Does your answer depend on composition?

(c) Obtain the expression for the composition dependence of the molar volumes for binary mixtures of 1 and 2. Then evaluate the partial molar volume for 1 at infinite dilution. Compare your result with that found in (b); what do you conclude?

(d) Rewrite your expression for $\overline{V}_1$ found in (a) in terms of the solute-free mole fractions for components 2 and 3. Now evaluate the partial molar volume for 1 at infinite dilution for solute-free mole fractions fixed at their values corresponding to $x_2 = x_3 = 0.2$. Repeat the evaluation for $x_2 = 2x_3 = 0.2$. On comparing these two results, what do you conclude?

(e) Using expressions from (d), evaluate $\overline{V}_1^{\infty}$ for binary mixtures of 1 and 2 and compare the result with those found in (b) and (c).

10.16 For the ternary mixture in the example of § 10.2.6, compute the component-1 activity coefficient γ_1^{+} as a function of x_1, using component 3 as the reference solvent.

(a) Plot your Henry's law, reference-solvent, ideal-solution fugacity as f_1 vs. x_1 and on the same plot show the corresponding fugacity curve for the Lewis-Randall ideal solution.

(b) Plot your activity coefficient γ_1^{+} vs. x_1 for the ternary, with $x_2/x_3 = 1$. On the same plot show the corresponding curve for the Lewis-Randall standard-state activity coefficient γ_1.

10.17 Combine the three equations (10.4.43)–(10.4.45) to eliminate the Lagrange multipliers and show that the result is (10.4.29) for the equilibrium constant. This shows that the stoichiometric and nonstoichiometric developments are two formulations of the same equations.

10.18 When the gas-phase reaction

$$CO + 2\,H_2 \rightarrow CH_3OH$$

reaches equilibrium at 700 K, the total pressure is found to be 3.1 bar and the composition of the mixture is $y_1 = 0.323$, $y_2 = 0.0323$, $y_3 = 0.645$, where 1 = CO, 2 = H_2, and 3 = CH_3OH. If this mixture is isothermally expanded to twice its original volume, determine the composition of the new equilibrium mixture.

10.19 Consider production of ethanol by the gas-phase reaction of ethylene with steam at 1 bar and 500 K. At 298 K the standard properties of formation are

	C_2H_4	H_2O	C_2H_5OH
Δg_f^o (kJ/mol)	68.46	–228.6	–168.5
Δh_f^o (kJ/mol)	52.51	–241.8	–235.1

The pure ideal-gas heat capacities can be represented by

$$c_p/R = A + BT + CT^2 + DT^3 + ET^4$$

with T in K and following parameter values from Poling et al. [14]

	A	B(10^3)	C(10^5)	D(10^8)	E(10^{11})
C_2H_4	4.221	–8.762	5.795	–6.729	2.511
H_2O	4.395	–4.186	1.405	–1.564	0.632
C_2H_5OH	4.396	0.628	5.546	–7.024	2.685

(a) Is the reaction endothermic or exothermic at 500 K?

(b) Determine the composition of the equilibrium mixture, assuming the standard heats of formation are constants at the values given above at 298 K.

(c) Repeat the calculation in (a), but use the given heat capacities to account for the effect of temperature on the heats of formation.

10.20 (a) Repeat the determination of the equilibrium composition in § 10.3.3 and § 10.4.4 for the stoichiometric method applied to ammonia synthesis, but evaluate the composition at 10 bar, 25°C rather than at 1 bar, 25°C.

(b) Repeat part (a) but use the nonstoichiometric method of § 10.3.7 and § 10.4.6.

10.21 (a) Repeat the calculation in § 10.3.3 and § 10.4.4 but at 200°C, 1 bar rather than at 25°C, 1 bar. Assume the heat of formation for gaseous ammonia is constant at –46.1 kJ/mol.

(b) Repeat part (a) but use the nonstoichiometric method.

10.22 (a) Repeat the calculation in § 10.3.3 and § 10.4.4 but at 200°C, 10 bar rather than at 25°C, 1 bar. You may assume the heat of formation for gaseous ammonia is constant at –46.1 kJ/mol.

(b) Compare and discuss your results from Problems 10.20(a), 10.21(a), and 10.22(a). In particular, how do changes in temperature and pressure affect the equilibrium conversion?

10.23 Ethyl acetate can be made by the esterification of ethanol with acetic acid according to

$$\text{HAc} + \text{EtOH} \rightarrow \text{EtAc} + \text{H}_2\text{O}$$

Values of Δg_f^o (kJ/mol) and Δh_{vap} (kJ/mol) are as follows:

	EtOH	**HAc**	**EtAc**	**H_2O**
Δg_f^o Liquid [16] at 298 K	–174.2	–389.6	–332.9	–237.3
Δg_f^o Gas [16] at 298 K	–168.4	–376.9	–327.6	–228.8
Δg_f^o Gas [17] at 355 K	–155.2	–365.3	–304.8	–226.0
Δh_{vap} [14] at 1 atm.	38.56	23.70	31.94	40.66

(a) At 298 K we could adjust the pressure in the reactor to obtain either a one-phase liquid mixture or a one-phase gas. Estimate the equilibrium constant K for both possibilities, and indicate whether the liquid or the gas would produce a higher equilibrium conversion of EtOH. In deciding whether a liquid or a gas phase should be used, are there other (nonthermodynamic) considerations that should be taken into account?

(b) At 355 K and 1.013 bar the four-component reaction mixture exists in vapor-liquid equilibrium. Using only the values given for Δg_f^o above, estimate the value of the liquid-phase equilibrium constant K for the VLE situation at 355 K.

10.24 Use the stoichiometric method to determine the equilibrium compositions resulting from gas-phase oxidation of nitric oxide (NO) to nitrogen dioxide (NO_2) performed at 1 bar and (a) 298.15 K, (b) 600 K, and (c) 1000 K. Assume the initial feed in each case contains one mole of oxygen for each mole of nitric oxide. The values of standard properties of formation at 25°C are as follows:

	Δg_f^o (kJ/mol)	Δh_f^o (kJ/mol)
NO (g)	86.55	90.25
NO_2 (g)	51.31	33.18

10.25 Repeat Problem 10.24 but use the nonstoichiometric method.

10.26 Repeat Problem 10.24 using the stoichiometric method, but evaluate the compositions at (a) 298.15 K, 30 bar and (b) 600 K, 30 bar. If necessary, assume the simple virial equation applies, $Z = 1 + BP/RT$ with values of the second virial coefficients provided by the Pitzer correlation in Problem 4.22. You may also assume that the gas-phase mixture forms a Lewis-Randall ideal solution.

10.27 Repeat Problem 10.26 but use the nonstoichiometric method.

11

ELEMENTARY COMPUTATIONAL PROCEDURES

Multiphase systems and chemical reactions pervade the chemical processing industries. For example, we routinely force the creation of a new phase to exploit the accompanying change in composition; consequently, phase changes are used in many separation processes, including distillation, crystallization, and solvent extraction. In addition, we routinely suppress the creation of a new phase, for example, to maintain inventory of liquids by controlling loss due to evaporation and to meet health and safety standards by controlling evaporation of flammable, hazardous, and toxic substances. Likewise, we often promote chemical reactions to convert inexpensive raw materials into valuable products. But we also try to prevent other reactions that convert valuable materials into costly wastes, and we try to prevent reactions that convert benign substances into hazardous or toxic chemicals. In all such situations, the design and operation of appropriate processes may hinge upon computing proper solutions to phase-equilibrium problems or reaction-equilibrium problems or both.

In previous chapters we developed the thermodynamics of phase and reaction equilibria, and we illustrated certain principles using straightforward computational procedures. We used only simple procedures so as not to detract from thermodynamic issues. In this chapter we consider more complex situations and therefore give more attention to computational techniques. No new thermodynamics is introduced in this chapter; instead, we try to show how the thermodynamics already developed can be used in multicomponent phase and reaction-equilibrium situations.

But while we devote more attention to computational issues here, our goals remain educational: mastery of the material in this chapter will not make you an expert in computational thermodynamics, though we hope this chapter can serve as a solid foundation for future study and practice. Of the many computational algorithms that are available, we have selected a small number of elementary ones that are representative. Moreover, just as we have found and exploited patterns that help organize thermodynamic concepts, so too can we find patterns in computational algorithms. The algorithms presented in this chapter build on certain fundamental patterns. Recognizing those patterns should help you understand the material and should help you appreciate the underlying unity that exists among superficially distinct topics.

Strategies for attacking phase and reaction equilibrium calculations divide into two basic types: (a) those that impose the equilibrium conditions through a coupled set of nonlinear algebraic equations (equality of fugacities for phase equilibria or relations between fugacities and equilibrium constants for reaction equilibria) and (b) those that impose the equilibrium conditions by minimizing the appropriate thermodynamic state function (usually the Gibbs energy at fixed T and P). These two strategies are thermodynamically equivalent, so the decision as to which to apply can be based on computational issues, such as convergence, computational resources required, and ease of computer-program preparation. But here we also have educational goals, and those goals appear to be best served by concentrating on strategies in class (a). In our opinion, the computational issues surrounding multivariable optimization are so technical and subtle that they detract too much from the thermodynamic issues we hope to illustrate and reinforce. As Acton has remarked [1], "... minimum-seeking methods are often used when a modicum of thought would disclose more appropriate techniques." In this chapter we ignore Gibbs-energy minimization.

In § 11.1 we present algorithms based on the strategies developed in Chapter 10—the phi-phi, gamma-phi, and gamma-gamma methods—to perform multicomponent vapor-liquid, liquid-liquid, and vapor-liquid-liquid computations. In § 11.2 we present a stoichiometric algorithm for solving single-phase, multireaction equilibrium problems. Before applying an algorithm to such problems, we must identify the number of independent reactions and assign values to the stoichiometric coefficients of those reactions. These rather troublesome preliminaries can be automated through a particular decomposition of the formula matrix **A**, as we show in § 11.2. Then in § 11.3 we briefly introduce and illustrate an elementary algorithm for performing coupled phase and reaction equilibrium calculations.

11.1 PHASE-EQUILIBRIUM CALCULATIONS

When $\mathcal{P}$ phases, each composed of C components, coexist in unconstrained equilibrium, the intensive state is identified by giving values for $\mathcal{F}$ independent intensive properties, where the number $\mathcal{F}$ is provided by the phase rule,

$$\mathcal{F} = C + 2 - \mathcal{P} \qquad (9.1.14)$$

So when two phases, α and β, are in equilibrium, the state is identified by values for C properties; those C properties could be the pressure plus $(C - 1)$ independent mole fractions $\{x^\alpha\}$ for the α phase. But although situations having $\mathcal{P} = 2$ require only C properties to identify the state, we often need to know more to satisfy the needs of a design or operational problem. For example, in addition to pressure and $\{x^\alpha\}$, we may also need to know temperature and the composition of the other phase $\{x^\beta\}$. Values for these additional properties can be computed by solving the equilibrium conditions

$$f_i^\alpha(T, P, \{x^\alpha\}) = f_i^\beta(T, P, \{x^\beta\}) \qquad i = 1, 2, \ldots, C \qquad (11.1.1)$$

These are C independent equations that can be solved for C unknowns. In this section we present representative algorithms for performing such calculations.

Table 11.1 Typical classes of problems encountered in multicomponent two-phase fluid-fluid equilibria[a]

Problem name in VLE situations	Knowns	Unknowns to find
Bubble-T	P and $\{x^\alpha\}$	T and $\{x^\beta\}$
Dew-T	P and $\{x^\beta\}$	T and $\{x^\alpha\}$
Bubble-P	T and $\{x^\alpha\}$	P and $\{x^\beta\}$
Dew-P	T and $\{x^\beta\}$	P and $\{x^\alpha\}$
Flash	T, P, and $\{z\}$	$\{x^\alpha\}$, $\{x^\beta\}$, N^β/N

[a] The knowns and unknowns apply to any two-phase fluid-fluid situation, including liquid-liquid, liquid-vapor, and gas-gas; however, the names apply only to liquid-vapor equilibria, with phase α = liquid and phase β = vapor. Names taken from Smith and van Ness [2].

From the measurables T, P, $\{x^\alpha\}$, and $\{x^\beta\}$, five common phase equilibrium problems can be contrived, depending on which quantities are known and which are unknown. For example, the problem introduced in the previous paragraph involves P and $\{x^\alpha\}$ as known and requires us to solve (11.1.1) for T and $\{x^\beta\}$. When phase α is liquid and phase β is vapor, this problem is called a *bubble-T* calculation, for we are to compute a point on the bubble curve of an isobaric Txy diagram. This along with the other four common problems are listed in Table 11.1.

In the table the second, third, and fourth problems each result from a permutation of the known and unknown quantities that occur in the bubble-T calculation. We refer to these as $\mathcal{F}$-problems, because each problem is well-posed when values are specified for $\mathcal{F}$ independent intensive properties, where the value of $\mathcal{F}$ is given by the phase rule (9.1.14). However, the flash problem in Table 11.1 differs from the others in that it is an $\mathcal{F}'$-problem; it is well-posed when values are specified for $\mathcal{F}'$ independent intensive properties, with the value of $\mathcal{F}'$ given by (9.1.12). Flash calculations pertain to separations by flash distillation in which a known amount N of one-phase fluid, having known composition $\{z\}$, is fed to a flash chamber. When T and P of the chamber are properly set, the feed partially flashes, producing a vapor phase of composition $\{x^\beta\}$ in equilibrium with a liquid of composition $\{x^\alpha\}$. The problem is to determine these compositions, as well as the fraction of feed that flashes N^β/N. Unlike the other problems in Table 11.1, the flash problem involves the relative amounts in the phases and therefore a solution procedure must invoke not only the equilibrium conditions (11.1.1) but also material balances.

To solve any of the problems cited in Table 11.1, we start with the equilibrium conditions (11.1.1) and reduce those equations to computational forms using one of the methods—phi-phi, gamma-phi, or gamma-gamma—presented in Chapter 10. With five kinds of problems and, in several cases, more than one possible solution strategy, many computational algorithms can be devised. We do not present all possible algorithms here; instead, we present one typical algorithm for each of the three strategies.

Problems at the end of the chapter give you opportunities to explore other algorithms. More technical discussions of these and other algorithms for phase-equilibrium calculations can be found in the book by Prausnitz et al. [3] and in the papers by Michelsen [4–6].

11.1.1 Phi-Phi Method Applied to Bubble-T Calculations

To illustrate an implementation of the phi-phi method (§ 10.1.1), let us consider the bubble-T calculation for vapor-liquid equilibria. Recall the phi-phi method uses FFF #1 for both liquid and vapor phases, and so, as a prerequisite to the calculation, we must choose an equation of state that reliably correlates the volumetric behavior of both phases. Typical candidates include the Peng-Robinson [7] and Redlich-Kwong-Soave [8] equations. For VLE calculations, the phi-phi equations (10.1.3) can be posed in terms of K-factors,

$$K_i = \frac{y_i}{x_i} = \frac{\varphi_i^{\ell}(T, v^{\ell}, \{x\})}{\varphi_i^{v}(T, v^{v}, \{y\})} \qquad i = 1, 2, \ldots, C \tag{11.1.2}$$

This represents an $\mathcal{F}$ problem in which C nonlinear algebraic equations are to be solved for C unknowns: T and $\{y\}$. In general, these equations must be solved by trial, and the standard method of attack is the Newton-Raphson scheme [9]. However, in particular problems, we hope to find alternative algorithms, for the Newton-Raphson method is computationally expensive and slow to converge.

For the bubble-T calculation in the phi-phi form, a viable alternative to Newton-Raphson is presented in Figure 11.1. This algorithm is composed of three principal parts: an initialization, an outer loop that searches for the unknown T, and an inner loop that searches for the vapor-phase mole fractions $\{y\}$. The algorithm can be used for any number of components, but it is restricted to equilibrium between two phases. In the special case of a single component, the algorithm is equivalent to the Maxwell equal-area construction given in (8.2.22).

Initialization. In this first segment of the algorithm, we set values for all parameters appearing in the equation of state and its associated mixing rules. We set the known values for the pressure and the liquid composition $\{x\}$. To test for convergence, we set a value for the tolerance ε; it will be applied to sums of computed vapor-phase mole fractions (Σy_i), so values in the range $10^{-6} \le \varepsilon \le 10^{-4}$ are appropriate. The initialization continues with first guesses for the unknown temperature T and the unknown mole fractions, but usually with the mole fractions posed in the form of K-factors. In most trial-and-error searches of this kind, success is heavily influenced by the initial guess, so some care should be taken in assigning initial values to T and $\{K\}$. The best sources for initial guesses are values obtained from computations at nearby states of the same system. Lacking those, estimates can be obtained by solving the bubble-T problem, assuming the system obeys a simple ideality, such as Raoult's law ($\varphi_i^v = 1$ and $\varphi_i^{\ell} = P_i^s / P$). This completes the initialization segment, and we now enter the main portion of the algorithm.

Inner loop. The search begins by computing the liquid phase and vapor phase fugacity coefficients. The equations of state used in these problems are explicit in pressure, so the φs are determined from (4.4.23), which involves an integration over the volume. This requires us to compute the molar volumes v^ℓ and v^v from the equation of state. If the equation is cubic in v, then it should be solved analytically using Cardan's method (Appendix C). However, if the equation is fifth order or higher, then it will have to be solved by trial for v. With the φs known, we compute the K-factors from (11.1.2) and hence get calculated values for the vapor mole fractions $\{y'\}$. Typi-

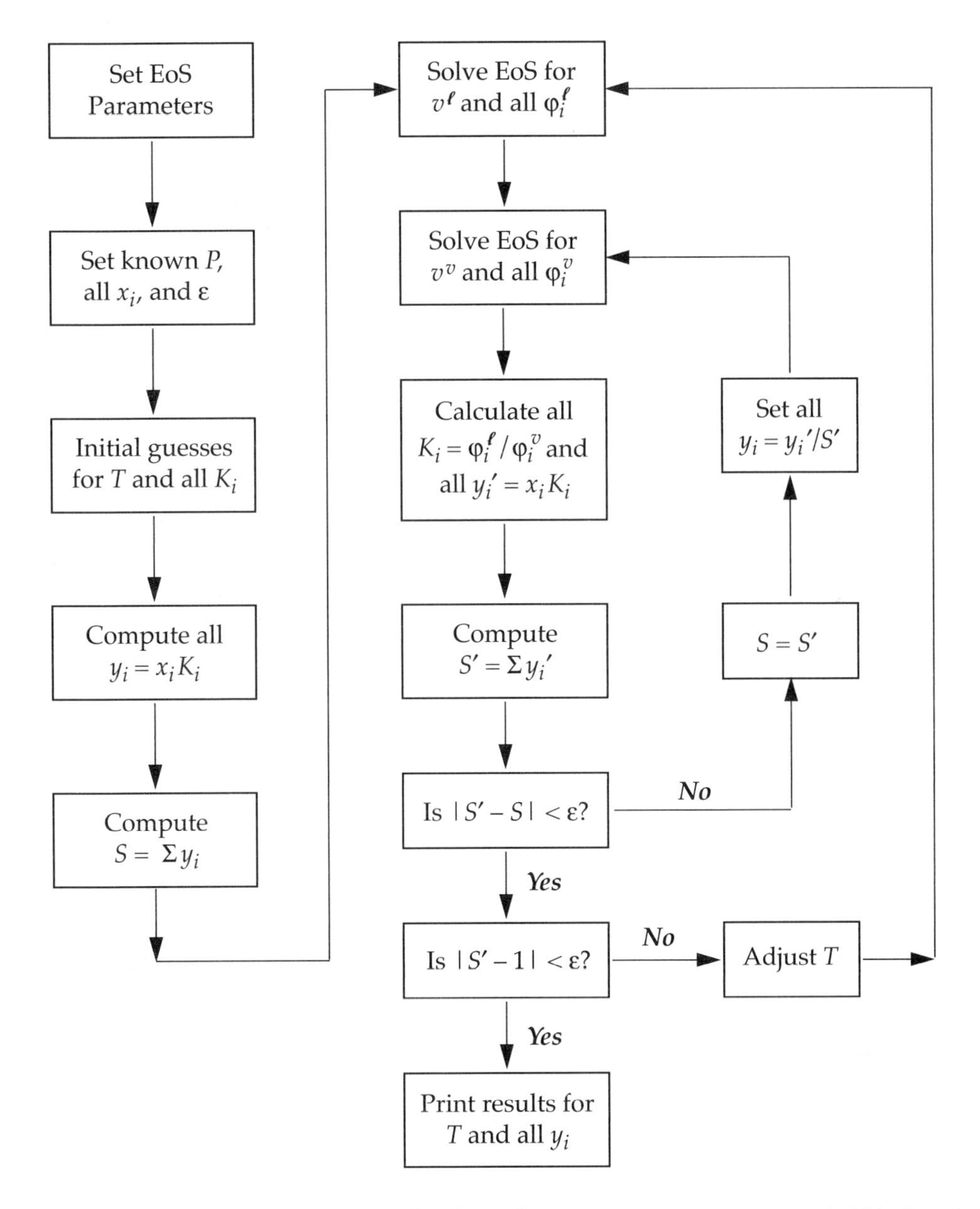

Figure 11.1 Algorithm for using the phi-phi method to solve multicomponent bubble-T problems in vapor-liquid equilibrium situations [3]

cally, the sum of these $\{y'\}$ differs from unity and it differs from the sum of the previous guesses for $\{y\}$; therefore, we normalize each element in $\{y'\}$ according to

$$y_i = \frac{y_i'}{\sum_i y_i'} = \frac{y_i'}{S'} \tag{11.1.3}$$

These normalized vapor-phase mole fractions become the new guesses, which we use to compute new values for the vapor-phase fugacity coefficients. The normalization (11.1.3) provides a self-consistent set of mole fractions; in general, it is good practice to use a self-consistent set of mole fractions when computing any property of a mixture. This step closes the inner search over the vapor composition.

Outer loop. When the sum S' stops changing within the inner loop, we test whether that sum equals unity (conservation of mass). If it does not, we adjust the temperature and compute new values for the liquid-phase fugacity coefficients. This step closes the outer search over temperature. In many cases each K-factor responds to a change in temperature in a sufficiently well-behaved way that T can be adjusted by the simple secant method: at the end of the kth iteration of the outer loop, the next guess $(k + 1)$ for T is taken to be

$$T_{k+1} = T_k - \frac{E_k}{\Delta E/\Delta T} \tag{11.1.4}$$

where $E = (S' - 1)$ is the error and each $\Delta w = w_k - w_{k-1}$ is a difference in values from the two previous iterations. However if (11.1.4) leads to instabilities, then a more conservative approach is required, such as bracketing plus bisection [9].

As with all algorithms for solving phase-equilibrium problems, the phi-phi method in Figure 11.1 is sensitive to the initial guesses made for T and $\{K\}$. With poor guesses, the algorithm tends to find the trivial solution, $y_i = x_i$ for all components. Also like most other methods, this algorithm performs poorly near vapor-liquid critical lines. Otherwise, the algorithm performs reasonably well; for example, for binaries with judicious initial guesses, it often converges in less than ten iterations over T.

11.1.2 Example

How do we use the phi-phi method to compute an isobaric *Txy* diagram for a binary mixture?

A typical isobaric Txy diagram appears in Figure 9.6. Here we outline how that diagram was computed, using $P = 30$ bar. The diagram is for an alkane(1)-aromatic(2) mixture modeled by the Redlich-Kwong equation of state,

$$P = \frac{RT}{v - b} - \frac{a}{\sqrt{T}v(v + b)} \tag{8.2.1}$$

Table 11.2 Values of Redlich-Kwong parameters for a certain alkane(1)-aromatic(2) mixture, computed using (8.4.18)–(8.4.23)

ij	T_{cij} (K)	P_{cij} (bar)	v_{cij} (cc/mol)	Z_{cij}	a_{ij} (cc/mol)^2 × bar K$^{0.5}$	b_i cc/mol
11	369.8	42.5	203.	0.281	18.28(10^7)	62.68
22	562.1	48.9	259.	0.271	45.26(10^7)	82.80
12	455.9	45.5	229.9	0.276	28.82(10^7)	

Mixing rules used here for a and b are given in (8.4.16) and (8.4.17). Pure-component critical properties, together with the mixing rules, provide the parameter values listed in Table 11.2. As a representative point, we take $P = 30$ bar and $x_1 = 0.6$, for which we are to compute T and y_1 using the bubble-T algorithm in Figure 11.1.

As the initial guess for the temperature, we might average the pure-component boiling points,

$$T \approx \sum_i x_i T_i^{\text{boil}} = 416 \text{ K} \tag{11.1.5}$$

with the pure boiling points at 30 bar computed from the Redlich-Kwong equation: 348 K for component 1 and 517 K for component 2. As initial guesses for the K-factors, we used $K_1 = 1.5$ and $K_2 = (1 - x_1K_1)/(1 - x_1) = 0.25$. At each iteration, molar volumes of both phases were computed analytically using Cardan's method (see Appendix C or [9]).

With a tolerance of $\varepsilon = 10^{-7}$, the algorithm converged in six iterations to $T = 387.68$ K and $y_1 = 0.8869$. The sum of the computed vapor-phase mole fractions gave $(\Sigma y_i - 1) = 7.4(10^{-9})$. Results are summarized in Table 11.3. The Tx_1 and Ty_1 points in Table 11.3 are plotted in Figure 9.6. By repeating this procedure for other x_1 values, we generate the complete Txy diagram shown in Figure 9.6.

Table 11.3 Results from phi-phi method for an alkane(1) and aromatic(2) in VLE at 30 bar with $x_1 = 0.6$; computed using algorithm in Figure 11.1 with Redlich-Kwong equation

	Liquid	Vapor
x_1 or y_1	0.6	0.8869
v (cc/mol)	122.30	742.54
φ_1	1.1836	0.8007
φ_2	0.1486	0.5257
f_1 (bar)	21.305	21.305
f_2 (bar)	1.783	1.783

11.1.3 Gamma-Phi Method Applied to Bubble-T Calculations

To illustrate the gamma-phi method (§ 10.1.2), we reconsider the bubble-T calculation. Now we intend to use FFF #1 for the vapor phase and one of FFF #2–5 for the liquid. In most cases, we use FFF #5 for the liquid-phase fugacity, then the gamma-phi equations (10.1.4) take the form

$$x_i \gamma_i(T, \{x\}; P_i^o) f_i^o(T, P_i^o) \exp\left[\int_{P_i^o}^{P} \frac{\overline{V}_i(T, \pi)}{RT} d\pi\right] = y_i \varphi_i(T, P, \{y\}) P \qquad (11.1.6)$$

For the liquid-phase standard state we usually choose the pure liquid at the system temperature and its vapor pressure $P_i^s(T)$,

$$f_i^o = \varphi_i^s(T) P_i^s(T) \qquad (11.1.7)$$

If data are lacking for partial molar volumes, we might approximate them using the pure-component molar volumes; then (11.1.6) becomes

$$x_i \gamma_i(T, \{x\}; P_i^s) \varphi_i^s(T) P_i^s(T) \exp\left[\frac{v_{\text{pure i}}(P - P_i^s)}{RT}\right] = y_i \varphi_i(T, P, \{y\}) P \qquad (11.1.8)$$

At low pressures the Poynting factor can be safely neglected. The activity coefficients are to be extracted from a model for g^E. We caution that if the vapor phase has $\varphi_i \neq 1$, then setting $\varphi_i^s = 1$ introduces an inconsistency, even at low pressures. An equation like (11.1.8) applies to each component, so we again have an $\mathcal{F}$-problem: the C nonlinear equations (11.1.8) are to be solved for C unknowns.

Those equations can be used to solve any of the problems cited in Table 11.1; for example, an algorithm for solving the bubble-T problem is presented in Figure 11.2. The gamma-phi algorithm in Figure 11.2 is structurally analogous to that for the phi-phi algorithm shown in Figure 11.1. Both algorithms are composed of three parts: an initialization stage, an outer search for the unknown temperature, and an inner search for the unknown $\{y\}$.

The initialization stage in Figure 11.2 is much like that in Figure 11.1; the principal difference is that now we initially set all vapor-phase fugacity coefficients to unity (i.e., we start by assuming the vapor is an ideal gas); this allows us to avoid making initial guesses for the $\{y\}$. However, we must still make an initial guess for the temperature, and the performance of the algorithm will be sensitive to that guess. The same considerations for initial guesses apply here as apply to phi-phi calculations (§ 11.1.1).

The search for T begins with the evaluation of the pure-liquid vapor pressures and fugacity coefficients. The temperature dependence of each vapor pressure is invariably represented by some relative of an Antoine equation. Values for fugacity coefficients are computed from a $PvTx$ equation of state; the same equation should be used for φ_i and for φ_i^s. Then, with values for the activity coefficients computed from the model chosen for g^E, the liquid-phase fugacity can be computed.

The algorithm in Figure 11.2 then calculates f_i^ℓ at low pressure, neglecting the Poynting factor that appears in (11.1.8). Including the Poynting factor introduces no complications, because at this point in the calculation, P and $\{x\}$ are known at the current value of T. The decision to include the Poynting factor should take into account the accuracy with which the equation of state estimates φ_i: there is no advantage to including a Poynting factor if doing so improves the liquid fugacity by 1% while the equation of state provides the vapor fugacity only to within 3%.

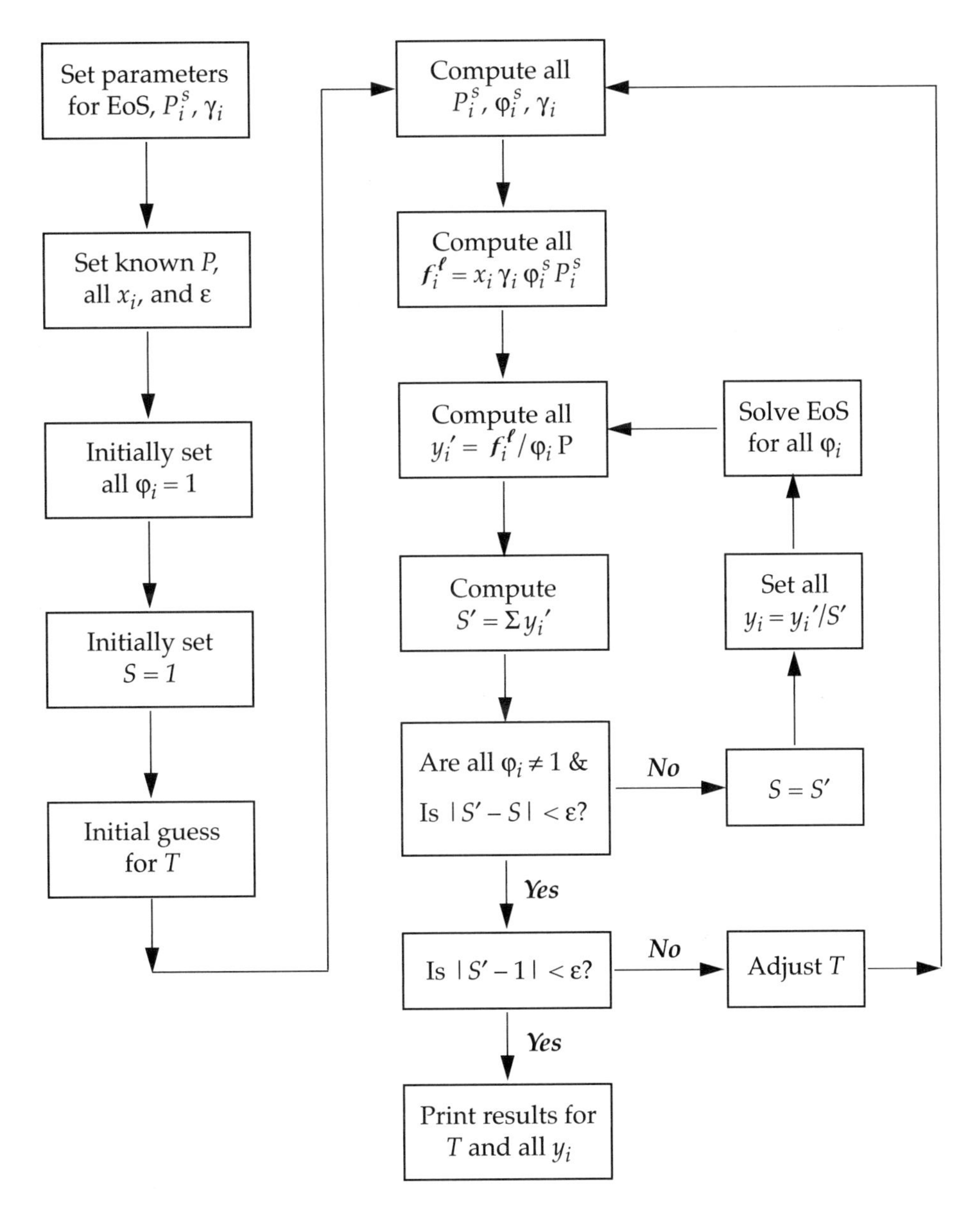

Figure 11.2 Algorithm for using the gamma-phi method to solve multicomponent bubble-T problems in vapor-liquid equilibrium situations at low pressures [3]

The remaining parts of the gamma-phi algorithm in Figure 11.2 parallel those already discussed in § 11.1.1 for the phi-phi algorithm of Figure 11.1. The low-pressure algorithm in Figure 11.2 is relatively robust compared to other algorithms discussed in this section: for most systems with reasonable initial guesses for T, convergence is usually quick and dependable. In many situations, convergence might also be attained at high pressures, but the results will be misleading if the models used for g^E and $\overline{V}_i$ do not include pressure and composition effects. In critical regions, inconsistencies *must* arise in the gamma-phi approach, and those inconsistencies will *not* be removed by simply changing from one g^E model to another or from one equation of state to another. To obtain consistent results in critical regions, we must use the phi-phi method with the same equation of state and same mixing rules for both phases.

The gamma-phi method seems to be most problematic when it is applied to high-pressure systems in which we use one of the dilute-solution standard states for activity coefficients. In these situations it is not unusual for the activity coefficient ($\gamma_i^* < 1$) to compete with the Poynting factor (> 1); in such cases, it is better to use FFF #4 rather than FFF #5, so these competing effects are all collected in the activity coefficient. However, these situations are further aggravated by slow convergence because the standard-state fugacity (now the Henry's constant H_i) changes relatively slowly with temperature; in some systems H_i may even pass through a maximum with T. This differs from the behavior encountered when a pure-component standard state is used, for then the standard-state fugacity, such as in (11.1.7), changes quickly with T.

11.1.4 Example

How do we use the bubble-T algorithm in Figure 11.2 to compute an isobaric *Txy* diagram for a binary mixture?

Consider binary mixtures of ethanol(1) and benzene(2), for which an isobaric *Txy* diagram at 2.5 bar is shown in Figure 9.9. There the liquid-phase activity coefficients were modeled using the Margules equations (5.6.12) and (5.6.13) with parameter values taken from Appendix E. Pure-component vapor pressures were taken from Appendix D. To compute a point on the diagram at 2.5 bar, we specify a liquid composition x_1, then apply the bubble-T algorithm to obtain the corresponding values for T and y_1. As a representative point, we choose $x_1 = 0.2$ at $P = 2.5$ bar, and set the convergence tolerance to $\varepsilon = 10^{-5}$. At 2.5 bar we expect the vapor to be nearly an ideal gas, but we test that expectation here.

Assume the vapor is an ideal gas. This means we set all $\varphi_i = \varphi_i^s = 1$. Then, from an initial guess of $T = 300$ K, the algorithm converges in 19 iterations with the final error in the sum of the vapor-phase mole fractions given by $(S - 1) = -0.3(10^{-5})$. The final values were $T = 368.4$ K and $y_1 = 0.451$, which are plotted on Figure 9.9. To obtain the complete diagram, we repeat this calculation using other values for x_1.

Correct for vapor-phase nonidealities. To account for the effects of nonideal-gas behavior, we compute fugacity coefficients using the simple virial equation

$$Z = 1 + \frac{BP}{RT} \tag{11.1.9}$$

Table 11.4 Values of the parameters in (11.1.12) for mixtures of ethanol(1) and benzene(2) at $340 \le T \le 380$ K[a]

ij	11	22	12
$\ln a_{ij}$	35.4488	21.0651	13.8196
n_{ij}	4.8594	2.4190	1.2883

[a] These apply for T in K and B in cc/mol. Original data from [10].

For a binary, expressions for fugacity coefficients are given in Problem 4.11,

$$\ln \varphi_i = \frac{P}{RT}(-B + 2y_i B_{ii} + 2y_j B_{ij}) \tag{11.1.10}$$

where the mixture second virial coefficient B is given by

$$B = \sum_i \sum_j y_i y_j B_{ij} \tag{11.1.11}$$

Values for the second virial coefficients of ethanol-benzene mixtures are given in [10]; here we use the following empiricism to represent their temperature dependencies between 340 and 380 K:

$$B_{ij} = \frac{-a_{ij}}{T^{n_{ij}}} \qquad 340 \le T \le 380 \text{ K} \tag{11.1.12}$$

Least-squares fits give the values for parameters a and n contained in Table 11.4.

We now repeat the bubble-T calculation for P = 2.5 bar, x_1 = 0.2. From an initial guess of 300 K, the algorithm converges in 16 iterations to T = 368.15 K and y_1 = 0.447. The corresponding values for the fugacity coefficients were φ_1 = 0.950 and φ_2 = 0.942; for the fugacity coefficients of the pure saturated vapors, φ_1^s = 0.949 and φ_2^s = 0.956.

Comparing the results for T and y_1 with those found above, we see that using the ideal-gas law introduces errors of less than 1% in both T and y_1. Whether these discrepancies are important depends on the use to be made of the results; for showing qualitative trends in *Txy* diagrams, discrepancies of less than 1% are generally insignificant, and so we used the ideal-gas assumption in computing the diagrams shown in Figure 9.9. But your application may differ. Judgements about nonidealities can be made by comparing φ_i / φ_i^s with unity.

These results suggest that the vapor-phase composition is more sensitive to the ideal-gas assumption than is the temperature. For example, at 10 bar and x_1 = 0.5, the ideal gas produces deviations of 2.7% in y_1 and 0.2% in T compared to results from the virial-equation. This difference in sensitivity is common; and, in fact, the uncertainties in $\{y\}$ often grow with increasing pressure more quickly than those in T.

11.1.5 Gamma-Gamma Method Applied to LLE Calculations

The gamma-gamma method (§ 10.1.3) is commonly applied to low-pressure, liquid-liquid equilibrium calculations. Recall that at low pressures, FFF #2–5 are all equivalent, so the gamma-gamma method reduces to (10.1.5). Then if we choose the same standard state for both liquid phases, (10.1.5) simplifies to

$$x_i^{\alpha} \gamma_i^{\alpha} (T, \{x^{\alpha}\}) = x_i^{\beta} \gamma_i^{\beta} (T, \{x^{\beta}\}) \qquad i = 1, 2, \ldots, C \qquad (10.1.6)$$

Since the system contains a total of C components, this represents C nonlinear algebraic equations that are to be solved for C unknowns.

But when the same g^E-model is used to obtain γ_i^{α} and γ_i^{β}, numerical procedures for solving (10.1.6) converge erratically, if at all. We therefore seek a procedure that is more reliable than a direct attack on (10.1.6). For example, note that if the system temperature and pressure are known (as they usually are for LLE situations), then the problem can be posed as an analogy to isothermal flash calculations. In such an approach, we take the known quantities to be T, P, and the set of overall system mole fractions $\{z\}$. These last are defined by

$$z_i = \frac{N_i}{N} = \frac{N_i^{\alpha} + N_i^{\beta}}{N^{\alpha} + N^{\beta}} \qquad (11.1.13)$$

Then the quantities to be computed would be the mole fractions in each phase, $\{x^{\alpha}\}$ and $\{x^{\beta}\}$, and the fraction of total material in one phase, $R = N^{\beta}/N$.

Before going further, let us pause to ask whether this problem is well-posed. Since the relative amount in a phase appears explicitly (R), the number of independent intensive properties needed to identify the state is not given by the Gibbs phase rule (9.1.14). Instead, we need

$$\mathcal{F}' = C + 1 - S \qquad (9.1.12)$$

Here we have imposed no external constraints, so $S = 0$, and therefore we must specify values for $(C + 1)$ properties to close the problem. This requirement is satisfied, for we know T, P, and $(C - 1)$ independent mole fractions z_i; the problem is well-posed. In fact, we have $(2C - 1)$ unknowns: $(C - 1)$ independent mole fractions x_i^{α}, $(C - 1)$ independent mole fractions x_i^{β}, plus R. And we have $(2C - 1)$ independent equations: C phase-equilibrium relations (10.1.6) plus $(C - 1)$ independent material balances.

The motivation for posing the LLE problem in this way is that it allows us to take advantage of the Rachford-Rice procedure [11], which is a robust algorithm traditionally applied to isothermal flash calculations. To develop that procedure, we introduce a distribution coefficient C_i for each component; this quantity is defined by

$$C_i \equiv \frac{x_i^{\beta}}{x_i^{\alpha}} \qquad (11.1.14)$$

In general, each C_i depends on the complete equilibrium state of the two-phase system (T, P, and compositions); however in many systems, distribution coefficients are less sensitive to changes of state than are mole fractions. The distribution coefficients play a role in LLE analogous to that played by K-factors in VLE; they are discussed further in § 12.1. The gamma-gamma form for the phase-equilibrium equations (10.1.6) can be expressed in terms of the distribution coefficients as

$$C_i = \frac{x_i^\beta}{x_i^\alpha} = \frac{\gamma_i^\alpha}{\gamma_i^\beta} \qquad i = 1, 2, \ldots, C \tag{11.1.15}$$

For each component i in the liquid-liquid system, we write a material balance

$$x_i^\alpha(1-R) + x_i^\beta R = z_i \tag{11.1.16}$$

where $R = N^\beta/N$. Then we use the definition (11.1.14) to eliminate x_i^β in favor of x_i^α,

$$x_i^\alpha(1-R) + x_i^\alpha C_i R = z_i \tag{11.1.17}$$

Solving for x_i^α, we obtain

$$x_i^\alpha = \frac{z_i}{1 + R(C_i - 1)} \tag{11.1.18}$$

We can also use (11.1.14) in (11.1.18) to obtain an expression for x_i^β,

$$x_i^\beta = \frac{z_i C_i}{1 + R(C_i - 1)} \tag{11.1.19}$$

But the mole fractions in each phase must sum to unity, so we define the Rachford-Rice function F by

$$F \equiv \sum_i^C x_i^\beta - \sum_i^C x_i^\alpha = \sum_i^C \frac{z_i(C_i - 1)}{1 + R(C_i - 1)} = 0 \tag{11.1.20}$$

By summing over the unknown mole fractions, we have reduced the problem from $(2C - 1)$ equations in $(2C - 1)$ unknowns to a problem of $(C + 1)$ equations in $(C + 1)$ unknowns. The unknowns are the C distribution coefficients plus the fraction R; the equations are the C equilibrium relations (11.1.15) plus (11.1.20). Recall that the overall mole fractions $\{z\}$ are known, so with guesses for all the C_i, (11.1.20) represents one equation in the remaining unknown R. The form (11.1.20) is effectively a $(C - 1)$-order polynomial in R and is generally solved by trial; in the special case of a binary, (11.1.20) is linear in R and it can be solved analytically.

This Rachford-Rice approach offers two principal advantages over other formulations of the LLE problem: (i) Equation (11.1.20) is one equation in the unknown R, independent of the number of components present. (The corresponding disadvantage is that we must make initial guesses for all C distribution coefficients.) (ii) Equation (11.1.20) readily lends itself to a solution by Newton's method. (See § A.6 in Appendix A.) In Newton's method the value of $R(k)$ at the end of the kth iteration is replaced by the next guess $R(k+1)$ by applying

$$R^{(k+1)} = R^{(k)} - \frac{F^{(k)}}{(\partial F/\partial R)_{\{C\}}} \tag{11.1.21}$$

where

$$\left(\frac{\partial F}{\partial R}\right)_{\{C\}} = -\sum_{i}^{C} \frac{z_i(C_i - 1)^2}{[1 + R(C_i - 1)]^2} < 0 \tag{11.1.22}$$

Since this derivative is *always* negative, F must be monotone in R for fixed values of the distribution coefficients $\{C\}$. This means that if the C_i were independent of composition, they would be constants, the curve $F(R)$ would cross $F = 0$ only at one value of R, and Newton's method would always converge. Further, it would converge quickly.

Unfortunately, the distribution coefficients C_i are not constants, and the Rachford-Rice function $F(R)$ actually represents a family of curves, as in Figure 11.3. At each iteration of the calculation, the C_i values change, moving the search from one curve to another. Nevertheless, each curve in the family is monotone in R, so the computation often converges.

A flow diagram for the Rachford-Rice method is shown in Figure 11.4. The algorithm divides into two parts: an initialization stage followed by a single search loop over R, the fraction of material in one phase. During initialization, we set values for all

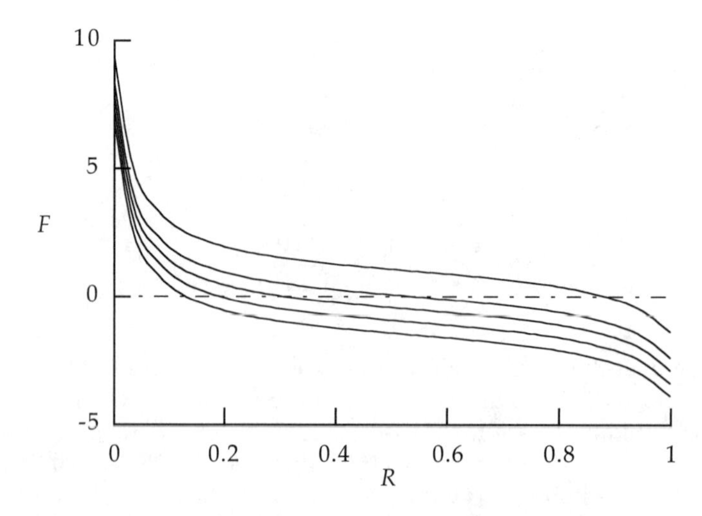

Figure 11.3 Schematic of a few members of the family of Rachford-Rice curves *F(R)* (11.1.20) for a ternary mixture in LLE. All curves here are for one set of overall mole fractions $\{z\}$ at the same T and P; however, each curve is for a different set of values for the distribution coefficients $\{C\}$.

parameters that appear in the models for activity coefficients. We also set the state by specifying values for T, P, and the overall mole fractions $\{z\}$, and we assign a value to the tolerance ε used to test whether the search for R has converged. The crucial initialization step is providing initial guesses for all distribution coefficients $\{C\}$; these initial guesses determine the success of the method. In most situations we have some rough idea as to how each component is distributed between two phases. Since we have defined $C_i = x_i^\beta / x_i^\alpha$, components expected to be predominantly in the β phase should have $C_i \gg 1$, while components predominantly in the α phase should have $C_i \ll 1$. Finally, initialization is completed by making a guess for R. Since $0 \le R \le 1$, we usually guess $R = 0.5$, unless we have information to the contrary.

The search for R is straightforward. We first solve the combined material balance-equilibrium equations (11.1.18) for all mole fractions $\{x^\alpha\}$ in one phase. Then we use the distribution coefficients to compute the mole fractions $\{x^\beta\}$ in the other phase.

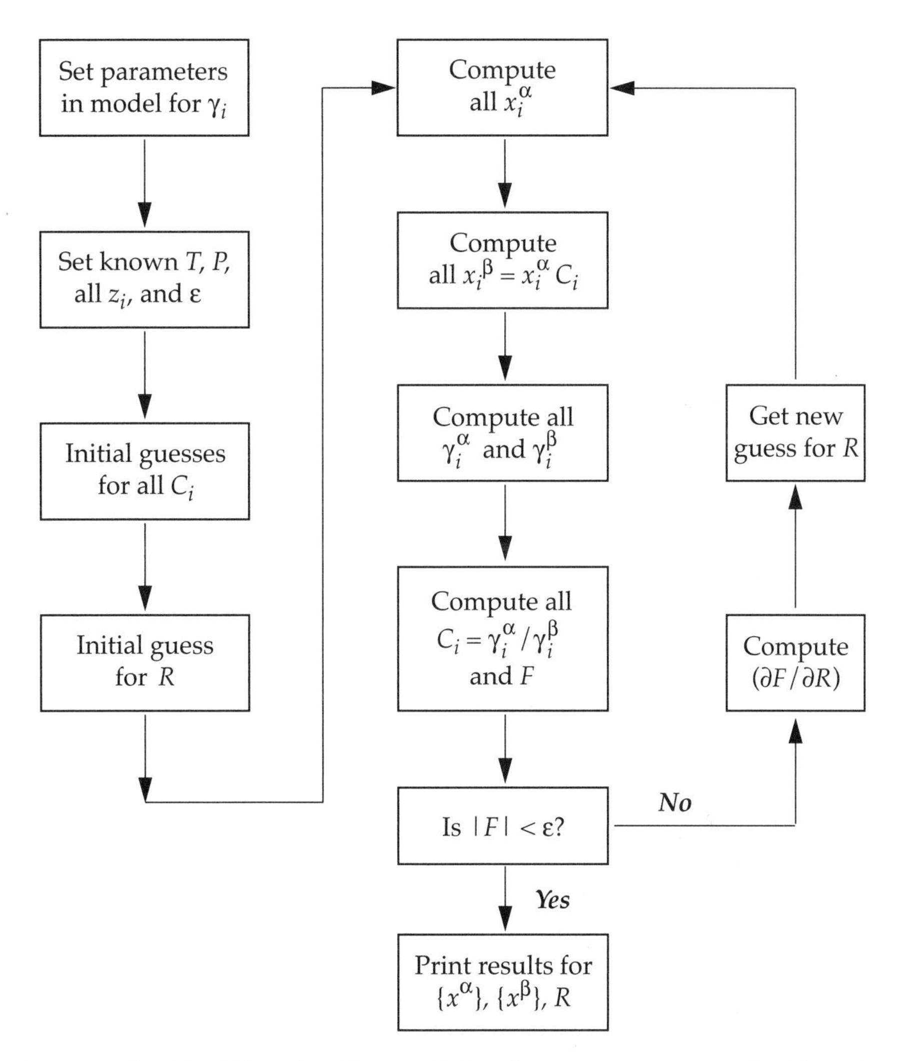

Figure 11.4 Rachford-Rice algorithm applied to the gamma-gamma method for solving multicomponent liquid-liquid equilibrium problems

With the compositions, we solve our selected model equations for all activity coefficients, and use (11.1.15) to obtain new values for the C_i. Then we can compute the Rachford-Rice function F from (11.1.20) and test for convergence. If convergence is lacking, we apply Newton's method (11.1.21) to get a new guess for R and iterate.

This procedure can fail to converge because, so long as the C_i are changing, each iteration moves us from one curve to another on the family of $F(R)$ curves shown in Figure 11.3. Therefore the slope of a curve, as computed by (11.1.22), is not necessarily the slope of the solution curve. Near the solution this discrepancy is often unimportant because nearby curves in $F(R)$ usually have similar slopes. But poor initial guesses will place the search point far from the solution, and curves at such points may have slopes that differ significantly from those along the true solution curve.

The Rachford-Rice algorithm can find nonphysical values for R. If the procedure gives $R > 1$ (or $R < 0$), then the system is a single liquid phase, rich in those components that dominate the β (or α) phase. However, we caution that solutions having $R >$ 1 and $R < 0$ may be false roots; that is, there may actually exist two-phase roots ($0 < R$ < 1) that solve the problem. These can usually be found by changing the initial guesses made for the distribution coefficients C_i.

In the special case of a binary mixture, we can obtain useful information from the Rachford-Rice procedure, even if we do not know the overall mole fractions $\{z\}$; to do so, we assume values for the z_i. Then, if a two-phase solution is found ($0 < R < 1$), the computed values for the compositions, $\{x^\alpha\}$ and $\{x^\beta\}$, will be correct regardless of the values assumed for $\{z\}$. However, the value for R changes with z_i. So if we have a binary and if we need only the compositions of the two liquid phases, and do not need R, then the Rachford-Rice remains a viable approach. Unfortunately, for two-phase situations with $C \geq 3$, the compositions, as well as R, depend on the z_i.

11.1.6 Example

How do we use the Rachford-Rice algorithm to compute a triangular diagram for liquid-liquid equilibria in a ternary mixture?

Consider mixtures of benzene(1), acetonitrile(2), and water(3) at 1.0133 bar, 333 K. Binary liquid mixtures of benzene and water are almost completely immiscible, so we expect the ternary to have a water-rich phase and an organic-rich phase with acetonitrile distributed between them. To model the activity coefficients, we choose the NRTL equation (see Appendix J) and use parameter values from Table J.1 [12].

To compute a point on the liquid-liquid saturation curve at the specified T and P, we choose an overall composition and apply the Rachford-Rice algorithm. For example, consider the overall composition having $z_1 = 0.3436$ and $z_2 = 0.3092$. We set the convergence tolerance to $\varepsilon = 10^{-5}$ (or 10^{-6} near the critical point). Then we guess the fraction of material in the water-rich phase, $R = 0.4$, and guess the component distribution coefficients: $C_1 = 7$, $C_2 = 2$, and $C_3 = 0.5$. Here we use $C_i = x_i^\beta / x_i^\alpha$, where α indicates the water-rich phase and β indicates the organic-rich phase.

From these initial values, the algorithm in Figure 11.4 converges in 23 iterations to $R = 0.6659$ with the phase compositions given in Table 11.5. Note that the guesses for the distribution coefficients were not particularly close to their final values ($C_1 = 214$, $C_2 = 6$, $C_3 = 0.06$); nevertheless, the equilibrium mole fractions agree with results com-

Table 11.5 Results from Rachford-Rice algorithm for two liquid phases in equilibrium at 333 K and 1.0133 bar

i	**Species**	z_i	x_i^α	x_i^β
1	C_6H_6	0.3436	0.0024	0.5147
2	C_2H_3N	0.3092	0.0711	0.4286
3	H_2O	0.3472	0.9265	0.0566

puted by McDonald and Floudas using a Gibbs-energy minimization technique [13]. The compositions in Table 11.5 represent the end points of one tie line, as shown in Figure 11.5. By repeating the calculation for other values of $\{z\}$, we obtain other tie lines and hence the two-phase line shown in the figure.

11.1.7 Calculation of Vapor-Liquid-Liquid Equilibrium

The methods presented in previous sections can be combined to attack multiphase equilibrium problems. To illustrate, we combine the gamma-phi method with the gamma-gamma method to solve three-phase, vapor-liquid-liquid problems. We again choose to pose these problems as analogies to isothermal flash calculations, as in § 11.1.5. Then such problems are well-posed when we have specified values for $\mathcal{F}'$ independent properties, where $\mathcal{F}'$ is given by (9.1.12) with $S = 0$,

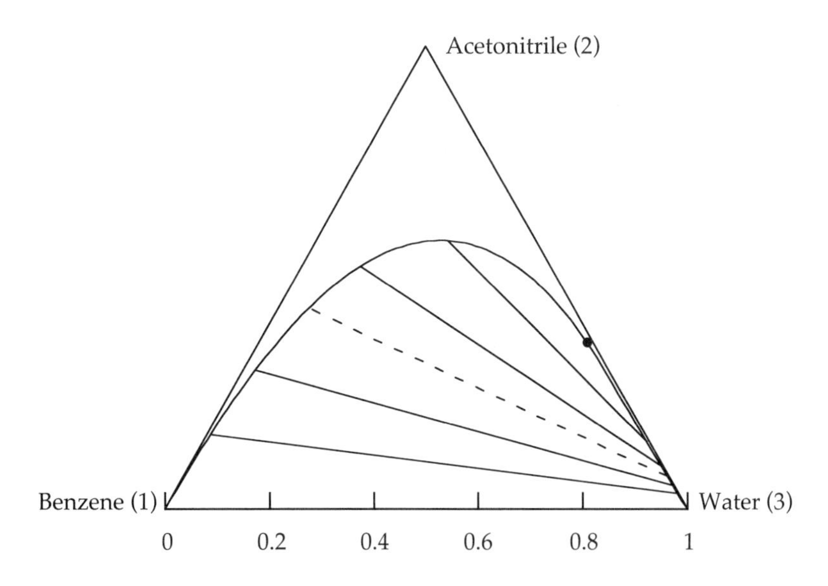

Figure 11.5 Liquid-liquid equilibria in mixtures of benzene(1), acetonitrile(2), and water(3) at 1.0133 bar and 333 K, computed from the Rachford-Rice algorithm in Figure 11.4 using the NRTL model for activity coefficients. Filled circle is an estimate of the liquid-liquid critical point; the estimate lies near $x_1 = 0.012$, $x_2 = 0.36$, $x_3 = 0.628$. Dashed line is the tie line used in the example to illustrate the computational procedure.

$$\mathcal{F}' = C + 1 \tag{11.1.23}$$

For $C \geq 3$, this condition is satisfied by specifying values for T, P, and $(C - 1)$ overall mole fractions $\{z\}$, with the z_i defined in (11.1.13). However for binaries ($C = 2$), these properties do not satisfy (11.1.23), and in fact the procedure presented in this section fails for binary mixtures in three-phase equilibria. The special case of a binary in VLLE is addressed in Appendix K.

With values known for $(C + 1)$ properties, we aim to compute values for $(3C - 1)$ other properties: $(C - 1)$ independent mole fractions $\{x^\alpha\}$ for liquid phase α, $(C - 1)$ fractions $\{x^\beta\}$ for liquid phase β, $(C - 1)$ fractions $\{y\}$ for the vapor, the fraction of material L in liquid α, and the fraction of material V in the vapor. To compute these quantities, we use a double implementation of the Rachford-Rice scheme. That is, we combine phase-equilibrium and material balance equations to obtain two equations that can be solved for the fractions L and V.

Those equations are derived by a procedure that parallels the one presented in § 11.1.5. We use K-factors to relate each mole fraction in the vapor to that of the same component in one liquid phase,

$$K_i = \frac{y_i}{x_i^\alpha} = \frac{\gamma_i^\alpha(T, v^\alpha, \{x^\alpha\})\, f_i^{o\alpha}}{\varphi_i(T, v^v, \{y\})\, P} \qquad i = 1, 2, \ldots, C \tag{11.1.24}$$

and we use distribution coefficients to relate the mole fractions of each component in the two liquid phases,

$$C_i = \frac{x_i^\beta}{x_i^\alpha} = \frac{\gamma_i^\alpha}{\gamma_i^\beta} \qquad i = 1, 2, \ldots, C \tag{11.1.15}$$

Let N be the total number of moles of material in the system, let N^α be the number of moles in phase α, and let N^v be the number in the vapor. Then the fractions L and V are given by $L = N^\alpha/N$ and $V = N^v/N$. A material balance on any one component can be written as

$$x_i^\alpha L + x_i^\beta (1 - L - V) + y_i V = z_i \tag{11.1.25}$$

Using (11.1.24) to eliminate y_i and (11.1.15) to eliminate x_i^β, we have

$$x_i^\alpha L + x_i^\alpha C_i(1 - L - V) + x_i^\alpha K_i V = z_i \tag{11.1.26}$$

Hence,

$$x_i^\alpha = \frac{z_i}{D_i} \tag{11.1.27}$$

Now we can combine (11.1.27) with the definition of C_i in (11.1.15) to obtain

$$x_i^{\beta} = x_i^{\alpha} C_i = \frac{z_i C_i}{D_i} \tag{11.1.28}$$

and we can combine (11.1.27) with the definition of K_i in (11.1.24) to obtain

$$y_i = x_i^{\alpha} K_i = \frac{z_i K_i}{D_i} \tag{11.1.29}$$

The denominators in (11.1.27)–(11.1.29) are given by

$$D_i = C_i + L(1 - C_i) + V(K_i - C_i) \tag{11.1.30}$$

As in the traditional Rachford-Rice approach, our strategy at this point is to reduce the number of unknowns by summing over the unknown mole fractions. Then we define two functions, analogous to the Rachford-Rice function in (11.1.20),

$$F_1 \equiv \sum_i^C x_i^{\alpha} - \sum_i^C x_i^{\beta} = \sum_i^C \frac{z_i(1 - C_i)}{D_i} = 0 \tag{11.1.31}$$

$$F_2 \equiv \sum_i^C y_i - \sum_i^C x_i^{\beta} = \sum_i^C \frac{z_i(K_i - C_i)}{D_i} = 0 \tag{11.1.32}$$

Our problem is to solve $(2C + 2)$ equations for $(2C + 2)$ unknowns. The unknowns are C K-factors, C distribution coefficients, plus L and V; the equations are the $2C$ equilibrium relations (11.1.15) and (11.1.24) plus the two equations (11.1.31) and (11.1.32). With guesses for all the K_i and C_i, (11.1.31) and (11.1.32) represent two equations to be solved for L and V. These are nonlinear algebraic equations, and to preserve the analogy with the simple Rachford-Rice procedure, we solve the two equations simultaneously via the Newton-Raphson method.

Newton-Raphson method. This is a trial-and-error method for solving simultaneous, nonlinear, algebraic equations. For our VLLE problem we would guess the two unknowns, L and V, use (11.1.31) and (11.1.32) to calculate values for the Rachford-Rice functions, F_1 and F_2, and then test for convergence. If our convergence criteria are not met at iteration k, we estimate values for the unknown L and V at the next iteration $(k + 1)$ by

$$L^{(k+1)} = L^{(k)} + \Delta L \tag{11.1.33}$$

$$V^{(k+1)} = V^{(k)} + \Delta V \tag{11.1.34}$$

To obtain values for the steps ΔL and ΔV, expand each of F_1 and F_2 at $(k + 1)$ in a Taylor series in L and V about their values at k, and truncate after the linear terms,

$$F_1^{(k+1)} = F_1^{(k)} + \Delta L\left(\frac{\partial F_1}{\partial L}\right)_V^{(k)} + \Delta V\left(\frac{\partial F_1}{\partial V}\right)_L^{(k)} \tag{11.1.35}$$

$$F_2^{(k+1)} = F_2^{(k)} + \Delta L\left(\frac{\partial F_2}{\partial L}\right)_V^{(k)} + \Delta V\left(\frac{\partial F_2}{\partial V}\right)_L^{(k)} \tag{11.1.36}$$

Set each lhs to zero and rearrange. The result is a pair of linear equations that can be solved for the steps ΔL and ΔV,

$$\begin{bmatrix} F_{1L} & F_{1V} \\ F_{2L} & F_{2V} \end{bmatrix} \begin{bmatrix} \Delta L \\ \Delta V \end{bmatrix} = \begin{bmatrix} -F_1 \\ -F_2 \end{bmatrix} \tag{11.1.37}$$

where F_1 and F_2 are evaluated at iteration k and the elements in the coefficient matrix are values of the derivatives at iteration k,

$$F_{1L} \equiv \left(\frac{\partial F_1}{\partial L}\right)_V = -\sum_i^C \frac{z_i(1-C_i)^2}{D_i^2} < 0 \tag{11.1.38}$$

$$F_{1V} \equiv \left(\frac{\partial F_1}{\partial V}\right)_L = -\sum_i^C \frac{z_i(1-C_i)(K_i-C_i)}{D_i^2} \tag{11.1.39}$$

$$F_{2L} \equiv \left(\frac{\partial F_2}{\partial L}\right)_V = F_{1V} \tag{11.1.40}$$

$$F_{2V} \equiv \left(\frac{\partial F_2}{\partial V}\right)_L = -\sum_i^C \frac{z_i(K_i-C_i)^2}{D_i^2} < 0 \tag{11.1.41}$$

Note that the coefficient matrix in (11.1.37) is symmetric. Once we have evaluated the derivatives (11.1.38)–(11.1.41), we solve (11.1.37) for ΔL and ΔV. This constitutes the Newton-Raphson method for solving sets of nonlinear algebraic equations [9].

VLLE Algorithm. The algorithm is summarized as a flow diagram in Figure 11.6. It divides into two parts: an initialization stage, followed by a single loop that searches for L and V. Note that this algorithm bears a strong analogy to the single Rachford-Rice scheme presented in Figure 11.4. As in all the algorithms discussed here, the crucial step in initialization is that in which initial guesses are made for all the K-factors and distribution coefficients. In many cases we have some knowledge as to how most components will be distributed among the three phases, so at least rough estimates can be contrived for the *K*s and *C*s. If such knowledge is lacking, estimates for the *K*s can be obtained from a two-phase vapor-liquid flash calculation (Problem 11.7), and estimates for the Cs can be obtained from a two-phase liquid-liquid calculation (§ 11.1.5). Finally, initial values are guessed for L and V; these are fractions $(0 \leq L \leq 1)$ and $(0 \leq V \leq 1)$ with $L + V < 1$.

The central portion of the algorithm in Figure 11.6 exactly parallels the standard Rachford-Rice procedure. First, we use (11.1.27)–(11.1.29) to compute the mole fractions for all phases, then we compute all fugacity coefficients and all activity coefficients. With those quantities we can obtain new estimates for the Cs and Ks from the phase-equilibrium relations (11.1.15) and (11.1.24). Now we use (11.1.31) and (11.1.32) to calculate values for the Rachford-Rice functions, F_1 and F_2, and test for convergence. If our convergence criteria are not met at iteration k, then we use the Newton-Raphson method to estimate the unknown L and V at the next iteration $(k + 1)$.

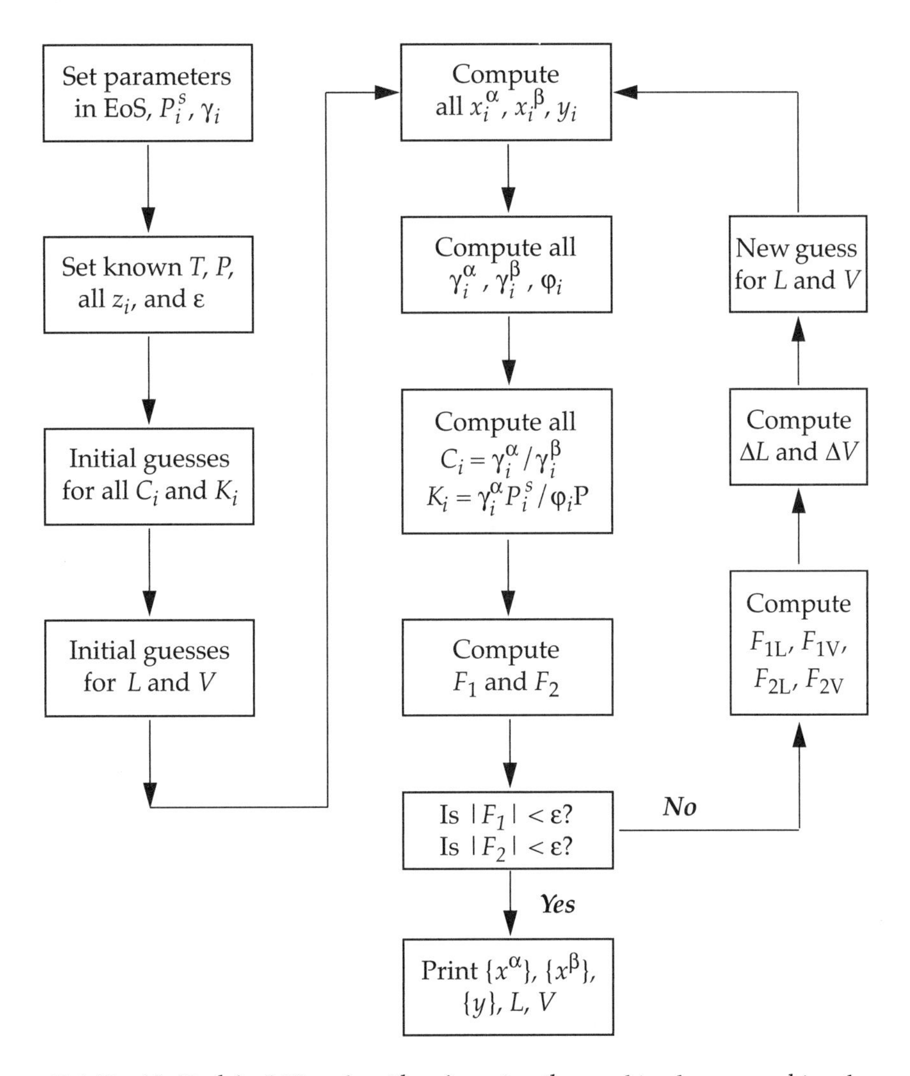

Figure 11.6 Double Rachford-Rice algorithm for using the combined gamma-phi and gamma-gamma methods to solve multicomponent $(C \geq 3)$ vapor-liquid-liquid equilibrium problems. This method fails for binary mixtures $(C = 2)$.

The Newton-Raphson procedure is notoriously slow to converge; for example, a ternary mixture may easily require 50 or more iterations. However, as implemented here, the method appears to be stable when used with reliable initial guesses for the *C*s and *K*s. If stability problems are encountered, then we may try to suppress amplification of the instabilities by replacing (11.1.33) and (11.1.34) with

$$L^{(k+1)} = L^{(k)} + \zeta \, \Delta L \tag{11.1.42}$$

$$V^{(k+1)} = V^{(k)} + \zeta \, \Delta V \tag{11.1.43}$$

where $0 \le \zeta \le 1$. This strategy often works, at the price of slowing convergence even further; however, slow convergence is better than no convergence at all.

This VLLE algorithm is prone to the same kinds of problems discussed in § 11.1.5 for the two-phase Rachford-Rice procedure: the algorithm is sensitive to the initial guesses made for the *C*s and *K*s, and nonphysical results for L and V may be false roots, or they may indicate that three phases do not form at the given conditions. The latter interpretation may hinge on the models chosen for the equation of state and for the activity coefficients. In addition, the absence of three phases can cause the coefficient matrix in (11.1.37) to become singular.

A principal advantage of this algorithm is that it applies to any number of components $C \ge 3$, though in every case we solve only the two equations (11.1.31) and (11.1.32). However, this method fails for binary mixtures. To see why, note that for binaries in three-phase equilibrium, (11.1.23) requires us to specify values for $\mathcal{F}' = 3$ variables. We then have five equations that can be solved for five unknowns. The five equations are four phase-equilibrium relations (11.1.15) and (11.1.24) plus the one Rachford-Rice function (11.1.31). In the Rachford-Rice approach, the five unknowns would be x_1^{α}, x_1^{β}, y_1, plus the fractions L and V. However, L and V appear in only one of our five equations, namely (11.1.31), and no second material balance, independent of (11.1.31), can be written for a binary mixture. Consequently, even though we set values for the three quantities (T, P, z_1), we have not closed the problem: instead, we have created an *indifferent* situation. However, if we can set values for (T, L, z_1) or (T, V, z_1), then we can solve our five equations for the remaining variables. But such problems could not be solved by an algorithm of the Rachford-Rice form.

Alternatively, when we pose an isothermal VLLE problem for binaries and use a g^E-model that is independent of pressure, the VLE and LLE problems decouple. Then we can apply the Rachford-Rice algorithm to the LLE problem and follow that with a bubble-P calculation to solve the VLE problem. This is described in Appendix K.

11.1.8 Example

How do we use the double Rachford-Rice method to compute the compositions of the phases of a ternary mixture in vapor-liquid-liquid equilibrium?

First we consider a mixture that exhibits two-phase liquid-liquid equilibria. In such mixtures, we can induce a vapor phase by raising the temperature or lowering the pressure or both. To illustrate, we use the ternary studied in § 11.1.6: benzene(1), ace-

Table 11.6 Mole fractions from double Rachford-Rice algorithm for two liquid phases in equilibrium with a vapor phase at 333 K and 0.7792 bar

i	Species	x_i^α	x_i^β	y_i
1	C_6H_6	0.0026	0.4786	0.4784
2	C_2H_3N	0.0762	0.4541	0.2819
3	H_2O	0.9212	0.0674	0.2397

tonitrile(2), and water(3). We showed in Figure 11.5 the triangular diagram containing LLE states for this mixture at 1.0133 bar, 333 K. That diagram was computed using the NRTL model for activity coefficients; we continue to use the NRTL model here, with parameters values from Table J.1. If we lower the pressure isothermally, we form a vapor phase, and since the pressure is low, we assume the vapor is an ideal gas. For pure vapor pressures, we use the Antoine model in Appendix D.

We consider the state at 0.7792 bar and 333 K, with the same overall mole fractions used in § 11.1.6: $z_1 = 0.3436$ and $z_2 = 0.3092$. We set the tolerance for convergence to $\varepsilon = 10^{-5}$. Let N^α be the total number of moles in the water-rich liquid phase, N^β the number in the organic-rich liquid, and N^v the number in the vapor; then we define the ratios $L = N^\alpha/N$ and $V = N^v/N$, where $N = (N^\alpha + N^\beta + N^v)$. To start the algorithm in Figure 11.6, we guess $L = 0.6$ and $V = 0.4$; for distribution coefficients, $C_i = x_i^\beta/x_i^\alpha$, we guess $C_1 = 7$, $C_2 = 2$, $C_3 = 0.5$; and for the K-factors, $K_i = y_i/x_i^\alpha$, we start with $K_1 = 10$, $K_2 = 4$, and $K_3 = 0.4$. From these initial values, the algorithm converged in 35 iterations to $L = 0.283$ and $V = 0.219$, with compositions as in Table 11.6. These results agree with those obtained by McDonald and Floudas [13] using a Gibbs energy minimization technique.

Note that for three components in three-phase equilibrium, the phase rule (9.1.14) requires only $\mathcal{F} = 2$ properties to identify the state. So with T and P fixed, any set of overall compositions $\{z\}$ that leads to a three-phase situation will produce the same compositions as given in Table 11.6. However, different sets $\{z\}$ will produce different distributions of material (L and V) among the three phases; some sets $\{z\}$ will produce only two equilibrium phases, and others will yield only a single phase. When only one or two phases are found, the compositions will differ from those in Table 11.6.

11.2 ONE-PHASE REACTION-EQUILIBRIUM CALCULATIONS

To solve reaction equilibrium problems, we must combine material balances with the criteria for reaction equilibria. Consequently, such problems bear a superficial resemblance to isothermal flash calculations, though in the case of reaction equilibria the material balances are applied to elements, not species. For $\mathcal{R}$ independent reactions involving C species in a single phase at fixed T and P, the criteria for equilibrium were given in § 7.6.1,

$$\mathcal{A}_j = -\sum_i^C \nu_{ij}\, \bar{G}_i = 0 \qquad j = 1, 2, \ldots, \mathcal{R} \tag{11.2.1}$$

where $\mathcal{A}_j$ is the affinity for reaction j and ν_{ij} is the stoichiometric coefficient for species i in reaction j. In § 10.3 we developed two strategies for solving the problems represented by (11.2.1); the strategies differ in how the elemental balances are imposed. In the *stoichiometric method* we use the elemental balances to obtain the stoichiometric coefficients ν_{ij}, and then we solve the $\mathcal{R}$ equations (11.2.1) for $\mathcal{R}$ extents of reaction ξ_j. In the *nonstoichiometric method* we impose the elemental balances as constraints on the minimization of G. This allows us to avoid evaluating any stoichiometric coefficients, but it requires us to solve for Lagrange multipliers as well as for C unknown mole numbers. In § 10.3 we illustrated both methods using simple situations involving a single reaction.

In this section we reformulate the reaction-equilibrium problem to obtain a computational algorithm that is particularly useful when many components are involved in multiple reactions; however, we continue to restrict attention to reactions occurring in a single phase. The reformulated algorithm offers the advantage of the stoichiometric method in that we only solve $\mathcal{R}$ equations for $\mathcal{R}$ unknowns, and it offers the advantage of the nonstoichiometric method in that we do not explicitly balance reactions or explicitly assign values to stoichiometric coefficients. The algorithm is based on a singular value decomposition of the formula matrix **A**, as discussed in § 11.2.1. From the decomposed formula matrix we can extract values for the stoichiometric coefficients, as in § 11.2.2; then, the equilibrium composition of the reaction mixture can be computed, as in § 11.2.3. Early developments in algorithms for reaction-equilibrium calculations are reviewed in the book by van Zeggeren and Storey [14]. More recent developments are discussed in a review paper by Seider et al. [15] and in the book by Smith and Missen [16].

11.2.1 Singular Value Decomposition of the Formula Matrix

Consider a closed reacting system containing C species in m_e elements with the elemental balances given by

$$\mathbf{AN} = \mathbf{b} = \text{constants} \tag{7.4.2}$$

Here **A** is an $(m_e \times C)$ formula matrix, **N** is a $(C \times 1)$ vector of mole numbers, and **b** is an $(m_e \times 1)$ vector of constant elemental abundances. (Basics of linear algebra are reviewed in Appendix B.) The matrix **A** is known from the chemical formulae of the species present, and the abundances **b** are known from the amounts initially loaded into the reactor. But the mole numbers **N** are unknown. Moreover, the sets **N** that satisfy the balances (7.4.2) are *not* unique: many different combinations of amounts of the given species (**N**) can produce the same elemental balances (**b**). This means that the formula matrix **A** is singular.

The singularity of **A** can also be deduced in another way. Recall that the rank of **A** is related to the number of independent reactions by

$$\mathcal{R} = C - \text{rank}(\mathbf{A}) \tag{7.4.5}$$

But **A** is generally not square ($m_e \neq C$) and, in any case, rank(**A**) $< C$ (else no reactions are occurring), so **A** must be singular.

In the language of linear algebra, **N** and **b** define vector spaces, and the dimension of a vector space corresponds to the number of linearly independent vectors, called *basis vectors*, that are needed to define the space. Then the multiplication in (7.4.2) can be interpreted as a transformation in which **A** maps a certain subspace of **N** into a subspace of **b**. In other words, only certain sets of mole numbers satisfy the elemental balances (7.4.2), and the possible sets of mole numbers depend on the chemical formulae for the species present in the system. That subspace of **b**, which is accessible to some **N**, is called the *range* of **A**; the dimension of the range equals rank(**A**). According to (7.4.2), any basis vectors for the range automatically satisfy the elemental balances. For example, if we let $\widehat{\mathbf{N}}$ represent one particular basis vector for the range, then

$$\mathbf{A}\widehat{\mathbf{N}} = \mathbf{b} \tag{11.2.2}$$

But for chemical-reaction problems, **A** is singular; consequently, in addition to the range, there must be another subspace of **N** that maps to zero under the transformation **A**,

$$\mathbf{A}\mathbf{N} = \mathbf{A}\mathbf{v}_j = \mathbf{0} \qquad j = 1, 2, \ldots, \mathcal{R} \tag{11.2.3}$$

The sets of mole numbers that satisfy (11.2.3) are sets of stoichiometric coefficients, $\mathbf{v}_j$, and the subspace of **N** that satisfies (11.2.4) is called the *nullspace* of **A**. The dimension of the nullspace is the number of independent vectors $\mathbf{v}_j$ (basis vectors) that satisfy (11.2.3); that is, the nullspace has dimension $\mathcal{R}$, which is the number of independent chemical reactions.

Since each vector $\mathbf{v}_j$ in (11.2.3) represents a set of stoichiometric coefficients and since stoichiometric coefficients are not unique, any linear combination of the basis vectors for the nullspace provides a legitimate set of stoichiometric coefficients for one reaction. Further, since the nullspace transforms to zero under **A**, *any* arbitrary linear combination of basis vectors for the nullspace, added to our particular range-space basis vector $\widehat{\mathbf{N}}$, satisfies the elemental balances,

$$\mathbf{A}\widehat{\mathbf{N}} + \mathbf{A}\mathbf{v}_j = \mathbf{b} + \mathbf{0} = \mathbf{b} \tag{11.2.4}$$

The mole numbers in $\widehat{\mathbf{N}}$ represent a *particular* solution to the elemental balances (7.4.2), while the stoichiometric coefficients given by (11.2.3) represent *general* solutions. We obtain all possible solutions by adding the general solution to a particular solution, as indicated by (11.2.4). Hence, all sets of mole numbers that satisfy the elemental balances can be obtained from

$$\mathbf{N} = \widehat{\mathbf{N}} + \sum_j^{\mathcal{R}} \mathbf{v}_j \tag{11.2.5}$$

Therefore, to find all possible sets of mole numbers that satisfy the elemental balances (7.4.2) for a reacting system, we need the basis vectors for both the range and the nullspace of the formula matrix **A**. This part of the problem is solved in the remainder of this section. With all possible **N** known, we would then search among those **N** for the particular set that satisfies the reaction-equilibrium criteria (11.2.1). One algorithm for solving this part of the problem is presented in § 11.2.3.

To start the determination of the basis vectors, first note that, since **A** is singular, we do not disturb the elemental balances if we augment **A** with rows of zeroes below row m_e, producing a square matrix **A′** of dimension $(C \times C)$. Then, to find the basis vectors for the range and nullspace, we perform a singular value decomposition of **A′**. In a singular value decomposition, **A′** is replaced by a product of three matrices [17]

$$\mathbf{A}' = \mathbf{U}\mathbf{W}\mathbf{V}^{\mathsf{T}} \tag{11.2.6}$$

Here **A′**, **U**, **W**, and **V** are each square of dimension $(C \times C)$. In addition, **W** is a diagonal matrix, and since **A′** is singular, some diagonal elements in **W** are zero. The number of zero elements equals the dimension of the nullspace, which for our problem is $\mathcal{R}$, the number of independent chemical reactions.

Further, for each zero diagonal element w_{kk}, the corresponding column $\mathbf{V}_k$ in **V** is an orthonormal basis vector for the nullspace. Therefore, any linear combination of the $\mathbf{V}_k$ provides a set of stoichiometric coefficients for one reaction. To simplify subsequent notation, we use the basis vectors from **V** to form a $(C \times \mathcal{R})$ matrix **P**; each column in **P** is one of the basis vectors $\mathbf{V}_k$. Then a set of stoichiometric coefficients can be obtained by

$$\mathbf{v}_j = \mathbf{P}\boldsymbol{\lambda}_j \tag{11.2.7}$$

where $\boldsymbol{\lambda}_j$ is an $(\mathcal{R} \times 1)$ vector of arbitrarily selected scale factors (or weights) for reaction j. For each species i in reaction j, (11.2.7) becomes

$$\nu_{ij} = \sum_{k}^{\mathcal{R}} p_{ik}\, \lambda_{kj} \tag{11.2.8}$$

While the number of zero diagonal elements in **W** provides the dimension of the nullspace, the number of nonzero elements provides the dimension of the range. Consequently, the number of nonzero elements on the diagonal in **W** equals rank(**A**). Further, for each nonzero diagonal element w_{kk}, the corresponding column $\mathbf{U}_k$ in **U** is an orthonormal basis vector for the range. We can find one particular vector $\widehat{\mathbf{N}}$ in the range that satisfies the elemental balances by

$$\widehat{\mathrm{N}} = \mathbf{A}^{-1}\mathrm{b} = (\mathbf{U}\mathbf{W}\mathbf{V}^{\mathsf{T}})^{-1}\mathrm{b} \tag{11.2.9}$$

But **U** and $\mathbf{V}^{\mathsf{T}}$ are orthogonal, so their inverses are merely their transposes, and **W** is diagonal, so its inverse is a matrix of diagonal elements $(1/w_{kk})$. Therefore, (11.2.9) becomes

$$\widehat{\mathrm{N}} = \mathbf{V}\,\mathrm{diag}[1/w_{kk}]\,\mathbf{U}^{\mathsf{T}}\,\mathrm{b} \tag{11.2.10}$$

The elements $w_{kk} = 0$ correspond to the nullspace. But we only want solutions for the range, so we remove the nullspace from (11.2.10) by setting $(1/w_{kk}) = 0$ if $w_{kk} = 0$. (As Press et al. remark, It isn't every day you get to set infinity equal to zero [9].)

To summarize, we perform a singular value decomposition of the augmented formula matrix to obtain the matrices **U**, **W**, and **V**. With these, we use (11.2.10) to obtain a particular basis vector $\widehat{\mathbf{N}}$ for the range. From **V**, we form **P** and then use (11.2.7) to obtain all sets of stoichiometric coefficients $\mathbf{v}_j$. Then we combine $\widehat{\mathbf{N}}$ and $\mathbf{v}_j$ into (11.2.5) to determine all sets of mole numbers that satisfy the elemental balances. Therefore, a singular value decomposition provides the number of independent reactions $\mathcal{R}$, all sets of $\mathcal{R}$ independent stoichiometric coefficients $\mathbf{v}_j$, and all possible combinations of mole numbers **N** that satisfy the elemental balances. A computer program for performing the decomposition is contained in the book by Press et al. [9]; routines for performing the decomposition are also available in *MATLAB* and in *Mathematica*™.

11.2.2 Example

How can stoichiometric coefficients be obtained from a singular value decomposition of the formula matrix?

Reconsider the problem posed in § 7.4.3: formation of synthesis gas (CO and H_2) by incomplete combustion of methane in oxygen. The products are CO_2, H_2O, CO, and H_2. So we have $C = 6$ species and $m_e = 3$ elements. The procedure involves these steps:

(1) Build the $(m_e \times C)$ formula matrix,

$$\mathbf{A} = \overbrace{\begin{bmatrix} 1 & 0 & 1 & 0 & 1 & 0 \\ 4 & 0 & 0 & 2 & 0 & 2 \\ 0 & 2 & 2 & 1 & 1 & 0 \end{bmatrix}}^{CH_4\ O_2\ CO_2\ H_2O\ CO\ H_2} \begin{cases} C \\ H \\ O \end{cases} \tag{7.4.31}$$

(2) Create the augmented matrix **A′** by adding rows of zeroes to **A** below its last row. The new matrix **A′** should be square, of dimension $(C \times C) = (6 \times 6)$.

(3) Perform the singular value decomposition on **A′**. A routine for doing so is listed in Press et al. [9]. The decomposition yields

$$\mathbf{U} = \begin{bmatrix} -0.197 & -0.265 & 0 & -0.944 & 0 & 0 \\ -0.966 & -0.216 & 0 & 0.141 & 0 & 0 \\ -0.166 & -0.940 & 0 & 0.298 & 0 & 0 \\ 0 & 0 & -1 & 0 & 0 & 0 \\ 0 & 0 & 0 & 0 & 1 & 0 \\ 0 & 0 & 0 & 0 & 0 & 1 \end{bmatrix} \tag{11.2.11}$$

$$\mathbf{W} = \begin{bmatrix} 5.016 & 0 & 0 & 0 & 0 & 0 \\ 0 & 3.223 & 0 & 0 & 0 & 0 \\ 0 & 0 & 0 & 0 & 0 & 0 \\ 0 & 0 & 0 & 1.206 & 0 & 0 \\ 0 & 0 & 0 & 0 & 0 & 0 \\ 0 & 0 & 0 & 0 & 0 & 0 \end{bmatrix} \tag{11.2.12}$$

$$\mathbf{V} = \begin{bmatrix} -0.810 & 0.186 & 0.459 & -0.315 & 0 & 0 \\ -0.066 & -0.583 & 0.459 & 0.495 & 0.249 & 0.371 \\ -0.106 & -0.665 & -0.115 & -0.288 & -0.664 & -0.093 \\ -0.418 & -0.158 & -0.344 & 0.481 & 0.166 & -0.650 \\ -0.072 & -0.374 & -0.344 & -0.535 & -0.664 & 0.093 \\ -0.385 & 0.134 & -0.574 & 0.234 & -0.166 & 0.650 \end{bmatrix} \tag{11.2.13}$$

Note that this is **V**, not its transpose $\mathbf{V}^\mathbf{T}$. We also caution that, to save space, we display only three significant figures for the elements in **V**; in practice, we generally need five significant figures for reliable results.

(4) The number of independent reactions $\mathcal{R}$ is the number of zero elements on the diagonal in **W**. Inspecting (11.2.12), we find $\mathcal{R} = 3$.

(5) For the zero elements w_{33}, w_{55}, and w_{66}, the corresponding columns $\mathbf{V}_3$, $\mathbf{V}_5$, and $\mathbf{V}_6$ in **V** are basis vectors for the nullspace. With these basis vectors we form a $(C \times \mathcal{R})$ matrix **P**,

$$\mathbf{P} = \begin{bmatrix} 0.459 & 0 & 0 \\ 0.459 & 0.249 & 0.371 \\ -0.115 & -0.664 & -0.093 \\ -0.344 & 0.166 & -0.650 \\ -0.344 & -0.664 & 0.093 \\ -0.574 & -0.166 & 0.650 \end{bmatrix} \tag{11.2.14}$$

(6) If we wanted a vector of stoichiometric coefficients for any one reaction j, we would now compute

$$\nu_{ij} = \sum_k^{\mathcal{R}} p_{ik} \lambda_{kj} \qquad i = 1, \ldots, C; \; j = 1, \ldots, \mathcal{R} \tag{11.2.15}$$

using an arbitrarily selected set of values for the scale factors λ_k. For example, if we happen to choose $\boldsymbol{\lambda}^\mathbf{T} = (-2.17945, -0.83166, -2.1350)$, then (11.2.15) gives

$$\boldsymbol{\nu}_1^\mathbf{T} = (-1, -2, 1, 2, 0, 0) \tag{11.2.16}$$

So one of the three independent reactions could be

$$CH_4 + 2O_2 \leftrightarrow CO_2 + 2H_2O \tag{11.2.17}$$

However, we need not solve explicitly for particular sets of stoichiometric coefficients; instead, we can leave the stoichiometric coefficients implicit, as described in § 11.2.2. This choice leads to the computational algorithm presented in § 11.2.3.

11.2.3 Implicit Stoichiometric Coefficients

Once the singular value decomposition has been performed on $\mathbf{A}'$, we can obtain a set of stoichiometric coefficients from (11.2.15), as shown above. This differs from a traditional balancing of reactions. In a traditional balancing, we do not select values for the scale factors λ. Instead, we select a value for one stoichiometric coefficient in each reaction, compute (in effect) the corresponding scale factors, and then use (11.2.15) to obtain all remaining stoichiometric coefficients $\mathbf{v}_j$ for the reaction j. We refer to this as an *explicit* evaluation of stoichiometric coefficients. We prefer this procedure when performing calculations by hand, for it can ensure that the stoichiometric coefficients are integers, or at least simple fractions.

However, when doing calculations by computer, there is no advantage to having simple numbers for the ν_{ij}, and in fact, in a computer environment, the traditional balancing procedure is less systematic than is a direct implementation of (11.2.15). Therefore we merely select values for the scale factors that simplify subsequent calculations; in most situations, simplifications are achieved by these choices:

$$\lambda_{kj} = 1 \qquad \textit{for } j = k \tag{11.2.18}$$

$$\lambda_{kj} = 0 \qquad \textit{for } j \neq k \tag{11.2.19}$$

These λ_{kj} can be used to form an $(\mathcal{R} \times \mathcal{R})$ identity matrix of scale factors,

$$\mathbf{\Lambda} = \mathbf{I} \tag{11.2.20}$$

Then (11.2.15) gives the stoichiometric coefficients as merely the elements of $\mathbf{P}$,

$$\nu_{ij} = p_{ij} \qquad \textit{for all } i \textit{ and } j \tag{11.2.21}$$

We refer to (11.2.21) as an *implicit* evaluation of the stoichiometric coefficients because we make no explicit choice of a value for any ν_{ij}; we merely take the p_{ij} values provided by the decomposition of $\mathbf{A}'$.

Substituting (11.2.21) into (7.4.13), we obtain an expression for the mole number of any species i in terms of the $\mathcal{R}$ extents of reaction,

$$N_i = N_i^o + \sum_j^{\mathcal{R}} \nu_{ij} \xi_j \tag{11.2.22}$$

where N_i^o is the amount of species i present at the start of the reactions. Our problem now is to find values for the $\mathcal{R}$ extents ξ_j that provide equilibrium values for the C mole numbers.

11.2.4 Computational Algorithm

In § 10.3 we presented two strategies for solving reaction-equilibrium problems: stoichiometric methods and nonstoichiometric methods. These methods can also be classified in terms of the linear algebra that surrounds the singular value decomposition discussed in § 11.2.1. Stoichiometric methods seek solutions by searching over those combinations of mole numbers that satisfy the stoichiometric constraints $\mathbf{Av} = \mathbf{0}$; that is, they search only in the nullspace of $\mathbf{A}$ and therefore stoichiometric methods are *nullspace methods* [18]. In contrast, nonstoichiometric methods seek solutions by searching over those combinations of mole numbers that satisfy the elemental balances $\mathbf{AN} = \mathbf{b}$; that is, such methods search in the range of $\mathbf{A}$ and consequently those are *range-space methods* [18]. In reaction-equilibrium problems null-space and range-space methods are thermodynamically equivalent, so the decision as to which to use can be based on computational efficiency and on the ease with which computer programs can be prepared.

But at this introductory level, our intent is to emphasize thermodynamic principles, so we prefer to discuss only elementary methods that can be implemented relatively easily, even at the price of computational efficiency. For reaction-equilibrium problems, this criterion is met by stoichiometric methods rather than by nonstoichiometric methods. Nonstoichiometric algorithms generally involve Gibbs-energy minimization with constraints. But considering the variety of (nonthermodynamic) problems that can arise in multivariate-optimization calculations and the likelihood that those problems will detract attention from relevant thermodynamic issues, we consider here only a version of a stoichiometric method. The algorithm presented here is a modification of the stoichiometric algorithm given in Chapter 4 of Smith and Missen [16].

For $\mathcal{R}$ independent chemical reactions taking place in a single phase at fixed temperature and pressure, the criteria for reaction equilibrium is (11.2.1): the affinities of all reactions become zero. Using the implicit stoichiometric coefficients from (11.2.21), we can write (11.2.1) as

$$\sum_i^C \nu_{ij} \overline{G}_i = 0 \qquad j = 1, 2, \ldots, \mathcal{R} \tag{11.2.23}$$

We use (5.4.3) to express the chemical potentials in terms of activities, and then (11.2.23) can be written as

$$\sum_i^C p_{ij}\, \Delta g_{if}^o + RT \sum_i^C p_{ij} \ln a_i = 0 \qquad j = 1, 2, \ldots, \mathcal{R} \qquad (11.2.24)$$

Here Δg_{if}^o is the Gibbs energy of formation for species i in its standard state (§ 10.4.2). These are $\mathcal{R}$ equations that can be solved for the equilibrium values of the $\mathcal{R}$ extents of reaction ξ_j. As usual, the calculation is done by trial.

A typical trial-and-error procedure is given in Figure 11.7. The procedure divides into two parts: an initialization stage, followed by a search for the mole numbers via the unknown extents of reaction. Note that this algorithm is structurally similar to the phase-equilibrium algorithms presented in § 11.1. For C species participating in $\mathcal{R}$ independent reactions, we have C unknown mole numbers, where $C > \mathcal{R}$. However, recall the mole numbers are not all independent; rather, they are coupled through the stoichiometry of the reactions. So the algorithm in Figure 11.7 actually solves for $\mathcal{R}$ independent extents of reaction ξ_j, from which the mole numbers can be obtained by applying (11.2.22).

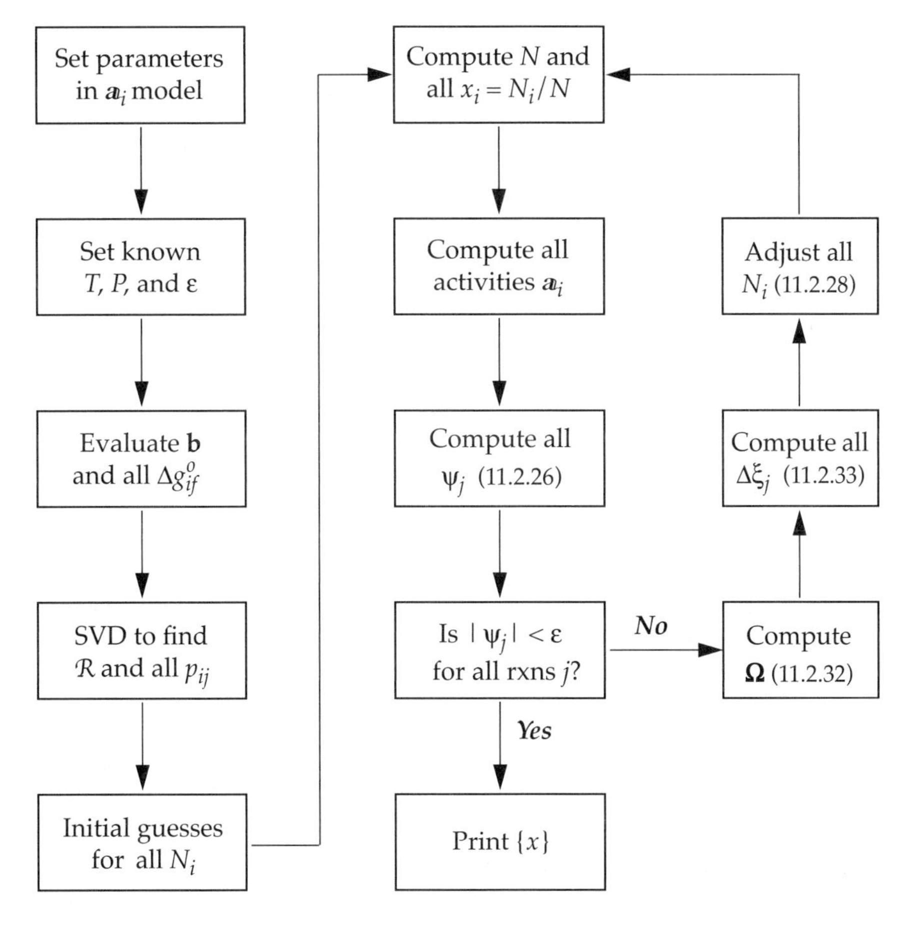

Figure 11.7 Algorithm for computing equilibrium mole fractions that result from multiple chemical reactions occurring in a single phase at fixed T and P. SVD means the singular value decomposition described in § 11.2.1 and 11.2.2.

Initialization. During the initialization of the algorithm, we first choose appropriate models for the activity of each species and assign values to any parameters that appear in those models. Typically, values for the activities are based on one of the FFF, and in Chapter 10 we have discussed the issues surrounding the selection of one of those formulae. We then set the fixed temperature and pressure, and we also assign a value to ε, which is the tolerance used to test for convergence. Next we obtain values for the Gibbs energies of formation for each species in its standard state; see § 10.4.2. Then we can perform a singular value decomposition of the augmented formula matrix $\mathbf{A}'$ to determine the number of independent reactions $\mathcal{R}$ and to obtain values for the stoichiometric coefficients p_{ij}.

Finally, we make initial guesses for the equilibrium values of all mole numbers N_i. Note that these guesses cannot be made arbitrarily, because although the mole numbers change during the reaction, they must *always* conserve the total number of atoms for each element,

$$\sum_i^C a_{ki} N_i^o = \sum_i^C a_{ki} N_i = b_k = \text{constant} \tag{11.2.25}$$

Here the a_{ki} are elements in the formula matrix $\mathbf{A}$: each a_{ki} represents the number of atoms of element k on a molecule of species i. The b_k are elemental abundances: each b_k is the total number of atoms of element k in the system. From the known initial amounts of species loaded into the reactor, we can determine each b_k; then the guesses for the final mole numbers must satisfy (11.2.25) for each element k. This completes the initialization stage of the algorithm.

Trial-and-error search. The search for the equilibrium mole numbers begins by computing the mole fractions x_i for each species. Then we use the model equations, together with appropriate FFF, to obtain values for the activities. With these, we compute the lhs of (11.2.24), which we now write as

$$\psi_j \equiv \sum_i^C p_{ij}\, \Delta g_{if}^o + RT \sum_i^C p_{ij} \ln \mathbf{a}_i \qquad j = 1, 2, \ldots, \mathcal{R} \tag{11.2.26}$$

and we test for convergence. If convergence is not attained at iteration k, we obtain new guesses for the mole numbers via the Newton-Raphson procedure, just as we did for the VLLE problem in § 11.1.7.

To develop the Newton-Raphson equations, we start with (7.4.15), in which a differential change in any mole number N_i is related to all $\mathcal{R}$ extents of reaction by

$$dN_i = \sum_j^{\mathcal{R}} \nu_{ij}\, d\xi_j = \sum_j^{\mathcal{R}} p_{ij}\, d\xi_j \tag{11.2.27}$$

Integrating this over a small increment, from iteration k to iteration $(k+1)$, we have

$$N_i^{(k+1)} = N_i^{(k)} + \zeta \sum_j^{\mathcal{R}} p_{ij}\, \Delta\xi_j^{(k)} \tag{11.2.28}$$

where ζ is a positive fraction used to control the stability of the algorithm, as in (11.1.42). To get the steps, $\Delta\xi_j$, we expand each ψ_j at iteration $(k+1)$ in a Taylor series in the ξs, about the ψ_j values at iteration k. Truncating after the linear term, we have

$$\psi_j^{(k+1)} = \psi_j^{(k)} + \sum_{\ell}^{\mathcal{R}} \sum_i^C \sum_m^C p_{ij} \left(\frac{\partial \ln a_i}{\partial N_m}\right)^{(k)} \left(\frac{\partial N_m}{\partial \xi_\ell}\right) \Delta\xi_\ell^{(k)} \tag{11.2.29}$$

But from (11.2.27) we have

$$\left(\frac{\partial N_m}{\partial \xi_\ell}\right) = p_{m\ell} \tag{11.2.30}$$

So we can write (11.2.29) as

$$\psi_j^{(k+1)} = \psi_j^{(k)} + \sum_{\ell}^{\mathcal{R}} \Omega_{j\ell}\, \Delta\xi_\ell^{(k)} \qquad j = 1, 2, \ldots, \mathcal{R} \tag{11.2.31}$$

where

$$\Omega_{j\ell} \equiv \sum_i^C \sum_m^C p_{ij} \left(\frac{\partial \ln a_i}{\partial N_m}\right)^{(k)} p_{m\ell} \tag{11.2.32}$$

Now we set the lhs of (11.2.31) to zero and rearrange; the result is a set of $\mathcal{R}$ linear equations

$$\mathbf{\Omega}\, \mathbf{\Delta\xi} = -\boldsymbol{\psi} \tag{11.2.33}$$

that can be solved for $\mathcal{R}$ increments $\Delta\xi_j$. Here $\mathbf{\Omega}$ is a symmetric $(\mathcal{R} \times \mathcal{R})$ matrix of elements $\Omega_{j\ell}$ given in (11.2.32), $\mathbf{\Delta\xi}$ is an $(\mathcal{R} \times 1)$ vector of increments in the extents of reactions, and $\boldsymbol{\psi}$ is an $(\mathcal{R} \times 1)$ vector of elements defined by (11.2.26). With the increments $\Delta\xi_j$ determined from (11.2.33), new guesses for the mole numbers can be computed from (11.2.28). This closes the search loop over the mole numbers.

Convergence. In many reaction-equilibrium problems, performance of this Newton-Raphson procedure is sensitive to the value used for the stability-control parameter ζ that appears in (11.2.28). We generally expect $0 < \zeta \le 1$. Small values of ζ tend to improve stability at the expense of slow convergence; inversely, large values of ζ can speed convergence but they can also promote growth of instabilities. In many problems, performance of the algorithm can be improved by changing the value of ζ as the calculation proceeds.

At the start of a calculation, ζ may have to be small ($\zeta \le 0.2$) to suppress amplification of large errors associated with the initial guess. But as the calculation proceeds, the search moves closer to a solution, and the algorithm may tolerate larger values for ζ. A simple, though crude, way to change ζ is to merely increase ζ at regular intervals during the calculation; for example, we might start with $\zeta = 0.1$ and increase ζ by a factor of 1.5 every five iterations during the search. Some experimentation is usually needed to find values for the initial ζ and for the scale factor that are optimal for a particular class of problems; nevertheless, it is not unusual for this simple procedure to reduce the number of iterations by factors of 3 to 5. Unfortunately, this procedure will not work for all problems; for example, in some cases, values of $\zeta > 0.5$ produce negative mole numbers. In these situations, we reduce ζ, perhaps iteratively using factors of 0.9, until we attain positive mole numbers for use in the next iteration. In short, manipulation of ζ is something of an art, and for some problems, a bit of finesse may be required to obtain convergence. Other schemes for changing ζ can be found in the book by Smith and Missen [16].

11.2.5 Example

How do we compute the equilibrium composition from a reaction carried out in a single phase at fixed T and P?

Let us determine the equilibrium composition resulting from the production of synthesis gas ($CO + H_2$) by oxidation of methane at 1500 K, 30 atm. The feed contains two moles of methane per mole of O_2 (from Smith and Missen [16]).

We have $C = 6$ species (CH_4, O_2, CO_2, H_2O, CO, H_2) and $m_e = 3$ elements (C, H, O). In § 11.2.2 the formula matrix **A** for this situation was constructed and the singular value decomposition performed. That decomposition gave $\mathcal{R} = 3$ independent reactions, with implicit stoichiometric coefficients contained in matrix **P** of (11.2.14). Choosing a basis of 1 mole of O_2 fed, the elemental abundances are

$$\mathbf{b}^{\mathbf{T}} = (b_C, b_H, b_O) = (2, 8, 2) \tag{11.2.34}$$

Table 11.7 gives the Gibbs energies of formation for each species in its standard state, taking the standard states to be ideal gases at 1500 K and 1 atm. At 30 atm, we assume the species are still ideal gases (since T is high). Then the activities are simply

$$a_i = (x_i P)/P_i^o \qquad \textit{ideal gas} \tag{11.2.35}$$

Table 11.7 Standard Gibbs energies of formation [16], initial guesses for mole numbers, and computed values for equilibrium mole fractions from production of synthesis gas at 1500 K and 30 atm.

i	**Species**	Δg_{if}^o **(kJ/mol)**	**1st guess for N_i**	**Final N_i**	**Final x_i**
1	CH_4	74.72	0.4	0.1142	0.0198
2	O_2	0	0.1	0.	0.
3	CO_2	–396.34	0.1	0.01885	0.0033
4	H_2O	–164.42	0.1	0.09532	0.0165
5	CO	–243.68	1.5	1.8670	0.3235
6	H_2	0	3.1	3.6763	0.6370

where $P = 30$ atm. and $P_i^o = 1$ atm. The mole number derivatives of the activities, needed for the quantity $\Omega_{j\ell}$ in (11.2.32), are

$$\left(\frac{\partial \ln a_i}{\partial N_m}\right)_{TP} = \frac{\delta_{ik} - x_i}{N_i} \tag{11.2.36}$$

where $\delta_{ik} = 1$ if $i = k$, but $\delta_{ik} = 0$ if $i \neq k$.

With the initial guesses for the mole numbers N_i given in Table 11.7 and a value of the stability control parameter $\zeta = 1$, the algorithm converges in 17 iterations to the mole numbers in Table 11.7. Final values for the affinities were $\mathcal{A}_1/RT = -2.0(10^{-8})$, $\mathcal{A}_2/RT = 1.6(10^{-8})$, and $\mathcal{A}_3/RT = 2.1(10^{-8})$. Note that the elemental balances $\mathbf{AN} = \mathbf{b}$ are satisfied by the initial guesses and by the final values of the mole numbers;

$$\text{for C:} \qquad N_1 + N_3 + N_5 = 2 \tag{11.2.37}$$

$$\text{for H:} \qquad 4N_1 + 2N_4 + 2N_6 = 8 \tag{11.2.38}$$

$$\text{for O:} \qquad 2N_2 + 2N_3 + N_4 + N_5 = 2 \tag{11.2.39}$$

Also note that although this problem involves three independent reactions, we need not identify explicitly any three reactions. Particular reactions and their stoichiometric coefficients remain implicit in the matrix $\mathbf{P}$ (11.2.14).

11.3 MULTIPHASE REACTION-EQUILIBRIUM CALCULATIONS

We now briefly introduce the problem of reaction equilibria in multiphase systems. In such problems, the difficulties that arise are more computational than thermodynamic, and since this is a book on thermodynamics, we do not delve deeply into the computational issues. The overriding theme here is that multiphase reaction problems combine the salient features of phase equilibria and reaction equilibria, and therefore such problems adhere to the general patterns established earlier in this chapter. We first consider computational difficulties that can be posed by indifferent situations (§ 11.3.1), then we present and illustrate one elementary algorithm (§ 11.3.2).

11.3.1 Computational Consequences of Indifferent Situations

In discussing states of multiphase, nonreacting systems in § 9.1, we presented two ways for identifying an intensive state: $\mathcal{F}$ and $\mathcal{F}'$ specifications. That discussion was extended to reacting systems in § 10.3.1. For both reacting and nonreacting systems we found that the difference between $\mathcal{F}$ and $\mathcal{F}'$ is the number of internal constraints: in an $\mathcal{F}$-specification we implicitly rely on internal constraints to complete an identification of state, but in an $\mathcal{F}'$-specification we explicitly include the consequences of internal constraints. In multiphase reacting systems, the difference is

$$\mathcal{F}' - \mathcal{F} = \mathcal{P} - 1 + \mathcal{R} \tag{11.3.1}$$

where $\mathcal{P}$ is the number of phases and $\mathcal{R}$ is the number of independent chemical reactions. One way to interpret the distinction between $\mathcal{F}$ and $\mathcal{F}'$ is that an $\mathcal{F}$-specification can only identify a class of indifferent states, while an $\mathcal{F}'$-specification may identify a member of such a class. For reacting systems the distinction between $\mathcal{F}$ and $\mathcal{F}'$ assumes more importance than it does in most phase-equilibrium situations [19, 20].

Indifferent situations can create problems in the trial-and-error procedures routinely used in calculations for phase and reaction equilibrium. In such calculations, we may start with an $\mathcal{F}$ or $\mathcal{F}'$ specification that properly closes the problem, but during the course of the trial-and-error search, the algorithm may enter an indifferent situation that couples properties that are otherwise independent. This may occur not only when azeotropes and critical points are encountered, but also when algorithms enter metastable and unstable regions of phase diagrams [20].

The response of a particular algorithm to indifferent situations may be unpredictable; for example, some algorithms may continue to search indefinitely because indifferent situations represent an infinite sequence of roots that are in many ways equivalent. Other algorithms, more dangerously, may arbitrarily select one member of the class of indifferent states and return that member as the solution. This response is not uncommon when a minimization method finds a local minimum (a metastable or unstable state) rather than the global minimum (the true equilibrium state). The possibilities for encountering indifferent situations increase when reaction-equilibria are coupled with phase-equilibria.

When we write algorithms for computing phase and reaction equilibria, we should try to implement guards that reduce the chances of search routines entering indifferent situations. When we use those algorithms, we should be aware that indifferent situations exist, that no guard is likely to protect against all eventualities, and therefore when a particular solution is found, it should not be accepted blindly.

11.3.2 Computational Algorithms for Multiphase, Multireaction Systems

In a typical problem, multiple reactions are taking place in a multiphase system at fixed T and P, and we are to compute the equilibrium compositions of all phases. At this point, such calculations raise no new thermodynamic issues; for example, for $\mathcal{R}$ independent reactions occurring among C species distributed between phases α and β, the problem is to solve the phase-equilibrium criteria

$$f_i^{\alpha} = f_i^{\beta} \qquad i = 1, 2, \ldots, C \qquad (7.3.12)$$

together with the reaction-equilibrium criteria,

$$\mathcal{A}_j = 0 \qquad j = 1, 2, \ldots, \mathcal{R} \qquad (7.6.3)$$

These reaction-equilibrium criteria apply to each phase in which reactions are occurring. But we can often simplify a calculation by assuming reactions occur in only one phase. Such an assumption is legitimate because the affinities in (7.6.3) are merely particular combinations of fugacities, and the phase-equilibrium criteria (7.3.12) require the same value of the fugacity for each component in all phases. This means that when phases α and β are in equilibrium, a reaction occurring in phase α must have the same value for the reaction equilibrium constant as it has for phase β, provided the standard-states used for the two phases are the same.

Coupled phase-reaction equilibrium problems not only raise no new thermodynamic issues, but they also raise few new computational issues. By building on the phase and reaction-equilibrium algorithms presented earlier in this chapter, we can devise an elementary algorithm. Reaction-equilibrium problems typically start with known values for T, P, and initial mole numbers N_i^o; in a phase-equilibrium context, these variables identify an $\mathcal{F}'$ problem, such as an isothermal flash calculation. Therefore we can combine the Rachford-Rice method with the reaction-equilibrium calculation given in § 11.2; an example is provided in Figure 11.8 for a vapor-liquid situation. This is a traditional way for attacking multiphase-multireaction problems [21, 22];

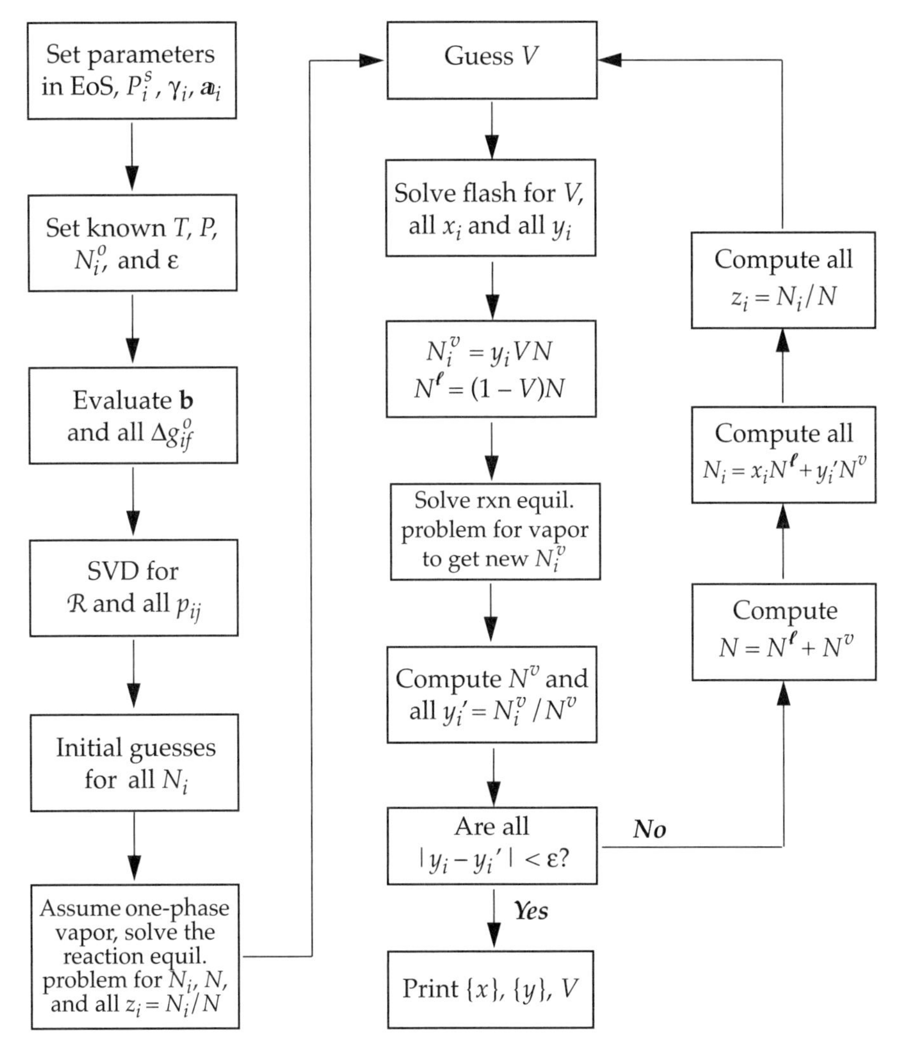

Figure 11.8 Elementary algorithm for computing equilibrium compositions from multiple reactions occurring in isothermal-isobaric, vapor-liquid situations. This algorithm combines the Rachford-Rice method for isothermal flash with the reaction-equilibrium method in Figure 11.7.

other methods have been reviewed by Seider and Widagdo [23]. Recent developments in computational algorithms include methods that minimize the Gibbs energy with guaranteed convergence to the global minimum [24, 25]. These are not elementary methods, and at present convergence is guaranteed only when fugacities are obtained from certain g^E-models (such as the Wilson equation); however, we can expect more developments in these methods in the future [26].

The algorithm in Figure 11.8 can be extended to liquid-liquid situations by combining the liquid-liquid form of the Rachford-Rice method (§ 11.1.5) with the reaction equilibrium algorithm from § 11.2.4. Again we would have only $(\mathcal{R} + 1)$ equations to solve for $(\mathcal{R} + 1)$ unknowns, no matter how many species are present. However, since such algorithms combine the Rachford-Rice and reaction-equilibrium procedures, they are susceptible to the same convergence problems as encountered in those methods. These include sensitivity to initial guesses and, in some problems, sensitivity to the parameter ζ used in (11.2.28) to control the stability of the reaction-equilibrium calculation. But if computer codes have already been written for the Rachford-Rice and reaction-equilibrium algorithms, then it is a relatively simple task to combine them into the algorithm in Figure 11.8. So this algorithm can serve as a first line of attack on a multiphase-multireaction problem; however, if the algorithm fails, then more sophisticated approaches will have to be tried [23].

11.3.3 Example

What are the effects of two-phase equilibria on equilibrium compositions obtained from one or more chemical reactions?

We consider esterification of ethanol with acetic acid to form ethyl acetate and water. This reaction has been much used for testing algorithms that perform simultaneous phase and reaction-equilibrium calculations. At ambient pressures, we assume the reaction occurs in a vapor phase; but depending on the exact values for T and P, the mixture may exist as one-phase vapor, one-phase liquid, or a two-phase vapor-liquid system. The feed contains equimolar amounts of ethanol and acetic acid. The problem is to determine the equilibrium state: the phases present and their compositions at 1.0133 bar and temperatures near 355 K.

Model parameter values. We estimate pure-component vapor pressures from an Antoine equation,

$$\ln P^s(\text{bar}) = A - \frac{B}{T(K) - C} \tag{11.3.2}$$

with parameters in Table 11.8 taken from Xiao et al. [22]. For the calculation, we assign the reaction to the vapor phase; therefore, we choose the standard states to be pure ideal gases at 1 bar. Values for the standard Gibbs energies of formation at 355 K and 358 K were taken from McDonald and Floudas [25] and are given in Table 11.8.

If a liquid phase appears, it will definitely be nonideal, and we adopt the Wilson equation (§ 5.6.5) as the model for liquid-phase activity coefficients. Values of the parameters Λ_{ij} in the Wilson model were taken from McDonald and Floudas [25], who in turn abstracted them from Suzuki et al. [27]; the values are given in Table 11.9. We assume the vapor at 1.0133 bar is an ideal gas, so the activities are merely

Table 11.8 Values for Antoine constants in (11.3.2) and standard Gibbs energies of formation (kJ/mol) at two temperatures [25]

Species	**A**	**B**	**C**	Δg^o_{if} **355 K**	Δg^o_{if} **358 K**
EtOH	11.41193	3316.920	60.44	–155.1994	–154.4923
HAc	10.78646	3785.565	39.63	–365.2934	–364.6783
EtAc	9.72377	2852.235	56.15	–304.7794	–303.5610
H_2O	11.72410	3841.196	45.14	–226.0446	–225.9061

$$a_i = \frac{y_i P}{P_i^o} \tag{11.3.3}$$

But carboxylic acids, such as acetic acid, are strongly associating [28], and even in vapors at low pressures, acetic acid will dimerize, producing a fugacity coefficient substantially less than unity. For vapor-liquid equilibrium situations, this has been taken into account in the values of the Wilson parameters fit by Suzuki et al. [27]. But we return to this issue later in this example.

Phases and compositions at 355 K, 1.0133 bar. For this situation we have $C = 4$ species and $m_e = 3$ elements. Using a feed containing 0.5 moles of EtOH and 0.5 moles of HAc, the elemental abundances for (C, H, O) are $\mathbf{b}^\mathsf{T} = (2, 5, 1.5)$. The formula matrix is

$$\mathbf{A} = \overbrace{\begin{bmatrix} 2 & 2 & 4 & 0 \\ 6 & 4 & 8 & 2 \\ 1 & 2 & 2 & 1 \end{bmatrix}}^{EtOH\ \ HAc\ \ EtAc\ \ HOH} \begin{cases} C \\ H \\ O \end{cases} \tag{11.3.4}$$

Table 11.9 Values of parameters Λ_{ij} in Wilson model at 355 K [25, 27]

i/*j*	**EtOH**	**HAc**	**EtAc**	$\mathbf{H_2O}$
EtOH	1	2.28180	0.77670	0.15347
HAc	0.27558	1	0.61790	0.26838
EtAc	0.55046	0.89277	1	0.12353
H_2O	0.92038	1.22642	0.14907	1

Table 11.10 Equilibrium compositions for esterification of ethanol (11.3.5) at 355 K and 1.0133 bar; the system is in vapor-liquid equilibrium with liquid-phase $\{x\}$, vapor phase $\{y\}$, and overall composition $\{z\}$. $V = 90$ mol%

i	**Species**	x_i	y_i	z_i
1	EtOH	0.0491	0.0829	0.0795
2	HAc	0.2042	0.0657	0.0795
3	EtAc	0.1174	0.4541	0.4205
4	H_2O	0.6294	0.3973	0.4205

A singular value decomposition of **A** (§ 11.2.1) identifies one independent reaction,

$$\text{EtOH} + \text{HAc} \rightarrow \text{EtAc} + \text{H}_2\text{O} \tag{11.3.5}$$

and since $\sigma = \Sigma \nu_i = 0$, the total number of moles ($N = 0.5 + 0.5 = 1$) is conserved throughout this one reaction.

We set the convergence tolerance $\varepsilon = 10^{-7}$ and the stability-control parameter $\zeta = 0.9$; then, with an initial guess for the final numbers of moles of $N = (0.1, 0.1, 0.4, 0.4)$, the algorithm in Figure 11.8 converges in 12 iterations. The result is a two-phase vapor-liquid system with $V = 90$ mole % of the material in the vapor phase. The computed equilibrium compositions are contained in Table 11.10. The final value of the affinity was $\mathcal{A}/RT = 4.3(10^{-8})$. A consistency test may be applied by checking whether the final compositions reproduce the known value of the equilibrium constant. Using the vapor-phase compositions and assuming an ideal gas, we find

$$K = \frac{y_3\, y_4}{y_1\, y_2} = 33.12 \tag{11.3.6}$$

which is within 2% of the value ($K = 33.72$) that we have computed from data in Stull et al. [29]. The above compositions agree almost exactly with results obtained by McDonald and Floudas [25]. From these results, the fractional conversion of ethanol is

$$\alpha_{EtOH} = \frac{N^o_{EtOH} - N^{eq}_{EtOH}}{N^o_{EtOH}} = 0.841 \tag{11.3.7}$$

Phases and compositions at 358 K, 1.0133 bar. With a slight increase in temperature, from 355 K to 358 K, the algorithm in Figure 11.8 finds no two-phase equilibrium; the system is a one-phase gas. Otherwise, with all temperature-independent parameters the same as in the previous calculation, and with the standard Gibbs energies of formation from Table 11.8, we now find the following equilibrium compositions:

$$y_{\mathrm{EtOH}} = y_{\mathrm{HAc}} = 0.0753 \tag{11.3.8}$$

and

$$y_{\mathrm{EtAc}} = y_{\mathrm{H_2O}} = 0.4247 \tag{11.3.9}$$

The final value of the affinity was $\mathcal{A}/RT = 5.7(10^{-8})$, and the equilibrium constant was computed to be

$$K = \frac{y_3\, y_4}{y_1\, y_2} = 31.81 \tag{11.3.10}$$

which is within 2% of the value ($K = 32.36$) that we computed from data in Stull et al. [29]. From these results, the fractional conversion of ethanol is now 84.9%, which differs only slightly from the conversion found for the two-phase situation in (11.3.7). This one-phase result at 358 K and the above compositions have been found by several workers [12, 13, 25, 30, 31]; however, note that since the mixture is now one-phase vapor, the liquid-phase activity coefficients are not applied, and therefore these results do not account for dimerization of acetic acid in the vapor.

Effects of dimerization on results at 358 K, 1.0133 bar. Water-acetic acid solutions in vapor-liquid equilibria have been studied by Sebastiani and Lacquaniti [32], who give the following for the equilibrium constant for low-pressure, vapor-phase dimerization of acetic acid:

$$\log K = \log\left[\frac{y_d}{y_a^2 (P/P^o)}\right] = \frac{3164}{T} - 7.5433 \tag{11.3.11}$$

where log is the base-10 logarithm, y_d is the mole fraction of dimer, y_a is the mole fraction of monomeric acetic acid, P^o is the standard state pressure in bar, and T must be in K. The constants in this expression are nearly the same as those given by Gmehling and Onken [33]. At 358 K, (11.3.10) gives

$$K = 19.71 \tag{11.3.12}$$

so vapor-phase dimerization is not negligible.

Since the algorithms in this chapter use Gibbs energies of formation, rather than equilibrium constants, we use the value of K in (11.3.11) to compute a Gibbs energy of formation for the dimer at 358 K. From (10.3.14) we have

$$\ln K = \frac{-\Delta g^o}{RT} \tag{11.3.13}$$

and from (10.4.18) we can write Δg^o in terms of properties of formation,

$$\Delta g^o = \sum_i^C \nu_i \, \Delta g^o_{if} \tag{11.3.14}$$

For the dimerization of acetic acid this becomes

$$\Delta g^o = -2\Delta g^o_{f\,\mathrm{HAc}} + \Delta g^o_{f\,\mathrm{dimer}} \tag{11.3.15}$$

Combining (11.3.11), (11.3.12), with (11.3.14), and using the value of Δg^o_f for HAc from Table 11.9, we obtain at 358 K,

$$\Delta g^o_{f\,\mathrm{dimer}} = -738.23 \text{ kJ/mol} \tag{11.3.16}$$

We have $C = 5$ species composed of $m_e = 3$ elements, and the formula matrix is

$$\mathbf{A} = \overbrace{\begin{bmatrix} 2 & 2 & 4 & 0 & 4 \\ 6 & 4 & 8 & 2 & 8 \\ 1 & 2 & 2 & 1 & 4 \end{bmatrix}}^{EtOH \;\; HAc \;\; EtAc \;\; HOH \;\; (HAc)_2} \left\{ \begin{matrix} C \\ H \\ O \end{matrix} \right. \tag{11.3.17}$$

A singular value decomposition of **A** (§ 11.2.1) finds two independent reactions: the esterification (11.3.5) and the dimerization,

$$2\mathrm{HAc} \rightarrow (\mathrm{HAc})_2 \tag{11.3.18}$$

Now $\sigma = \Sigma\nu_i \neq 0$, so the total number of moles will change during these two reactions. We still use a feed containing 0.5 moles of EtOH and 0.5 moles of HAc, so the elemental abundances remain $\mathbf{b}^\mathsf{T} = (2, 5, 1.5)$.

Dimerization effectively decreases the fugacity coefficient of HAc, because the mole fraction of monomeric HAc decreases; therefore, we do not expect dimerization to cause any condensation of the one-phase vapor found in part (2). So we perform the reaction-equilibrium calculation using the one-phase algorithm in Figure 11.7. We use the same values for the algorithmic parameters ε and ζ as used in part (1), with an initial guess for the final numbers of moles of $\mathbf{N} = \{0.1, 0.0333, 0.4, 0.4, 0.0333\}$. At 358 K, 1.0133 bar, the algorithm converges in 10 iterations to the values in Table 11.11.

The total amount of material at equilibrium is 0.9642 moles for each mole of feed. The final values of the affinities were $\mathcal{A}_e/RT = -2.1(10^{-8})$ and $\mathcal{A}_d/RT = -3.8(10^{-8})$, where subscript e refers to the esterification and subscript d refers to the dimerization. The computed values for the equilibrium constants are $K_e = 31.78$ which differs by less than 2% from the value (32.36) extracted from data in Stull et al. [29], and $K_d = 19.69$, which differs by only 0.1% from the value in (11.3.11). (Recall we use $P^o = 1$ bar.)

The fractional conversion of ethanol is now $\alpha_{\mathrm{EtOH}} = 0.774$. So the effect of acetic acid dimerization is to decrease the fractional conversion of ethanol by 7.5% from the

Table 11.11 Equilibrium compositions for the esterification of ethanol (11.3.5) at 358 K and 1.0133 bar, including effects of dimerization of acetic acid (11.3.17); at these conditions, the system is one-phase vapor.

i	**Species**	N_i	y_i
1	EtOH	0.11319	0.11739
2	HAc	0.04159	0.04314
3	EtAc	0.38681	0.40117
4	H_2O	0.38681	0.40117
5	$(HAc)_2$	0.03580	0.03713

value found by ignoring dimerization. This change is not insignificant. We can contemplate including additional reactions to account for formation of acetic acid -mers of higher order (see Problem 11.25), but dimer formation dominates at 358 K.

11.4 SUMMARY

In this chapter we have presented a collection of elementary algorithms for solving multicomponent phase-equilibrium problems, reaction-equilibrium problems, and phase-equilibrium problems coupled to reaction-equilibrium problems. The algorithms are particular implementations of the problem-solving strategies introduced in Chapter 10: phi-phi, gamma-phi, and gamma-gamma methods for phase equilibrium, plus the stoichiometric method for reaction equilibrium. The algorithms were selected and presented in ways that are intended to emphasize underlying structural similarities; for example, we were able to base several algorithms on the Rachford-Rice procedure traditionally applied to isothermal flash calculations.

However, none of these algorithms will serve in all problems situations, for our goals are primarily educational, not computational. Our intent has been to illustrate thermodynamic and computational principles that can foster development of sound engineering judgement. In that light, we enumerate here those issues that should be addressed in translating any problem statement into a computational procedure.

In setting up and carrying out any engineering computation, we have three fundamental issues to resolve before a calculation should be attempted: (1) Is the problem well-posed? (2) How should the problem be formulated mathematically? (3) What computational techniques can be applied? These are not three separate issues; rather, they are coupled. For example, the way a problem is posed influences problem formulation, and problem formulation influences computational technique.

Is the problem well-posed? This issue concerns whether we have enough information to compute the required unknowns. In phase and reaction-equilibrium computations, this issue is resolved by a proper application of the generalized phase rule; it might not be properly resolved by a routine application of the Gibbs phase rule. In particular, we have discussed two kinds of subtleties that are often overlooked.

First, we distinguished between an $\mathcal{F}$ and an $\mathcal{F}'$ identification of state. Less information is provided by an $\mathcal{F}$-specification than by an $\mathcal{F}'$-specification, but in particular situations one or the other may be more appropriate. For example, in vapor-liquid equilibrium calculations, an $\mathcal{F}$-specification is sufficient to close a bubble-T problem, but an $\mathcal{F}$-specification fails to close an isothermal flash problem. Furthermore, most reaction-equilibrium problems are not closed by $\mathcal{F}$-specifications; they require $\mathcal{F}'$-specifications. We have also illustrated that in some situations an $\mathcal{F}$-specification may be sufficient, but an $\mathcal{F}'$-specification may lead to a more advantageous problem formulation and solution technique. The principal pitfall is to apply an $\mathcal{F}$-specification to a problem that demands an $\mathcal{F}'$-specification, for then the problem is ill-posed.

Second, we raised the specter of indifferent situations. These occur when either the number or type of knowns does not allow computation of all unknowns. Most troublesome are those situations in which the problem is initially well-posed but during the computation, variables that were initially independent become coupled. When such coupling occurs, the behavior of trial-and-error search algorithms may be erratic or erroneous, so we should not accept results as reliable merely because they were generated on a computer using a sophisticated algorithm.

How should the problem be formulated mathematically? In the introduction to this chapter we noted that a phase or reaction-equilibrium problem can be formulated in two general ways: (a) as a set of coupled algebraic equations or (b) as a multivariable minimization problem. For pedagogical reasons, we have presented only formulations of type (a). In phase-equilibrium calculations the algebraic equations originate in the phi-phi, gamma-phi, and gamma-gamma methods presented in Chapter 10. In § 10.5 we discussed the issues that lead us to choose one of these strategies over the others. For phase-equilibrium conditions supplemented with material balances, we choose the Rachford-Rice formulation.

For reaction-equilibrium computations, we have discussed only stoichiometric methods, in which the elemental balances are imposed explicitly through $\mathcal{R}$ sets of stoichiometric coefficients. For one-phase systems, these formulations require us to solve only $\mathcal{R}$ algebraic equations for $\mathcal{R}$ extents of reaction; therefore, they require us to identify $\mathcal{R}$ independent reactions. Such stoichiometric methods appear to be most effective when the number of species C is not much greater than the number of elements ($C \approx m_e$). Otherwise, when $C \gg m_e$, nonstoichiometric methods may be more computationally efficient [16, 18], though this comment probably depends on the particular algorithms being compared.

What computational techniques can be applied? For sets of nonlinear algebraic equations, the traditional approach is the Newton-Raphson method. This method should be familiar to those who regularly perform numerical computations. Like any trial-and-error procedure, it has advantages and limitations. But note that we have not hesitated to avoid Newton-Raphson when a clearly better method is available. We should not be surprised when the Newton-Raphson procedure fails, either for certain sets of initial guesses or even for all guesses of certain problem formulations. Nor should we expect that convergence of the Newton-Raphson procedure implies correctness; the procedure can indiscriminately find nonphysical solutions, false solutions, and indifferent solutions. But the Newton-Raphson algorithm is a fundamental computational tool, and the development of competent engineers begins with mastery of basic tools, coupled to a strong desire to move beyond the basics.

LITERATURE CITED

[1] F. S. Acton, *Numerical Methods That Work,* The Mathematical Association of America, Washington, D.C., 1990.

[2] J. M. Smith and H. C. Van Ness, *Introduction to Chemical Engineering Thermodynamics,* 4th ed., McGraw-Hill, New York, 1987.

[3] J. M. Prausnitz, E. A. Grens, T. F. Anderson, C. A. Eckert, R. Hsieh, and J. P. O'Connell, *Computer Calculations for Multicomponent Vapor-Liquid and Liquid-Liquid Equilibria,* Prentice-Hall, Englewood Cliffs, NJ, 1980.

[4] M. L. Michelsen, "The Isothermal Flash Problem. Part I. Stability Analysis," *Fluid Phase Equil.,* **9**, 1 (1982).

[5] M. L. Michelsen, "The Isothermal Flash Problem. Part II. Phase-Split Calculation," *Fluid Phase Equil.,* **9**, 21 (1982).

[6] M. L. Michelsen, "Some Aspects of Multiphase Calculations," *Fluid Phase Equil.,* **30**, 15 (1986).

[7] D. -Y. Peng and D. B. Robinson, "A New Two-Constant Equation of State," *IEC Fund.,* **15**, 59 (1976).

[8] G. Soave, "Equilibrium Constants from a Modified Redlich-Kwong Equation of State," *Chem. Eng. Sci.,* **27**, 1197 (1972).

[9] W. H. Press, B. P. Flannery, S. A. Teukolsky, and W. T. Vetterling, *Numerical Recipes,* Cambridge University Press, Cambridge, 1986.

[10] J. G. Hayden and J. P. O'Connell, "A Generalized Method for Predicting Second Virial Coefficients," *IEC Process Dev. Des.,* **14**, 209 (1975); supplements.

[11] H. H. Rachford, Jr. and J. D. Rice, "Procedure for Use of Electronic Digital Computers in Calculating Flash Vaporization Hydrocarbon Equilibrium," *J. Petrol. Techno.,* **4**(10), Sec. 1, p. 19 (October, 1952).

[12] J. Castillo and I. E. Grossman, "Computation of Phase and Chemical Equilibria," *Computers Chem. Eng.,* **5**, 99 (1981).

[13] C. M. McDonald and C. A. Floudas, "Global Optimization for the Phase and Chemical Equilibrium Problem: Application to the NRTL Equation," *Computers Chem. Eng.,* **19**(11), 1111 (1995).

[14] F. van Zeggeren and S. H. Storey, *The Computation of Chemical Equilibria,* Cambridge University Press, Cambridge, 1970.

[15] W. D. Seider, R. Gautam, and C. W. White III, "Computation of Phase and Chemical Equilibria: A Review," *ACS Symp. Series,* R. G. Squires and G. V. Reklaitis (eds.), **124**(5), 115 (1980).

[16] W. R. Smith and R. W. Missen, *Chemical Reaction Equilibrium Analysis,* Wiley and Sons, New York, 1982; reprinted by Kreiger Publishing, Malabar, FL, 1991.

[17] J. L. Goldberg, *Matrix Theory with Applications,* McGraw-Hill, New York, 1991.

[18] H. Greiner, "An Efficient Implementation of Newton's Method for Complex Nonideal Chemical Equilibria," *Computers Chem. Engr.,* **15**(2), 115 (1991).

[19] I. Prigogine and R. Defay, *Chemical Thermodynamics*, D. H. Everett (transl.), Longmans, Green and Co., London, 1954.

[20] R. A. Heidemann, "Non Uniqueness in Phase and Reaction Equilibrium Computations," *Chem. Engr. Sci.*, **33**, 1517 (1978).

[21] R. V. Sanderson and H. H. Y. Chien, "Simultaneous Chemical and Phase Equilibrium Calculation," *IEC Process Des. Dev.*, **12**(1), 81 (1973).

[22] W. Xiao, K. Zhu, W. Yuan, and H. H. Y. Chien, "An Algorithm for Simultaneous Chemical and Phase Equilibrium Calculation," *A. I. Ch. E. J.*, **35**(11), 1813 (1989).

[23] W. D. Seider and S. Widagdo, "Multiphase Equilibria of Reactive Systems," *Fluid Phase Equil.*, **123**, 283 (1996).

[24] C. A. Floudas, A. Aggarwal, and A. R. Ciric, "A Global Optimum Search for Nonconvex NLP and MINLP Problems," *Computers Chem. Eng.* **13**(10), 1117 (1989).

[25] C. M. McDonald and C. A. Floudas, "Global Optimization and Analysis for the Gibbs Free Energy Function Using the UNIFAC, Wilson, and ASOG Equations," *IEC Res.*, **34**, 1674 (1995).

[26] E. S. Pérez Cisneros, R. Gani, and M. L. Michelsen, "Reactive Separation Systems. I. Composition of Physical and Chemical Equilibrium," *Chem. Engr. Sci.*, **52**, 527 (1970).

[27] I. Suzuki, H. Komatu, and M. Hirata, "Formulation and Prediction of Quaternary Vapor-Liquid Equilibria Accompanied by Esterification," *J. Chem. Eng. Japan*, **3**(2), 152 (1969).

[28] J. M. Prausnitz, R. N. Lichtenthaler, and E. Gomes de Azevedo, *Molecular Thermodynamics of Fluid-Phase Equilibria*, 2nd ed., Prentice-Hall, Englewood Cliffs, NJ, 1986, p. 138f.

[29] D. R. Stull, E. F. Westrum, Jr., and G. C. Sinke, *The Chemical Thermodynamics of Organic Compounds*, Wiley, New York, 1969.

[30] G. Lantagne, B. Marcos, and B. Cayrol, "Computation of Complex Equilibria by Nonlinear Optimization," *Computers Chem. Eng.*, **12**(6), 589 (1988).

[31] B. George, L. P. Brown, C. H. Farmer, P. Buthod, and F. S. Manning, "Computation of Multicomponent, Multiphase Equilibrium," *IEC Process Des. Dev.*, **15**(3), 372 (1976).

[32] E. Sebastiani and L. Lacquaniti, "Acetic Acid-Water System Thermodynamic Correlation of Vapor-Liquid Equilibrium Data," *Chem. Eng. Sci.*, **22**, 1155 (1967).

[33] J. Gmehling and U. Onken, *Vapor-Liquid Equilibrium Data Collection*, DECHEMA Chemistry Data Series, DECHEMA, Frankfurt, 1977f, vol. 1, part 1, p. 687.

PROBLEMS

11.1 Develop an algorithm that applies the phi-phi method, with a volumetric equation of state, to perform bubble-P calculations for multicomponent mixtures. Your development should include (a) a flow diagram for the algorithm and (b) a list of all data you would need before the algorithm could be implemented.

11.2 Write a computer program that implements the bubble-P algorithm developed in Problem 11.1. Use the Redlich-Kwong equation with mixing rules from § 8.4.4. To compute phase volumes, the cubic is best solved analytically via Cardan's method (Appendix C).

(a) Use your program to compute the isothermal Pxy diagram for mixtures of methane and propane at 278 K. Critical properties of the pures are given in Table 8.1.

(b) In presenting your diagram, include a qualitative description of its main features and discuss to what extent the behavior of your algorithm differs in different regions of the diagram.

11.3 Write a computer program that implements the phi-phi method for bubble-T calculations. Use the Redlich-Kwong equation of state (8.2.1) with mixing rules given in § 8.4.4. To obtain the volumes of the phases, the cubic is best solved analytically using Cardan's method (Appendix C). Apply your program to binary mixtures of methane(1) and propane(2), whose critical properties are in Table 8.1. For each of the following states, use your program to determine T, y_1, v^ℓ, and v^v, where the last two are the molar volumes of the liquid and vapor, respectively. (a) $P = 12.5$ bar, $x_1 = 0.05$; (b) $P = 30.8$ bar, $x_1 = 0.453$; (c) $P = 50$ bar, $x_1 = 0.5$; (d) $P = 83.2$ bar, $x_1 = 0.6$.

11.4 Determine the bubble temperature for an equimolar mixture of diethyl ether(1) and acetone(2) at 1 bar. The mixture obeys $g^E/RT = 0.712\, x_1 x_2$. Pure-component vapor pressures are given in Appendix D.

11.5 At 50°C a binary liquid mixture of components 1 and 2 exists in liquid-liquid equilibrium. The mixture obeys $g^E/RT = 2.25\, x_1 x_2$. Determine the compositions of the two phases.

11.6 Determine the dew-point pressure for an equimolar vapor composed of ethanol(1) and water(2) at 90°C. Pure-component vapor pressures can be estimated from the correlation in Appendix D. If necessary, assume the activity coefficients obey the Wilson model, with parameter values in Appendix E.

11.7 (a) Following the derivation given in § 11.1.5, develop the equation, analogous to (11.1.20), for solving multicomponent isothermal flash problems, which take the following form: given T, P, and $\{z\}$, find $\{x\}$, $\{y\}$, and V, where V is the fraction of feed that flashes. Your result should contain K-factors in place of distribution coefficients. Also obtain the expression for $(\partial F/\partial V)$.

(b) Prepare a flow diagram, analogous to that appearing in Figure 11.4, for the Rachford-Rice procedure applied to isothermal flash calculations.

11.8 A liquid mixture of 25 mole % benzene and 75% toluene, initially at 100°C, 1.5 bar, is fed to an isothermal flash chamber. What pressure should be maintained in the chamber to produce an equimolar vapor product? The pure-component vapor pressures are given in Appendix D. Clearly state and justify any assumptions made.

11.9 (a) Write a computer program that performs the Rachford-Rice isothermal flash calculation, as developed in Problem 11.7. Assume the vapor mixtures are ideal gases and that the liquid mixtures obey the multicomponent version of the Porter equation given in § 5.6.4.

(b) Consider a ternary mixture at 50°C, 3 bar, having Porter parameters $A_{12} = 3$, $A_{13} = 1.4$, and $A_{23} = 0.7$. The vapor pressures of the pure components are $P_1^s = 2.5$ bar, $P_2^s = 1.5$ bar, and $P_3^s = 2.1$ bar. Let V represent the fraction of the total mixture that is in the vapor phase. Use your program to determine V and the composition of each phase for mixtures at 50°C, 3 bar, and each of the following values for the overall mole fraction,

(i) $z_1 = 0.2, \quad z_2 = 0.5$
(ii) $z_1 = 0.3, \quad z_2 = 0.4$
(iii) $z_1 = 0.35, z_2 = 0.35$
(iv) $z_1 = 0.4, \quad z_2 = 0.3$

11.10 The excess Gibbs energy for a certain binary mixture obeys the Porter equation $g^E = \beta x_1 x_2$ with the temperature dependence of β given by (T in K)

$$\beta/RT = -0.002T^2 + 1.22T - 183.6$$

(a) Construct plots of the change in Gibbs energy on mixing g^m/RT vs. x_1 for two isotherms: $T = 300$ K and $T = 330$ K. Plot both curves on the same axes.

(b) If the system splits into two liquid phases at either temperature, determine the compositions of the two phases.

11.11 Consider a binary mixture of components 1 and 2 in LLE at fixed T and P. Show that the Rachford-Rice function F (11.1.20) is linear in R and can be written in terms of overall mole fractions z_i and distribution coefficients C_i as

$$R = \frac{z_1(1 - C_1) + z_2(1 - C_2)}{(1 - C_1)(1 - C_2)}$$

11.12 (a) Write a computer program that applies the Rachford-Rice procedure, developed in § 11.1.5, to multicomponent liquid-liquid equilibrium calculations. Assume the liquid mixtures obey the multicomponent version of the Porter equation given in § 5.6.4.

(b) Consider a ternary mixture at 50°C, 3 bar, with Porter parameters $A_{12} = 3$, $A_{13} = 1.4$, and $A_{23} = 0.7$. Let L represent the fraction of the total mixture that is in phase α, where phase α is rich in component 1. Use your program to determine L and the composition of each phase for mixtures at 50°C, 3 bar, and the following values of the overall mole fraction,

(i) $z_1 = 0.2, z_2 = 0.5$
(ii) $z_1 = 0.4, z_2 = 0.4$
(iii) $z_1 = 0.5, z_2 = 0.2$

11.13 Ten moles of benzene, twenty moles of toluene, and ten moles of water completely fill a closed vessel. The mixture is brought into vapor-liquid-liquid equilibrium at 90°C and 1.5 bar. Pure-component vapor pressures are given in Appendix D.

(a) Estimate the composition of the vapor phase. Clearly state and justify any assumptions made.

(b) Determine the total number of moles in each of the three phases.

11.14 Consider ternary mixtures of benzene(1), acetonitrile(2), and water(3) at 300 K, 0.10133 bar. The liquid-phase activity coefficients can be modeled by the NRTL equation (see Appendix J); values of the NRTL parameters at 300 K are contained in Table J.1. Pure-component vapor pressures are in Appendix D.

(a) For overall mole fractions $\{z\} = \{0.3436, 0.3092, 0.3472\}$ determine whether the mixture at this T and P can exist in VLE by using the Rachford-Rice algorithm (Problem 11.7) to perform an isothermal flash calculation.

(b) For the same T, P, and overall compositions as in part (a), determine whether the mixture can exist in LLE by applying the Rachford-Rice LLE algorithm of Figure 11.4.

(c) How do you explain the results found in (a) and (b)? Can they both be correct? If not, which is the correct answer?

11.15 At 1.0133 bar, 343.15 K, ternary mixtures of ethanol(1), ethyl acetate(2), and water(3) exhibit liquid-liquid immiscibility. These mixtures have been modeled using the NRTL equations (Appendix J) with parameters given in Table J.1.

(a) For overall mole fractions $\{z\} = \{0.04, 0.3, 0.66\}$, use the Rachford-Rice LLE algorithm in Figure 11.4 to determine the compositions of the two phases and fraction of the total system that forms the water-rich phase.

(b) Repeat part (1) for $\{z\} = \{0.11, 0.4, 0.49\}$.

(c) Compute the complete triangular diagram for this mixture at 1.0133 bar, 343.15 K.

11.16 (a) Write a computer program that applies the Rachford-Rice procedure, developed in § 11.1.4, to multicomponent vapor-liquid-liquid equilibrium calculations. Assume the vapor mixtures are ideal gases and that the liquid mixtures obey the multicomponent Porter model given in § 5.6.4.

(b) Consider a ternary mixture at 50°C, 3.7 bar, with Porter parameters $A_{12} = 3.5$, $A_{13} = 1.4$, and $A_{23} = 0.7$. The pure vapor pressures are $P_1^s = 1.5$ bar, $P_2^s = 2.5$ bar, and $P_3^s = 2.1$ bar. Let L represent the fraction of the total that is in phase α, where phase α is rich in component 1, and let V be the fraction in the vapor. Use your program to determine L, V, and the composition of each of the three phases for mixtures at 50°C, 3.7 bar, and the following values of the overall mole fraction,

(i) $z_1 = 0.5$, $z_2 = 0.2$
(ii) $z_1 = 0.3$, $z_2 = 0.4$
(iii) $z_1 = 0.4$, $z_2 = 0.3$

11.17 Three moles of benzene, four moles of acetonitrile, and four moles of water are placed in a closed vessel and brought into VLLE at 333 K, 0.70 bar. Assume the liquid phases obey the NRTL equations (Appendix J), with parameter values given in Table J.1. Appendix D contains Antoine parameters for vapor pressures of the pure-components. Use the double Rachford-Rice algorithm of Figure 11.6 to determine the following:

(a) the total number of moles in each phase,

(b) the number of moles of acetonitrile in each phase,

(c) the mole fraction of acetonitrile in each phase.

11.18 Repeat Problem 11.17 for an initial loading that contains 2.5 moles of benzene, five moles of acetonitrile, and 3.5 moles of water. What quantities in your answers here differ from their values in Problem 11.17? In comparing the results here with those in Problem 11.17, why do the values of some properties change, while others do not?

11.19 In § 11.1.7 we noted that the double Rachford-Rice algorithm for VLLE does not apply to binary mixtures. However, in Appendix K we present simple alternatives that usually apply to isothermal VLLE calculations for binary mixtures. To practice the procedure in Appendix K, consider liquid mixtures of toluene(1) and water(2), which are almost completely immiscible at ambient conditions. At 10°C we need the pressure at which this mixture exhibits VLLE, and we need to know the compositions of the three phases. The pure-component vapor pressures can be found from the correlation given in Appendix D. Perform the calculation twice:

(a) Do a hand calculation assuming the liquids are completely immiscible.

(b) Write a computer program that implements the procedure described in Appendix K. For the liquid-phase activity coefficients, use the NRTL model (Appendix J) with parameter values from Table J.1. What are the % deviations in the values for P and y_1 from (a) and (b)?

11.20 Write a computer program that uses the singular value decomposition procedure in § 11.2 to find the number of independent reactions $\mathcal{R}$ and the values of the stoichiometric coefficients. (A listing of a program for performing the decomposition is given in Press et al. [9].) You may want to be able to interact with the program, so you can select values for the stoichiometric coefficients of some species and then let the program compute values for the remaining coefficients. Apply your program to the following sets of reactants and products:

(a) Formation of methanol: (CO_2, H_2, CH_3OH, CO, H_2O).

(b) Chlorination of methane: (CH_4, Cl_2, CH_3Cl, CH_2Cl_2, $CHCl_3$, CCl_4, HCl).

(c) Combustion of a gas in air: (CH_4, C_2H_6, O_2, N_2, CO, CO_2, NO, H_2O).

(d) Formation of ethylene by dehydrogenation of ethane: (C_2H_6, H_2, CH_4, C_2H_2, C_2H_4, C_3H_6, C_3H_8, C_4H_8, C_6H_6).

(e) Formation of diethylsulfide by reacting ethanol with hydrogen and gaseous sulfur: (C_2H_5OH, H_2, S_2, C_2H_5SH, $(C_2H_5)_2S$, H_2S, $(C_2H_5)_2O$, C_2H_6, CH_3CHO, C_2H_4, H_2O).

11.21 At ambient conditions, binary liquid mixtures of ethyl acetate(1) and water(2) are partially miscible. At 343.15 K determine the pressure at which this binary exhibits three-phase VLLE and find the compositions of the three phases. The pure-component vapor pressures can be modeled by an Antoine equation with parameters in Table 11.9. Perform the calculation twice:

(a) Do a hand calculation, assuming the liquids are completely immiscible.

(b) Apply the algorithm described in Appendix K. For the liquid-phase activity coefficients, use the NRTL equations (Appendix J) with parameter values in Table J.1. What are the % deviations in the values obtained for P and y_1 from (a) and (b)?

11.22 At 25°C and 1 atm. a certain binary liquid mixture of components 1 and 2 deviates from ideal-solution behavior according to

$$\frac{g^E}{RT} = x_1 x_2 (2x_1 + 3x_2)$$

(a) Show whether at any composition, this mixture violates the diffusional stability criterion and therefore splits into two liquid phases.

(b) If a phase split does occur, determine the compositions of the two phases.

11.23 Write a computer program that implements the algorithm in Figure 11.7 for computing multireaction equilibria. Test your program by computing the equilibrium compositions resulting from the production of synthesis gas, as illustrated in § 11.2.5. Then use your program to compute the compositions for the synthesis-gas reactions at the conditions of § 11.2.5, except with the following modifications:

(a) Take the feed to be 2 moles of methane for every 1.5 moles of O_2.

(b) Take the feed to be 3 moles of methane for every mole of O_2.

(c) Assume oxygen is supplied as air and that the accompanying nitrogen is inert; hence, the feed ratios are 2 moles CH_4/mole O_2 and 4 moles N_2/mole O_2. Note that inerts do not change the formula matrix; they merely change the species activities by changing the mole fractions.

11.24 In some cases, a given set of reactants and products can have more than one independent set of stoichiometric coefficients. For example, consider formation of methane from synthesis gas: the reactants are carbon monoxide and hydrogen; the products are methane, carbon dioxide, and water.

(a) Balance this reaction by hand. Can you find more than one independent set of stoichiometric coefficients for reactions involving all five species? How can you test whether you have found all independent sets?

(b) How do you expect a singular value decomposition of the formula matrix **A** to behave for this collection of reactants and products? Perform the decomposition to test your expectation. Is there any advantage to the decomposition over the hand balancing done in part (a)?

11.25 Use the one-phase multireaction algorithm in Figure 11.6 to determine the extent to which formation of tetramers of acetic acid affect the fractional conversion during esterification of ethanol. That is, repeat the vapor-phase calculation at 358 K, 1.0133 bar illustrated in the last part of § 11.3.3, but now include not only dimers but also tetramers. (Spectroscopic evidence suggests that formation of trimers is unfavored [32].) Sebastiani and Lacquaniti give the equilibrium constant for formation of tetramers as [32]

$$\log K = \log\left[\frac{y_t}{y_a^4 (P/P^o)^3}\right] = \frac{5884}{T} - 14.8572 \qquad \text{(P11.25.1)}$$

where log is the base-10 logarithm, y_t is the mole fraction of tetramer, y_a is the mole fraction of monomeric acetic acid, T is in K, and P^o = 1 bar. As a basis for the calculation, use a feed that contains 0.5 moles of ethanol and 0.5 moles of acetic acid. Compare your results for the equilibrium composition and the fractional conversion of ethanol with those given in the last part of § 11.3.3.

11.26 Consider the following two simultaneous reactions taking place in a gas phase,

$$A + B \rightarrow 2C$$

$$B + C \rightarrow D$$

The reference state is taken to be 300 K and 0.1 MPa, and at this state the pure components have the following values for properties of formation.

	A	B	C	D
$(\Delta g_f^o)/R$ (K)	$4. \times 10^4$	$-1. \times 10^4$	$2. \times 10^4$	$-3. \times 10^4$
$(\Delta h_f^o)/R$ (K)	$3. \times 10^4$	$5. \times 10^4$	$-1. \times 10^4$	$4. \times 10^4$
c_p^o/R	5	4	4.5	8.5

A reactor is initially loaded with four moles of A and eight moles of B, then the reaction is allowed to proceed, reaching equilibrium at 300 K and 0.1 MPa. If, from the equilibrium condition, the temperature is increased isobarically to 350 K, would the equilibrium mole fraction of component C increase, decrease, or remain constant? Justify your answer.

12

SELECTED APPLICATIONS

When we apply thermodynamics to industrial and research problems, we should draw fundamental ideas from Parts I and II, devise an appropriate solution strategy, as in Chapter 10, and combine those with a computational technique, as in Chapter 11. Such a procedure provides values for measurables that can be used to interpret novel phenomena, to design new processes, and to improve existing processes. The procedure is illustrated in this chapter for several well-developed situations. They include conventional phase-equilibrium calculations for vapor-liquid, liquid-liquid, and solid-solid equilibria (§ 12.1); solubility calculations for gases in liquids, solids in liquids, and solutes in near-critical solvents (§ 12.2); independent variables in steady-flow processes (§ 12.3); heat effects for flash separators, absorbers, and chemical rectors (§ 12.4); and effects of changes of state on selected properties (§ 12.5).

12.1 PHASE EQUILIBRIA

When two or more bulk phases are in contact and at equilibrium, the measurables of interest are usually temperature, pressure, and the compositions of the phases. Of these measurables, the most important are often the compositions; for example, in the design and operation of separation processes, we routinely need the composition of a particular phase, or when the temperature and pressure change, we need to know the extent to which the compositions also change. When engineering applications involve fluid-fluid equilibria, we often find that, besides *absolute* compositions, *relative* compositions can be informative and important. We identify such relative measures as members of a class of *engineering quantities*: certain variables or combinations of variables that facilitate thermodynamic analyses of practical problems but whose definitions invoke no new thermodynamic fundamentals. The engineering quantities considered in this section include K-factors (§ 12.1.1) and relative volatilities (§ 12.1.2), which are used in VLE, together with distribution coefficients and selectivities, which are used in LLE (§ 12.1.3). However, relative compositions are less useful in analyzing solid-solid equilibria (SSE), so in § 12.1.4 we are content to show how methods from Chapter 10 can be used to correlate SSE data.

12.1.1 K-Factors in Vapor-Liquid Equilibria

Consider a multicomponent mixture in vapor-liquid equilibrium; let $\{x\}$ represent the set of mole fractions for the liquid and let $\{y\}$ be the same for the vapor. In a closed system, the compositions $\{x\}$ and $\{y\}$ will change, often drastically, with changes in T and P. However, in many systems the ratio y_i/x_i, for each component i, is less sensitive to changes of state than is either x_i or y_i by itself. This observation is exploited by introducing two quantities: the K-factor and the relative volatility (§12.1.2). We have already encountered the K-factor in the Rachford-Rice method for flash calculations; see (11.1.24) and Problem 11.7.

For each component i in a multicomponent VLE situation, define a K-factor as the ratio of the vapor-phase mole fraction y_i to the liquid-phase mole fraction x_i,

$$K_i \equiv \frac{y_i}{x_i} \tag{12.1.1}$$

The K-factor is an intensive measurable property (state function); it may be greater than unity or less than unity. A mixture has a K-factor for each of its C components, but only $(C-1)$ are independent: if we know the composition of one phase together with values for $(C-1)$ K-factors, then the last may be computed by

$$K_i = \frac{1-\sum_{j\neq i}^{C} y_j}{1-\sum_{j\neq i}^{C} x_j} = \frac{1-\sum_{j\neq i}^{C} x_j K_j}{1-\sum_{j\neq i}^{C} x_j} \tag{12.1.2}$$

For example, a binary has $K_2 = (1 - x_1K_1)/(1 - x_1)$. In general, a K-factor depends on temperature, pressure, and the mole fractions $\{x\}$ and $\{y\}$, but in many systems, the K-factors respond weakly and regularly to changes of state.

To illustrate, let us consider a Lewis-Randall ideal solution in equilibrium with an ideal gas, so the system obeys Raoult's law (§ 10.2.2),

$$x_i P_i^s = y_i P \qquad \textit{Raoult's law} \tag{12.1.3}$$

Then each K-factor is merely

$$K_i = \frac{P_i^s}{P} \qquad \textit{Raoult's law} \tag{12.1.4}$$

where P_i^s is the vapor pressure of pure i at the mixture temperature T. So in this special case, the K-factors are constants when T and P are fixed; they depend on the phase compositions implicitly through T and P. Along isotherms, the Raoult's law K-factor varies as $1/P$; this is illustrated in Figure 12.1. The Henry's law K-factor also varies as $1/P$. Of course, most liquid mixtures are not ideal, so (12.1.4) does not generally apply. For nonideal solutions, we can identify two general forms for K-factors, depending on whether the equality of fugacities (10.2.11) is posed in terms of phi-phi or gamma-phi.

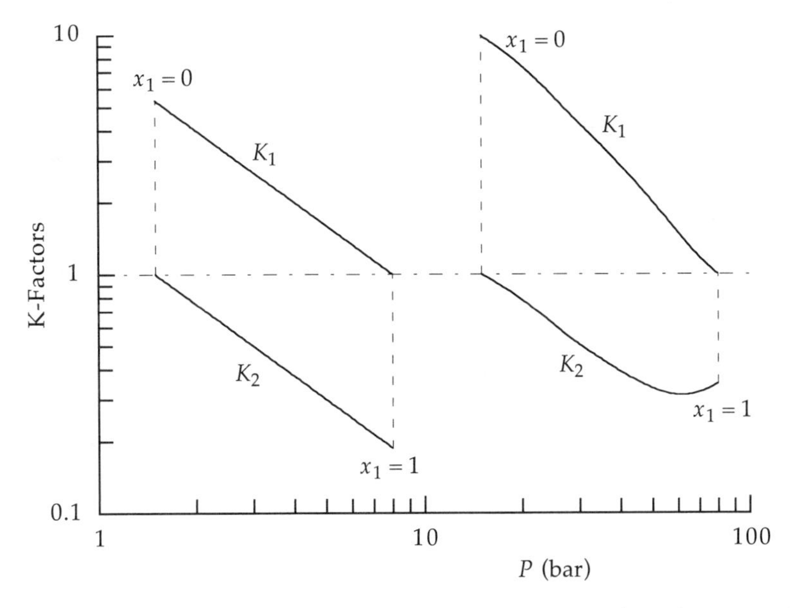

Figure 12.1 (*left*) Isothermal plot of K-factors in a binary mixture that obeys Raoult's law. In such cases, each K varies as $1/P$. (*right*) Same as at left except for a nonideal mixture; the nonlinearities in K_1 and K_2 may be caused by nonidealities in the liquid or in the vapor or in both.

Phi-phi form. If we choose FFF #1 for both vapor and liquid, then the equality of fugacities takes the phi-phi form (10.1.3), and the K-factor becomes

$$K_i = \frac{\varphi_i^{\ell}(T, v^{\ell}, \{x\})}{\varphi_i^{v}(T, v^{v}, \{y\})} \tag{12.1.5}$$

Values for the fugacity coefficients would, as usual, be obtained from a $PvTx$ equation of state using (4.4.23). In writing (12.1.5) note that in each argument list we have included the molar volume of the phase; by so doing, we emphasize that values for those volumes must be computed from the equation of state, even when the pressure is used to specify the state.

At low pressures the vapor will be ideal, so $\varphi_i^v \approx 1$; then, on using (5.5.8) for the liquid fugacity coefficient, neglecting the Poynting factor at low pressures, and taking the pure-i vapor pressure as the standard-state pressure, (12.1.5) reduces to

$$K_i = \varphi_i^{\ell}(T, P, \{x\}) = \varphi_i^{\ell}(T, P_i^s, \{x\}) \frac{P_i^s(T)}{P} \tag{12.1.6}$$

If the liquid is an ideal solution, then (5.1.3) gives $\varphi_{\text{pure i}}^{\ell}(T, P_i^s, \{x\}) = \varphi_{\text{pure i}}^{\ell}(T, P_i^s) = \varphi_{\text{pure i}}^{v}(T, P_i^s) \approx 1$, and (12.1.6) reduces to the Raoult's law form (12.1.4). Even when φ_i^{ℓ} is independent of composition, but not equal to unity, the K-factors still adhere to the functional form of Raoult's law (12.1.4) but the numerical values for the K-factors will differ from the Raoult's law values.

For binary mixtures, the effect of pressure on a K-factor can be obtained by combining the pressure derivatives of x_i and y_i that appear in (9.3.13) and (9.3.14). But for many systems, φ_i^ℓ is a weak function of $\{x\}$ and then along low-pressure isotherms, (12.1.6) suggests that K_i decreases as $1/P$ when the pressure is increased. This behavior is shown in Figure 12.1. Each plot in the figure shows ($\ln K_i$) plotted at fixed T against ($\ln P$), and in each case the curve is linear at low pressures, as required by (12.1.6). If the mixture has φ_i^ℓ independent of $\{x\}$, then $K_i \propto 1/P$ and the linear relation is preserved at all pressures at fixed T, as on the left in Figure 12.1. If the mixture is weakly nonideal ($\gamma_i \neq 1$ or $\varphi_i \neq 1$ or both), then the linear relation is disrupted, but only at high pressures, as on the right in Figure 12.1.

If the mixture is sufficiently nonideal that an azeotrope forms, then the curves for K_1 and K_2 cross; their intersection occurs at $K_1 = K_2 = 1$, which identifies the azeotrope. This possibility is shown on the left in Figure 12.2. Finally, if one component is supercritical, then the mixture may have a critical point at the fixed T; if this occurs, then $K_1(P)$ and $K_2(P)$ are two branches of the same curve. Those branches coincide at $K_1 = K_2 = 1$, which identifies the mixture critical point, as on the right in Figure 12.2.

Effects of temperature are shown in Figure 12.3 for binary mixtures of methane(1) and propane(2), with K_1 and K_2 computed from (12.1.5) using the Redlich-Kwong equation of state, as in § 8.4.4. At the temperatures used in Figure 12.3, methane is supercritical, so the isotherms appear as on the right in Figure 12.2. At low pressures both K_1 and K_2 decrease linearly on this logarithmic plot, again confirming that $K_i \propto 1/P$ along isotherms. At low pressures, $K_1 = H_1/P$, while K_2 is equal to unity at the pure propane vapor pressure; at the high-pressure mixture critical points $K_1 = K_2 = 1$.

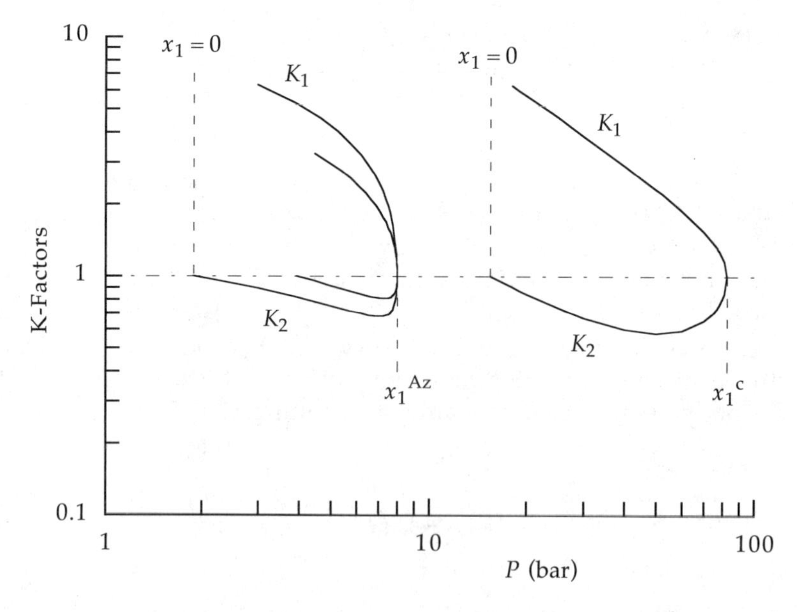

Figure 12.2 (*left*) Isothermal plot of K-factors in a binary mixture that forms an azeotrope. At this azeotrope the pressure is a maximum and the two curves intersect with $K_1 = K_2 = 1$. (*right*) Isothermal plot of K-factors in a binary mixture that passes through a mixture critical point. In these cases, K_1 and K_2 are two branches of the same curve; the two coincide at the critical point with $K_1 = K_2 = 1$.

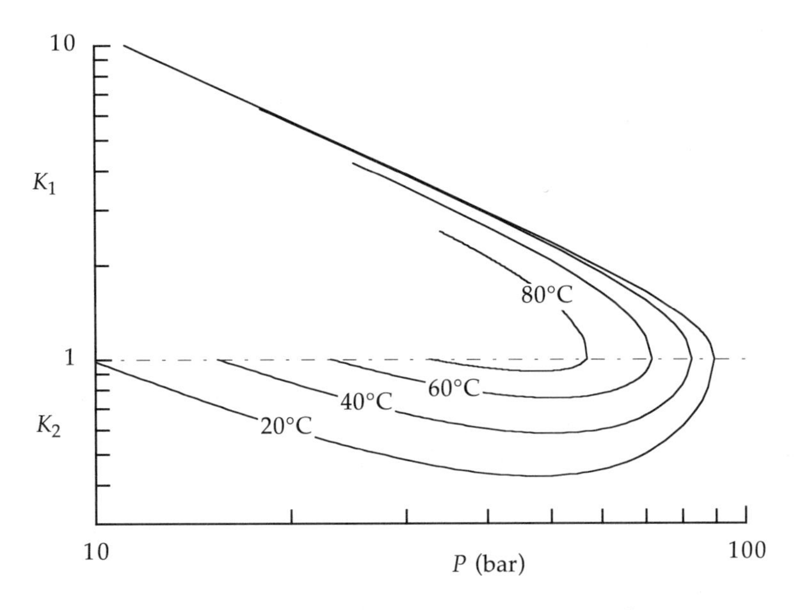

Figure 12.3 Effect of temperature on K-factors in binary mixtures. These K-factors were computed from the phi-phi form (12.1.5) using the Redlich-Kwong equation of state applied to vapor-liquid equilibria in binary mixtures of methane(1) and propane(2). At these temperatures methane is supercritical and the mixtures exhibit critical points at $K_1 = K_2 = 1$. This is an example of the class of binaries shown on the right in Figure 12.2.

Gamma-phi form. If we choose FFF #1 for the vapor and one of FFF #2–5 for the liquid, then at low pressure the equality of fugacities takes the gamma-phi form (10.1.4), and the resulting expression for the K-factor (12.1.1) is

$$K_i = \frac{\gamma_i(T, \{x\}) f_i^o(T)}{\varphi_i(T, v^v, \{y\}) P} \tag{12.1.7}$$

If we use FFF #5 for the liquid-phase fugacity, take the standard state to be pure saturated liquid i at T, and assume the low-pressure vapor is an ideal gas, then (12.1.7) becomes

$$K_i = \frac{\gamma_i(T, \{x\})\, P_i^s(T)}{P} \tag{12.1.8}$$

which is equivalent to (12.1.6). This gives $K_i \propto 1/P$, as it should along low-pressure isotherms.

Using the gamma-phi form (12.1.7) with the activity coefficient based on a pure-component standard state, we can determine the limiting behavior of K_i. In the pure limit,

$$\lim_{x_i \to 1} K_i = \frac{f_i^s(T)}{\varphi_i^s(T)\, P_i^s(T)} = 1 \qquad \textit{fixed T; keep VLE} \tag{12.1.9}$$

Note that this limit is to be taken along a vapor-liquid saturation curve with temperature held constant. If pure-component i is supercritical at the mixture T, then pure i does not exist in VLE at T, and the limit in (12.1.9) is undefined. This occurs for supercritical methane in Figure 12.3.

For the infinite-dilution limit, we choose component i to be a solute, and use the solute-free, dilute-solution standard state, so $f_i^o = H_{is}$. Then we have

$$\lim_{x_i \to 0} K_i = \frac{\gamma_i^\infty f_i^o}{\varphi_i^{v,\infty} P} = \frac{H_{is}}{\varphi_i^{v,\infty} P} \qquad \textit{fixed } T, \{x^{sf}\}; \textit{ keep VLE} \qquad (12.1.10)$$

This limit is taken along a vapor-liquid saturation curve with fixed T and fixed liquid-phase solute-free mole fractions $\{x^{sf}\}$. Here, φ_i^∞ is the vapor-phase fugacity coefficient at infinite dilution, and we have used (10.2.41) to identify the solute-free Henry's constant. At low pressures, the fugacity coefficient will be unity, and (12.1.10) will reduce to the ratio H_{is}/P; so along isotherms, we again have $K_i \propto 1/P$. In general, a K-factor changes continuously with x_i between the limits given by (12.1.9) and (12.1.10); it will often, though not always, be monotone in x_i. This discussion illustrates that it is sometimes easier to conceptualize property behavior using one formulation (gamma-phi), although calculations may be better done using another formulation (phi-phi).

12.1.2 Relative Volatilities in Vapor-Liquid Equilibria

For each pair of components i and j in a multicomponent vapor-liquid equilibrium situation, define a relative volatility α_{ij} as the ratio of the two K-factors,

$$\alpha_{ij} \equiv \frac{K_i}{K_j} = \frac{y_i/x_i}{y_j/x_j} \qquad (12.1.11)$$

The relative volatility is an intensive measurable property (state function). It measures the ease of separating i from j by distillation. For example, α_{ij} can be used to estimate the minimum number of ideal stages needed for a distillation column. If $\alpha_{ij} = 1$, no separation is possible by simple distillation. It is also used to estimate residue curves in multicomponent distillation designs.

In general, the relative volatility depends on $(T, P, \{x\}, \{y\})$, but in some mixtures α_{ij} is only weakly affected by changes of state. For example, in the special case of a Lewis-Randall ideal solution in equilibrium with an ideal gas, Raoult's law for the K-factors (12.1.4) produces

$$\alpha_{ij} = \frac{P_i^s(T)}{P_j^s(T)} \qquad \textit{Raoult's law} \qquad (12.1.12)$$

So in this special case, the relative volatility is independent of pressure and the compositions of both phases. It may depend on temperature, but often the temperature dependence is weak. Otherwise, general expressions for α_{ij} can be obtained from both the phi-phi and gamma-phi forms for VLE.

Phi-phi form. Substituting the phi-phi expression for the K-factors (12.1.5) into the definition (12.1.12), we obtain

$$\alpha_{ij} = \frac{\varphi_i^\ell(T, v^\ell, \{x\})}{\varphi_i^v(T, v^v, \{y\})} \frac{\varphi_j^v(T, v^v, \{y\})}{\varphi_j^\ell(T, v^\ell, \{x\})} \qquad (12.1.13)$$

As usual, the fugacity coefficients would be evaluated from a *PvTx* equation of state using (4.4.23). At low pressures we can substitute the approximation (12.1.6) for the K-factors; then (12.1.13) simplifies to

$$\alpha_{ij} = \frac{\varphi_i^\ell(T, P_i^s, \{x\})}{\varphi_j^\ell(T, P_j^s, \{x\})} \frac{P_i^s(T)}{P_j^s(T)} \qquad \text{low } P \qquad (12.1.14)$$

Although this approximation contains no explicit pressure term, it still depends implicitly on P; for example, if we fix T and change P while maintaining VLE, then the liquid compositions change and therefore α_{ij} changes.

Curves for relative volatilities α_{12} are shown in Figure 12.4 for the methane(1) + propane(2) mixtures whose K-factors were shown in Figure 12.3. At low pressures, each isotherm in Figure 12.4 starts at the vapor pressure of pure propane, and as P increases, α_{12} decreases almost linearly, reaching unity at the mixture critical line. The simple Raoult's law form for α_{12} (12.1.12) does not apply to the mixtures in Figure 12.4, even at low pressures, because all the temperatures in the figure are above the pure methane critical point; so no value of P_1^s exists at any of the states shown in the figure. Note that α_{12} increases as T is decreased isobarically; that is, at low temperatures and low pressures, the vapor phase is dominated by a relatively large fraction of methane.

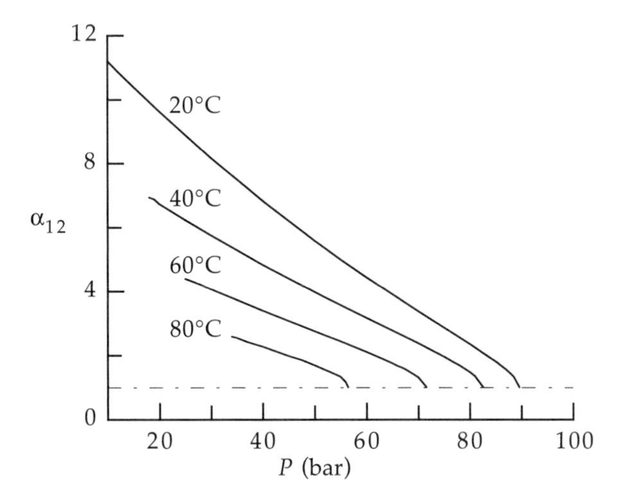

Figure 12.4 Relative volatilities for vapor-liquid equilibria in binary mixtures of methane(1) and propane(2), computed from the K-factors appearing in Figure 12.3. For these mixtures, each isotherm terminates at a mixture critical point, where $\alpha_{12} = 1$.

Gamma-phi form. If we substitute the low-pressure gamma-phi expression for the K-factors (12.1.7) into the definition of the relative volatility (12.1.12), we obtain

$$\alpha_{ij} = \frac{\gamma_i(T, \{x\})\, f_i^o(T)\; \varphi_j(T, v^v, \{y\})}{\gamma_j(T, \{x\})\, f_j^o(T)\; \varphi_i(T, v^v, \{y\})} \tag{12.1.15}$$

The explicit P in the K-factor (12.1.7) cancels from (12.1.15). The liquid-phase activity coefficients would be obtained from one of FFF #2–5; often, FFF #5 would be used, as in (12.1.8). If we can substitute the K-factor approximation (12.1.8) into (12.1.15), we find

$$\alpha_{ij} = \frac{\gamma_i(T, \{x\})\, P_i^s(T)}{\gamma_j(T, \{x\})\, P_j^s(T)} \tag{12.1.16}$$

Equation (12.1.16) is often valid, not only at low pressures, but also at moderate pressures, for in many situations the ratios of φs and Poynting factors may be close to unity, even though individual φs and Poynting factors are not. Furthermore, if the liquid is a Lewis-Randall ideal solution, then the activity coefficients are unity and (12.1.16) reduces to the Raoult's law form (12.1.12); but generally, the activity coefficients are not unity and their values must be obtained from a model for g^E.

Typical examples of how α_{ij} changes with composition are shown in Figure 12.5 for three binary mixtures: acetone(1) + methanol(2), acetone(1) + chloroform(3), and methanol(2) + chloroform(3). These α_{ij} were computed from the low-pressure form (12.1.16) using the Margules model (5.6.11); the corresponding activity coefficients for these three binaries appear in Figure 5.7. All three α_{ij} are monotone in the mole frac-

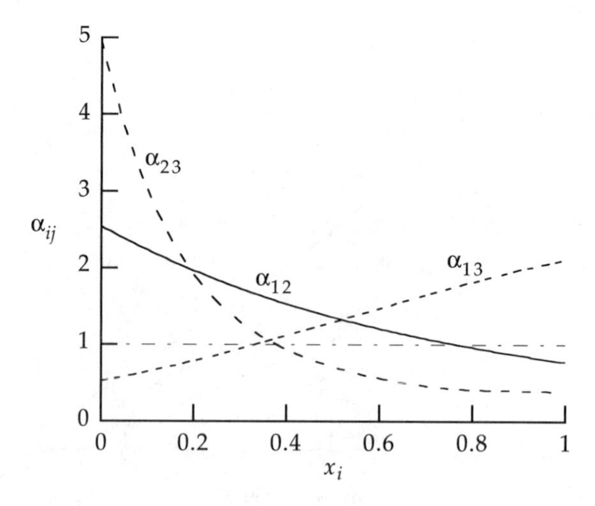

Figure 12.5 Relative volatilities for vapor-liquid equilibria in three binary mixtures: acetone(1) + methanol(2), acetone(1) + chloroform(3), and methanol(2) + chloroform(3). All computed at 60°C from the gamma-phi form (12.1.16) using the Margules equations (5.6.12) and (5.6.13) with parameters from Appendix E.

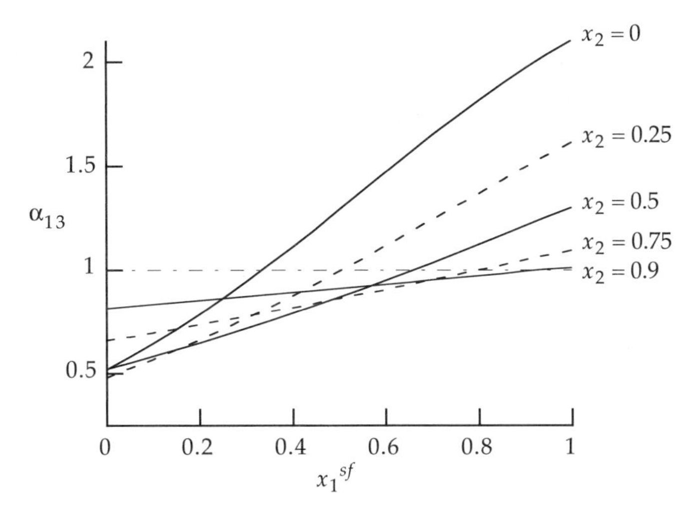

Figure 12.6 Relative volatilities α_{13} for vapor-liquid equilibria in ternary mixtures of acetone(1) + methanol(2) + chloroform(3). All computed at 60°C from the gamma-phi form (12.1.16) using the Margules equations (5.6.23) with parameters from Appendix E. The quantity on the abscissa is the methanol-free mole fraction for acetone; that is, $x_1^{sf} = x_1/(x_1 + x_3)$.

tion (the usual case for binaries) and all three are less than 1 for some compositions, but greater than 1 for others (often, but not always, the case). When binary mixtures have $\alpha_{ij} = 1$, the system has the same composition in both phases; this identifies either a homogeneous azeotrope (§ 9.3.4) or a gas-liquid critical point (§ 9.3.5). But those g^E models that contain no pressure effects cannot predict vapor-liquid critical points; so in Figure 12.5, those points having $\alpha_{ij} = 1$ all represent azeotropes and the components could not be separated by a simple distillation.

Representative behavior for one relative volatility in a ternary mixture is shown in Figure 12.6. The components for the ternaries used in this figure are the same as those used for the binaries in the previous figure. Figure 12.6 shows how α_{13} responds when the methanol-free mole fraction of acetone x_1^{sf} is changed and the methanol mole fraction x_2 is held fixed. These relative volatilities were computed from the low-pressure form (12.1.16) using the multicomponent version of the Margules equations (5.6.23). The figure shows that α_{13} is well-behaved, being roughly linear in x_1^{sf}. Note that when x_2 is large enough, $\alpha_{13} < 1$ for all x_2; this suggests that extractive distillation could be used to separate components 1 and 3. Also note that the major effect of methanol(2) is to reduce α_{13} more when the amount of acetone is small ($x_1^{sf} \to 1$) rather than when the amount of chloroform is small ($x_1^{sf} \to 0$).

Limiting behaviors. The gamma-phi form (12.1.15) is convenient for determining limiting behaviors of relative volatilities. In the following we use activity coefficients in a pure-component standard state. First consider the pure-1 limit of α_{12} in a multicomponent mixture, taking the limit with T fixed, VLE maintained (so $P \to P_1^s$), and all liquid mole fractions except x_1 driven to zero. This limit, applied to (12.1.15), yields

$$\lim_{x_1 \to 1} \alpha_{12} = \frac{\varphi_1^s(T)\, P_1^s(T)}{\varphi_{\text{pure 1}}^v} \frac{\varphi_2^{v,\infty}}{\gamma_{21}^\infty\, P_2^s(T)} = P_1^s(T) \frac{\varphi_2^{v,\infty}}{H_{21}(T)} \tag{12.1.17}$$

The limiting activity coefficient is defined by (10.2.40) and we have used (10.2.47) to introduce the reference solvent Henry's constant H_{21}. For a binary at low pressures the fugacity coefficient is unity, while the reference-solvent and solute-free Henry's constants are the same, so (12.1.17) reduces to

$$\lim_{x_1 \to 1} \alpha_{12} = \frac{P_1^s(T)}{H_{21}(T)} = \frac{P_1^s(T)}{\gamma_{21}^{\infty} P_2^s(T)} \qquad \textit{binary, low P} \qquad (12.1.18)$$

where the activity coefficient is that at infinite dilution of 2 in pure 1.

We may also consider solute-free infinite-dilution limits, taken at fixed T, with fixed solute-free mole fractions, and VLE preserved. For a multicomponent mixture, this limit is

$$\lim_{x_1 \to 0} \alpha_{12} = \frac{\gamma_{12}^{\infty} f_1^o}{\varphi_1^{v,\infty}} \frac{\varphi_2^v(T, v^v, \{y^{sf}\})}{\gamma_2(T, v^\ell, \{x^{sf}\}) f_2^o} \qquad \textit{fixed T, \{x^{sf}\}} \qquad (12.1.19)$$

Here the component-2 fugacity and activity coefficients depend on the solute-free mole fractions. However, for a binary $x_2^{sf} = 1$, $\gamma_2 = 1$, and at low pressures the fugacity coefficients are unity, so (12.1.19) reduces to

$$\lim_{x_1 \to 0} \alpha_{12} = \frac{\gamma_{12}^{\infty} P_1^s(T)}{P_2^s(T)} \qquad \textit{binary, low P} \qquad (12.1.20)$$

where the activity coefficient is that at infinite dilution of 1 in pure 2. Note that (12.1.20) is a permuted form of (12.1.18). For binaries at low pressures, a plot of relative volatility vs. mole fraction forms a continuous curve between the limits in (12.1.18) and (12.1.20); it is nearly always monotone. Therefore, if one limit has $\alpha_{12} < 1$ while the other limit has $\alpha_{12} > 1$, then $\alpha_{12} = 1$ at some intermediate mole fraction; that is, an azeotrope occurs. So the limiting values of α_{12} provide a simple test for the existence of azeotropes in binary mixtures.

12.1.3 Distribution Coefficients for Liquid-Liquid Equilibria

Consider liquid phases α and β in equilibrium. For each component i, we may define a distribution coefficient C_i, as the ratio of the equilibrium mole fractions of i in the two phases,

$$C_i \equiv \frac{x_i^\beta}{x_i^\alpha} \qquad (12.1.21)$$

The ratio C_i is an intensive measurable property (state function); it depends on temperature, pressure, and the composition of each phase. We may find $C_i > 1$ or $C_i < 1$.

The distribution coefficient was first introduced in (11.1.14) as an aid in performing LLE calculations; it is analogous to the K-factor in vapor-liquid equilibrium. Just as for the K-factor, different forms for C_i can be obtained, depending on whether we choose the phi-phi, gamma-phi, or gamma-gamma form to represent the equality of fugacities. However, the gamma-phi approach is little used for LLE, so here we consider only the phi-phi and gamma-gamma forms for C_i.

Phi-phi form. If the phi-phi form is used (§ 10.1.1), then the definition (12.1.21) reduces to a ratio of fugacity coefficients

$$C_i = \frac{\varphi_i^{\alpha}(T, v^{\alpha}, \{x^{\alpha}\})}{\varphi_i^{\beta}(T, v^{\beta}, \{x^{\beta}\})} \tag{12.1.22}$$

and values for the φs would be computed from a *PvTx* equation of state that applies to each liquid phase. Usually we use the same equation of state for both phases, but at states well away from any consolute point, different equations might be used.

At low pressures we can insert the approximation (12.1.6) for liquid-phase fugacity coefficients, obtaining

$$C_i = \frac{\varphi_i^{\alpha}(T, P_i^{o\alpha}, \{x^{\alpha}\})\, P_i^{o\alpha}}{\varphi_i^{\beta}(T, P_i^{o\beta}, \{x^{\beta}\})\, P_i^{o\beta}} \qquad \textit{low } P \tag{12.1.23}$$

where P_i^o is any convenient pressure for the standard state. Often the same standard-state pressure is chosen for both phases ($P_i^{o\alpha} = P_i^{o\beta}$) and then (12.1.23) simplifies further. But in any event, (12.1.23) shows that at low pressures, the distribution coefficient is independent of system pressure P.

Gamma-gamma form. If the gamma-gamma form (§ 10.1.3) is used for the equality of the fugacities, then the distribution coefficient (12.1.21) becomes

$$C_i = \frac{\gamma_i^{\alpha}(T, P, \{x^{\alpha}\})\, f_i^{o\alpha}(T, P_i^{o\alpha})}{\gamma_i^{\beta}(T, P, \{x^{\beta}\})\, f_i^{o\beta}(T, P_i^{o\beta})} \tag{12.1.24}$$

We would extract an appropriate expression for the activity coefficients from FFF #2–5. Usually we use the same FFF for the same component in both phases, but this is not necessary, and we usually choose the same standard-state fugacity for the same component in both phases, then (12.1.24) reduces to a simple ratio of activity coefficients,

$$C_i = \frac{\gamma_i^{\alpha}(T, P, \{x^{\alpha}\})}{\gamma_i^{\beta}(T, P, \{x^{\beta}\})} \tag{12.1.25}$$

When a pure-component standard state is used for both activity coefficients, then the pure-component and infinite-dilution limits are straightforward. Taking the pure-component limit, with T held fixed, (12.1.24) becomes

$$\lim_{x_i^\alpha \to 1} C_i = \frac{f_i^{o\alpha}(T, P_i^{o\alpha})}{f_i^{o\beta}(T, P_i^{o\beta})} \tag{12.1.26}$$

and if the same standard state is used for component i in both phases, then the rhs of (12.1.26) reduces to unity. On taking the dilute-solution limit, with T and the solute-free mole fractions $\{x^{sf}\}$ fixed, (12.1.24) becomes

$$\lim_{x_i^\alpha \to 0} C_i = \frac{\gamma_i^{\alpha,\infty}(T, P, \{x^{\alpha,sf}\}) f_i^{o\alpha}(T, P_i^{o\alpha})}{\gamma_i^{\beta,\infty}(T, P, \{x^{\beta,sf}\}) f_i^{o\beta}(T, P_i^{o\beta})} \qquad \textit{fixed } \{x^{sf}\} \tag{12.1.27}$$

If the same standard state is used for i in both phases, then the rhs of (12.1.27) reduces to a ratio of infinite-dilution activity coefficients.

For ternary mixtures, the limits (12.1.27) define a tie line at the edge of a triangular diagram for two-phase equilibria; see, for example, Figure 9.25. But we note that an infinite-dilution tie line is not parallel to the edge, for if it were, the infinite-dilution distribution coefficient would be unity. This means that C_i is discontinuous when the infinite-dilution limit passes over to the value of C_i for zero concentration.

When one phase (say α) is rich in component i, while the other phase is dilute in i, then it is often advantageous to select different standard states for component i in the two phases. Then instead of (12.1.24), we would have

$$C_i = \frac{\gamma_i^{\alpha}(T, P, \{x^{\alpha}\}) f_i^{o\alpha}(T, P_i^{o\alpha})}{\gamma_i^{*}(T, P, \{x^{\beta}\}) H_{is}(T, P_i^{o\beta}, \{x^{\beta sf}\})} \tag{12.1.28}$$

In the limit of extreme immiscibility, (12.1.26) applies to phase α while (12.1.27) applies to phase β, so (12.1.28) becomes

$$\lim_{\substack{x_i^\alpha \to 1 \\ x_i^\beta \to 0}} C_i = \frac{f_i^{o}(T, P_i^{o})}{H_{is}(T, P_i^{o}, \{x^{sf}\})} \qquad \textit{fixed } \{x^{sf}\} \tag{12.1.29}$$

Sample values for a distribution coefficient are shown in Figure 12.7 for a solute (component 1) distributed between two completely immiscible solvents (components 2 and 3). We let phase β contain solvent 2 and let phase α contain solvent 3. The curves were computed from the gamma-gamma form (12.1.25) using Porter's equation (5.6.4) for activity coefficients. The figure shows how the distribution of solute changes when the identity of one solvent (component 2) is changed. This kind of calculation is routinely done when screening candidate solvents for use in a separation process by solvent extraction. In the calculations for Figure 12.7, the Porter parameter $A_{13} = 1.9$ remained unchanged throughout the study, while the value of A_{12} was changed systematically, as indicated in the figure. (Recall that in the Porter model, A_{12} and A_{13} must be < 2, else additional phase splits will occur.)

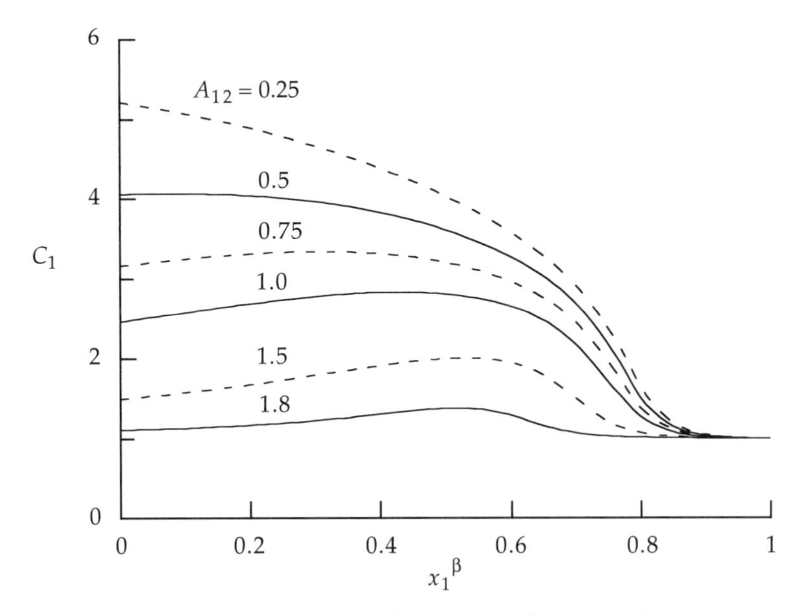

Figure 12.7 Response of the distribution coefficient C_1 to changes in the identity of one solvent (component 2 = phase β) when a solute (component 1) is distributed between two completely immiscible solvents (2 and 3). These curves were computed from the gamma-gamma form (12.1.25) using Porter's equation for the activity coefficients. For all calculations, the Porter parameter $A_{13} = 1.9$, while the value of A_{12} was as shown.

If solvent 2 had $A_{12} = A_{13}$, then the solute would attain the same mole fraction in both phases and C_1 would be unity for all compositions. Figure 12.7 shows that as A_{12} is decreased away from $A_{13} = 1.9$, C_1 increases; that is, the mole fraction of solute increases in the solvent-2 phase. As a general rule, the larger the disparity in intermolecular forces between solute 1 and each solvent, 2 and 3, the easier it is to extract solute from one phase into the other. Note that each curve in the figure obeys the pure-component (12.1.26) and dilute-solution (12.1.27) limits (with the same standard states used for both phases). Also note that at high concentrations ($x_1^\beta > 0.85$) both phases are dominated by solute molecules and no separation occurs. Finally, note that a weak maximum occurs in C_1 when $A_{12} > 1$.

For completeness, we mention that in describing liquid-liquid equilibria, some authors define a *selectivity* β_{ij} to be the ratio of two distribution coefficients,

$$\beta_{ij} \equiv \frac{C_i}{C_j} \tag{12.1.30}$$

So the selectivity used in liquid-liquid equilibria is analogous to the relative volatility α_{ij} used in vapor-liquid equilibria. In general, β_{ij} depends on temperature, pressure, and the compositions of both liquid phases; however, since both phases are liquids, β_{ij} is often little affected by moderate changes in pressure. At consolute points, the compositions are the same in the two phases, so $C_i = C_j = 1$, and therefore $\beta_{ij} = 1$. Otherwise, when values of β_{ij} are very different from unity, then it is feasible to separate components by liquid extraction.

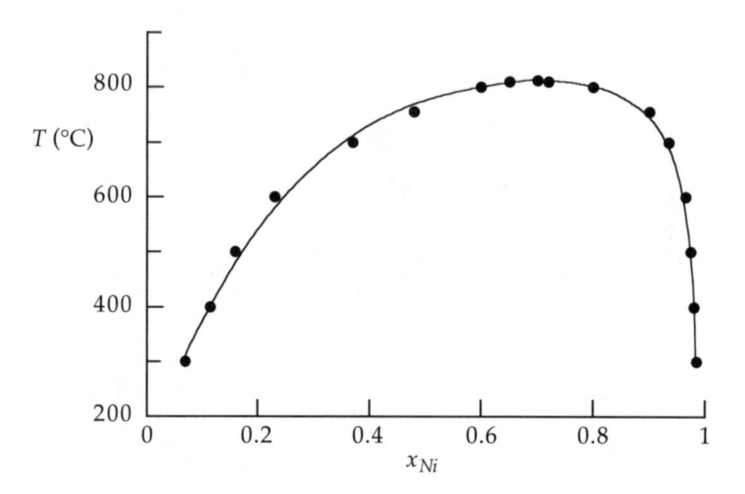

Figure 12.8 Correlation of solid-solid equilibria for binary alloys of nickel + gold. Points are experimental data from the collection of Hansen [1]. Line was computed via the gamma-gamma method using the Margules equation with each Margules parameter fit to the data via a quadratic in temperature.

12.1.4 Correlating Solid-Solid Equilibria

Thermodynamically, solid-solid equilibria are isomorphic to liquid-liquid equilibria, so a thermodynamic description of SSE poses the same problems as does a description of LLE. Those descriptions usually take the form of Txx diagrams: unless pressure is very high, P has negligible effects on the properties of solids. Correlations of Txx data for SSE are based on gamma-gamma expressions for fugacities, and often a simple Porter or Margules equation is sufficient to correlate available data [2]. In correlating SSE data, the challenge is usually not in identifying an appropriate model for g^E, but in correlating the temperature dependence of the parameters in the model.

A typical Txx diagram is shown in Figure 12.8, which applies to binary alloys of nickel + gold. These alloys exhibit a solid-solid UCST near 812°C. The line in the figure is a correlation of experimental data (points) and was computed via the gamma-gamma method together with the Rachford-Rice algorithm from § 11.1.5. The activity coefficients were modeled using the Margules equations (5.6.12) with the parameters A_1 and A_2 fit to the data using quadratics in temperature (see Problem 12.9).

12.2 SOLUBILITIES

Solubility is an oft ill-defined term, used rather indiscriminately to refer to small amounts of a solute of one phase dissolved in a solvent of another phase. Invariably, the solvent is a liquid or dense fluid, though it may contain any number of components, while the solute may be gas, liquid, or solid. Solubility problems are really phase-equilibrium problems and are attacked using the general strategies presented in Chapter 10. In this section we describe the three common solubility problems: *gas solubility,* which refers to supercritical gases dissolved in liquids (§ 12.2.1); *solid solubility,* which refers to solids dissolved in liquids (§ 12.2.2); and solubilities in *near-critical systems,* which usually involve a liquid or solid dissolved in a near-critical fluid (§ 12.2.3).

12.2.1 Gases in Liquids

Gas solubility usually refers to the liquid-phase mole fraction x_i that occurs in a VLE situation in which the system temperature T is above the critical temperature of the pure solute ($T > T_{ci}$), but below the critical temperature of at least one other component ($T < T_{cj}$). An example is CO_2 dissolved in a carbonated beverage. These situations can be described using either a phi-phi or a gamma-phi approach; the gamma-phi method is the traditional approach.

Phi-phi. If a reliable $PvTx$ equation of state is available, then we may use the phi-phi method to compute gas solubilities. Thermodynamically, this is merely phi-phi applied to VLE and the general approach has been discussed in § 10.1.1 and § 12.1.1. But in practice, this is a relatively recent development because reliable equations of states have only recently been devised for supercritical solutes in subcritical solvents. When the phi-phi method is used, computed solubilities are found to be sensitive to the temperature dependence of parameters in the equation of state; they are also sensitive to the mixing rules used for those parameters. In particular, when cubic equations are used, the temperature dependence and mixing rule for the parameter a must be chosen with care. However, we judge this to be a modeling problem, not a thermodynamic problem.

Gamma-phi. In the gamma-phi approach to gas solubilities, FFF #1 is always used for the vapor, so the issues center on appropriate expressions for the fugacities of components in the liquid phase. Let component 1 be the supercritical solute and 2 be the subcritical solvent. For liquid fugacities we often use FFF #5 and for the solvent we would likely choose a pure-component standard state with the standard-state pressure equal to the vapor pressure of the pure liquid ($P_2^o = P_2^s$). Then FFF #5 becomes

$$f_2(T, P, x_2) = x_2\, \gamma_2(T, P_2^s, x_2)\, \varphi_2^s(T)\, P_2^s(T) \exp\left[\frac{1}{RT}\int_{P_2^s}^{P} \bar{V}_2\, d\pi\right] \tag{12.2.1}$$

At low pressures, the Poynting factor is negligible, and if $x_2 \approx 1$, then $\gamma_2 \approx 1$; otherwise, the activity coefficient would be obtained from an appropriate model for g^E.

We might also select FFF #5 for the solute in the liquid phase. Since we expect solubility problems to have $x_1 < 0.05$, we would choose a Henry's law standard state, with the standard-state pressure taken to be the vapor pressure of pure solvent (i.e., $P_1^o = P_2^s$). Now FFF #5 takes the form

$$f_1(T, P, x_1) = x_1\, \gamma_1^*(T, P_2^s, x_1)\, H_1(T, P_2^s) \exp\left[\frac{1}{RT}\int_{P_2^s}^{P} \bar{V}_1\, d\pi\right] \tag{12.2.2}$$

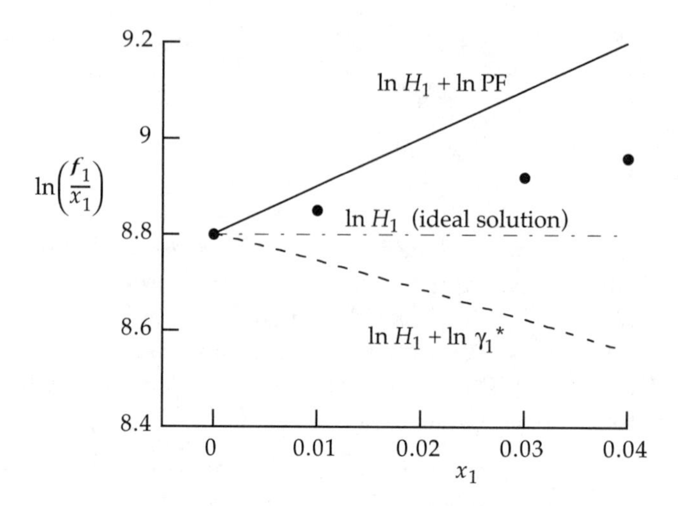

Figure 12.9 For the fugacity of a supercritical solute(1) in a liquid solvent(2), the Poynting factor (PF) in FFF #5 (12.2.2) tends to compensate for the effects of the activity coefficient. This plot shows contributions to fugacities of hydrogen(1) in methanol(2) at 294.15 K. Points are the experimental data of Krichevskii et al. [3]. Horizontal line is for a Henry's law ideal solution. Upper line includes only the Poynting factor, while the lower line includes only the activity coefficient. Adapted from a figure in Campanella et al. [4].

Usually, neither the activity coefficient nor the Poynting factor in (12.2.2) is negligible; further, while the Poynting factor is > 1, the activity coefficient γ^* is < 1 (see § 10.2.3). Therefore, these two terms tend to compete in correcting for nonidealities. Such competing effects are illustrated in Figure 12.9 for the fugacities of hydrogen dissolved in methanol.

We prefer to avoid these competing effects by using FFF #4 rather than FFF #5; then the effects of all nonidealities combine into a single activity coefficient, which we evaluate at the system pressure rather than a standard-state pressure. With this choice, the solute fugacity takes the form

$$f_1(T, P, x_1) = x_1 \gamma_1^*(T, P, x_1) H_1(T, P_2^s) \tag{12.2.3}$$

The value of $\gamma_1^*(T, P, x_1)$ may be > 1 or < 1 (Since the points lie above the horizontal line in Figure 12.9, we have $\gamma_1^* > 1$ in Figure 12.9.), but it is often near unity because it combines the competing effects that appear in FFF #5. Although use of FFF #4 avoids the need for values or estimates of partial molar volumes, which appear in FFF #5 (12.2.2), in practice, models for γ_1^* require values for the solution density [4].

Temperature effects. To illustrate how a solute mole fraction x_1 responds to changes in temperature, we consider a binary mixture and use gamma-phi along with FFF #4,

$$x_1 \gamma_1^*(T, P, x_1) H_1(T, P_2^s) = y_1 \varphi_1 P \tag{12.2.4}$$

When T is changed, at fixed P and y_1, the response of x_1 is dominated by the response of the Henry's constant; so, to a good approximation,

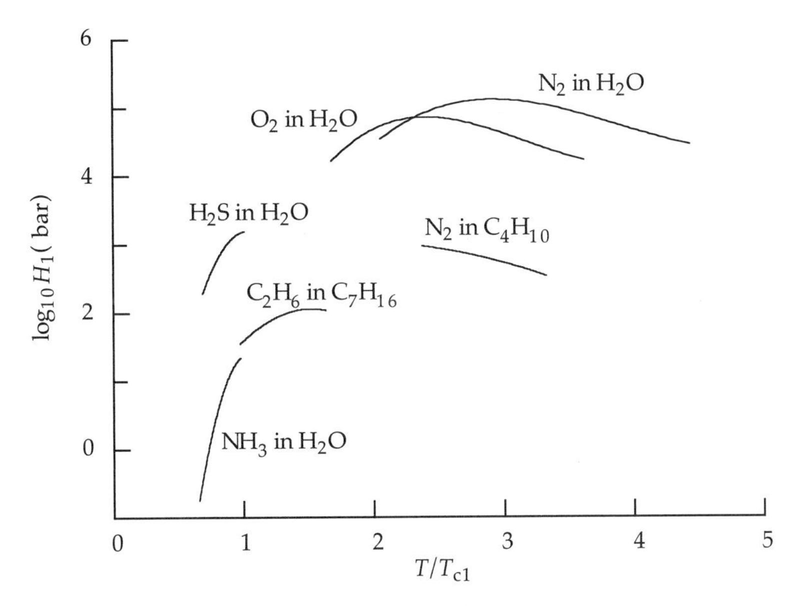

Figure 12.10 Response of the Henry's constant to changes in temperature for binary mixtures of a gas solute(1) in a liquid solvent(2). For temperatures well above the critical temperature of the solute ($T \gg T_{c1}$), H_1 decreases with increasing T. But for temperatures near and below T_{c1}, H_1 increases with T. Here, each Henry's constant is evaluated at the vapor pressure of pure solvent, $H_1(T, P_2^s)$. Adapted from a figure in Cysewski and Prausnitz [5].

$$\left(\frac{\partial \ln x_1}{\partial T}\right)_{Py} = -\left(\frac{\partial \ln H_1(T, P_2^s)}{\partial T}\right)_{Py} \tag{12.2.5}$$

Experimental data for $H_1(T)$ are shown in Figure 12.10 for several binary systems. The observed behavior generally divides into two groups:

(a) For temperatures near and below the solute critical temperature ($T \le T_{c1}$), H_1 increases with T.

(b) But at some $T > T_{c1}$, H_1 passes through a maximum, and thereafter H_1 decreases with increasing T.

These observations, combined with (12.2.5), indicate that, when ($T \gg T_{c1}$), a gas solubility generally increases with T, but when T is near or below T_{c1}, the solubility decreases with increasing T:

$$\left.\left(\frac{\partial \ln x_1}{\partial T}\right)_{Py}\right\} \begin{array}{ll} > 0 & \text{for } (T \gg T_{c1}) \\ < 0 & \text{for } (T \le T_{c1}) \end{array} \tag{12.2.6}$$

Note in Figure 12.10 that the behavior of the Henry's constant $H_1(T)$ differs substantially from that of pure-component vapor pressures $P^s(T)$, such as in Figure 9.2. This has implications for the use of $H_1(T)$ or $P^s(T)$ as standard-state fugacities.

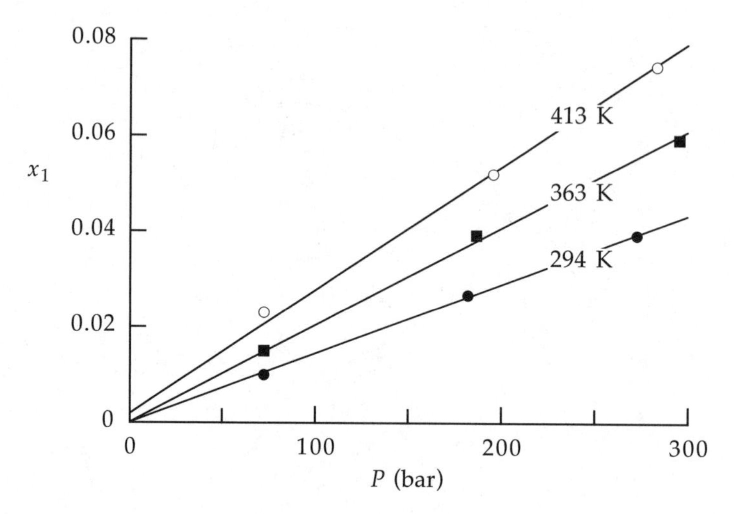

Figure 12.11 Effect of pressure on the solubility of hydrogen(1) in methanol(2). Points are experimental data from Krichevskii et al. [3]. Lines are least squares fits to the data; so, within experimental error, these values for the solubility x_1 are linear in P. At high temperatures, note that the straight-line intercept at $P = 0$ does not occur at $x_1 = 0$.

Pressure effects. The response of gas solubility to a change in pressure can also be deduced from the gamma-phi form; to do so, we use (12.2.4) to write

$$\left(\frac{\partial \ln x_1}{\partial P}\right)_{Ty} = \frac{1}{P} + \left(\frac{\partial \ln \varphi_1}{\partial P}\right)_{Ty} - \left(\frac{\partial \ln \gamma_1^*}{\partial P}\right)_{Ty} \tag{12.2.7}$$

At moderate pressures the last two terms on the right are small and we expect the solubility to increase linearly with pressure: $x_1 \propto P$. At high pressures the effects of the fugacity and activity coefficients cannot be neglected from (12.2.7) and the linear behavior of $x_1(P)$ will be disrupted. Figure 12.11 illustrates this behavior for hydrogen in methanol. In the figure, note that the slopes of the lines are not merely $1/H_1$, which would arise from a naïve use of Henry's law ($x_1 = y_1 P/H_1$); instead, the straight lines occur because the nonideal term φ_1/γ_1^* in (12.2.4) is also apparently linear in P.

12.2.2 Solids in Liquids

Solid solubility usually refers to the liquid-phase mole fraction x_i that occurs in an LSE situation when the system temperature is below the melting temperature of the pure solute ($T < T_m$), but above the melting temperature of at least one other component ($T_{mj} < T < T_{cj}$) or when the solution is above its eutectic temperature. An example is the equilibrium concentration of salt in water, which pertains to our attempts to prevent icing of winter roads. These situations are best described using the gamma-gamma form,

$$x_i^\ell \gamma_i^\ell f_i^{o\ell} = x_i^s \gamma_i^s f_i^{os} \tag{12.2.8}$$

with the standard state of solid solute taken to be pure solid at T and that for liquid solute taken to be pure liquid at T. The solubility can then be expressed in the form

$$\ln x_i^\ell = \ln\left(\frac{x_i^s \gamma_i^s}{\gamma_i^\ell}\right) + \ln\left(\frac{f_i^{os}(T)}{f_i^{o\ell}(T)}\right) \tag{12.2.9}$$

To obtain expressions for the standard-state fugacities appearing in (12.2.9), we follow the strategy suggested in § 10.1.4. There we noted that, rather than evaluate the individual standard state fugacities, it is easier to evaluate the ratio that appears on the right in (12.2.9). The expression for that ratio is contained in (10.1.10); so, using (10.1.10) in (12.2.9) yields

$$\ln x_i^\ell = \ln\left(\frac{x_i^s \gamma_i^s}{\gamma_i^\ell}\right) - \frac{\Delta h_i^m(T_m)}{R}\left(\frac{1}{T} - \frac{1}{T_m}\right) + \int_{T_m}^{T} \frac{\Delta c_p^o(\tau)}{R}\left(\frac{1}{\tau} - \frac{1}{T}\right) d\tau \tag{12.2.10}$$

Here Δh^m is the latent heat of melting for pure i and Δc_p^o is the difference in isobaric heat capacities for pure liquid i and pure solid i. Common approximations for Δc_p^o are described in § 10.1.4. Typically, the last term in (12.2.10) is small and can be neglected.

A common use of (12.2.10) is to determine freezing point depressions ($\Delta T = T_m - T$) that occur when small amounts of solute(1) are dissolved in an otherwise pure solvent(2) (e.g., salt in water). In such problems (12.2.10) can be applied only to the solvent(2). Several forms for ΔT can be obtained from (12.2.10), depending on the particular assumptions made; however, it is usual to assume that the solid is pure, so $x_2^s = 1$ and $\gamma_2^s = 1$, and to assume $\gamma_2^\ell \approx 1$ because the liquid is also nearly pure. Then if we neglect the last two terms altogether, (12.2.10) reduces to

$$\ln x_2^\ell \approx -\frac{\Delta h_2^m(T_m)}{RT}\left(\frac{\Delta T}{T_m}\right) \tag{12.2.11}$$

Therefore for a known concentration ($x_1^\ell = 1 - x_2^\ell$) of solute dissolved in the liquid, (12.2.11) can be used to estimate the amount by which the pure-solvent freezing point decreases (ΔT). (In contrast, boiling point elevations are usually determined from the gamma-phi method, using the Lewis-Randall rule for the solvent and the solute-free Henry's law for the solute.)

Expressions deduced from (12.2.10), such as (12.2.11), can also be used to determine molecular weights. For example, we could dissolve a known mass of solute in a known mass of solvent and then measure the freezing-point depression ΔT of the solvent. If the molecular weight of the solvent is known, then using the measured value of ΔT in (12.2.11) gives the mole fraction of the solute; then, the molecular weight of the solute can be computed from the known mass. Further discussions of solid solubilities, including descriptions of the effects of nonidealities and dissociation, have been given by Tsonopoulos [6].

12.2.3 Near-Critical Systems

Now we consider solubility problems in which the solute is a condensed phase (liquid or solid), but the pressure is high and the temperature is near or above the critical temperature of the solvent; in such cases, the solvent is not a liquid but a dense fluid. These situations are important because the solubility (i.e., the solute mole fraction in the fluid x_i^f) can be much larger than its value at low pressures. The solubility increases for two reasons:

(a) The effect of P on the condensed phase fugacity, which is quantified by a Poynting factor for the solute; as P increases, this Poynting factor might become greater than unity.

(b) The nonideality of the fluid, which is quantified by the solute fugacity coefficient $\varphi_i(T, v^f, \{x^f\})$; as P increases, this fugacity coefficient can become less than unity.

In addition to an amplification of the solubility, small changes of fluid state in the critical region can cause drastic changes in the solubility x_i^f. Such changes reflect large changes in the molar volume v^f because, in the critical region of a pure fluid, $(\partial v/\partial P)$ is large.

As a typical example, consider the solubility of a solid in a supercritical fluid. For the equality of fugacities we choose the gamma-phi form and use FFF #5 for the condensed phase; then for solute i we write

$$x_i^f \varphi_i(T, v^f, \{x^f\})\, P = x_i^s \gamma_i^s(T, P_i^o, \{x^s\}) f_i^{os}(T) \exp\left[\frac{1}{RT}\int_{P_i^{os}}^{P} \overline{V}_i^s \, d\pi\right] \tag{12.2.12}$$

For the solid-phase standard state we choose the pure solid on its sublimation curve at the system temperature T; hence, $f_i^{os} = P_i^{sub}(T) \ll P$. To simplify (12.2.12) we first assume the solid phase is essentially pure solute, so $x_i^s = 1$, $\gamma_i^s = 1$, and the partial molar volume is the pure-solid molar volume, v_i^{os}. If we also assume the solid is incompressible, then (12.2.12) reduces to

$$x_i^f \approx \frac{P_i^{sub}(T)}{\varphi_i(T, v^f, \{x^f\})\, P} \exp\left[\frac{v_i^{os} P}{RT}\right] \tag{12.2.13}$$

Although the solubility may be enhanced by the increased pressure, the value is still small; generally $x_i^f < 0.01$. So for a binary, the supercritical fluid is nearly pure solvent, and the fugacity coefficient in (12.2.13) is approximately its value at infinite dilution. Then we are left with

$$x_i^f \approx \frac{P_i^{sub}(T)}{\varphi_i^{\infty}(T, v^f)\, P} \exp\left[\frac{v_i^{os} P}{RT}\right] \tag{12.2.14}$$

At low pressures, the Poynting factor is unity and the fluid phase is essentially an ideal gas, so (12.2.14) takes the ideal-gas form P_i^{sub}/P and the solubility decreases as P increases. But at high pressures, the Poynting factor is large, the fluid is no longer ideal, and the fugacity coefficient is small (< 1). At high P these effects can combine to make the solubility increase with increasing P. Between these extremes, the fluid changes from ideal-gas to nonideal-gas behavior, and x_i^f passes through a minimum with pressure.

The approximation (12.2.14) is illustrated in Figure 12.12 for solid methane in hydrogen. The points in the figure are experimental data, which pass through a minimum as expected. The solid line is the ideal-gas result, decreasing linearly with P on this logarithmic plot. The dashed line is the ideal-gas result corrected with the Poynting factor [i.e., $\varphi_i = 1$ in (12.2.14)]. The dash-dot line is the complete approximation (12.2.14), with the fugacity coefficient computed from the simple virial equation,

$$Z = 1 + \frac{B}{v} \tag{12.2.15}$$

with

$$\ln\varphi_1^{\infty} = \frac{2B_{12}}{v_{\text{pure 2}}} - \ln\left(\frac{Pv_{\text{pure 2}}}{RT}\right) \tag{12.2.16}$$

where methane is component 1 and $B_{12} = -102$ cc/mol is the unlike-interaction, second virial coefficient [7]. Figure 12.12 is typical in that the solubility enhancement is primarily caused by fluid-phase nonidealities, while the condensed-phase Poynting factor makes only a small contribution to the enhancement. At the highest pressures in Figure 12.12, agreement between experiment and (12.2.14) can be improved by including the third virial coefficient (or an empirical equivalent) in the equation of

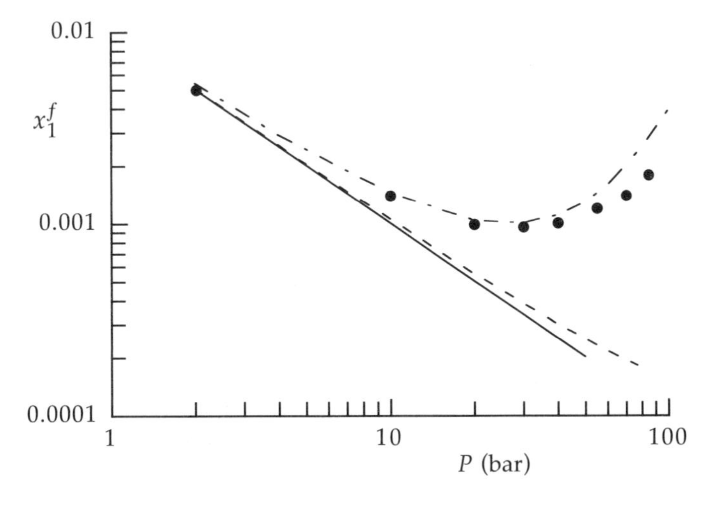

Figure 12.12 Supercritical enhancement of the solubility of solid methane(1) in fluid hydrogen(2) at 76 K. Points are experimental data of Hiza and Herring [8]. Solid line is from the ideal-gas law; dashed line is the ideal-gas result corrected by a Poynting factor; dash-dot line is the approximation (12.2.14) with the fugacity coefficient computed from the simple virial equation via (12.2.16). Figure after Chueh and Prausnitz [7].

state (12.2.15) [9]. At 76 K, $B_{22} = -12$ cc/mol [10], so unlike attractions between solute and solvent are stronger than those between solvent molecules. As pressure increases, the effect of this attraction on φ_1 overcomes the slight tendency of CH_4 to volatilize, so solubility increases. At higher pressures, where the third virial coefficient C becomes important, this decrease in φ with increasing P is reduced, because C is positive while B is negative.

For liquids in near-critical fluids the analysis is generally more complicated than the above development for solids because the equilibrium liquid phase is not pure. To the degree that the solvent dissolves in the liquid, the solubility of the solute can decrease. But such solutions may be nonideal, and if the corresponding activity coefficients are greater than unity, then they may partially compensate for the decrease in solubility.

Finally, we note that solubilities in near-critical fluids can often be enhanced by adding other components—so called *entrainers* or *cosolvents*. These components have B_{13} values that are more negative than B_{12}, further decreasing φ_i^{∞}, and thereby producing even larger values of x_1^f at high pressures.

12.3 INDEPENDENT VARIABLES IN STEADY-FLOW PROCESSES

Many kinds of material and chemical processes are carried out under steady flow conditions, and in analyzing such situations, just as in analyzing any situation, we must ensure that we have enough data to perform a proper analysis. This requires us to determine the number of independent variables required to close the problem. The presentation here builds on material presented in § 3.1, 3.6, 9.1, and 10.3.1.

12.3.1 Thermodynamic Stuff Equations

Analysis of steady-flow situations invoke the thermodynamic stuff equations, which apply to both equilibrium and nonequilibrium situations. The stuff equations include material balances, the energy balance, and the entropy balance. Material balances are essential when the number of inlets to a system differs from the number of outlets, when stream compositions change, and when chemical reactions occur. The energy balance expresses the first law for open systems and can be used to determine heat effects in workfree processes or work effects in adiabatic processes. The entropy balance expresses the second law for open systems and contains the heat but not the work. However, the entropy balance can be used in calculations only when we can quantify the entropy generation term. This can rarely be done in chemical processing situations, so we usually compute heat effects from just material and energy balances. After a process is fully evaluated, we may determine the value for the entropy generated and use it to compare efficiencies of alternative processes.

The differential forms of the stuff equations were given in § 3.6. For a system of C components, the material balances take the form

$$dN_i = \sum_{\alpha} dN_{\alpha i} - \sum_{\beta} dN_{\beta i} \qquad i = 1, 2, \ldots, C \tag{12.3.1}$$

where the lhs is the change in number of moles of component i accumulated in the system. Similarly, the energy balance can be written as

$$d\left(\sum_i N_i \overline{U}_i\right) = \Delta H_{\alpha\beta} + \delta Q + \delta W_{sh} \tag{12.3.2}$$

where the lhs is the change in total internal energy in the system and

$$\Delta H_{\alpha\beta} \equiv \sum_\alpha \sum_i \overline{H}_{\alpha i}\, dN_{\alpha i} - \sum_\beta \sum_i \overline{H}_{\beta i}\, dN_{\beta i} \tag{12.3.3}$$

In these equations, index α runs over all inlets to the system while β runs over all outlets. The energy balance (12.3.2) is a special form of (3.6.3) assuming negligible mass for the boundary and negligible kinetic and potential energy changes across the system. Note we have C material balances (12.3.1) but only one energy balance (12.3.2).

In the chemical processing industries, steady-flow systems are common, so the accumulation terms on the left sides of (12.3.1) and (12.3.2) are normally zero, and the rate forms of the balance equations can be used. Then the material and energy balances can be expressed as

$$\sum_\alpha \dot{N}_{\alpha i} - \sum_\beta \dot{N}_{\beta i} = 0 \qquad i = 1, 2, \ldots, C \tag{12.3.4}$$

$$\dot{Q} + \dot{W}_{sh} + \sum_\alpha \sum_i \overline{H}_{\alpha i}\, \dot{N}_{\alpha i} - \sum_\beta \sum_i \overline{H}_{\beta i}\, \dot{N}_{\beta i} = 0 \tag{12.3.5}$$

These constitute $(C + 1)$ equations that can be solved for $(C + 1)$ unknowns. In particular situations we may have additional equations available as a result of other constraints that apply, such as phase equilibrium.

12.3.2 Counting Numbers of Independent Variables

In § 3.6.2 we developed general expressions for determining the number of independent variables that apply to steady-flow processes involving only one phase and no reactions. We now extend those results to any number of phases and reactions. As in § 3.6.2, we are usually concerned with one of two quantities: either $\mathcal{V}$, the number of interactions available for changing the state, or $\mathcal{F}_{ex}$, the number of independent variables required to identify the final extensive state. The following discussion is obviously related to those in § 9.1 and § 10.3.1, but those presentations did not consider streams flowing through open systems.

As in § 3.6.2, we select the system to be a control volume that is open to steady-state energy and mass transfers. The system contains $\mathcal{P}$ homogeneous phases in which $\mathcal{R}$ independent chemical reactions are occurring. Material crosses system boundaries via N_p inlet and outlet streams; each stream is a one-phase mixture of C components. For

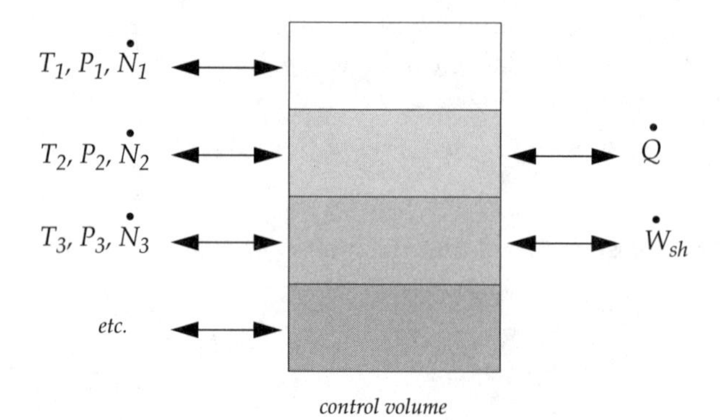

Figure 12.13 Schematic of an open system (a control volume) containing $\mathcal{P}$ phases in which are occurring $\mathcal{R}$ chemical reactions. The system has a total of N_p input and output streams through which material is flowing at steady state. It also has energy conduits to provide for steady-state shaft work and steady-state heat transfer.

steady flow we must have $N_p \geq 2$. Energy crosses system boundaries via thermal interactions and shaft-work modes; a schematic appears in Figure 12.13.

Interactions between system and surroundings are unaffected by the content or behavior of material within system boundaries. Therefore, the number of interactions available for changing the state, $\mathcal{V}$, is the same number as found in § 3.6.2,

$$\mathcal{V} = N_p(C+2) + 2 - S_{ext} \tag{3.6.12}$$

The state of each stream can be manipulated through C mole numbers, a thermal interaction, and a shaft-work mode; hence, the term $(C + 2)$. In addition, energy can enter or leave the control volume through a thermal interaction and a shaft-work mode; hence, the 2 on the rhs of (3.6.12). And we might impose external constraints by blocking S_{ext} interactions. If no external constraints are imposed, then we have the maximum number of available interactions,

$$\mathcal{V}_{max} = N_p(C+2) + 2 \tag{12.3.6}$$

But neither the number of phases nor the number of reactions in the control volume affect the number of material and energy conduits. Recall, this insensitivity of $\mathcal{V}$ to internal constraints also occurs for closed systems: (9.1.1) for $\mathcal{V}$ in closed multiphase systems is the *same* as (3.1.3) for closed one-phase systems.

To obtain the number of independent variables needed to identify the final state, $\mathcal{F}_{ex}$, we apply (3.6.14); that is, we subtract from $\mathcal{V}$ the number of internal constraints. Internal constraints include C material balances on the control volume, one energy balance on the control volume, and $(C + 2)(\mathcal{P} - 1)$ phase equilibrium relations. As discussed in § 10.3.1, chemical reactions introduce no additional constraints because each new constraint (a reaction equilibrium condition) is accompanied by a new variable (an extent of reaction). Therefore,

$$\mathcal{F}_{ex} = N_p(C+2) + 2 - S_{ext} - [C + 1 + (C+2)(\mathcal{P} - 1)] \tag{12.3.7}$$

and simplifying leaves us with

$$\mathcal{F}_{ex} = (N_p - \mathcal{P})(C+2) + 3 - S_{ext} \qquad \textit{equilibrium among } \mathcal{P} \textit{ phases} \qquad (12.3.8)$$

We emphasize that (12.3.8) assumes equilibrium is reached among the $\mathcal{P}$ phases. If equilibrium is not attained, then the phase-equilibrium relations do not contribute to (12.3.7), and instead of (12.3.8), we have

$$\mathcal{F}_{ex} = N_p(C+2) - (C-1) - S_{ext} \qquad \mathcal{P} \textit{ phases not in equilibrium} \qquad (12.3.9)$$

In either case, if we have values for $\mathcal{F}_{ex}$ process variables, then we can apply the constraint relations to compute values for the remaining variables contributing to the total $\mathcal{V}$. In the special case that the control volume contains only a single phase ($\mathcal{P} = 1$), then (12.3.8) reduces to (12.3.9), which is the same as (3.6.14).

Note that all quantities counted in $\mathcal{V}$, $\mathcal{F}_{ex}$, and S_{ext} are process variables: properties of streams and energy conduits that cross system boundaries—none are system properties. For example, the quantities counted in the external constraints S_{ext} are typically compositions, flow rates, or temperatures imposed on some streams; the absence of any shaft work applied to the control volume; or the absence of a thermal interaction between the control volume and its surroundings (adiabatic process). When chemical reactions occur, all compounds may not be present in all streams; for example, feed streams may contain only reactants. In such cases S_{ext} is increased by unity for each compound missing from each stream.

The number of properties $\mathcal{F}_{ex}$ specifies an extensive state. For example, the compositions of streams might be given by setting values for sets of mole numbers, such as $\{N_1\}$ for stream 1, $\{N_2\}$ for stream 2, etc. Alternatively, we can satisfy the number required by $\mathcal{F}_{ex}$ by giving an $\mathcal{F}'$-specification of intensive states plus an extensive *basis*. For example, we could set the mole fractions of the streams $\{z_1\}$ and $\{z_2\}$; set the relative amounts in streams, N_1/N_2; and choose a basis, such as $N_1 = 1$ mole.

12.3.3 Example

How many variables must be known to design or analyze an isothermal flash unit?

A binary mixture of components 1 and 2 is to be separated by an isothermal flash, as shown schematically in Figure 12.14. The chamber has a single feed composed of one-phase liquid and two product streams: an overhead vapor product and a bottoms liquid product. We choose the system to be the fluid in the flash chamber, so we have $C = 2$, $\mathcal{P} = 2$, and $N_p = 3$.

The maximum number of interactions available for changing the state is given by (12.3.6),

$$\mathcal{V}_{max} = N_p(C+2) + 2 = 3(2+2) + 2 = 14 \qquad (12.3.10)$$

These fourteen would be T, P, and mole numbers N_1 and N_2 for the three streams (total of twelve), plus any heat and shaft work that cross the boundary (two more).

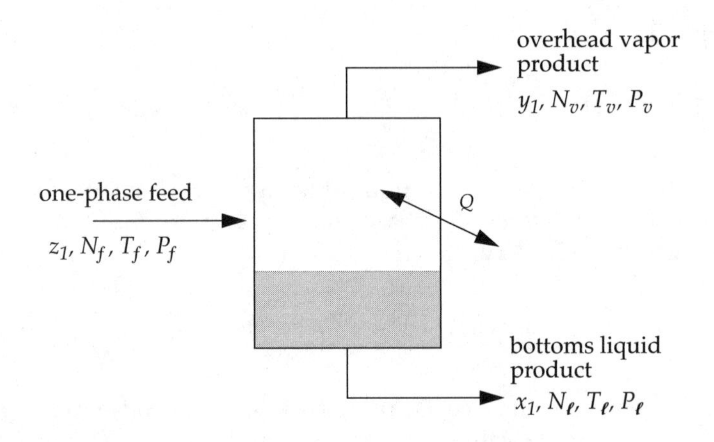

Figure 12.14 Schematic of an isothermal flash unit. Note that heat transfer is required to maintain constant temperature in the unit.

However, these fourteen are not all independent, so a smaller number of properties is sufficient to identify the state. That number is given by (12.3.8)

$$\mathcal{F}_{ex} = (N_p - \mathcal{P})(C+2) + 3 - S_{ext} = (2+2)(3-2) + 3 - S_{ext} = 7 - S_{ext} \quad (12.3.11)$$

We have three external constraints: (1) no shaft work ($W_s = 0$), (2) isothermal operation (outlet temperatures are same as temperature in the unit), and (3) isobaric operation (outlet pressures are same as in the unit). So (12.3.11) gives

$$\mathcal{F}_{ex} = 7 - 3 = 4 \quad (12.3.12)$$

Therefore, if we have values for four of the 14 variables and use the three external constraints, we can apply seven internal constraint relations to compute values for the remaining seven quantities. The seven internal constraint relations are (a) a material balance on component 1, (b) a material balance on component 2, (c) an energy balance on the system, (d) equality of liquid and vapor fugacities for component 1 in the unit, (e) equality of liquid and vapor fugacities for component 2 in the unit, (f) equality of liquid and vapor temperatures in the unit, and (g) equality of liquid and vapor pressures in the unit. The last four express chemical, thermal, and mechanical equilibrium between the vapor and liquid phases in the chamber.

Variations on this basic problem occur; for example, if the feed were a two-phase fluid, then the specification would require the relative amounts in the two phases. In some applications, feed and product streams may be split into multiple inlets and outlets, changing the value of N_p. Further, the problem situation may change depending on the particular four properties used to satisfy (12.3.12). Many such variations are possible, but two are common: process analysis and process design.

In a *process analysis* we are given values for the inputs to a process, and we must compute values for the process outputs. For example, in a process analysis of our flash situation, we might know values for properties of the feed stream (T_f, P_f, z_1, N_f) and we would need to compute property values for the product streams (x_1, y_1, N_v, N_ℓ) plus the heat duty Q. To solve this problem, we would solve the phase-equilibrium relations for x_1 and y_1, the material balances for N_v and N_ℓ, and the energy balance for the heat duty Q.

In a *process design* we are given values for the outputs from a process, and we must compute values for the process inputs. In the flash situation, we might know the product properties (x_1, y_1, N_v, and N_ℓ) and we would compute the feed properties (T_f, P_f, z_1, and N_f) plus the heat duty Q. But note that in a process design the equations to be solved—phase equilibrium plus material and energy balances—are exactly the same as those to be solved in a process analysis. Moreover, the value of $\mathcal{F}_{ex}$, the number of variables needed to close each problem, is also the same. Analysis differs from design only in the identities of knowns and unknowns.

Another type of design problem is *optimization*, in which we seek to adjust operating conditions to maximize or minimize an additional variable. For example, we might seek the feed composition that optimizes energy efficiency, where the efficiency is measured by the heat duty per mole of feed; that is, we seek to minimize Q/N_f. In these kinds of problems, additional nonthermodynamic equations and models may be involved. Nevertheless, although numerical values for computed results could differ, the thermodynamic description would be unaltered.

12.4 HEAT EFFECTS IN STEADY-FLOW PROCESSES

We now turn to processing situations in which heat effects are of primary importance; examples include chemical reactors and separators that exploit phase partitioning. Thermodynamic analysis of these situations invoke the stuff equations; in particular, steady-state heat effects are computed from (12.3.5). To obtain the partial molar enthalpies that appear in (12.3.5), we need enthalpies as functions of composition; so in § 12.4.1 we show how enthalpy-concentration diagrams can be constructed from volumetric equations of state applied to binary mixtures in phase equilibrium. Then we apply the energy balance (12.3.5) to multicomponent flash separators (§ 12.4.2), binary absorbers (§ 12.4.3), and chemical reactors (§ 12.4.4).

12.4.1 Enthalpy-Concentration Diagrams

Before energy balances can be used in the analysis or design of multicomponent flow processes, we must have data or correlations for mixture enthalpies as functions of composition. Such correlations can be developed from models for volumetric equations of state or from models for g^E; in the latter case, we would need a form for the temperature-dependence of the parameters in the g^E model. In this section we discuss how an equation of state can be used to compute an enthalpy-concentration diagram for a two-phase equilibrium situation; such diagrams contribute to the analysis or design of distillation columns.

To have a simple example, we consider an alkane(1) + aromatic(2) mixture, modeled by the Redlich-Kwong equation (8.2.1). Certain vapor-liquid phase diagrams for this mixture were displayed and discussed in § 9.3. Here our objective is to compute residual enthalpies for vapor and liquid that coexist in equilibrium; in particular, we want to construct an isothermal plot of h^{res} vs. x and y. (We will call this an *hxy* diagram, even though it is h^{res} that is actually plotted.) To do so, we set the temperature, pick a liquid composition x_1, and then perform a bubble-P calculation to obtain values

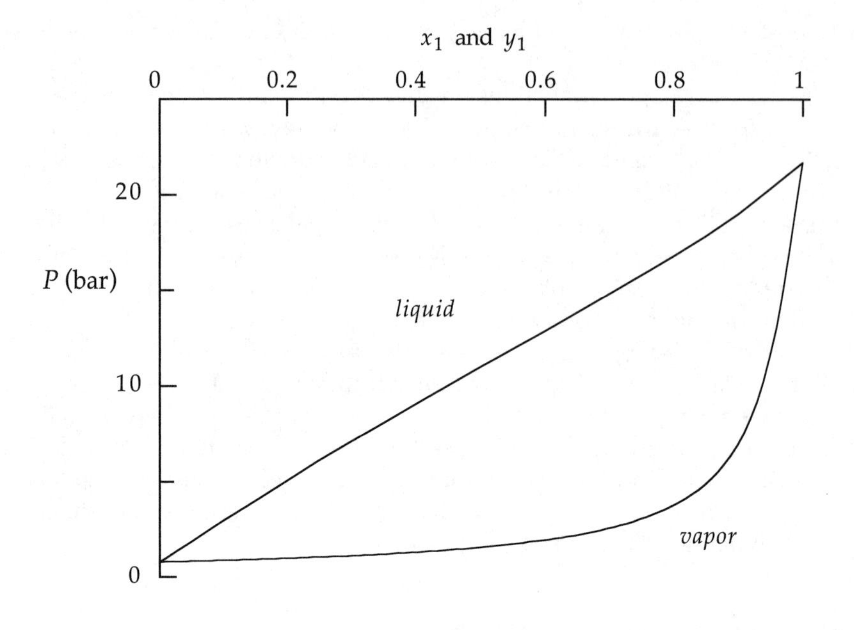

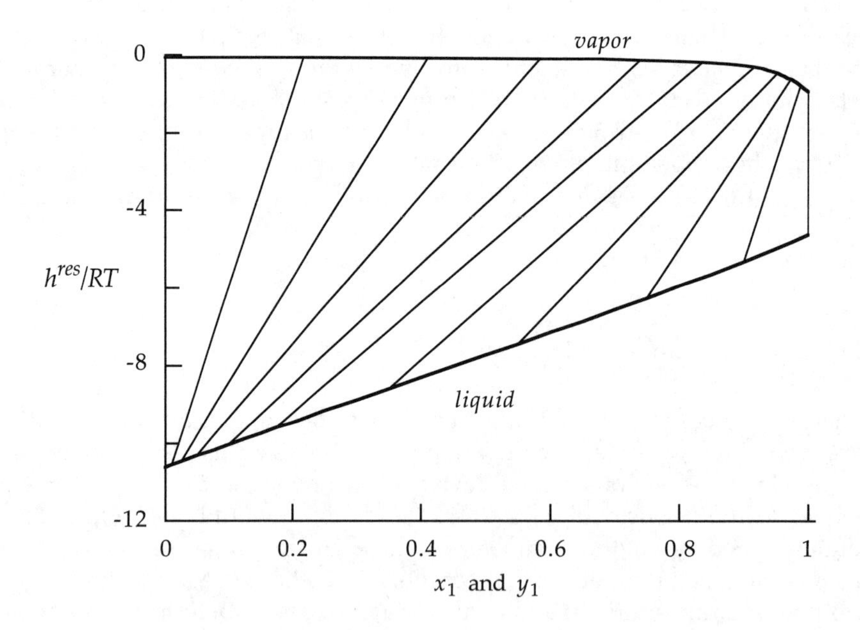

Figure 12.15 *Top*: Isothermal *Pxy* diagram for a binary mixture of an alkane(1) + an aromatic(2) at 330 K, computed from the Redlich-Kwong equation of state using the phi-phi method. This diagram is the same as in Figure 9.5. *Bottom*: The corresponding isothermal (residual) enthalpy-concentration diagram for the same mixture as at top, also computed from the Redlich-Kwong equation using (12.4.1). Note that differences in liquid and vapor h^{res} values at $x_1 = 0$ and at $x_1 = 1$ estimate the pure-component latent heats of vaporization.

for P and y_1. The calculation is done via the phi-phi method (§ 10.1.1) and involves only a slight modification of the bubble-T logic diagram that appears in Figure 11.1. With T, P, x_1, and y_1 known, we can compute h^{res} for each phase. A general expression for h^{res} can be obtained by using the Redlich-Kwong equation to evaluate the integral in (4.4.14) for u^{res}; then we use $h^{res} = u^{res} - Pv^{res}$. The result can be written as

$$\frac{h^{res}}{RT} = \frac{b}{v-b} - \frac{a}{RT^{1.5}}\left[\frac{3}{2b}\ln\left(\frac{v+b}{v}\right) + \frac{1}{v+b}\right] \qquad \textit{Redlich-Kwong} \qquad (12.4.1)$$

The mixing rules used here for a and b are those simple ones given in § 8.4.4. The molar volumes for each phase were computed from the equation of state at the known T and P using Cardan's method to solve the cubic (see Appendix C).

The resulting diagram is shown in Figure 12.15 for $T = 330$ K. Also included on the figure is the corresponding Pxy diagram. The computed hxy diagram shows that values of h^{res} for saturated liquid are roughly linear in x_1; likewise, the saturated vapor values are linear in y_1, except at high pressures when the vapor is rich in the alkane component. Tie lines connect two equilibrium phases, and the difference in h^{res} values at the ends of a tie line gives the latent heat of vaporization for a particular equilibrium situation. Similarly, the differences in h^{res} at $x_1 = 0$ and $x_1 = 1$ give the latent heats for the pure components. For these mixtures, more energy must be provided to flash an aromatic-rich liquid than to flash the same amount of alkane-rich liquid.

For completeness, we show in Figure 12.16 the residual Gibbs energies for the same mixtures and states appearing on the hxy diagram in Figure 12.15. These Gibbs energies were computed from the fugacity coefficients, with the help of (6.2.13),

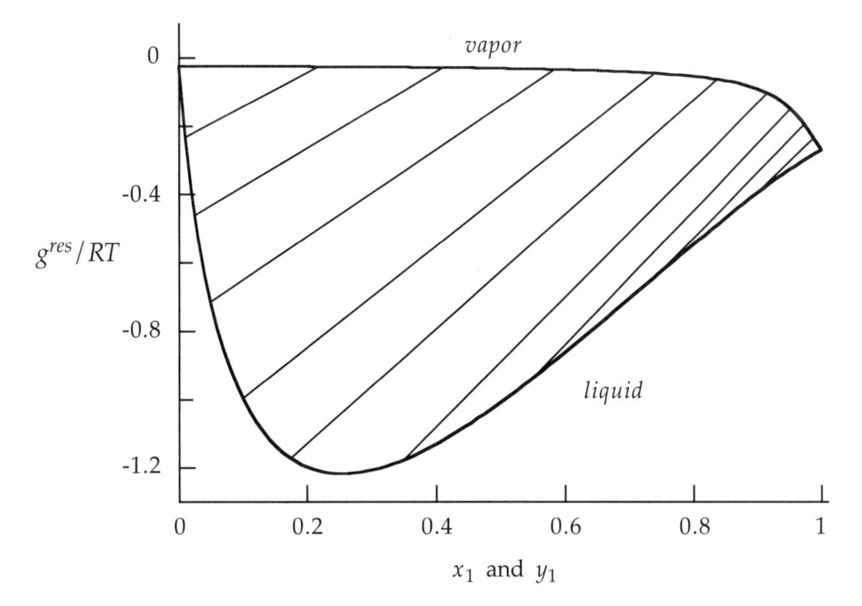

Figure 12.16 Isothermal residual Gibbs energies for the alkane(1) + aromatic(2) mixtures used in Figure 12.15. These curves were computed from the Redlich-Kwong equation of state at 330 K. Tie lines are the same as those on the hxy diagram in Figure 12.15.

$$\frac{g^{res}}{RT} = \frac{1}{RT}\sum_i x_i \bar{G}_i^{res} = \sum_i x_i \ln\varphi_i \tag{12.4.2}$$

Values for the fugacity coefficients were obtained from the Redlich-Kwong equation when the phi-phi method was being used to solve the bubble-P problem. Since P is not constant for the diagram in Figure 12.16 (P must change with x_1 if VLE is to be maintained at fixed T), there is no simple representation for the excess Gibbs energy g^E that corresponds to the diagram.

The tie lines on the g^{res} plot in Figure 12.16 are the same tie lines shown on the h^{res} plot in Figure 12.15. Note that each pure component has the same value of g^{res} for its vapor and liquid phases, but the two phases in VLE have different values for g^{res} (that is, the tie lines are not horizontal). For these mixtures, each liquid has a much more negative value of g^{res} than does the vapor at the same T and P; that is, compared to the vapor phases, much more work must be done to convert these liquids into ideal gases (see § 6.3.2). Note that since $P_1^s \approx 22$ bar at 330 K, the pure liquid alkane (component 1) is a hypothetical state at the lower pressures of the mixtures at this T; the phi-phi approach can readily handle this situation, but use of gamma-phi with a g^E model would be awkward.

12.4.2 Flash Separators

In many separation processes, heat effects are important because heat serves as the "separation agent" [11, 12]; that is, heat transfer to or from the system promotes creation of a new phase in which the compositions differ from those in the original phase. Examples include distillation and crystallization. In the analysis and design of such processes, thermodynamics can identify the direction of diffusion across phase boundaries, it can give us the equilibrium compositions of phases, and it can provide the direction and amount of heat to be transferred; but thermodynamics cannot give us any information about the rates of any of these processes.

In a multicomponent flash process, a single-phase feed is split into a vapor product and a liquid product, as in Figure 12.14. The three streams generally have different compositions, and the two product streams may or may not be in equilibrium with one another. Let the feed have temperature T_f, pressure P_f, composition $\{z\}$, and flow rate N_f. Similarly, let the vapor product have T_v, P_v, $\{y\}$, and N_v, and let the liquid product have T_ℓ, P_ℓ, $\{x\}$, and N_ℓ. Then the steady-state material balance (12.3.4) can be written for each component as

$$N_{if} = N_{iv} + N_{i\ell} \qquad i = 1, 2, \dots, C \tag{12.4.3}$$

while the steady-state energy balance (12.3.5) becomes

$$Q = h_v(T_v, P_v, \{y\})N_v + h_\ell(T_\ell, P_\ell, \{x\})N_\ell - h_f(T_f, P_f, \{z\})N_f \tag{12.4.4}$$

As a basis for the heat analysis, we choose one mole of feed, for then the material balances (12.4.3) can be expressed in terms of mole fractions. We also introduce $V = N_v/N_f$, the fraction of feed that flashes; then the material balances (12.4.3) become

$$z_i = y_i V + x_i(1-V) \qquad i = 1, 2, \ldots, C \tag{12.4.5}$$

and the energy balance (12.4.4) becomes

$$q \equiv \frac{Q}{N_f} = h_v V + h_\ell(1-V) - h_f \tag{12.4.6}$$

or

$$q = V(h_v - h_\ell) + (h_\ell - h_f) \tag{12.4.7}$$

To solve (12.4.5) and (12.4.7), we must choose how the enthalpies will be evaluated. We may use either residual enthalpies relative to the ideal gas, as in Chapter 4, or excess enthalpies relative to the ideal solution, as in Chapter 5. For consistency, we should use the same approach for computing enthalpies as we use to solve the phase-equilibrium problem:

(a) If we compute product compositions using the phi-phi method, then we should use residual enthalpies for both phases.

(b) But if we use a gamma-phi method, then we should use residual enthalpies for the vapor and excess enthalpies for the liquid.

First, we discuss the phi-phi approach to enthalpies, then we describe the gamma-phi method.

Phi-phi method. In this approach, we compute residual enthalpies for both the liquid and vapor using a volumetric equation of state. A form of the energy balance containing residual enthalpies can be obtained by combining (4.2.4) for the residual enthalpy, (3.3.22) for the response of h^{ig} to a change in T, and the energy balance (12.4.6); the result is

$$q = V\left[h_v^{res}(T_v, P_v, \{y\}) + \int_{T_f}^{T_v}\left(\sum_i y_i c_{p,\,\text{pure i}}^{ig}(T)\right)dT\right] \tag{12.4.8}$$

$$+ (1-V)\left[h_\ell^{res}(T_\ell, P_\ell, \{x\}) + \int_{T_f}^{T_\ell}\left(\sum_i x_i c_{p,\,\text{pure i}}^{ig}(T)\right)dT\right]$$

$$- h_f^{res}(T_f, P_f, \{z\})$$

In applying (12.4.8), three important cases arise, depending on the temperatures of the feed and product streams.

Case 1: $T_v = T_\ell = T_f$. When all three streams are at the same temperature, then the ideal-gas integrals in (12.4.8) vanish, and we are left with terms containing only residual enthalpies,

$$q = V(h_v^{res} - h_\ell^{res}) + (h_\ell^{res} - h_f^{res}) \tag{12.4.9}$$

This situation typically occurs in *isothermal flash* processes. Further, when equilibrium is achieved in an isothermal flash chamber, the product pressures are equal, $P_v = P_\ell$.

Case 2: $T_v = T_\ell \neq T_f$. In these processes, the two product streams are at the same temperature ($T \equiv T_v = T_\ell$), but the feed is at some other temperature. Now we have one less constraint than we had for isothermal operation, and therefore, we have one more independent variable than in case 1. In analyzing case-2 situations, we usually know both T_f and T; then the ideal-gas terms in the energy balance (12.4.8) combine to yield

$$q = Vh_v^{res}(T, P_v, \{y\}) + (1-V)h_\ell^{res}(T, P_\ell, \{x\}) - h_f^{res}(T_f, P_f, \{z\}) \tag{12.4.10}$$

$$+ \int_{T_f}^{T} \left(\sum_i [Vy_i + (1-V)x_i]\, c_{p,\,\text{pure i}}^{ig}(\tau) \right) d\tau$$

Using the material balance (12.4.5) in the ideal-gas term, (12.4.10) reduces to

$$q = Vh_v^{res} + (1-V)h_\ell^{res} - h_f^{res} + \Delta h^{ig} \tag{12.4.11}$$

where the ideal-gas term does not depend on the product compositions,

$$\Delta h^{ig} = \int_{T_f}^{T} \left(\sum_i z_i c_{p,\,\text{pure i}}^{ig}(\tau) \right) d\tau \tag{12.4.12}$$

The energy balance (12.4.11) is most often used in analyses of *adiabatic flash* units ($q = 0$); then we have

$$Vh_v^{res} + (1-V)h_\ell^{res} - h_f^{res} + \Delta h^{ig} = 0 \tag{12.4.13}$$

That is, the adiabatic flash problem is isenthalpic. Analyses of adiabatic flash problems are characterized by the constraint $q = 0$, which replaces the constraint T = constant of isothermal flash problems. Hence, the value for $\mathcal{F}_{ex}$ is the same for both adiabatic and isothermal flash problems.

The adiabatic energy balance (12.4.13) can be expressed in the form of a lever rule in enthalpies,

$$V = \frac{h_f^{res} - h_\ell^{res} - \Delta h^{ig}}{h_v^{res} - h_\ell^{res}} \tag{12.4.14}$$

This form of the energy balance, together with phase-equilibrium relations and material balances, can be applied to adiabatic flash units to obtain the fraction of feed that flashes V, the compositions of the products $\{x\}$ and $\{y\}$, and the temperature of the product streams T.

Case 3: $T_v \neq T_\ell \neq T_f$. When the feed and product streams all have different temperatures, then no further simplification of the rhs of the energy balance occurs, and the full form (12.4.8) must be applied to flash calculations. These situations typically occur when equilibrium is not attained in the flash chamber. This means an additional variable must be known to close the problem, usually a temperature of a product.

Gamma-phi method. Instead of phi-phi, we might choose to solve the VLE problem for a flash unit using the gamma-phi method. Then the energy balance (12.4.6) should be written in terms of residual enthalpies for the vapor and excess enthalpies for the liquid. The vapor product enthalpies would still be obtained from the form used in (12.4.8), but liquid phase enthalpies would now be computed by

$$h_\ell = h_\ell^E(T_\ell, \{x\}) + \sum_i x_i[\Delta h_{vap,i}(T_\ell) + h_{svi}^{res}(T_\ell) + h_{\text{pure i}}^{ig}(T_\ell)] \tag{12.4.15}$$

Here $\Delta h_{vap,i}$ is the latent heat of vaporization of pure i, h_{svi}^{res} is the residual enthalpy of pure saturated vapor i, and we have neglected any effects of pressure on liquid enthalpies. Assuming the feed is also one-phase liquid, then the energy balance is

$$\begin{aligned} q = {} & Vh_v^{res}(T_v, P_v, \{y\}) + (1-V)h_\ell^E(T_\ell, \{x\}) - h_f^E(T_f, \{z\}) \\ & + V\int_{T_f}^{T_v}\left(\sum_i y_i c_{p,\text{ pure i}}^{ig}(T)\right)dT + (1-V)\int_{T_f}^{T_\ell}\left(\sum_i x_i c_{p,\text{ pure i}}^{ig}(T)\right)dT \\ & + \sum_i [(1-V)x_i \Delta h_{vap,i}(T_\ell) - z_i \Delta h_{vap,i}(T_f)] \\ & + \sum_i [(1-V)x_i h_{svi}^{res}(T_\ell) - z_i h_{svi}^{res}(T_f)] \end{aligned} \tag{12.4.16}$$

This is a general form of the steady-state energy balance for flash calculations based on the gamma-phi method. Simplifications may occur, depending on the temperatures of the feed and product streams. For example, for an isothermal flash (12.4.16) reduces to

$$\begin{aligned} q = {} & Vh_v^{res}(T, P_v, \{y\}) + (1-V)h_\ell^E(T, \{x\}) - h_f^E(T, \{z\}) \\ & - V\sum_i y_i[\Delta h_{vap,i}(T) + h_{svi}^{res}(T)] \end{aligned} \tag{12.4.17}$$

Calculational procedures for using these gamma-phi forms are the same as those for using phi-phi forms.

12.4.3 Example

How do we determine the heat effect for a steady-state isothermal flash?

An isothermal flash unit is to separate a mixture of the alkane(1) and aromatic(2) components used in preparing the *hxy* diagram in Figure 12.15. Equimolar saturated liquid is fed to the unit at a steady rate of 2500 mole/min; the unit operates at $T_v = T_\ell \equiv T$ = 330 K and $P_v = P_\ell \equiv P = 5$ bar. According to § 12.3.3, these four values (z_1, N_f, T, P) are sufficient to close the problem.

The flash problem is workfree, has negligible kinetic and potential energy changes, and is a steady-state process, so the heat effect is given by the energy balance (12.4.7). In addition, the phase-equilibrium diagrams in Figure 12.15 were determined from the Redlich-Kwong equation of state using the phi-phi method; therefore, we choose to solve the energy balance using residual enthalpies, as in (12.4.9). For saturated liquid feed and the same pressure in the two product streams, the energy balance (12.4.9) becomes

$$q = V[h_v^{res}(T, P, \{y\}) - h_\ell^{res}(T, P, \{x\})] + [h_\ell^{res}(T, P, \{x\}) - h_f^{res}(T, P_f^s, \{z\})] \tag{12.4.18}$$

Here P_f^s is the saturation pressure for the liquid feed at T. To obtain the residual enthalpies of the products, we must obtain values for the product compositions by solving the phase-equilibrium problem at 330 K and 5 bar. To obtain the fraction of feed that flashes, we must solve a material balance on the flash unit.

Solutions to the phase-equilibrium problem are contained on the *Pxy* diagram in Figure 12.15. At 330 K and 5 bar, that diagram gives $x_1 = 0.2$ and $y_1 = 0.85$. With these mole fractions we apply a lever rule to obtain the fraction that flashes,

$$V = \frac{z_1 - x_1}{y_1 - x_1} = \frac{0.5 - 0.2}{0.85 - 0.2} = 0.545 \tag{12.4.19}$$

Taking the mole fractions $x_1 = 0.2$ and $y_1 = 0.85$ to the *hxy* diagram in Figure 12.15, we find the residual enthalpies of the vapor and liquid product streams:

$$h_v^{res}/RT = -0.2 \quad \text{and} \quad h_\ell^{res}/RT = -9.5 \tag{12.4.20}$$

Similarly, the enthalpy of the feed is obtained from the *hxy* diagram; for saturated one-phase liquid at $z_1 = 0.5$, we read

$$h_f^{res}/RT = -7.7 \tag{12.4.21}$$

Substituting all these results into (12.4.18), we find

$$q = 9.0 \text{ kJ/mol of feed} \tag{12.4.22}$$

The positive value means that this amount of heat must be *added* to the flash unit to maintain the temperature at 330 K. Since the process is workfree, this is the actual heat effect, regardless of any irreversibilities.

12.4.4 Binary Absorbers

Distillation is typical of those separation processes in which heat effects are large because heat transfer is used to create a new phase. But even in separations that do not use heat in this way, other heat effects, such as heats of mixing, can be significant. An example is a gas-liquid absorption column in which a gas-phase component is strongly absorbed into a liquid phase. In such separations, it is not unusual for the liquid to have a large negative value for h^E, so if the column is operated isothermally, then a large amount of heat must be removed, or if the column is operated adiabatically, then the liquid leaves the column at a much higher temperature than it entered. In isothermal columns, the problem is aggravated because large absorbers are difficult to cool; in adiabatic columns, the problem is aggravated when high liquid temperatures decrease the gas solubility in the liquid. Consequently, an absorber design that ignores heat effects tends to underdesign the column [13].

A typical gas-liquid absorber is shown schematically in Figure 12.17. The corresponding form for the steady-state energy balance (12.3.5) is

$$Q + h^a(T^a, \{y^a\})N^a + h^c(T^c, \{x^c\})N^c$$

$$- h^b(T^b, \{y^b\})N^b - h^d(T^d, \{x^d\})N^d = 0 \qquad (12.4.23)$$

In practice, differences in pressure across absorbers have small effects on stream enthalpies, compared to the effects of differences in temperature and composition, so pressure effects are ignored in (12.4.23). If the column is sufficiently tall and operated isothermally, then the inlet gas and outlet liquid may be assumed to be in equilibrium; but the energy balance (12.4.23) applies regardless of whether phase equilibrium is attained.

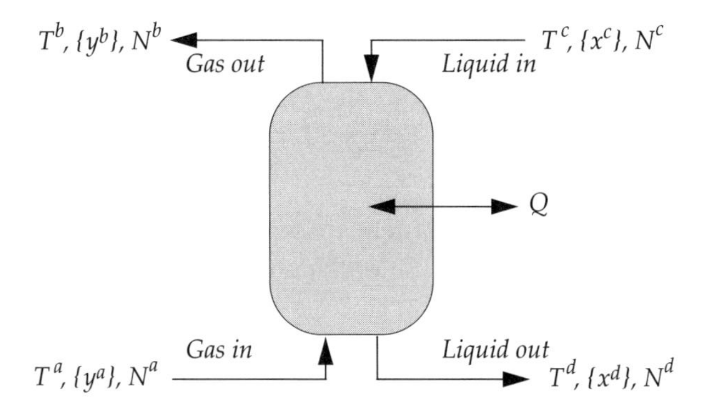

Figure 12.17 Schematic diagram for a typical gas-liquid absorption column

If the column is to be operated isothermally, then the typical problem is to use the energy balance (12.4.23), together with material balances and phase-equilibrium relations, to compute Q and the composition $\{x^d\}$ for the liquid leaving the column. If the column is to be operated adiabatically, then the typical problem is to determine both the temperature T^d and the composition $\{x^d\}$ at the liquid outlet. We illustrate both problems using absorption of ammonia from air into water; the following problem was originally analyzed by Sherwood and Pigford [13].

Problem statement. Ammonia(1) is to be removed from air(2) by contacting with water(3) in an absorption column. The column is to be operated adiabatically and at steady state; it can be represented schematically as in Figure 12.17, with superscripts a–d denoting the following streams: a ⇒ gas inlet, b ⇒ gas outlet, c ⇒ liquid inlet, and d ⇒ liquid outlet. The gas phase enters the column at 20°C and has $y_1^a = 0.416$. The column removes 99 mole % of the ammonia from the gas, and the gas phase leaves the column at 20°C. Pure water ($x_3^c = 1$) is fed to the column at 20°C.

We must determine the minimum flow rate needed for the water (N^c), as well as the temperature (T^d) and composition $\{x^d\}$ of the liquid leaving the column. We choose a basis of one mole of air entering ($N_2^a = 1$ mole).

Additional data. The gas and liquid do not reach equilibrium in the absorber, but the water flow rate is a minimum when the inlet gas and outlet liquid are in equilibrium. This means y_1^a and x_1^d are related by a K-factor, $K_1 = y_1^a / x_1^d$. For ammonia distributed between air and water the K-factor depends on T according to [13]

$$\ln K_1 = A - \frac{B}{T(K)} \tag{12.4.24}$$

where A = 14.481 and B = 4315.1. If the liquid temperature differs from that of the gas, then T in (12.4.24) would be the liquid value, because the mole fraction of air in the gas, y_2, is not sensitive to T; $y_2 \approx 1$ until the liquid approaches its boiling point.

A "heat of solution" q^s for N moles of water absorbing one mole of ammonia from air can be defined by

$$q^s \equiv h^d(T, x_1^d) - h^a(T) \tag{12.4.25}$$

At 20°C and $x_1 < 0.3$, this heat of solution depends on the mole fraction in the liquid according to [13]

$$q^s = (-35.2 + 8.1x_1^d) \text{ kJ/mol NH}_3 \tag{12.4.26}$$

Check whether we have a closed problem. We have $N_p = 4$ streams and $C = 3$ components (ammonia, water, and "air"). Since the phases do not reach equilibrium in the unit, $\mathcal{F}_{ex}$ is given by (12.3.9)

$$\mathcal{F}_{ex} = N_p(C+2) - (C-1) - S_{ext} = 4(3+1) - (-1) - S_{ext} = 18 - S_{ext} \tag{12.4.27}$$

We have the following external constraints:

(a)	Adiabatic and workfree	2
(b)	No pressure effect in any stream	4
(c)	No ammonia in liquid inlet	1
(d)	No air in liquid inlet or outlet	2
(e)	No water in gas inlet or outlet	2
(f)	Equilibrium between gas in and liquid out	1

So we have $\mathcal{S}_{ext} = 12$, and (12.4.27) gives $\mathcal{F}_{ex} = 6$: we need values for six quantities to close the problem. Here we have values for six: (1) composition of the gas inlet, $y_1^a = 0.416$; (2) amount of ammonia in the gas outlet, $N_1{}^b = 0.01\ N_1^a$; (3) temperatures of three streams, $T^a = T^b = T^c = 20°C$; (4) basis of $N_2^a = 1$ mole of air fed to the column.

Material balance on ammonia. A balance on ammonia around the column gives

$$N_1^a = N_1^b + N_1^d \tag{12.4.28}$$

Therefore the amount of ammonia removed from the gas stream is

$$N_1^d = N_1^a - N_1^b = N^a - N^b \tag{12.4.29}$$

This amount must be absorbed by the liquid, so

$$N_1^d = N^d - N^c \tag{12.4.30}$$

Dividing this by the total amount of liquid leaving the column, we have

$$x_1^d = 1 - \frac{N^c}{N^d} \tag{12.4.31}$$

This relates the unknown mole fraction x_1^d to the unknown liquid feed rate N^c, but we do not yet know the liquid discharge rate N^d.

However, we can solve for N_1^d, because we are given $y_1^a = 0.416$. From the definition of a mole fraction and our basis of $N_2^a = 1$, we can write

$$N_1^a = \frac{y_1^a}{1 - y_1^a} \tag{12.4.32}$$

We are also given the recovery, $N_1^b = 0.01 N_1^a$; therefore,

$$N_1^d = 0.99 N_1^a = 0.99\left(\frac{y_1^a}{1 - y_1^a}\right) = 0.7052 \text{ mol } NH_3 \tag{12.4.33}$$

Phase equilibrium. To get the minimum liquid flow rate, we are assuming the inlet gas stream is in equilibrium with the liquid outlet stream; this means the ammonia fractions in the two streams are related by a K-factor,

$$x_1^d = \frac{y_1^a}{K_1(T^d)} \tag{12.4.34}$$

where the temperature dependence of K_1 is given by (12.4.24). We have the value for y_1^a, but (12.4.34) is still one equation in two unknowns: x_1^d and T^d. As a second equation, we write the energy balance.

Energy balance. For our adiabatic column, the steady-state energy balance (12.4.23) becomes

$$h^a(T, y_1^a)N^a + h^c(T)N^c - h^b(T, y_1^b)N^b - h^d(T^d, x_1^d)N^d = 0 \tag{12.4.35}$$

where $T = T^a = T^b = T^c = 20°\text{C}$. The liquid feed is pure water, so $h^c = h_3(T)$, and with little error, we can assume that enthalpies for air-ammonia mixtures are independent of composition, $h^a \approx h^b$. Then (12.4.35) is

$$h^a(T)[N^a - N^b] + h_3(T)N^c - h^d(T^d, x_1^d)N^d = 0 \tag{12.4.36}$$

Now substitute (12.4.29) for the term in brackets and use (12.4.30) to eliminate N^d,

$$N_1^d[h^a(T) - h^d(T^d, x_1^d)] + N^c[h_3(T) - h^d(T^d, x_1^d)] = 0 \tag{12.4.37}$$

In the second term, we assume the presence of ammonia has a negligible effect on the enthalpy of the liquid, so we write that term as

$$h_3(T) - h^d(T^d, x_1^d) \approx h_3(T) - h_3(T^d) = c_p(T - T^d) \tag{12.4.38}$$

Here c_p is the heat capacity of liquid water, which is nearly constant over modest temperature changes. The term in the first bracket in (12.4.37) is (approximately) the negative heat of solution; so, using (12.4.25) and (12.4.38) in (12.4.37), the energy balance reduces to

$$-N_1^d q^s + N^c c_p(T - T^d) = 0 \tag{12.4.39}$$

Dividing through by N^d and using (12.4.31), we have

$$-x_1^d q^s + (1 - x_1^d)c_p(T - T^d) = 0 \tag{12.4.40}$$

The phase equilibrium relation (12.4.34) and the energy balance (12.4.40) constitute two equations that can be solved for the two unknowns: x_1^d and T^d. We can combine these two to eliminate T^d, leaving a single nonlinear question in one unknown:

$$T + \frac{xq^s}{c_p(1-x)} - \frac{B}{\ln(x) + A - \ln y} = 0 \tag{12.4.41}$$

where $x \equiv x_1^d$, $y \equiv y_1^a = 0.416$, $T = 293.15$ K, $c_p = 0.0754$ kJ/mol, A and B are given under (12.4.24). Solving (12.4.41) for x via Newton's method, we find $x \equiv x_1^d = 0.0838$. Then (12.4.33) gives the required water feed rate,

$$N^c = N_3^d = N_1^d\left(\frac{1-x_1^d}{x_1^d}\right) = 7.7 \text{ mol water} \tag{12.4.42}$$

Finally, we can solve (12.4.40) for the liquid discharge temperature: $T^d = 62°\text{C}$. The increase in water temperature means the process generates heat; since this process is adiabatic, that heat cannot cross system boundaries, so some of it warms the water.

Compare with isothermal operation. If this column were operated isothermally at 20°C, then (12.4.24) for the K-factor would immediately give the mole fraction for ammonia in the liquid outlet as $x_1^d = 0.528$ (cf. $x_1^d = 0.0838$ for adiabatic operation). Then instead of (12.4.42), the minimum water flow rate would be

$$N^c = N_3^d = N_1^d\left(\frac{1-x_1^d}{x_1^d}\right) = 0.63 \text{ mol water} \tag{12.4.43}$$

So, for the *same* rate of ammonia removed from the gas (0.7052 moles per mole of air in), the differences between isothermal and adiabatic operation include (i) an order of magnitude smaller flow rate for water in the isothermal column and (ii) a factor of six larger mole fraction of ammonia at the liquid outlet from the isothermal column. The 42° difference in temperature causes a much greater solubility of ammonia in the liquid from the isothermal column; however, a significant amount of heat must be removed to keep the temperature constant. This heat can be estimated from (12.4.26),

$$q = N_1^d q^s = -31 \text{ kJ/mol air in} \tag{12.4.44}$$

12.4.5 Chemical Reactors

The control of temperature in chemical reactors is important, not only because temperature affects conversion, but also because temperature strongly affects many reaction *rates*. In most reactor designs, attaining an economically viable reaction rate is more important than attaining the equilibrium conversion. Many reactions are exothermic, and heat may have to be removed to maintain the desired temperature; but some are endothermic, and then heat may have to be supplied. In either case, an industrial reaction often involves a substantial energy effect, and such an effect impacts both the design and operation of chemical reactors.

We consider here a steady-flow reactor that is supplied reactants through one inlet stream and that discharges products through a single outlet stream. In the stoichiometric approach, we have $\mathcal{R}$ independent reactions occurring among C species. Because of the reactions, a material balance on each species must now include generation and consumption terms,

$$N_i^{\alpha} - N_i^{\beta} + \sum_j^{\mathcal{R}} \nu_{ij}\, \xi_j = 0 \qquad i = 1, 2, \ldots, C \qquad (12.4.45)$$

Here superscript α indicates the inlet and β indicates the outlet. The stoichiometric coefficients ν_{ij} are < 0 for reactants (consumption) and > 0 for products (generation). In (12.4.45) the extent for reaction j can take any legitimate value; here, the ξ_j are not limited to their equilibrium values. Note that the number of moles for any component i is not necessarily conserved.

The corresponding steady-state energy balance can be written as

$$Q + N^{\alpha} h^{\alpha}(T^{\alpha}, P^{\alpha}, \{x^{\alpha}\}) - N^{\beta} h^{\beta}(T^{\beta}, P^{\beta}, \{x^{\beta}\}) = 0 \qquad (12.4.46)$$

This means the heat effect Q is simply the change in enthalpy across the reactor. However, we cannot obtain numerical values for the stream enthalpies h^{α} and h^{β} that appear in (12.4.46); we can only obtain values for enthalpy differences or values relative to some reference state. Further, we cannot simply form a difference $(h^{\alpha} - h^{\beta})$ in (12.4.46) because mole numbers are not necessarily conserved in chemical reactors; that is, we expect to have $N^{\beta} \neq N^{\alpha}$ [see (12.4.45)]. Therefore, we evaluate the enthalpy difference in (12.4.46) by creating a hypothetical process that connects the reactant state to the product state. Since heats of reaction are typically tabulated and correlated for reactions carried out in ideal-gas states at a reference temperature (see Chapter 10), we use a multistep hypothetical process involving ideal-gas reactions. Many different process paths can be legitimately constructed for this problem; the choice from among them is usually dictated by computational convenience and available data.

Our hypothetical process involves the five steps shown schematically in Figure 12.18. Since enthalpy is a state function, ΔH across the entire reactor is given by the sum of the enthalpy changes across the five stages of our process,

$$N^{\beta} h^{\beta} - N^{\alpha} h^{\alpha} = N^{\alpha} \Delta h_{12} + N^{\alpha} \Delta h_{23} + (N^{\beta} - N^{\alpha}) \Delta h_{34} + N^{\beta} \Delta h_{45} + N^{\beta} \Delta h_{56} \qquad (12.4.47)$$

We now identify computational forms for each of these five enthalpy changes.

Step 1-2. In this first step, the reactant mixture at $(T^{\alpha}, P^{\alpha}, \{x^{\alpha}\})$ is converted into an ideal-gas mixture at the same state. The enthalpy change is the negative residual enthalpy for the reactant mixture,

$$N^{\alpha} \Delta h_{12} = -N^{\alpha} h^{res}(T^{\alpha}, P^{\alpha}, \{x^{\alpha}\}) \qquad (12.4.48)$$

Step 2-3. In the second step, the temperature of the ideal-gas mixture is changed from T^α to a reference temperature T^o. The enthalpy change for this step is therefore merely a sensible heat effect,

$$N^\alpha \Delta h_{23} = \sum_i^C N_i^\alpha \int_{T^\alpha}^{T^o} c_{pi}^{ig}(T)\, dT \tag{12.4.49}$$

The value used for T^o is that at which values for ideal-gas heats of reaction are available, either from experiment or as computed from heats of formation. (See next step.)

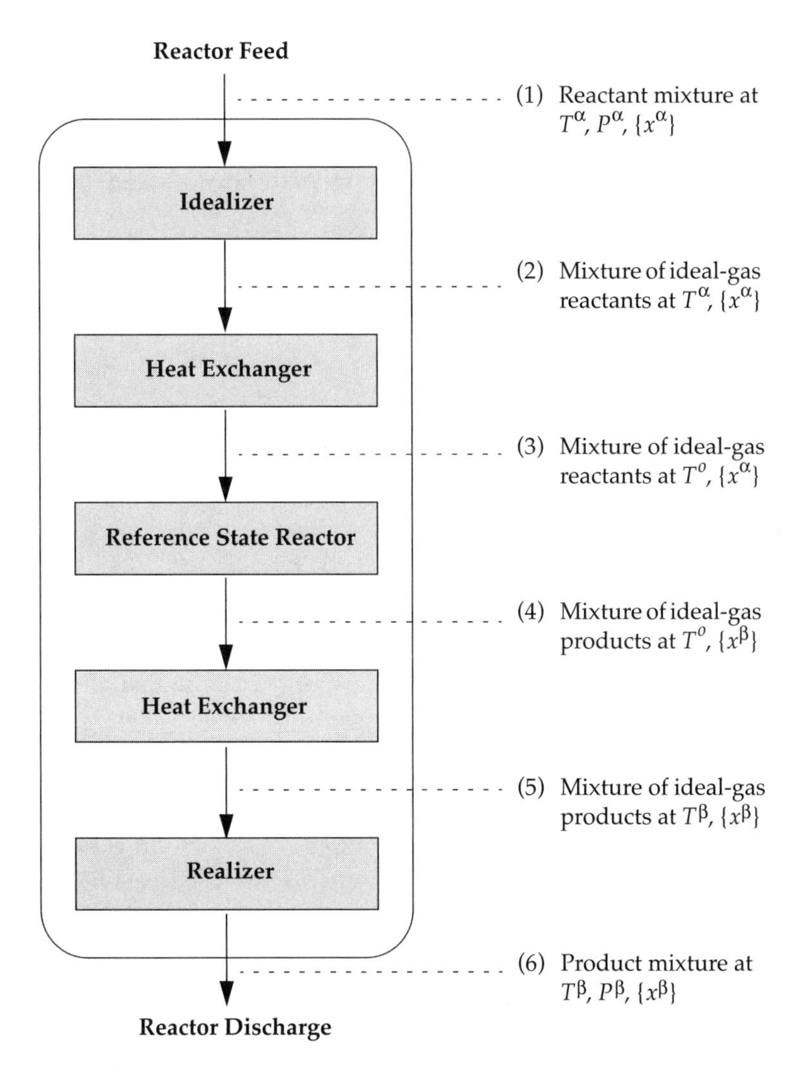

Figure 12.18 Schematic of a hypothetical thermodynamic process for determining heat effects associated with a chemical reactor. Since ideal-gas enthalpies are independent of pressure, we do not need to specify pressures for the ideal-gas states (2)–(5).

Step 3-4. At this point the reactions are allowed to proceed, converting the ideal-gas reactants into a mixture of ideal-gas products. The enthalpy change is the combined heats for all $\mathcal{R}$ reactions,

$$(N^{\beta} - N^{\alpha})\Delta h_{34} = \sum_{j}^{\mathcal{R}} \xi_j \, \Delta h_j^{ig}(T^o) \tag{12.4.50}$$

Often these heats of reaction are computed from heats of formation, as described in § 10.4.2,

$$(N^{\beta} - N^{\alpha})\Delta h_{34} = \sum_{j}^{\mathcal{R}} \xi_j \sum_{i}^{C} \nu_{ij} \, \Delta h_{if}^{ig}(T^o) \tag{12.4.51}$$

Step 4-5. Now the temperature of the ideal-gas product is changed from the reference temperature T^o to the final product temperature T^{β}; so again we have a sensible heat effect,

$$N^{\beta} \Delta h_{45} = \sum_{i}^{C} N_i^{\beta} \int_{T^o}^{T^{\beta}} c_{pi}^{ig}(T) \, dT \tag{12.4.52}$$

Step 5-6. Finally, the ideal-gas mixture is converted to a real mixture at the product state. The corresponding enthalpy change is the residual enthalpy for the product mixture,

$$N^{\beta} \Delta h_{56} = N^{\beta} h^{res}(T^{\beta}, P^{\beta}, \{x^{\beta}\}) \tag{12.4.53}$$

By combining (12.4.48)–(12.4.50) and (12.4.52)–(12.4.53) with the energy balance (12.4.47), we obtain a form of the steady-state energy balance that allows us to compute heat effects for chemical reactors. To obtain numerical values for quantities on the rhs in (12.4.48)–(12.4.53), we need an equation of state to obtain residual enthalpies, along with ideal-gas heat capacities and ideal-gas heats of reaction.

Note that the hypothetical process in Figure 12.18 makes no assumptions about the phases of the feed or discharge streams. If some species are in condensed phases, either in the reactant or product streams, the process in the figure still applies because the residual enthalpies would contain the energy effects associated with any phase changes that occur in going to and returning from the ideal-gas states. Note also that the process in Figure 12.18 can include inert components. Inerts merely have $\nu_{ij} = 0$ for all reactions j, so the heat effect for an inert is merely the sensible effect that changes its temperature from that of the reactants T^{α} to that of the products T^{β}. This comment also applies to unreacted portions of reactants, because they behave as inerts.

The form of the energy balance (12.4.47) is a general one that can be used for any number of species and reactions; it applies in a number of special cases, including the following important ones.

(a) For an isothermal reactor ($T^\alpha = T^\beta$) with feed composition and extents of reaction ξ_j known, we can solve the energy balance (12.4.47) to obtain the total heat effect Q. This would tell us whether heat must be supplied or removed to control the temperature to the desired value.

(b) For an adiabatic reactor ($Q = 0$) with a known feed state (T^α, P^α, $\{x^\alpha\}$) and known extents of reaction, we can solve the energy balance for the product temperature T^β.

(c) However, if equilibrium is attained in an adiabatic reactor, then the calculation is complicated because we must solve the reaction equilibrium equations (Chapter 11) simultaneously with the energy balance (12.4.47) to obtain the extents of reaction and the product temperature T^β.

12.5 RESPONSE OF SELECTED PROPERTIES

Thermodynamics is particularly useful in reducing the amount of experimental data needed for determining how properties respond to changes of state. Such changes could be illustrated using many thermodynamic properties; however, we will confine the discussion here to two important classes of quantities: standard-state fugacities (§ 12.5.1) and yields from chemical reactions (§ 12.5.2).

12.5.1 Standard-State Fugacities

In § 5.1 we defined a standard state to be a well-defined state of a real or hypothetical pure substance; therefore, we need only consider how the standard-state fugacity f_i^o responds to changes in temperature and pressure. General expressions for the temperature and pressure derivatives of f_i^o can be written immediately from (4.3.13) and (4.3.14) (or see Table 6.2); hence,

$$\left(\frac{\partial \ln f_i^o}{\partial P}\right)_T = \frac{\overline{V}_i^o}{RT} \tag{12.5.1}$$

and

$$\left(\frac{\partial \ln f_i^o}{\partial T}\right)_P = \frac{-\overline{H}_i^{o,\,res}}{RT^2} = \frac{-(\overline{H}_i^o - h_{\text{pure i}}^{ig})}{RT^2} \tag{12.5.2}$$

The pressure effect in (12.5.1) can arise only when we use FFF #2; in FFF #3–5, f_i^o is always evaluated at the standard-state pressure P_i^o. Further, on integrating (12.5.1) over a change of pressure, we obtain the Poynting factor, which appears in FFF #3. So our emphasis here is on how temperature affects f_i^o. In § 10.2 we identified two common classes of standard states: those based on the pure component and those based on infinitely dilute solutions. We consider those two choices here.

Pure-component standard states. When we take the standard state to be based on a pure component, then $f_i^o = f_{\text{pure i}}$, and the derivatives in (12.5.1) and (12.5.2) become

$$\left(\frac{\partial \ln f_i^o}{\partial P}\right)_T = \frac{v_{\text{pure i}}}{RT} \tag{12.5.3}$$

and

$$\left(\frac{\partial \ln f_i^o}{\partial T}\right)_P = \frac{-h^{res}_{\text{pure i}}}{RT^2} \tag{12.5.4}$$

Since $v/RT > 0$, f_i^o must always increase with an isothermal increase in pressure, no matter which pure-component standard state we use.

Turning to the temperature derivative in (12.5.4), we recall that the residual enthalpy can, in general, be either positive or negative. But for condensed phases it is invariably negative, and then f_i^o must increase with isobaric increases in T. For a liquid mixture at T and P, we usually take the standard state for component i to be pure saturated liquid i at T (hence, $P_i^o = P_i^s$). Then, if the mixture temperature changes, the standard state pressure also changes. In such cases, we need, not the isobaric derivative in (12.5.4), but that along the saturation curve (indicated by a subscript σ),

$$\left(\frac{\partial \ln f_i^o}{\partial T}\right)_\sigma = \left(\frac{\partial \ln \varphi_i^s}{\partial T}\right)_\sigma + \left(\frac{\partial \ln P_i^s}{\partial T}\right)_\sigma \tag{12.5.5}$$

$$= \frac{-(h^v_{\text{pure i}} - h^{ig}_{\text{pure i}}) + (h^v_{\text{pure i}} - h^\ell_{\text{pure i}})}{RT^2} \tag{12.5.6}$$

$$= \frac{-h^{res,\ell}_{\text{pure i}}}{RT^2} \tag{12.5.7}$$

So in this special case the derivatives in (12.5.4) and (12.5.7) are numerically equal.

At low temperatures (hence low vapor pressures), the saturated vapor is nearly an ideal gas ($\varphi_i^s \approx 1$), so our first approximation would be that f_i^o for liquids changes with T in the same way as the vapor pressure,

$$\left(\frac{\partial \ln f_i^o}{\partial T}\right)_\sigma \approx \left(\frac{\partial \ln P_i^s}{\partial T}\right)_\sigma \tag{12.5.8}$$

But at high temperatures (hence high vapor pressures), we need both φ_i^s and P_i^s in (12.5.5). Nevertheless, $\ln f_i^o(T)$ often remains nearly linear in $1/T$, as shown in Figure 12.19 for liquid water. Along a pure saturation curve we find $\varphi_i^s < 1$, so in general, we expect

$$f_i^o(T, P_i^s) < P_i^s(T) \tag{12.5.9}$$

If the mixture temperature T is above the critical temperature of pure i ($T > T_{ci}$), then no vapor pressure exists for i. However, if T is within about 10% of T_{ci}, then we

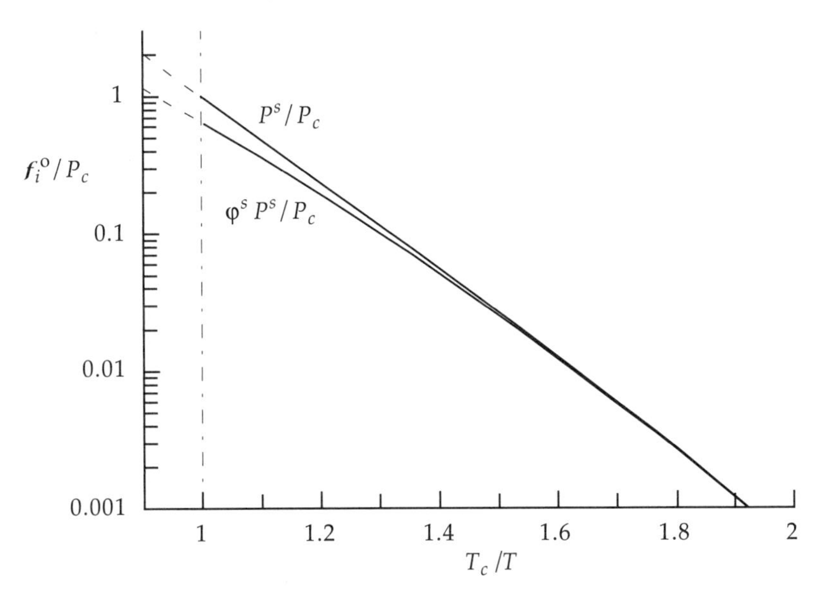

Figure 12.19 Effect of temperature on fugacity of a pure saturated liquid. Vapor-phase nonidealities (φ_i^s) lower f_i^o from the pure vapor-pressure curve, but the variation of f_i^o with $1/T$ remains roughly linear. At supercritical temperatures, pure vapor pressures do not exist; nevertheless, for ($0.9 < T_{ci}/T < 1$), we may choose the hypothetical pure liquid for the standard state and obtain a value of f_i^o by extrapolation. These values were computed for pure water using data from steam tables [14].

can still estimate f_i^o. To do so, we use the hypothetical pure liquid at T as the standard state and obtain a value for the standard-state fugacity by extrapolation. This procedure is indicated in Figure 12.19 for states having $0.9 \le T_{ci}/T < 1$.

Infinite-dilution standard states. When the standard state is based on an infinitely dilute solution, then the standard-state fugacity is a Henry's constant. In § 10.2 we introduced two kinds of Henry's constants for multicomponent mixtures: the solute-free form H_{is} and the reference-solvent form H_{ir}. For binary mixtures these two are the same. Here we use H_{is} to illustrate the response to changes in T and P; analogous expressions apply for H_{ir}.

When the solute-free Henry's constant is used as the standard-state fugacity, the derivatives in (12.5.1) and (12.5.2) become

$$\left(\frac{\partial \ln f_i^o}{\partial P}\right)_T = \left(\frac{\partial \ln H_{is}}{\partial P}\right)_T = \frac{\overline{V}_i^\infty}{RT} \tag{12.5.10}$$

and

$$\left(\frac{\partial \ln f_i^o}{\partial T}\right)_P = \left(\frac{\partial \ln H_{is}}{\partial T}\right)_P = \frac{-(\overline{H}_i^\infty - h_{\text{pure i}}^{ig})}{RT^2} \tag{12.5.11}$$

Don't confuse the Henry's constant, which appears on the lhs of (12.5.10) and (12.5.11), with the partial molar enthalpy, which appears on the rhs of (12.5.11).

Integrating (12.5.10) over a change in pressure yields a Poynting factor, which has already been displayed in (10.2.22) for H_{is} and in (10.2.30) for H_{ir}. In those Poynting factors, $\overline{V}_i^\infty$ may be positive or negative. For gases at temperatures below the solvent critical point, it is usually positive, and the Henry's constant increases with an isothermal increase in pressure. But as the solvent critical point is approached, $\overline{V}_i^\infty$ becomes very large, forcing f_i^o to diverge. In these situations, the divergence can be avoided by using FFF #4.

The effect of T is complex, because the difference in enthalpies on the rhs of (12.5.11) can be of either sign. Examples are shown in Figure 12.20, using Henry's constants from Figure 12.10, but now plotted versus $1/T$. The slopes of the lines in Figure 12.20 are given by

$$\left(\frac{\partial \ln H_{is}}{\partial(1/T)}\right)_P = \frac{\overline{H}_i^\infty - h_{\text{pure i}}^{ig}}{R} \tag{12.5.12}$$

At low temperatures ($T < T_{c1}$, where 1 = the gas solute), $1/T$ is large, and the numerator in (12.5.12) is negative, so the Henry's constant decreases with increasing $1/T$. At high temperatures ($T \gg T_{c1}$), $1/T$ is small, and the numerator is positive, so the Henry's constant increases with increasing $1/T$. At some intermediate temperature, generally above the solute critical temperature (T_{c1}), H_{is} passes through a maximum. Near solvent critical points, f_i^o may diverge because of its temperature dependence, similar to the divergence discussed under (12.5.11).

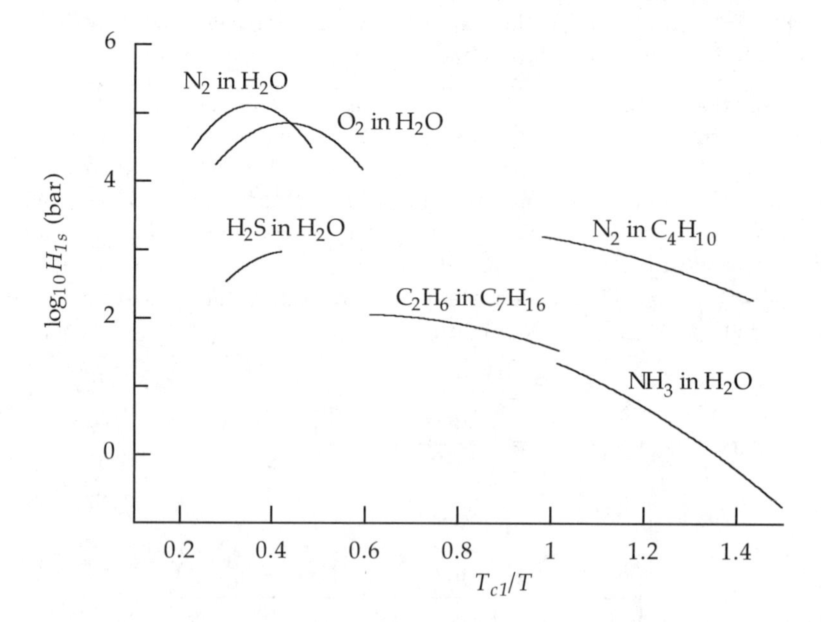

Figure 12.20 Effect of temperature on Henry's constants for several gases(1) in liquids(2). These are the same data as plotted in Figure 12.10, but here we plot $1/T$ on the abscissa to emphasize that the temperature dependence of a dilute-solution standard-state fugacity differs from that for a pure-component standard state; cf. this with Figure 12.19.

Note that the behavior of the standard-state fugacity in an infinite-dilution standard state differs qualitatively from that in a pure-component standard state (cf. Figure 12.19 with 12.20). Unlike (12.5.12), in which the rhs changes sign with T, the slope of the fugacity in the pure-component standard state (Figure 12.19) remains negative and roughly constant.

12.5.2 Response of Yields from Chemical Reactions

By controlling such operating variables as T, P, and amount of inerts, we can influence the rate and yield attained in equilibrium chemical reactors. Although thermodynamics cannot address issues of rates, it can indicate how such operating variables might be manipulated to improve product yields. To keep the following presentation simple, we consider only a single reaction written in the "forward" direction, so the equilibrium constant $K > 1$.

The equilibrium constant for one reaction was defined in (10.3.13) as

$$K = \prod_i^C a_i^{\nu_i} = \prod_i^C \left(\frac{f_i}{f_i^o}\right)^{\nu_i} \tag{12.5.13}$$

where the ν_i are stoichiometric coefficients: $\nu_i < 0$ for reactants and $\nu_i > 0$ for products. Here we use FFF #1 for the fugacities along with the gas-phase standard state ($f_i^o = 1$ bar) from § 10.4.1. Then (12.5.13) becomes

$$K(T) = \prod_i^C [x_i \varphi_i(T, P, \{x\}) P]^{\nu_i} = K_x K_\varphi P^\sigma \tag{12.5.14}$$

where $\sigma \equiv \sum_i^C \nu_i$, while

$$K_x \equiv \prod_i^C x_i^{\nu_i} \quad \text{and} \quad K_\varphi \equiv \prod_i^C \varphi_i^{\nu_i} \tag{12.5.15}$$

Rearranging (12.5.14), we have

$$K_x = K(T)\, K_\varphi^{-1}(T, P, \{x\})\, P^{-\sigma} \tag{12.5.16}$$

As the reaction proceeds, the extent ξ increases toward its equilibrium value, the product mole fractions increase, so we expect K_x to increase. Therefore, if we can manipulate operating variables to increase the equilibrium value of the extent, then we expect to also increase K_x, which measures conversion of reactants to products.

Pressure effects. For reactions completely carried out in condensed phases, changes in pressure have negligible effects on yields; that is, for liquids, φP in (12.5.14) is roughly constant unless the pressure change is very large. But for gas-phase reactions, pressure changes can be important. Recall that the equilibrium constant K does not depend on the reactor pressure P, so the pressure derivative of (12.5.15) gives

$$\left(\frac{\partial \ln K_x}{\partial P}\right)_T = \frac{-\sigma}{P} - \left(\frac{\partial \ln K_\varphi}{\partial P}\right)_T \tag{12.5.17}$$

Unless P is very high, the first term dominates the rhs and we have

$$\left(\frac{\partial \ln K_x}{\partial P}\right)_T \approx \frac{-\sigma}{P} \tag{12.5.18}$$

Since $P > 0$, the response of K_x to a change in pressure is determined by σ; that is, by whether the reaction increases or decreases the number of moles. If $\sigma < 0$, then the number of moles decreases during the reaction, and we increase conversion by increasing P. Inversely, if $\sigma > 0$, then the number of moles increases during the reaction, and we increase conversion by decreasing P. If $\sigma = 0$, then the number of moles is conserved by the reaction, and a change in pressure has little effect on conversion.

Temperature effects. At fixed pressure, changes in temperature affect both K and K_φ,

$$\left(\frac{\partial \ln K_x}{\partial T}\right)_P = \left(\frac{\partial \ln K}{\partial T}\right)_P - \left(\frac{\partial \ln K_\varphi}{\partial T}\right)_P \tag{12.5.19}$$

but the equilibrium constant K dominates the response. For example, for ideal gases the second term on the rhs is identically zero. In addition, K changes with T according to a Gibbs-Helmholtz equation (10.3.19), so (12.5.19) leads to

$$\left(\frac{\partial \ln K_x}{\partial T}\right)_P \approx \left(\frac{\partial \ln K}{\partial T}\right)_P = \frac{\Delta h^o}{RT^2} \tag{12.5.20}$$

Therefore, the response to a temperature change depends on the heat of reaction. For exothermic reactions, $\Delta h^o < 0$, so we increase K_x by decreasing T; but for endothermic reactions, $\Delta h^o > 0$, and we increase K_x by increasing T.

However, changes in T affect not only the equilibrium conversion, but also the reaction rate, and at least for elementary reactions, the rate increases with T (the law of Arrhenius). So in a proper design and operation of many reactors, the effects of temperature on equilibrium conversion must be balanced against the effects on rate. For example, to achieve economically viable rates, exothermic reactions are often performed at high temperatures, even though this decreases the equilibrium conversion.

Heats of reaction also influence the choice between adiabatic and isothermal reactors. When a reactant mixture of the same composition and temperature is fed to both an adiabatic reactor and an isothermal reactor, the equilibrium conversion is almost always less in the adiabatic reactor; this is true for both endothermic and exothermic reactions. Endothermic reactions performed in adiabatic reactors are accompanied by a fall in temperature, decreasing conversion. In such situations, we try to improve both the rate and the conversion by feeding reactants at high temperatures. But if a high temperature cannot be maintained in an adiabatic reactor, then we should consider adding heat and operating the reactor isothermally.

Similarly, exothermic reactions performed in an adiabatic reactor are accompanied by a rise in temperature, decreasing conversion but increasing the rate. In such situations we seek an economic balance between these competing effects, often by adjusting operating variables so that the heat generated is sufficient to sustain the desired rate. But if the high temperature reduces product yield too much, then we consider removing heat and operating isothermally. In some situations, a reaction may generate heat more quickly than it can be removed, causing localized hot spots to form in the reactor and leading to catastrophic failure of the reactor vessel.

Effect of inerts. The first-order effect of adding inerts to a reactor is to decrease the mole fractions of reactants and products by increasing the total number of moles N. But during a reaction, N may increase, decrease, or remain constant. So we separate N from the other terms in the definition of K_x (12.5.15),

$$K_x \equiv \prod_i^C x_i^{\nu_i} = \prod_i^C \left(N_i^{\nu_i} N^{-\nu_i} \right) = N^{-\sigma} K_n \tag{12.5.21}$$

where

$$K_n \equiv \prod_i^C N_i^{\nu_i} \tag{12.5.22}$$

Combining (12.5.21) with (12.5.16), we find

$$K_n = N^{\sigma} K_x = N^{\sigma} K K_{\varphi}^{-1} P^{-\sigma} \tag{12.5.23}$$

At fixed T and P, the response to a change in N is given by

$$\left(\frac{\partial \ln K_n}{\partial N} \right)_{TP} = \frac{\sigma}{N} - \left(\frac{\partial \ln K_{\varphi}}{\partial N} \right)_{TP} \tag{12.5.24}$$

which is usually dominated by the first term on the rhs,

$$\left(\frac{\partial \ln K_n}{\partial N} \right)_{TP} \approx \frac{\sigma}{N} \tag{12.5.25}$$

Since $N > 0$, the response is determined by σ. When $\sigma < 0$, the amount of product produced is decreased by adding inerts. But when $\sigma > 0$, the amount of product is increased by adding inerts. If $\sigma = 0$, adding inerts has little effect.

12.6 SUMMARY

We have used this chapter to illustrate how thermodynamics can contribute to the analysis and design of selected engineering processes. The applications considered here included calculations for phase equilibria, solubilities, heat effects in steady-flow processes, and the response of certain variables to changes of state.

Phase-equilibrium calculations were discussed for vapor-liquid equilibria (VLE), liquid-liquid equilibria (LLE), and solid-solid equilibria (SSE). Results from VLE calculations often take the form of K-factors and relative volatilities, especially when thermodynamic calculations serve as intermediate steps in computer-aided process-design programs. In those situations, K-factors are routinely provided to subprograms that size distillation columns and gas-liquid absorbers. Similarly, the distribution coefficients computed for LLE serve as bases for sizing solvent-extraction columns; moreover, liquid-liquid distribution coefficients may be helpful in screening candidate solvents for use in an extraction.

Solubility calculations are merely phase-equilibrium calculations applied to supercritical gases in liquids, solids in liquids, and solutes in near-critical fluids. The last application has drawn substantial attention, for near-critical extraction processes are being applied, not only in the chemical and energy industries, but also in food processing, purification of biological products, and clean-up of hazardous wastes.

In the section on heat effects, we emphasized how the steady-state energy balance can be used to design and analyze flash separators, absorption columns, and chemical reactors. In each application we developed a general form for the energy balance, and then we showed how it simplifies when it is applied to adiabatic and isothermal operations. We also noted that engineering calculations for process *design* involve the same quantities and the same equations as those for process *analysis*. Process design differs from process analysis only in the identities of the knowns and unknowns.

Finally, we ended the chapter by discussing how changes of state affect standard-state fugacities and yields from chemical reactions. These are important issues, but they also illustrate how thermodynamics can be used to answer such questions. For example, equilibrium yields from chemical reactions might be improved by changing temperature, changing pressure, or adding inerts; the considerations are as follows.

(a) For exothermic reactions, decrease T to increase equilibrium yield of product, but for endothermic reactions, increase T.

(b) If the total number of moles increases during a reaction, then the yield is increased by decreasing P, or by adding inerts, or both.

(c) If the total number of moles decreases during a reaction, then product yield is increased by increasing P, or by removing product as it forms, or both.

(d) If the total number of moles remains constant during the reaction, then yield is little affected either by changes in P or by addition of inerts.

A principal objective of this chapter has been to illustrate the kinds of problems that thermodynamics can address and the kinds of industrial situations to which it can be applied. But more importantly, in this chapter we have tried to show *how* thermodynamics should be applied. For example, our goal has not been merely to show that the energy balance can be applied to flash separators and chemical reactors, but rather we have tried to develop a procedure for applying the energy balance, so that you can use that procedure for any processing equipment—not just separators and reactors. Henri Poincaré once remarked that, in a logical development, the order in which elements are placed is much more important than the identities of the elements themselves [15]. In other words, the examples used here are not nearly so important as the *pattern* we followed in developing the examples. To reap full benefit from this chapter, study the patterns.

LITERATURE CITED

[1] M. Hansen, *Constitution of Binary Alloys*, McGraw-Hill, New York, 1958, p. 220.

[2] R. T. DeHoff, *Thermodynamics in Materials Science*, McGraw-Hill, New York, 1993.

[3] I. R. Krichevskii, N. M. Zhavoronkov, and D. S. Tsiklis, "Solubility of Hydrogen, Carbon Monoxide, and of Their Mixtures in Methanol Under Pressure," *Zhur. Fiz. Khim.*, **9**, 317 (1937).

[4] E. A. Campanella, P. M. Mathias, and J. P. O'Connell, "Equilibrium Properties of Liquids Containing Supercritical Substances," *A. I. Ch. E. J.*, **33**, 2057 (1987).

[5] G. R. Cysewski and J. M. Prausnitz, "Estimation of Gas Solubilities in Polar and Nonpolar Solvents," *IEC Fund.*, **15**, 304 (1976).

[6] C. Tsonopoulos, *Properties of Dilute Aqueous Solutions of Organic Solutes*, Ph.D. Dissertation, University of California at Berkeley, 1970; C. Tsonopoulos and J. M. Prausnitz, "Activity Coefficients of Aromatic Solutes in Dilute Aqueous Solutions," *IEC Fund.*, **10**, 593 (1971).

[7] P. L. Cheuh and J. M. Prausnitz, "Third Virial Coefficients of Nonpolar Gases and Their Mixtures," *A. I. Ch. E. J.*, **13**, 896 (1967).

[8] M. J. Hiza and R. N. Herring, "Solid-Vapor Equilibrium in the System Hydrogen-Methane," *Advan. Cryog. Eng.*, **10**, 182 (1965).

[9] D. J. Quiram, J. P. O'Connell, and H. D. Cochran, Jr., "The Solubility of Solids in Compressed Gases," *J. Supercrit. Fluids*, **7**, 159 (1994).

[10] J. H. Dymond and E. B. Smith, *The Virial Coefficients of Pure Gases and Mixtures*, Clarendon Press, Oxford, 1980.

[11] D. F. Rudd, G. J. Powers, and J. J. Sirola, *Process Synthesis*, Prentice-Hall, Englewood Cliffs, NJ, 1973.

[12] C. J. King, *Separation Processes*, 2nd ed., McGraw-Hill, New York, 1980.

[13] T. K. Sherwood and R. L. Pigford, *Absorption and Extraction*, McGraw-Hill, New York, 1952.

[14] J. H. Keenan and F. G. Keyes, *Thermodynamic Properties of Steam*, Wiley and Sons, New York, 1936; 21st printing, 1950.

[15] H. Poincaré, "Le Raisonnement Mathématique," in *Science et Méthod*, E. Flammarion, Paris, 1908; reprinted as "Mathematical Creation," in *The Creative Process*, B. Ghiselin (ed.), The New American Library, Inc., New York, 1952.

[16] W. B. Kay and F. M. Warzel, "Phase Relations of Binary Systems that Form Azeotropes. II. The Ammonia-Isooctane System," *A. I. Ch. E. J.* **4**, 296 (1958).

PROBLEMS

12.1 Consider a mixture of subcritical components in VLE. If FFF #1 is used for the vapor and FFF #4 is used for the liquid, write a completely general expression for the relative volatility α_{12} in terms of T, P, $\{x\}$, and $\{y\}$. List the quantities in your expression that must be modeled or approximated.

12.2 At 30°C and 1 bar, a certain binary liquid mixture has the following values for the activity coefficients at infinite dilution: $\gamma_1^\infty = 12.5$ and $\gamma_2^\infty = 2.5$. Do you expect this mixture to exhibit an azeotrope at 30°C? At this temperature the pure component vapor pressures are $P_1^s = 2.3$ bar and $P_2^s = 1.5$ bar.

12.3 The activity coefficients at infinite dilution for both components of a binary system can be described by the equation

$$\ln\gamma_1^\infty = \ln\gamma_2^\infty = 0.9 - 70/T(K)$$

At 350 K the ratio of saturation pressures is 1.3. Do you expect the mixture to have an azeotrope at 350 K? If so, estimate its composition. If not, give the bounds on the relative volatility α_{12} over the entire range of compositions.

12.4 Mixtures of diethyl ether ($C_4H_{10}O$)(1) and ethanol(2) are to be separated into essentially pure components by a distillation column operating at low pressure. Estimate the bounds on the relative volatility α_{12}. Assume the liquid mixtures obey the Margules correlation (5.6.11) with $A_1 = 0.1665 + 233.74/T(\text{K})$ and $A_2 = 0.5908 + 197.55/T(\text{K})$. Pure-component vapor pressures are in Appendix D.

12.5 (a) Making reasonable assumptions, estimate the mole fraction for oxygen dissolved in Lake Huron when the ambient temperature is 20°C.

(b) Early on a summer morning, a dense fog covers Lake Huron and the air temperature is 17°C. Estimate the mole fraction of water in the air immediately above the lake.

12.6 Of the three liquid materials, sedentone(1), rasaline(2) and thermolide(3), component 1 is miscible in 2 and 3, but components 2 and 3 are partially immiscible. At $P = 1$ bar the three binaries are quadratic mixtures with

$$\frac{g_{ij}^E}{RT} = x_i x_j \left[A_{ij} + \frac{B_{ij}}{T(K)}\right] \quad \text{and} \quad v_{ij}^E = x_i x_j C_{ij}$$

The parameter values are as follows:

Binary	A_{ij}	B_{ij} (K)	C_{ij} (cc/mol)
12	–0.1	100.	–10.
23	150.	5.	20.
13	–1.0	600.	15.

Determine the distribution coefficient at infinite dilution for component (1) distributed between the phases rich in (2) and rich in (3). Do the calculation twice: (a) at $T = 300$ K, $P = 0.1$ MPa and (b) at $T = 350$ K, $P = 10.0$ MPa. The pure component volumes are $v_1 = 15$, $v_2 = 120$, and $v_3 = 150$, all in cm^3/mol.

12.7 Accurately estimate the solubility of hydrogen(1) in benzene(2) at 450 K and 150 bar. The following data may be of some use.

(a) The molar volume of liquid benzene is $v_2 = 113.6$ cm^3/mol and its isothermal compressibility is $\kappa_T = 0.66$/MPa.

(b) Values of virial coefficients include $B_{11} = 10$ cm^3/mol, $B_{12} = 13$ cm^3/mol, $B_{22} = -535$ cm^3/mol, $C_{111} = 300$ cm^6/mol^2, and $C_{222} = 30{,}000$ cm^6/mol^2.

(c) Expressions for excess properties relative to Henry's law include

$$g^E/RT = 1.24x_1^2 \quad \text{and} \quad v^E = [172 - 310\,P\,(\text{MPa})]x_1^2 \text{ cm}^3/\text{mol}.$$

(d) At infinite dilution, the partial molar volume (in cm^3/mol) obeys

$$1/\overline{V}_1^{\infty} = 0.00943 + 0.0411P(\text{MPa})$$

(e) At 473 K the Henry's constant is $H_1(P_2^s) = 1.17$ MPa. At 450 K and P_2^s, the residual partial molar enthalpy at infinite dilution is

$$\overline{H}_1^{\infty} - h^{ig} = 3.227RT$$

(f) Although they could be computed, assume $y_1 = 0.09081$ and take the vapor volume to be 256.3 cm^3/mol.

(g) Vapor pressures for pure benzene can be obtained from Appendix D.

12.8 For a solid that does not dissociate in water, estimate the number of moles of the solid that must be added to 100 liters of water to lower the freezing point to –3°C. At 0°C the heat of fusion for pure water is about 6 kJ/mol.

12.9 A five kilogram sample composed of an equimolar mixture of nickel and gold is brought to equilibrium at 350°C.

(a) Use Figure 12.8 to estimate the amount of gold in the nickel-rich solid phase at 350°C.

(b) Calculate the amount of gold in the nickel-rich phase, assuming the Margules equations apply. Using T in Kelvin and nickel as component 1, the Margules parameters obey

$$A_1 = 3.1381 + 5.1269(10^{-3})T - 5.5015(10^{-6})T^2$$

$$A_2 = 7.1331 - 8.9209(10^{-3})T + 2.7109(10^{-6})T^2$$

12.10 Consider mixtures of ethane(1) in water(2) at 350 K. Show that the vapor mole fraction of water goes through a minimum when the pressure of the system is increased at constant temperature. Clearly state and justify all assumptions. The following data may be helpful (all at 350 K): $P_2^s = 0.0416$ MPa, $v_{\text{pure }2} = 18.5$ cm^3/mol, while the second virial coefficients take these values, $B_{11} = -129$, $B_{12} = -77$, and $B_{22} = -560$, all in cm^3/mol.

12.11 Derive (12.2.16), which gives the fugacity coefficient at infinite dilution based on the simple virial equation (12.2.15).

12.12 Estimate the infinite-dilution distribution coefficient C_1^∞ for naphthalene(1) distributed between immiscible phases of water(2) and benzene(3) at 300 K. Assume $\Delta C_p = 0.6\,\Delta S_m$ and that the binaries with naphthalene are described by the Porter equation. The following data may help; T_m is the melting point and values of the solubilities are given here in gm/100 gm solvent at 300 K.

Species	T_m (K)	Mol wt	$\frac{\Delta h_m}{RT}$	Solubility in H_2O	Solubility in C_6H_6
1	353	128	6.5	0.003	46.
2	273	18	2.6	...	7.0
3	279	78	4.3	0.17	...

12.13 (a) Consider a binary mixture in VLE at an azeotrope. Sketch a schematic *hxy* diagram for this system, as in the lower portion of Figure 12.15, and show the azeotropic tie-line.

(b) If instead of an azeotrope, the mixture had a critical point at $x_1 = 0.75$ and at the temperature of the diagram, sketch the corresponding isothermal *hxy* diagram.

12.14 Lipids are long chain (C_{12}–C_{30}) biological hydrocarbons that have a single functional group, such as a carboxylic acid, ester, or alcohol, at one end of the chain. Several methods can be considered for separating lipids from the plant or animal cells in which they are found. Candidate processes include these:

(i) Solvent extraction by mixing with an organic, such as dichloromethane or n-propanol, followed by evaporation of the solvent;

(ii) "Supercritical extraction" by contacting with high pressure CO_2, followed by decreasing the pressure to separate the CO_2; and

(iii) Distillation at very low pressures.

As usual, the choice of method depends on separability, which requires phase-equilibrium information. For each of the above methods (i)–(iii),

(a) Use material in this chapter to describe the basic phase-equilibrium problem involved and list the fundamental equation(s) governing the mole fraction of lipid (x_1) in the phase external to the cells.

(b) Give FFF and appropriate approximations for evaluating x_1.

(c) Indicate which of the quantities in (b) would be readily accessible (found in handbooks, etc.) and which would have to be obtained from models, with parameters either estimated or obtained by fitting data.

(d) For the quantities in your answer for (c), give the methods you would use to predict or correlate data. (For example, if you propose some measurements for obtaining activity coefficients, list the specific quantities to be measured and how you would obtain the model parameters.)

12.15 Twenty moles of saturated liquid ammonia at 1.38 MPa are to be mixed with 80 moles of compressed iso-octane, originally at 1.38 MPa and 328 K. The final mixture is to be one-phase liquid at 1.38 MPa. Determine the direction and amount of heat transferred in this process. For data, see Kay and Warzel [16].

12.16 (a) Use the Redlich-Kwong equation with the mixing rules in § 8.4.4 to compute an isothermal *hxy* diagram for binary mixtures of ethane(1) + propane(2) in VLE at 290 K.

(b) An isothermal flash unit is fed an equimolar mixture of liquid ethane and propane. If the unit operates at 18 bar, how much heat must be supplied (per mole of feed) to keep the temperature at 290 K?

12.17 The Rachford-Rice procedure, described in § 11.1.5 for LLE and in Problem 11.7 for VLE, is an example of a process *analysis* calculation. Using the equations from the Rachford-Rice procedure, develop an algorithm that could be used for the *design* of an isothermal VL flash process. That is, for given values of the product compositions $\{x\}$ and $\{y\}$ and vapor fraction V, your algorithm should determine the required feed composition $\{z\}$, together with the T and P to be maintained in the flash chamber.

12.18 Calculate the heat that must be removed from the isothermal reaction of iron and oxygen to form Fe_2O_3 at 291 K. The following data are available:

(a) The heat removed when 1 mole of iron is isothermally dissolved in dilute HCl to give dilute aqueous $FeCl_3$ plus H_2 (gas) at 291 K is 47.7 kJ.

(b) The heat removed when 1 mole of Fe_2O_3 is dissolved in dilute HCl to give dilute aqueous $2FeCl_3$ and H_2O is 155.9 kJ.

(c) The standard enthalpy of formation of liquid water at 291 K is –286.14 kJ.

12.19 How much heat per mole is required for continuous vaporization of the atmospheric azeotrope formed from jaypocus ogreate(1) and wahooic aggravate(2)? Use all and only these data:

$$\ln P_i^s = a_i + b_i/T \quad \text{and} \quad g^E/RT = x_1 x_2(0.5 + 500/T)$$

where P_i^s is in bar, T in K, $a_1 = 12.710$, $b_1 = -5289$ K, $a_2 = 9.29$, and $b_2 = -4289$ K.

12.20 A certain plant produces methanol by reacting H_2 with CO. In the process, stoichiometric portions of the reactants are compressed isothermally and reversibly at 400 K from 1 to 30 MPa. The compressor discharges the gas to a well-insulated catalytic reactor where methanol is formed to the equilibrium extent.

(a) If the surroundings are at 300 K, determine the amounts and directions of work and heat transferred during the compression step; use a basis of one mole of CO fed.

(b) Calculate the outlet temperature of the reactor as a function of the extent of reaction.

Ignore any pressure drop in the reactor and justify any assumptions about fluid ideality.

12.21 The rearrangement reaction A ↔ B occurs in both liquid and gas phases. If those phases are ideal, then show that computation of the equilibrium K-factor, $K_A = y_A/x_A$, requires only the equilibrium constants in the two phases. What if the phases were nonideal?

12.22 Ethylene glycol(1), $(CH_2OH)_2$, can be made from the reaction of carbon monoxide(2) with hydrogen(3) if a liquid-phase catalyst is available. One form of reactor could be a downflow column in which liquid glycol, containing the catalyst, contacts a cocurrent gas stream. The column should be long enough to allow equilibrium to be reached, with one of the reactant gases depleted.

(a) To maximize the extent of reaction at a given pressure $P > P_1^s$, should CO or H_2 be the limiting reactant?

(b) Estimate the outlet ratio of CO to H_2 in the vapor when the conditions are 0.507 MPa, 350 K.

(c) Would the equilibrium extent decrease or increase if N_2 were present in the gas stream?

12.23 Estimate the maximum amount of ethyl acetate(3) that could be obtained from an isothermal reactor that is fed 2 moles of ethanol(1) and 2 moles of acetic acid(2) at 370 K. Pure vapor pressures are given by expressions in Table 11.8. Properties of formation for pure saturated liquids at 298 K are as follows.

Species	$\Delta g_f^o/RT$	$\Delta h_f^o/RT$
EtOH	–67.8	–109
HAc	–160.	–201
EtAc	–131.	–189
H_2O	–95.6	–115

12.24 For a large insulated, countercurrent heat exchanger, estimate the number of moles of N_2 that can be continuously heated from 298 to 353 K by two moles of NO_2 entering at 373 K and leaving at 298 K if

(a) No reaction is assumed.

(b) Equilibrium association always occurs: $2NO_2 \rightarrow N_2O_4$ (Reference states are ideal gases at 298 K, 0.1013 MPa.).

Species	$\Delta g_f^o/RT$	$\Delta h_f^o/RT$	c_p^{ig}/R
NO_2	20.7	13.4	4.5
N_2O_4	39.5	3.70	9.5
N_2	0	0	3.5

AFTERWORD

AFTERWORD

You are part of a development group assigned to determine the properties and phase behavior of certain mixtures that are to be used in a new process for your company. Your supervisor is relying on the group to provide a quick and thorough assessment of the proposed process: each day of production delay costs the company one million dollars.

You begin by asking how the information will be used: Is it for exploratory research, conceptual design, process development, equipment sizing, troubleshooting? You next ask what processing steps are involved: reactions, separations, heating, cooling, pumping, expansions, recycles? And which steps could affect business decisions for commercialization: Are reaction yields limited by rates or by equilibrium conversions? Are separations hindered by formation of azeotropes or solutropes? If additional solvents are introduced, how will they be removed, so the product is not contaminated? Can any solvents be recycled to avoid disposal and waste? Finally, you ask precisely what properties are being requested. Are they compositions of phases in equilibrium? Densities and enthalpies of single phase liquids, gases, or solids? Reaction rate constants? In short, you must decide what properties are to be quantified and then decide how those values will be used: in appropriate hand calculations or in a process simulator.

At this preliminary stage, you may be tempted to skimp on the quality of property data, but then you remember that inadequate thermodynamic information can lead to improper designs and process failures. Material and energy stuff equations have always been used in the analysis and design of such unit operations as heat exchangers, reactors, separators, and compressors. But in earlier times, values for energies and enthalpies were obtained from approximate models. Similarly, in staged processes such as distillation, extraction, and chromatography, equilibrium is assumed at each stage, so models had to be used to obtain estimates for fugacities. Often those early models were limited by incomplete knowledge of conditions actually pertaining in the process, by lack of insight into how properties should be modeled, and by our inability to solve coupled sets of nonlinear algebraic and differential equations.

Furthermore, early models had to be simple because hand calculation was necessarily the mode, but now computer-based "process simulators" readily solve complex, multiparameter equations. Such simulators enable us to generate alternative "what-if" scenarios to study feasibility and optimization; they also allow us to probe the smallest details of process facilities and conditions. However, this powerful capability is limited by the approximations we provide to the simulator and by our interpretations of the output that the simulator provides to us. Casual, uncritical use of process simulators can obscure the significance of results and lead to process designs that are physically unrealizable. Therefore, you must give some attention to the accuracy with which property values will be needed and to the computational resources that will be required to achieve the required accuracy.

You and your coworkers begin to organize your plan of attack. You conclude that the process is not far enough along for detailed design; there are still issues about catalyst life and raw materials that are to be solved by others. Those problems could cause the project to be delayed or canceled. So at this point, feasibility is the most important concern. That is, in identifying the steps by which raw materials are to be converted into valuable products, two aspects come crucially into play: *feasibility,* in terms of Nature's possibilities, and *economic opportunity,* in terms of alternatives and human valuation. Inevitably a new product can be made in many different ways, but infeasible and uneconomic paths should be eliminated early in a design process, and thermodynamics can play an invaluable role in this endeavor.

Now you begin to ask about the conditions and species of the process: What are the temperatures and pressures relative to the melting and critical values? Are the compositions nearly pure or very dilute? Are the molecules of the substances likely to associate with themselves or solvate with others? Are there to be additional solvents and what are their properties? You also formulate relevant "always true" relations and you begin to consider appropriate models for nonideal gases and nonideal solutions. These questions can usually be answered by elementary analysis, shrewd selection of property formulation, reasonable estimates of property values, and thoughtful evaluation of the results. In this way you should be able to address and prioritize the overwhelming number of questions that always arise in a new undertaking.

For many decades, situations of this kind have been attacked by engineers in the energy and petrochemical industries. Those decades of experience teach us that such situations can be successfully resolved by appealing to fundamental thermodynamic relations among systems, properties, and interactions. But in contemporary society, even more complex situations are being posed by evolving technologies. For example, nanoscale systems involve such small numbers of molecules that we cannot ignore the effects of size distributions or of surface and edge effects. Biomolecular and other polymeric substances often respond to interactions by changing internal conformations. Electromagnetic fields are being used to manipulate the structure and dynamics of many systems. With modern instruments, the richness of Nature is becoming more obvious, stimulating searches for innovative means of exploiting energy, creating new materials, and modifying living processes.

The systems treated in detail in this text were limited in three ways. First, the systems were large enough that fluctuations of properties were not noticeable and that extensive properties were homogeneous to degree unity. Second, the interactions of a system with its surroundings were limited to the thermal interaction, the *PV* work mode, and material transfers. Third, formulations for modeling substances and mixtures implied that any nonideality of the system, whether relative to the ideal gas or the ideal solution, could be separated from that ideality, or else it could be accounted for by interactions among different molecules, rather than by changes in the molecules themselves.

These assumptions may be inadequate for many engineering situations of contemporary interest; however, this does not mean that the thermodynamic laws are invalid or that the basic methodology must be modified. For example, all of the operations in Chapter 3 remain valid, but the specifics will need to be adapted to treat complex cases. In particular, extensions must be made to include the effects of system size, additional work modes and their variables, and effects of molecular configuration, especially as density and composition change.

Although contemporary systems and processes may be complex, the techniques and the content of this book still apply. But to maximize the value of our approach, you may need to create new definitions, characterize other properties, consider additional interactions that influence complex systems, implement connections to molecular theory and statistical mechanics, and derive appropriate relations that are amenable to reliable modeling. In the past such characterizations were commonly done in terms of macroscopic measurables, but now molecular structure is being used to describe complex systems, including alternative-energy systems, biochemicals, colloids and interfaces, electrolytes, polymers, and exotic materials.

In this book you have been confronted with fundamental thermodynamics, and you have seen how that thermodynamics can be used to analyze traditional applications. Although you may encounter few traditional applications in your future professional life, the fundamentals still apply, so you should be able to deal successfully with contemporary applications as they arise. The fundamentals are permanent and universal, it is only the applications that go in and out of style.

APPENDICES

A

TOOLS FROM THE CALCULUS

Classical thermodynamics makes extensive use of the calculus; in fact, thermodynamics employs calculus so extensively that it is worthwhile to have a summary of the most important concepts. That summary is provided here. Throughout this appendix, as in all thermodynamics, we presume that functions such as $f(x)$ and $f(x, y)$ satisfy the required conditions of continuity and differentiability.

A.1 PARTIAL DERIVATIVES

Basic relations among thermodynamic variables are routinely stated in terms of partial derivatives; these relations include the fundamental equations from the first and second laws, as well as innumerable relations among properties. Here we define the partial derivative and give a graphical interpretation. Consider a variable z that depends on two independent variables, x and y,

$$z = f(x, y) \tag{A.1.1}$$

At a specified y-value the derivative of z with respect to (wrt) x, if it exists, is called the *partial derivative* of z wrt x; it is defined by

$$\left(\frac{\partial z}{\partial x}\right)_y = \lim_{\Delta x \to 0} \frac{f(x + \Delta x, y) - f(x, y)}{\Delta x} \tag{A.1.2}$$

where Δx represents a change in the value of x. The subscript on the lhs of (A.1.2) indicates that, during the operation, y remains fixed at a particular value. Likewise, we define the partial derivative of z wrt y, at a fixed value of x, by

$$\left(\frac{\partial z}{\partial y}\right)_x = \lim_{\Delta y \to 0} \frac{f(x, y + \Delta y) - f(x, y)}{\Delta y} \tag{A.1.3}$$

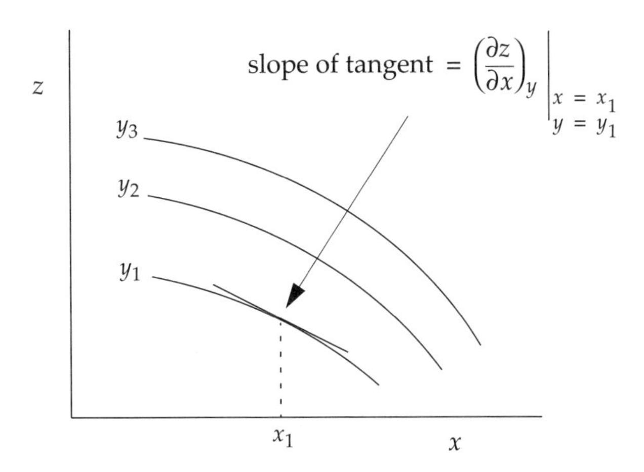

Figure A.1 The partial derivative in (A.1.2) can be interpreted as the slope of a y-level curve on a plot of z vs. x.

The quantity $z = f(x, y)$ can be represented on a plot of z vs. x, with y treated as a parameter. Then the partial derivative of z wrt x, evaluated at y_1, is the slope of the tangent to the y_1-level curve at any point x; this is illustrated in Figure A.1. In general the numerical value of the partial derivative changes when either x or y changes. Similarly, we could plot z vs. y with x treated as a parameter; then the partial derivative of z wrt y, evaluated at a particular x_1, is the slope of the tangent to the x_1-level curve at any point y.

A.2 TOTAL DIFFERENTIALS

General forms for property changes and balance equations are usually posed as total differentials. Consider a quantity z that depends on two independent variables,

$$z = f(x, y) \tag{A.1.1}$$

We would like to know how z responds when we change either x or y or both. We have already found that the partial derivative $(\partial z/\partial x)_y$ tells how z responds to a change in x at a fixed value of y; likewise, $(\partial z/\partial y)_x$ tells how z responds to a change in y at a fixed value of x. Then, if these partial derivatives are continuous, the total differential dz tells how z responds when we simultaneously change both x and y:

$$dz = \left(\frac{\partial z}{\partial x}\right)_y dx + \left(\frac{\partial z}{\partial y}\right)_x dy \tag{A.2.1}$$

If z depends on more than two independent variables, an additional term is added to (A.2.1) for each additional variable. The total differential (A.2.1) can be used to form other partial derivatives; for example, to express how z responds to changes in x with another quantity w held fixed, we use (A.2.1) to write

$$\left(\frac{\partial z}{\partial x}\right)_w = \left(\frac{\partial z}{\partial x}\right)_y + \left(\frac{\partial z}{\partial y}\right)_x \left(\frac{\partial y}{\partial x}\right)_w \tag{A.2.2}$$

Note that w might be a third independent variable or it might be some function of x and y.

A.3 IMPLICIT FUNCTION THEOREM

Thermodynamic descriptions of systems are often reformulated by exploiting connections among properties. One reformulation among differential properties can be obtained by applying the implicit function theorem. Consider some function

$$y = f(x) \tag{A.3.1}$$

then there must be some other function F such that

$$F(x, y) = 0 \tag{A.3.2}$$

that is, y can be considered as an implicit function of x. The total differential of F must vanish,

$$dF = \left(\frac{\partial F}{\partial x}\right)_y dx + \left(\frac{\partial F}{\partial y}\right)_x dy = 0 \tag{A.3.3}$$

That is, changes in x and y can occur only in ways that preserve $F = 0$, which in turn preserves the original functionality $y = f(x)$. Now, we could also consider

$$y = y(F, x) \tag{A.3.4}$$

so that

$$dy = \left(\frac{\partial y}{\partial F}\right)_x dF + \left(\frac{\partial y}{\partial x}\right)_F dx = \left(\frac{\partial y}{\partial x}\right)_F dx \tag{A.3.5}$$

Then (A.3.3) for dF becomes

$$dF = \left(\frac{\partial F}{\partial x}\right)_y dx + \left(\frac{\partial F}{\partial y}\right)_x \left(\frac{\partial y}{\partial x}\right)_F dx = 0 \tag{A.3.6}$$

and we obtain the implicit function theorem,

$$\left(\frac{\partial y}{\partial x}\right)_F = \frac{-\left(\frac{\partial F}{\partial x}\right)_y}{\left(\frac{\partial F}{\partial y}\right)_x} \quad \text{provided} \quad \left(\frac{\partial F}{\partial y}\right)_x \neq 0 \tag{A.3.7}$$

In thermodynamics the implicit function theorem is usually written in the form of a *triple product rule*:

$$\left(\frac{\partial y}{\partial x}\right)_F \left(\frac{\partial F}{\partial y}\right)_x \left(\frac{\partial x}{\partial F}\right)_y = -1 \tag{A.3.8}$$

Example A.3.1. For the function $y = x^2$ we can write

$$F(x, y) = x^2 - y = 0 \tag{A.3.9}$$

Then

$$\left(\frac{\partial F}{\partial y}\right)_x = -1 \qquad \text{and} \qquad \left(\frac{\partial F}{\partial x}\right)_y = 2x \tag{A.3.10}$$

Therefore,

$$\left(\frac{\partial y}{\partial x}\right)_F = \frac{-\left(\frac{\partial F}{\partial x}\right)_y}{\left(\frac{\partial F}{\partial y}\right)_x} = \frac{-2x}{-1} = 2x \tag{A.3.11}$$

which we can verify by explicit differentiation of the original function. ☒

A.4 EXACT DIFFERENTIAL EQUATIONS

In thermodynamics the concept of *state* is important because state functions have the special characteristics of exactness. In this section we show how to test whether a quantity is exact, hence, whether it is a state function. Consider a differential equation in two variables, x and y,

$$M(x, y)dx + N(x, y)dy = 0 \tag{A.4.1}$$

This equation is said to be exact, if there exists some function $u(x, y)$ such that

$$M = \left(\frac{\partial u}{\partial x}\right)_y \qquad \text{and} \qquad N = \left(\frac{\partial u}{\partial y}\right)_x \tag{A.4.2}$$

For then the original equation is the total differential of u,

$$du = M(x, y)dx + N(x, y)dy = 0 \tag{A.4.3}$$

and the solution of the differential equation (A.4.1) can immediately be written as

$$u(x, y) = \text{constant} \tag{A.4.4}$$

This solution identifies a family of level curves on a plot of y vs. x. If the partial derivatives M and N are continuous, then no two level curves can intersect; i.e., through every point (x, y) there passes only one level curve. There is, for example, no point (x, y) where $u = 2$ and simultaneously $u = 3$.

Changes in x and y produce

$$\Delta u = u(x_2, y_2) - u(x_1, y_1) \tag{A.4.5}$$

So a change in u depends on only the initial and final values of x and y, and it is independent of any intermediate x or y-values that might be visited during the change. The necessary and sufficient condition for exactness is that the second cross-partial derivatives be equal:

$$\left(\frac{\partial M}{\partial y}\right)_x = \left(\frac{\partial N}{\partial x}\right)_y \tag{A.4.6}$$

If a differential equation contains more than two independent variables, the necessary and sufficient conditions for exactness are that relations analogous to (A.4.6) must be satisfied by *every* pair of independent variables.

Example A.4.1. The differential equation

$$y\,dx + x\,dy = 0 \tag{A.4.7}$$

is exact because

$$\left(\frac{\partial M}{\partial y}\right)_x = \left(\frac{\partial y}{\partial y}\right)_x = 1 \tag{A.4.8}$$

and

$$\left(\frac{\partial N}{\partial x}\right)_y = \left(\frac{\partial x}{\partial x}\right)_y = 1 \tag{A.4.9}$$

The solution to the differential equation is therefore

$$u(x, y) = xy = \text{constant} \tag{A.4.10}$$

We may check this solution by forming the total differential and appealing to (A.4.7),

$$du = d(xy) = y\,dx + x\,dy = 0 \tag{A.4.11}$$

So $u = xy$ is indeed a constant; that is, to satisfy (A.4.7), changes in x and y can occur only in ways that preserve their product,

$$\Delta u = \int du = u(x_2, y_2) - u(x_1, y_1) = x_2 y_2 - x_1 y_1 = 0 \tag{A.4.12}$$

Example A.4.2. However, the differential equation

$$y dx - x dy = 0 \tag{A.4.13}$$

is not exact, as it stands, because

$$\left(\frac{\partial M}{\partial y}\right)_x = \left(\frac{\partial y}{\partial y}\right)_x = 1 \tag{A.4.14}$$

while

$$\left(\frac{\partial N}{\partial x}\right)_y = -\left(\frac{\partial x}{\partial x}\right)_y = -1 \tag{A.4.15}$$

But the equation (A.4.13) can be made exact by finding an integrating factor. ▣

A.5 INTEGRATING FACTORS

Most differential equations are not exact; however, in some cases, a multiplicative factor (called an *integrating factor*) can be found that transforms the equation into one that is exact. In thermodynamics this transformation can sometimes be used to convert heat or work into state functions.

Consider a linear, first-order ordinary differential equation. All such equations can be written in the form

$$\frac{dy}{dx} + yP(x) = Q(x) \tag{A.5.1}$$

In the form used in § A.4, this equation would appear as

$$[yP(x) - Q(x)]dx + dy = 0 \tag{A.5.2}$$

We seek an integrating factor $F(x)$ that makes this equation exact. Applying the factor we would have

$$F(x)[yP(x) - Q(x)]dx + F(x)dy = 0 \tag{A.5.3}$$

or

$$M(x, y)dx + N(x, y)dy = 0 \tag{A.5.4}$$

where $M = F(x)[y\,P(x) - Q(x)]$ and $N = F(x)$. For exactness we must have

$$\left(\frac{\partial N}{\partial x}\right)_y = \left(\frac{\partial M}{\partial y}\right)_x \tag{A.4.6}$$

Using our expressions for M and N, this becomes

$$\left(\frac{\partial F}{\partial x}\right)_y = F(x)P(x) \tag{A.5.5}$$

Separating variables,

$$\frac{dF(x)}{F(x)} = P(x)dx \tag{A.5.6}$$

or

$$F(x) = \exp[\int P(x)dx] \tag{A.5.7}$$

This identifies the functional form for the integrating factor. In practice the integration can be done using definite limits or indefinitely with the addition of an integration constant. If the integration is done indefinitely, then either the integration constant can be evaluated from a known reference condition or it can be absorbed into other constants that will appear in the solution to the differential equation.

With the above choice for $F(x)$, our scaled equation (A.5.3) is exact; i.e., it can be written in the form

$$du(x, y) = Q(x)\,F(x)\,dx \tag{A.5.8}$$

where

$$\left(\frac{\partial u}{\partial x}\right)_y = yP(x)\exp[\int P(x)dx] \tag{A.5.9}$$

and

$$\left(\frac{\partial u}{\partial y}\right)_x = \exp[\int P(x)dx] \tag{A.5.10}$$

So

$$u(x, y) = y\exp[\int P(x)dx] \tag{A.5.11}$$

and the solution to the original differential equation (A.5.1) is

$$y\exp[\int P(x)dx] = \int F(x)Q(x)\,dx = \text{constant} \tag{A.5.12}$$

Note that integrating factors are *not* unique. For example, let v be some function of x and y. If $F(x)$ is an integrating factor that reduces a differential equation to the form

$$dv(x, y) = 0 \tag{A.5.13}$$

then, $[f(v)F(x)]$ is also an integrating factor, where $f(v)$ can be chosen arbitrarily.

Example A.5.1. Consider the differential equation in Example A.4.2,

$$ydx - xdy = 0 \tag{A.5.14}$$

which is not exact as it stands. Writing it in the form

$$\frac{dy}{dx} - \frac{y}{x} = 0 \tag{A.5.15}$$

and comparing with (A.5.1), we identify

$$P(x) = -\frac{1}{x} \qquad \text{and} \qquad Q(x) = 0 \tag{A.5.16}$$

Then from (A.5.7), we have

$$F(x) = \exp\left[-\int \frac{dx}{x}\right] = \exp[\ln x^{-1}] \tag{A.5.17}$$

So the integrating factor is

$$F(x) = \frac{1}{x} \tag{A.5.18}$$

Applying this to (A.5.14), we find

$$\frac{y}{x}dx - dy = 0 \tag{A.5.19}$$

So (A.5.11) leads to

$$du = d\left(\frac{y}{x}\right) = 0 \tag{A.5.20}$$

and the solution is

$$\frac{y}{x} = \text{constant} \tag{A.5.21}$$

Further, not only is $(1/x)$ an integrating factor for (A.5.1), but so too is $[f(y/x)\,(1/x)]$, where f is any function. You may wish to verify this claim. ▣

A.6 LEGENDRE TRANSFORMATIONS

One way for obtaining new relations among thermodynamic variables is by applying a Legendre transform. Assume we have a curve represented by a set of points, $y = f(x)$, where the curve is convex for all x; i.e.,

$$\frac{d^2f}{dx^2} > 0 \tag{A.6.1}$$

Instead of the set of points, we could represent the same data by a set of tangents; we use s for a tangent to $f(x)$,

$$s \equiv \frac{df}{dx} \tag{A.6.2}$$

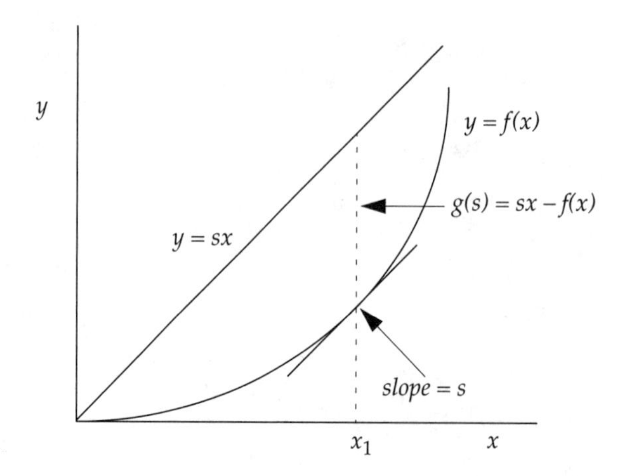

Figure A.2 Illustration of a Legendre transform from $f(x)$ to $g(s)$, where s is the slope of $f(x)$ at x_1 and $g(s)$ is the vertical distance from the point $f(x_1)$ to the line sx

The Legendre transform connects the point-representation of the data to the tangent-representation. To construct the transform, note first that we want to replace x with s as the independent variable. Since f is convex, each value of the slope s is unique. Further note that if we introduce a new independent variable, then we must also introduce a new function, call it $g(s)$. The Legendre transformation performs the mapping from the old function to the new one,

$$f(x) \rightarrow g(s) \tag{A.6.3}$$

To select the form for $g(s)$, we construct on the plot of $f(x)$ vs. x the line $y = sx$, where s is the slope of the tangent at some point x_1 on $y = f(x)$; see Figure A.2. Since $f(x)$ is convex, the vertical distance between the line and the curve is a maximum at x_1; i.e., that distance is single-valued. So we use that distance as the new dependent variable $g(s)$,

$$g(s) = sx - f(x) \tag{A.6.4}$$

This form for $g(s)$ is the Legendre transform commonly used in classical mechanics.

Note that we could as easily choose to define the new function, call it $h(s)$, by

$$h(s) \equiv f(x) - sx = -g(s) \tag{A.6.5}$$

This is the Legendre transform used in thermodynamics. Note that the total differentials dg and dh are simply related,

$$dg = sdx + xds - df = sdx + xds - sdx = xds = -dh \tag{A.6.6}$$

So

$$x = \frac{dg}{ds} = -\frac{dh}{ds} \tag{A.6.7}$$

That is, in the new space of g vs. s, the slope of the curve is the old independent variable x, while in the new space of h vs. s, the negative slope of the curve is x.

Example A.6.1. Say we have $f(x) = \exp[x]$, and we want to form a Legendre transform. Using (A.6.2), we define

$$s = \frac{df}{dx} = \exp[x] \tag{A.6.8}$$

So

$$x = \ln s \tag{A.6.9}$$

and therefore from (A.6.4) the Legendre transform is

$$h(s) = f(x) - sx = s(1 - \ln s) \tag{A.6.10}$$

▣

Example A.6.2. Newton's method as a Legendre transform. When we must solve a nonlinear algebraic equation of the form

$$f(x) = 0 \tag{A.6.11}$$

we often apply Newton's method: (a) guess a value for x, call it x_{old}, (b) test whether $f(x_{old})$ is tolerably close to zero, (c) if it is not, evaluate the slope at x_{old},

$$f' = \left.\frac{df}{dx}\right|_{x_{old}} \tag{A.6.12}$$

and (d) obtain a new guess, x_{new}. The new guess is formed by constructing the straight line through (x_{old}, f) that has slope f' and extrapolating that line to $f = 0$,

$$x_{new} = x_{old} - \frac{f}{f'} \tag{A.6.13}$$

If we rewrite this as

$$x_{new} = x_{old} - f\,\frac{dx}{df} \tag{A.6.14}$$

then we see that Newton's prescription has the structure of a Legendre transform. That is, we can interpret Newton's method as a transformation from the space (x_{old}, f) to the space $(x_{new}, 1/f')$.

Provided the slope f' is nonzero and finite, then it is clear from the transform that at the roots (where $f = 0$), the values of x_{old} coincide with those of x_{new}: the roots occur at fixed points of the Legendre transform. So in Newton's method, the original problem, "find values of x that make $f = 0$" is replaced with the equivalent problem, "find values of x that are invariant under a Legendre transformation."

Newton's method is most reliable when f is monotone in x; that is, when the slope f' is nonzero and has the same sign for all values of x. In such cases the Legendre transform produces a unique x_{new} for each x_{old}. Conversely, when the transformation is not unique, because $f(x)$ has an extremum or worse because $f(x)$ is oscillatory, then Newton's method is susceptible to convergence problems. ▣

A.7 EULER'S THEOREM ON HOMOGENEOUS FUNCTIONS

Many extensive thermodynamic properties are homogeneous in the mole numbers. This fact can be exploited through Euler's theorem to immediately integrate total differentials of such extensive properties. A function $f(x, y)$ is said to be homogeneous of order n if

$$f(\lambda x, \lambda y) = \lambda^n f(x, y) \tag{A.7.1}$$

where λ is an arbitrary constant scale factor. Then Euler's theorem states that for such a function, if its partial derivatives exist,

$$x\left(\frac{\partial f}{\partial x}\right)_y + y\left(\frac{\partial f}{\partial y}\right)_x = nf(x, y) \tag{A.7.2}$$

We can prove this by differentiating (A.7.1) wrt λ and setting $\lambda = 1$, since λ is arbitrary.

A.8 GIBBS-DUHEM EQUATION

In the thermodynamic description of multicomponent systems, a principal relation is the Gibbs-Duhem equation. Astarita [1] has shown that the Gibbs-Duhem equation is not merely a thermodynamic relation; it is a general repercussion of the properties of homogeneous functions. Consider a multivariant function, such as

$$F = F(\{N\}, \{Y\}) \tag{A.8.1}$$

such that F is homogeneous of degree one in the N_i and homogeneous of degree zero in the Y_i. In thermodynamics, the N_i could be mole numbers, and $\{Y\}$ could represent temperature and pressure. Since F is homogeneous, if all the N_i are changed by the same factor $\lambda \neq 1$ (while keeping all the Y_i fixed), then F changes by that same factor,

$$F(\{\lambda N\}, \{Y\}) = \lambda F(\{N\}, \{Y\}) \tag{A.8.2}$$

The change in F can be expressed as

$$\Delta F = \lambda F(\{N\}, \{Y\}) - F(\{N\}, \{Y\}) = (\lambda - 1)F(\{N\}, \{Y\}) \tag{A.8.3}$$

We now seek another expression for ΔF. Note that we can write ΔF as

$$\Delta F = \int_{F(\{N\})}^{F(\{\lambda N\})} dF \tag{A.8.4}$$

where

$$dF = \sum_i \bar{F}_i \, dN_i \tag{A.8.5}$$

and the $\bar{F}_i$ (the partial molar properties) are defined by

$$\bar{F}_i \equiv \left(\frac{\partial F}{\partial N_i}\right)_{\{Y\}, N_{j \neq i}} \tag{A.8.6}$$

Substituting (A.8.5) into (A.8.4), we have

$$\Delta F = \int_{N_i}^{\lambda N_i} \sum_i \bar{F}_i dN_i \tag{A.8.7}$$

Reverse the order of summation and integration, and note that the $\bar{F}_i$ are homogeneous of degree zero in the N_i. Then we can write

$$\Delta F = \sum_i \bar{F}_i \int_{N_i}^{\lambda N_i} dN_i = \sum_i \bar{F}_i (\lambda - 1) N_i \tag{A.8.8}$$

Equating (A.8.3) with (A.8.8) leaves

$$F = \sum_i \bar{F}_i N_i \tag{A.8.9}$$

which is the fundamental relation between an extensive property and its partial molar derivatives. The total differential of F can now be written as

$$dF = \sum_i \bar{F}_i dN_i + \sum_i N_i d\bar{F}_i \tag{A.8.10}$$

Comparing (A.8.5) with (A.8.10), we obtain the Gibbs-Duhem equation

$$\sum_i N_i d\bar{F}_i = 0 \qquad \text{fixed } \{Y\} \tag{A.8.11}$$

When the $\{Y\}$ change, (A.8.11) becomes

$$\sum_i N_i d\bar{F}_i = \sum_i N_i \sum_j \left(\frac{\partial \bar{F}_i}{\partial Y_j}\right)_{\{N\}, Y_{k \neq j}} dY_j \tag{A.8.12}$$

Reverse the order of summations on the rhs and take the N_i inside the derivative,

$$\sum_i N_i d\bar{F}_i = \sum_j \frac{\partial}{\partial Y_j}\left(\sum_i N_i \bar{F}_i\right)_{\{N\}, Y_{k \neq j}} dY_j = \sum_j \left(\frac{\partial F}{\partial Y_j}\right)_{\{N\}, Y_{k \neq j}} dY_j \tag{A.8.13}$$

This can be written as the generalized form of the Gibbs-Duhem equation,

$$\sum_i N_i d\bar{F}_i - \sum_j \left(\frac{\partial F}{\partial Y_j}\right)_{\{N\}, Y_{k \neq j}} dY_j = 0 \tag{A.8.14}$$

A.9 MEAN VALUE THEOREM

Changes in thermodynamic properties are computed by performing integrations; in some situations, the integration amounts to the evaluation of the mean for a continuous function. This mean is defined by the mean value theorem. If $f(x)$ is piecewise continuous on $[a, b]$, then there is some value of f, designate it by f_m, such that

$$f_m = \frac{1}{b-a}\int_a^b f(x)\,dx \tag{A.9.1}$$

The quantity f_m is called the *mean value* of f on the interval $[a, b]$.

This theorem simply says that the area under the curve $f(x)$ on the interval $[a, b]$ is identically the area of some rectangle of height h and width w:

$$hw = \int_a^b f(x)\,dx \tag{A.9.2}$$

If the area is known, then this is one equation containing two unknowns, h and w. Therefore, we can choose one arbitrarily. Let's choose the width w to be the length of the interval in x, as in Figure A.3,

$$w = |b-a| = b-a \qquad \textit{for } b > a \tag{A.9.3}$$

So (A.9.2) can be written

$$h[b-a] = \int_a^b f(x)\,dx \tag{A.9.4}$$

Then combining (A.9.4) with (A.9.1) identifies the height h as the mean value f_m.

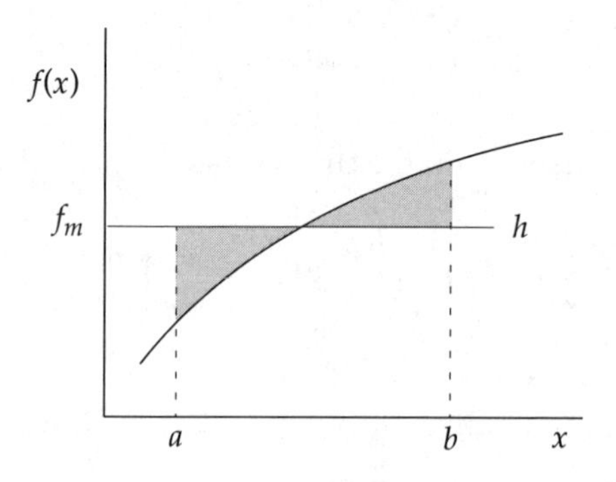

Figure A.3 The mean-value theorem for integrals states that the area under the curve $f(x)$ from a to b is equal to the area of a rectangle of width $w = (b-a)$ and height f_m. This requires that the two shaded areas be equal. The height f_m is called the *mean value* of $f(x)$ on the interval $[a, b]$.

A.10 TAYLOR SERIES

Many approximations used in modeling thermodynamic properties are based on the Taylor series. Examples are the virial expansions for the equation of state and the Redlich-Kister expansion of the excess Gibbs energy. Let $f(x)$ and all its derivatives be continuous and single-valued on $[a, b]$. Then the Taylor series provides an approximation to $f(b)$ if we know f at a nearby point $x = a$ and if we can evaluate derivatives of f at $x = a$,

$$f(b) = f(a) + (b-a)\left(\frac{df}{dx}\right)\bigg|_a + \ldots + \frac{(b-a)^n}{n!}\left(\frac{d^n f}{dx^n}\right)\bigg|_a \tag{A.10.1}$$

The first-order approximation represents a straight line through $f(a)$; this linear extrapolation is used in Newton's method, discussed in Example A.6.2. The second-order approximation represents a parabola through $f(a)$, and in general the nth-order approximation represents an nth-order polynomial through $f(a)$.

For a function that depends on two variables, $f(x, y)$, we may (under the same restrictions as above) perform a double Taylor expansion about a known point $f(a, c)$. For $a \le x \le b$ and $c \le y \le e$, the double expansion is

$$f(b, e) = f(a, c) + (b-a)\left(\frac{\partial f}{\partial x}\right)_y\bigg|_{a,\,c} + (e-c)\left(\frac{\partial f}{\partial y}\right)_x\bigg|_{a,\,c} \tag{A.10.2}$$

$$+ \frac{(b-a)^2}{2!}\left(\frac{\partial^2 f}{\partial x^2}\right)_y\bigg|_{a,\,c} + \frac{2(b-a)(e-c)}{2!}\left(\frac{\partial^2 f}{\partial x \partial y}\right)\bigg|_{a,\,c} + \frac{(e-c)^2}{2!}\left(\frac{\partial^2 f}{\partial y^2}\right)_x\bigg|_{a,\,c} + \ldots$$

A.11 LEIBNIZ RULE FOR DIFFERENTIATING INTEGRALS

Evaluation of some thermodynamic derivatives may require us to differentiate definite integrals. The general prescription for so doing was given by Leibniz. The problem is to find the general expression for dF/dx, when $F(x)$ is given by

$$F(x) = \int_{a(x)}^{b(x)} f(x, z)\, dz \tag{A.11.1}$$

In the notation of (A.11.1), the definition of the derivative in (A.1.3) appears as

$$\frac{dF}{dx} = \lim_{\Delta x \to 0} \frac{F(x + \Delta x) - F(x)}{\Delta x} \tag{A.11.2}$$

Substituting (A.11.1) into (A.11.2), we have

$$\frac{dF}{dx} = \lim_{\Delta x \to 0} \frac{1}{\Delta x}\left[\int_{a(x+\Delta x)}^{b(x+\Delta x)} f(x+\Delta x, z)dz - \int_{a(x)}^{b(x)} f(x, z)dz\right] \tag{A.11.3}$$

We divide the first integral into three parts,

$$\frac{dF}{dx} = \lim_{\Delta x \to 0} \frac{1}{\Delta x}\left[\int_{b(x)}^{b(x+\Delta x)} f(x+\Delta x, z)dz + \int_{a(x)}^{b(x)} f(x+\Delta x, z)dz \right. \tag{A.11.4}$$

$$\left. - \int_{a(x)}^{a(x+\Delta x)} f(x+\Delta x, z)dz - \int_{a(x)}^{b(x)} f(x, z)dz\right]$$

The second and fourth terms can be combined to give

$$\frac{dF}{dx} = \lim_{\Delta x \to 0} \frac{1}{\Delta x}\left[\int_{b(x)}^{b(x+\Delta x)} f(x+\Delta x, z)dz - \int_{a(x)}^{a(x+\Delta x)} f(x+\Delta x, z)dz \right. \tag{A.11.5}$$

$$\left. + \int_{a(x)}^{b(x)} [f(x+\Delta x, z) - f(x, z)]dz\right]$$

The first two integrals can each be simplified with the help of the mean value theorem: let $f(x, b)$ be the mean of the first integrand and $f(x, a)$ be the mean of the second. The third integral can be simplified by interchanging the limit with the integral and applying the definition of a derivative. Hence, we are left with

$$\frac{dF}{dx} = f(x, b) \lim_{\Delta x \to 0} \frac{1}{\Delta x}[b(x+\Delta x) - b(x)] \tag{A.11.6}$$

$$- f(x, a) \lim_{\Delta x \to 0} \frac{1}{\Delta x}[a(x+\Delta x) - a(x)] + \int_{a(x)}^{b(x)} \left(\frac{\partial f}{\partial x}\right)_z dz$$

Using the definition of the derivative for the first two terms, we obtain

$$\frac{dF}{dx} = f(x, b)\frac{db(x)}{dx} - f(x, a)\frac{da(x)}{dx} + \int_{a(x)}^{b(x)} \left(\frac{\partial f}{\partial x}\right)_z dz \tag{A.11.7}$$

This is the Leibniz rule for differentiating definite integrals. In those special cases in which one or both limits (a and b) are constants, independent of x, then (A.11.7) simplifies accordingly.

A.12 L'HOSPITAL'S RULE

Often we need to determine limiting values of thermodynamic properties. But sometimes, those limits appear indeterminate because the property is a ratio and both the numerator $f(x)$ and the denominator $g(x)$ either vanish or diverge as x reaches its limiting value. Such limits can still be obtained by applying l'Hospital's rule.

Consider two functions $f(x)$ and $g(x)$ that both vanish when $x = x_o$; that is, $f(x_o) = 0$ and $g(x_o) = 0$. We want to evaluate the ratio of these two functions in the limit as x approaches x_o,

$$\lim_{x \to x_o} \frac{f(x)}{g(x)} = ? \tag{A.12.1}$$

When x is near x_o, f and g can be reliably approximated by their Taylor expansions,

$$f(x) = f(x_o) + (x - x_o)\left(\frac{df}{dx}\right)\bigg|_{x = x_o} + O[(x - x_o)^2] \tag{A.12.2}$$

and

$$g(x) = g(x_o) + (x - x_o)\left(\frac{dg}{dx}\right)\bigg|_{x = x_o} + O[(x - x_o)^2] \tag{A.12.3}$$

We have $f(x_o) = g(x_o) = 0$; further, the higher-order terms in each expansion vanish more rapidly than the linear term, as x approaches x_o. Therefore, we obtain l'Hospital's rule,

$$\lim_{x \to x_o} \frac{f(x)}{g(x)} = \frac{\left(\frac{df}{dx}\right)\Big|_{x = x_o}}{\left(\frac{dg}{dx}\right)\Big|_{x = x_o}} \tag{A.12.4}$$

This states that if f and g both reach zero simultaneously, then the limit is determined by the *rates* at which f and g each approach zero.

If two other functions $u(x)$ and $v(x)$ each diverge as x approaches x_o, then to obtain the limiting value of their ratio, we apply l'Hospital's rule to the inverse ratio of their inverses,

$$\lim_{x \to x_o} \frac{u(x)}{v(x)} = \lim_{x \to x_o} \frac{(1/v(x))}{(1/u(x))} = \frac{\left(\frac{dv^{-1}}{dx}\right)\Big|_{x = x_o}}{\left(\frac{du^{-1}}{dx}\right)\Big|_{x = x_o}} \tag{A.12.5}$$

LITERATURE CITED

[1] G. Astarita, "Thermodynamics: A View from Outside," *Fluid Phase Equil.*, **82**, 1 (1993).

B

ELEMENTS OF LINEAR ALGEBRA

Linear algebra provides a notation for concisely representing linear algebraic equations and it provides a set of operations by which those equations can be solved. Linear equations arise in many common situations, such as taking inventories by material balances. As a particular example, consider a closed system initially loaded with N total moles of three liquid components: 1, 2, and 3. That initial mixture had mole fractions z_1, z_2, and z_3. When equilibrium is attained, the system is found to have divided into three phases: α, β, and γ. The mole fractions for phase α are $\{w_i\}$, those in phase β are $\{x_i\}$, and those in phase γ are $\{y_i\}$. These mole fractions are related to the original overall mole fractions z_i by material balances:

$$w_1 N^{\alpha} + x_1 N^{\beta} + y_1 N^{\gamma} = z_1 N \tag{B.0.1}$$

$$w_2 N^{\alpha} + x_2 N^{\beta} + y_2 N^{\gamma} = z_2 N \tag{B.0.2}$$

$$N^{\alpha} + N^{\beta} + N^{\gamma} = N \tag{B.0.3}$$

Here N^{α}, N^{β}, and N^{γ} are the total amounts in each phase. A typical problem is to use known values of the mole fractions and the overall amount N to solve (B.0.1)–(B.0.3) for the values of N^{α}, N^{β}, and N^{γ}. For three equations in three unknowns, a primitive calculational procedure may be adequate, but for more—say 20 equations in 20 unknowns—we seek a systematic calculational procedure that can be implemented on a computer. In what follows, we first develop the notation, then we introduce operations for solving a system of equations like that in (B.0.1)–(B.0.3).

B.1 MATRICES

A matrix $\mathbf{A}$ is a two-dimensional array of elements a_{ij} arranged in a particular order. In general, the array has m rows ($i = 1, 2, \ldots, m$) and n columns ($j = 1, 2, \ldots, n$), with the elements ordered as follows:

$$\mathbf{A} = \begin{bmatrix} a_{11} & a_{12} & a_{13} & \cdots & a_{1n} \\ a_{21} & a_{22} & a_{23} & \cdots & a_{2n} \\ \cdots & \cdots & \cdots & \cdots & \cdots \\ \cdots & \cdots & \cdots & \cdots & \cdots \\ a_{m1} & a_{m2} & a_{m3} & \cdots & a_{mn} \end{bmatrix} \tag{B.1.1}$$

For any element a_{ij}, the first subscript i indicates the row, while the second subscript j indicates the column. The *dimension* of **A** is identified by the total number of its rows and columns; that is, a general matrix **A** has dimension $(m \times n)$. If the number of rows is the same as the number of columns $(m = n)$, then the matrix is *square*, and n is said to be the *order* of the square matrix. A matrix having more than one row and one column will be indicated by an upper-case, bold face, sans serif character, such as **A**.

If an array has only one column $(m \times 1)$ or one row $(1 \times n)$, then it is a *vector*. We will use an unembellished, bold face, lower case character to represent a *column vector*,

$$\mathbf{x} = \begin{bmatrix} x_1 \\ x_2 \\ \cdots \\ x_m \end{bmatrix} \tag{B.1.2}$$

An object that has only one row and one column (1×1) is a *scalar*.

Elements of a matrix **A** having $i = j$ lie on the diagonal of **A**; they are called the *diagonal* elements of **A**. Those having $i \neq j$ are *off-diagonal* elements, as shown in Figure B.1. Elements having $i > j$ lie below the diagonal in the *lower triangular* part of **A**, while those having $i < j$ lie above the diagonal in the *upper triangular* part of **A**. If all elements above the diagonal are zero, the matrix is lower triangular; if all elements below the diagonal are zero, it is upper triangular. A square matrix whose off-diagonal elements are all zero is said to be a *diagonal* matrix.

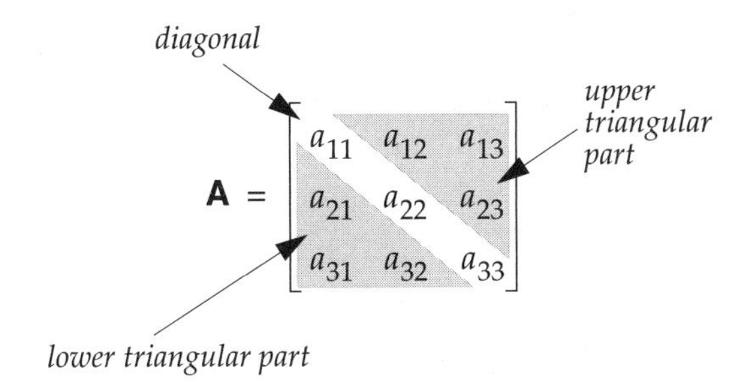

Figure B.1 The diagonal of any matrix **A** is composed of elements a_{ii}; elements above the diagonal lie in the upper triangular part of **A**, while those below are in the lower triangular part.

B.2 SPECIAL MATRICES

A matrix whose elements are all zero is called a *zero matrix* $\mathbf{0}$; it can have any dimension, for example,

$$\mathbf{0} = \begin{bmatrix} 0 & 0 \\ 0 & 0 \end{bmatrix} \quad \text{and} \quad \mathbf{0} = \begin{bmatrix} 0 & 0 & 0 \\ 0 & 0 & 0 \\ 0 & 0 & 0 \end{bmatrix} \tag{B.2.1}$$

If the diagonal elements of a diagonal matrix are all unity, then the matrix is an *identity matrix* $\mathbf{I}$; it can have any order, for example,

$$\mathbf{I} = \begin{bmatrix} 1 & 0 \\ 0 & 1 \end{bmatrix} \quad \text{and} \quad \mathbf{I} = \begin{bmatrix} 1 & 0 & 0 \\ 0 & 1 & 0 \\ 0 & 0 & 1 \end{bmatrix} \tag{B.2.2}$$

Any square matrix whose off-diagonal elements all satisfy $a_{ji} = a_{ij}$ is *symmetric*; for example, the identity matrix is symmetric, as are these:

$$\begin{bmatrix} 1 & 2 \\ 2 & 5 \end{bmatrix} \quad \text{and} \quad \begin{bmatrix} 2 & 1 & 3 \\ 1 & 1 & 4 \\ 3 & 4 & 6 \end{bmatrix} \tag{B.2.3}$$

Any square matrix whose off-diagonal elements all satisfy $a_{ji} = -a_{ij}$ is *skew-symmetric*; examples include

$$\begin{bmatrix} 1 & -2 \\ 2 & 5 \end{bmatrix} \quad \text{and} \quad \begin{bmatrix} 2 & 1 & -3 \\ -1 & 1 & 4 \\ 3 & -4 & 6 \end{bmatrix} \tag{B.2.4}$$

Every matrix $\mathbf{A}$ has a transpose $\mathbf{A}^T$ obtained by interchanging the rows and columns in $\mathbf{A}$. So the transpose of $\mathbf{A}$ has

$$a_{ji} \text{ in } \mathbf{A}^T = a_{ij} \text{ in } \mathbf{A} \qquad \textit{for all i and all j} \tag{B.2.5}$$

For example, if

$$\mathbf{A} = \begin{bmatrix} 2 & 1 \\ 3 & 4 \end{bmatrix} \quad \text{then} \quad \mathbf{A}^T = \begin{bmatrix} 2 & 3 \\ 1 & 4 \end{bmatrix} \tag{B.2.6}$$

Therefore, if $\mathbf{A}$ is of dimension $(m \times n)$, then $\mathbf{A}^T$ is of dimension $(n \times m)$. If $\mathbf{A}$ is symmetric, then $\mathbf{A} = \mathbf{A}^T$. The transpose of a column vector is a *row vector*; so, if

$$\mathbf{x} = \begin{bmatrix} 1 \\ 2 \end{bmatrix} \qquad \text{then} \qquad \mathbf{x}^{T} = \begin{bmatrix} 1 & 2 \end{bmatrix} \tag{B.2.7}$$

B.3 OPERATIONS

Two matrices **A** and **B** are equal only if they have the same dimensions and only if each element in **A** is the same as the corresponding element in **B**. That is, to have

$$\mathbf{A} = \mathbf{B} \tag{B.3.1}$$

we must have

$$a_{ij} = b_{ij} \qquad \textit{for all } i \textit{ and all } j \tag{B.3.2}$$

B.3.1 Addition

Matrix addition is defined only for matrices that have the same dimension: if **A** and **B** both have dimension $(m \times n)$, then their sum is also an $(m \times n)$ matrix given by

$$\mathbf{A} + \mathbf{B} = \begin{bmatrix} (a_{11} + b_{11}) & (a_{12} + b_{12}) & \cdots & \cdots \\ (a_{21} + b_{21}) & (a_{22} + b_{22}) & \cdots & \cdots \\ \cdots & \cdots & \cdots & \cdots \\ (a_{m1} + b_{m1}) & \cdots & \cdots & (a_{mn} + b_{mn}) \end{bmatrix} \tag{B.3.3}$$

Similarly, if **A** and **B** have dimension $(m \times n)$, then their difference is another $(m \times n)$ matrix, given by

$$\mathbf{A} - \mathbf{B} = \begin{bmatrix} (a_{11} - b_{11}) & (a_{12} - b_{12}) & \cdots & \cdots \\ (a_{21} - b_{21}) & (a_{22} - b_{22}) & \cdots & \cdots \\ \cdots & \cdots & \cdots & \cdots \\ (a_{m1} - b_{m1}) & \cdots & \cdots & (a_{mn} - b_{mn}) \end{bmatrix} \tag{B.3.4}$$

The addition of **A** to itself λ times is the same as multiplying **A** by a scalar λ; in such cases, every element in **A** is multiplied by λ,

$$\lambda\mathbf{A} = \begin{bmatrix} \lambda a_{11} & \lambda a_{12} & \lambda a_{13} & \cdots & \lambda a_{1n} \\ \lambda a_{21} & \lambda a_{22} & \lambda a_{23} & \cdots & \lambda a_{2n} \\ \cdots & \cdots & \cdots & \cdots & \cdots \\ \cdots & \cdots & \cdots & \cdots & \cdots \\ \lambda a_{m1} & \lambda a_{m2} & \lambda a_{m3} & \cdots & \lambda a_{mn} \end{bmatrix} \tag{B.3.5}$$

B.3.2 Vector Multiplication

The inner or *dot product* of two vectors is defined only between a row vector $\mathbf{x}^T$ and a column vector $\mathbf{y}$, and only when the number of columns in $\mathbf{x}^T$ is the same as the number of rows in $\mathbf{y}$. The dot product is a scalar given by

$$\mathbf{x}^T\mathbf{y} = \begin{bmatrix} x_1 & \dots & x_m \end{bmatrix} \begin{bmatrix} y_1 \\ \dots \\ y_m \end{bmatrix} = \sum_i^m x_i\, y_i \tag{B.3.6}$$

The dot product of a vector with itself gives the sum of the squares of its elements,

$$\mathbf{x}^T\mathbf{x} = \sum_i^m x_i^2 \tag{B.3.7}$$

Because of the restriction on the dimensions of the two multipliers, the dot product does not commute; that is,

$$\mathbf{x}^T\mathbf{y} \neq \mathbf{y}\mathbf{x}^T \tag{B.3.8}$$

so, on the lhs of (B.3.6), we distinguish the *premultiplier* $\mathbf{x}^T$ from the *postmultiplier* $\mathbf{y}$.

B.3.3 Matrix Multiplication

Multiplication of matrices **A** and **B** is defined only when the number of columns in the premultiplier **A** is the same as the number of rows in the postmultiplier **B**. For example, if **A** has dimension $(m \times n)$ and **B** has dimension $(n \times p)$, then their product **C** is an $(m \times p)$ matrix,

$$\underbrace{\mathbf{A}}_{(m\times n)}\underbrace{\mathbf{B}}_{(n\times p)} = \underbrace{\mathbf{C}}_{(m\times p)} \tag{B.3.9}$$

In the product matrix **C**, each element c_{ij} is the dot product of the ith row in **A** with the jth column in **B**,

$$c_{ij} = \sum_k^n a_{ik}\, b_{kj} \qquad \textit{for all i and all j} \tag{B.3.10}$$

For example, let **A** and **B** be the following matrices,

$$\mathbf{A} = \begin{bmatrix} 1 & 2 & 3 \\ 4 & 5 & 6 \end{bmatrix} \quad \text{and} \quad \mathbf{B} = \begin{bmatrix} 7 & 10 \\ 8 & 11 \\ 9 & 12 \end{bmatrix} \tag{B.3.11}$$

Then their product, $\mathbf{AB} = \mathbf{C}$, is a (2×2) matrix given by

$$\mathbf{C} = \begin{bmatrix} [1(7) + 2(8) + 3(9)] & [1(10) + 2(11) + 3(12)] \\ [4(7) + 5(8) + 6(9)] & [4(10) + 5(11) + 6(12)] \end{bmatrix} = \begin{bmatrix} 50 & 68 \\ 112 & 167 \end{bmatrix} \tag{B.3.12}$$

Note that matrix multiplication does not commute; that is, in general,

$$\mathbf{AB} \neq \mathbf{BA} \tag{B.3.13}$$

In fact, it often happens that even when the operation **AB** is defined, the operation **BA** may not be defined; this is especially true for nonsquare matrices. Also note that any matrix multiplied by the zero matrix yields another zero matrix,

$$\mathbf{0B} = \mathbf{B0} = \mathbf{0} \tag{B.3.14}$$

and multiplying any matrix by the identity matrix merely reproduces the original matrix,

$$\mathbf{AI} = \mathbf{IA} = \mathbf{A} \tag{B.3.15}$$

Finally, note that it is possible for the product of two nonzero matrices to yield the zero matrix; for example,

$$\begin{bmatrix} 1 & -2 \\ -2 & 4 \end{bmatrix} \begin{bmatrix} 2 & 2 \\ 1 & 1 \end{bmatrix} = \begin{bmatrix} 0 & 0 \\ 0 & 0 \end{bmatrix} \tag{B.3.16}$$

Consequently, division by matrices is not defined.

B.3.4 Matrix Multiplied by a Vector

At this point we have discussed multiplying vectors by vectors and matrices by matrices; we can also multiply matrices by vectors. An $(m \times n)$ matrix **A** can be multiplied by a vector in two situations: (i) It can be postmultiplied by an $(n \times 1)$ column vector, $\mathbf{A}\mathbf{x}$, and (ii) it can be premultiplied by a $(1 \times m)$ row vector, $\mathbf{y}^{\mathrm{T}}\mathbf{A}$. One use of matrix-vector multiplication is to economically represent sets of linear algebraic equations. For example, the three material balance equations given in (B.0.1)–(B.0.3) can be expressed simply as

$$\mathbf{Ax} = \mathbf{b} \tag{B.3.17}$$

where **A** is a (3×3) coefficient matrix,

$$A = \begin{bmatrix} w_1 & x_1 & y_1 \\ w_2 & x_2 & y_2 \\ 1 & 1 & 1 \end{bmatrix} \tag{B.3.18}$$

x is a (3 × 1) vector of amounts in each phase, and **b** is a (3 × 1) vector of total amounts in the system,

$$\mathbf{x} = \begin{bmatrix} N^\alpha \\ N^\beta \\ N^\gamma \end{bmatrix} \quad \text{and} \quad \mathbf{b} = \begin{bmatrix} Nz_1 \\ Nz_2 \\ N \end{bmatrix} \tag{B.3.19}$$

B.4 DETERMINANTS

Every square matrix **A** of any order n has associated with it a scalar, represented by $|\mathbf{A}|$, called the *determinant* of **A**. The value of $|\mathbf{A}|$ is defined in terms of the elements of **A** by

$$|\mathbf{A}| = \sum_{k_1}^{n}\sum_{k_2}^{n}\cdots\sum_{k_n}^{n} e_{k_1k_2\ldots k_n} a_{1k_1}a_{2k_2}\ldots a_{nk_n} \tag{B.4.1}$$

The coefficient e takes one of three values,

$$e_{k_1k_2\ldots k_n} = \begin{cases} 0 & \text{if any two subscripts are the same} \\ 1 & \text{if } (k_1, k_2, \ldots, k_n) \text{ is an even permutation of } 1, 2, \ldots, n \\ -1 & \text{if } (k_1, k_2, \ldots, k_n) \text{ is an odd permutation of } 1, 2, \ldots, n \end{cases} \tag{B.4.2}$$

A permutation results from an exchange of any two indices in the sequence $(k_1\ k_2 \ldots k_n)$. If the number of exchanges needed to restore the sequence to the natural order (1, 2, … , n) is even, then the original sequence is an *even permutation*; otherwise, it is an *odd permutation*. For a square matrix of order n, the sum in (B.4.1) contains $n!$ terms.

The determinant of a (2 × 2) matrix

$$\mathbf{A} = \begin{bmatrix} a_{11} & a_{12} \\ a_{21} & a_{22} \end{bmatrix} \tag{B.4.3}$$

is given by

$$|\mathbf{A}| = a_{11}a_{22} - a_{12}a_{21} \tag{B.4.4}$$

The determinant of a (3×3) matrix

$$\mathbf{B} = \begin{bmatrix} b_{11} & b_{12} & b_{13} \\ b_{21} & b_{22} & b_{23} \\ b_{31} & b_{32} & b_{33} \end{bmatrix} \tag{B.4.5}$$

is given by

$$|\mathbf{B}| = b_{11}b_{22}b_{33} + b_{12}b_{23}b_{31} + b_{13}b_{32}b_{21} - b_{31}b_{22}b_{13} - b_{21}b_{12}b_{33} - b_{11}b_{23}b_{32} \tag{B.4.6}$$

Note that the determinant of the transpose is the same as the determinant of the original matrix,

$$|\mathbf{A}^{\mathrm{T}}| = |\mathbf{A}| \tag{B.4.7}$$

Determinants are not usually evaluated from the definition (B.4.1); instead, their values may be obtained by Laplace's expansion or by reduction to triangular form. We do not review those methods here, except to note that the determinant of a triangular matrix is merely the product of its diagonal elements. For example, if **A** is lower triangular,

$$\mathbf{A} = \begin{bmatrix} a_{11} & 0 & 0 \\ a_{21} & a_{22} & 0 \\ a_{31} & a_{32} & a_{33} \end{bmatrix} \tag{B.4.8}$$

Then

$$|\mathbf{A}| = a_{11}\, a_{22}\, a_{33} \tag{B.4.9}$$

And if **B** is upper triangular,

$$\mathbf{B} = \begin{bmatrix} b_{11} & b_{12} & b_{13} \\ 0 & b_{22} & b_{23} \\ 0 & 0 & b_{33} \end{bmatrix} \tag{B.4.10}$$

Then

$$|\mathbf{B}| = b_{11}\, b_{22}\, b_{33} \tag{B.4.11}$$

Algorithms and computer programs for reducing any square matrix to triangular form (**LU** decomposition) are contained in the book by Press et al. [1].

Determinants enjoy many special characteristics; for example, the value of a determinant is invariant under several kinds of operations applied to any of its rows or columns. We do not review those many attributes here, except to note the following important one: If two rows (or columns) of a square matrix are the same, or if one row (or column) is a scalar multiple of any other row (or column), then the value of the determinant is zero. You may want to test this using some simple (2×2) matrices.

B.5 LINEAR INDEPENDENCE

Consider two equations that are linear in two unknowns; for example,

$$2x + 3y = 7 \tag{B.5.1}$$

$$4x + 6y = 10 \tag{B.5.2}$$

If we plot each of these equations in the two-dimensional space of the unknowns, we obtain a straight line from each. In general, the intersection of those straight lines represents the solution to the equations. However, for the two equations given above, the straight lines are parallel: they do not intersect and therefore no unique solutions exist for x and y. Such equations are said to be *linearly dependent*. On writing (B.5.1) and (B.5.2) in matrix form, we have

$$\begin{bmatrix} 2 & 3 \\ 4 & 6 \end{bmatrix} \begin{bmatrix} x \\ y \end{bmatrix} = \begin{bmatrix} 7 \\ 10 \end{bmatrix} \tag{B.5.3}$$

Note that the elements in the second row of the coefficient matrix are a scalar multiple of those in the first row; consequently, the determinant of the coefficient matrix is zero,

$$\begin{vmatrix} 2 & 3 \\ 4 & 6 \end{vmatrix} = 2(6) - 3(4) = 0 \tag{B.5.4}$$

Extending these observations to any number of equations and unknowns, we have the following: For any set of n linear equations in n unknowns, if the determinant of the coefficient matrix is zero, then the equations are not all linearly independent and the equations have no unique solution. Any square matrix $\mathbf{A}$ having $|\mathbf{A}| = 0$ is said to be *singular*. Inversely, if the coefficient matrix is *nonsingular*, so $|\mathbf{A}| \neq 0$, then the equations are linearly independent and a unique solution exists.

Recall that for any $(n \times n)$ matrix, n is called the *order* of the matrix. For any $(m \times n)$ rectangular matrix $\mathbf{A}$, the rank of $\mathbf{A}$ is defined to be the order of the largest nonsingular square matrix that can be formed from $\mathbf{A}$ by crossing out entire rows or columns or both. This means that the rank can be no larger than the smaller of m and n

$$0 \leq \text{rank}(\mathbf{A}) \leq \min[m, n] \tag{B.5.5}$$

If $\mathbf{A}$ is square and of order n, then (B.5.5) becomes

$$0 \le \text{rank}(\mathbf{A}) \le n \qquad \textit{square matrix} \qquad \text{(B.5.6)}$$

and if **A** is square and nonsingular, then its rank is the same as its order,

$$\text{rank}(\mathbf{A}) = n \qquad \textit{square nonsingular matrix} \qquad \text{(B.5.7)}$$

The above statements regarding linear independence and the existence of solutions can now be restated in terms of the rank of the coefficient matrix: For a set of n linear equations in n unknowns, a unique solution exists only if the rank of the coefficient matrix **A** is the same as the order of **A**; that is, a unique solution exists only if (B.5.7) is satisfied.

B.6 SOLVING SYSTEMS OF LINEAR EQUATIONS

Any system of n linear equations in n unknowns can be represented in matrix form as

$$\mathbf{A}\mathbf{x} = \mathbf{b} \qquad \text{(B.6.1)}$$

and we have found that unique solutions exist for the unknowns **x**, provided **A** is nonsingular.

B.6.1 Matrix Inverse

One way to find solutions to (B.6.1) is as follows: Imagine that we could find another $(n \times n)$ matrix $\mathbf{A}^{-1}$ such that

$$\mathbf{A}^{-1}\mathbf{A} = \mathbf{A}\mathbf{A}^{-1} = \mathbf{I} \qquad \text{(B.6.2)}$$

Then we could solve (B.6.1) by premultiplying both sides by $\mathbf{A}^{-1}$,

$$\mathbf{A}^{-1}\mathbf{A}\mathbf{x} = \mathbf{A}^{-1}\mathbf{b} \qquad \text{(B.6.3)}$$

Then

$$\mathbf{I}\mathbf{x} = \mathbf{A}^{-1}\mathbf{b} \qquad \text{(B.6.4)}$$

Hence

$$\mathbf{x} = \mathbf{A}^{-1}\mathbf{b} \qquad \text{(B.6.5)}$$

The matrix $\mathbf{A}^{-1}$ is the *inverse* of **A**; it exists and is unique, so long as **A** is nonsingular.

One way to obtain the inverse $\mathbf{A}^{-1}$ from a known (square) matrix **A** is as follows. For each element a_{ij} of **A**, define the *ij-minor* $|\mathbf{M}_{ij}|$ to be the determinant of the matrix remaining when we cross out the ith row and the jth column of **A**. Now attach a sign to each minor according to the rule

$$c_{ij} = (-1)^{i+j}\,|\mathbf{M}_{ij}| \qquad \text{(B.6.6)}$$

Each signed minor c_{ij} is called a *cofactor,* and each is an element in an $(n \times n)$ matrix of cofactors $\mathbf{C}$. Then the inverse of $\mathbf{A}$ can be obtained by

$$\mathbf{A}^{-1} = \frac{\mathbf{C}^{\mathrm{T}}}{|\mathbf{A}|} \tag{B.6.7}$$

The inverse of a product of matrices is the product of their inverses in reverse order,

$$(\mathbf{ABC})^{-1} = \mathbf{C}^{-1}\,\mathbf{B}^{-1}\,\mathbf{A}^{-1} \tag{B.6.8}$$

The inverse of a diagonal matrix is merely another diagonal matrix in which each diagonal element is the inverse of the corresponding diagonal element in the original matrix; for example,

$$\begin{bmatrix} a_{11} & 0 \\ 0 & a_{22} \end{bmatrix}^{-1} = \begin{bmatrix} 1/a_{11} & 0 \\ 0 & 1/a_{22} \end{bmatrix} \tag{B.6.9}$$

If one (or more) diagonal elements in a diagonal matrix is zero, then the matrix is singular and no inverse exists. Finally, if the inverse of a matrix $\mathbf{A}$ equals its transpose $\mathbf{A}^{\mathrm{T}}$ then $\mathbf{A}$ is said to be *orthogonal.*

B.6.2 Cramer's Rule

To illustrate the use of an inverse for solving systems of linear equations, consider two equations in two unknowns,

$$\begin{bmatrix} a_{11} & a_{12} \\ a_{21} & a_{22} \end{bmatrix} \begin{bmatrix} x_1 \\ x_2 \end{bmatrix} = \begin{bmatrix} b_1 \\ b_2 \end{bmatrix} \tag{B.6.10}$$

The minors of the coefficient matrix are $|\mathbf{M}_{11}| = a_{22}$, $|\mathbf{M}_{12}| = a_{21}$, $|\mathbf{M}_{21}| = a_{12}$, and $|\mathbf{M}_{22}| = a_{11}$. So the matrix of cofactors is

$$\mathbf{C} = \begin{bmatrix} a_{22} & -a_{21} \\ -a_{12} & a_{11} \end{bmatrix} \tag{B.6.11}$$

and its transpose is

$$\mathbf{C}^{\mathrm{T}} = \begin{bmatrix} a_{22} & -a_{12} \\ -a_{21} & a_{11} \end{bmatrix} \tag{B.6.12}$$

Then the vector of unknowns can be found by

$$\mathbf{x} = \frac{\mathbf{C}^T\mathbf{b}}{|\mathbf{A}|} = \frac{1}{|\mathbf{A}|}\begin{bmatrix} a_{22} & -a_{12} \\ -a_{21} & a_{11} \end{bmatrix}\begin{bmatrix} b_1 \\ b_2 \end{bmatrix} \tag{B.6.13}$$

So

$$\mathbf{x} = \frac{1}{|\mathbf{A}|}\begin{bmatrix} (a_{22}\,b_1 - a_{12}\,b_2) \\ (-a_{21}\,b_1 + a_{11}\,b_2) \end{bmatrix} \tag{B.6.14}$$

Hence,

$$x_1 = \frac{(a_{22}\,b_1 - a_{12}\,b_2)}{|\mathbf{A}|} = \frac{\begin{vmatrix} b_1 & a_{12} \\ b_2 & a_{22} \end{vmatrix}}{|\mathbf{A}|} \tag{B.6.15}$$

and

$$x_2 = \frac{(-a_{21}\,b_1 + a_{11}\,b_2)}{|\mathbf{A}|} = \frac{\begin{vmatrix} a_{11} & b_1 \\ a_{21} & b_2 \end{vmatrix}}{|\mathbf{A}|} \tag{B.6.16}$$

where $|\mathbf{A}|$ for a 2×2 is given by (B.4.4). These last two equations are known as *Cramer's rule*; it extends to any number of linear equations and unknowns. But note that if the original coefficient matrix $\mathbf{A}$ is singular, then $|\mathbf{A}| = 0$, the equations are linearly dependent, and no solutions exist for x_1 and x_2.

We can now indicate the solution to the three-phase material balance problem (B.0.1)–(B.0.3). On applying Cramer's rule to the matrix form (B.3.17) for the three material balances, we find an expression for the amount in phase α, relative to the total amount in the system,

$$\frac{N^\alpha}{N} = \frac{\begin{vmatrix} z_1 & x_1 & y_1 \\ z_2 & x_2 & y_2 \\ 1 & 1 & 1 \end{vmatrix}}{\begin{vmatrix} w_1 & x_1 & y_1 \\ w_2 & x_2 & y_2 \\ 1 & 1 & 1 \end{vmatrix}} = \frac{x_2(z_1 - y_1) + y_2(x_1 - z_1) + z_2(y_1 - x_1)}{x_2(w_1 - y_1) + y_2(x_1 - w_1) + w_2(y_1 - x_1)} \tag{B.6.17}$$

Analogous results are obtained for N^β/N and N^γ/N. The result (B.6.17) is a form of the tie-triangle rule for three components in three-phase equilibrium (see § H.2).

Cramer's rule is usually sufficient for solving two equations in two unknowns or three equations in three unknowns. However, for larger sets of equations, other solution procedures are preferred, such as Gauss-Jordan reduction and the Gauss-Seidel method. But in most cases, the best method is **LU** decomposition, in which the coefficient matrix **A** is decomposed into a product of a lower triangular matrix **L** and an upper triangular matrix **U**. The procedure is contained in the book by Press et al. [1].

B.7 QUADRATIC FORMS

Consider a nonlinear equation of the form

$$a_{11}x^2 + 2a_{12}xy + a_{22}y^2 = 0 \tag{B.7.1}$$

Such equations can be written in matrix form like this,

$$\begin{bmatrix} x & y \end{bmatrix} \begin{bmatrix} a_{11} & a_{12} \\ a_{21} & a_{22} \end{bmatrix} \begin{bmatrix} x \\ y \end{bmatrix} = 0 \tag{B.7.2}$$

or

$$\mathbf{x}^T \mathbf{A} \mathbf{x} = 0 \tag{B.7.3}$$

where **A** is square and symmetric ($a_{21} = a_{12}$). The scalar $\mathbf{x}^T\mathbf{A}\mathbf{x}$ is called a *quadratic form*. If **A** is not symmetric ($a_{21} \neq a_{12}$), (B.7.3) represents no loss of generality, because any square matrix can be written as the sum of symmetric and skew-symmetric parts. For example,

$$\mathbf{A} = \frac{1}{2}(\mathbf{A} + \mathbf{A}^T) + \frac{1}{2}(\mathbf{A} - \mathbf{A}^T) \tag{B.7.4}$$

But the skew-symmetric part always obeys

$$\mathbf{x}^T(\mathbf{A} - \mathbf{A}^T)\,\mathbf{x} = 0 \tag{B.7.5}$$

So we have

$$\mathbf{x}^T \mathbf{A} \mathbf{x} = \frac{1}{2}\mathbf{x}^T(\mathbf{A} + \mathbf{A}^T)\,\mathbf{x} \tag{B.7.6}$$

where $(\mathbf{A} + \mathbf{A}^T)$ is always symmetric.

If a quadratic form $p = \mathbf{x}^T\mathbf{A}\mathbf{x}$ has $p > 0$ for all $\mathbf{x} \neq 0$, then the quadratic form is said to be *positive definite*; if it has $p \geq 0$ for all $\mathbf{x} \neq 0$, then it is *positive semidefinite*. Similarly, if $p = \mathbf{x}^T\mathbf{A}\mathbf{x}$ has $p < 0$ for all $\mathbf{x} \neq 0$, then the quadratic form is said to be *negative definite*; if it has $p \leq 0$ for all $\mathbf{x} \neq 0$, then it is *negative semidefinite*.

One way to determine the definiteness of a quadratic form is to determine the signs of its principal minors. In any square matrix **A**, the principal minors are the determinants $|\mathbf{M}_i|$ formed from the first i rows and i columns of **A**. For example, for the 2×2 matrix in (B.7.2),

$$|\mathbf{M}_1| = a_{11} \tag{B.7.7}$$

and

$$|\mathbf{M}_2| = \begin{vmatrix} a_{11} & a_{12} \\ a_{21} & a_{22} \end{vmatrix} \tag{B.7.8}$$

A square matrix of order n has n principal minors. Then, a quadratic form is positive definite if *all* its principal minors are positive,

$$|\mathbf{M}_i| > 0 \qquad \textit{for all } i \tag{B.7.9}$$

and it is positive semidefinite if all its principal minors are positive or zero,

$$|\mathbf{M}_i| \geq 0 \qquad \textit{for all } i \tag{B.7.10}$$

Similarly, a quadratic form $\mathbf{x}^T\mathbf{A}\mathbf{x}$ is negative definite if all its odd-order principal minors are positive,

$$|\mathbf{M}_i| > 0 \qquad i = 1, 3, 5, \textit{etc.} \tag{B.7.11}$$

and all its even-order principal minors are negative,

$$|\mathbf{M}_i| < 0 \qquad i = 2, 4, 6, \textit{etc.} \tag{B.7.12}$$

A quadratic form is negative semidefinite if all its odd-order principal minors are positive or zero and all its even-order principal minors are negative or zero.

LITERATURE CITED

[1] W. H. Press, B. P. Flannery, S. A. Teukolsky, and W. T. Vetterling, *Numerical Recipes*, Cambridge University Press, Cambridge, 1986.

C

SOLUTIONS TO CUBIC EQUATIONS

It was at the University of Bologna during the Italian Renaissance that Scipione del Ferro (1465–1526) first discovered how to solve any depressed cubic equation for its real roots. A depressed cubic lacks a quadratic term,

$$ax^3 + cx + d = 0 \tag{C.0.1}$$

In 1535 the procedure was discovered independently by Niccolo Fontana (1497–1557), a.k.a. Tartaglia (the "Stammerer"). In keeping with the practice of the times, both Scipione and Tartaglia kept the solution secret, for men have always understood that doors may open to those who know what remains hidden to others. But eventually Tartaglia succumbed to relentless cajoling and revealed the solution to Gerolamo Cardano [in English, Jerome Cardan (1501–1576)]. Thereafter, Cardan discovered how to depress any cubic and then he could find the real roots of them all; he even had some appreciation for the existence of imaginary roots. Continuing along similar lines, Cardan's pupil Ludovico Ferrari (1522–1565) found a way to reduce the general quartic to a cubic and thereby he was able to solve any quartic equation for its real roots.

Such are the bare outlines of a remarkable story of discovery—an outline stripped of the personalities involved and therefore missing the tragicomical blend of bluster, chicanery, and brilliance that makes the story unique. In 1545 Cardan published the cubic's analytic solution in his book, *Artis magnae sive de regulis algebraicis liber unus,* which is now usually known as *Ars Magna.* But Cardan's was a complex personality and his character so far from endearing that he has been unevenly treated by historians of mathematics (compare [1] and [2]). For example, as the first published solution to the cubic, some cite *De Aequationum Recognitione et Emendatione,* by François Viète (1540–1603). But this was not published until 1615—seventy years after the *Ars Magna* appeared and twelve years after Viète's death—and it is not clear whether Viète's is an independent discovery, or whether he is reciting Cardan's method in a more congenial notation, or whether that portion of the manuscript was inserted after Viète's death. Cardan's story is sympathetically told by Dunham [1]; excerpts from the *Ars Magna,* with additional commentary, are contained in the book edited by Struik [3].

Here is Cardan's method for obtaining the real roots to any cubic having real coefficients. First write your cubic in the form

$$x^3 + bx^2 + cx + d = 0 \tag{C.0.2}$$

and then compute

$$p = \frac{2b^3 - 9bc + 27d}{54} \tag{C.0.3}$$

and

$$q = \frac{b^2 - 3c}{9} \tag{C.0.4}$$

If $(p^2 - q^3) > 0$, then the cubic has only one real root,

$$x_1 = -\text{sgn}(p)\left(\frac{r^2 + q}{r}\right) - \frac{b}{3} \tag{C.0.5}$$

where $\text{sgn}(p) = p/|p|$ and

$$r = \left(\sqrt{p^2 - q^3} + |p|\right)^{1/3} \tag{C.0.6}$$

But if $(p^2 - q^3) \le 0$, then the cubic has three real roots,

$$x_1 = (-2\sqrt{q})\cos(\theta/3) - b/3 \tag{C.0.7}$$

$$x_2 = (-2\sqrt{q})\cos((\theta + 2\pi)/3) - b/3 \tag{C.0.8}$$

$$x_3 = (-2\sqrt{q})\cos((\theta + 4\pi)/3) - b/3 \tag{C.0.9}$$

where

$$\theta = \text{acos}(p/\sqrt{q^3}) \tag{C.0.10}$$

LITERATURE CITED

[1] W. Dunham, *Journey Through Genius: The Great Theorems of Mathematics,* Wiley, New York, 1990.

[2] W. W. Rouse Ball, *A Short Account of the History of Mathematics,* Macmillan, London, 1927; reprinted by Dover, New York, 1960.

[3] D. J. Struik (ed.), *A Source Book of Mathematics 1200–1800,* Harvard University Press, Cambridge, MA, 1969.

D

VAPOR PRESSURES OF SELECTED FLUIDS

Vapor pressures of most pure fluids can be adequately correlated by the Antoine equation (8.2.31). The following table provides values of the Antoine parameters for a few fluids. These apply for the vapor pressure P^s in bar and temperature T in Kelvin [1].

Fluid	Formula	A	B	C
Acetone	C_3H_6O	10.031	2940.5	35.93
Acetonitrile	C_2H_3N	9.667	2945.5	49.15
Benzene	C_6H_6	9.281	2788.5	52.36
Carbon tetrachloride	CCl_4	9.254	2808.2	46.0
Chloroform	$CHCl_3$	9.353	2696.8	46.16
Ethanol	C_2H_5OH	12.292	3804.0	41.68
Diethyl ether	$C_4H_{10}O$	9.463	2511.3	41.95
Methanol	CH_3OH	11.967	3626.6	34.29
Methyl acetate	$C_3H_6O_2$	9.509	2601.9	56.15
Toluene	$C_6H_5CH_3$	9.394	3096.5	53.67
Water	H_2O	11.683	3816.4	46.13

LITERATURE CITED

[1] R. C. Reid, J. M. Prausnitz, and T. K. Sherwood, *The Properties of Gases and Liquids*, 3rd ed., McGraw-Hill, New York, 1977.

E

PARAMETERS IN MODELS FOR G EXCESS

The following tables provide values for parameters in models for the excess Gibbs energy of selected binary liquid mixtures. Table E.1 contains values for the Porter equation (§ 5.6.2), Table E.2 for the Margules equation (§ 5.6.3), and Table E.3 for Wilson's equation (§ 5.6.5).

LITERATURE CITED IN TABLES E.1–E.3

[1] R. H. Perry, C. H. Chilton, and S. D. Kirkpatrick (eds.), *Chemical Engineer's Handbook*, 4th ed., McGraw-Hill, New York, 1963, p. 13–17.

[2] O. A. Hougen, K. M. Watson, and R. A. Ragatz, *Chemical Process Principles*, Part II, Thermodynamics, 2nd ed., Wiley, New York, 1959, p. 930.

[3] H. V. Kehiaian, J. P. E. Grolier, M. R. Kechavarz, G. C. Benson, O. Kiyohara, and Y. P. Handa, "Thermodynamic Properties of Binary Mixtures Containing Ketones VII," *Fluid Phase Equil.*, **7**, 95 (1981).

[4] M. L. McGlashan, J. E. Prue, and I. E. J. Sainsbury, "Equilibrium Properties of Mixtures of Carbon Tetrachloride and Chloroform," *Trans. Faraday Soc.*, **50**, 1284 (1954).

[5] D. S. Adcock and M. L. McGlashan, "Heats of Mixing," *Proc. Roy. Soc. A*, **226**, 266 (1954).

[6] O. Redlich, *Thermodynamics: Fundamentals, Applications*, Elsevier, Amsterdam, 1976.

[7] M. J. Holmes and M. van Winkle, "Prediction of Ternary Vapor-Liquid Equilibria from Binary Data," *Ind. Eng. Chem.*, **62**(1), 21 (1970); L. H. Ballard and M. van Winkle, "Vapor-Liquid Equilibria at 760 mm Pressure," *Ind. Eng. Chem.*, **44**, 2450 (1952).

[8] J. Gmehling and U. Onken, *Vapor-Liquid Equilibrium Data Collection*, Chemistry Data Series (in several volumes), DECHEMA, Frankfurt am Main, Germany, 1977.

Table E.1 Selected binary liquid mixtures in which the excess Gibbs energy approximately obeys Porter's equation (5.6.1)

Component 1	Component 2	T(°C)	A	Ref.
Acetone	Ethyl ether	35–56	0.741	[1]
	Methanol	56–64	0.560	[1]
	Benzene	56–80	0.405	[1]
Benzene	Cyclohexane	80	0.335	[2]
2-Butanone	n-Hexane	50	1.280	[3]
		60	1.220	
		70	1.166	
Carbon tetrachloride	Chloroform[a]	25	0.172	[4]
		40	0.154	
		55	0.138	
Carbon tetrachloride	Cyclohexane[b]	30	0.108	[5]
		40	0.101	
		50	0.094	
		60	0.088	
		70	0.083	
Ethanol	Ethyl acetate	72–78	0.896	[1]
	Toluene	77–110	1.757	[1]
Ethylbenzene	o-Xylene	136–144	0.0081	[6]
	m-Xylene	136–139	0.0083	[6]
	p-Xylene	136–138	0.0071	[6]
Methanol	Ethyl acetate	62–77	1.16	[1]
	Methyl acetate	57–64	1.064	[1]
	Trichloroethylene	65–87	1.946	[1]
	2-Propanol	65–82	–0.0754	[7]

[a] $A = -0.2034 + 111.86/T(K)$

[b] $A = -1.7056 + 141.45/T(K) + 0.2357 \ln T(K)$

Table E.2 Selected binary liquid mixtures in which the excess Gibbs energy approximately obeys the Margules equation (5.6.11)[a]

Component 1	Component 2	T(°C)	A_1	A_2
Acetone	Benzene	57.7–76.5	0.316	0.461
	Carbon tetrachloride	56.–70.8	0.764	0.918
	Chloroform	57.5–64.4	–0.561	–0.840
	Methanol	55.3–64.6	0.579	0.618
Benzene	Carbon tetrachloride	76.6–79.9	0.0855	0.121
	Chloroform	62.–79.2	–0.167	–0.236
	Methanol	58.–78.6	1.710	2.293
	n-Hexane	68.6–77.9	0.516	0.365
2-Butanone	Chloroform	62.9–79.7	–0.686	–0.85
Ethanol	Benzene	67.9–76.9	1.472	1.836
	Cyclohexane	65.–74	1.726	2.473
	n-Hexane	58.1–78.3	1.940	2.705
	Toluene	77.–110.6	1.571	1.648
Methanol	Chloroform	53.5–63.	0.832	1.736
n-Octane	Ethylbenzene	125.7–136.2	0.201	0.188
1-Propanol	Benzene	77.–97.2	1.336	1.596
	Ethylbenzene	97.–118.9	1.330	1.239
	n-Hexane	66.2–89.6	1.867	1.536
	Ethyl acetate	78.–96.	0.519	0.641
2-Propanol	Acetone	56.8–79.8	0.514	0.632
	Benzene	71.8–82.4	1.269	1.520
	Ethyl acetate	75.9–80.3	0.517	0.476
Toluene	Phenol	110.5–172.7	1.034	0.714

[a] All mixtures here are at 760 mm Hg. Values of parameters taken from a larger collection given in [8].

Table E.3 Selected binary liquid mixtures in which the excess Gibbs energy can be approximately represented by Wilson equations (5.6.24) and (5.6.30)[a]

Component 1	Component 2	ρ_1 (mol/l)	ρ_2 (mol/l)	$\Delta\lambda_{12}$ (kJ/mol)	$\Delta\lambda_{21}$ (kJ/mol)
Acetone	Chloroform	13.50	12.40	0.486	–2.120
	Ethanol		17.04	0.730	1.060
	Methanol		24.55	–0.479	2.281
	Water		55.34	1.441	6.201
Benzene	Ethanol	11.18	17.04	1.115	5.290
	Methanol		24.55	0.475	7.753
	1-Propanol		13.31	1.494	4.269
2-Butanone	Chloroform	11.09	12.40	–0.954	–1.048
Ethanol	Cyclohexane	17.04	9.20	8.041	1.520
	n-Heptane		6.78	7.980	1.935
	Toluene		9.36	5.317	0.973
	Water		55.34	1.754	3.812
Methanol	Chloroform	24.55	12.40	7.087	–1.514
	Water		55.34	0.347	2.178
n-Octane	Ethylbenzene	6.11	8.13	–0.722	1.391
	Phenol		12.03	3.524	6.755
1-Propanol	n-Heptane		6.78	6.180	1.105
	Water		55.34	3.793	5.843
Toluene	Phenol	9.36	12.03	0.138	3.283

[a] All mixtures here are at 760 mm Hg. Values of parameters taken from a larger collection given in [8].

F

A STABILITY CONDITION FOR BINARIES

In this appendix we prove that a stable, one-phase, binary mixture must have values for component fugacities that are less than the corresponding pure-component values; that is, we prove that a stable, one-phase, binary mixture must have

$$f_1(T, P, \{x\}) < f_{\text{pure }1}(T, P) \tag{F.0.1}$$

where either component can be labeled 1. We start with the one-phase stability criterion for mixtures (8.3.14), written in terms of the chemical potential,

$$\bar{G}_{11} \equiv \left(\frac{\partial \bar{G}_1}{\partial N_1}\right)_{TPN_2} > 0 \qquad \textit{not unstable} \tag{8.3.14}$$

This can be written in terms of the fugacity as

$$\left(\frac{\partial \ln f_1}{\partial x_1}\right)_{TP} > 0 \qquad \textit{not unstable} \tag{F.0.2}$$

We apply this to two situations.

Situation 1. If (F.0.2) is satisfied for all x_1 between 0 and 1, then f_1 increases monotonically from 0 at $x_1 = 0$ to $f_{\text{pure }1}$ at $x_1 = 1$. The mixture remains a stable single phase at all compositions, and at every x_1 (F.0.1) is obeyed.

Situation 2. If the condition (F.0.2) is violated over some range of x_1, then the mixture is not stable over some compositions and it may split into two phases α and β. The curve for $f_1(x_1)$ either oscillates or it separates into distinct branches. The phase equilibrium conditions require

$$f_1^{\alpha}(T, P, \{x^{\alpha}\}) = f_1^{\beta}(T, P, \{x^{\beta}\}) \tag{F.0.3}$$

Further, each of these phases is stable, so they each satisfy the stability requirement (F.0.2).

Let β designate the phase that is rich in component 1. We presume pure 1 is a stable phase, so by continuity, mixtures from x_1^{β} to $x_1 = 1$ are stable single phases, and because of (F.0.2) their fugacities must be less than $f_{\text{pure}\,1}$. Therefore they approach $f_{\text{pure}\,1}$ from below; hence,

$$f_1^{\beta}(T, P, \{x^{\beta}\}) < f_{\text{pure}\,1}(T, P) \tag{F.0.4}$$

In fact, the mixture at x_1^{β} has the smallest fugacity of any stable, one-phase mixture that is rich in component 1.

Combining (F.0.3) with (F.0.4), the component-2 rich phase must obey

$$f_1^{\alpha}(T, P, \{x^{\alpha}\}) < f_{\text{pure}\,1}(T, P) \tag{F.0.5}$$

We also assume pure 2 is a stable phase, so again by continuity, mixtures from $x_1 = 0$ to x_1^{α} must be stable single phases, and by (F.0.2) their fugacities must increase monotonically from 0 to f_1^{α}. In fact, the mixture at x_1^{α} has the largest fugacity of any stable, one-phase mixture that is rich in component 2. Hence,

$$0 < f_1^{\alpha}(T, P, \{x^{\alpha}\}) = f_1^{\beta}(T, P, \{x^{\beta}\}) < f_{\text{pure}\,1}(T, P) \tag{F.0.6}$$

If only the two phases, α and β, form as a result of the phase split, then one-phase mixtures at compositions between x_1^{α} and x_1^{β} can only be metastable or unstable. Therefore, all stable one-phase mixtures are bounded by $[0, x_1^{\alpha}]$ or by $[x_1^{\beta}, 1]$, so they all are described by (F.0.6). Hence at the given T and P, all stable one-phase mixtures must satisfy (F.0.1). *QED*

If a third phase γ forms (such as in VLLE), then the above argument still holds; we just must be careful to identify phase α as that richest in component 2 and phase β as the one richest in component 1. Then the third phase has some composition between x_1^{α} and x_1^{β}, but it must have the *same* values for fugacities as phases α and β; so, analogous to (F.0.6) we would have

$$0 < f_1^{\alpha}(T, P, \{x^{\alpha}\}) = f_1^{\beta}(T, P, x^{\beta}) = f_1^{\gamma}(T, P, \{x^{\gamma}\}) < f_{\text{pure}\,1}(T, P) \tag{F.0.7}$$

Therefore all stable one-phase mixtures still obey (F.0.1).

Mixtures having $f_1 > f_{\text{pure}\,1}$ are either metastable (they satisfy (F.0.2)) or unstable (they violate (F.0.2)). We caution that while stable one-phase mixtures must obey (F.0.1), the converse is not true: mixtures satisfying (F.0.1) are not necessarily stable. They could be stable, unstable, or metastable.

G

NOTATION IN VARIATIONAL CALCULUS

Variational calculus is concerned with finding extrema; for example, what is the shortest distance between two points on the surface of a parabolic cylinder? In ordinary, garden-variety calculus, we deal with *functions*, which are objects whose values depend on the values of numerical quantities. But in the variational calculus, the focus of attention is on *functionals*, which are objects whose values depend on functions. For example, we may interpret the entropy as a functional because its value depends on other thermodynamic functions, such as temperature, pressure, and composition. Since the functionals differ from functions, we sometimes find it convenient to use a notation for operators on functionals that differs somewhat from the notation for operators on functions. For our purposes, the most important notational distinction occurs for differential operators.

Let f be a functional that depends on C functions x_i, i = 1, 2, ... , C. When f is at a stationary point (a maximum or minimum), the functions $\{x\}$ have values $\{x_o\}$. The variation of any x_i about its stationary value can be represented by

$$x_i = x_{io} + \delta x_i \tag{G.0.1}$$

where δx is read as the "variation of x." In our situations, the stationary point may be an equilibrium state of a system and the variations might be caused by natural fluctuations. When the $\{x\}$ all fluctuate, we are interested in the total response of the functional f,

$$\Delta f = f(\{x\}) - f(\{x_o\}) \tag{G.0.2}$$

where Δf represents the total response. Natural fluctuations about equilibrium states are small, so the total response can be estimated by a Taylor expansion about the stationary point:

$$\Delta f = \delta f + \delta^2 f + \ldots \tag{G.0.3}$$

Then the first-order variation of f merely means the total differential, evaluated at the stationary point,

$$\delta f = \sum_i^C \left(\frac{\partial f}{\partial x_i}\right)_{\{x_o\}} \delta x_i \tag{G.0.4}$$

Similarly, the second-order variation of f is given by

$$\delta^2 f = \sum_i^C \sum_j^C \left[\frac{\partial}{\partial x_i}\left(\frac{\partial f}{\partial x_j}\right)_{\{x_o\}}\right]_{\{x_o\}} \delta x_i \, \delta x_j \tag{G.0.5}$$

Since the quantities f of interest to us form exact differentials, the second-order variation in (G.0.5) is invariant under an exchange of the indices i and j. We need not proceed further into the variational calculus here because in this book we use the variations δf and $\delta^2 f$ merely as a notational convenience; relations (G.0.4) and (G.0.5) define the notation we use.

H

TRIANGULAR DIAGRAMS

The phase behavior of a ternary mixture is conventionally presented on an equilateral triangular diagram, such as in Figure H.1. Any point on the diagram represents a ternary mixture of a particular composition. The compositions are usually given in mole fractions, but weight fractions can also be used. We first review the basic principles of triangular diagrams (§ H.1), then we give the tie-triangle rule for three components in a three-phase equilibrium situation (§ H.2).

H.1 BASIC FEATURES OF TRIANGULAR DIAGRAMS

On a triangular diagram, the vertices represent pure components; in Figure H.1 we have called the pure components A, B, and C. Then each edge of the triangle represents all binary mixtures formed by two of the three components; for example, the edge $\overline{AC}$ represents all mixtures of components A and C. A particular point on an edge divides the edge into two segments, and the lengths of those segments are simply related to the composition of the binary represented by the point. For example, in Figure H.1 the point v on edge $\overline{AC}$ represents the binary mixture that has mole fractions $x_A = \overline{vC}/\overline{AC}$ and $x_C = \overline{vA}/\overline{AC}$.

Any point on the interior of the triangle, such as point P in Figure H.1, represents a ternary mixture. To obtain the composition of the mixture at P, drop perpendiculars $\overline{aP}$, $\overline{bP}$, and $\overline{cP}$ to each of the three edges. In an equilateral triangle the lengths of these perpendiculars always sum to the altitude of the triangle, $\overline{hA}$,

$$\overline{aP} + \overline{bP} + \overline{cP} = \overline{hA} \tag{H.1.1}$$

This geometric statement is equivalent to a material balance on the mixture, and therefore it is true regardless of the identities of the components or their intermolecular forces. In fact, the mole fractions of the mixture represented by P are given by the ratios $x_A = \overline{aP}/\overline{hA}$, $x_B = \overline{bP}/\overline{hA}$, and $x_C = \overline{cP}/\overline{hA}$. Since the mole fractions must always sum to unity, the composition is determined by giving values for any two indepen-

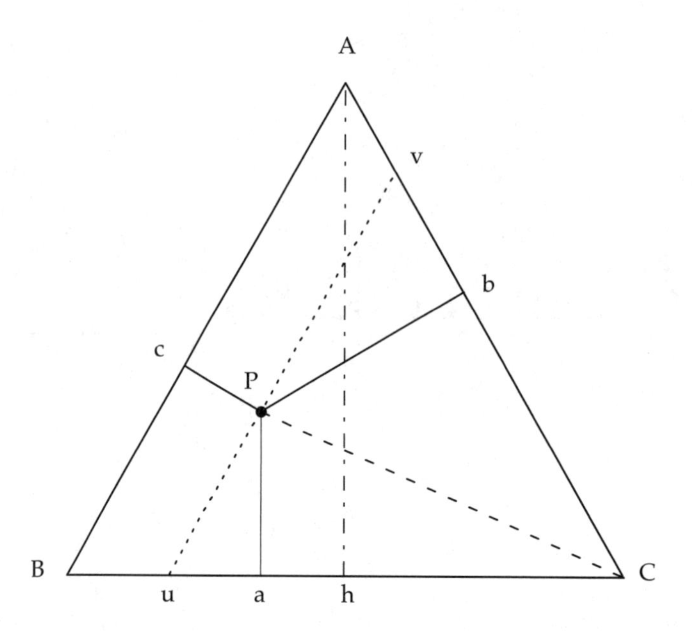

Figure H.1 Standard equilateral triangle for representing ternary mixtures

dent mole fractions; that is, to locate any point, such as *P*, we need values for only two of its three perpendiculars.

The symmetry of equilateral triangles imposes invariants on certain lines that represent particular classes of ternary mixtures. We identify two such invariant classes here. One set of invariants is composed of lines that are parallel to an edge. Since every point on such a line is the same perpendicular distance from the edge, every mixture on that line has the same fraction in the component represented by the opposite vertex. For example, the line $\overline{uv}$ in Figure H.1 is parallel to edge $\overline{AB}$, and therefore *every* mixture on $\overline{uv}$ has the same value for the C-component mole fraction: $x_C = \overline{cP}/\overline{hA}$.

A second set of invariants contains lines that pass through a vertex. On such a line, every point has the two perpendiculars to adjacent edges in the same ratio, and therefore every mixture on the line has the same relative amounts in those two components. For example, the line $\overline{PC}$ in Figure H.1 passes through vertex C, and therefore *every* mixture on $\overline{PC}$ has the same ratio of mole fractions for components A and B. This means that, if we designate component C as the "solute", then all along line $\overline{PC}$ the solute-free mole fractions for components A and B are both constants, x_A^{sf} = constant and x_B^{sf} = constant.

H.2 TIE-TRIANGLE RULE

When ternary mixtures exhibit three-phase equilibria, a tie-triangle rule can be used to obtain the relative amounts in the three phases. This is analogous to the lever rule for binaries in two-phase equilibria. Consider a ternary mixture of components 1, 2, and 3 in three-phase equilibrium at the overall composition represented by a point *P*, as in Figure H.2. In the figure, the tie-triangle ABC bounds the three-phase region, the

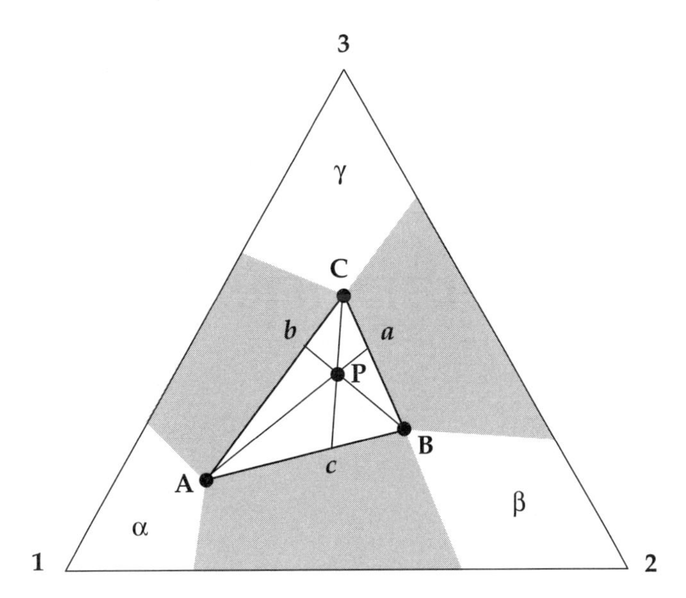

Figure H.2 For a ternary mixture of components 1, 2, and 3 in three-phase equilibrium, the relative amounts in the three phases can be obtained from the tie-triangle rule (H.2.1)–(H.2.3) applied to an isothermal-isobaric triangular diagram, such as this. Here, α, β, and γ are the three phases. Point P represents the overall composition of the three-phase system. The three-phase equilibrium region is bounded by the tie-triangle ABC; shaded regions are two-phase situations, and the areas α, β, and γ are single phase situations.

shaded areas are two-phase regions, and the areas α, β, γ are each single phases. To obtain the fraction in each phase, draw a straight line from each vertex (A, B, C) of the tie triangle, through P, and extend each line to intersect the opposite side (at points *a*, *b*, *c*). Then the fraction of material in each phase is given by a ratio of line segments,

$$R_\alpha = \begin{pmatrix}\text{fraction of material} \\ \text{in phase } \alpha\end{pmatrix} = \frac{\overline{Pa}}{\overline{Aa}} \tag{H.2.1}$$

$$R_\beta = \begin{pmatrix}\text{fraction of material} \\ \text{in phase } \beta\end{pmatrix} = \frac{\overline{Pb}}{\overline{Bb}} \tag{H.2.2}$$

$$R_\gamma = \begin{pmatrix}\text{fraction of material} \\ \text{in phase } \gamma\end{pmatrix} = \frac{\overline{Pc}}{\overline{Cc}} \tag{H.2.3}$$

This tie-triangle rule is a consequence of material balances on the system (see (B.6.17) in Appendix B) and therefore it applies to *any* three-phase equilibrium situation involving ternary mixtures.

I

LAGRANGE MULTIPLIERS

The problem addressed in this appendix is how to optimize a function when constraints apply. Optimize means find an extremum—a maximum or a minimum. When there are no constraints, we generally optimize by taking derivatives and setting them equal to zero. The function to be differentiated is called the *objective function*. But constraints couple variables that would otherwise have been independent, preventing our taking simple derivatives. The method of Lagrange multipliers provides a way to circumvent this problem; essentially, this is done by introducing additional degrees of freedom, one for each constraint. Rather than develop the theory for this approach, we illustrate its use by a simple example.

Our problem is to maximize the area of a rectangle under the constraint that the perimeter must be 100. Let A be the area, with b and h the lengths of two sides, so

$$A = bh \tag{I.0.1}$$

In this problem b and h are coupled through the constraint,

$$2b + 2h = 100 \tag{I.0.2}$$

This is called an equality constraint; it is formally written as

$$2b + 2h - 100 = 0 \tag{I.0.3}$$

The constraint establishes a relation between b and h, so we cannot, for example, take the derivative of A wrt b while holding h fixed. Instead, we create a new objective function F from the original one (I.0.1) and the constraint (I.0.3). This new function is defined by

$$F = bh + \lambda(2b + 2h - 100) \tag{I.0.4}$$

where λ is the *Lagrange multiplier*. Note that when we satisfy the constraint, then (I.0.4) reduces to $F = bh = A$. The presence of the second term in F enforces the constraint, so we can treat b and h as independent when taking derivatives of F; however, the multiplier λ is a new unknown that usually must be found to solve the problem.

Forming the derivatives, we have

$$\left(\frac{\partial F}{\partial b}\right)_{h\lambda} = h + 2\lambda = 0 \tag{I.0.5}$$

$$\left(\frac{\partial F}{\partial h}\right)_{b\lambda} = b + 2\lambda = 0 \tag{I.0.6}$$

$$\left(\frac{\partial F}{\partial \lambda}\right)_{bh} = 2b + 2h - 100 = 0 \tag{I.0.7}$$

Note (I.0.7): the derivative wrt the multiplier always recovers the constraint. These are three algebraic equations that can be solved for the three unknowns: b, h, and λ. It is clear from (I.0.5) and (I.0.6) that the sides of the rectangle must be equal; that is, to satisfy the constraint, the rectangle must be a square. Then from (I.0.7), we find

$$b = h = 25 \tag{I.0.8}$$

Of all rectangles having perimeter = 100, that having the largest area is the square of side = 25.

When several equality constraints are to be applied, we introduce one multiplier for each. But in general, we should not introduce more constraints than there are initially independent variables. Doing so creates an over-constrained problem that usually has no solution. In some problems the Lagrange multiplier has a physical significance, but none appears to apply to the λ in the simple problem above.

J

NRTL MODEL

The introductory discussion of models for liquid-phase activity coefficients, presented in Chapter 5, included a description of the Wilson equation, which is appropriate for many nonelectrolyte mixtures that exhibit large deviations from ideality. However, the Wilson model cannot correlate liquid-liquid equilibrium data, and therefore it cannot be used in LLE and VLLE calculations. To overcome this deficiency, Renon and Prausnitz [1] devised the NRTL model for g^E (NonRandom, Two-Liquid).

This model is similar to the Wilson model, but for binaries the Wilson equations involve only two adjustable parameters, while the NRTL equations involve three. For binary mixtures of components 1 and 2, the NRTL equation for g^E takes this form:

$$\frac{g^E}{RT} = x_1 x_2 \left(\frac{\tau_{21} \mathcal{G}_{21}}{x_1 + x_2 \mathcal{G}_{21}} + \frac{\tau_{12} \mathcal{G}_{12}}{x_1 \mathcal{G}_{12} + x_2} \right) \tag{J.0.1}$$

where

$$\tau_{ij} = \frac{\Delta g_{ij}}{RT} \qquad \text{and} \qquad \mathcal{G}_{ij} = \exp(-\alpha \tau_{ij}) \tag{J.0.2}$$

The three adjustable parameters are α, Δg_{12}, and Δg_{21} (note that $\Delta g_{21} \neq \Delta g_{12}$). The quantities Δg_{ij} have some characteristics in common with the Wilson parameters $\Delta\lambda_{ij}$: they are independent of composition and, usually, they are assumed to be either constants or linear in temperature. The value of the parameter α usually lies between 0.2 and 0.5. For many mixtures, the model is not particularly sensitive to α, so if a value cannot be found, arbitrarily setting α (say to 0.3) often proves satisfactory.

For binary mixtures, (J.0.1) leads to the these expressions for activity coefficients:

$$\ln \gamma_1 = x_2^2 (\tau_{21} \mathcal{G}_{21} \Omega_{21} + \tau_{12} \Omega_{12}) \tag{J.0.3}$$

and

$$\ln\gamma_2 = x_1^2(\tau_{21}\Omega_{21} + \tau_{12}\mathcal{G}_{12}\Omega_{12}) \tag{J.0.4}$$

where

$$\Omega_{ij} \equiv \frac{\mathcal{G}_{ij}}{(x_i\mathcal{G}_{ij} + x_j)^2} \tag{J.0.5}$$

For multicomponent mixtures, the NRTL equations generalize to

$$\frac{g^E}{RT} = \sum_i^C \frac{x_i L_i}{M_i} \tag{J.0.6}$$

$$\ln\gamma_i = \frac{L_i}{M_i} + \sum_j^C \frac{x_j \mathcal{G}_{ij}}{M_j}\left(\tau_{ij} - \frac{L_j}{M_j}\right) \tag{J.0.7}$$

where

$$L_i = \sum_k^C x_k \tau_{ki} \mathcal{G}_{ki} \tag{J.0.8}$$

$$M_i = \sum_k^C x_k \mathcal{G}_{ki} \tag{J.0.9}$$

$$\mathcal{G}_{ij} = \exp(-\alpha_{ij}\tau_{ij}) \tag{J.0.10}$$

$$\left.\begin{array}{ll} \tau_{ij} = \Delta g_{ij}/RT & \text{for } i \neq j \\ \tau_{ij} = 0 & \text{for } i = j \end{array}\right\} \tag{J.0.11}$$

Here C is the number of components, and note we have

$$\alpha_{ji} = \alpha_{ij} \tag{J.0.12}$$

but

$$\tau_{ji} \neq \tau_{ij} \qquad \text{because} \qquad \Delta g_{ji} \neq \Delta g_{ij} \tag{J.0.13}$$

For a binary mixture, (J.0.6) reduces to (J.0.1) while (J.0.7) reduces to (J.0.3) and (J.0.4). Like the Wilson equations, one advantage to the NRTL model is that all adjustable parameters are binary parameters, so no multicomponent data are needed to obtain their values. Parameter values for a few selected binaries are given in Table J.1.

Table J.1 Values of NRTL parameters for selected binary liquid mixtures

Component 1	Component 2	T (K)	Δg_{12} (cal/mol)	Δg_{21} (cal/mol)	$\alpha_{12} = \alpha_{21}$	Ref.
Acetonitrile	Water	300	415.38	1016.28	0.20202	[2]
	Water	333	363.57	1262.4	0.3565	[2]
Benzene	Acetonitrile	300	693.61	92.47	0.67094	[2]
	Acetonitrile	333	998.2	65.74	0.88577	[2]
	Water	300	3892.44	3952.2	0.23906	[2]
	Water	333	3883.2	3849.57	0.24698	[2]
Ethanol	Ethyl acetate	343	–480.377	1148.848	0.1	[3]
	Water	343	–53.732	1166.524	0.3	[3]
Ethyl acetate	Water	343	611.817	1869.890	0.3	[3]
Toluene	Water	283	2101.4	3265.0	0.2	[4]

LITERATURE CITED

[1] H. Renon and J. M. Prausnitz, "Local Compositions in Thermodynamic Excess Functions for Liquid Mixtures," *A. I. Ch. E. J*, **14**, 135 (1968).

[2] J. Castillo and I. E. Grossman, "Computation of Phase and Chemical Equilibria," *Computers Chem. Eng.*, **5**, 99 (1981).

[3] F. van Zandijcke and L. Verhoeye, "Vapor-Liquid Equilibrium of Ternary Systems with Limited Miscibility at Atmospheric Pressure," *J. Appl. Chem. Biotechnol.*, **24**, 709 (1974).

[4] J. M. Sorensen and W. Arlt, *Liquid-Liquid Equilibrium Data Collection*, vol. 5, part 2, DECHEMA, Frankfurt am Main, Germany, 1980.

K

SIMPLE ALGORITHMS FOR BINARY VLLE

In § 11.1.7 we noted that the double Rachford-Rice algorithm for VLLE does not apply to binary mixtures. Here we develop simple alternatives that often can be used for isothermal VLLE calculations of binary mixtures.

The problem is this: we have a binary of components 1 and 2 in VLLE at a known temperature. We are to find the pressure and the compositions of the three phases. We have four unknowns, but we also have four independent phase-equilibrium relations:

$$f_i^{\alpha} = f_i^{v} \qquad i = 1, 2 \tag{K.0.1}$$

$$f_i^{\alpha} = f_i^{\beta} \qquad i = 1, 2 \tag{K.0.2}$$

where superscripts α and β indicate liquid phases and superscript v indicates the vapor. We choose the gamma-phi method for the VLE problem (K.0.1) and the gamma-gamma method for the LLE problem (K.0.2). Then our four equations become

$$x_i^{\alpha} \gamma_i^{\alpha} \varphi_i^{s} P_i^{s} = y_i \varphi_i P \qquad i = 1, 2 \tag{K.0.3}$$

$$x_i^{\alpha} \gamma_i^{\alpha} = x_i^{\beta} \gamma_i^{\beta} \qquad i = 1, 2 \tag{K.0.4}$$

Here we consider low to moderate pressures, so we ignore the Poynting factor that would otherwise appear in (K.0.3). In general, the four equations (K.0.3)–(K.0.4) must be solved simultaneously.

Pressure independent γ-model. But many γ-models contain no pressure dependence, and if we use such a model, then our four equations decouple: that is, we can solve the LLE problem separately from the VLE problem. In such cases, our strategy is to first solve the LLE problem (K.0.4) by applying the Rachford-Rice LLE algorithm in Figure 11.4. For that calculation, any overall composition z_1 can be used, so long as it provides two-phase roots. This gives the liquid-phase mole fractions. Then with T and x_1^α known, we can solve the VLE problem (K.0.3), which now is merely a bubble-P calculation for y_1 and P.

Ideal-gas vapor phase. In the special case that the pressure is low enough for the vapor to be an ideal gas, the bubble-P calculation can be done analytically. Setting $\varphi_i = 1$ and $\varphi_i^s = 1$, the sum of the two equations in (K.0.3) gives the pressure,

$$P = x_1^\alpha \gamma_1^\alpha P_1^s + x_2^\alpha \gamma_2^\alpha P_2^s \tag{K.0.5}$$

then

$$y_1 = x_1^\alpha \gamma_1^\alpha P_1^s / P \tag{K.0.6}$$

Completely immiscible liquids. In the very special case that we know that the components are (essentially) immiscible as liquids, then the low-pressure problem simplifies further. For example, say phase α is essentially pure component 1, so

$$x_1^\alpha \approx 1 \tag{K.0.7}$$

and then phase β must be essentially pure component 2, so

$$x_2^\beta \approx 1 \tag{K.0.8}$$

Now the two equations in (K.0.3) become

$$P_1^s \approx y_1 P \tag{K.0.9}$$

and

$$P_2^s \approx y_2 P \tag{K.0.10}$$

Their sum gives the total pressure,

$$P_1^s + P_2^s = P \tag{K.0.11}$$

and with P known, the vapor-phase composition can be obtained from either (K.0.9) or (K.0.10).

❧

NOTATION

In the following lists, parentheses hold equation numbers, table numbers, figure numbers, or problem numbers where the symbol is defined or first introduced.

ROMAN LOWER CASE

a	Helmholtz energy, intensive (3.7.19)
a	Parameter in cubic equation of state (4.5.54)
a_{ki}	Number of atoms of element k on a molecule of species i (7.4.1)
$\mathfrak{a}_i$	Activity of component i (5.4.2)
b	Parameter in cubic equation of state (4.5.54)
b_k	Total number of atoms of element k (7.4.1)
c_p	Isobaric heat capacity, intensive (Table 3.2)
c_v	Isometric heat capacity, intensive (Table 3.2)
e	Total energy, intensive (2.4.7)
f	Generic property, intensive (3.4.2)
g	Gravitational acceleration (1.2.1)
g	Gibbs energy, intensive (3.2.24)
h	Enthalpy, intensive (2.4.15)
k	Boltzmann constant [Problem 1.5 and (2.3.6)]
k_{ij}	Binary interaction parameter in equation of state (4.5.80)
m	Mass of object or system (1.2.1)
m_e	Number of elements (7.4.1)
q	Heat, intensive (P2.17.1)
s	Entropy, intensive (2.4.21)
t	Time (Problem 1.15)
u	Internal energy, intensive (2.2.11)
v	Volume, intensive (1.2.4)
$\mathbf{v}_j$	Rate for reaction j (7.4.50)
w	Sonic velocity (P3.15.1)
w	Work, intensive (P2.10.1)

x	Horizontal distance (2.1.1)
x_i	Mole fraction of component i (1.2.7)
y_i	Mole fraction of component i in vapor phase (§ 9.3.1)
z	Vertical distance (1.2.1)
z_i	Overall mole fraction for component i (11.1.13)

ROMAN UPPER CASE

A	Area (2.1.9)
A	Helmholtz energy, extensive (3.2.11)
$\mathbf{A}$	Formula matrix (7.4.2)
A, B, C	Parameters in models (§ 5.6.1)
B	Second virial coefficient (4.5.9)
B'	Pressure second virial coefficient (4.5.21)
C	Third virial coefficient (4.5.10)
C_i	Generic conceptual property (Table 3.1)
C_i	Distribution coefficient for component i (11.1.14)
C_p	Isobaric heat capacity, extensive (3.3.8)
C_v	Isometric heat capacity, extensive (3.3.7)
C'	Pressure third virial coefficient (4.5.22)
E	Total energy [(2.1.8) and (2.2.9)]
F	Generic thermodynamic property, extensive (3.3.1)
F	Force (2.1.1)
F	Rachford-Rice function (11.1.20)
G	Gibbs energy, extensive (3.2.13)
H	Enthalpy, extensive (2.4.1)
H	Henry's constant (10.2.20), (10.2.27)
K_i	K-factor for component i (11.1.2), (12.1.1)
K_j	Equilibrium constant for reaction j (10.3.13)
L	Fraction of feed in liquid product (Problem 9.4)
M_i	Generic measurable property (Table 3.1)
N	Total number of moles (1.2.4)
N_{A}	Avogadro's number [Problem 1.5 and under (2.3.6)]
N_i	Number of moles of component i (1.2.7)
N_p	Number of mass ports to and from a system (3.6.11)
P	Absolute pressure (1.2.2)
Q	Heat, extensive (2.2.1)
R	Gas constant (Problem 1.5 and Problem 3.2)
R	Fraction of material in one of two liquid phases [below (11.1.13)]
S	Entropy, extensive (2.3.5)
T	Absolute temperature (2.3.6)
U	Internal energy, extensive (2.1.27)
$\mathbf{U}, \mathbf{W}, \mathbf{V}$	Matrices in singular value decomposition (11.2.6)
V	Volume, extensive (1.2.5)
V	Fraction of material in vapor phase (Problem 9.4)
W	Work, extensive (2.1.1)
Z	Compressibility factor (4.3.1)

ROMAN SCRIPT

$\mathcal{A}_j$	Affinity for reaction j (7.4.41)
$\mathcal{C}$	Number of components (3.1.2); number of species (7.4.1)
$\mathcal{F}$	Dissipative components of driving forces (1.3.4)
$\mathcal{F}$	Number of independent properties for intensive state (3.1.8)
$\mathcal{F}_{ex}$	Number of independent properties for extensive state (3.1.6)
$\mathcal{F}'$	Number of independent properties (9.1.11)
f_i	Fugacity of component i (4.3.8)
$\mathcal{N}$	Number of molecules (2.2.11)
$\mathcal{P}$	Pressure to overcome dissipative forces (2.1.11)
$\mathcal{P}$	Number of phases (9.1.5)
$\mathcal{R}$	Number of independent chemical reactions (7.4.5)
$\mathcal{S}$	Number of internal constraints (3.1.6)
$\mathcal{S}_{ext}$	Number of external constraints (3.1.3)
$\mathcal{V}$	Number of orthogonal interactions (3.1.3)
$\mathcal{V}_{max}$	Maximum number of interactions (3.1.1)

GREEK LOWER CASE

α	Volume expansivity (3.3.6)
α_{ij}	Relative volatility (12.1.11)
β	Reciprocal thermal energy, $\beta = 1/RT$ (4.5.45)
β	Dimensionless group in equation of state (8.2.12)
β_{ij}	Selectivity (12.1.30)
γ	Ratio of heat capacities (P3.23.1)
γ_i	Activity coefficient for component i (5.4.5)
γ_v	Thermal pressure coefficient (3.3.5)
δ	Differential driving force (1.3.3)
δ	Small amount of path function (2.2.5)
δ	Variational operator (G.0.4)
δ_{12}	Combination of second virial coefficients (5.3.8)
$\Delta\lambda_{ij}$	Parameters in Wilson model (5.6.30)
ε	Tolerance in trial-and-error searches (§ 11.1.1)
ζ	Convergence parameter in trial-and-error searches (11.1.42)
η	Packing fraction (4.5.2)
κ_s	Adiabatic compressibility (3.3.26)
κ_T	Isothermal compressibility (3.3.25)
λ	Integrating factor (2.3.3)
λ_k	Lagrange multiplier (10.3.33)
ν	Velocity [(2.1.5) and (2.3.6)]
ν_{ij}	Stoichiometric coefficient for species i in reaction j (7.4.10)
$\mathbf{\nu}_j$	Vector of stoichiometric coefficients for reaction j (7.4.17)
ξ_j	Extent of reaction j (7.4.12)
π	Dummy integration variable corresponding to pressure (4.4.2)
ρ	Density, mass (Table 3.2) or molar (4.5.8)
σ	Diameter of hard sphere (4.5.2)
σ_j	Algebraic sum of stoichiometric coefficients in reaction j (7.4.21)
τ	Dummy integration variable for temperature (10.1.10)

υ	Number of degrees of freedom (4.1.3)
φ_i	Fugacity coefficient of component i (4.3.18)
φ_i	Apparent volume fraction for component i (P5.12.3)
ψ	Dummy integration variable for intensive volume (4.4.13)
ω	Acentric factor (P4.22.2)

GREEK UPPER CASE

Δ	Delta operator: $\Delta x = x_2 - x_1$ (1.2.3)
Δ	Net total driving force (1.3.1)
Λ_{ij}	Parameters in Wilson model (5.6.24)
Ψ	Dummy integration variable for extensive volume (4.4.17)
Ω	Term in Wilson model for activity coefficients (5.6.27)

SUBSCRIPTS AND SUPERSCRIPTS; ROMAN

Az	Azeotrope (9.3.21)
acc	Accumulation (7.5.1)
ad	Adiabatic (2.1.27)
B	Boyle (4.5.12)
b	Boundary (2.2.10)
b	Boiling (Problem 1.10)
c	Configurational (2.2.13)
c	Critical (Problem 1.10)
con	Consumption (7.5.1)
dev	Deviation (4.0.1)
dif	Diffusion (7.5.6)
E	Excess property (5.2.1)
ext	External to a system (1.3.1)
f	Formation property (10.4.15)
f	Property of feed stream (Figure 12.14)
gen	Generated [(2.3.8) and (7.5.1)]
hs	Hard sphere (4.5.14)
I	Interface (7.2.1)
i	Index over components (3.4.2)
ig	Ideal gas (4.1.2)
irr	Irreversible (2.1.16)
is	Ideal solution (5.1.1)
j	Index over reactions (7.4.6)
k	Kinetic [(2.1.7) and (2.2.11)]
k	Index over phases or system parts (7.1.2)
ℓ	Liquid (8.2.18)
m	Change of property on mixing (3.7.38)
m	Melting (Problem 1.10)
mix	Mixture property (3.7.37)
o	Standard state property (5.1.3)
o	Initial value (7.4.6)
p	Potential [(2.1.4) and (2.2.11)]

R	Reduced by critical property (P4.31.1)
ref	Reference (4.0.1)
res	Residual (4.2.1)
rev	Reversible (2.1.14)
rxn	Reaction (7.5.2)
s	Saturation [Problem 6.5 and (8.2.18)]
s	Solid (10.1.7)
sf	Solute free (10.2.19)
sh	Shaft [under (2.4.2)]
sp	Spinodal (8.2.17)
sub	Sublimation [below (9.2.1)]
sur	Surroundings (7.1.3)
T	Transpose of a matrix or a vector (B.2.5)
t	Total (3.7.9)
ub	Upper bound (7.4.23)
v	Vapor (8.2.18)
vap	Vaporization (8.2.23)
wf	Workfree (2.2.7)

SUBSCRIPTS AND SUPERSCRIPTS; GREEK AND OTHER SYMBOLS

α	Feed stream (3.6.2); bulk phase (7.2.4)
β	Discharge stream (3.6.2); bulk phase (7.2.4)
σ	Saturation (9.2.1)
®	Reference-state property (4.3.12)
∞	Infinite dilution (5.4.14)
$+$	Referred to reference-solvent standard state (10.2.32)
$*$	Referred to solute-free standard state (10.2.24)

ABBREVIATIONS

cc	Cubic centimeter, cm^3 (7.1.45)
EoS	Equation of state (Figure 4.7)
FFF	Famous fugacity formulae (6.4.1)
GGE	Gas-gas equilibrium (§ 9.5.2)
LCEP	Lower critical end point (Figure 9.21)
LCST	Lower critical solution temperature (§ 9.3.6)
lhs	Left-hand side [under (2.3.8)]
LLE	Liquid-liquid equilibrium (start of § 9.3)
LSE	Liquid-solid equilibrium (Figure 9.27 and Table 9.2)
NRTL	Nonrandom, two-liquid model (Appendix J)
rhs	Right-hand side [under (2.2.2)]
UCEP	Upper critical end point (Figure 9.21)
UCST	Upper critical solution temperature (§ 9.3.6)
VLE	Vapor-liquid equilibrium [below (8.2.10)]
VLLE	Vapor-liquid-liquid equilibrium (start of § 9.3)
wrt	With respect to [above (3.2.22)]

INDEX